Epigenetic Drug Discovery

Giordanetto, F. (Ed.)

Early Drug Development
Bringing a Preclinical Candidate to the Clinic

2018

ISBN: 978-3-527-34149-8

Vol. 73

Holenz, Jörg (Ed.)

Lead Generation
Methods and Strategies

2016

ISBN: 978-3-527-33329-5

Vol. 68

Handler, N. and Buschmann, H. (Eds.)

Drug Selectivity
An Evolving Concept in Medicinal Chemistry

2018

ISBN: 978-3-527-33538-1

Vol. 72

Erlanson, D. A. and Jahnke, W. (Eds.)

Fragment-based Drug Discovery
Lessons and Outlook

2015

ISBN: 978-3-527-33775-0

Vol. 67

Vaughan, T., Osbourn, J., and Jalla, B. (Eds.)

Protein Therapeutics

2017

ISBN: 978-3-527-34086-6

Vol. 71

Urbán, L., Patel, V. F., and Vaz, R. J. (Eds.)

Antitargets and Drug Safety

2015

ISBN: 978-3-527-33511-4

Vol. 66

Ecker, G. F., Clausen, R. P., and Sitte, H. H. (Eds.)

Transporters as Drug Targets

2017

ISBN: 978-3-527-33384-4

Vol. 70

Keserü, G. M. and Swinney, D. C. (Eds.)

Kinetics and Thermodynamics of Drug Binding

«UNTERTITEL»2015

ISBN: 978-3-527-33582-4

Vol. 65

Martic-Kehl, M. I. and Schubiger, P.A. (Eds.)

Animal Models for Human Cancer
Discovery and Development of Novel Therapeutics

2017

ISBN: 978-3-527-33997-6

Vol. 69

Pfannkuch, F. and Suter-Dick, L. (Eds.)

Predictive Toxicology

From Vision to Reality

2014

ISBN: 978-3-527-33608-1

Vol. 64

Epigenetic Drug Discovery

Edited by
Wolfgang Sippl and Manfred Jung

WILEY-VCH

Volume Editors

Wolfgang Sippl
Martin-Luther-Universität
Institut für Pharmazie
Wolfgang-Langenbeck-Straße 4
06120 Halle
Germany

Manfred Jung
Albert-Ludwigs-Universität
Institut für Pharmazeutische
Wissenschaften
Albertstraße 25
Albert-Ludwigs-Universität
79104 Freiburg
Germany

Series Editors

Dr. Raimund Mannhold
Rosenweg 7
40489 Düsseldorf
Germany
mannhold@uni-duesseldorf.de

Dr. Helmut Buschmann
Aachen, Germany
Sperberweg 15
52076 Aachen
Germany
hbuschmann@gmail.com

Dr. Jörg Holenz
Head of Neuroscience Virtual
Proof-of-Concept DPU
R&D Neurosciences TAU
GSK
1250 S. Collegeville Road
Collegeville, PA 19426-0989
USA
joerg.x.holenz@gsk.com

Library of Congress Card No.:
applied for

British Library Cataloguing-in-Publication Data
A catalogue record for this book is available from the British Library.

Bibliographic information published by the Deutsche Nationalbibliothek
The Deutsche Nationalbibliothek lists this publication in the Deutsche Nationalbibliografie; detailed bibliographic data are available on the Internet at <http://dnb.d-nb.de>.

Print ISBN: 978-3-527-34314-0
ePDF ISBN: 978-3-527-80928-8
ePub ISBN: 978-3-527-80926-4
oBook ISBN: 978-3-527-80925-7

Cover Design SCHULZ Grafik-Design, Fußgönheim
Typesetting SPi Global, Chennai, India
Printing and Binding C.O.S. Printers Pte Ltd, Singapore

Printed on acid-free paper

Preface

Target-based drug discovery continues to be a dominating paradigm in industrial research, and current strategies for epigenetic therapy are no exception. Already in 1942 Waddington introduced the term epigenetics [1]. He defined it as "the branch of biology which studies the causal interactions between genes and their products which bring the phenotype into being" [2]. Originally, epigenetics referred to all molecular pathways modulating the expression of a genotype into a particular phenotype. Later on, with the rapid growth of genetics, its meaning has gradually narrowed. Epigenetics has been defined as "the study of changes in gene function that are mitotically and/or meiotically heritable and that do not entail a change in DNA sequence" [3, 4].

Epigenetic changes such as DNA methylation and histone methylation and acetylation alter gene expression at the level of transcription by upregulating, downregulating, or silencing genes completely. At the molecular level, epigenetic regulation involves hierarchical covalent modification of DNA and the proteins that package DNA, such as histones. Dysregulation of epigenetic events can be pathological, leading to cardio-vascular disease, neurological disorders, metabolic disorders, and cancer development, whereas the main focus of epigenetic drug discovery efforts has been on cancer. Thus, identifying drugs that inhibit these epigenetic changes are of great clinical interest [5].

Considering this prominent role in drug development, nine years ago a volume on "Epigenetic targets in drug discovery" [6] was published, covering in detail the available knowledge on methodology, epigenetic target classes, and inhibitor development. Since that period, novel findings in this field accumulated very fast. Thus, we felt time is mature to organize a second edition focusing on these recent developments. The editors of the initial book, Wolfgang Sippl and Manfred Jung, agreed to organize also the updating task. The new volume, presented here, focuses on medicinal chemistry applied to epigenetic targets, one of the fastest growing areas of drug discovery in recent years and comprises three parts. The introduction describes the gain of knowledge within the last decade. The second part concerns current methods including structural biology of epigenetic targets, computer-based technologies, mass spectrometry, peptide microarrays, chemical probe development, and epigenetic multi-targeting. The third part focuses on epigenetic targets like HDAC, SirT, HAT, methyltransferase and demethylase modulators, DNA modifiers, bromodomain and methyl-lysine reader proteins, and parasitic epigenetic targets.

We thank Wolfgang Sippl and Manfred Jung for organizing this volume and to work with such excellent authors. Last, but not least we thank Frank Weinreich and Waltraud Wüst from Wiley-VCH for their valuable contributions to this project and to the entire book series.

Raimund Mannhold
Düsseldorf
Helmut Buschmann
Aachen
Jörg Holenz
May 2018 *Boston, USA*

References

1 Waddington, C.H. (1942). The epigenotype. *Endeavour* 1: 18–20.
2 Waddington, C.H. (1968). Towards a theoretical biology. In: *The Basic Ideas of Biology*, 1–32. Edinburgh University Press.
3 Wu, C. and Morris, J.R. (2001). Genes, genetics, and epigenetics: a correspondence. *Science* 293 (5532): 1103–1105.
4 Arrowsmith, C.H., Bountra, C., Fish, P.V. et al. (2012). Epigenetic protein families: a new frontier for drug discovery. *Nat. Rev. Drug Discovery* 11: 384.
5 Heerboth, S., Lapinska, K., Snyder, N. et al. (2014). Use of epigenetic drugs in disease: an overview. *Genet. Epigenet.* 6: 9–19.
6 Sippl, W. and Jung, M. (2008). *Epigenetic Targets in Drug Discovery*, Methods and Principles in Medicinal Chemistry, vol. 42. Wiley-VCH.

Foreword

The role of histone becomes, thus, part of the problem of how the environment affects gene activity. Biology has by now outgrown the abstract and rigid limitations of classical genetics; for now it is clear that the chromosome, like other centres of vital activity, is subject to regulation by feed-back of the periphery.

A. E. Mirsky, 1965 (sic !) [1]

We had already used this quotation in the foreword for the first edition of this book in 2009. Still it is fascinating to read these prophetic words from more than 50 years ago. Already in 1950 Stedman had discussed the role of histones in differentiation [2] and in 1964 Allfrey reported on the acetylation on histones [3]. The words of Mirsky are from a Ciba Foundation symposium on histones and already then, a "functional correlation between histone acetylation and the RNA-synthetic capacity of the chromatin" was suggested.

Since the first edition of our book in 2009 the field of Epigenetics and the related drug discovery and development efforts have made amazing progress. By now, two inhibitors of DNA methyltransferases and five inhibitors of histone deacetylases (HDACs) have been approved for cancer treatment. Among the latter, tucidinostat (Chidamide) is of special interest as it is has been developed in China by a Chinese drug discovery company (Chipscreen Biosciences) [4] and we expect to see an increasing importance of China not only in science but also in drug development in the upcoming years. In addition, it is the first class-I selective orally available HDAC inhibitor approved and it will be interesting to see if resp. for which patients the class selectivity will have an advantage.

For other targets, like histone methyltransferase and demethylases, the therapeutic potential was already clearly visible nine years ago but now several of them have indeed moved into clinical trials [5] and the results of these are awaited with great anticipation. For acetyltransferases, the progress is still very limited but just recently the first clearly drug like HAT inhibitor has been presented [6] and it will be interesting to see if this "oldest class" of epigenetic target will still find its way to patients.

One target class that had not been covered in our first edition of 2009 are the so-called reader proteins that specifically bind to acetylated resp. methylated lysines and mediate the signals that have been set and maintained by the equilibrium of acetyltransferases and deacetylase resp. methyltransferase and demethylases. The inhibitors of the acetyl-lysine readers (bromodomain containing proteins) have developed at an amazing pace from the groundbreaking publication in 2010 [7] to several candidates in clinical trials already. The methyl lysine readers still lag a bit behind in their development but first inhibitors have emerged and there is also a perspective to target recognition of non-histone client proteins for these readers [8].

The technique of chemically induced proteolysis (PROTAC approach) [9] has received specific interest due to its recent application in the field of epigenetics [10] and shows great potential as an approach that is conceptually different from standard chemical target inhibition. Thus, we can expect more exciting years for biology and drug discovery and development in the field of epigenetics.

We apologize to all scientists whose efforts in the field were not duly cited in this book. We thank our authors, the editors and publishers from Wiley-VCH and our families for support.

Wolfgang Sippl
Manfred Jung

References

1 Mirsky, A.E. (1966). *Histones – Their Role in the Transfer of Genetic Information* (Ciba Foundation Study Group, Nr. 24), 112. London: Churchill Ltd.
2 Stedman, E. (1950). Cell specificity of histones. *Nature* 166: 780–781.
3 Allfrey, V.G., Faulkner, R., and Mirsky, A.E. (1964). Acetylation and methylation of histones and their possible role in the regulation of RNA synthesis. *Proc. Natl. Acad. Sci. U.S.A.* 51: 786–794.
4 Lu, X., Ning, Z., Li, Z. et al. (2016). Development of chidamide for peripheral T-cell lymphoma, the first orphan drug approved in China. *Intractable Rare Dis. Res.* 5: 185–191.
5 For an overview see e.g.Morera, L., Lübbert, M., and Jung, M. (2016). Targeting histone methyltransferases and demethylases in clinical trials for cancer therapy. *Clin. Epigenetics* 8: 57.
6 Lasko, L.M., Jakob, C.G., Edalji, R.P. et al. (2017). Discovery of a selective catalytic p300/CBP inhibitor that targets lineage-specific tumours. *Nature* 550: 128–132.
7 Filippakopoulos, P., Qi, J., Picaud, S. et al. (2010). *Nature* 468: 1067–1073.
8 Metzger, E., Willmann, D., McMillan, J. et al. (2016). Assembly of methylated LSD1 and CHD1 drives AR dependent transcription and translocation. *Nat. Struct. Mol. Biol.* 23: 132–139.
9 Cromm, P.M. and Crews, C.M. (2017). Targeted protein degradation: from chemical biology to drug discovery. *Cell Chem. Biol.* 24: 1181–1190.
10 Winter, G.E., Buckley, D.L., Paulk, J. et al. (2015). Phthalimide conjugation as a strategy for in vivo target protein degradation. *Science* 348: 1376–1381.

Contents

Part I

Introduction – Epigenetics

1

Epigenetics: Moving Forward

Lucia Altucci

Università degli Studi della Campania 'Luigi Vanvitelli', Dipartimento di Medicina di Precisione,
Vico L. De Crecchio 7, 80138 Napoli, Italy

Both the focus on epigenetics and the simple use of the term "epigenetic" have significantly augmented since the 1940s, when Sir Conrad Waddington opened the ground to this field. Since then, the definition of epigenetics became more inclusive, often defined as "*stably heritable phenotype resulting from changes in a chromosome without alterations in the DNA sequence*" (2008 Cold Spring Harbor Epigenetics meeting). In more common words, the term epigenetics derives from επί-genetics, which literally means "on top of" genetics, referring to the modifications of chromatin that are able to switch genes "on" or "off" affecting the cell's "interpretation" of genes and consequently function, specialization, phenotype, and cell fate [1].

Recently, the importance of the epigenetics has become evident from the plethora of articles, conferences, and consortia on the topic over the past decade. All over the world, research was intensified more and more on basic as well as biomedical-oriented epigenetic-based methodologies, targets, and applications. Funders initiated concerted actions to promote standardization and collaboration of the worldwide efforts aiming to unveil the role of transcriptional and epigenetic mechanisms in specification of cell fates and functions, such as the "American Association for Cancer Research Human Epigenome Task Force" and the "European Union Network of Excellence." The "International Human Epigenome Consortium" (IHEC) was founded to coordinate and standardize the production of reference epigenomes with a focus on cell states relevant to health and diseases, thereby accelerating translation of new knowledge to improve therapy [2]. IHEC has also coordinated the international efforts by bringing together the European Commission that funded "Blueprint consortium" (http://www.blueprint-epigenome.eu/) with, as a mere example, the NIH that funded "Roadmap on Epigenomics" (http://www.roadmapepigenomics.org). In addition, IHEC introduced common bioinformatics standards, models, and tools to analyze and interpret epigenomic data in a uniform and interoperable manner [3].

Epigenetic Drug Discovery, First Edition. Edited by Wolfgang Sippl and Manfred Jung.

1.1 Why This Enormously Increased Interest?

One reason is the need to address fundamental questions to understand the way the genome and environment interact in development and aging and how the epigenome affects or is affected by health and disease.

In addition, there is an urgent need to develop new ways to "drug" the epigenome and to translate discoveries into improvements of human health. Despite being quite stable and hereditable, epigenome modifications can be easily changed within the cell, affecting cell fate and functions. This epigenome plasticity opens the way to the pharmacological exploitation and to the identification and characterization of chromatin-targeting drugs. The identification of increasing numbers of new players acting as "*writers, erasers, or readers*" of the epigenome suggests that an intricate and very well-defined epi-modulated setting is responsible for maintaining the plasticity potential, ultimately guarantying cell identity and cell heterogeneity of otherwise similar tissues. Given that new modifications/new players are being uncovered, additional complexity arises, and a better understanding and frequent revisiting of the mechanism(s) of chromatin regulation and plasticity – ultimately at the single-cell level – are needed. The potential of this emerging knowledge toward its translation into biomedical applications is breathtaking. For example, a huge number of studies (many of which using high-throughput approaches) have unveiled the significance of certain histone marks, epi-enzymes, and chromatin-regulating factors in different human pathologies such as cancer, neurological disorders, diabetes, immunological pathologies, etc. [4]. Translating this basic knowledge to bedside practice has triggered investments in the identification and development of new drugs able to re-equilibrate deregulated epigenome areas acting by inhibiting or (currently more rarely) activating chromatin enzymes and/or by interfering the function of chromatin readers.

In addition to the rapidly accumulating knowledge on the mechanisms of action of chromatin-targeting "(epi)drugs," we have only beginning to unravel the different substrates of the epi-enzymes. "Epi-drugs" are designed to inhibit (or activate) histone-modifying enzymes or DNA methyltransferases or to interfere with readers of the resulting chromatin modifications. However, these chromatin modifiers (and the respective "epi-drugs") affect various substrates, including proteins in signal transduction pathways and cell structure. Such insights will turn out to be crucial to develop a better rational design of drugs treatment (and combination thereof), further exploiting and expanding the promise of epigenetically acting drugs.

It is still debated whether selective or broad chromatin modulators will be more effective [5]. As has been demonstrated in some cancer types harboring mutated enzymes, a selective "epi-drug" approach (active exclusively or preferentially on the mutant) may be preferred. On the other hand, a broad modulator might become more useful when concomitant alterations of different epi-targets are playing a role. This might also include hybrid molecules acting contextually on one epi-target and one non-epi-target.

Among the best studied chromatin-targeting drugs, HDAC inhibitors [6] and DNA-demethylating agents [7] have entered the clinic for anticancer treatment and prevention. Despite that HDAC inhibitors mostly induce hyperacetylation, this cannot be considered as a parameter of response. This issue highlights the need for a detailed understanding and development of markers of treatment response along with (epi)drug development. This will become a challenging task considering that epigenetic-based approaches have been proposed for very different diseases. In cancer patients, the altered expression of epi-players (overexpression or silencing) or a qualitative deregulation such as the mutation in one of the epi-enzymes has been one of the parameters of choice although patient's stratification on the basis of HDAC expression levels appears not always predictive of a better response. The presence of a well-characterized target mutation may instead prove to be more useful for patient stratification. Small molecules able to selectively modulate the mutated enzymes/targets may display tumor-specific action.

Interestingly, different groups of enzymes display diverse ways of deregulation; for example, HDACs are generally quantitatively overexpressed in cancer [8] (with the exception of HDAC2 mutations [9], for example, in colon tumorigenesis), whereas HATs appears more frequently mutated [10, 11]. Furthermore, the direct and indirect deregulation of methylation control through mutations in DNA methyltransferases and isocitrate dehydrogenases (IDH) genes appears to go along with abnormal histone and DNA methylation as a common feature of tumors with IDH1 and IDH2 mutations and altered stem cell differentiation and eventual tumorigenesis [12]. Description of inactivating mutations in TET2 suggests that cellular transformation is in part caused by the deregulation of 5-mC conversion. The TET enzymes have particular relevance in hematological cancers and solid tumors with mutations causing TET inactivation [13].

1.2 Looking Forward to New Avenues of Epigenetics

The constant flow of discoveries in the epigenetic field adds new layers of complexity and may lead to novel approaches for treatment. Novel chromatin marks are identified, and insight from mining of these targets (alone and within the context of others) may rapidly change our view. For example, hydroxymethyl cytosine and its modulation is at present a focus of discussions aimed at unraveling its mode of action and its potential role in cancer as well as other human diseases [14] [15]. The levels of 5hmC in the brain of patients with neurodegenerative disorders have been reported to be highly compromised, indicating a potential role of 5hmC in neurodegenerative diseases including Alzheimer's disease, Parkinson's disease, and Huntington's disease. It has yet to be established whether this is the cause or the consequence of the onset and progression of these diseases [16].

The burst in acquisition of scientific knowledge and in evolving new technologies will also pave the way to new concepts in the regulation and deregulation of the epigenome. Emerging single-cell epigenomic methods [17] are being developed with the exciting potential to transform our knowledge of gene regulation [18]. Until recently, our epigenetic modifications have been studied in bulk measurements in populations of cells.

The development of single-cell technologies is likely to cause a profound transformation of epigenome studies and their interpretation, in particular, in cases where (epi)genetic heterogeneity is overriding. In recent years, many of the high-throughput sequencing technologies hither to assaying population have been adapted and became assayable at the single-cell level. Combined single-cell methods such as simultaneous assessment of the transcriptome and DNA methylome may provide deeper insight in epigenetic–transcriptional correlations, allowing analyses on the causal relationships between phenome and the epigenome state. Furthermore, combined genome and epigenome analyses will likely open up new avenues to dissect the complex contribution of genomic and epigenomic heterogeneities [19].

A better integration of high-throughput data, bioinformatics interpretation, novel epi-marks, and chromatin players has the potential to bridge basic knowledge with the clinics both for epi-marks mining for diagnosis of disease treatment and outcome prediction and for disease prevention. Furthermore, many chromatin-targeting drugs have been identified and characterized in the past decade for their beneficial action against different human diseases. Even though the beneficial effect and link to the selective chromatin-regulating action has to be better corroborated and strengthened, their clinical potential is clear. In agreement, HDACi have been approved for the therapy of cutaneous T-cell lymphoma (CTCL) and recently for the treatment of multiple myeloma [20], as are DNA-demethylating agents for the treatment of myelodysplastic syndrome (MDS). In addition, the action of HDAC inhibitors against cancer might also be linked to the modulation on the immune system, potentially shedding a different light for their clinical use [21]. That histone methylation is also altered in cancer that led to the identification of lysine methyltransferases and demethylases as promising targets for new anticancer drugs. Inhibitors (targeting the histone methyltransferases DOT1L and EZH2 as well as the demethylase LSD1) have already reached the first stages of clinical trials in cancer therapy [22].

Also pharmacological inhibition of BET proteins shows therapeutic action in a variety of different pathologies, particularly in models of cancer and inflammation [23]. Such effects have been attributed to subsets of downstream targets. While it is clear that the therapeutic potential is huge, the current understanding of molecular mechanisms that underlie the therapeutic effects of pharmacological BET bromodomain inhibition still need better understanding [24].

Drug discovery efforts in the epigenetic field are not only focused on cancer but also on more chronic diseases opening the way to new opportunities for the epi-targeted treatments. For example, I-BET151 has been reported to effectively prevent type 1 diabetes in a mouse model for this disease [25, 26], suggesting that an epigenetic treatment of diabetes might be at our doorstep. Along these lines, different classes of "epi-drugs" that have been suggested to decrease obesity and clinical trials at different stages are ongoing, aiming to a better definition of their potential [27]. Recent studies have identified SIRT1 activators that may delay multiple diseases of aging and extend lifespan *in vivo* [28]. In theory, such molecules could act against diseases, potentially extending healthy years of life. Potential roles of SIRT1 and SIRT2 modulation in neurodegenerative diseases have been proposed [29, 30] and an SIRT1 inhibitor (Selisistat) is in clinical trial against Huntington's disease [31].

These are only examples of the critical need to illuminate the drug discovery efforts in the identification and characterization of the novel epi-drugs [32]. Thus, in this volume an overview of state-of-the-art knowledge and development in drug design for epi-targets, their mechanisms of actions, and the increasing spectrum of applications is presented. Furthermore, current methodologies are discussed including the structural biology of epigenetic targets, computer-based technologies, mass spectrometry, peptide microarrays, chemical probe development, and epigenetic multi-targeting. In addition, the "epi-drug" classes such as HDAC, SirT, HAT, methyltransferase and demethylase modulators, DNA modifiers, bromodomain, and methyl-lysine reader proteins are examined. Finally this volume will also address challenges and promises of parasitic epigenetic targets. A new promising approach is chemically induced proteolysis by so-called PROTACs (proteolysis targeting chimeras), where a ligand to the target of interest is fused to a moiety that leads, e.g. to ubiquitinylation and subsequent proteolytic degradation. This will phenocopy knockdowns, resp. knockout studies, and is promising prolonged target inactivation and might become a new paradigm in drug discovery and hence also in epigenetics [33–35].

Acknowledgments

Blueprint (282510), EPIGEN (MIUR-CNR); MIUR (20152TE5PK), AIRC (17217); COST EPICHEMBIO CM1406.

References

1. Deans, C. and Maggert, K.A. (2015). What do you mean, "epigenetic"? *Genetics* 199: 887–896.
2. International Cancer Genome, C, Hudson, T.J., Anderson, W. et al. (2010). International network of cancer genome projects. *Nature* 464: 993–998.
3. Stunnenberg, H.G., International Human Epigenome, C, and Hirst, M. (2016). The International Human Epigenome Consortium: a blueprint for scientific collaboration and discovery. *Cell* 167: 1145–1149.
4. Stunnenberg, H.G., International Human Epigenome, C, and Hirst, M. (2016). The International Human Epigenome Consortium: a blueprint for scientific collaboration and discovery. *Cell* 167: 1897.
5. Benedetti, R., Conte, M., Iside, C., and Altucci, L. (2015). Epigenetic-based therapy: from single- to multi-target approaches. *Int. J. Biochem. Cell Biol.* 69: 121–131.
6. Benedetti, R., Conte, M., and Altucci, L. (2015). Targeting histone deacetylases in diseases: where are we? *Antioxid. Redox Signaling* 23: 99–126.
7. Pechalrieu, D., Etievant, C., and Arimondo, P.B. (2017). DNA methyltransferase inhibitors in cancer: from pharmacology to translational studies. *Biochem. Pharmacol.* 129: 1–13.

8 Conte, M. and Altucci, L. (2012). Molecular pathways: the complexity of the epigenome in cancer and recent clinical advances. *Clin. Cancer Res.* 18: 5526–5534.
9 Ropero, S., Fraga, M.F., Ballestar, E. et al. (2006). A truncating mutation of HDAC2 in human cancers confers resistance to histone deacetylase inhibition. *Nat. Genet.* 38: 566–569.
10 Di Cerbo, V. and Schneider, R. (2013). Cancers with wrong HATs: the impact of acetylation. *Briefings Funct. Genomics* 12: 231–243.
11 Pasqualucci, L., Dominguez-Sola, D., Chiarenza, A. et al. (2011). Inactivating mutations of acetyltransferase genes in B-cell lymphoma. *Nature* 471: 189–195.
12 Yang, H., Ye, D., Guan, K.L., and Xiong, Y. (2012). IDH1 and IDH2 mutations in tumorigenesis: mechanistic insights and clinical perspectives. *Clin. Cancer Res.* 18: 5562–5571.
13 Scourzic, L., Mouly, E., and Bernard, O.A. (2015). TET proteins and the control of cytosine demethylation in cancer. *Genome Med.* 7: 9.
14 Branco, M.R., Ficz, G., and Reik, W. (2011). Uncovering the role of 5-hydroxymethylcytosine in the epigenome. *Nat. Rev. Genet.* 13: 7–13.
15 Pfeifer, G.P., Kadam, S., and Jin, S.G. (2013). 5-hydroxymethylcytosine and its potential roles in development and cancer. *Epigenetics Chromatin* 6: 10.
16 Sherwani, S.I. and Khan, H.A. (2015). Role of 5-hydroxymethylcytosine in neurodegeneration. *Gene* 570: 17–24.
17 Clark, S.J., Lee, H.J., Smallwood, S.A. et al. (2016). Single-cell epigenomics: powerful new methods for understanding gene regulation and cell identity. *Genome Biol.* 17: 72.
18 Bintu, L., Yong, J., Antebi, Y.E. et al. (2016). Dynamics of epigenetic regulation at the single-cell level. *Science* 351: 720–724.
19 Hou, Y., Guo, H., Cao, C. et al. (2016). Single-cell triple omics sequencing reveals genetic, epigenetic, and transcriptomic heterogeneity in hepatocellular carcinomas. *Cell Res.* 26: 304–319.
20 Imai, Y., Maru, Y., and Tanaka, J. (2016). Action mechanisms of histone deacetylase inhibitors in the treatment of hematological malignancies. *Cancer Sci.* 107: 1543–1549.
21 Kroesen, M., Gielen, P., Brok, I.C. et al. (2014). HDAC inhibitors and immunotherapy; a double edged sword? *Oncotarget* 5: 6558–6572.
22 Morera, L., Lubbert, M., and Jung, M. (2016). Targeting histone methyltransferases and demethylases in clinical trials for cancer therapy. *Clin. Epigenetics* 8: 57.
23 Shu, S., Lin, C.Y., He, H.H. et al. (2016). Response and resistance to BET bromodomain inhibitors in triple-negative breast cancer. *Nature* 529: 413–417.
24 Shi, J. and Vakoc, C.R. (2014). The mechanisms behind the therapeutic activity of BET bromodomain inhibition. *Mol. Cell* 54: 728–736.
25 Kitagawa, Y. and Ohkura, N. (2014). Treating type-1 diabetes with an epigenetic drug. *elife* 3: e05720.
26 Fu, W., Farache, J., Clardy, S.M. et al. (2014). Epigenetic modulation of type-1 diabetes via a dual effect on pancreatic macrophages and beta cells. *elife* 3: e04631.

27 Arguelles, A.O., Meruvu, S., Bowman, J.D., and Choudhury, M. (2016). Are epigenetic drugs for diabetes and obesity at our door step? *Drug Discovery Today* 21: 499–509.
28 Hubbard, B.P. and Sinclair, D.A. (2014). Small molecule SIRT1 activators for the treatment of aging and age-related diseases. *Trends Pharmacol. Sci.* 35: 146–154.
29 Donmez, G. (2013). Sirtuins as possible targets in neurodegenerative diseases. *Curr. Drug Targets* 14: 644–647.
30 Donmez, G. and Outeiro, T.F. (2013). SIRT1 and SIRT2: emerging targets in neurodegeneration. *EMBO Mol. Med.* 5: 344–352.
31 Carafa, V., Rotili, D., Forgione, M. et al. (2016). Sirtuin functions and modulation: from chemistry to the clinic. *Clin. Epigenetics* 8: 61.
32 Altucci, L. and Rots, M.G. (2016). Epigenetic drugs: from chemistry via biology to medicine and back. *Clin. Epigenetics* 8: 56.
33 Toure, M. and Crews, C.M. (2016). Small-molecule PROTACS: new approaches to protein degradation. *Angew. Chem. Int. Ed.* 55: 1966–1973.
34 Lu, J., Qian, Y., Altieri, M. et al. (2015). Hijacking the E3 ubiquitin ligase cereblon to efficiently target BRD4. *Chem. Biol.* 22: 755–763.
35 Winter, G.E., Buckley, D.L., Paulk, J. et al. (2015). DRUG DEVELOPMENT. Phthalimide conjugation as a strategy for in vivo target protein degradation. *Science* 348: 1376–1381.

Part II

General Aspects/Methodologies

2

Structural Biology of Epigenetic Targets: Exploiting Complexity

Martin Marek, Tajith B. Shaik, and Christophe Romier

Institut de Génétique et de Biologie Moléculaire et Cellulaire (IGBMC), Département de Biologie Structurale Intégrative, Université de Strasbourg, CNRS, INSERM, 1 rue Laurent Fries, 67404 Illkirch Cedex, France

2.1 Introduction

In the last two decades, epigenetic effectors have increasingly been shown to be major regulators of nuclear processes, with direct implications for cell homeostasis, response to external stimuli, development, and onset and progression of many diseases [1]. As a consequence, both fundamental research in epigenetics and the development of epigenetic drugs (epi-drugs) for therapy have become major fields of investigation.

Initial studies focused on epigenetic enzymes involved in the deposition and removal of epigenetic marks and on the reader domains responsible for the specific recognition of these marks [1, 2]. Yet, the discovery that other epigenetic effectors such as histone variants, histone chaperones, and ATP-dependent chromatin remodelers are also implicated in diseases further broadens the number of targets for epi-drug design [1, 3].

A few epi-drugs are already approved for the treatment of diseases, notably cancer [2c, 3c]. Their clinical use is often accompanied by serious undesirable side effects due to the fact that many epigenetic effectors belong to families whose members are often functionally different but structurally similar. This makes selective inhibition a major issue for the design of next-generation epi-drugs. In this respect, structural information is invaluable in helping deciphering precisely in molecular terms the mechanisms governing epigenetic processes and in aiding next-generation epi-drug design.

Another important reason for the reduced usage of epi-drugs is the strong interplay between epigenetic effectors. Notably, many epigenetic effectors act within large macromolecular complexes that represent the bona fide functional epigenetic units and that bear different epigenetic activities. This organization has two major consequences for the design and the use of epi-drugs. First, these complexes are physically and functionally linking epigenetic activities. Thus, modulating one activity with small molecules is likely to affect the other activities. Second, regulatory subunits can change partner/substrate recognition, enzymatic activity/kinetics, and inhibitor binding. Here again, deciphering the

Epigenetic Drug Discovery, First Edition. Edited by Wolfgang Sippl and Manfred Jung.

structures of these large molecular assemblies, or at least those of their active sub-complexes, is of paramount importance for understanding epigenetic mechanisms and for aiding epi-drug design.

A wealth of structural data has already been obtained on epigenetic effectors and their interactions with inhibitors, substrates, and protein partners, unraveling the diversity and complexity of these interactions. The huge amount of published structural data prevents an exhaustive description of all these results. Chapter 2 on epigenetic enzymes [2b] and specific chapters of this book are providing precise structural information on epigenetic enzymes and readers. In this chapter, we have chosen to focus primarily on epigenetic macromolecular complexes from the various classes of epigenetic effectors whose structures have enlarged our understanding of epigenetic mechanisms and pave the way for designing next-generation selective epi-drugs. Specifically, macromolecular interactions as well as mechanisms leading to structural rearrangements are described, highlighting ways of modulating the activity of epigenetic effectors.

2.2 DNA Methylases: The DNMT3A–DNMT3L–H3 and DNMT1–USP7 Complexes

In human, DNA methylation occurs predominantly on cytosines (5-methylcytosine) in CpG motifs that often form clusters known as CpG islands [4]. The initial view that DNA methylation is a rather stable epigenetic mark has been completely revisited in the last decade as new demethylation pathways have been characterized [5]. It is now commonly accepted that DNA methylation is a highly dynamic mark that is important in developmental processes. Specifically, methylation patterns are strongly perturbed in diseases, notably in cancers [4, 5].

De novo DNA methylation is carried out by the DNA methyltransferases DNMT3A and DNMT3B, whereas DNMT1 is required for the maintenance of the methyl mark by methylating hemimethylated DNA. DNMT3L, an inactive paralogue of DNMT3A/B, binds to and stimulates the activity of DNMT3A. DNMT3A activity is also stimulated in a DNMT3L-independent manner by histone H3 when its lysine 4 is not methylated. DNMT3A and DNMT3L both have an ADD (ATRX-DNMT3-DNMT3L) domain followed by a methyltransferase domain. Yet, DNMT3A catalytic domain (CD) is active, whereas the one of DNMT3L (CD-like) is inactive.

The ADD domains of DNMT3A and DNMT3L are able to bind to the N-terminus of histone H3 when it is not methylated on lysine 4. Methylation prevents binding due to steric hindrance [6]. The 3.8 Å resolution crystal structure of DNMT3A (ADD-CD) bound to DNMT3L (CD-like) shows that DNMT3L (CD-like) forms an extensive interaction with DNMT3A CD. This suggests that stimulation of DNMT3A activity by DNMT3L comes from a stabilization of the DNMT3A CD (Figure 2.1a) [7]. However, DNMT3A ADD domain and the linker region that connects it to DNMT3A (CD) pack against DNMT3A CD at a position where substrate DNA would be expected to bind, indicating

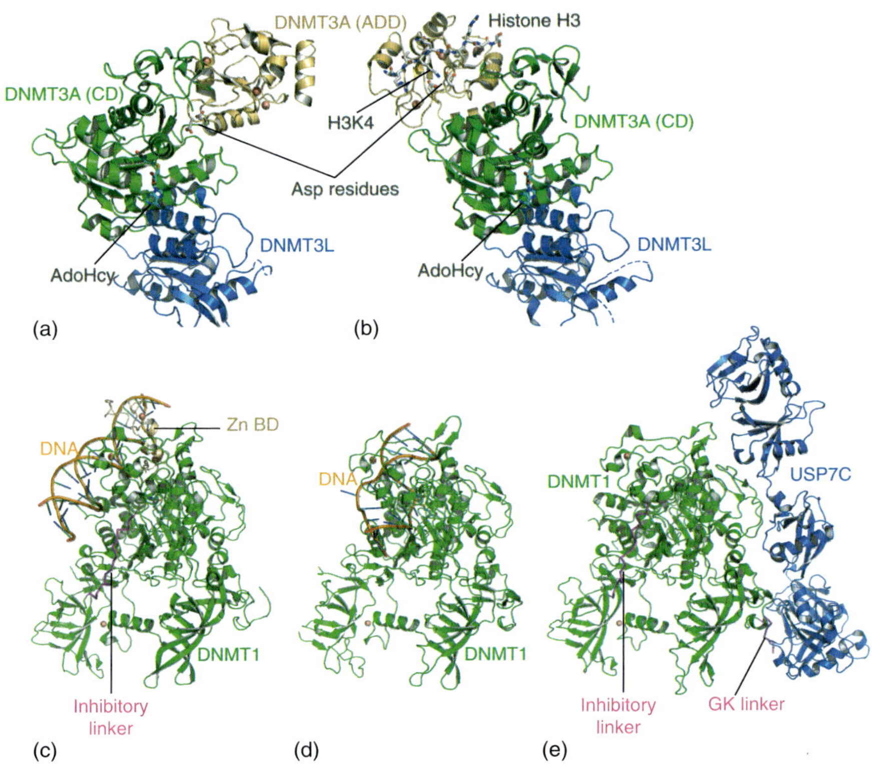

Figure 2.1 Structures of DNA methyltransferases. Ribbon representation of DNA methyltransferases (a, b) Structures of the *de novo* DNA methylation complex DNMT3A–DNMT3L in inhibitory (a) and activated (b) states. DNMT3A catalytic domain (CD) is colored green and its ADD (ATRX–DNMT3–DNMT3L) domain is colored wheat. DNMT3L inactive CD is colored blue. Zinc ions are shown as red spheres. Histone H3-activating peptide and important DNMT3A aspartate residues are represented as sticks with gray carbons. An AdoHcy (S-adenosyl-L-homocysteine) molecule is represented as sticks with cyan carbons. The coloring is identical in all figures unless stated. (c–e) Structures of methyl mark maintenance DNA methyltransferase DNMT1 in inhibitory (a) and active (b) states and in a stabilizing complex with USP7 (c). DNMT1 is shown in green, DNA in orange, and USP7 in blue. DNMT1 zinc binding domain (Zn BD; wheat) recognizing unmethylated DNA is shown as well as the inhibitory and regulatory GK linkers (magenta ribbons).

that this structure represents an inhibitory form of the DNMT3A–DNMT3L complex.

Upon binding of histone H3 N-terminus unmethylated on lysine K4 to DNMT3A ADD domain, the ADD domain makes a large movement, interacting with another surface of DNMT3A (CD), thus freeing the DNA binding surface of this CD (Figure 2.1b) [7]. Specifically, H3K4 binds to DNMT3A ADD domain aspartate residues that are otherwise involved in the formation of the inactive DNMT3A conformation.

DNMT1 is also an essential DNA methylase and is the target of two of the few FDA-approved epi-drugs [2c, 3c]. In contrast to DNMT3A/B enzymes, DNMT1 can only methylate hemimethylated DNA. The 3.0 and 2.6 Å crystal structures of

DNMT1 bound to non-methylated DNA and to hemimethylated DNA suggest a mechanism by which this enzyme carries out this discrimination [8]. Specifically, unmethylated DNA is recognized by a zinc finger of DNMT1 [8a]. This recognition positions the linker that connects the zinc finger to the first bromo-adjacent homology 1 (BAH1) domain of DNMT1 between the DNA and the active site of the enzyme, leading to an inactive complex (Figure 2.1c).

In the structure of the productive complex [8b], the DNA is found inserted into the active site of the CD (Figure 2.1d). Actually, in this structure the major conformational change observed concerns the catalytic loop that adopts a conformation compatible with catalysis. Yet, this structure was obtained with a shorter construct of DNMT1 that does not encompass the zinc finger and the following linker that are playing a major role in DNMT1 autoinhibition in the presence of unmethylated DNA. It remains therefore to be understood whether the presence of hemimethylated DNA prevents zinc finger binding and autoinhibition or whether the removal of the inhibition is due to an active mechanism.

DNMT1 has been shown to be regulated through various pathways and partner proteins. One of them is the ubiquitin-specific protease 7 (USP7) that stabilizes DNMT1. The 2.9 Å crystal structure of USP7 C-terminus (USP7C) in complex with DNMT1 has been solved [9]. The overall structure of DNMT1 in this complex is highly similar to the one in the autoinhibited form, including the positioning of the DNMT1 inhibitory N-terminal linker in DNMT1 DNA binding site (Figure 2.1e).

The DNMT1–USP7C complex structure reveals that USP7C, which is composed of several ubiquitin-like domains, binds to DNMT1 on the side opposite to the methylase active site. A critical interaction is made with DNMT1 KG linker that contains several Lysine–Glycine repeats. Specifically, the lysines of this linker are forming multiple interactions with residues of USP7C, and acetylation of these lysines precludes interaction between USP7 and DNMT1, favoring the *in vivo* degradation of DNMT1 [9].

2.3 Histone Arginine Methyltransferases: The PRMT5–MEP50 Complex

Protein arginine methyltransferases (PRMTs) are monomethylating and symmetrically or asymmetrically dimethylating arginine residues in histones and other cellular effectors [10]. The role and mode of action of PRMTs have long remained poorly understood. This picture is however changing as more data is obtained on this class of enzymes, showing that they are also involved in a wide range of diseases. Specifically, to develop therapeutic strategies targeting these enzymes, the deciphering in molecular terms of the specific recognition by PRMTs of their substrates and of the influence of partner proteins on PRMTs activity and substrate recognition has to be addressed.

The 2.0 and 3.0 Å crystal structures of human and *Xenopus laevis* PRMT5 in complex with one of its partners, MEP50, have provided novel information on these issues [11]. PRMT5 monomethylates and symmetrically dimethylates different substrates [10]. PRMT5 is composed of two domains: an N-terminal TIM barrel and a C-terminal CD that adopts a canonical arginine methyltransferase fold. The structures of the PRMT5–MEP50 complex reveal the formation of a tetramer of PRMT5–MEP50 dimers where PRMT5 forms the core of the octamer and MEP50 is located on the outside of the complex.

The 2.0 Å crystal structure of the human PRMT5–MEP50 complex in the presence of a AdoMet analogue and an H4 N-terminal tail peptide shows how the substrate is recognized in the active site of PRMT5 and suggests how active site residues participate to the methylation process [11a] (Figure 2.2a). Interestingly, the crystal structures of PRMT5–MEP50 bound to selective PRMT5 inhibitors show how these inhibitors can bind directly to these active site residues, leading to selective inhibition [12]. In addition, these different structures also reveals the molecular basis by which phosphorylation of tyrosine residues in the substrate binding groove can diminish catalytic activity by opposing to substrate binding.

Yet, these structures do not reveal the role of MEP50 in the complex. This information is provided by a lower resolution electron microscopy (EM) structure of the PRMT5–MEP50 complex bound to one of its substrate, nucleoplasmin. This structure reveals that nucleoplasmin interacts predominantly with MEP50 that serves as a docking platform for the substrate to be presented to PRMT5 [11b].

2.4 Histone Lysine Methyltransferases: The MLL3–RBBP5–ASH2L and the PRC2 Complexes

Proteins of the MLL family play major roles in development and are mainly responsible for the methylation of lysine 4 of H3 (H3K4), an epigenetic mark associated with activation of transcription [13]. MLL1 has been most studied

Figure 2.2 Structures of arginine and lysine methyltransferases. Ribbon representation of arginine and lysine methyltransferases. (a). Structure of the PRMT5–MEP50 dimer. PRMT5 arginine methyltransferase catalytic domain (CD) is colored green, its TIM-barrel domain is colored wheat, and MEP50 is colored blue. Histone H4 peptide binding to PRMT5 CD is shown as sticks with gray carbon as well as PRMT5 important active site residues and regulatory tyrosines. An AdoMet (S-adenosyl-L-methionine) analogue bound to PRMT5 CD is shown as sticks with cyan carbons. (b). Structure of lysine methyltransferase MLL3 (SET domain; green) in complex with a RBBP5 peptide (yellow) and ASH2L C-terminal domain (blue). Histone H3 N-terminal peptide and an AdoHcy molecule are shown as sticks with gray and cyan carbons, respectively. Residues at the interface of the three proteins that form a hydrogen bond network (MLL3 Arg network) are also shown as sticks. (c). Structure of the polycomb repressive complex 2 (PRC2) (EZH2/EED/SUZ12). Methyltransferase EZH2 is colored green except its catalytic SET domain that is colored wheat. EED is colored blue and the SUZ12 VEFS domain is colored yellow. An AdoHcy molecule bound to EZH2 SET domain is shown as sticks with cyan ribbons. Two H3 N-terminal peptides are shown as sticks with gray carbon: a H3K27me3 peptide bound to EED and a mutated H3M27 peptide bound to EZH2 SET domain. Two regulatory elements are shown in magenta: the SET activation loop (SAL) (shown as ribbon) and the stimulation-responsive motif (SRM). This latter motif is suggested to transmit the signal of H3K27me3 binding to the EED subunit to the catalytic SET domain.

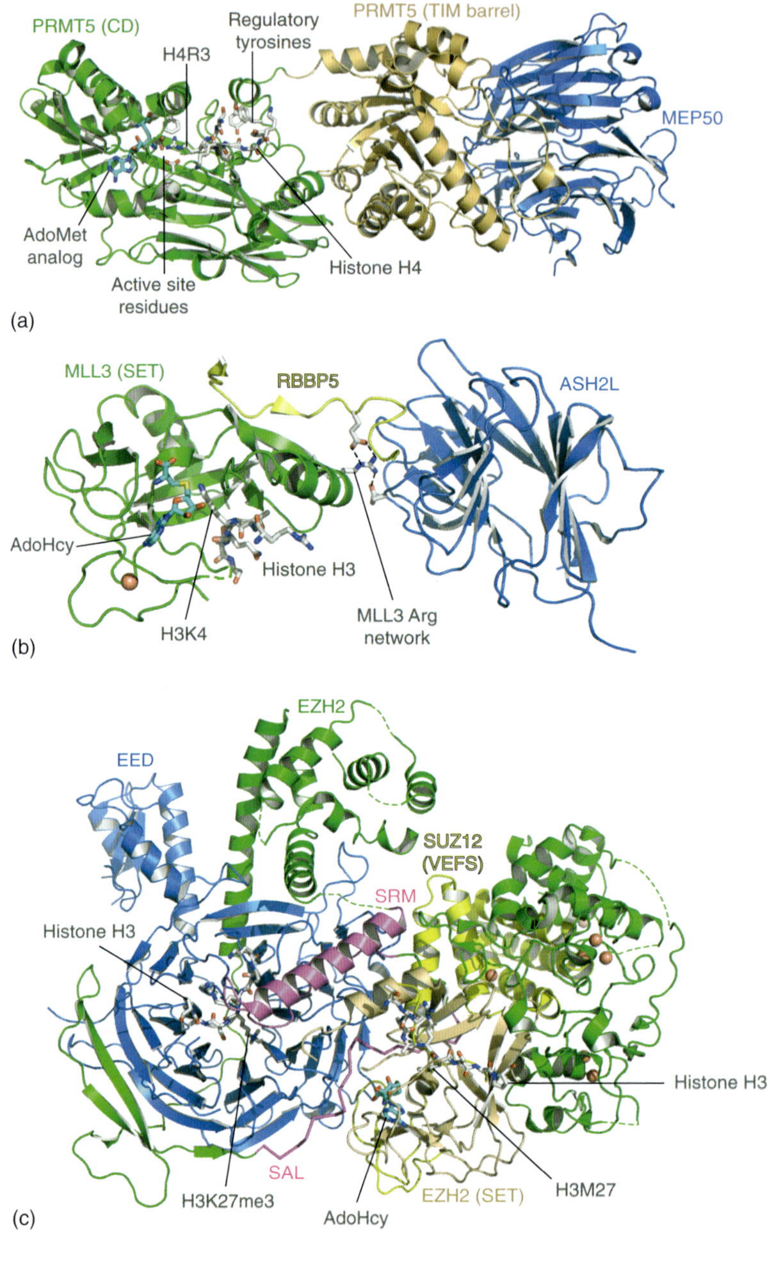

due to its implication in leukemia, but other MLL proteins have also been shown to be involved in cancers [14]. Proteins of the MLL family are part of large complexes that all share the ASH2L and RBBP5 subunits. These two subunits are sufficient to stimulate the methyltransferase activity of the MLL family members that otherwise display poor activity. Only MLL1 requires a third partner, WDR5, for full activation.

The molecular basis for MLL protein activity stimulation has long remained poorly understood. The recent crystallographic structure at 2.4 Å resolution of the human complex formed between the SET methyltransferase domain of MLL3, the C-terminal domain of ASH2L, and a long peptide of RBBP5 in the presence of S-adenosyl-L-homocysteine (AdoHcy; product of the demethylation reaction of the S-adenosyl-L-methionine (SAM) MLL3 cofactor) and of a H3 peptide substrate has shed light on this issue (Figure 2.2b) [15].

In this structure, the RBBP5 peptide is shown forming a link between the MLL3 SET domain and the ASH2L C-terminal domain. Specifically, RBBP5 N-terminus interacts with MLL3, whereas RBBP5 C-terminus binds to ASH2L. Importantly, all three proteins interact at one precise location with an arginine residue of MLL3 binding to different residues from ASH2L and RBBP5. Although this interface is limited, it appears essential for the stability of the complex and its conformation and for the stimulation of the methyltransferase activity. The stability of the MLL3–ASH2L–RBBP5 complex is further reinforced by residues neighboring the MLL3 arginine. These neighboring residues, in contrast to the arginine, are not conserved in MLL1. This potentially explains the requirement for WDR5 for the stabilization and the full stimulation of the MLL1–ASH2L–RBBP5 complex activity [15].

Surprisingly, the structure of MLL3 SET domain alone is not much different from the one of MLL3 bound to ASH2L–RBBP5, raising the question of the requirement of the ASH2L–RBBP5 complex for MLL3 activity stimulation. Binding of a H3 peptide to this latter complex only induces local conformational changes, but these changes cannot explain the poor activity of free MLL3. Measurements of the structural dynamics of MLL3 alone and in complex with ASH2L–RBBP5 by NMR and molecular simulation techniques suggest that some subdomains of MLL3 are intrinsically dynamic and that the binding of ASH2L–RBBP5 confers the sufficient stability to the SET domain to be able to bind stably its histone H3 target and to perform its methyltransferase activity.

Probably, one of the most awaited achievements of the last years has been the structural characterization of the polycomb repressive complex 2 (PRC2) [16]. PRC2 trimethylates histone H3 at lysine 27, and H3K27me3 is a major epigenetic mark of facultative heterochromatin that is associated with gene silencing [17]. Perturbation of PRC2 activity has been linked to multiple diseases, notably cancers, which explains that this complex currently represents a major target in epi-drug development, epi-drugs targeting PRC2 being currently in preclinical and clinical trials [18].

PRC2 is composed of a core complex formed by subunits EZH2 (catalytic subunit), EED (embryonic ectoderm development), SUZ12 (suppressor of zeste 12), and RBAP46/48 [17b]. Specifically, a sub-complex formed by EZH2, EED, and SUZ12 is sufficient for activity, EZH2 being inactive by itself. Interestingly, binding of H3K27me3 to PRC2 has been shown to allosterically stimulate the activity of the complex, whereas a H3K27M mutation yields to the inhibition of PRC2 [19]. In addition, PRC2 is composed of several facultative subunits that are further responsible for the modulation of the activity of this complex [17a, 17b].

Our understanding in molecular terms of the function of PRC2 has long remained obscure. The 1.9 Å crystal structure of the WD40 domain of human

EED in complex with a H3K27me3 peptide shed the first light on PRC2 structure/function relationships [19]. Specifically, the trimethylated lysine is recognized within a central aromatic cage at the surface of EED WD40 domain. Another essential information came from the 2.0 Å crystal structures of the human EZH2 C-terminal SET domain that is responsible for the H3K27 methylation activity of EZH2 [20]. This structure revealed that the SET domain alone adopts an inactive conformation, with the substrate and SAM cofactor binding sites being occluded by different parts of the SET domain itself.

A first step in our structural understanding of the PRC2 complex initially came from the 21 Å EM structure of a human EZH2/EED/SUZ12/RBAP48/AEBP2 sub-complex [21]. This low resolution structure, complemented by labeling, cross-linking, and mass spectrometry techniques, enabled the location of the different proteins composing the sub-complex as well as specific domains within the EM map. This structure revealed a four lobe (a–d) organization, with SUZ12 and AEBP2 forming a physical link between lobes a, b and lobes c, d. Interestingly, EZH2, EED, and a SUZ12 C-terminal domain (VEFS), which are sufficient for activity, appear to form the major components of the A, B lobes.

This initial structural information was complemented and refined by the 2.3–2.9 Å resolution crystal structures of the EZH2/EED/SUZ12 VEFS sub-complex from human and the thermophilic yeast *Chaetomium thermophilum* (Figure 2.2c) [16]. The organization of the complex appears very similar to the one derived from the EM structure. The high resolution data however provides unprecedented detailed molecular information on the PRC2 core sub-complex assembly and its function. The complex can be divided in two lobes. The regulatory lobe is composed of EED and the N-terminal region of EZH2 (corresponding to lobe A in the EM structure). In this lobe, EZH2 is encircling EED through multiple domains, ensuring a very tight interaction between the two subunits. The second catalytic lobe (lobe B in the EM structure) is composed primarily of the C-terminal region of EZH2, including its SET methylation domain. The SUZ12 VEFS domain is found at the interface between these two moieties, interacting with and stabilizing the interaction between both lobes.

Importantly, in the complex the EZH2 SET domain adopts an active conformation, the AdoHcy product being bound at the SAM binding site and the peptide binding groove being correctly formed. Actually, a H3M27 mutant (where H3K27 lysine is replaced by a methionine) peptide, which is absolutely required for crystallization and is known to be involved in tumorigenesis, is found bound to the SET domain where H3M27 occupies the H3K27 binding pocket. A surprising aspect of the EZH2 CD is that it is bipartite, the C-terminal SET domain being complemented by a functionally essential SET activation loop (SAL) located in the N-terminal region of EZH2 (regulatory lobe).

The EZH2/EED/SUZ12 VEFS sub-complex could be crystallized in the absence (basal state) and in the presence (activated state) of a H3K27me3 peptide. In the activated state, the peptide is found bound to EED as observed in the EED–H3K27me3 structure (Figure 2.2c). Strikingly, upon H3K27me3 binding to EED, the N-terminal EZH2 region, termed stimulation-responsive motif (SRM), which directly follows the SAL, becomes ordered and visible in the electron density, interacting with the H3K27me3 peptide. The SRM also makes direct interactions with the SET domain, suggesting an explanation for the allosteric activation of PRC2 by the H3K27me3 epigenetic mark.

2.5 Histone Lysine Ubiquitinylases: The PRC1 Complex

Polycomb repressive complex 1 (PRC1) is also involved in transcriptional repression and, like PCR2, has been linked with various cancers [18b]. However, unlike PRC2 that is a lysine methyltransferase, PRC1 mono-ubiquitinylates histone H2A at lysine 119. Ubiquitinylation requires the activity of three enzymes called E1, E2, and E3. First, an E1 enzyme transfers ubiquitin to an E2-conjugating enzyme. Then, the E2 enzyme transfers the ubiquitin to a lysine side chain of a target protein that is specifically recognized by the E3 enzyme [22].

PRC1 acts as a complex that is minimally composed of three proteins. These three proteins have various homologues yielding different PRC1 complexes that have distinct gene targets [17b–d]. Two of these proteins are forming the E3 enzyme, whereas the third one is the E2 enzyme. Various unrelated E2/E3 structures have shown how E2 and E3 enzymes are acting in concert, but how the PRC1 specifically recognizes its nucleosome target has long remained poorly understood.

The 3.3 Å resolution crystallographic structure of a human PRC1 (RING1B–BMI1–UBCH5C)–nucleosome complex has enabled the deciphering in molecular terms of this recognition (Figure 2.3a) [23]. RING1B and BMI1 are the two proteins that compose the E3 enzyme that is supposed to recognize specifically the nucleosome. Accordingly, RING1B makes the most extensive interactions with the nucleosome from the three PRC1 proteins. Specifically, it interacts primarily with the so-called “acidic patch” of the nucleosome, a set of acidic residues of H2A and H2B that are exposed at the surface of the histone octamer [24]. Actually, many different nuclear effectors that interact with histone pairs and the nucleosome have been shown to target the acidic patch through an “arginine-anchor” mechanism [25]. RING1B forms a particularly large number of interactions with the acidic patch, not only with arginines but also with lysine residues.

BMI1 is also interacting with the histone octamer, albeit mostly with the H3 and H4 histones. This interaction is less extensive than in the case of RING1B but still contributes to the recognition of the nucleosome by PRC1. E2 enzyme UBCH5C interacts mostly with RING1B that is centrally positioned in the RING1B–BMI1–UBCH5C complex and that anchors it on the H2A–H2B acidic patch. This positions UBCH5C active site directly over the H2A C-terminal tail that harbors K119 (Figure 2.3a). Interestingly, UBCH5C does not interact with any histone. Rather, it makes contacts with the entry/exit and dyad nucleosomal DNA. Thus, through its interactions with all histones and with DNA, PRC1 ensures that it is engaged with the nucleosome.

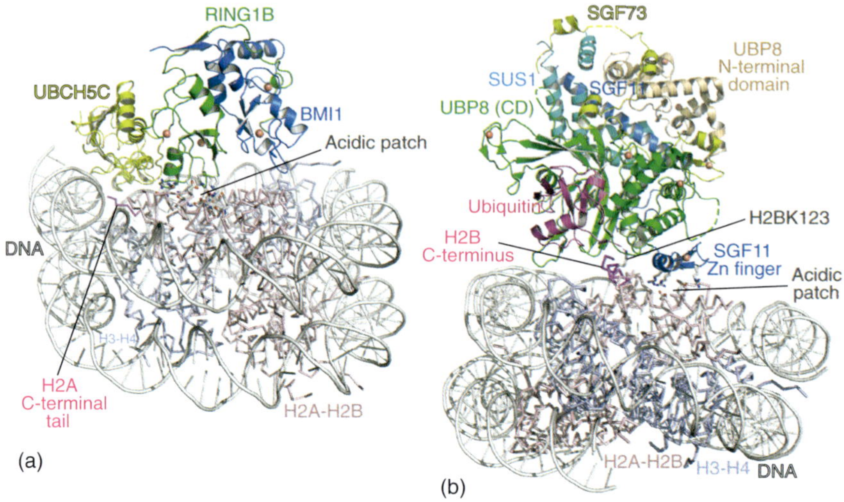

Figure 2.3 Structures of (de)ubiquitinylases. Ribbon representation of (de)ubiquitinylases. (a) Structure of the polycomb repressive complex 1 (PRC1) (RING1B–BMI1–UBCH5C) ubiquitinylase bound to the nucleosome. RING1B is colored green, BMI1 blue, and UBCH5C yellow. The DNA is colored white and the H3–H4 and H2A–H2B pairs are colored light blue and light pink, respectively. The nucleosome acidic patch residues as well as RING1B positively charged residues that form a strong interaction network are shown as sticks with gray carbons. The H2A C-terminal tail targeted by UBCH5C is shown as magenta ribbon. (b) Structure of the deubiquitination module (DUBm) (UBP8–SGF11–SUS1–SGF73) of the SAGA co-activator bound to the nucleosome. UBP8 N-terminal domain is colored wheat; its catalytic domain (CD) is colored green. SGF11 is colored blue for its N-terminal domain and dark blue for its zinc finger C-terminal domain (Zn finger). SUS1 is colored cyan and SGF73 N-terminal region is colored yellow. SGF11 positively charged residues and nucleosome active site residues are shown as stick with gray carbons. H2B C-terminal helix is shown in magenta as well as the ubiquitin molecule bound to the H2BK123 residue (shown as sticks with gray carbons). The nucleosome is displayed as in (a).

2.6 Histone Lysine Deubiquitinylases: The SAGA Deubiquitination Module

SAGA (Spt-Ada-GCN5-acetyl transferase) is a 1.8 MDa transcriptional co-activator complex that acts during RNA polymerase II transcription activation and that also couples transcription elongation with RNA export [26]. SAGA bears two enzymatic activities: a histone acetyltransferase (HAT) activity, through its GCN5 subunit, and a histone deubiquitination activity, through its yeast UBP8/human USP22 subunit. Both enzymes require additional subunits for functional activity on their cognate substrate, the nucleosome.

SAGA deubiquitination module (DUBm) is composed of four proteins that are all required for activity: yeast UBP8, SGF11, SGF73, and SUS1, which deubiquitinylate yeast H2B K123, and human USP22, ATXN7L3, ATXN7, and ENY2, which deubiquitinylate human H2B K120 [26]. Only the first hundred residues of SGF73/ATXN7 are required for complex formation and the deubiquitination activity, its C-terminal region being involved in the attachment of the DUBm to the rest of the SAGA complex. Importantly, polyglutamine extensions in human ATXN7 N-terminal domain are responsible for the SCA7 neurodegenerative disease [27].

The requirement for the four proteins and their role in the deubiquitination activity have remained poorly understood. Initial complex reconstitution experiments with the yeast DUBm have shown that the C-terminal zinc finger of SGF11 and the C-terminal deubiquitination domain of UBP8 are not required for complex assembly [28]. The same study showed that SGF73 not only requires the three other subunits to be incorporated within the complex but also stabilizes UBP8 in this complex.

The 1.9–2.7 Å crystal structures of the yeast SAGA DUBm in the absence and in the presence of ubiquitin have revealed the intricacy of the DUBm formation and the role of all four subunits in forming a functionally active deubiquitination complex (Figure 2.3b) [29]. The SAGA DUBm is composed of an assembly lobe and a catalytic lobe. The assembly lobe is formed by a long N-terminal α-helix of SGF11 around which SUS1 is wrapped. The N-terminal zinc-containing domain of UBP8 docks itself onto this SGF11–SUS1 complex, which then serves as a platform for binding for the first half of SGF73 N-terminal domain.

The UBP8 C-terminal deubiquitination CD makes only little interactions with the assembly lobe, being separated from its N-terminal domain by a linker. In fact, it is the second half of the SGF73 N-terminal domain that forms the interface between the assembly lobe and the UBP8 deubiquitination CD, locking this domain into a position that makes the DUBm catalytically active (Figure 2.3b). Importantly, polyglutamine extensions in ATXN7, the human homologue of yeast SGF73, which are responsible of the SCA7 disease, are found in the first half of the ATXN7 N-terminal domain that is expected to be part of the assembly lobe [27, 29b, 30]. Although the mechanism by which the disease occurs remains obscure, this highlights the importance of SGF73–ATXN7 N-terminal part in the DUBm assembly and function.

Finally, the SGF11 C-terminal zinc finger, which is not required for DUBm assembly, is found bound to the deubiquitination domain of UBP8, close to the active site, and is separated from SGF11 N-terminal α-helix by a long linker. This zinc finger has initially been thought to recognize DNA in the nucleosome, but the 3.9 Å crystal structure of the DUBm in complex with an ubiquitinated nucleosome revealed that in fact SGF11 zinc finger recognizes the acidic patch of the nucleosome with several arginine residues using an arginine-anchor mechanism (Figure 2.3b) [25b].

The DUBm does not make extensive interactions with the rest of the nucleosome. Only a small part of the UBP8 deubiquitination domain appears to contact H2B and the DNA (Figure 2.3b). Interestingly, the ubiquitin appears not to make any contacts with the nucleosome and binds to the UBP8 deubiquitination domain in a canonical manner. Comparison with the nucleosome-free DUBm structure reveals some structural rearrangements upon ubiquitin binding, but the DUBm by itself already seems to be in an active conformation. Since UBP8 alone is very poorly active, it is the formation of the DUBm module that is sufficient to lock UBP8 into this active conformation.

The DUBm is part of the larger SAGA complex that also bears a HAT activity. The low resolution structure of a SAGA-nucleosome complex by EM suggests that the same nucleosome could be engaged by the two enzymatic activities of SAGA at the same time [31]. Structure determination of the GCN5–ADA2–ADA3–SGF29 HAT module in complex with the nucleosome is awaited to understand how SAGA HAT activity is carried out. Actually, a structure combining both HAT and DUB modules bound to a mononucleosome would give unprecedented information on how two different epigenetic enzymatic functions can collaborate.

2.7 Histone Acetyltransferases: The MSL1 and NUA4 Complexes

Lysine acetylation is one of the best characterized covalent modifications of histone tails [32]. HAT enzymes have been classified into five major subfamilies based on sequence homology and substrate acetylation properties: HAT1, GCN5/PCAF, MYST, P300/CBP, and RTT109 [33]. All HAT enzymes share a structurally conserved acetyl-CoA binding domain. However, activity of most HATs is regulated by tethering into multisubunit complexes. The enzymatic specificities of different HAT complexes are a matter of intensive research as a majority of isolated HAT CDs exhibit no or little substrate selectivity.

MOF (males absent on the first) is a HAT of the MYST family that has been shown to be an essential component of the *Drosophila* dosage compensation male-specific lethal (MSL) complex, which also harbors proteins MSL1, MSL2, MSL3, MLE, and noncoding RNAs roX1 and roX2 [34]. In human, four orthologues of the *Drosophila* proteins were identified: MOF, MSL1, MSL2, and MSL3. Specifically, while isolated MOF HAT domain is able to acetylate free histones H3 and H4, the entire multisubunit complex can only acetylate histone H4 in a nucleosomal context. Both MSL1 and MSL3 are important for stimulating MOF activity.

MSL1 is predicted to be mostly unstructured but has a scaffolding role within the MSL complex, binding to MOF, MSL2, and MSL3. Specifically, the C-terminal region of MSL1 (PEHE domain) binds consecutively MOF and MSL3, whereas its central region binds to MSL2 N-terminal ring domain. The crystal structures of mammalian MOF HAT domain in complex with MSL1 PEHE N-terminus, of MSL1 PEHE C-terminus in complex with the MRG domain of MSL3, and of MSL1 central domain in complex with MSL2 ring domain have provided first insights into how the MSL complex assembles [35].

Specifically, the 2.0–2.7 Å crystal structures of MOF (HAT)–MSL1 (PELE-N) complex show that MSL1 binding does not induce major conformational changes in MOF, raising the question of the role of MSL1 and MSL3 in stimulating MOF activity (Figure 2.4a). Yet, both published structures do not agree on the location of the very N-terminus of the MSL1 construct used. In one structure, this region is close to the active site, where it could influence substrate binding. In addition, MSL3 binding site on MSL1 is close to the MOF binding site and MSL3 could also influence substrate recognition.

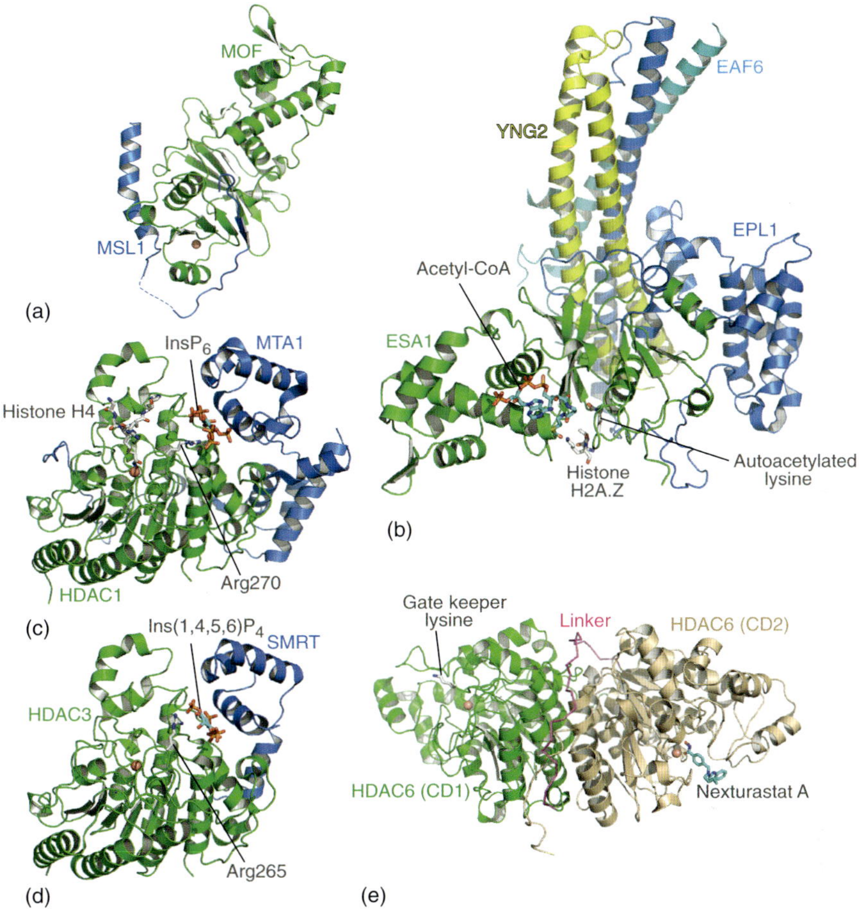

Figure 2.4 Structures of histone acetyltransferases (HATs) and histone deacetylases (HDACs). Ribbon representation of HATs and HDACs. (a). Structure of MOF HAT domain (green) bound to MSL1 (blue). (b). Structure of the NUA4 HAT complex (ESA1–EPL1–YNG2–EAF6). ESA1 is colored green, EPL1 blue, YNG2 yellow, and EAF6 cyan. Acetyl-CoA is shown as sticks with cyan carbons and histone H2A.Z peptide as well as an autoacetylated lysine as sticks with gray carbons. (c, d) Structures of HDAC1 (c) and HDAC3 (d) in complex with corepressors MTA1 and SMRT, respectively, and with inositol phosphate molecules. HDACs are colored green and corepressors blue, with inositol phosphate molecules shown as sticks with cyan carbons. Histone H4 peptide analogue and important arginine residues are shown as sticks with gray carbons. (e) Structure of HDAC6 tandem CD1–CD2 HDAC domains. CD1 is shown in green and CD2 in wheat. The linker connecting both HDAC domains is shown as magenta ribbon. CD1 gatekeeper lysine and HDAC6 selective inhibitor Nexturastat A molecule bound to CD2 are shown as sticks with gray and cyan carbons, respectively.

Another structural analysis, on the NuA4 acetyltransferase complex, has brought further light on how HAT specificity toward a particular histone tail in a nucleosomal context can be achieved [36]. NUA4 is a large HAT complex comprising 13 subunits that preferentially acetylates histone H4 in the nucleosomes [37]. The NUA4 HAT module consists of four subunits: in yeast the catalytic subunit ESA1, EPL1, YNG2, and EAF6 and in mammals TIP60, EPC1, ING3, and EAF6. This core complex can recapitulate most of the enzymatic activities of the holoenzyme.

The 2.7–3.2 Å crystal structures of the yeast NUA4 HAT core module (ESA1 HAT domain, EPL1 N-terminal and central domains, YNG2 N-terminal domain, and EAF6) alone and in complex with acetyl-CoA or a H2A.Z histone N-terminal tail reveal an organization in two lobes that are interacting weakly and are relatively independent from each other (Figure 2.4b) [36]. EPL1 N-terminal region wraps and interacts extensively with ESA1 HAT domain, forming the catalytic lobe. The central domain of EPL1 forms a long coiled coil structure that bundles with the helices formed by the N-terminal domain of YNG2 and by EAF6 to form the assembly lobe.

Importantly, binding of EPL1 to ESA1 HAT induces conformational changes in the active site of the CD by reorganizing some important loops. The presence of an autoacetylated lysine in ESA1 is observed that stabilizes the new conformation of an active site loop. The structure of the complex in the presence of a histone tail peptide shows that these conformational changes observed are required for the accommodation of the peptide in a productive way. Interestingly, the complex by itself reveals very little binding specificity toward the peptide, raising the question on how specificity is achieved.

The 8 Å resolution EM structures of the NUA4 HAT core complex bound to a nucleosome are providing answers to this question [36, 38]. Specifically, the crystal structure of the NUA4 core complex and of the nucleosome can be unambiguously fitted in the electron density, revealing that the catalytic HAT subunit is oriented in close vicinity to the histone H4 tail. In addition, ESA1 N-terminal Tudor domain, which was included in the EM analysis, is also found bound to the nucleosome. The complex also appears to contact the acidic patch and DNA elements. This is the first structural evidence on how an otherwise poorly selective HAT can be specifically positioned to catalyze acetylation of its cognate histone substrate.

2.8 Histone Deacetylases: HDAC1–MTA1 and HDAC3–SMRT Complexes and HDAC6

The family of histone deacetylases (HDACs) has been divided into four classes depending on their folds and their sequence similarities. Class I, II (IIa and IIb), and IV deacetylases (HDACs) adopt an arginase–deacetylase α/β-fold and rely on a zinc ion for activity [39]. Class III deacetylases are referred as sirtuins and adopt a Rossmann fold, relying on NAD^+ for activity [40].

Zinc-dependent HDAC enzymes catalyze the removal of the acetyl group from the Nε atom of lysines and are the most pharmacologically investigated epigenetic targets so far. Accordingly, most currently approved epigenetic drugs (Vorinostat (SAHA), Romidepsin, Belinostat, Panobinostat, and Tucidinostat (Chidamide)) are inhibitors of HDACs [41]. The HDAC inhibitors have been proven to be effective anticancer agents, mostly in hematological malignancies, but recent discoveries suggested that they can also be therapeutically beneficial in the treatments of multiple other human diseases including neurodegenerative and immune disorders as well as viral and parasitic infections [41b, 42].

All HDAC inhibitors approved so far are targeting the active site of these enzymes, binding to the catalytic zinc ion, and thus preventing recognition and deacetylation of cognate cellular substrates. The major issue with clinical usage of these approved HDAC inhibitors is the fact that they have no or little selectivity and target the structurally similar but functionally different 11 human HDAC isozymes (HDAC1–11). Therefore, the treatments with currently approved HDAC inhibitors are often accompanied by serious undesirable side effects, which hamper their clinical usage. Several isozyme-selective inhibitors have been developed [41b], but the approved inhibitors mostly show limited selectivity. The hydroxamates target mostly class I and HDAC6 (class IIb), while Tucidinostat is selective for class I.

Selective targeting of HDAC–protein interactions is emerging as an alternative for inhibition studies [41a, 43]. Class I HDAC1, HDAC2, and HDAC3 are typically recruited into multisubunit macromolecular complexes. Importantly, the recruitment of these HDACs to corepressor complexes triggers maximal HDAC activity. The structures of HDAC1 and HDAC3 in complex with cognate corepressors metastasis-associated protein 1 (MTA1) and silencing mediator for retinoid and thyroid hormone receptor (SMRT) provide the molecular explanation for this enhanced deacetylase activity.

In the 3.0, 3.3, and 2.0 crystal structures of HDAC1–MTA1, HDAC1–MTA1–H4 peptide, and HDAC3–SMRT, the corepressor domains wrap around HDAC CDs, and, in the case of HDAC1, this domain also mediates dimerization of the complex (Figure 2.4c,d) [44]. Strikingly, the structures of HDAC1–MTA1 and HDAC3–SMRT complexes reveal an unexpected regulation mechanism of HDAC activity mediated by inositol phosphates. In HDAC3–SMRT complex, endogenous Ins(1,4,5,6)P_4 is found in a basic pocket formed between the HDAC and the corepressor domain, and that is close to the active site pocket. The 3.3 Å crystal structure of HDAC1–MTA1 with exogenous $InsP_6$ also revealed an inositol molecule bound in the basic pocket, but it shows certain structural differences when compared with HDAC3–SMRT complex [44c].

The binding of an inhibitor in the HDAC active site potentiates inositol polyphosphate binding in the basic pocket, which indicates mutual communication between the active site pocket and the inositol binding pocket. This cross-talk is likely mediated by HDAC1 residue Arg270 and HDAC3 Arg265 that interconnect the active site pocket with the inositol binding pocket. Interestingly, while mutation of HDAC3 Arg265 led to the complete loss of deacetylase activity, the mutation of HDAC1 Arg270 is less functionally important. Yet, HDAC1 kinetics is regulated upon inositol polyphosphate binding, leading to the hypothesis of HDAC1 activation by entropic allostery [44c].

In contrast to other HDACs, HDAC6 is unique as it harbors two functional CDs (CD1 and CD2). The crystal structures of human HDAC6 CD2 and zebrafish HDAC6 CD1, CD2, and tandem CD1–CD2 brought important structural and mechanistic insights into HDAC6 catalytic mechanism and its inhibitions by small-molecule compounds [45]. The 2.9 Å crystal structure of zebrafish HDAC6 tandem CD1–CD2 reveals that the tandem forms together with the interdomain linker an ellipsoid-shaped complex with pseudo-twofold symmetry (Figure 2.4e) [45b]. The two CDs interact strongly with each other, and the linker domain connecting CD1 and CD2 also functions as a domain–domain glue to further stabilize the CD1–CD2 complex. Importantly, biochemical experiments show that maximal HDAC6 activity is achieved when both CD1 and CD2 are physically interacting with each other.

Both HDAC6 CD1 and CD2 are structurally very similar. However, biochemical data shows that the CD2 domain exhibits relatively broad substrate specificity, while CD1 appears specific for substrates bearing a carboxy-terminal acetyl-lysine residue [45a]. This selectivity seems achieved by the fact that CD1 active site pocket is more constricted due to the presence of a lysine residue (a leucine in CD2) that protrudes into the catalytic pocket. This lysine plays the role of a gatekeeper that confers CD1 a specificity toward carboxy-terminal acetyl-lysine substrates by hydrogen bonding with the α-carboxylate group of the acetyl-lysine. In addition, the hydroxyl group of a CD1 tyrosine (a phenylalanine in CD2 and in other HDACs) could also contributes to the observed specificity via an additional hydrogen bonding with the carboxy-terminal acetyl-lysine residue.

The structure of HDAC6 was solved in complex with a HDAC6-selective inhibitor, Nexturastat A, providing specific information on HDAC selective inhibition. Other works on the HDAC8 isozyme have also contributed to address this issue that is essential for developing the next generation of HDAC epi-drugs [42a, 42c, 46].

2.9 Histone Variants and Histone Chaperones: A Complex and Modular Interplay

The sequence differences between histone variants and canonical histones can concern a few residues to full structural domains [3d]. Besides their common DNA packaging role, canonical and variant histones can have very different functions. Specifically, whereas canonical histones are deposited onto the chromatin during replication, histone variants are deposited throughout the cell cycle, where they play specific functional roles at precise genomic loci. Due to their broad implication in many nuclear processes, histone variants are involved in many diseases, notably in cancers.

This importance of histone variants explains that the transfer of histones between the different cellular compartments and their deposition onto and removal from the chromatin is a highly regulated process. This requires the action of a large number of dedicated histone chaperones that are ensuring that the various histones are present at their correct genomic loci [3e]. Depending on their role, some of these chaperones can recognize any histone pair, whereas others bind only to histone pairs of the H3–H4 or H2A–H2B family. Other chaperones are binding to very specific variant pairs.

Importantly, not only the sequence differences but also the epigenetic marks borne by histones can affect their interaction with histone chaperones. In addition, histone chaperones are often members of large epigenetic complexes, where they collaborate with different epigenetic functions. In the last decade, a large amount of structural data has been gathered on the recognition of histone pairs, canonical and variants, by histone chaperones [3e]. A few of them are described here that highlight the interplay between histones, histone chaperones, and other epigenetic effectors.

During replication, many different effectors are required to disassemble and then reassemble chromatin. Maintenance of the epigenetic state encoded by specific epigenetic marks is essential during this process. This is rendered complicated by the fact that both old histones and newly synthetized histones are used for reassembly. Specifically, new histones need to be identified and modified according to the previous epigenetic state encoded by the old histones. The histone chaperone anti-silencing function 1 (ASF1) participates to this process by binding H3–H4 pairs. ASF1 was the first chaperone whose structure in complex with a histone pair was solved and that was shown to interact with the H3–H4 pair in a way that prevents this pair to homodimerize and to interact with H2A–H2B [47].

The replicative helicase plays an important role in the replication process. Its MCM2 subunit has been shown to bind H3–H4, and the structure of an N-terminal fragment of MCM2 in complex with H3–H4 has revealed that MCM2 binds to a H3–H4 dimer, almost completely shielding the histone tetramer DNA binding surface [48]. Interestingly, ASF1 and MCM2 were shown to act in concert, and the 2.3 and 3.5 Å crystal structures of an ASF1–MCM2–H3–H4 complex reveals that both chaperones can bind simultaneously to the H3–H4 pair (Figure 2.5a) [48a, 49]. In this complex, ASF1 breaks the dimerization interface between the H3 and H4 pairs, and the complex displays a 1 : 1 : 1 : 1 stoichiometry. Yet, the two chaperones interact minimally with each other, showing the modularity of the chaperone–histone pair interactions.

Prior and after replication, it is essential that newly synthetized histone pairs are recognized for further processing. The TONSL–MMS22L complex has been shown to interact and cooperate with the free and chromatin-bound H3–H4 pair, and with ASF1 and MCM2. The 2.4 Å crystal structure of TONSL ankyrin repeat domain in complex with H3–H4 and MCM2 reveals that TONSL binds minimally to the globular domain of the histones but interacts extensively with H4 N-terminal tail (Figure 2.5b) [50]. Both TONSL and MCM2 chaperones do not interact directly, and modeling shows that ASF1 could also bind to the TONSL–MCM2–H3–H4 complex, reinforcing the concept of modularity. Strikingly, TONSL–H4 interactions are only possible when H4K20 is unmethylated, and this epigenetic mark appears specific to newly synthesized H4 histone. Modeling shows that TONSL, which has been shown to be associated with chromatin, could remain bound to the H4 tail in a nucleosomal context, thus marking nucleosomes that have incorporated newly synthesized histones.

None of the ASF1, MCM2, and TONSL chaperones can distinguish between the different members of the H3 family. Yet, many chaperones bear this capacity in order to deposit histone variants at specific loci. One specific H3 variant is CENPA that is found exclusively at centromeres. The human histone chaperone HJURP has been shown to recognize specifically the CENPA–H4 pair, and the 2.3–2.6 Å crystal structures of the human and yeast complexes show how HJURP binds to CENPA–H4, preventing CENPA–H4 dimerization and shielding part of the DNA binding surface of this pair (Figure 2.5c) [51]. The determinants for the specific recognition of CENPA over H3 were found to be minimal.

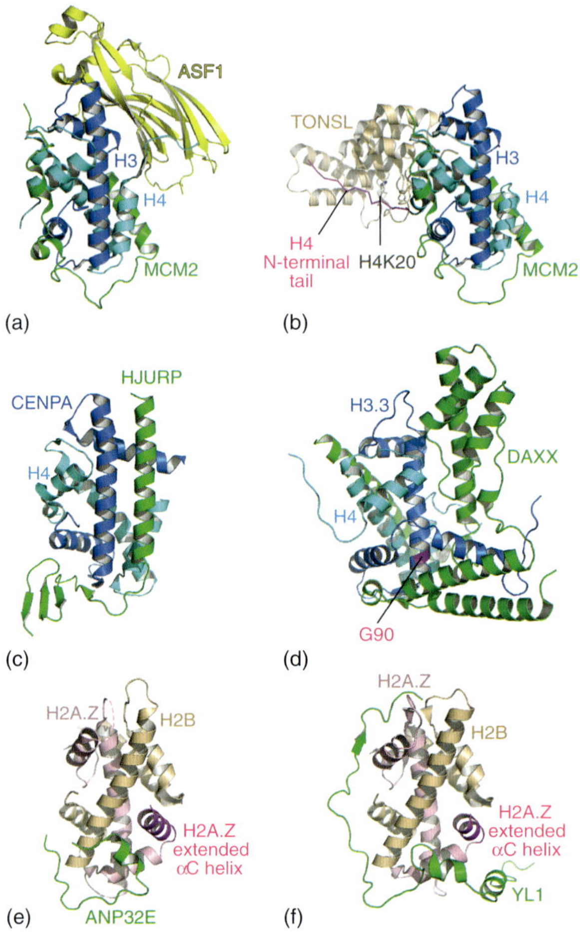

Figure 2.5 Structures of histone chaperone–histone pair complexes. Ribbon representation of histone chaperone–histone pair complexes. H3 (and related H3 histone variants) and H4 are colored blue and cyan, respectively. H2A (and related H2A histone variants) and H2B are colored light pink and wheat, respectively. (a) Structure of the MCM2–ASF1–H3–H4 complex. MCM2 is colored green and ASF1 yellow. (b) Structure of the MCM2–TONSL–H3–H4 complex. MCM2 is colored green and TONSL wheat. H4 N-terminal tail is shown as magenta ribbon and H4K20 as sticks with gray carbons. (c) Structure of HJURP in complex with the CENPA-H4 variant pair. HJURP is colored green. (d) Structure of the DAXX–H3.3–H4 complex. DAXX is colored green. G90 that conveys H3.3-specific recognition is colored magenta. (e, f) Structures of H2A.Z–H2B variant pair in complex with ANP32E removal (e) and YL1 deposition (f) histone chaperones. Both chaperones are colored green. The extended αC helix from H2A.Z is colored magenta.

H3.3 is a variant that is deposited in gene bodies as well as in pericentric and telomeric heterochromatin. H3.3 varies by only five residues from the canonical H3.1 histone. Deposition at heterochromatin loci is carried out by the histone chaperone death domain-associated protein (DAXX). The 2.8 Å crystal structures of the DAXX–H3.3–H4 complex shows that DAXX interacts extensively with the H3.3–H4 pair (Figure 2.5d) [52]. Despite the large interface between DAXX and H3.3–H4, as for CENPA the specific recognition of H3.3 relies on very few amino acid changes, notably on one glycine residue (G90) that is a methionine in H3.1. Interestingly, the 2.2 Å crystal structure of another unrelated H3.3 histone chaperone, UBN1, in complex with ASF1 and H3.3–H4 reveals the same mechanism of specific H3.3 recognition by UBN1 through this glycine [53]. In addition, the DAXX–H3.3–H4 complex can be targeted by the viral protein BNRF1 from the Epstein–Barr virus to hijack the deposition machinery. BNRF1 interacts not only with DAXX but also with H3.3–H4 [54].

Not only H3 but also H2A and H2B have various members in their families and dedicated histone chaperones. New data has shed light on the recognition and processing of the H2A.Z variant, which acts in both transcription and DNA repair. Specifically, two large functionally homologous human ATP-dependent chromatin remodelers were shown to act on H2A.Z: P400/TIP60 and SRCAP [55]. The P400/TIP60 complex is involved not only in chromatin remodeling but also in acetylation through its TIP60 subunit that corresponds to yeast ESA1 from the NUA4 complex described above. The histone chaperone ANP32E has been shown to belong to P400/TIP60 and to evict H2A.Z from the chromatin. The 1.5–2.6 Å crystal structure of the ANP32E–H2A.Z–H2B complex shows a minimal interface between the chaperone and the histone pair (Figure 2.5e) [55, 56]. ANP32E binding to H2A.Z–H2B causes the doubling in size of H2A.Z αC helix, with direct implication for H2A.Z–H2B eviction from the nucleosome. This extension of H2A.Z αC helix is only enabled by the absence, compared with H2A, of a glycine residue in this helix. Strikingly, this glycine is also the only determinant for H2A.Z-specific recognition by ANP32E.

The second remodeling complex involved in H2A.Z biology is SRCAP. This complex has been shown to be involved in H2A.Z deposition, and the histone chaperone YL1 is involved in this mechanism. The 1.9–2.7 crystal structures of the YL1–H2A.Z–H2B complex reveal that this chaperone also doubles in size H2A.Z αC helix, but it also interacts more extensively with the histone pair, covering its entire DNA binding surface (Figure 2.5f) [25c, 25d]. Even if the H2A.Z αC helix glycine plays a role in the specific H2A.Z recognition by YL1, other determinants are required for full recognition, in contrast with what has been observed with ANP32E. In this case, the ATP-dependent remodeling activity of SRCAP is probably needed to prior evict the H2A–H2B pair from the nucleosome. For ANP32E, nucleosome remodeling and recognition of H2A.Z by a single determinant appear much more relevant for accessing H2A.Z αC helix buried at the heart of the nucleosome, prior to H2A.Z-H2B eviction through H2A.Z αC helix extension.

2.10 ATP-Dependent Remodelers: CHD1, ISWI, SNF2, and the SNF2-Nucleosome Complex

ATP-dependent chromatin remodelers are large multisubunit complexes that use the energy of ATP to partially or fully disassemble, slide, or change the conformation and composition of nucleosomes [3a, 57]. Their ATPase subunits all have a CD composed of two RecA-like (DExx and HELICc) subdomains. Based on the sequence similarity of their CDs and of the presence or absence of additional auxiliary domains, the remodelers have been classified in four large families: SWI/SNF, IWSI, CHD, and INO80.

The increasing reports of the implication of remodelers in many different diseases make these epigenetic effectors important targets for therapeutic developments. Specifically, the remodelers of the SWI/SNF family are found mutated in 20% of all human tumors [3b]. Yet, inhibition of the ATPase CD might cause strong selectivity problems. It is therefore essential to understand in molecular terms not only how these CDs cooperate with the auxiliary domains of the catalytic subunits and with their partner subunits in their respective complexes but also how they recognize and act on their nucleosomal substrates. The intrinsic flexibility and the large size of the remodelers have slowed down the structural characterization of these epigenetic machines. Yet, our understanding on how remodelers are acting and are being regulated has made important progress.

The CHD family is characterized by a tandem of chromodomains in the N-terminal region of the ATPase subunit. The 3.7 Å crystal structure of CHD1 core encompassing both chromodomains and the ATPase CD reveals that these different domains are interacting extensively, the two chromodomains bridging both RecA-like lobes of the ATPase (Figure 2.6a) [58]. The observed positioning of lobe 1 compared with lobe 2 implies that CHD1 ATPase domain is in an inactive conformation. In addition, part of the linker bridging the two chromodomains and that forms two long antiparallel α-helices is found binding to lobe 2 at a position where DNA is supposed to bind. Binding of methylated proteins to CHD1 chromodomains, like in the case of KDM1A [59], should possibly relieve inhibition. Interestingly, the histone H4 N-terminal tail has also been shown to stimulate CHD1 activity even in absence of the chromodomains, demonstrating of an intricate regulatory mechanism of CHD1 activity.

In the ISWI family, two domains (AutoN and NegC) flanking the ATPase CD have been shown to regulate the ATPase activity. The 2.4 Å crystal structure of the core IWSI ATPase subunit encompassing the AutoN, ATPase, and NegC domains shows, as for CHD1, a nonproductive positioning of the two RecA-like lobes (Figure 2.6b) [60]. Here, the AutoN domain is shown interacting with both lobes of the ATPase domain, possibly stabilizing, as the two chromodomains of CHD1, the nonproductive conformation of the enzyme. Interestingly, the N-terminal tail of H4 is also stimulating the activity of ISWI. The 3.0 Å crystal structure of an N-terminal H4 peptide bound to ISWI lobe 2 shows that H4 binds to one of the anchoring surface of AutoN on lobe 2, suggesting an explanation to the stimulating activity of H4 by preventing the interaction between the ISWI AutoN and CDs [60].

In contrast, the role of ISWI NegC domain is less clear from the structural analysis since this domain projects out of the structure. In the IWSI family, the NegC domain is followed by a Hand-Slant-Slide (HSS) domain that contacts naked DNA. The 3.2–3.6 Å crystal structure of the HSS domain from the remodeler subunit ISW1a in complex with another subunit of this complex, LOC3, in the absence and in the presence of DNA showed that the HSS–LOC3 complex is able to bind two DNA molecules [61]. The low resolution EM structure

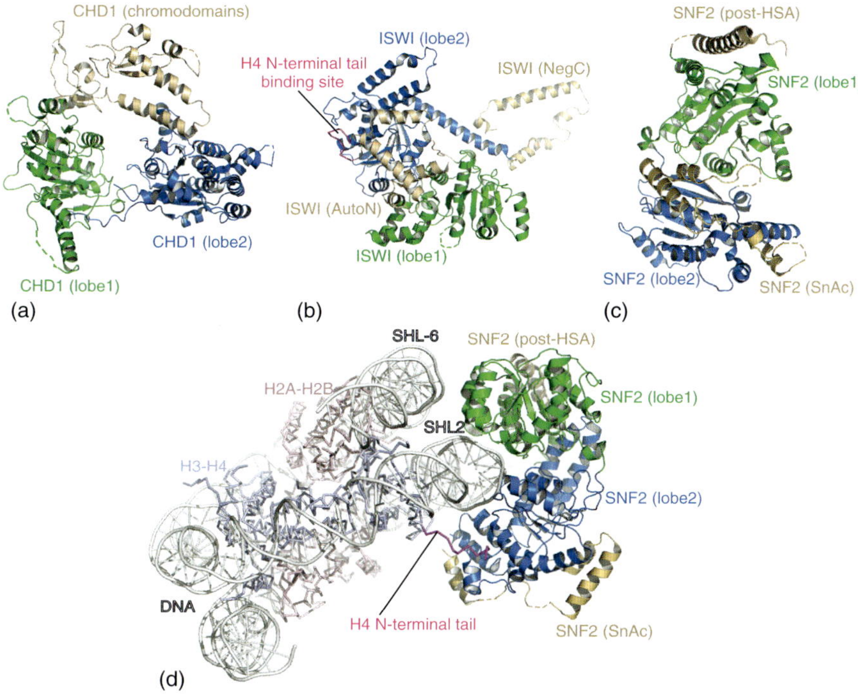

Figure 2.6 Structures of ATP-dependent chromatin remodeling subunits. Ribbon representation of ATP-dependent chromatin remodeling subunits. The RecA-like lobe 1 and lobe 2 of the ATPase domain are colored green and blue, respectively, and the auxiliary domains are colored wheat. (a) Structure of CHD1. The double chromodomains of CHD1 link the two ATPase lobes, keeping them in an inactive conformation. (b) Structure of ISWI. The AutoN domain also links the two ATPase lobes and keep them in an inactive conformation. The part of the AutoN domain that binds to the H4 binding pocket of lobe 2 is shown as magenta ribbon. (c) Structure of SNF2 in an inactive conformation. (d) Structure of SNF2 bound to the nucleosome. The color coding for the nucleosome is as in Figure 2.3. The H4 N-terminal tail that binds to SNF2 lobe 2 is shown as magenta ribbon. SNF2 is bound to the nucleosome in an active conformation and interacts with the nucleosome DNA at SHL2 and SHL-6.

of the HSS–LOC3 complex with a nucleosome having free DNA extensions at the nucleosome entry and exit points further shows that the HSS–LOC3 complex can interact with both the DNA entering and exiting the nucleosome. This study also suggests that the substrate of ISWI family members could be a dinucleosome rather than a mononucleosome and that the HSS domain could define the minimal spacing between two nucleosomes [61]. Interestingly, the NegC domain could possibly help with this process, functionally linking the ATPase CD and the HSS domain.

In the SNF family, the ATPase core domains are flanked on the N-terminus by a helicase/SANT-associated (HSA) domain and a post-HSA domain and on the C-terminus by a SnAC domain. The HSA domain forms a long α-helix that binds to other proteins, such as actin or actin-related proteins (ARPs), and regulates the remodeling activity by an unknown mechanism [62]. The role of the other auxiliary domains is less well understood. The 2.3 Å crystal structure of a yeast SNF2 encompassing a small part of the HSA, the post-HSA, the ATPase, and SnAC domains has shed light on the interactions between these different domains [63].

Although the part of the HSA domain included in the analysis is not seen in density, the post-HSA is found forming a long helix interacting with the RecA-like lobe 1 of the ATPase (Figure 2.6c). The SnAC forms extensive interactions with lobe 2, having little contacts with the first lobe. Interestingly, the two core domains of the ATPase interact strongly with each other, but their relative positioning, as for CHD1 and ISWI, is not compatible with a productive mode. In contrast with CHD1 and ISWI, however, this nonproductive conformation is not stabilized by auxiliary domains.

The EM structure at 4.0–4.7 Å resolution of a SNF2-nucleosome complex has provided a long-awaited structural view at sufficient resolution on how the catalytic subunit of a remodeler engages its nucleosome substrate (Figure 2.6d) [64]. Upon binding, the two RecA-like lobes are reorienting themselves to now form a productive ATPase CD. This requires not only the movement relative to each other of the different structural elements observed in the unbound structure but also the structuration of several disordered elements.

SNF2 is found bound at superhelical location 2 (SHL2) of the nucleosome, a position already known to bind preferentially several remodelers and where the H4 N-terminal tail protrudes from the nucleosome. Accordingly, the H4 tail is found bound at the same position on lobe 2 as was observed for the ISWI lobe 2–H4 tail complex [60]. Since H4 is not required to stimulate SNF2 activity, this interaction may in this case only position correctly SNF2 on the nucleosome. In the case of ISWI, it may both remove the inhibition by the AutoN domain and position ISWI on the nucleosome.

Besides the limited SNF2–H4 interaction, SNF2 interacts strongly with the nucleosomal DNA at the SHL2. This interaction is mediated within the cleft formed between lobes 1 and 2 of the ATPase domain and involves primarily contacts with the phosphate backbone. In addition, SNF2 also interacts with its lobe 1 at SHL-6, most likely to help position the enzyme in a correct orientation onto the nucleosome. Importantly, binding of SNF2 to the nucleosome in the absence of any nucleotide appears already sufficient to distort the DNA at SHL2, lifting the DNA off from its canonical path on the histone octamer surface, thus probably priming the remodeling reaction to come.

In the last decade, our mechanistic understanding of ATP-dependent chromatin remodelers has made a lot of progress. It remains to be understood how remodelers from the INO80 family, which have a long insertion between their RecA-like domains, are structurally organized and what are the changes in the remodelers upon nucleotide binding. Importantly, many features observed in the SNF2-nucleosome structure might be conserved in other remodelers due to sequence similarity of mechanistically important residues, suggesting analogous modes of action. Yet, existing sequence and structural differences between the remodelers are also likely to convey specific functional outputs. The same applies for the associated subunits forming the full remodeler complexes.

2.11 Epigenetic Readers: Histone Crotonylation Readers and the 53BP1-Nucleosome (H2AK15Ub–H4K20me2) Complex

Epigenetic readers are becoming more and more attractive targets for epi-drug development. This is exemplified by the case of the bromodomains that recognize the acetyl-lysine epigenetic marks, and other known readers are currently also targets for epi-drug development (see Chapters 14 and 15 in this book). Interestingly, recent progresses in mass spectrometry instrumentation and high resolution proteomic approaches are identifying so-far unknown histone modifications and their readers, which should define future targets in epi-drug development.

Among the epigenetic marks recently identified are lysine propionylation (Kpr), butyrylation (Kbu), and crotonylation (Kcr) [65]. These findings suggest that not only there are corresponding enzymes that lay and erase these marks but also reader modules that transduce this information into functional outcomes. Among the first structurally characterized readout modules of non-acetyl modifications are readers of lysine crotonylation such as PHD fingers and YEATS domains [66]. Crotonylation differs from other acylations in its rigidity and planar configuration due to the π-electron conjugation. Biophysical measurements reveal that the double PHD finger (DPF) domains of human MOZ and DPF2 bind a range of histone lysine acylations but display the strongest affinity for crotonylated lysine residues [66b].

Specifically, the 1.4–1.6 Å crystal structures of the MOZ DPF domain in complex with histones H3K14cr, H3K14bu, and H3K14pr revealed that these non-acetyl acylations are accommodated in a hydrophobic "dead-end" pocket with selectivity for crotonylation (Figure 2.7a). The observed selectivity toward crotonyl-lysine is achieved by intimate encapsulation and an amide-sensing hydrogen bonding network. Interestingly, sequence and structural comparison revealed that a glycine residue is critical for the pocket formation owing to its side-chain-free feature. In most classical histone H3K4-binding PHD fingers, this glycine is replaced by a bulkier residue such as tyrosine or phenylalanine, which fills the pocket and prevents Kcr accommodation.

A different mode of specific recognition of crotonyl-lysine was described in AF9 YEATS domain [66c, 67]. The NMR and 2.7 crystal structures of AF9 YEATS complexed with histone H3K9cr and H3K18cr peptides provided molecular insight into specific recognition of crotonylated substrates (Figure 2.7b). In the structure, Kcr is anchored in a specific pocket, and the Kcr-flanking H3 residues are recognized by extensive polar or hydrophobic contacts. Careful inspection of the crystal structure revealed an extended aromatic sandwiching cage with crotonyl specificity arising from π-aromatic and hydrophobic interactions between the planar crotonylamide group and aromatic rings of a phenylalanine and tyrosine residues. Interestingly, these structural features are conserved within the YEATS family members, but not within the bromodomains, providing additional evidence for why bromodomains do not bind crotonyl-lysine substrates.

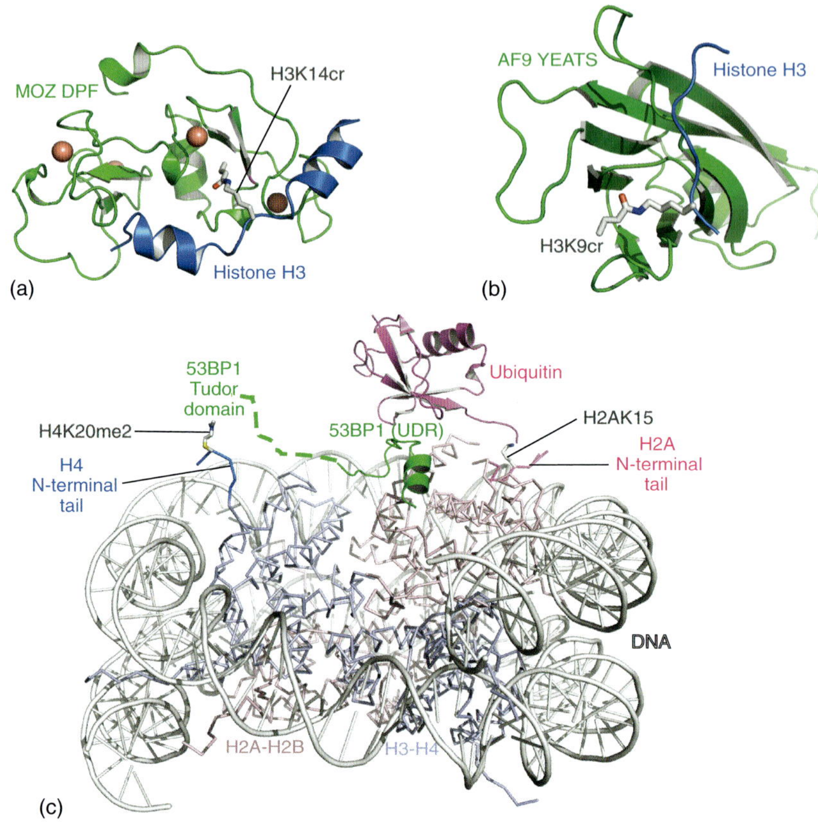

Figure 2.7 Structures of epigenetic readers. Ribbon representation of epigenetic readers. (a, b) Structures of crotonylation readers: MOZ double PHD fingers bound to H3K14cr and AF9 YEATS domain bound to H3K9cr. The readers are colored green and histone H3 is colored blue. The crotonylated residues are shown as sticks with gray carbons. (c) Structure of 53BP1 bound to the H4K20 dimethylated and H2AK15 ubiquitinylated nucleosome. The nucleosome is colored as in Figure 2.3. 53BP1 is colored green. 53BP1 ubiquitin-dependent recruitment motif (UDR) is shown as ribbon. 53BP1 Tudor domain was not provided in the model deposited and is just indicated. The H4 N-terminal tail is shown as blue ribbon and the H2A N-terminal tail as magenta ribbon. H4K20me2 and H2AK15 are shown as sticks with gray carbons. Ubiquitin is colored magenta.

Readers can bind to small peptides, but *in vivo* this recognition is often carried out in a nucleosomal context. How this is done remains poorly understood. The 4.5 Å EM structure of the tumor suppressor 53BP1, which acts in DNA repair and binds to nucleosomes simultaneously ubiquitinylated on H2AK15 and dimethylated on H4K20, provides an interesting view on modified nucleosome recognition (Figure 2.7c) [68]. Specifically, a small region of 53BP1 composed of a Tudor domain followed by a small ubiquitin-dependent recruitment motif (UDR) is sufficient to bind to this doubly modified nucleosome. Although the medium resolution of the structure renders the structural interpretation somewhat difficult, it appears that the Tudor domain lays over the H4K20me2 epigenetic mark, whereas the following small UDR motif meanders on the nucleosome surface, interacting with specific motifs, including the acidic patch.

An important aspect of this mode of binding is the positioning of a part of this UDR between the histone octamer and the ubiquitin. This binding of the UDR enables the bridging between the histone octamer and the ubiquitin, in agreement with the observation made with the SAGA DUBm that nucleosome-bound ubiquitin does not interact with the nucleosome. This ubiquitin–UDR–nucleosome interaction stabilizes the ubiquitin in a specific position and enables the recognition of the H2A–K15 ubiquitinylated nucleosome.

Interestingly, the previously described 8 Å resolution structure of the core NUA4 HAT complex bound to a nucleosome through a Tudor domain, an interaction with the acidic patch and DNA elements [36], corroborates the 53BP1-nucleosome structure. Yet, better resolution will be required in the future for the structures of such complexes to enable the design of next-generation epi-drugs. Nevertheless, the complexity of the recognition of modified nucleosome by larger epigenetic effectors, as the ones described in this chapter, indicates that selective modulation of epigenetic effector–nucleosome interactions could be achieved.

2.12 Conclusions

Structural analyses provide an essential information when it comes to designing more selective and more potent drugs. The results presented here highlight the intricacy of the assembly, the interactions, and the mechanisms of epigenetic effectors to achieve precise epigenetic regulation. Specifically, the embedding of epigenetic enzymes within large macromolecular complexes yields a high degree of modulation of their activity through structural changes and allosteric mechanisms. This has major implications for the design of novel, more selective, and more potent drugs targeting the active site of epigenetic enzymes.

This also opens the way to design drugs that will modulate (i) the intramolecular interactions within complexes and (ii) the interactions of these complexes with their substrates. Importantly, regulation of these mechanisms through small molecules could help select between processes that yield different biological outcomes. Another important aspect concerns the fact that a drug targeting the same enzyme can have different effects depending on the macromolecular environment of the enzyme (i.e. free or embedded in various complexes). In addition, some complexes are bearing different epigenetic activities. It will be interesting for these complexes to design dual inhibitors that can modulate these different activities at the same time.

A surprising observation made from the different epigenetic effector–nucleosome complexes whose structures have been solved so far is that a major interaction of the epigenetic players with the nucleosome involves targeting the H2A–H2B acidic patch. The other interactions appear less strong but serve to position correctly the effectors on the nucleosome. It remains to be investigated whether targeting other interaction regions with the nucleosome could help selectively modulate the chromatin accessibility of specific complexes. Yet, epigenetic complexes are often larger than the sub-complexes currently being structurally characterized when bound to the nucleosome. This should open many more possibilities for selective modulation of epigenetic effector–nucleosome interactions.

The structural results on epigenetic players obtained in the last decade represent the beginnings of a new era where epigenetic complexes will be at the heart of integrated structural, chemical, and medicinal biology. Notably, the ongoing major technological advances in EM that are enabling a "resolution revolution" [69] are expected to further ease the high resolution structure determination of epigenetic complexes, with direct implications in epi-drug discovery. Importantly, the combination of new-generation epi-drugs with the new revolutionary technologies currently developed in genome editing and targeting [70] should provide unprecedented means to specifically intervene at specific genomic loci to correct genetic and epigenetic aberrations to cure human diseases.

Note: Recent publications on the structures of CHD1 and INO80 bound to nucleosomes have enlarged our understanding of remodeler/nucleosome interactions.

Acknowledgments

This work and the authors of this manuscript have been supported by funding from the European Union's Seventh Framework Programme for research, technological development, and demonstration under grant agreement nos. 241865 (SEtTReND) and 602080 (A-ParaDDisE); by institutional funds from the Centre National de la Recherche Scientifique (CNRS), the Institut National de la Santé et de la Recherche Médicale (INSERM), and the Université de Strasbourg; by la Fondation ARC pour la Recherche sur le Cancer (contract PJA 20141201960); by the IdEX Attractivité Université de Strasbourg; by the French Infrastructure for Integrated Structural Biology (FRISBI) (ANR-10-INSB-05-01); and by Instruct as part of the European Strategy Forum on Research Infrastructures (ESFRI).

References

1 Allis, C.D. and Jenuwein, T. (2016). The molecular hallmarks of epigenetic control. *Nat. Rev. Genet.* 17 (8): 487–500.

2 (a) Kouzarides, T. (2007). Chromatin modifications and their function. *Cell* 128 (4): 693–705. (b) Romier, C., Wurtz, J.M., Renaud, J.P., and Cavarelli, J. (2009). Structural biology of epigenetic targets. In: *Epigenetic Targets in Drug Discovery*, vol. 42 (ed. W. Sippl and M. Jung), 23–56. Weinheim, Germany: Wiley-VCH Verlag GmbH & Co. KGaA. (c) Cai, S.F., Chen, C.W., and Armstrong, S.A. (2015). Drugging chromatin in cancer: recent advances and novel approaches. *Mol. Cell* 60 (4): 561–570.

3 (a) Clapier, C.R. and Cairns, B.R. (2009). The biology of chromatin remodeling complexes. *Annu. Rev. Biochem.* 78: 273–304. (b) Kadoch, C., Hargreaves, D.C., Hodges, C. et al. (2013). Proteomic and bioinformatic analysis of mammalian SWI/SNF complexes identifies extensive roles in human malignancy. *Nat. Genet.* 45 (6): 592–601. (c) Brien, G.L., Valerio, D.G., and Armstrong, S.A. (2016). Exploiting the Epigenome to control cancer-promoting gene-expression programs. *Cancer Cell* 29 (4): 464–476. (d) Buschbeck, M. and Hake, S.B. (2017). Variants of core histones and their roles in cell fate decisions, development and cancer. *Nat. Rev. Mol. Cell Biol.* 18 (5): 299–314. (e) Hammond, C.M., Stromme, C.B., Huang, H. et al. (2017). Histone chaperone networks shaping chromatin function. *Nat. Rev. Mol. Cell Biol.* 18 (3): 141–158.

4 Schubeler, D. (2015). Function and information content of DNA methylation. *Nature* 517 (7534): 321–326.

5 Jeschke, J., Collignon, E., and Fuks, F. (2016). Portraits of TET-mediated DNA hydroxymethylation in cancer. *Curr. Opin. Genet. Dev.* 36: 16–26.

6 (a) Ooi, S.K., Qiu, C., Bernstein, E. et al. (2007). DNMT3L connects unmethylated lysine 4 of histone H3 to de novo methylation of DNA. *Nature* 448 (7154): 714–717. (b) Otani, J., Nankumo, T., Arita, K. et al. (2009). Structural basis for recognition of H3K4 methylation status by the DNA methyltransferase 3A ATRX-DNMT3-DNMT3L domain. *EMBO Rep.* 10 (11): 1235–1241.

7 Guo, X., Wang, L., Li, J. et al. (2015). Structural insight into autoinhibition and histone H3-induced activation of DNMT3A. *Nature* 517 (7536): 640–644.

8 (a) Song, J., Rechkoblit, O., Bestor, T.H., and Patel, D.J. (2011). Structure of DNMT1-DNA complex reveals a role for autoinhibition in maintenance DNA methylation. *Science* 331 (6020): 1036–1040. (b) Song, J., Teplova, M., Ishibe-Murakami, S., and Patel, D.J. (2012). Structure-based mechanistic insights into DNMT1-mediated maintenance DNA methylation. *Science* 335 (6069): 709–712.

9 Cheng, J., Yang, H., Fang, J. et al. (2015). Molecular mechanism for USP7-mediated DNMT1 stabilization by acetylation. *Nat. Commun.* 6: 7023.

10 Blanc, R.S. and Richard, S. (2017). Arginine methylation: the coming of age. *Mol. Cell* 65 (1): 8–24.

11 (a) Antonysamy, S., Bonday, Z., Campbell, R.M. et al. (2012). Crystal structure of the human PRMT5:MEP50 complex. *Proc. Natl. Acad. Sci. U.S.A.* 109 (44): 17960–17965. (b) Ho, M.C., Wilczek, C., Bonanno, J.B. et al. (2013). Structure of the arginine methyltransferase PRMT5-MEP50 reveals a mechanism for substrate specificity. *PLoS One* 8 (2): e57008.

12 (a) Chan-Penebre, E., Kuplast, K.G., Majer, C.R. et al. (2015). A selective inhibitor of PRMT5 with in vivo and in vitro potency in MCL models. *Nat. Chem. Biol.* 11 (6): 432–437. (b) Mavrakis, K.J., McDonald, E.R. III,, Schlabach, M.R. et al. (2016). Disordered methionine metabolism in MTAP/CDKN2A-deleted cancers leads to dependence on PRMT5. *Science* 351 (6278): 1208–1213.

13 Smith, E., Lin, C., and Shilatifard, A. (2011). The super elongation complex (SEC) and MLL in development and disease. *Genes Dev.* 25 (7): 661–672.
14 Ford, D.J. and Dingwall, A.K. (2015). The cancer COMPASS: navigating the functions of MLL complexes in cancer. *Cancer Genet.* 208 (5): 178–191.
15 Li, Y., Han, J., Zhang, Y. et al. (2016). Structural basis for activity regulation of MLL family methyltransferases. *Nature* 530 (7591): 447–452.
16 (a) Jiao, L. and Liu, X. (2015). Structural basis of histone H3K27 trimethylation by an active polycomb repressive complex 2. *Science* 350 (6258): aac4383. (b) Justin, N., Zhang, Y., Tarricone, C. et al. (2016). Structural basis of oncogenic histone H3K27M inhibition of human polycomb repressive complex 2. *Nat. Commun.* 7: 11316.
17 (a) Holoch, D. and Margueron, R. (2017). Mechanisms regulating PRC2 recruitment and enzymatic activity. *Trends Biochem. Sci* 42 (7): 531–542. (b) Margueron, R. and Reinberg, D. (2011). The Polycomb complex PRC2 and its mark in life. *Nature* 469 (7330): 343–349. (c) Blackledge, N.P., Rose, N.R., and Klose, R.J. (2015). Targeting Polycomb systems to regulate gene expression: modifications to a complex story. *Nat. Rev. Mol. Cell Biol.* 16 (11): 643–649. (d) Simon, J.A. and Kingston, R.E. (2013). Occupying chromatin: Polycomb mechanisms for getting to genomic targets, stopping transcriptional traffic, and staying put. *Mol. Cell* 49 (5): 808–824.
18 (a) Comet, I., Riising, E.M., Leblanc, B., and Helin, K. (2016). Maintaining cell identity: PRC2-mediated regulation of transcription and cancer. *Nat. Rev. Cancer* 16 (12): 803–810. (b) Koppens, M. and van Lohuizen, M. (2016). Context-dependent actions of Polycomb repressors in cancer. *Oncogene* 35 (11): 1341–1352.
19 Margueron, R., Justin, N., Ohno, K. et al. (2009). Role of the polycomb protein EED in the propagation of repressive histone marks. *Nature* 461 (7265): 762–767.
20 (a) Antonysamy, S., Condon, B., Druzina, Z. et al. (2013). Structural context of disease-associated mutations and putative mechanism of autoinhibition revealed by X-ray crystallographic analysis of the EZH2-SET domain. *PLoS One* 8 (12): e84147. (b) Wu, H., Zeng, H., Dong, A. et al. (2013). Structure of the catalytic domain of EZH2 reveals conformational plasticity in cofactor and substrate binding sites and explains oncogenic mutations. *PLoS One* 8 (12): e83737.
21 Ciferri, C., Lander, G.C., Maiolica, A. et al. (2012). Molecular architecture of human polycomb repressive complex 2. *Elife* 1: e00005.
22 Streich, F.C. Jr., and Lima, C.D. (2014). Structural and functional insights to ubiquitin-like protein conjugation. *Annu. Rev. Biophys.* 43: 357–379.
23 McGinty, R.K., Henrici, R.C., and Tan, S. (2014). Crystal structure of the PRC1 ubiquitylation module bound to the nucleosome. *Nature* 514 (7524): 591–596.
24 (a) Luger, K., Mader, A.W., Richmond, R.K. et al. (1997). Crystal structure of the nucleosome core particle at 2.8 a resolution. *Nature* 389 (6648): 251–260. (b) Suto, R.K., Clarkson, M.J., Tremethick, D.J., and Luger, K. (2000). Crystal structure of a nucleosome core particle containing the variant histone H2A.Z. *Nat. Struct. Biol.* 7 (12): 1121–1124.

25 (a) McGinty, R.K. and Tan, S. (2015). Nucleosome structure and function. *Chem. Rev.* 115 (6): 2255–2273. (b) Morgan, M.T., Haj-Yahya, M., Ringel, A.E. et al. (2016). Structural basis for histone H2B deubiquitination by the SAGA DUB module. *Science* 351 (6274): 725–728. (c) Latrick, C.M., Marek, M., Ouararhni, K. et al. (2016). Molecular basis and specificity of H2A.Z-H2B recognition and deposition by the histone chaperone YL1. *Nat. Struct. Mol. Biol.* 23 (4): 309–316. (d) Liang, X., Shan, S., Pan, L. et al. (2016). Structural basis of H2A.Z recognition by SRCAP chromatin-remodeling subunit YL1. *Nat. Struct. Mol. Biol.* 23 (4): 317–323.

26 Spedale, G., Timmers, H.T., and Pijnappel, W.W. (2012). ATAC-king the complexity of SAGA during evolution. *Genes Dev.* 26 (6): 527–541.

27 Helmlinger, D., Hardy, S., Sasorith, S. et al. (2004). Ataxin-7 is a subunit of GCN5 histone acetyltransferase-containing complexes. *Hum. Mol. Genet.* 13 (12): 1257–1265.

28 Diebold, M.L., Fribourg, S., Koch, M. et al. (2011). Deciphering correct strategies for multiprotein complex assembly by co-expression: application to complexes as large as the histone octamer. *J. Struct. Biol.* 175 (2): 178–188.

29 (a) Kohler, A., Zimmerman, E., Schneider, M. et al. (2010). Structural basis for assembly and activation of the heterotetrameric SAGA histone H2B deubiquitinase module. *Cell* 141 (4): 606–617. (b) Samara, N.L., Datta, A.B., Berndsen, C.E. et al. (2010). Structural insights into the assembly and function of the SAGA deubiquitinating module. *Science* 328 (5981): 1025–1029.

30 David, G., Abbas, N., Stevanin, G. et al. (1997). Cloning of the SCA7 gene reveals a highly unstable CAG repeat expansion. *Nat. Genet.* 17 (1): 65–70.

31 Durand, A., Bonnet, J., Fournier, M. et al. (2014). Mapping the deubiquitination module within the SAGA complex. *Structure* 22 (11): 1553–1559.

32 Choudhary, C., Weinert, B.T., Nishida, Y. et al. (2014). The growing landscape of lysine acetylation links metabolism and cell signalling. *Nat. Rev. Mol. Cell Biol.* 15 (8): 536–550.

33 Marmorstein, R. and Zhou, M.M. (2014). Writers and readers of histone acetylation: structure, mechanism, and inhibition. *Cold Spring Harbor Perspect. Biol.* 6 (7): a018762.

34 Keller, C.I. and Akhtar, A. (2015). The MSL complex: juggling RNA-protein interactions for dosage compensation and beyond. *Curr. Opin. Genet. Dev.* 31: 1–11.

35 (a) Hallacli, E., Lipp, M., Georgiev, P. et al. (2012). Msl1-mediated dimerization of the dosage compensation complex is essential for male X-chromosome regulation in drosophila. *Mol. Cell* 48 (4): 587–600. (b) Huang, J., Wan, B., Wu, L. et al. (2012). Structural insight into the regulation of MOF in the male-specific lethal complex and the non-specific lethal complex. *Cell Res.* 22 (6): 1078–1081. (c) Kadlec, J., Hallacli, E., Lipp, M. et al. (2011). Structural basis for MOF and MSL3 recruitment into the dosage compensation complex by MSL1. *Nat. Struct. Mol. Biol.* 18 (2): 142–149.

36 Xu, P., Li, C., Chen, Z. et al. (2016). The NuA4 Core complex acetylates Nucleosomal histone H4 through a double recognition mechanism. *Mol. Cell* 63 (6): 965–975.

37 (a) Doyon, Y. and Cote, J. (2004). The highly conserved and multifunctional NuA4 HAT complex. *Curr. Opin. Genet. Dev.* 14 (2): 147–154. (b) Selleck, W., Fortin, I., Sermwittayawong, D. et al. (2005). The *Saccharomyces cerevisiae* piccolo NuA4 histone acetyltransferase complex requires the enhancer of Polycomb a domain and chromodomain to acetylate nucleosomes. *Mol. Cell. Biol.* 25 (13): 5535–5542.
38 Chittuluru, J.R., Chaban, Y., Monnet-Saksouk, J. et al. (2011). Structure and nucleosome interaction of the yeast NuA4 and piccolo-NuA4 histone acetyltransferase complexes. *Nat. Struct. Mol. Biol.* 18 (11): 1196–1203.
39 Lombardi, P.M., Cole, K.E., Dowling, D.P., and Christianson, D.W. (2011). Structure, mechanism, and inhibition of histone deacetylases and related metalloenzymes. *Curr. Opin. Struct. Biol.* 21 (6): 735–743.
40 Moniot, S., Weyand, M., and Steegborn, C. (2012). Structures, substrates, and regulators of mammalian sirtuins – opportunities and challenges for drug development. *Front. Pharmacol.* 3: 16.
41 (a) Millard, C.J., Watson, P.J., Fairall, L., and Schwabe, J.W. (2017). Targeting class I histone deacetylases in a "complex" environment. *Trends Pharmacol. Sci.* 38 (4): 363–377. (b) West, A.C. and Johnstone, R.W. (2014). New and emerging HDAC inhibitors for cancer treatment. *J. Clin. Invest.* 124 (1): 30–39.
42 (a) Marek, M., Oliveira, G., Pierce, R.J. et al. (2015). Drugging the schistosome zinc-dependent HDACs: current progress and future perspectives. *Future Med. Chem.* 7 (6): 783–800. (b) Shuttleworth, S.J., Bailey, S.G., and Townsend, P.A. (2010). Histone deacetylase inhibitors: new promise in the treatment of immune and inflammatory diseases. *Curr. Drug Targets* 11 (11): 1430–1438. (c) Chakrabarti, A., Oehme, I., Witt, O. et al. (2015). HDAC8: a multifaceted target for therapeutic interventions. *Trends Pharmacol. Sci.* 36 (7): 481–492.
43 Maolanon, A.R., Kristensen, H.M., Leman, L.J. et al. (2017). Natural and synthetic macrocyclic inhibitors of the histone deacetylase enzymes. *ChemBioChem* 18 (1): 5–49.
44 (a) Millard, C.J., Watson, P.J., Celardo, I. et al. (2013). Class I HDACs share a common mechanism of regulation by inositol phosphates. *Mol. Cell* 51 (1): 57–67. (b) Watson, P.J., Fairall, L., Santos, G.M., and Schwabe, J.W. (2012). Structure of HDAC3 bound to co-repressor and inositol tetraphosphate. *Nature* 481 (7381): 335–340. (c) Watson, P.J., Millard, C.J., Riley, A.M. et al. (2016). Insights into the activation mechanism of class I HDAC complexes by inositol phosphates. *Nat. Commun.* 7: 11262.
45 (a) Hai, Y. and Christianson, D.W. (2016). Histone deacetylase 6 structure and molecular basis of catalysis and inhibition. *Nat. Chem. Biol.* 12 (9): 741–747. (b) Miyake, Y., Keusch, J.J., Wang, L. et al. (2016). Structural insights into HDAC6 tubulin deacetylation and its selective inhibition. *Nat. Chem. Biol.* 12 (9): 748–754.
46 (a) Heimburg, T., Chakrabarti, A., Lancelot, J. et al. (2016). Structure-based design and synthesis of novel inhibitors targeting HDAC8 from *Schistosoma mansoni* for the treatment of Schistosomiasis. *J. Med. Chem.* 59 (6): 2423–2435. (b) Kannan, S., Melesina, J., Hauser, A.T. et al. (2014). Discovery

of inhibitors of *Schistosoma mansoni* HDAC8 by combining homology modeling, virtual screening, and in vitro validation. *J. Chem. Inf. Model.* 54 (10): 3005–3019. (c) Marek, M., Kannan, S., Hauser, A.T. et al. (2013). Structural basis for the inhibition of histone deacetylase 8 (HDAC8), a key epigenetic player in the blood fluke *Schistosoma mansoni*. *PLoS Pathog.* 9 (9): e1003645. (d) Stolfa, D.A., Marek, M., Lancelot, J. et al. (2014). Molecular basis for the antiparasitic activity of a mercaptoacetamide derivative that inhibits histone deacetylase 8 (HDAC8) from the human pathogen *Schistosoma mansoni*. *J. Mol. Biol.* 426 (20): 3442–3453.

47 (a) English, C.M., Adkins, M.W., Carson, J.J. et al. (2006). Structural basis for the histone chaperone activity of Asf1. *Cell* 127 (3): 495–508. (b) Natsume, R., Eitoku, M., Akai, Y. et al. (2007). Structure and function of the histone chaperone CIA/ASF1 complexed with histones H3 and H4. *Nature* 446 (7133): 338–341.

48 (a) Huang, H., Stromme, C.B., Saredi, G. et al. (2015). A unique binding mode enables MCM2 to chaperone histones H3-H4 at replication forks. *Nat. Struct. Mol. Biol.* 22 (8): 618–626. (b) Richet, N., Liu, D., Legrand, P. et al. (2015). Structural insight into how the human helicase subunit MCM2 may act as a histone chaperone together with ASF1 at the replication fork. *Nucleic Acids Res.* 43 (3): 1905–1917.

49 Wang, H., Wang, M., Yang, N., and Xu, R.M. (2015). Structure of the quaternary complex of histone H3-H4 heterodimer with chaperone ASF1 and the replicative helicase subunit MCM2. *Protein Cell* 6 (9): 693–697.

50 Saredi, G., Huang, H., Hammond, C.M. et al. (2016). H4K20me0 marks post-replicative chromatin and recruits the TONSL-MMS22L DNA repair complex. *Nature* 534 (7609): 714–718.

51 (a) Cho, U.S. and Harrison, S.C. (2011). Recognition of the centromere-specific histone Cse4 by the chaperone Scm3. *Proc. Natl. Acad. Sci. U.S.A.* 108 (23): 9367–9371. (b) Hu, H., Liu, Y., Wang, M. et al. (2011). Structure of a CENP-A-histone H4 heterodimer in complex with chaperone HJURP. *Genes Dev.* 25 (9): 901–906. (c) Zhou, Z., Feng, H., Zhou, B.R. et al. (2011). Structural basis for recognition of centromere histone variant CenH3 by the chaperone Scm3. *Nature* 472 (7342): 234–237.

52 (a) Elsasser, S.J., Huang, H., Lewis, P.W. et al. (2012). DAXX envelops a histone H3.3-H4 dimer for H3.3-specific recognition. *Nature* 491 (7425): 560–565. (b) Liu, C.P., Xiong, C., Wang, M. et al. (2012). Structure of the variant histone H3.3-H4 heterodimer in complex with its chaperone DAXX. *Nat. Struct. Mol. Biol.* 19 (12): 1287–1292.

53 Ricketts, M.D., Frederick, B., Hoff, H. et al. (2015). Ubinuclein-1 confers histone H3.3-specific-binding by the HIRA histone chaperone complex. *Nat. Commun.* 6: 7711.

54 Huang, H., Deng, Z., Vladimirova, O. et al. (2016). Structural basis underlying viral hijacking of a histone chaperone complex. *Nat. Commun.* 7: 12707.

55 Obri, A., Ouararhni, K., Papin, C. et al. (2014). ANP32E is a histone chaperone that removes H2A.Z from chromatin. *Nature* 505 (7485): 648–653.

56 Mao, Z., Pan, L., Wang, W. et al. (2014). Anp32e, a higher eukaryotic histone chaperone directs preferential recognition for H2A.Z. *Cell Res.* 24 (4): 389–399.

57 Bartholomew, B. (2014). Regulating the chromatin landscape: structural and mechanistic perspectives. *Annu. Rev. Biochem.* 83: 671–696.

58 Hauk, G., McKnight, J.N., Nodelman, I.M., and Bowman, G.D. (2010). The chromodomains of the Chd1 chromatin remodeler regulate DNA access to the ATPase motor. *Mol. Cell* 39 (5): 711–723.

59 Metzger, E., Willmann, D., McMillan, J. et al. (2016). Assembly of methylated KDM1A and CHD1 drives androgen receptor-dependent transcription and translocation. *Nat. Struct. Mol. Biol.* 23 (2): 132–139.

60 Yan, L., Wang, L., Tian, Y. et al. (2016). Structure and regulation of the chromatin remodeller ISWI. *Nature* 540 (7633): 466–469.

61 Yamada, K., Frouws, T.D., Angst, B. et al. (2011). Structure and mechanism of the chromatin remodelling factor ISW1a. *Nature* 472 (7344): 448–453.

62 Schubert, H.L., Wittmeyer, J., Kasten, M.M. et al. (2013). Structure of an actin-related subcomplex of the SWI/SNF chromatin remodeler. *Proc. Natl. Acad. Sci. U.S.A.* 110 (9): 3345–3350.

63 Xia, X., Liu, X., Li, T. et al. (2016). Structure of chromatin remodeler Swi2/Snf2 in the resting state. *Nat. Struct. Mol. Biol.* 23 (8): 722–729.

64 Liu, X., Li, M., Xia, X. et al. (2017). Mechanism of chromatin remodelling revealed by the Snf2-nucleosome structure. *Nature* 544 (7651): 440–445.

65 Sabari, B.R., Zhang, D., Allis, C.D., and Zhao, Y. (2017). Metabolic regulation of gene expression through histone acylations. *Nat. Rev. Mol. Cell Biol.* 18 (2): 90–101.

66 (a) Andrews, F.H., Shinsky, S.A., Shanle, E.K. et al. (2016). The Taf14 YEATS domain is a reader of histone crotonylation. *Nat. Chem. Biol.* 12 (6): 396–398. (b) Xiong, X., Panchenko, T., Yang, S. et al. (2016). Selective recognition of histone crotonylation by double PHD fingers of MOZ and DPF2. *Nat. Chem. Biol.* 12 (12): 1111–1118. (c) Zhang, Q., Zeng, L., Zhao, C. et al. (2016). Structural insights into histone Crotonyl-lysine recognition by the AF9 YEATS domain. *Structure* 24 (9): 1606–1612.

67 Li, Y., Sabari, B.R., Panchenko, T. et al. (2016). Molecular coupling of histone Crotonylation and active transcription by AF9 YEATS domain. *Mol. Cell* 62 (2): 181–193.

68 Wilson, M.D., Benlekbir, S., Fradet-Turcotte, A. et al. (2016). The structural basis of modified nucleosome recognition by 53BP1. *Nature* 536 (7614): 100–103.

69 Kuhlbrandt, W. (2014). Biochemistry. The resolution revolution. *Science* 343 (6178): 1443–1444.

70 Doudna, J.A. and Charpentier, E. (2014). Genome editing. The new frontier of genome engineering with CRISPR-Cas9. *Science* 346 (6213): 1258096.

3

Computer-based Lead Identification for Epigenetic Targets

Chiara Luise, Tino Heimburg, Berin Karaman, Dina Robaa, and Wolfgang Sippl

Martin-Luther University of Halle-Wittenberg, Institute of Pharmacy, Department of Pharmaceutical Chemistry, Wolfgang-Langenbeck-Straße 4, 06120 Halle/Saale, Germany

3.1 Introduction

The two major biochemical pathways of epigenetic regulation are DNA methylation and posttranslational modifications of amino acids in histone proteins. Histone modifications interact with each other and constitute a particular pattern of alterations in the chromatin structure, the so-called histone code [1]. Posttranslational modifications of histones have been shown to participate in a wide array of cellular processes, including regulation of gene transcription, gene repression, and mitosis. Regulations of these covalent modifications and their implications are currently of great interest in the scientific community [2]. Although histone proteins are under the control of various posttranslational modifications, such as acetylation, acylation, methylation, phosphorylation, ubiquitination, or glycosylation, the focus of most studies has been mainly on lysine acetylation/deacetylation and methylation/demethylation. Over the last decade, many of the enzymes that regulate these histone modifications have been identified and characterized on a molecular level. With large effort, a wealth of three-dimensional (3D) structures of epigenetic targets has been determined, which helped to understand epigenetic targets and their regulation on a molecular level (see Chapter 2). From the early discovery of histone deacetylase (HDAC) inhibitors, such as trichostatin A, to the more recent discovery of novel histone-modifying enzyme inhibitors, structure- and computer-based approaches were applied to analyze target–ligand interactions and to rationalize the development of small-molecule modulators [3]. This chapter will summarize the most commonly used computer-based approaches to identify and optimize small-molecule inhibitors and modulators. In addition, we will discuss the progress on small-molecule inhibitors of epigenetic targets that were identified by computer-based approaches. We will focus on histone-modifying enzymes (writers and erasers), whereas examples of successful applications of computer-based ligand design for reader proteins can be found in Chapters 14 and 15.

Epigenetic Drug Discovery, First Edition. Edited by Wolfgang Sippl and Manfred Jung.

3.2 Computer-based Methods in Drug Discovery

It has become common sense in drug discovery to use computer-based *in silico* design techniques to identify and optimize novel molecules with a desired biological activity [4]. As a practical matter, computer-based methods are frequently split into disciplines that focus on either structure-based or ligand-based approaches. When the 3D structure of a target protein is known, then it is possible to apply structure-based methods. New candidate molecules may be docked into a particular binding pocket in order to study if they can interact with the protein in an optimal way. In an ideal situation, the target protein has already been crystallized in complex with a first lead structure. However, in particular for novel targets, frequently structural information is not available at the time of the first experiments (e.g. high-throughput screening (HTS)). When no 3D structure is available, but a sufficient number of active analogs have already been synthesized, then pharmacophore-based methods can be applied to identify novel molecules that fit a particular pharmacophore model [5, 6]. It may seem straightforward to develop new molecules for known proteins by applying structure-based approaches, but there are significant problems involved. Protein flexibility, multiple ligand binding modes, solvation, and entropic effects are some of the problems that must be solved to end up with reliable models. There are also available studies that demonstrate the efficiency of ligand- and structure-based approaches in direct comparison [6], as well as synergies obtained by using both approaches in combination [7]. As the costs for computational power keep dropping and parallel computing is now available, *in silico* modeling techniques are no longer a limiting problem to analyze large databases comprising millions of compounds [4]. In general, *in silico* methods gained influence on the validation of targets, hit finding, hit-to-lead expansion, lead optimization, and also the prediction of suitable ADMET and off-target activity profiles [8].

3.2.1 Pharmacophore-based Methods

The classical definition of a pharmacophore [9], which is the spatial arrangement of functional groups in a molecule necessary to mediate a biological effect, is based on a 3D point of view of molecules. It reflects the way medicinal chemists characterize the interaction pattern of molecules for a given target protein. However, depending on the different research areas, scientists have different views. Computational chemists often use the term pharmacophore in a more abstract way. Influenced by the structural representation of molecules, a set of topological connections is used to define the properties and dimension of a molecule in 2D. Here, the spatial and topological distribution of pharmacophoric features is converted to a lower dimensional representation, e.g. vectors. Such vectors, which represent pharmacophore descriptors, are referred to as "fingerprints," "keys," "bitstrings," or "correlation vectors," depending on the type of information stored. The pharmacophore descriptors or fingerprints can be regarded as a transformed molecular representation instead of an explicit 3D structure [10].

Both fingerprint and pharmacophore models are often used to rapidly screen large compound collections to identify novel potential hits [11].

3.2.2 QSAR

Quantitative structure–activity relationship (QSAR) techniques are the most prominent computational means to support chemistry within ligand-based drug design projects. The primary aim of these techniques is to establish a correlation of the biological activities of a series of structurally and biologically characterized compounds with the spatial fingerprints of numerous field properties of each molecule, such as steric demand, lipophilicity, and electrostatic interactions. Typically, a QSAR study allows identifying a correlation between molecular structure and biological effect and provides guidelines for the design of next-generation compounds with enhanced biological activity. Frequently applied methods include the comparative molecular field analysis (CoMFA) [12], the comparative molecular similarity indices analysis (CoMSIA) [13], and the GRID/Generating Optimal Linear PLS Estimations (GOLPEs) program [14]. Several reviews have been published in the last few years focusing on the basic theory, the pitfalls, and the application of 3D QSAR approaches [15–18]. Apart from the commercial distribution, a major factor behind the ongoing enthusiasm for CoMFA-related approaches comes from the proven ability of several of these methods to correctly estimate the biological activity of novel compounds [15].

3.2.3 Docking

The increase of available structural data of protein–ligand complexes has accelerated the development of structure-based methods. Today, protein–ligand docking is considered the most important tool for structure-based drug design. The docking process can be divided into two major parts: (i) the correct placement of a molecule in a protein binding pocket and (ii) the prediction of ligand affinities by a so-called scoring function [6]. Considering the complexity of the underlying thermodynamics of the ligand binding process, this is evidently a challenging task. After more than twenty years of development, a large number of docking programs and even more scoring functions are available. In contrast to pharmacophore screening approaches, the performance of docking methods is a trade-off between computational demand and accuracy. A major problem of current docking programs is that they do not take protein flexibility into account or do not consider protein–ligand desolvation processes. Only very recently, docking programs were developed (e.g. AutoDock4 [19], GOLD4 [20], Glide [21], or FlexE [22]), which consider protein flexibility to some extent [23]. Another problem that often occurs in ligand docking is the correct placement of water molecules within the binding site, which represent putative ligand binding partners. Structure-based approaches are able to provide important information about the interaction between a ligand and a macromolecule, but the accurate prediction of the binding affinity is still challenging. A detailed discussion about the limitations of docking and scoring methods can be found in recently published reviews [24, 25].

3.2.4 Virtual Screening

HTS of chemical libraries is a well-established method for finding new lead structures in early drug discovery. However, the increasing number of purchasable compounds makes it necessary to have fast and reliable prefiltering methods to reduce the number of compounds selected for experimental testing. In contrast to HTS, virtual screening (VS) hit compounds are selected on the basis of computer-based predictions. In this way, VS approaches try to select the most promising compounds from an electronic database for biological testing. VS can be carried out by applying different kinds of approaches: by searching databases for molecules with chemical similarity to a set of known actives [11], by identifying compounds that match a given pharmacophore [4], or by fitting molecules into the 3D structure of a macromolecular target [26]. In addition, prescreening filters are usually applied to filter a compound library before docking for compounds with favorable pharmacokinetics, optimum oral bioavailability, compatibility with some types of metabolism, and consequently low toxicity [27]. The secret to success in VS lies in the choice of an appropriate combination of methods [24].

To evaluate the performance of a given VS technique, validation studies are usually carried out [28]. For this purpose, known binders are pooled with a set of nonbinding or assumedly nonbinding compounds (decoys) to assess how well the known binders "enrich" during the course of the screening process. Such considerations are relevant for the development of new methods; however, in actual drug discovery projects, at the end of the day, a newly discovered chemotype or a drug-like lead structure represents the ultimate goal. Since the early VS attempts, a plethora of application studies have been performed, and the list of success stories is steadily growing. Despite the fact that VS is still a developing discipline, it has frequently been reviewed [4, 8, 24–26, 29, 30].

A major issue concerning VS results is the "quality" of the derived hits [31]. There are numerous VS studies that reported weakly active compounds, promiscuous binders, or assay interfering molecules and often a stringent biochemical or cellular validation is missing. To increase the "quality" of VS hits, a couple of approaches have been developed in the last years. In 2010, Baell and Holloway [32] analyzed compounds that showed activity in multiple assays and suggested that these compounds may interfere with the bioactivity detection technology. Such compounds were named pan-assay interference compounds (PAINS). Since then, the PAINS concept has been widely accepted by the scientific community, and several PAINS filters have been developed to triage VS hits flagged with these alerts prior to experimental validation. Also chemical databases, such as ZINC (http://zinc15.docking.org/) and ChEMBL (www.ebi.ac.uk/chembl/), flag now compounds containing PAINS alerts. Today, some scientific journals have started to recommend that all VS hit compounds should be passed through one of the available PAINS filters. However, several authors have noted that the application of the PAINS filtering could discard viable drug candidates because such alerts have actually been found in approved drugs [33].

3.2.5 Binding Free Energy Calculation

An accurate prediction of binding affinities for a diverse set of ligands using scoring functions turns out to be genuinely difficult [26]. In general, this is a problem of subtraction of large numbers, inaccurately calculated, to arrive at a small number. The large numbers are the protein–ligand interaction energy on one hand and the cost of bringing protein and ligand out of the solvent and into a complex on the other hand. The result of this subtraction is the free energy of binding, the small number we want to know. As long as full protein–ligand flexibility and the complexity of protein–ligand–water interactions are not covered in docking and scoring, it cannot be expected to end up with accurate affinity prediction. Of course, more accurate methods may be considered. Among the most accurate today are thermodynamic integration/free energy calculation methods, which can sometimes calculate the differences in affinities between related molecules within 1 kcal mol^{-1} [34]. The problem with these methods is that they demand more computation time as to be infeasible for a large compounds selection usually used in VS. Recent reports suggest that progress is being made in calculating absolute binding affinities; nevertheless, these methods still remain too slow for VS, though they may be useful in rescoring docking hit lists [35, 36]. For recent reviews on binding free energy calculation methods, see Refs. [37–39].

3.3 Histone Deacetylases

3.3.1 Zinc-Dependent HDACs

So far, 18 different members of HDACs have been discovered in humans and have been classified into four classes based on their homology to yeast HDACs [40]. Class I includes four different subtypes (HDAC1, HDAC2, HDAC3, and HDAC8), and class II contains six subtypes, which are divided into two subclasses: class IIa with subtypes HDAC4, HDAC5, HDAC7, and HDAC9, and class IIb with subtypes HDAC6 and HDAC10. Class I and class II HDACs share significant structural homology, especially within the highly conserved catalytic domain. HDACs 6 and 10 are unique as they have two catalytic domains. HDAC11 is referred to as class IV. While the activity of class I, II, and IV HDACs depends on a zinc-based catalysis mechanism, the class III enzymes, generally referred to as sirtuins, require nicotinamide adenine dinucleotide as a cofactor for their catalysis.

Co-crystal structures of HDACs in complex with hydroxamic acid-containing ligands have been useful in dissecting the molecular details of HDAC inhibition [41]. The first of these structures was obtained with histone deacetylase-like protein (HDLP) complexed with two different inhibitors [3]. Later, crystal structures of HDAC1 [42], HDAC2 [43], HDAC3 [44], HDAC4 [45], HDAC6 [46, 47], HDAC7 [48], HDAC8 [49], and HDAC10 [50] have been resolved (see Chapter 2 for more details). Crystal structures of HDAC5, HDAC9, and HDAC11 have not yet been reported. The architecture of the solved HDACs is very similar, with the residues in the active site being conserved between bacterial and human HDACs.

However, the organization of the loops and distal helices is different in the known HDAC structures, e.g. HDAC8 shows a more open and accessible substrate binding channel. The deacetylase active site with the zinc ion is located at the end of an approximately 12 Å long tunnel. The zinc ion is penta-coordinated by two aspartic acids and one histidine residue in addition to ligand and/or water molecules. The remainder of the substrate channel is made up of lipophilic amino acids that are highly conserved across the isoforms. Especially, two aromatic side chains that restrict the width of the tunnel can be found at the same position in the different HDAC structures. At the top of the binding cavity is a rim leading to the surface, which is formed by several loop regions that differ between subtypes. As an example, the binding modes of the co-crystallized inhibitors Nexturastat A (**1**) (PDB ID 5G0I) and Panobinostat (**2**) (PDB ID 5EF8) at HDAC6 are shown in Figure 3.1 (chemical structures of the inhibitors are shown in Figure 3.2).

In the crystal structures, the inhibitors coordinate either in a bidentate (Figure 3.1a) or monodentate way (Figure 3.1b) to the catalytic zinc ion and make a series of hydrogen bonds via their hydroxamic acid moiety. In the monodentate binding mode, a water molecule is bound to the zinc ion and is bridging the interaction with the hydroxamic acid. The hydroxamic acids are linked by a spacer with bulky cap groups. The spacer can be aromatic or aliphatic and participates in van der Waals interactions throughout the long acetyl-lysine channel, whereas the terminal part of the inhibitor interacts with residues at the rim of HDACs. In general, the binding mode of the co-crystallized pan-HDAC inhibitor trichostatin is conserved among the different species and subtypes [51].

Ligand-bound structures of the target protein provide a suitable starting point for structure-based drug design. Since the first of the solved crystal structure was that of HDLP, molecular modeling studies started with the analysis of the inhibitor binding mode for this bacterial enzyme. A variety of early studies were reported where structure-based optimization of HDAC inhibitors was successfully guided by docking the compounds into the active site [52–60] (see also Chapter 9). Homology models of HDAC1 and HDAC6 complexed with hydroxamate-based inhibitors were generated by several groups [61–72]. The refined models were generally checked for structural integrity using molecular dynamics (MD) simulations and various protein structure checking tools. Later the crystal structures of HDAC1 and HDAC6 could be solved, which confirmed the predicted binding mode of the reported inhibitors [42, 47].

Numerous VS studies have been reported using either pharmacophore-based modeling or docking to identify novel HDAC inhibitors [73]. However, often the described hits were either weakly active (IC_{50} value above 10 μM) or categorized as PAINS, or a clear cellular characterization has not been carried out [73]. Therefore, such VS hits are not suitable compounds for further lead optimization. In the following we will highlight a few successful VS projects with biologically validated hits.

At Argenta, a VS approach was initiated based on the published crystal structure (PDB ID 1C3R) of the HDLP [74]. A virtual library of 644 hydroxamic acids was generated from a database of available carboxylic acids. The designed compounds were docked and scored using the FlexX program [22]. Based on

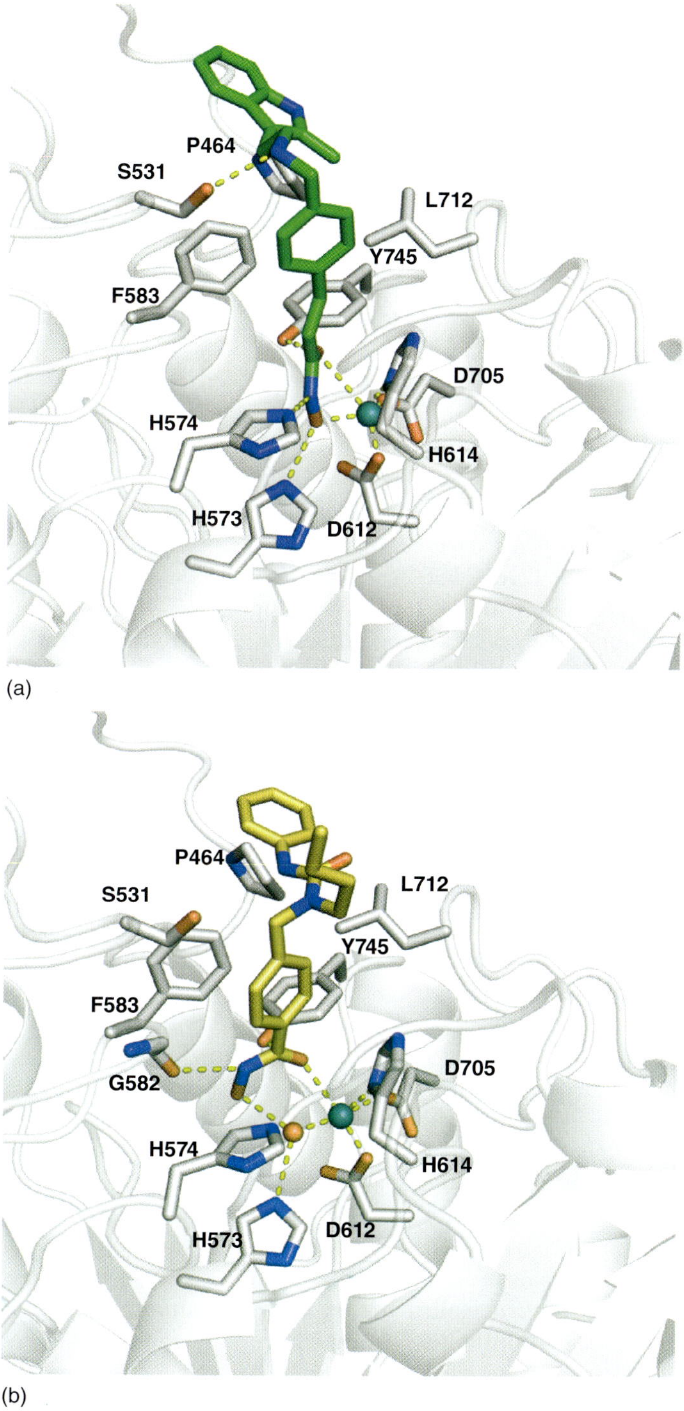

Figure 3.1 (a, b) X-ray structure of HDAC6 in complex with the inhibitors panobinostat (a) and nexturastat A (b). Only the residues in the binding pocket are shown. Hydrogen bonds are displayed as dashed line.

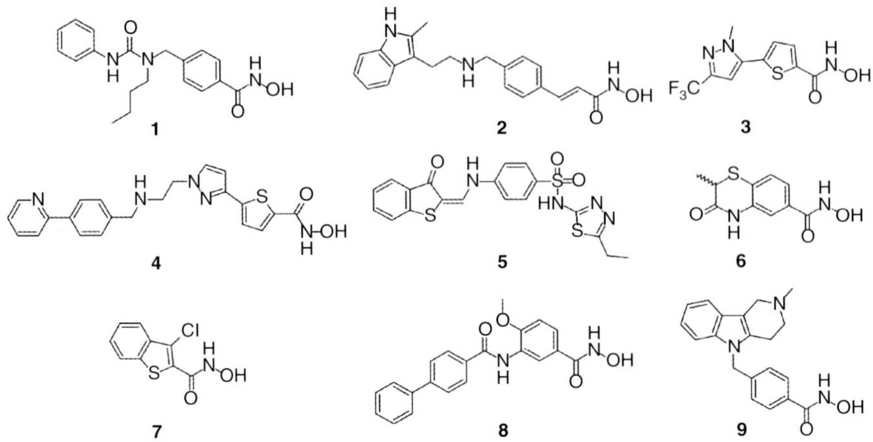

Figure 3.2 Molecular structures of HDAC inhibitors mentioned in the text.

the VS results, 75 compounds were selected and tested in an HDAC (subtype) enzyme assay. ADS100380 (**3**; Figure 3.2) was identified as a submicromolar (IC_{50} = 0.75 μM) HDAC inhibitor. Based on the docking results, optimization of ADS100380 was designed to make additional interactions at the entrance to the HDAC active site by tethering hydrophobic aromatic groups to the pyrazole nitrogen. Several nanomolar HDAC inhibitors were obtained by the structure-guided optimization process (**4**; Figure 3.2).

Tang et al. [75] applied a combination of QSAR and VS to identify inhibitors of HDAC1. Based on 59 known HDAC inhibitors and calculated MolConnZ and MOE chemical descriptors, they generated and validated a QSAR model. The model was used to predict the biological activity of 45 hits derived from a virtual screen of about 9.5 million purchasable compounds. Four molecules were tested *in vitro* against HDAC1 and HDAC6, resulting in 3 μM hits. However, all three hits were hydroxamic acids and structurally related to the known HDAC inhibitors SAHA and nexturastat A underlying the challenge in identifying novel non-hydroxamate HDAC inhibitors.

A structure-based design approach to identify non-hydroxamate-based HDAC1 inhibitors has been conducted by Park et al. [76]. A homology model of HDAC1 was generated using the available crystal structure of HDAC8 in complex with a hydroxamic acid, and, as chemical library for VS, a chemical database distributed by InterBioScreen Ltd. (www.ibscreen.com) comprising 460 000 synthetic and natural compounds was used. The compounds were docked using the docking program AutoDock [77] and a modified scoring function that included a novel solvation model. The 100 top-ranked compounds were purchased and tested *in vitro* against the enzyme. Six compounds could be identified as active HDAC1 inhibitors from which four molecules were non-hydroxamic acids. The most active hit (Cpd. **5**, IC_{50} = 4.2 μM; Figure 3.2) contained as putative zinc-chelating group a *N*-[1,3,4]-thiadiazol-2-yl sulfonamide that was not reported before for HDAC inhibitors. Even though the authors stated that the inhibitor might show better pharmacokinetic properties compared with hydroxamate derivatives, no cellular effects of **5** or related thiadiazole sulfonamides have been reported so far.

A successful VS campaign has also been published by Kannan et al. [78]. They screened for inhibitors of the HDAC8 orthologue from the human parasite *Schistosoma mansoni* in order to discover novel antiparasitics. The VS was carried out using a homology model of smHDAC8 and the GOLD docking program. Among 75 tested VS hits, about 35% showed activity against the parasitic enzyme, and nine compounds showed an IC_{50} in the micromolar range. The most promising hits were fragment-like compounds with a hydroxamate group such as T5979345 (**6**) and T6072858 (**7**; Figure 3.2). To prove the predicted binding mode, the compounds were later co-crystallized with smHDAC8 and were used for structure-guided optimization [79]. Based on the crystal structure of J1038 with smHDAC8, a series of benzhydroxamic acids were synthesized, which resulted in potent smHDAC8 inhibitors able to kill the parasite *ex vivo* (e.g. **8** $IC_{50} = 75$ nM; Figure 3.2) [80]. The crystal structure of one of the designed benzhydroxamates could nicely show the favorable interactions with the parasitic smHDAC8. Two hydrogen bonds between the amide linker and Lys20/His292 stabilize the inhibitor in the active site (Figure 3.3). This interaction is not possible in human HDAC1 and HDAC6 resulting in the selectivity for smHDAC8.

Besides the low number of successful VS campaigns for HDACs, numerous studies have been published where structure-based design approaches, such as

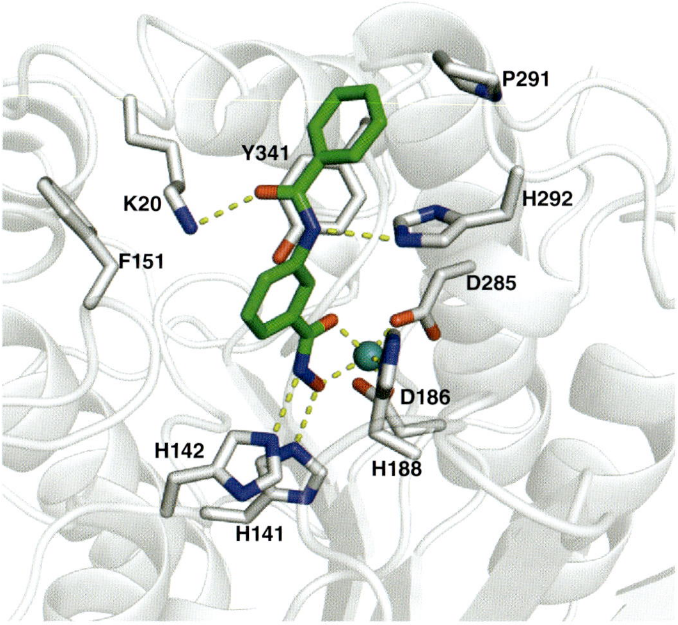

Figure 3.3 X-ray structure of smHDAC8 in complex with a benzhydroxamate-based inhibitor (PDB ID 5FUE) showing the interaction with the lysine channel. Only the residues in the binding pocket are shown. Hydrogen bonds are displayed as dashed line.

docking and binding free energy calculation, resulted in potent and selective HDAC inhibitors. Kozikowski and coworkers discovered in 2010 the HDAC6 inhibitor tubastatin A (**9**) via structure-based drug design combined with homology modeling [81]. The comparison of two homology models of HDAC1 and HDAC6 revealed that the catalytic channel rim differs greatly between both isoforms and suggested that this channel is wider and shallower in HDAC6 than in HDAC1. Therefore, a shorter aromatic moiety was proposed to fit this channel to possibly enhance the selectivity of the inhibitor for HDAC6. *In vitro* testing of Tubastatin A confirmed the predicted high selectivity for HDAC6.

Further examples of successful applications of computer- and structure-based approaches for the design of HDAC inhibitors can be found in two recently published reviews [82, 83].

3.3.2 Sirtuins

The class III deacetylases, named sirtuins, are structurally and functionally different from other HDACs. In contrast to the zinc-dependent deacetylation of classical HDACs, sirtuins depend on NAD^+ to carry out the catalytic reaction. A variety of sirtuin crystal structures have been published over the last few years. Crystal structures of sirtuins derived from prokaryotes [84–91] as well as human sirtuin subtypes, such as Sirt1 [92, 93], Sirt2 [94–97], Sirt3 [98–106], Sirt5 [101, 105, 107–109], and Sirt6 [110], have provided insights into the overall structure of sirtuins. So far no 3D structures of human Sirt4 and Sirt7 isoforms are available. For a detailed review, the interested reader is referred to [111]. All solved sirtuin structures contain a conserved 275 amino acid catalytic domain with variable N- and C-termini. The structure of the catalytic domain consists of a large classical Rossmann fold and a small zinc binding domain. The interface between the large and the small subdomain is commonly subdivided into A, B, and C pocket. This division is based on the interaction of adenine (A), ribose (B), and nicotinamide (C), which are parts of the NAD^+ cofactor (Figure 3.4a). The acetyl-lysine peptide substrate and NAD^+ cofactor bind to the active tunnel at the interface of two domains. Upon binding of the substrate or a pseudo-substrate, the zinc binding domain rotates toward the Rossmann-fold domain and induces a "closure" of the active cleft. This closure of both domains helps to correctly orientate the acetyl-lysine in the hydrophobic tunnel, allowing the formation of an hydrogen bond between the Nε atom of lysine and the backbone carbonyl of a conserved valine residue (Val233, Sirt2 numbering) as well as further van der Waals interactions with hydrophobic residues. Recently, Rumpf et al. [112] showed that sirtuins can also adopt a "locked-open conformation." It was shown that upon binding of the highly selective Sirt2 inhibitor SirReal2 (Figure 3.4b, the zinc binding domain rotates toward the Rossmann-fold domain; however, no closure of the substrate channel was observed even in the presence of a substrate. This newly discovered conformation of Sirt2 resembles an intermediate state between the apo, "open" conformation, and substrate bound, "closed" conformation.

Several inhibitors are available for Sirt2 that have been discovered by applying computer-based and VS approaches [113–115]. For example, one of the first

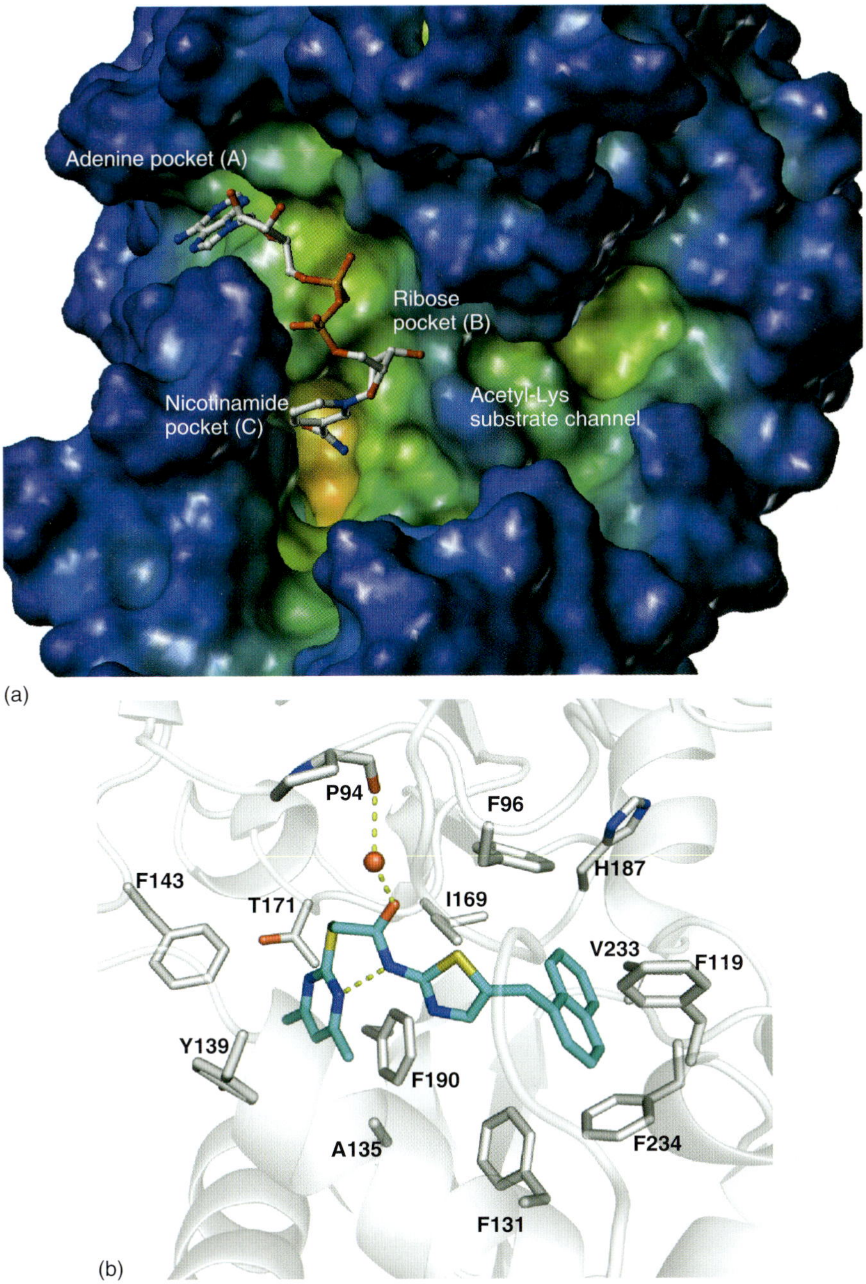

Figure 3.4 (a) X-ray structure of human Sirt2 in complex with NAD^+ (predicted by docking). The molecular surface of the protein is colored according to the cavity depth. (b) X-ray structure of human Sirt2 in complex with the selective inhibitor SirReal2 (PDB ID 4RMH). Only the residues in the binding pocket are shown. Hydrogen bonds are displayed as dashed line.

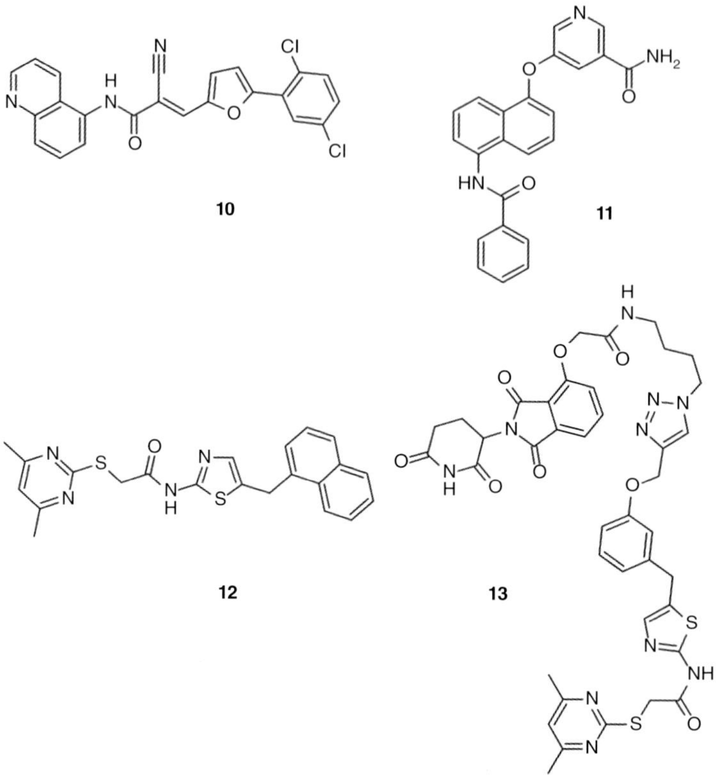

Figure 3.5 Molecular structures of sirtuin inhibitors mentioned in the text.

potent Sirt2 inhibitors, the vinylnitrile AGK2 (**10**; Figure 3.5), has been identified by focused library screening [116]. Outeiro et al. developed 3D models of human Sirt2 in different conformations by combining available human and yeast X-ray structures. The high flexibility of the active site loop made this strategy necessary. The different Sirt2 conformations were subsequently used for docking the inhibitors using the ICM program [117]. Comparative analysis of the low energy ligand conformations confirmed that the preferred site for ligand binding is pocket C where the nicotinamide part of NAD^+ is interacting. A hydrogen bonding pattern for AKG2 was observed, which is similar to that of other Sirt2 inhibitors. AGK2 inhibits Sirt2 with an IC_{50} value of 3.5 μM and shows more than 10-fold selectivity over Sirt1 and Sirt3. In further biological experiments, AGK2 displayed the ability to block α-synuclein-mediated toxicity in a Parkinson's disease model, possibly by modulating tubulin acetylation.

Fragment-based drug design has been used as an alternative computational tool as opposed to traditional VS approaches for the identification of novel sirtuin modulators.

Cui et al. [118] carried out a fragment-based approach to identify selective and potent Sirt2 inhibitors. The authors could successfully combine the chemical fragments of suramin and nicotinamide, which resulted in the design of

5-benzamidonaphthalen-1/2-yloxy- nicotinamide derivatives. Among these, compound **11** (Figure 3.5) showed nanomolar inhibitory activity against Sirt2 (IC_{50} = 0.0483 μM) and 200–900-fold selectivity over Sirt1 (IC_{50} = 12.0 μM) and Sirt3 (IC_{50} = 44.2 μM). The authors presented two putative binding modes based on docking studies of compound **11** into the Sirt2 crystal structure. While the nicotinamide moiety binds into the nicotinamide binding pocket and forms hydrogen bonds with the conserved residues Ile169 and Asp170 in both docking poses, the naphthalene group makes π–π interactions either with Phe96 and Phe119 or Phe119 and His187 in model 1 and model 2, respectively.

Rumpf et al. described the X-ray structures of human Sirt2 with a series of compounds bearing an aminothiazole scaffold [112]. These inhibitors are among the most potent and selective Sirt2 inhibitors. Interestingly, these inhibitors – called sirtuin rearranging ligands (SirReals) – were found to rearrange the Sirt2 binding pocket and behave as allosteric inhibitors. In addition, the original lead structure, SirReal2 (**12**; Figure 3.5), was found to be highly selective over other sirtuin isotypes. By solving the X-ray structures of several SirReals with Sirt2, it was shown that the isoform selective inhibition by these inhibitors is derived from a ligand-induced structural rearrangement of the active site. SirReal2 binds to a highly hydrophobic pocket adjacent to the zinc binding domain and does not prevent the binding of NAD^+ in its productive conformation. Guided by the structural insights obtained from the hSirt2–SirReal2 crystal structure, the authors aimed to systematically probe the limits of variation within the scaffold of the SirReals [119]. About 50 synthesized aminothiazoles were docked into the Sirt2 crystal structure and were used for binding free energy calculations. A good correlation was observed between the experimental inhibition data and the calculated MM-GB/SA protein-inhibitor binding energies. Moreover, the authors were able to rationalize their results with a crystal structure of hSirt2 in complex with the most potent inhibitor from this series.

To probe the cellular effects of Sirt2, not only by small-molecule enzyme inhibitors but also by eliminating the protein with all its scaffolding functions from the cellular network, Schiedel et al. [120] used the derived crystal structures of SirReals in complex with Sirt2 to develop so-called proteolysis targeting chimeras (PROTAC) as Sirt2 inhibitors. The PROTAC concept is based on bifunctional small molecules that are capable of hijacking the cellular quality control by recruiting the protein of interest (POI) to E3 ubiquitin ligases for polyubiquitinylation and thus to induce its proteasomal degradation. Since its first description by Crews and coworkers in the year 2001 [121], the PROTAC concept has been widely applied to induce the degradation of various proteins such as kinases, transcription factors, and epigenetic reader proteins [122]. Based on the crystal structure of a SirReal/Sirt2 and thalidomide/cereblon complex and a protein–protein docking approach, Schiedel et al. designed a SirReal-PROTAC (**13**; Figure 3.5) with an appropriate linker able to interact with both binding pockets of the Sirt2/cereblon complex (Figure 3.6). In HeLa cells, the SirReal-PROTAC induced isotype-selective Sirt2 degradation that results in the hyperacetylation of the microtubule network coupled with

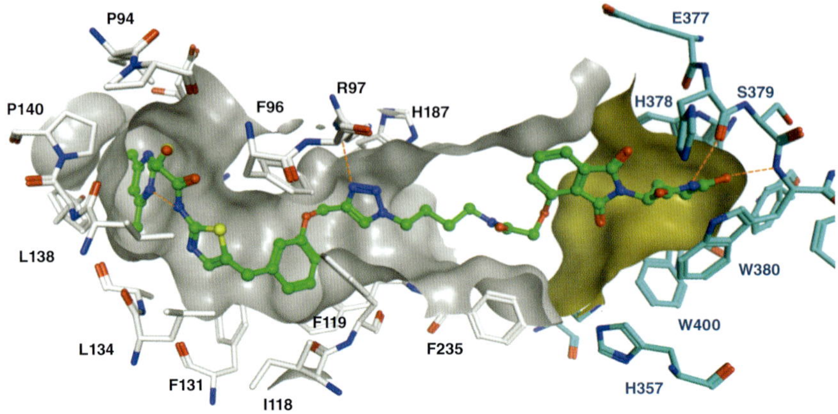

Figure 3.6 PROTAC-Sirt2/cereblon complex. Only the residues of both binding pockets are shown. Hydrogen bonds are displayed as dashed orange lines.

enhanced process elongation. The SirReal-PROTAC **13** was the first example of a probe that is able to chemically induce the degradation of an epigenetic eraser protein.

3.4 Histone Methyltransferases

Histone methyltransferases can be subdivided into three classes: SET-domain-containing lysine methyltransferases, non-SET-domain lysine methyltransferases, and protein arginine methyltransferases (PRMTs). All of them use S-adenosyl-L-methionine (SAM) as a co-substrate for methylation, and S-adenosylhomocysteine (SAH) is formed as a by-product [123]. Similar to other druggable target classes like protein kinases, PKMTs and PRMTs also share a common mechanism of catalysis, whereby the cofactor SAM is the donor for the transfer of a methyl group to a nitrogen atom of either a lysine or arginine side chain. Unlike histone acetylation, histone methylation does not alter the charge of the lysine residue in question. It affects the basicity, hydrophobicity, and the size of the amino acid side-chain group, which affects the affinity of proteins that recognize such side chains. Although these changes are subtle, there are proteins that are able to bind selectively to certain methylated residues.

So far, more than 20 lysine methyltransferases [124] and 11 protein arginine methyltransferases (PRMT1–1) have been identified, but enzymatic activity could not be demonstrated for all of the members [125–127]. Lysine methyltransferases have been designated historically with a variety of names. Eventually, a common nomenclature for chromatin-modifying enzymes has been proposed instead of these. The human lysine methyltransferases have been renamed KMTs and are subdivided into eight classes [128]. Many of the methyltransferases have been linked to cancer, and, therefore, the development of inhibitors is desirable [129].

By applying a combination of structure-based VS and biochemical characterization, two drug-like substrate-competitive PRMT1 inhibitors, namely, allantodapsone (**14**) and stilbamidine (**15**; Figure 3.7), have been reported [130]. Based on a generated homology model of human PRMT1 and subsequent docking into the substrate binding site, key interactions between the identified inhibitors and PRMT1 could be derived. A common feature of the actively tested inhibitors is the hydrogen bonding of a basic or polar group with the active site Glu152. A basic or polar moiety of the inhibitors mimics the guanidine group of the arginine side chain of the peptide substrate. In addition, the inhibitors show van der Waals interactions with several aromatic residues of the binding pocket (Tyr47, Tyr156, Trp302). The inhibitors induced hypomethylation at H4R3 in HepG2 cells. Functional activity was confirmed in a reporter gene assay with MCF7a cells. Both inhibitors showed a reduction of estrogen receptor activation by estradiol in a dose-dependent manner. Competition studies using histone substrate and cofactor SAM showed that the identified inhibitors are substrate-competitive inhibitors, which is in agreement with the obtained docking results. In the further course of research, larger databases were screened for new arginine methyltransferase inhibitors with a modified screening procedure including a pharmacophore prefilter. This approach led to the discovery of inhibitors **16**, **17**, and **18** (see Figure 3.7) from the ChemBridge database with inhibitory effects down to 13 μM [131].

Schapira and coworkers reported on the successful application of a structure-based VS to identify inhibitors for the PRMT4 isoform that is also named CARM1 [132]. They started their VS with a structural analysis of available PRMTs and inhibitors. A common feature that they identified was a basic amine tail that anchored in the PRMT substrate arginine binding channel. Deconstructing the already known inhibitors showed that fragments preserving the basic tail bind with exceptionally high ligand efficiency, probably through interaction with a conserved catalytic glutamate side chain, conserved in PRMTs. Then they built a PRMT-focused virtual library composed of commercially available compounds where diverse scaffolds were appended to the basic amine tails from known PRMT inhibitors. The virtual library was derived by filtering the ZINC database (~22 million compounds), resulting in ~132 000 compounds that were subsequently used for docking studies. The top-ranked docking poses were rescored using a MM-GB/SA protocol (AMBER software package) and applying a short 1ns MD simulation for the most promising hits. This stepwise reduction of hits resulted in 51 promising compounds that were tested *in vitro* against CARM1. A valuable hit with an IC_{50} of 1.9 μM could be derived, and after two rounds of structure-based optimization, a nanomolar lead **19** (IC_{50} CARM1 50 nM; Figure 3.7), with excellent selectivity against other PRMT isoforms, could be developed.

Another VS on PRMTs was recently reported by Zheng and coworkers [133]. The authors used the crystal structure of PRMT5 and docked the filtered SPECS library to the substrate binding pocket of this PRMT isoform. Among the purchased and tested 42 hits, a low micromolar hit could be obtained (IC_{50} = 35.6 μM) that was further optimized to compound DC_C01 (**20**; Figure 3.7) with an IC_{50} of 2.8 μM. Selectivity was tested and confirmed against

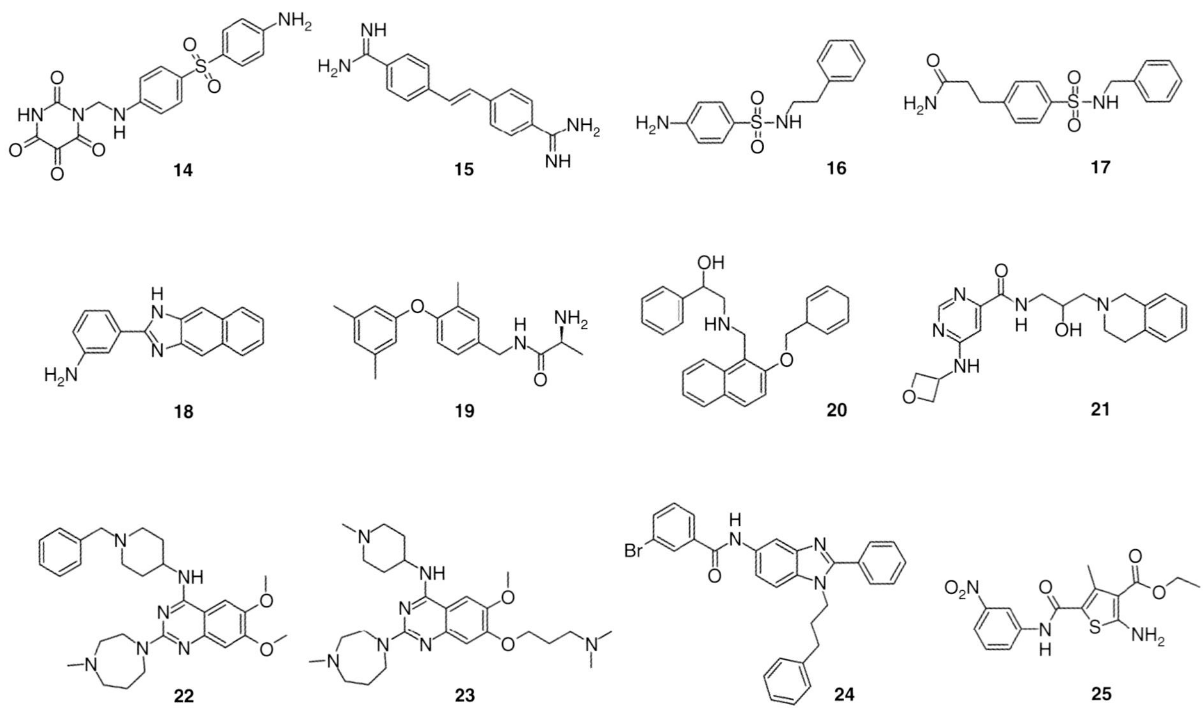

Figure 3.7 Molecular structures of histone methyltransferase inhibitors mentioned in the text.

PRMT1 and the lysine methyltransferase EZH2. However, compared with the potency of the reported reference inhibitor for PRMT5, EPZ015666 with an IC_{50} of 22 nM (**21**; Figure 3.7), the optimized hit is still ~100 times less potent.

Jenuwein et al. applied a combination of VS and HTS to screen the Boehringer Ingelheim chemical compound library for inhibitors of the lysine methyltransferase G9a. A similarity searching approach was carried out using nine known inhibitors of arginine and lysine methyltransferases [134, 135]. The approach was used both for a ligand-based and protein-based VS. For the protein-based screening, a homology model based on the crystal structure of DIM-5 [136], a lysine methyltransferase with 30% identity in the SET domain to G9a, was developed. For both protein- and ligand-based screening, top-scored compounds were selected. Additional compounds were randomly selected, and a total of 125 000 substances were subsequently evaluated in HTS. The experimental *in vitro* testing on G9a revealed seven hits with activities in the low micromolar range. Besides an unselective inhibitor, the selective inhibitor BIX-01294 (**22**; Figure 3.7) was identified. BIX-01294 inhibited G9a at 2.7 μM (IC_{50}) and showed no activity against SUV39H1 or PRMT1. Cell lines treated with BIX-01294 showed a reduction in histone H3 lysine 9 dimethylation, and the mono- or trimethyl stages appeared unaffected. In addition, the mode of action for two inhibitors was examined by measuring the dependence of the reaction kinetics on the concentration of the cofactor SAM. The quinazoline-based scaffold of BIX-01294 has been further optimized by structure–activity relationship (SAR) exploration with the support of X-ray co-crystal structures, leading to the discovery of the highly potent inhibitor UNC0224 (**23**; Figure 3.7) [137].

Benzoxazole and benzimidazole derivatives were discovered by VS using the available G9a crystal structure and the ChemBridge CORE library. A structure-guided chemical optimization resulted in compound GA001 (**24**; Figure 3.7) with moderate G9a inhibition and cellular activity (*in vitro* $IC_{50} = 1.32$ μM, MCF7 cells $IC_{50} = 5.73$ μM) [138].

Also for the lysine methyltransferase SET7 (also called KMT7, SETD7, SET9), a target-based VS resulted in the identification of micromolar hits [139]. The SPECS library was filtered and subsequently docked into the SAM cofactor binding site of SET7. Pharmacophore filtering and visual analysis of the putative binding mode paved the way for structure-guided optimization of the most promising hit. Of note, compound DC-S239 (**25**; Figure 3.7), with IC_{50} value of 4.59 μM, displayed selectivity for SET7 against the methyltransferases DOT1L, EZH2, NSD1, SETD8, and G9a.

3.5 Histone Demethylases

In 2004, Shi and coworkers discovered the first demethylase that specifically removes methyl groups from histone H3K4 and named it lysine-specific demethylase 1 (LSD1) [140]. LSD1 (also named KDM1A) shows a high sequence homology with FAD-dependent amine oxidases. Two years later, another family of histone demethylases was identified and termed JmjC protein [141]. These demethylases use Fe(II) and α-ketoglutarate (α-KG) in an oxygenation reaction to remove the methyl groups from lysine residues. Since the expression levels of several of these demethylases are increased in primary tumors, histone demethylases are considered as potential drug targets [142]. Therefore, it is not surprising that a huge effort is made to explore the structural requirements of modulating these targets and to identify small-molecule inhibitors. A variety of crystal structures have been solved in the last four years, which helped in understanding the binding of substrates and first available unselective inhibitors (for more details see Chapter 2) [143].

3.5.1 LSD1 (KDM1A)

As revealed by its crystal structure, LSD1 is composed of three domains: the tower domain (aa 419–520), the SWIRM domain (aa 166–260), and the amine oxidase domain (AOD) (aa 520–852) [144, 145]. The catalytic site is located in the AOD, where both the cofactor FAD and the substrate bind [146]. The AOD of LSD1 shares a significant sequence similarity with other FAD-dependent amine oxidases. It shows a 26% homology with plant amine oxidase (PAO) and 20% homology with MAO-A and MAO-B enzymes [147]. While the FAD binding pocket shares high similarity with the other amine oxidases, substantial differences are found in the subdomain, which accommodates the natural substrate [148]. This subdomain is much more spacious and open in LSD1, and it harbors a deep polar pocket in order to accommodate the peptide substrate. Moreover, the classical aromatic cage found in other amine oxidases, which is responsible for the recognition of the positively charged methylated amine through cation–π interactions, is missing in LSD1. Only one aromatic amino acid (Tyr761) is preserved, whereas the second one is replaced by Thr810. There are currently more than 60 crystal structures of LSD1 available in the Protein Data Bank (PDB). For a detailed description of the structural features, see Chapter 11. The available crystal structures have already been used for structure-based optimization of LSD1 inhibitors and for VS of new chemotypes. Docking and VS of LSD1 small-molecule inhibitors is challenging due to the huge size of the binding pocket (more than 1700 $Å^3$) as well as the fact that many known inhibitors are irreversible inhibitors. Structurally related proteins that should not be inhibited by LSD1 inhibitors are the amine oxidases MAO-A and MAO-B.

The first successful VS of LSD1 inhibitors was reported by Schmitt et al. in 2013 [149]. They used the combination of similarity-based screening and docking to the LSD1 substrate pocket to identify irreversible inhibitors. The Enamine compound collection (~750 000 cpds.) was first screened for compounds bearing a warhead known from irreversible amine oxidase inhibitors such as tranylcypromine or pargyline. The 3-aryloxy-2-hydroxypropylamine T5342129 (**26**; Figure 3.8) with an *N*-propargyl group was identified as an LSD1 inhibitor with a potency of 44 μM in a hydrogen peroxide-dependent assay and 34 μM in a formaldehyde dehydrogenase (FDH) assay. The irreversible inhibitor hypothesis was confirmed in an *in vitro* dilution assay. The cellular activity was confirmed using MCF7 cells and measuring the increased level of H3K4 dimethylation after inhibitor incubation.

Sorna et al. carried out a structure-based VS to identify reversible LSD1 inhibitors [150]. Computational docking and scoring was done using Glide and ICM [21, 117] by stepwise reducing the size of the compound collection (ZINC database ~13 million cpds.). About 10 000 molecules were docked to the FAD cofactor binding pocket and gave, after inspection of the docking poses, 121 hits that were tested *in vitro*. Six submicromolar hits could be identified that all share a *N*′-(1-phenylethylidene)benzohydrazide scaffold. Further chemical optimization and SAR studies aided in the discovery of compound **27** (Figure 3.8), with a K_i of 31 nM. Compound **27** was shown to be a reversible and specific inhibitor of LSD1 and showed a noncompetitve binding mode in kinetic experiments. Cellular effects such as an increase of the H3K9me2 mark and growth inhibition of several cancer cell lines were reported.

A combination of pharmacophore filtering and docking was applied by Zhou et al. [151] to identify novel scaffolds as putative LSD1 inhibitors. They filtered the SPECS database (about 170 000 compounds) by a pharmacophore model generated on the basis of 22 known LSD1 inhibitors. Even if different types of inhibitors (e.g. irreversible tranylcypromine-like compounds, FAD-competitive, and non-competitive inhibitors) were considered, the authors generated common pharmacophore hypotheses. The pharmacophore screening resulted in about 7000 molecules that were further filtered. Finally, 950 compounds were docked to the LSD1 substrate binding pocket. Among nine *in vitro* tested compounds, XZ09 (**28**; Figure 3.8) showed an IC_{50} value of 2.41 μM and selectivity for LSD1 compared with MAO-A/B. However, an experimental validation concerning target engagement and cellular effects has not been reported so far.

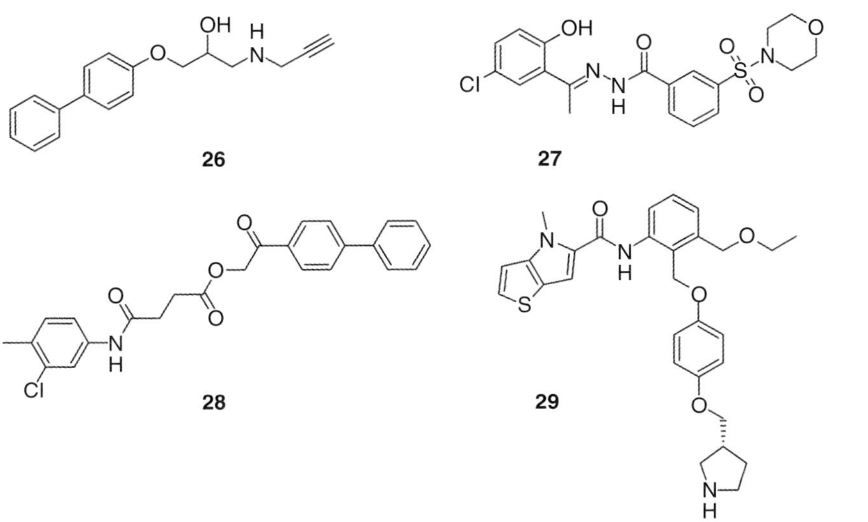

Figure 3.8 Molecular structures of LSD1 histone demethylase inhibitors mentioned in the text.

Recently, several groups combined the structural data available for LSD1 with medicinal chemistry approaches to guide the optimization of novel inhibitors [152–157]. These studies resulted in several potent and selective inhibitors belonging to novel chemotypes. Vianello et al. were also able to co-crystallize the optimized thieno[3,2-b]pyrrole-5-carboxamide **29** (IC_{50} = 7.8 nM; Figure 3.8) with LSD1, which shows a reversible binding mode in the H3-substrate binding pocket [158].

3.5.2 Jumonji Histone Demethylases

The Jumonji-type demethylases belong to the cupin superfamily of enzymes. They show a conserved β-barrel fold that harbors the binding site for Fe^{2+} and the cofactor α-KG. Coordination of Fe^{2+} is achieved through two histidines and an aspartate or glutamate. Besides the coordination with the two histidine residues and the acidic amino acid from the conserved motif, the octahedral coordination of Fe^{2+} is completed by the cofactor α-KG and a water molecule. The binding pocket of α-KG itself shows less conservation among the subfamilies and can hence be exploited for the development of selective inhibitors. Numerous crystal structures of JmjC-domain-containing proteins in complex with inhibitors have appeared in the last few years, which gives an important insight into the inhibitory mechanism of these compounds (for a review, the reader can refer to [159]). All inhibitors reported so far are metal chelators, which bind competitively in the α-KG binding pocket. This is the main reason why many of these inhibitors show an inhibitory activity at different JmjC-protein subtypes as well as at other Fe(II)/α-KG-dependent enzymes, such as prolyl hydroxylase domain-containing proteins (PHDs) and factor-inhibiting hypoxia-inducible factor (FIH). Crystal structure analysis showed that JMJDs possess a subpocket, adjacent to the α-KG binding pocket, which is larger and more open than in the related enzymes FIH and PHD in order to accommodate the peptide substrate.

In recent years, several lead compounds have been disclosed as inhibitors of KDMs. Although most of these compounds lack selectivity and are expected to act at related and undesired targets, some breakthroughs have been made toward finding selective inhibitors. While targeting the metal ion is obviously necessary for the activity of the inhibitors, the simultaneous targeting of the substrate binding pocket, which shows less conservation among α-KG-dependent oxygenases, seems to be profitable in achieving selectivity [160].

The first VS for JumonjiC proteins has been reported by Chu et al. [161]. The authors used available crystal structures of JMJD2A (KDM4A) and docked the NCI dataset into the binding pocket. From the top-ranked 3000 they selected 10 compounds for *in vitro* testing and identified compound **30** (Figure 3.9) as a micromolar inhibitor of JMJD2A (IC_{50} = 6.4 μM) and JMJD2B (IC_{50} = 9.3 μM). The compound showed selectivity against JMJD2D and JMJD2E and increased the level of the JMJD2A/B-depending mark H3K9me3 in LNCaP cells. The inhibitor does not contain a classical metal ion chelating moiety, and the binding mode of the molecule is currently not clear, even if an interaction between the

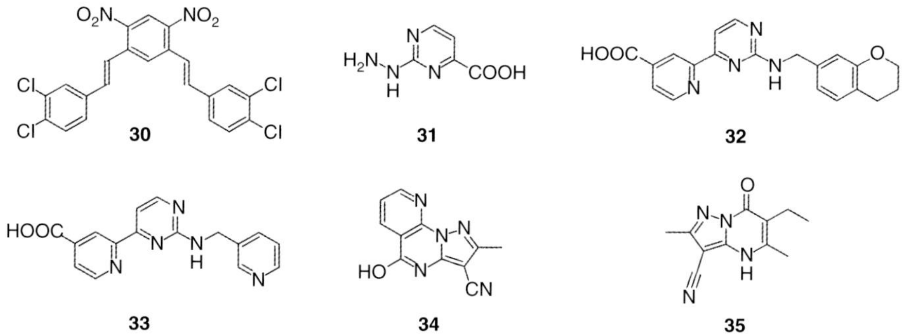

Figure 3.9 Molecular structures of JumonjiC histone demethylase inhibitors mentioned in the text.

nitro group and the metal ion of JMJD2A has been predicted by the docking study.

Wu et al. [162] generated a homology model of JARID1a (also known as KDM5a) using JMJD2 (KDM4) crystal structures as template. Taking the homology model, they set up a VS and identified a micromolar hit with an aminopyrimidine moiety as metal chelating moiety. The hit was further optimized to the submicromolar inhibitor **31** (Figure 3.9). The pyrimidinecarboxylic acid is mimicking the pyridinecarboxylic core structure known already from other JMJD2 inhibitors. Cellular testing showed that the compound is able to increase the level of H3K4me3 in a dose-dependent manner.

A potent JMD2A (KDM4A) inhibitor was identified in a recent VS study [163]. Roatsch et al. filtered several compound collections for molecules containing a metal ion chelating moiety. The retrieved hits were docked into the ketoglutarate binding pocket of JMJD2A, and the top-ranked compounds were tested *in vitro* against several KDMs. Compound **32** (Figure 3.9) was found to be a submicromolar inhibitor of KDM4A (IC_{50} = 370 nM). A co-crystal X-ray structure of one of the VS hits (**33**; Figure 3.9) revealed the mode of binding of these compounds as competitive to 2-oxoglutarate as predicted by the docking study (Figure 3.10). Selectivity studies revealed a preference for JMJD2A and JARID1A over JMJD3. Since the carboxylic acid derivatives showed weak cellular effects (due to penetration problems), ester prodrugs were synthesized that showed an inhibition of proliferation of KYSE-150 cells.

To retrieve new JMJD2D (KDM4D) inhibitors, Fang et al. performed molecular docking-based VS against various chemical libraries [164]. They used the GOLD docking program and docked the compounds into the ketoglutarate binding pocket of JMJD2D and used the GoldScore scoring function to rank the docking poses. Thirty compounds were tested *in vitro*, and a pyrazolo[1,5-a]pyrimidine-3-carbonitrile showed an inhibition in the primary screening. Further chemical modification resulted in compound **34** (Figure 3.9), which showed submicromolar (IC_{50} = 410 nM) inhibition of the enzyme. The docking study predicted the nitrile as metal chelating moiety, which is in

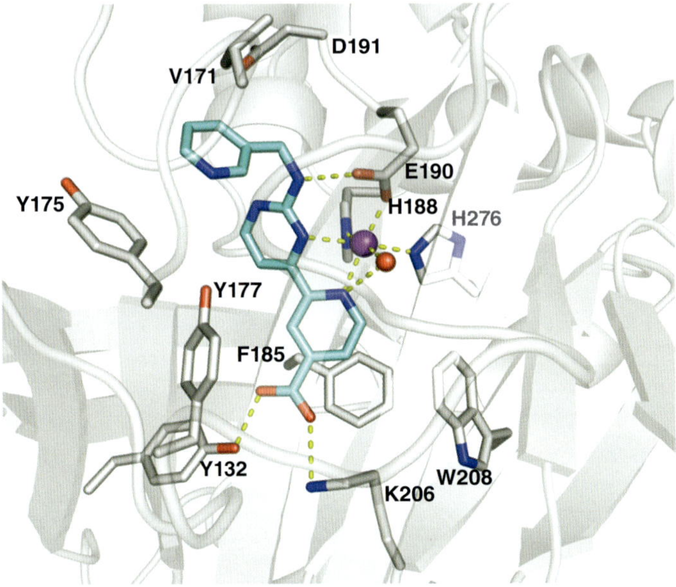

Figure 3.10 X-ray structure of the VS hit **35** with the histone demethylase JMJD2A (KDM4A, PDB ID 5ANQ).

agreement with a recently reported crystal structure of JARID1A (KDM5A) in complex with a pyrazolopyrimidine-3-carbonitrile (**35**; Figure 3.9) that shows high structural similarity to compound **34** [165]. However, no cellular activity of compound **34** has been reported so far.

3.6 Summary

Despite many technical challenges, structure-based VS has an important role in drug discovery. As a steadily growing number of epigenetic targets are characterized biologically and also structurally, structure-based methods are more and more applied to design specific inhibitors to elucidate their therapeutic potential. Ligand docking and scoring technologies have steadily improved, and the importance of the adequate validation of pragmatic VS protocols is now well recognized. While both ligand- and receptor-based approaches have already demonstrated their worth in the identification of novel lead compounds for epigenetic targets, there is still the challenge to improve the predictive accuracy of scoring functions, particularly to enable scoring-based methods to have a greater impact in guiding lead optimization.

Acknowledgments

C. L. is supported by the PhD program of Saxony-Anhalt. We also thank the CM1406 European COST action "Epigenetic Chemical Biology" for support.

References

1 Wolffe, A.P. and Matzke, M.A. (1999). Epigenetics: regulation through repression. *Science* 286: 481–486.
2 Kouzarides, T. (2007). Chromatin modifications and their function. *Cell* 128: 693–705.
3 Finnin, M.S., Donigian, J.R., Cohen, A. et al. (1999). Structures of a histone deacetylase homologue bound to the TSA and SAHA inhibitors. *Nature* 401: 188–193.
4 Kirchmair, J., Distinto, S., Schuster, D. et al. (2008). Enhancing drug discovery through in silico screening: strategies to increase true positives retrieval rates. *Curr. Med. Chem.* 15: 2040–2053.
5 Hawkins, P.C.D., Skillman, A.G., and Nicholls, A. (2007). Comparison of shape-matching and docking as virtual screening tools. *J. Med. Chem.* 50: 74–82.
6 Evers, A., Hessler, G., Matter, H., and Klabunde, T. (2005). Virtual screening of biogenic amine-binding G-protein coupled receptors: comparative evaluation of protein- and ligand-based virtual screening protocols. *J. Med. Chem.* 48: 5448–5465.
7 Kirchmair, J., Laggner, C., Wolber, G., and Langer, T. (2005). Comparative analysis of protein-bound ligand conformations with respect to catalyst's conformational space subsampling algorithms. *J. Chem. Inf. Model.* 45: 422–430.
8 Taboureau, O., Baell, J.B., Fernandez-Recio, J., and Villoutreix, B.O. (2012). Established and emerging trends in computational drug discovery in the structural genomics era. *Chem. Biol.* 19: 29–41.
9 Wermuth, G., Ganellin, C.R., Lindberg, P., and Mitscher, L.A. (1998). Glossary of terms used in medicinal chemistry (IUPAC recommendations 1998). *Pure Appl. Chem.* 70: 1129–1143.
10 Ahlstrom, M.M., Ridderstrom, M., Luthman, K., and Zamora, I. (2005). Virtual screening and scaffold hopping based on GRID molecular interaction fields. *J. Chem. Inf. Model.* 45: 1313–1323.
11 Muegge, I. and Mukherjee, P. (2016). An overview of molecular fingerprint similarity search in virtual screening. *Expert Opin. Drug Discovery* 11: 137–148.
12 Cramer, R.D., Patterson, D.E., and Bunce, J.D. (1988). Comparative molecular-field analysis (Comfa).1. Effect of shape on binding of steroids to carrier proteins. *J. Am. Chem. Soc.* 110: 5959–5967.
13 Klebe, G. and Abraham, U. (1993). On the prediction of binding-properties of drug molecules by comparative molecular-field analysis. *J. Med. Chem.* 36: 70–80.
14 Baroni, M., Costantino, G., Cruciani, G. et al. (1993). Generating optimal linear PLS estimations (GOLPE) – an advanced chemometric tool for handling 3D-QSAR problems. *Quant. Struct.-Act. Relat.* 12: 9–20.
15 Cramer, R.D. (2014). Template CoMFA applied to 116 biological targets. *J. Chem. Inf. Model.* 54: 2147–2156.

16 Cruciani, G.E. (2006). *Methods and Principles in Medicinal Chemistry – Molecular Interaction Fields*. New York: Wiley-VCH.
17 Speck-Planche, A. and Cordeiro, M.N. (2015). Multitasking models for quantitative structure-biological effect relationships: current status and future perspectives to speed up drug discovery. *Expert Opin. Drug Discovery* 10: 245–256.
18 Doweyko, A.M. (2004). 3D-QSAR illusions. *J. Comput.-Aided Mol. Des.* 18: 587–596.
19 Osterberg, F., Morris, G.M., Sanner, M.F. et al. (2002). Automated docking to multiple target structures: incorporation of protein mobility and structural water heterogeneity in AutoDock. *Proteins* 46: 34–40.
20 Jones, G., Willett, P., Glen, R.C. et al. (1997). Development and validation of a genetic algorithm for flexible docking. *J. Mol. Biol.* 267: 727–748.
21 Friesner, R.A., Banks, J.L., Murphy, R.B. et al. (2004). Glide: a new approach for rapid, accurate docking and scoring. 1. Method and assessment of docking accuracy. *J. Med. Chem.* 47: 1739–1749.
22 Rarey, M., Kramer, B., Lengauer, T., and Klebe, G. (1996). A fast flexible docking method using an incremental construction algorithm. *J. Mol. Biol.* 261 (3): 470–489.
23 Xu, M. and Lill, M.A. (2013). Induced fit docking, and the use of QM/MM methods in docking. *Drug Discov. Today Technol.* 10: e411–e418.
24 Chen, Y.C. (2015). Beware of docking. *Trends Pharmacol. Sci.* 36: 78–95.
25 Tame, J.R. (2005). Scoring functions – the first 100 years. *J. Comput.-Aided Mol. Des.* 19: 445–451.
26 Leach, A.R., Shoichet, B.K., and Peishoff, C.E. (2006). Prediction of protein–ligand interactions. Docking and scoring: successes and gaps. *J. Med. Chem.* 49: 5851–5855.
27 Ripphausen, P., Nisius, B., and Bajorath, J. (2011). State-of-the-art in ligand-based virtual screening. *Drug Discovery Today* 16: 372–376.
28 Kolb, P. and Irwin, J.J. (2009). Docking screens: right for the right reasons? *Curr. Top. Med. Chem.* 9: 755–770.
29 Lionta, E., Spyrou, G., Vassilatis, D.K., and Cournia, Z. (2014). Structure-based virtual screening for drug discovery: principles, applications and recent advances. *Curr. Top. Med. Chem.* 14: 1923–1938.
30 Stumpfe, D., Ripphausen, P., and Bajorath, J. (2012). Virtual compound screening in drug discovery. *Future Med. Chem.* 4: 593–602.
31 Ripphausen, P., Nisius, B., Peltason, L., and Bajorath, J. (2010). Quo vadis, virtual screening? A comprehensive survey of prospective applications. *J. Med. Chem.* 53: 8461–8467.
32 Baell, J.B. and Holloway, G.A. (2010). New substructure filters for removal of pan assay interference compounds (PAINS) from screening libraries and for their exclusion in bioassays. *J. Med. Chem.* 53: 2719–2740.
33 Capuzzi, S.J., Muratov, E.N., and Tropsha, A. (2017). Phantom PAINS: problems with the utility of alerts for pan-assay interference compounds. *J. Chem. Inf. Model.* 57: 417–427.

34 Pearlman, D.A. (2005). Evaluating the molecular mechanics Poisson–Boltzmann surface area free energy method using a congeneric series of ligands to p38 MAP kinase. *J. Med. Chem.* 48: 7796–7807.

35 Chang, C.E. and Gilson, M.K. (2004). Free energy, entropy, and induced fit in host-guest recognition: calculations with the second-generation mining minima algorithm. *J. Am. Chem. Soc.* 126: 13156–13164.

36 Deng, Y.Q. and Roux, B. (2006). Calculation of standard binding free energies: aromatic molecules in the T4 lysozyme L99A mutant. *J. Chem. Theory Comput.* 2: 1255–1273.

37 Genheden, S. and Ryde, U. (2015). The MM/PBSA and MM/GBSA methods to estimate ligand-binding affinities. *Expert Opin. Drug Discovery* 10: 449–461.

38 Broomhead, N.K. and Soliman, M.E. (2017). Can we rely on computational predictions to correctly identify ligand binding sites on novel protein drug targets? Assessment of binding site prediction methods and a protocol for validation of predicted binding sites. *Cell Biochem. Biophys.* 75: 15–23.

39 Ryde, U. and Soderhjelm, P. (2016). Ligand-binding affinity estimates supported by quantum-mechanical methods. *Chem. Rev.* 116: 5520–5566.

40 de Ruijter, A.J., van Gennip, A.H., Caron, H.N. et al. (2003). Histone deacetylases (HDACs): characterization of the classical HDAC family. *Biochem. J.* 370: 737–749.

41 Hodawadekar, S.C. and Marmorstein, R. (2007). Chemistry of acetyl transfer by histone modifying enzymes: structure, mechanism and implications for effector design. *Oncogene* 26: 5528–5540.

42 Millard, C.J., Watson, P.J., Celardo, I. et al. (2013). Class I HDACs share a common mechanism of regulation by inositol phosphates. *Mol. Cell* 51: 57–67.

43 Bressi, J.C., Jennings, A.J., Skene, R. et al. (2010). Exploration of the HDAC2 foot pocket: synthesis and SAR of substituted N-(2-aminophenyl)benzamides. *Bioorg. Med. Chem. Lett.* 20: 3142–3145.

44 Watson, P.J., Fairall, L., Santos, G.M., and Schwabe, J.W.R. (2012). Structure of HDAC3 bound to co-repressor and inositol tetraphosphate. *Nature* 481: 335–340.

45 Bottomley, M.J., Lo Surdo, P., Di Giovine, P. et al. (2008). Structural and functional analysis of the human HDAC4 catalytic domain reveals a regulatory structural zinc-binding domain. *J. Biol. Chem.* 283: 26694–26704.

46 Miyake, Y., Keusch, J.J., Wang, L.L. et al. (2016). Structural insights into HDAC6 tubulin deacetylation and its selective inhibition. *Nat. Chem. Biol.* 12: 748–754.

47 Hai, Y. and Christianson, D.W. (2016). Histone deacetylase 6 structure and molecular basis of catalysis and inhibition. *Nat. Chem. Biol.* 12: 741–747.

48 Schuetz, A., Min, J., Allali-Hassani, A. et al. (2008). Human HDAC7 harbors a class IIa histone deacetylase-specific zinc binding motif and cryptic deacetylase activity. *J. Biol. Chem.* 283: 11355–11363.

49 Somoza, J.R., Skene, R.J., Katz, B.A. et al. (2004). Structural snapshots of human HDAC8 provide insights into the class I histone deacetylases. *Structure* 12: 1325–1334.

50 Hai, Y., Shinsky, S.A., Porter, N.J., and Christianson, D.W. (2017). Histone deacetylase 10 structure and molecular function as a polyamine deacetylase. *Nat. Commun.* 8: 15368.
51 Zhang, Y., Fang, H., Jiao, J., and Xu, W. (2008). The structure and function of histone deacetylases: the target for anti-cancer therapy. *Curr. Med. Chem.* 15: 2840–2849.
52 Massa, S., Mai, A., Sbardella, G. et al. (2001). 3-(4-Aroyl-1H-pyrrol-2-yl)-*N*-hydroxy-2-propenamides, a new class of synthetic histone deacetylase inhibitors. *J. Med. Chem.* 44: 2069–2072.
53 Van Ommeslaeghe, K., Elaut, G., Brecx, V. et al. (2003). Amide analogues of TSA: synthesis, binding mode analysis and HDAC inhibition. *Bioorg. Med. Chem. Lett.* 13: 1861–1864.
54 Wang, D.F., Wiest, O., Helquist, P. et al. (2004). On the function of the 14 A long internal cavity of histone deacetylase-like protein: implications for the design of histone deacetylase inhibitors. *J. Med. Chem.* 47: 3409–3417.
55 Lu, Q., Wang, D.S., Chen, C.S. et al. (2005). Structure-based optimization of phenylbutyrate-derived histone deacetylase inhibitors. *J. Med. Chem.* 48: 5530–5535.
56 Rodriquez, M., Terracciano, S., Cini, E. et al. (2006). Total synthesis, NMR solution structure, and binding model of the potent histone deacetylase inhibitor FR235222. *Angew. Chem. Int. Ed.* 45: 423–427.
57 Maulucci, N., Chini, M.G., Micco, S.D. et al. (2007). Molecular insights into azumamide e histone deacetylases inhibitory activity. *J. Am. Chem. Soc.* 129: 3007–3012.
58 Liu, T., Kapustin, G., and Etzkorn, F.A. (2007). Design and synthesis of a potent histone deacetylase inhibitor. *J. Med. Chem.* 50: 2003–2006.
59 Di Micco, S., Terracciano, S., Bruno, I. et al. (2008). Molecular modeling studies toward the structural optimization of new cyclopeptide-based HDAC inhibitors modeled on the natural product FR235222. *Bioorg. Med. Chem.* 16: 8635–8642.
60 Chen, P.C., Patil, V., Guerrant, W. et al. (2008). Synthesis and structure-activity relationship of histone deacetylase (HDAC) inhibitors with triazole-linked cap group. *Bioorg. Med. Chem.* 16: 4839–4853.
61 Lavoie, R., Bouchain, G., Frechette, S. et al. (2001). Design and synthesis of a novel class of histone deacetylase inhibitors. *Bioorg. Med. Chem. Lett.* 11: 2847–2850.
62 Remiszewski, S.W., Sambucetti, L.C., Atadja, P. et al. (2002). Inhibitors of human histone deacetylase: synthesis and enzyme and cellular activity of straight chain hydroxamates. *J. Med. Chem.* 45: 753–757.
63 Mai, A., Massa, S., Ragno, R. et al. (2003). 3-(4-Aroyl-1-methyl-1H-2-pyrrolyl)-*N*-hydroxy-2-alkylamides as a new class of synthetic histone deacetylase inhibitors. 1. Design, synthesis, biological evaluation, and binding mode studies performed through three different docking procedures. *J. Med. Chem.* 46: 512–524.
64 Park, H. and Lee, S. (2004). Homology modeling, force field design, and free energy simulation studies to optimize the activities of histone deacetylase inhibitors. *J. Comput.-Aided Mol. Des.* 18: 375–388.

65 Wang, D.F., Helquist, P., Wiech, N.L., and Wiest, O. (2005). Toward selective histone deacetylase inhibitor design: homology modeling, docking studies, and molecular dynamics simulations of human class I histone deacetylases. *J. Med. Chem.* 48: 6936–6947.
66 Kim, H.M., Hong, S.H., Kim, M.S. et al. (2007). Modification of cap group in delta-lactam-based histone deacetylase (HDAC) inhibitors. *Bioorg. Med. Chem. Lett.* 17: 6234–6238.
67 Witter, D.J., Harrington, P., Wilson, K.J. et al. (2008). Optimization of biaryl selective HDAC1&2 inhibitors (SHI-1:2). *Bioorg. Med. Chem. Lett.* 18: 726–731.
68 Moradei, O.M., Mallais, T.C., Frechette, S. et al. (2007). Novel aminophenyl benzamide-type histone deacetylase inhibitors with enhanced potency and selectivity. *J. Med. Chem.* 50: 5543–5546.
69 Schäfer, S., Saunders, L., Eliseeva, E. et al. (2008). Phenylalanine-containing hydroxamic acids as selective inhibitors of class IIb histone deacetylases (HDACs). *Bioorg. Med. Chem.* 16: 2011–2033.
70 Yan, C., Xiu, Z., Li, X. et al. (2008). Comparative molecular dynamics simulations of histone deacetylase-like protein: binding modes and free energy analysis to hydroxamic acid inhibitors. *Proteins* 73: 134–149.
71 Mukherjee, P., Pradhan, A., Shah, F. et al. (2008). Structural insights into the *Plasmodium falciparum* histone deacetylase 1 (PfHDAC-1): a novel target for the development of antimalarial therapy. *Bioorg. Med. Chem.* 16: 5254–5265.
72 Weerasinghe, S.V., Estiu, G., Wiest, O., and Pflum, M.K. (2008). Residues in the 11 Å channel of histone deacetylase 1 promote catalytic activity: implications for designing isoform-selective histone deacetylase inhibitors. *J. Med. Chem.* 51: 5542–5551.
73 Krishna, S., Kumar, V., and Siddiqi, M.I. (2016). Recent advances in computer-assisted structure-based identification and design of histone deacetylases inhibitors. *Curr. Top. Med. Chem.* 16: 934–947.
74 Price, S., Bordogna, W., Bull, R.J. et al. (2007). Identification and optimisation of a series of substituted 5-(1H-pyrazol-3-yl)-thiophene-2-hydroxamic acids as potent histone deacetylase (HDAC) inhibitors. *Bioorg. Med. Chem. Lett.* 17: 370–375.
75 Tang, H., Wang, X.S., Huang, X.P. et al. (2009). Novel inhibitors of human histone deacetylase (HDAC) identified by QSAR modeling of known inhibitors, virtual screening, and experimental validation. *J. Chem. Inf. Model.* 49: 461–476.
76 Park, H., Kim, S., Kim, Y.E., and Lim, S.J. (2010). A structure-based virtual screening approach toward the discovery of histone deacetylase inhibitors: identification of promising zinc-chelating groups. *ChemMedChem* 5: 591–597.
77 Forli, S., Huey, R., Pique, M.E. et al. (2016). Computational protein-ligand docking and virtual drug screening with the AutoDock suite. *Nat. Protoc.* 11: 905–919.

78 Kannan, S., Melesina, J., Hauser, A.T. et al. (2014). Discovery of inhibitors of *Schistosoma mansoni* HDAC8 by combining homology modeling, virtual screening, and in vitro validation. *J. Chem. Inf. Model.* 54: 3005–3019.

79 Marek, M., Kannan, S., Hauser, A.T. et al. (2013). Structural basis for the inhibition of histone deacetylase 8 (HDAC8), a key epigenetic player in the blood fluke Schistosoma mansoni. *PLoS Pathog.* 9: e1003645.

80 Heimburg, T., Chakrabarti, A., Lancelot, J. et al. (2016). Structure-based design and synthesis of novel inhibitors targeting HDAC8 from *Schistosoma mansoni* for the treatment of Schistosomiasis. *J. Med. Chem.* 59: 2423–2435.

81 Butler, K.V., Kalin, J., Brochier, C. et al. (2010). Rational design and simple chemistry yield a superior, neuroprotective HDAC6 inhibitor, tubastatin A. *J. Am. Chem. Soc.* 132: 10842–10846.

82 De Vreese, R. and D'Hooghe, M. (2017). Synthesis and applications of benzohydroxamic acid-based histone deacetylase inhibitors. *Eur. J. Med. Chem.* 135: 174–195.

83 Micelli, C. and Rastelli, G. (2015). Histone deacetylases: structural determinants of inhibitor selectivity. *Drug Discovery Today* 20: 718–735.

84 Avalos, J.L., Boeke, J.D., and Wolberger, C. (2004). Structural basis for the mechanism and regulation of Sir2 enzymes. *Mol. Cell* 13: 639–648.

85 Avalos, J.L., Bever, K.M., and Wolberger, C. (2005). Mechanism of sirtuin inhibition by nicotinamide: altering the NAD(+) cosubstrate specificity of a Sir2 enzyme. *Mol. Cell* 17: 855–868.

86 Cosgrove, M.S., Bever, K., Avalos, J.L. et al. (2006). The structural basis of sirtuin substrate affinity. *Biochemistry* 45: 7511–7521.

87 Hoff, K.G., Avalos, J.L., Sens, K., and Wolberger, C. (2006). Insights into the sirtuin mechanism from ternary complexes containing NAD+ and acetylated peptide. *Structure* 14: 1231–1240.

88 Hawse, W.F., Hoff, K.G., Fatkins, D.G. et al. (2008). Structural insights into intermediate steps in the Sir2 deacetylation reaction. *Structure* 16: 1368–1377.

89 Min, J., Landry, J., Sternglanz, R., and Xu, R.M. (2001). Crystal structure of a SIR2 homolog-NAD complex. *Cell* 105: 269–279.

90 Zhao, K., Harshaw, R., Chai, X., and Marmorstein, R. (2004). Structural basis for nicotinamide cleavage and ADP-ribose transfer by NAD(+)-dependent Sir2 histone/protein deacetylases. *Proc. Natl. Acad. Sci. U.S.A.* 101: 8563–8568.

91 Zhao, K., Chai, X., and Marmorstein, R. (2003). Structure of a Sir2 substrate, Alba, reveals a mechanism for deacetylation-induced enhancement of DNA binding. *J. Biol. Chem.* 278: 26071–26077.

92 Davenport, A.M., Huber, F.M., and Hoelz, A. (2014). Structural and functional analysis of human SIRT1. *J. Mol. Biol.* 426: 526–541.

93 Zhao, X., Allison, D., Condon, B. et al. (2013). The 2.5 Å crystal structure of the SIRT1 catalytic domain bound to nicotinamide adenine dinucleotide (NAD(+)) and an indole (EX527 analogue) reveals a novel mechanism of histone deacetylase inhibition. *J. Med. Chem.* 56: 963–969.

94 Finnin, M.S., Donigian, J.R., and Pavletich, N.P. (2001). Structure of the histone deacetylase SIRT2. *Nat. Struct. Biol.* 8: 621–625.

95 Moniot, S., Schutkowski, M., and Steegborn, C. (2013). Crystal structure analysis of human Sirt2 and its ADP-ribose complex. *J. Struct. Biol.* 182: 136–143.

96 Rumpf, T., Gerhardt, S., Einsle, O., and Jung, M. (2015). Seeding for sirtuins: microseed matrix seeding to obtain crystals of human Sirt3 and Sirt2 suitable for soaking. *Acta Crystallogr. F Struct. Biol. Commun.* 71: 1498–1510.

97 Yamagata, K., Goto, Y., Nishimasu, H. et al. (2014). Structural basis for potent inhibition of SIRT2 deacetylase by a macrocyclic peptide inducing dynamic structural change. *Structure* 22: 345–352.

98 Bao, X.C., Wang, Y., Li, X. et al. (2014). Identification of 'erasers' for lysine crotonylated histone marks using a chemical proteomics approach. *eLife* 3: 1–18.

99 Disch, J.S., Evindar, G., Chiu, C.H. et al. (2013). Discovery of thieno[3,2-d]pyrimidine-6-carboxamides as potent inhibitors of SIRT1, SIRT2, and SIRT3. *J. Med. Chem.* 56: 3666–3679.

100 Gertz, M., Fischer, F., Nguyen, G.T.T. et al. (2013). Ex-527 inhibits sirtuins by exploiting their unique NAD(+)-dependent deacetylation mechanism. *Proc. Natl. Acad. Sci. U.S.A.* 110: E2772–E2781.

101 Gertz, M., Giang, T.T.N., Fischer, F. et al. (2012). A molecular mechanism for direct sirtuin activation by resveratrol. *PLoS One* 7.

102 Jin, L., Wei, W.T., Jiang, Y.B. et al. (2009). Crystal structures of human SIRT3 displaying substrate-induced conformational changes. *J. Biol. Chem.* 284: 24394–24405.

103 Nguyen, G.T.T., Gertz, M., and Steegborn, C. (2013). Crystal structures of Sirt3 complexes with 4′-bromo-resveratrol reveal binding sites and inhibition mechanism. *Chem. Biol.* 20: 1375–1385.

104 Nguyen, G.T.T., Schaefer, S., Gertz, M. et al. (2013). Structures of human sirtuin 3 complexes with ADP-ribose and with carba-NAD(+) and SRT1720: binding details and inhibition mechanism. *Acta Crystallogr. D Biol. Crystallogr.* 69: 1423–1432.

105 Szczepankiewicz, B.G., Dai, H., Koppetsch, K.J. et al. (2012). Synthesis of Carba-NAD and the structures of its ternary complexes with SIRT3 and SIRT5. *J. Org. Chem.* 77: 7319–7329.

106 Wu, J.H., Zhang, D.Y., Chen, L. et al. (2013). Discovery and mechanism study of SIRT1 activators that promote the deacetylation of fluorophore-labeled substrate. *J. Med. Chem.* 56: 761–780.

107 Du, J.T., Zhou, Y.Y., Su, X.Y. et al. (2011). Sirt5 Is a NAD-dependent protein lysine demalonylase and desuccinylase. *Science* 334: 806–809.

108 Schuetz, A., Min, J.R., Antoshenko, T. et al. (2007). Structural basis of inhibition of the human NAD(+)-dependent deacetylase SIRT5 by suramin. *Structure* 15: 377–389.

109 Zhou, Y.Y., Zhang, H.M., He, B. et al. (2012). The bicyclic intermediate structure provides insights into the desuccinylation mechanism of human sirtuin 5 (SIRT5). *J. Biol. Chem.* 287: 28307–28314.

110 Jiang, H., Khan, S., Wang, Y. et al. (2013). SIRT6 regulates TNF-alpha secretion through hydrolysis of long-chain fatty acyl lysine. *Nature* 496: 110–113.

111 Schiedel, M., Robaa, D., Rumpf, T. et al. (2018). The current state of NAD+-dependent histone deacetylases (sirtuins) as novel therapeutic targets. *Med. Res. Rev.* 38: 147–200.

112 Rumpf, T., Schiedel, M., Karaman, B. et al. (2015). Selective Sirt2 inhibition by ligand-induced rearrangement of the active site. *Nat. Commun.* 6: 1–13.

113 Neugebauer, R.C., Sippl, W., and Jung, M. (2008). Inhibitors of NAD+ dependent histone deacetylases (sirtuins). *Curr. Pharm. Des.* 14: 562–573.

114 Trapp, J., Jochum, A., Meier, R. et al. (2006). Adenosine mimetics as inhibitors of NAD(+)-dependent histone deacetylases, from kinase to sirtuin inhibition. *J. Med. Chem.* 49: 7307–7316.

115 Karaman, B., Jung, M., and Sippl, W. (2016). Structure-based design and computational studies of sirtuin inhibitors, Chapter 11. In: *Epi-Informatics*. Boston: Academic Press.

116 Outeiro, T.F., Kontopoulos, E., Altmann, S.M. et al. (2007). Sirtuin 2 inhibitors rescue alpha-synuclein-mediated toxicity in models of Parkinson's disease. *Science* 317: 516–519.

117 Abagyan, R., Totrov, M., and Kuznetsov, D. (1994). ICM – A new method for protein modeling and design: applications to docking and structure prediction from the distorted native conformation. *J. Comput. Chem.* 15: 488–506.

118 Cui, H.Q., Kamal, Z., Ai, T. et al. (2014). Discovery of potent and selective sirtuin 2 (SIRT2) inhibitors using a fragment-based approach. *J. Med. Chem.* 57: 8340–8357.

119 Schiedel, M., Rumpf, T., Karaman, B. et al. (2016). Aminothiazoles as potent and selective Sirt2 inhibitors: a structure-activity relationship study. *J. Med. Chem.* 59: 1599–1612.

120 Schiedel, M., Herp, D., Hammelmann, S. et al. (2018). Chemically induced degradation of sirtuin 2 (Sirt2) by a proteolysis targeting chimera (PROTAC) based on sirtuin rearranging ligands (SirReals). *J. Med. Chem.* 61: 482–491.

121 Toure, M. and Crews, C.M. (2016). Small-molecule PROTACS: new approaches to protein degradation. *Angew. Chem. Int. Ed.* 55: 1966–1973.

122 Neklesa, T.K., Winkler, J.D., and Crews, C.M. (2017). Targeted protein degradation by PROTACs. *Pharmacol. Ther.* 174: 138–144.

123 Smith, B.C. and Denu, J.M. (2009). Chemical mechanisms of histone lysine and arginine modifications. *Biochim. Biophys. Acta* 1789: 45–57.

124 Qian, C. and Zhou, M.M. (2006). SET domain protein lysine methyltransferases: structure, specificity and catalysis. *Cell. Mol. Life Sci.* 63: 2755–2763.

125 Boisvert, F.M., Chenard, C.A., and Richard, S. (2005). Protein interfaces in signaling regulated by arginine methylation. *Sci. STKE* 2005: re2.

126 Bedford, M.T. (2007). Arginine methylation at a glance. *J. Cell Sci.* 120: 4243–4246.

127 Pal, S. and Sif, S. (2007). Interplay between chromatin remodelers and protein arginine methyltransferases. *J. Cell. Physiol.* 213: 306–315.

128 Allis, C.D., Berger, S.L., Cote, J. et al. (2007). New nomenclature for chromatin-modifying enzymes. *Cell* 131: 633–636.

129 Schneider, R., Bannister, A.J., and Kouzarides, T. (2002). Unsafe SETs: histone lysine methyltransferases and cancer. *Trends Biochem. Sci* 27: 396–402.

130 Spannhoff, A., Heinke, R., Bauer, I. et al. (2007). Target-based approach to inhibitors of histone arginine methyltransferases. *J. Med. Chem.* 50: 2319–2325.
131 Heinke, R., Spannhoff, A., Meier, R. et al. (2009). Virtual screening and biological characterization of novel histone arginine methyltransferase PRMT1 inhibitors. *ChemMedChem* 4: 69–77.
132 de Freitas, R.F., Eram, M.S., Smil, D. et al. (2016). Discovery of a potent and selective coactivator associated arginine methyltransferase 1 (CARM1) inhibitor by virtual screening. *J. Med. Chem.* 59: 6838–6847.
133 Ye, Y., Zhang, B.D., Mao, R.F. et al. (2017). Discovery and optimization of selective inhibitors of protein arginine methyltransferase 5 by docking-based virtual screening. *Org. Biomol. Chem.* 15: 3648–3661.
134 Cheng, D., Yadav, N., King, R.W. et al. (2004). Small molecule regulators of protein arginine methyltransferases. *J. Biol. Chem.* 279: 23892–23899.
135 Mason, J.S., Morize, I., Menard, P.R. et al. (1999). New 4-point pharmacophore method for molecular similarity and diversity applications: overview of the method and applications, including a novel approach to the design of combinatorial libraries containing privileged substructures. *J. Med. Chem.* 42: 3251–3264.
136 Zhang, X., Yang, Z., Khan, S.I. et al. (2003). Structural basis for the product specificity of histone lysine methyltransferases. *Mol. Cell* 12: 177–185.
137 Liu, F., Chen, X., Allali-Hassani, A. et al. (2009). Discovery of a 2,4-diamino-7-aminoalkoxyquinazoline as a potent and selective inhibitor of histone lysine methyltransferase G9a. *J. Med. Chem.* 52: 7950–7953.
138 Zhang, J., Yao, D., Jiang, Y. et al. (2017). Synthesis and biological evaluation of benzimidazole derivatives as the G9a histone methyltransferase inhibitors that induce autophagy and apoptosis of breast cancer cells. *Bioorg. Chem.* 72: 168–181.
139 Meng, F., Cheng, S., Ding, H. et al. (2015). Discovery and optimization of novel, selective histone methyltransferase SET7 inhibitors by pharmacophore- and docking-based virtual screening. *J. Med. Chem.* 58: 8166–8181.
140 Shi, Y.J., Lan, F., Matson, C. et al. (2004). Histone demethylation mediated by the nuclear arnine oxidase homolog LSD1. *Cell* 119: 941–953.
141 Tsukada, Y., Fang, J., Erdjument-Bromage, H. et al. (2006). Histone demethylation by a family of JmjC domain-containing proteins. *Nature* 439: 811–816.
142 Wang, G.G., Allis, C.D., and Chi, P. (2007). Chromatin remodeling and cancer, part I: covalent histone modifications. *Trends Mol. Med.* 13: 363–372.
143 Rose, N.R., Ng, S.S., Mecinovic, J. et al. (2008). Inhibitor scaffolds for 2-oxoglutarate-dependent histone lysine demethylases. *J. Med. Chem.* 51: 7053–7056.
144 Chen, Y., Yang, Y., Wang, F. et al. (2006). Crystal structure of human histone lysine-specific demethylase 1 (LSD1). *Proc. Natl. Acad. Sci. U.S.A.* 103: 13956–13961.
145 Stavropoulos, P., Blobel, G., and Hoelz, A. (2006). Crystal structure and mechanism of human lysine-specific demethylase-1. *Nat. Struct. Mol. Biol.* 13: 626–632.

146 Yang, M., Culhane, J.C., Szewczuk, L.M. et al. (2007). Structural basis of histone demethylation by LSD1 revealed by suicide inactivation. *Nat. Struct. Mol. Biol.* 14: 535–539.

147 Forneris, F., Battaglioli, E., Mattevi, A., and Binda, C. (2009). New roles of flavoproteins in molecular cell biology: histone demethylase LSD1 and chromatin. *FEBS J.* 276: 4304–4312.

148 Hojfeldt, J.W., Agger, K., and Helin, K. (2013). Histone lysine demethylases as targets for anticancer therapy. *Nat. Rev. Drug Discovery* 12: 917–930.

149 Schmitt, M.L., Hauser, A.T., Carlino, L. et al. (2013). Nonpeptidic propargylamines as inhibitors of lysine specific demethylase 1 (LSD1) with cellular activity. *J. Med. Chem.* 56: 7334–7342.

150 Sorna, V., Theisen, E.R., Stephens, B. et al. (2013). High-throughput virtual screening identifies novel N′-(1-phenylethylidene)-benzohydrazides as potent, specific, and reversible lsd1 inhibitors. *J. Med. Chem.* 56: 9496–9508.

151 Zhou, C., Kang, D., Xu, Y.G. et al. (2015). Identification of novel selective lysine-specific demethylase 1 (LSD1) inhibitors using a pharmacophore-based virtual screening combined with docking. *Chem. Biol. Drug Des.* 85: 659–671.

152 Niwa, H. and Umehara, T. (2017). Structural insight into inhibitors of flavin adenine dinucleotide-dependent lysine demethylases. *Epigenetics* 12: 340–352.

153 Mould, D.P., Bremberg, U., Jordan, A.M. et al. (2017). Development of 5-hydroxypyrazole derivatives as reversible inhibitors of lysine specific demethylase 1. *Bioorg. Med. Chem. Lett.* 27: 3190–3195.

154 Xi, J., Xu, S., Wu, L. et al. (2017). Design, synthesis and biological activity of 3-oxoamino-benzenesulfonamides as selective and reversible LSD1 inhibitors. *Bioorg. Chem.* 72: 182–189.

155 Li, Z.H., Liu, X.Q., Geng, P.F. et al. (2017). Discovery of [1,2,3]triazolo [4,5-d]pyrimidine derivatives as novel LSD1 inhibitors. *ACS Med. Chem. Lett.* 8: 384–389.

156 Wang, S., Zhao, L.J., Zheng, Y.C. et al. (2017). Design, synthesis and biological evaluation of [1,2,4]triazolo[1,5-a]pyrimidines as potent lysine specific demethylase 1 (LSD1/KDM1A) inhibitors. *Eur. J. Med. Chem.* 125: 940–951.

157 Sartori, L., Mercurio, C., Amigoni, F. et al. (2017). Thieno[3,2-b]pyrrole-5-carboxamides as new reversible inhibitors of histone lysine demethylase KDM1A/LSD1. Part 1: high-throughput screening and preliminary exploration. *J. Med. Chem.* 60: 1673–1692.

158 Vianello, P., Sartori, L., Amigoni, F. et al. (2017). Thieno[3,2-b]pyrrole-5-carboxamides as new reversible inhibitors of histone lysine demethylase KDM1A/LSD1. Part 2: structure-based drug design and structure-activity relationship. *J. Med. Chem.* 60: 1693–1715.

159 Hoffmann, I., Roatsch, M., Schmitt, M.L. et al. (2012). The role of histone demethylases in cancer therapy. *Mol. Oncol.* 6: 683–703.

160 Hauser, A.-T., Roatsch, M., Schulz-Fincke, J. et al. (2015). Discovery of histone demethylase inhibitors Chapter 18. In: *Epigenetic Technological Applications*. Boston: Academic Press.

161 Chu, C.H., Wang, L.Y., Hsu, K.C. et al. (2014). KDM4B as a target for prostate cancer: structural analysis and selective inhibition by a novel inhibitor. *J. Med. Chem.* 57: 5975–5985.
162 Wu, X.A., Fang, Z., Yang, B. et al. (2016). Discovery of KDM5A inhibitors: homology modeling, virtual screening and structure-activity relationship analysis. *Bioorg. Med. Chem. Lett.* 26: 2284–2288.
163 Roatsch, M., Robaa, D., Pippel, M. et al. (2016). Substituted 2-(2-aminopyrimidin-4-yl)pyridine-4-carboxylates as potent inhibitors of JumonjiC domain-containing histone demethylases. *Future Med. Chem.* 8: 1553–1571.
164 Fang, Z., Wang, T.Q., Li, H. et al. (2017). Discovery of pyrazolo[1,5-a] pyrimidine-3-carbonitrile derivatives as a new class of histone lysine demethylase 4D (KDM4D) inhibitors. *Bioorg. Med. Chem. Lett.* 27: 3201–3204.
165 Gehling, V.S., Bellon, S.F., Harmange, J.C. et al. (2016). Identification of potent, selective KDM5 inhibitors. *Bioorg. Med. Chem. Lett.* 26: 4350–4354.

4

Mass Spectrometry and Chemical Biology in Epigenetics Drug Discovery

Christian Feller[1], *David Weigt*[2, 3], *and Carsten Hopf*[2, 3]

[1] *Institute of Molecular Systems Biology, ETH Zürich, Department of Biology, Auguste-Piccard-Hof 1, 8093 Zürich, Switzerland*
[2] *Heidelberg University, Hartmut Hoffmann-Berling International Graduate School of Molecular and Cellular Biology (HBIGS), Im Neuenheimer Feld 501, 69120 Heidelberg, Germany*
[3] *Mannheim University of Applied Sciences, Center for Mass Spectrometry and Optical Spectroscopy (CeMOS), Paul-Wittsack-Str. 10, 69163 Mannheim, Germany*

4.1 Introduction: Mass Spectrometry Technology Used in Epigenetic Drug Discovery

The state of chromatin is a major determinant of the cellular state that regulates all aspects of genome functions. The principles that govern chromatin functions – such as the organization of gene expression programs and the maintenance of genome stability – are based on proteins forming intricate interaction networks. Among them are hundreds of factors that write, read, and erase histone modifications. These posttranslational modifications (PTMs) function as central organization hubs by regulating chromatin structure and coordinating protein recruitment and function at chromatin.

Alteration in the composition of chromatin often leads to diseases including cancer and immuno-inflammatory and neurological disorders [1–3]. Consequently, the development of small-molecule drug-like regulators targeting chromatin-modifying enzymes has gained significant attention in academic research and industry. However, current drug development projects frequently fail because of a compound's lack of efficacy as well as off-target effects and other safety concerns. Therefore, although first epigenetic drug candidates are already being evaluated in the clinics and many more are in drug discovery pipelines, there is a strong need for new technologies to better understand the molecular mode of action (MoA) of these compounds within the cells, tissues, and whole organisms.

Contrasting with the complexity of protein interactions and their regulation by histone modifications, the prevalent methods to study chromatin functions using genomic and antibody-based strategies are limited to study a single protein or single modification at a time. Furthermore, recent studies alarmed that many histone modification-directed antibodies display substantial off-target reactivities. For example, antibodies raised against single acetylation sites preferentially

Epigenetic Drug Discovery, First Edition. Edited by Wolfgang Sippl and Manfred Jung.

bind to poly-acetylated peptides, and di- and trimethylation-directed antibodies often cross-react with lower methylation states [4]. In addition, modifications frequently co-occur in close proximity at the histone N-terminal "tail" domains. These form combinatorial modification motifs that may bear additional functions beyond the individual modification sites [5, 6]. Antibodies cannot directly and specifically assess these hyper-modified histone isoforms, thereby hindering our understanding of the histone code [7].

Current mass spectrometry (MS) technology provides an indispensable toolbox for epigenetic drug discovery [8, 9]. A substantial number of bottom-up proteomics methods, which analyze complex mixtures of tryptic peptides derived from digested sub-proteomes, are capable of identifying and quantifying thousands of proteins simultaneously. They have been custom-tailored in recent years to serve five main purposes in epigenetic drug discovery: (i) identification of putative drug targets of phenotypic screening hits or other compounds with unknown MoA, (ii) in-depth characterization of protein complexes containing the drug-binding protein, (iii) elucidation of a drug's MoA by systematic characterization of histone PTMs, (iv) MS-based compound screening and selectivity profiling, and (v) imaging the spatial distribution of epigenetic drugs and/or their pharmacodynamic action.

4.1.1 Mass Spectrometry Workflows for the Analysis of Proteins

Two complementary MS methods exist to analyze proteins. In the top-down proteomic strategy, which is mostly limited to low complexity mixtures containing a single or few proteins, the intact protein is directly analyzed by MS. Middle-down proteomics is a variant of the top-down strategy with increased sensitivity and specificity, where large fragments of proteins (>50 amino acids) are analyzed [10].

In bottom-up proteomics, peptides derived from digested proteins are analyzed with the mass spectrometer, and protein information is inferred by reassigning peptides back to proteins (Figure 4.1). Bottom-up proteomics in combination with high performance liquid chromatography (HPLC) is the most widely applied MS strategy, because chromatographic separation and the mass range and fragmentation process of peptides match closest current instrumentation capabilities than those using intact proteins or large protein fragments. Peptides are commonly generated with the serine endoprotease trypsin, which hydrolyzes peptide bonds after the basic amino acids lysine and arginine to generate peptides in a range of 5–25 amino acids. After enzymatic digestion and subsequent peptide purification (desalting), peptide mixtures are separated according to their hydrophobicity by reverse phase chromatography typically using C18 resins. The resulting reduced complexity and increased peptide concentration in the chromatographic space (retention time) is critical to improve selectivity and sensitivity, in particular for low abundant proteins and their modifications. Even though chromatographic separation of complex peptide mixtures has been enhanced by operating HPLC at higher pressures when using longer columns and smaller particle sizes of the chromatographic resins, typically more than tens to hundreds of peptides elute and ionize synchronously, which has profound implications on the MS acquisition and data analysis strategy.

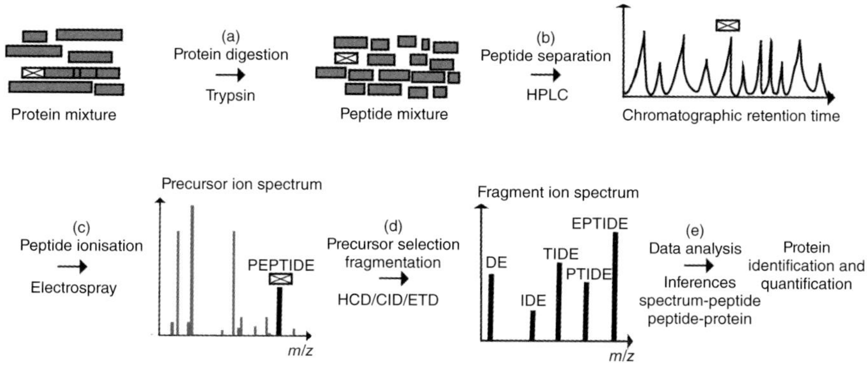

Figure 4.1 Bottom-up proteomic workflow. (a) Proteins are digested to peptides. (b) Peptides are separated by high performance liquid chromatography (HPLC). (c) Many peptides elute from the HPLC column and ionize at the same time. (d) Precursor ions are selected and fragmented to generate fragment ion spectra. (e) Computational proteomic workflows infer peptide sequence from aligning measured spectra to theoretical spectra derived from genomic/proteomic databases. Peptides are reassigned to proteins, and quantities are often inferred from integrating precursor (DDA) or fragment ion (SRM, DIA) signals over time (chromatogram).

Peptides eluting from the HPLC column are ionized by electrospray and then transferred to the vacuum of the mass spectrometer, where their mass-to-charge (*m/z*) ratio is measured. In the mass spectrometer, the ions are guided by the electric or magnetic fields of mass analyzers to a detector composed of electron multipliers, where the signal is converted to a MS spectrum defined by *m/z* of the ions and their signal intensities. The performance of the MS is determined by the scan speed of acquiring MS spectra per time unit, the sensitivity of ion selection and transmission, and the quality of the spectra characterized by the mass accuracy and resolution to distinguish close-by signals. These parameters critically depend on the type of mass analyzers and their combinations (ion trap, orbitrap, quadrupole, time of flight (ToF), and Fourier transform ion cyclotron resonance (FT-ICR) [11]) and their operating mode and need to be optimized and tailored to the needs of the specific application.

The identity of a peptide is typically determined by tandem mass spectrometry (MS/MS). After selecting a peptide ion with a mass analyzer, this precursor/parent ion is collided with an inert gas to yield product/fragment ions characterized by successive losses of amino acids. The fragment ion spectra are assigned to peptides using sequence alignment algorithms incorporated in proteomic database search engines. The quality and thus the confidence in a peptide-spectrum match (PSM) critically depends on several parameters, including the resolution and mass accuracy of the fragment ion spectrum, the ratio of observed to theoretical fragment ions, the fraction of fragment ions mapped to the peptide sequence, and the type of fragmentation (such as collision-induced dissociation or higher energy collisional dissociation).

Depending on the instrumentation, acquisition, and data analysis strategy, three data acquisition modes are used for bottom-up proteomics: data-dependent acquisition (DDA), data-independent acquisition (DIA), and targeted proteomic modes such as selected/parallel reaction monitoring (SRM/PRM). In DDA mode, also termed "shotgun proteomics," the mass spectrometer is operated in iterative cycles of recording parent ion spectra over a wide mass range (typically 300 to 1400 *m/z*) followed by acquiring fragment spectra for the most abundant precursors (typically 10–20). Peptides are often quantified by integrating their precursor signal along the chromatographic retention (chromatogram), whereas information from fragment ions are used for peptide identification. Shotgun proteomics is the most widely used bottom-up proteomic strategy that benefits from a combination of fast, high resolution, and high mass accuracy mass analyzers (such as the ones implemented in commercial quadrupole-orbitrap and linear ion trap-quadrupole-orbitrap setups [11]) and established bioinformatics pipelines. However, despite high scan speeds of 20 Hz for recent instruments and by reducing the re-selection of already sequenced precursors by on-the-flight decisions of the MS operating software, still only a minor fraction of the often more than 100 000 peptide features detected in complex whole cell lysates is identified [12]. This phenomenon, known as stochastic undersampling, reduces the reproducibility and sensitivity of shotgun proteomic approaches.

To overcome the limitations of DDA for stochastic undersampling, targeted proteomic methods were developed where tens of peptides are measured in a highly sensitive, specific, and reproducible manner. In an SRM experiment, these benefits are achieved by continuously measuring fragment ions defined by a spectral assay [13]. A spectral assay needs to be established before the SRM experiment by selecting the most intense 3–5 fragment ions from a set of 2–4 "best-flying" peptides for a protein of interest. Furthermore, synthetic reference peptides are used to optimize fragmentation conditions (collision energy) and transmission settings for every pair of precursor–fragment ions called a "transition." The SRM experiment has become the gold standard for sensitive and reproducible proteomics and is the method of choice for target validation following discovery by high-throughput shotgun proteomics and for clinical and industry applications where high confidence is required. The PRM assay is a recent variant of the SRM method, which measures all fragment ions for tens to hundreds of peptides. Implemented on a quadrupole-orbitrap instrument, PRM achieves high specificity by recording fragment ions with higher resolution and mass accuracy compared with the SRM experiments implemented on triple-quadrupole instruments [14].

DIA workflows were recently developed with the aim to close the gap between quantifying thousands of proteins in a single MS experiment (typical for DDA) while achieving high reproducibility (similar to SRM/PRM). In a SWATH-MS (sequential windowed acquisition of all theoretical fragment ion spectra) experiment, which presents the first high-throughput DIA method, a fast-scanning triple-ToF mass spectrometer rapidly scans through a wide range of product ions (typically 400 to 1200 *m/z*) generated by broad isolation windows (typically 25 Da, compared with 2–4 Da for DDA and SRM/PRM experiments) [15]. To deconvolute the resulting chimeric spectra – which violates the "one fragment ion spectrum–one peptide" dogma of DDA and SRM/PRM – sophisticated software tools were developed that require high-quality spectral libraries (similar to SRM/PRM) to extract chromatograms in a peptide-centric fashion [16].

Proteome-wide spectral libraries covering at least two peptides per protein were recently developed to quantify the proteomes of human cells and several model organisms [17]. Importantly, because DIA maps represent digital maps of all observable fragment ions (within a dynamic intrascan range of 4 orders of magnitude), these datasets can be re-mined with new spectral libraries, for example, once those become available for proteoforms such as the ones formed by PTMs and sequence variants.

The quantitative accuracy of an MS experiment is intrinsically limited by co-eluting and co-ionizing peptides and other ions that suppress the ionization efficiency ("matrix effect"). These effects are more pronounced for highly complex peptide mixtures, ion trap acquisitions with limited dynamic range, and samples that were insufficiently purified from contaminating salts. To minimize these distortions in label-free experiments, *in situ* metabolic labeling of proteins in cell culture (e.g. "stable isotope labeling with amino acids in cell culture" (SILAC)) or chemical labeling of peptides (e.g. "isobaric tags for relative and absolute quantitation" (iTRAQ) or "tandem mass tag" (TMT)) allows for more accurate side-by-side comparison of two or a few samples [11, 18]. In addition to signal distortions by ion suppression effects, peptides with different sequences and modification status evoke different signal intensities (MS response factor). Synthetic reference peptides can be used to determine and normalize for different MS response factors and thereby improve quantitative accuracy. Moreover, spiking isotopically labeled ("heavy") synthetic reference peptides into protein lysates allows to estimate absolute quantities of a protein and its modifications. However, this approach is labor intensive and expensive and thus limited to experiments that require very high quantitative accuracy [19].

4.1.2 Mass Spectrometry Imaging

Mass spectrometry imaging (MSI) has emerged as a versatile label-free technology for simultaneous analysis of the spatial distribution of hundreds or even thousands of molecules including proteins (top-down analysis up to 20–30 kDa), peptides (bottom-up or middle-down analysis), lipids, small-molecule compounds, and drugs [20–22]. The primary ionization methods used in MSI today are matrix-assisted laser desorption/ionization (MALDI) (Figure 4.2), desorption electrospray ionization (DESI), and secondary ion mass spectrometry (SIMS) [23]. Ion sources are combined with ToF, Fourier transform (FT) ion trap (such as orbitrap), ion cyclotron resonance (FT-ICR), or other mass analyzers. A typical workflow for the most frequently used MALDI-ToF MSI shall be outlined here (Figure 4.2).

Tissue of choice is harvested and snap frozen in liquid nitrogen or isopentane. Thin cryosections (often 8–12 μm) are cut, mounted onto conductive indium tin oxide-coated slides, delipidated (for protein imaging only), and desiccated. A suitable chemical MALDI matrix is applied to tissue by spraying or sublimation (for maximum spatial resolution). The chosen matrix determines the selectivity for various classes of analytes, e.g. 2,5-dihydroxybenzoic acid may be used for imaging of lipid and other small molecules in both positive and negative ion mode. A UV laser (355 nm) scans the tissue and generates one mass spectrum

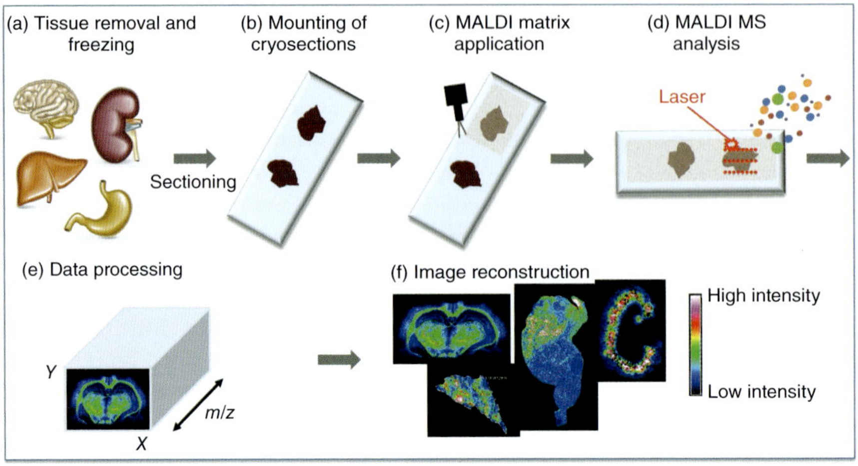

Figure 4.2 Schematic representation of MALDI mass spectrometry imaging workflow. (a) Tissue is harvested and immediately cryopreserved. (b) 8–12 μm sections are cut using a cryostat and mounted onto conductive indium tin oxide-coated slides. (c) A suitable chemical MALDI matrix is applied by spraying or sublimation. Choice of matrix determines selectivity for various classes of analytes, e.g. sinapinic acid may be used for imaging of intact proteins. (d) A UV laser scans across the tissue and generates one mass spectrum per pixel, typically with spatial resolutions of 5–100 μm and at speeds of up to 50 pixels per second. (e) A so-called data cube is generated, in which ion intensity values are mapped to *x*, *y*, and mass-to-charge (*m/z*) coordinates. One ion intensity image can be generated per *m/z*. (f) Examples of reconstructed images include normalized ion intensity images on a false color scale or co-localization images of two and more masses.

per pixel, typically with spatial resolutions of 5–100 μm and at speeds of up to 50 pixels per second. As a consequence, ion intensity values are mapped to *x*, *y*, and mass-to-charge (*m/z*) coordinates. One ion intensity image can be generated per *m/z* and displayed, for example, as normalized ion intensity images on a false color scale or co-localization images of two and more masses.

Since pre-fractionation (e.g. by HPLC) of the "sample" analyzed in every pixel is not possible, ion suppression effects are prevalent, and great care must be taken to normalize and properly process data, in order to obtain at least semiquantitative results [20–22, 24]. Despite this cautionary note, MSI is widely used in drug discovery and chemical biology today, as it is a label-free method that permits side-by-side analysis of parent compound, its metabolites, and putative (lipid) biomarkers all on the same slide [25–27]. Some MSI instruments do not offer sufficient resolving power for unanimous assignments of measured *m/z* to defined molecular entities. Therefore, ultrahigh resolution MSI in FT-ICR (resolving power $R > 1\,000\,000$) or orbitrap (R up to 100 000) mass spectrometers and/or MS/MS imaging of a selected fragment ion is required for signal verification. In MS/MS imaging, a selected few (but far from most) precursor ions can be selected (in ToF, quadrupole, or other selectors) for subsequent analysis of characteristic product ions. It has been known for some time that posttranslationally modified histones can be analyzed by MSI [28].

Histone PTMs in tissue do not only represent candidate diagnostic biomarkers but also pharmacodynamic biomarkers that can indicate the site of drug action [29]. Conceptually, therefore, MSI could evolve into a semiquantitative tool with utility in PK–PD studies.

4.2 Target Identification and Selectivity Profiling: Chemoproteomics

Chemical proteomics, aka chemoproteomics methods, systematically investigate a small-molecule compound's interactions with a sub-proteome, i.e. with its potential targets and off-targets [30–34]. The MS method of choice today is mostly quantitative bottom-up data-dependent proteomics (see Section 4.1.1). During the discovery phase, i.e. whenever unknown target proteins and their complex members are to be identified, precursor ions are selected in "shotgun mode," whereas targeted precursor selection (i.e. SRM or PRM) may be employed later on for selectivity profiling, when all relevant parameters (e.g. retention times, precursor ion mass, optimized transitions, etc.) are known.

The key to a successful chemical proteomics study is therefore a properly designed affinity enrichment or probing strategy with high affinity chemical tools of known selectivity (or lack thereof – depending on the aim of the study). In the most frequently used approach, the so-called compound pulldowns, derivatives of tool compounds with known target profile or a phenotypic screening hit or any other compound of interest with an insufficiently characterized MoA ("bait compounds"), are immobilized on a chromatography resin (Figure 4.3). Good understanding of structure–activity relationship (SAR) enables exploration of different linkage positions of that compound. SAR-preserving linkage facilitates binding of candidate efficacy targets, whereas linkage positions that are predicted to interfere with activity may nevertheless offer binding sites for potential off-targets. Cell or tissue lysates are incubated with the resin. Bound proteins are subsequently eluted from the chromatography column and identified (and often quantified as well) by LC–MS/MS bottom-up shotgun proteomics. To account for nonspecific protein binding to the resin, inactive bait compounds are used as controls. Preferably, however, competition-binding experiments are typically performed, where multiple parallel samples are treated with increasing concentrations of free test compound to determine EC_{50} values [30]. Even K_d values can be calculated, provided that protein depletion is accounted for [35].

Chemical proteomics investigates targets in their cellular state, i.e. as protein complexes, e.g. histone deacetylase (HDAC, also known as KDAC to acknowledge non-histone lysine acetylation substrates) complexes, consisting of catalytic and other subunits that may be posttranslationally modified. Disease-relevant cells or tissues such as the ones in which the initial phenotypic screen is performed can be employed. Moreover, chemoproteomics enables the identification of a comprehensive list of putative targets and off-targets in these cells or tissues, thus helping to generate hypotheses regarding MoA or safety concerns.

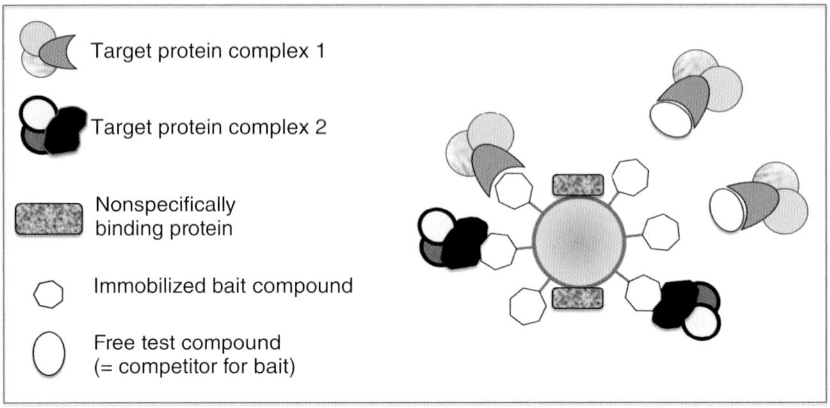

Figure 4.3 Schematic representation of a typical chemoproteomics workflow. A bait compound (white hexamer) immobilized on a resin may capture several target protein complexes as well as nonspecific binders. In a competition-binding experiment, a free test compound competes with binding of a subset of target protein complexes, whereas the bait compound itself competes with all specific binding. Chemical proteomics is therefore also useful for selectivity profiling of free test compounds against the sub-proteome captured by the bait. Note that all specifically bound target protein complexes are eluted when the free bait compound is used as competitor. In this situation complementary affinity purification approaches, e.g. immobilized antibodies directed against the "orange protein," would be required for successful deconvolution of individual target protein complexes.

LY-294002 may serve as a point in case: the phosphatidyl inositol-3-kinase (PI3K) inhibitor (as well as its PI3K-inactive structural analogue) is also a potent inhibitor of bromodomain proteins [36]. This example illustrates the emerging concept that drug selectivity must not only be profiled against members of the same target class.

Examples of phenotypic screening hits or other compounds with unknown MoA that serendipitously have been revealed to target epigenetic mechanisms are rare, but the case of putative immune modulators that were found to target BET (bromodomain and extraterminal domain) family proteins may serve as an example [37]. Following a phenotypic screen for inducers of an apoA1 reporter gene, Chung et al. used linkable derivatives of an active screening hit, I-BET762, and of its inactive enantiomer to identify proteins that selectively bound to the former. Following chromatography, targets were identified by MS as the BET proteins BRD2, BRD3, and BRD4. In most instances, however, analogues of known tool compounds are used as bait. In that capacity, they allow for selectivity profiling and in-depth characterization of mega-Dalton-sized target protein complexes (Table 4.1). Whereas the most relevant examples shall be presented here, a more detailed discussion can be found in recent reviews [8, 9].

A complementary chemoproteomics method, activity-based proteome profiling (ABPP), has been reviewed elsewhere [34]. Here, a non-immobilized chemical probe is incubated with cell or tissue extract. The probe consists of a binding group that mediates the initial noncovalent binding to targets, a reactive moiety that forms a covalent bond and an affinity "handle" such as biotin for

Table 4.1 Protein–small-molecule interactions. 1. Target identification and selectivity profiling using immobilized compounds. 2. Compound screening/profiling using modified histone peptides.

Targeted protein complex	Identification strategy: Target identification/ selectivity profiling with immobilized compound	Identification strategy: Compound screening/ profiling in LC–MS/MS or MALDI-ToF MS assays with modified histone peptides
Writer		
Histone/lysine acetyltransferase (HAT/KAT)	[38]	[39]
Protein arginine methyltransferase (PRMT)	[40, 41]	[42, 43]
Eraser		
Histone/lysine deacetylases (HDAC/KDAC, sirtuins)	[44–48] a)	[49–51]
Lysine demethylase (KDM)	[52, 53]	[54–58]
Reader		
BET type of bromodomains	[36, 37, 59–61]	

a) Thermal proteome profiling (TPP), where drug is not immobilized, is first described in 2014 [62].

subsequent analysis and purification. For example, ABPP approaches have been used for examination of HDAC complexes [63].

4.2.1 Histone Deacetylase and Acetyltransferase Chemoproteomics

Whereas cyclic tetrapeptides, trapoxins, have been used in Schreiber's seminal chemical biology work on HDACs [64], vorinostat derivatives are currently preferred in HDAC chemoproteomics. An immobilized *p*-aminomethyl analogue of the FDA-approved HDACi vorinostat (suberoylanilide hydroxamic acid; SAHA) captures multiple HDAC target protein complexes (HDAC1, 2, 3, 6, 8, and 10) including four separate HDAC1/2-containing complexes [44]. The non-selectivity of vorinostat enables the use of this matrix for selectivity profiling, as the target complexes are selectively displaced from the resin (see Figure 4.3). Proteomics-based selectivity profiling with cellular protein complexes reveals more molecular selectivity detail than that with recombinant proteins. For instance, an aminobenzamide HDACi targeted some HDAC complexes, but did not potently target the HDAC–Sir3 silencer complex [44]. In the alternative ABPP approach, a benzophenone analogue of vorinostat (SAHA-BPyne) in combination with bottom-up LC–MS/MS is useful for analysis of HDAC1, 2, 3, and 6 [47]. A cell-permeable vorinostat-chloroalkane probe is the latest chemical biology tool for HDAC target complex studies: it facilitates intracellular target binding, enrichment using commercially available magnetic beads, and proteomics-based target identification [65]. Selective Sirt2 inhibitors and their SAR have recently been described, opening up the possibility of chemoproteomics studies of class III HDACs [66, 67]. CoA cofactor-based bisubstrate inhibitors of lysine acetyltransferases, e.g. as clickable probes, can be used for affinity purification of protein complexes containing multiple such enzymes from three phylogenetically distinct families and including orphan transferases [38].

4.2.2 Bromodomain Chemoproteomics

The I-BET affinity matrix that was initially used for identification of BET proteins as targets of a phenotypic screening hit [37] can be used for EC_{50} determinations and more detailed studies of BET protein complexes [59]. A cautionary note is that non-selectivity of affinity matrices as well as test compounds together with the extraordinary molecular mass of some epigenetic target protein complexes can result in concentration responses of many (sometimes >100) proteins identified in competition-binding experiments. Therefore, chemoproteomics methods need to be complemented by orthogonal affinity-based approaches (see Section 4.3). There is an increasing number of options for affinity capture of BET proteins that includes dual bromodomain and kinase inhibitors [36, 68]. Interestingly, some bead-immobilized kinase inhibitors capture bromodomain proteins other than BETs, e.g. ATAD2 [61]. One limitation of single compounds as affinity capture ligands stems from the fact that the number of possible targets that a test compound can replace is limited by the choice of targets captured by the immobilized compound instead of by its own target repertoire. Since it would be too cumbersome to synthesize one or multiple affinity ligands for every compound to be investigated in detail, affinity resins with multiple non-selective compounds akin to "kinobeads" [30] that cover a large portion of the kinome would be desirable. Initial "bromobeads" for chemoproteomics selectivity profiling have been reported [60], and recent selective and non-selective bromodomain modulators discussed elsewhere in this book are bound to lead to new affinity matrices for chemoproteomics in the future.

4.2.3 Demethylase Chemoproteomics

GSK-J3, a linkable analogue of the selective JMJD3 lysine demethylase inhibitor GSK-J1, is useful for in-depth studies of this epigenetic target [52]. However, because of the exquisite selectivity of this tool compound, additional compounds would be necessary for investigations of other demethylase proteins or protein complexes. For chemoproteomics selectivity profiling, in analogy to similar tools described for deacetylases and bromodomain proteins (Sections 4.2.1. and 4.2.2.), an initial combination of non-selective 2-oxoglutarate-dependent dioxygenase binders and demethylase inhibitors that enable profiling against several KDM3 and 4 (JMJD1 and JMJD2) subfamily members has been reported [53]. This "jumonobead" affinity matrix can enrich 40 out of 60 known dioxygenase-dependent enzymes including 18 JmJC domain-containing proteins from human cells.

4.2.4 Methyltransferase Chemoproteomics

A biotinylated analogue of the lysine methyltransferase inhibitor UNC0638 can be used for studies of the lysine methyltransferases G9a and GLP [40]. Another UNC0638-based affinity matrix with covalent linkage enriches a G9a-dependent mega-Dalton repressome from primary macrophages [41]. Also in this case of compound pulldowns, as has been noted for mega-Dalton deacetylase and bromodomain protein complexes, many targets of epigenetic drugs or tool compounds are enriched together with the complexes they are part of. When performing competition-binding experiments with a dilution series of a small-molecule compound, not only concentration–response curves of the direct targets of the drug are recorded but also matching curves for the proteins, which are co-enriched as protein complex members. This feature of co-competition can be used in combination with other proteomic methods to unequivocally define the composition of epigenetic target complexes and reader complexes that contribute to a compound's MoA [44, 59].

4.3 Characterization of Epigenetic Drug Target Complexes and Reader Complexes Contributing to Drug's Mode of Action

Histones are epigenetically modified in a dynamic manner by multiprotein complexes, which can be classified as writers adding PTMs, erasers removing PTMs, and readers assembling PTM-dependent docking sites for additional protein complexes that may be relevant for a drug's MoA [3]. In epigenetics, more than in most other fields, it is protein complexes rather than single gene product enzymes that carry out the functions of writers, erasers, or readers. Because of this high level of complexity, proteomic data from several orthogonal enrichment methods typically need to be combined in order to elucidate protein complex composition or the MoA of drugs [44, 59]. These approaches include co-purifications via immobilized drugs (see Section 4.2), antibodies against native protein complex constituents, affinity matrices against epitope tags or other recombinant tags, and affinity enrichment using bait peptides.

4.3.1 Immunoaffinity Purification of Native Protein Complexes

For immunoaffinity purifications (IAP) of protein complexes, antibodies that target either one of the complex's native protein subunits or a recombinant epitope tag can be used (Figure 4.4). The advantage of the former strategy is the ability to analyze cellular complexes with correct stoichiometry and possibly PTMs that reflect the cellular functional state [69–71]. Combination of IAP with drug pulldowns is often required for successful deconvolution of target protein complex composition: especially in epigenetics, mixtures of protein complexes are typically co-eluted from the chemoproteomics affinity matrix (compare Figure 4.3). Therefore, an unequivocal mapping of proteins to complexes is impossible with chemoproteomics alone. For instance, immobilized pan-HDAC

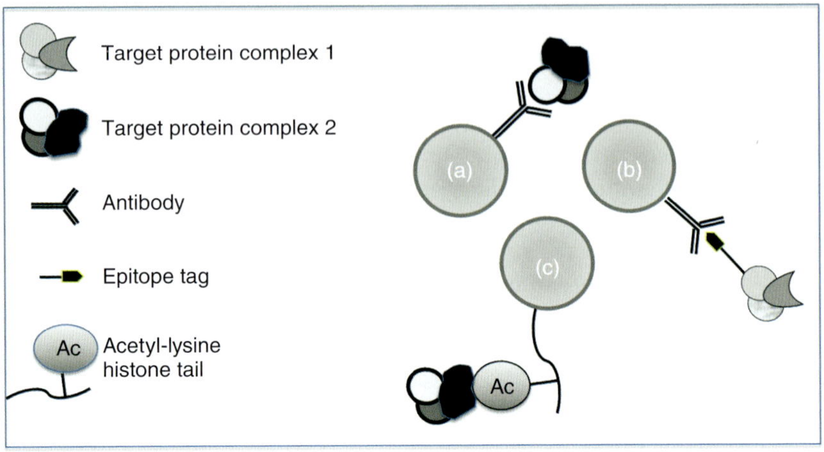

Figure 4.4 Strategies for affinity enrichment of epigenetic target and/or reader protein complexes. (a) A specific antibody directed against a known subunit of an epigenetic protein complex is used for immunoaffinity purification (IAP). When combined with chemoproteomics results (Figure 4.3), target protein complexes can be delineated. Here, endogenous protein complexes are analyzed. (b) Similar to (a), but an epitope tag on overexpressed "orange protein" is used for IAP. (c) Similar to (a), an immobilized histone N-terminal peptide with a defined set of PTMs is used for affinity enrichment.

inhibitors affinity-capture all HDAC1-, HDAC2-, or HDAC3-containing protein complexes simultaneously. However, when chemoproteomic studies are complemented with IAPs using antibodies against unique subunits of single HDAC1, HDAC2, and HDAC3 complexes, e.g. antibodies against LSD1, then integration of the two datasets enables characterization of individual HDAC protein complexes and identification of novel target complexes and of the target protein itself within the complex [30, 44] (Figure 4.4). A potentially new therapeutic option for mixed lineage leukemia, suggested by studies combining chemoproteomics, IAP, and additional affinity capture proteomic experiments using compounds targeting the acetyl-lysine recognizing BET proteins, may serve as another example [59]. Despite these successes and some large-scale projects [72, 73], IAP is still rarely used in many laboratories, perhaps reflecting the high cost and often insufficient specificity of available antibodies for proteomics studies.

4.3.2 Immunoaffinity Purification with Antibodies Against Epitope Tags

Epitope tags, short peptide sequences recognized by well-characterized monoclonal antibodies, are attached to proteins of interest by means of molecular cloning. They can be used for standardized IAP. Common epitope tags include FLAG, HA, and c-Myc tags [69, 71]. Whereas standardized IAP is a plus, the need for often poorly reproducible cellular (over)expression of the tagged protein in the presence of the untagged endogenous protein generally leads to variable complex composition. As endogenous protein complex subunits may have low abundance, high degrees of overexpression of the epitope-tagged bait protein can lead to a small number of correctly assembled complexes that would be available for analysis. Low level expression of the tagged protein is therefore required [69]. Bacterial artificial chromosomes (BAC) enable the gene expression from endogenous promoters in the presence of most regulatory genetic elements or even knock-ins of recombinant proteins. They have been used, for instance, for studies of trimethyl lysine reader protein complexes [74].

In epigenetics, epitope tagging is widely used for detailed studies of target protein complexes: for example, IAP of FLAG-tagged BRPF1 followed by LC–MS/MS has revealed that the exchange of associated factors directs a switch in histone tail specificity of the HBO1 acetyltransferase/HBO1 in a complex with JADE acetylates histone H4, whereas HBO1 in a complex with BRPF1 acetylates histone H3 [75]. The polycomb repressive complex 2 (PRC2), the EZH2-containing histone methyl transferase complex inducing transcriptional silencing, may serve as another example: EZH2 knockout does not abolish H3K27 methylation in mouse embryonic stem cells. Affinity purifications of FLAG- and biotin-tagged PRC2 subunit EED found the EZH2 homologue EZH1 in a noncanonical PRC complex. Apparently, EZH1 preserves H3K27 methylation mark on some genes in EZH2-deficient cells [76].

4.3.3 Affinity Enrichment Using Histone Tail Peptides as Bait

Immobilized histone tail peptides or even complete nucleosomes with a defined set of PTMs can be used for affinity enrichment of epigenetic reader complexes that serve as PTM-dependent molecular scaffold and thus contribute to a writer or eraser compound's MoA [59, 74, 77](Figure 4.4c). Used as a stand-alone affinity enrichment method in a SILAC quantitative proteomics experiment, protein binding to non-methylated H3 peptide can, for instance, be compared with that to H3K4me3 peptide, thus revealing differential binding of the transcription factor FHIID to the latter [74]. In some cases, it may be beneficial to combine compound pulldowns, IAP, and histone tail binding studies to assess a compound's MoA in maximum detail [59].

4.4 Elucidation of a Drug's Mode of Action: Analysis of Histone Posttranslational Modifications by MS-Based Proteomics

According to the histone code hypothesis, individual modifications functionally interact with each other to bring about new functions beyond those of the individual modification site [78]. Methods to study how a single modification functions in the context of others are limited. Antibodies directed against single modification sites are the most widely used reagents to study histone biology in the cellular context, but these reagents are only available for few of the more than 500 estimated histone modifications [79]. Moreover, antibodies are frequently of limited specificity, and they are less suited to study combinatorial modifications co-occurring at the same histone molecule [4].

MS-based proteomics recently matured as a technology that can provide complementary information to the chromatin toolbox by characterizing most, if not all, histone modifications and some of their combinatorial isoforms with high specificity and quantitative accuracy. Recent examples of MS-based histone modification profiling identified cancer-promoting changes in the epigenome [80, 81], characterized how diet and microbiome affects the modification status at chromatin [82], and helped to identify new chromatin regulators that mitigate the aging-associated decline of physiological integrity [83, 84]. Moreover, MS-based profiling of histone modification states can be used to characterize the specificity of antibodies, enzymes, and drugs *in vitro* and in the cellular context [85]. Surprisingly, however, the (histone) substrates for many putative histone-modifying enzymes are still not characterized, and only a few epigenetic drugs have been assessed through profiling across comprehensive panels of (histone) modifications.

4.4.1 Histone Modification MS Workflows

Modified histones provide a formidable challenge for MS-based proteomics because of (i) the high number of different modification types (PTM diversity), (ii) frequently co-occurring modifications at the same peptide (PTM combinations), (iii) very similar sequences across histone isoforms that often differ only by a single amino acid on an individual peptide (high sequence homology), and (iv) a skewed amino acid distribution toward a few frequently occurring amino acids (enriched short sequence motifs). These obstacles limit the use of conventional proteomic strategies (Feller, manuscript in preparation). However, some of these challenges have been met that led to robust MS workflows that enable the quantification of 20–60 histone modifications including some of their combinations in a single experiment.

Most histone proteomic workflows start by enriching these highly basic proteins by acid extraction either from purified nuclei or directly from crude cell or tissue lysates (Figure 4.5). Because histones are strongly enriched for lysines and arginines, trypsin generates many too short peptides that cannot be retained on C18 chromatographic resins. Therefore, in most protocols all unmodified and

Figure 4.5 Histone PTM workflow. Histone PTM workflow exemplified for H3 proteoform with four highly abundant modifications (H3K9me3K14acK23acK27me2). (a) Histones are extracted by diluted sulfuric acid. (b) Chemical acetylation by d6-acetic anhydride transfers a deuterated (d3)-acetyl group to free and monomethylated lysines. Subsequent trypsin digestion yields ArgC-like peptides that remain the connectivity between close-by lysines. (c) Lysine methylation induces characteristic shifts in chromatographic retention time, which allows MS^1-based quantification of methylation isoforms. Co-eluting isobaric positional isomers (H3K9ac, H3K14ac) require additional MS^2-based discrimination and quantification. (d) PTMs alter a peptide's response factor. To achieve quantitative accuracy, synthetic reference peptides to acetylation and methylation motifs were used to determine the LC–MS correction factor. A quantification tag (green hexamer), proteolytically released from the histone peptide, was used to normalize for variations in the quantities of the synthetic peptides. Source: Feller et al. 2015 [86]. Reproduced with permission of Elsevier.

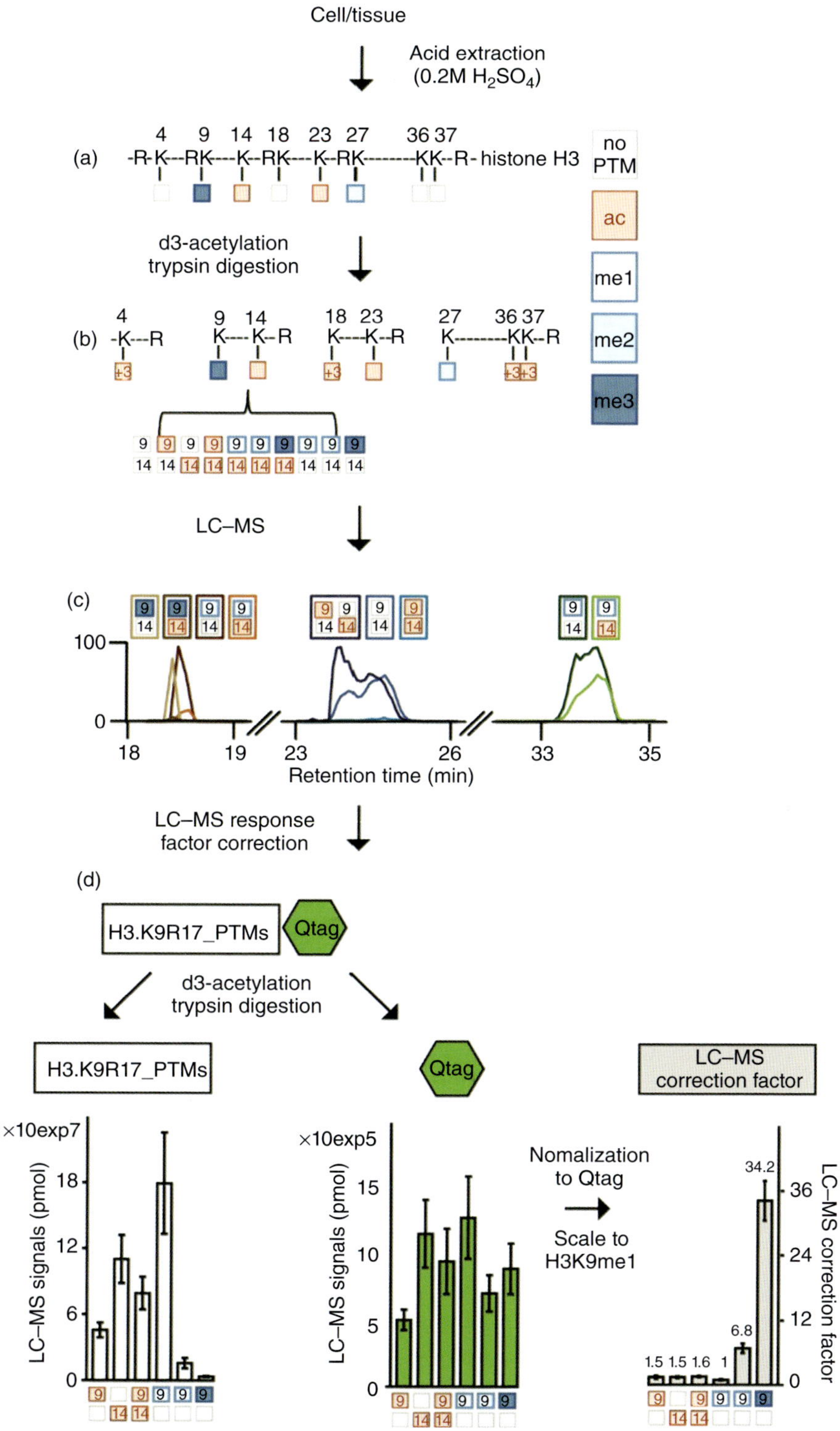

Cell/tissue
Acid extraction
(0.2M H_2SO_4)
(a)
4 9 14 18 23 27 36 37
-R-K--RK---K--RK---K-RK--------KK--R- histone H3
no PTM
ac
me1
me2
me3
d3-acetylation
trypsin digestion
(b)
+3
LC–MS
(c)
100
0
18 19 23 26 33 35
Retention time (min)
LC–MS response
factor correction
(d)
H3.K9R17_PTMs
Qtag
d3-acetylation
trypsin digestion
LC–MS
correction factor
×10exp7
LC–MS signals (pmol)
×10exp5
Nomalization
to Qtag
Scale to
H3K9me1
LC–MS correction factor
1.5 1.5 1.6 1 6.8 34.2

monomethylated lysines are first derivatized with propionic anhydride, which protects these lysines in the subsequent trypsinization reaction to yield ArgC-like peptides generated after arginine cleavage [87]. Alternatively, propionic anhydride can be replaced with d6-deuterated acetic anhydride [88]. Implemented on high resolution instruments, using d6-deuterated acetic anhydride has the advantage to harmonize the peptide properties among modified peptide isoforms sharing the same sequence but differing in their location of the lysine acetylation site [86]. The incorporation of three heavy hydrogens creates a mass shift that allows distinguishing the "heavy" chemical acetylation from cellular acetylation. After the trypsinization step, the resulting peptides are also derivatized with propionic anhydride or deuterated acetic anhydride to propionylate/acetylate the peptide's free N-termini. This installs additional hydrophobicity to increase the chromatographic retention of small peptides, such as the ones containing the di- and trimethylation states of lysine 4 on histone H3 (H3K4me2/3). In addition, this peptide-level derivatization eliminates the charge at the peptide's N-terminus, which increases sensitivity by distributing the peptide ions to fewer dominating charge states [87].

Another advantage of the protein-level derivatization is to yield longer peptides often containing several lysines. This peptide feature allows analyzing modification cross-talk occurring between close-by lysines (and other amino acid residues) within the same peptide sequence. In many cases, several positional isomer peptides are generated that contain the same modification type, which is located on different amino acid residues (sites). Identifying and quantifying these isobaric peptides – i.e. peptides that share the same precursor *m/z* – requires chromatographic separation or distinguishing fragment ions (Figure 4.5). Recording high-quality fragment ions over the chromatographic peak is often achieved by programming the mass spectrometer to specifically target these positional isomers using inclusion lists (while other modifications may be identified with DDA in the same MS experiment) or performing a PRM/SRM experiment [80, 87, 89]. Care should be taken when selecting the precursor isolation width, because similar peptides with similar precursor *m/z* and shared product ions may co-elute, which results in interfering spectra that compromises quantification accuracy. This phenomenon is particularly pronounced when applying standard DIA methods to measure histone PTMs. Moreover, for some combinatorial modification motifs, such as the doubly acetylated H4 isoforms (e.g. H4K5acK12ac), a standard tandem MS strategy is not sufficient. Identification and quantification of these co-eluting combinatorial modifications requires re-fragmenting specific fragment ions and acquiring MS3 ions [86] or following characteristic fragment ion patterns using specialized software [90].

Most quantitative histone MS studies report 20–60 histone modification motifs. These often include N-terminal acetylation on histones H2A (K5), H3 (K9, K14, K18, K23, K27), and H4 (K5, K8, K12, K16), mono-/di-/trimethylation on H3 (K4, K9, K27, K36, and K79) and H4 (K20), and phosphorylation on H3S10. The result of a histone modification MS experiment is often reported as "estimated" stoichiometry. For this analysis, the integrated signal intensities for a single modification motif are normalized relative to all modification motifs sharing the same sequence. For example, to estimate the site occupancy of the combinatorial H3K9me2S10pK14ac motif, its signal intensity is divided by the summed intensities of all peptides decorated with any permutation of H3K9ac, H3K9me1/2/3, H3S10p, and H3K14ac. To improve quantitative accuracy, synthetic peptides can be used to correct for varying response factors influenced by the peptide's sequence and modification type (Figure 4.4). For instance, a comparison of response factors across histone H3 methylation motifs indicated that some modifications, such as H3K4me2/3, show more than 10-fold lower detection efficiencies than their unmodified peptide siblings [86, 91].

4.4.2 Application of Histone MS Workflows to Characterize Epigenetic Drugs

The MS analysis of histone modification patterns in response to (epigenetic) drug treatment provides the most direct readout for a drug's MoA in the cell. So far, however, only few epigenetic inhibitors have been characterized against larger panels of cellular modifications using quantitative MS, including inhibitors against G9A/GLP (BIX01294, UNC0638, and UNC0646), EZH1/2 (GSK126 and UNC1999), SMYD2 and CBP/p300 (C646), and HDACs. These initial studies exemplify how MS-based histone profiling in conjunction with phenotypic assays allows evaluating a drug's potency and selectivity while also uncovering the cellular "off-target" responses to these perturbations.

The G9A/GLP inhibitors BIX01294 and UNC0638 specifically reduce H3K9me2 in mouse embryonic fibroblasts and human cell lines, whereas the mono- and trimethylation states at this site remain mostly unaffected [42, 92]. MS measurements further showed that the abundance of other methylation sites occurring on H3K4, H3K27, H3K36, H3K79, and H4K20 did not change in response to these inhibitors. Profiling the G9A/GLP inhibitor UNC0646 against a larger panel of 41 modification states on H3, another study reported reduction of all three methylation states on H3K9. Of note, the trimethylated state decreased only in peptides where the adjacent K14 residue was acetylated, (H3K9me3K14ac) [93]. In agreement, G9A knockout mouse ES cells also display reduced levels of the mono- and dimethylated forms of H3K9 [42] as well as H3K9me3K14ac [93]. These results highlight the importance of comparing the drug and target enzyme(s) side by side across extended panels of histone modification states, which allows a more accurate evaluation of a drug's potency and selectivity in the cellular system. Moreover, all three tested G9A/GLP inhibitors BIX01294, UNC0638, and UNC0646 led to elevated acetylation at H3K9. In a side-by-side comparison between BIX01294 and UNC0638 in human MDA-MB-231 breast cancer cells, reduction of H3K9me2 and increase of H3K14ac was more pronounced for UNC0638, suggesting a higher potency of this drug at least in this cell line [92]. Interestingly, phenotypic assays such as viability assays combined with immunoblotting experiments for H3K9me2 showed that the cellular potency and toxicity differed by more than fivefold across seven human cell lines [92]. These results demonstrate the impact of the cellular system on the activity of a chemical modulator and suggest that an accurate description of the MoA benefits from profiling drugs across different cellular systems. In-depth analysis of the steady-state and drug-responsive epigenome by histone profiling in conjunction with MS imaging of the drug and elucidation of the chromatin protein network engaging the drug should provide helpful information to understand the differences across these cellular systems.

Two studies recently characterized the EZH2 inhibitor GSK126 and the EZH1/EZH2 dual inhibitor UNC1999 using MS-based histone profiling [93, 94]. In agreement with the known targets of these enzymes, both studies identified a reduction of the di- and trimethylation states on H3K27. However, while the authors observed increased levels of H3K27me1 in mouse ES cells treated with GSK126 and in mouse cells lacking EZH2 or other PRC2 components [93], UNC1999 treatment in murine leukemia cells led to a more varied response of H3K27me1-containing motifs depending on the state of H3K36 [94]. Moreover, both studies detected elevated levels of most methylation states on H3K36. This may be the result of a conversion from doubly to singly methylated motifs (e.g. H3K27me2K36me2 to H3K36me2) and reflect a negative cross-talk between the enzymes methylating these two sites, as suggested before [95]. In comparison with combinatorial modifications at other histone regions, the function of these abundant co-occurring motifs containing different methylation states at lysines 27 and 36 on H3 is less well understood [6, 95].

While the methyltransferases G9A/GLP and EZH1/EZH2 discussed above presumably catalyze the methylation reaction only at a single lysine residue on histones, most acetyltransferases (KATs/HATs) and deacetylases (KDACs/HDACs) appear to have broader substrate specificity toward histone and non-histone proteins. Defining the precise cellular targets for these enzyme classes is challenging, because they often function on the context of large, mega-Dalton protein complexes, which complicates their detailed biochemical analysis. In a few cases where this challenge succeeded, it was shown that the histone substrate (free histones, octamers, and nucleosomes) and associated cofactors forming alternative protein complexes modulate the specificity of the acetylating and deacetylating enzymes [96]. Moreover, *in vivo* studies are complicated because many KATs and KDACs are encoded in multiple paralogue genes in mammalian cells, which often compensate function. Together, these challenges when studying KATs and KDACs highlight the needs for new strategies to investigate their functions and to improve drug development pipelines for this clinically relevant class of enzymes.

In order to gain understanding of the substrate specificity of lysine acetyltransferases and deacetylases, we recently developed an improved histone MS workflow and applied it to analyze the histone acetylation and methylation landscape after depleting all KATs and KDACs expressed in *Drosophila* KC cells [86]. Drosophila melanogaster was chosen as a model because its genome contains less paralogues for these chromatin enzymes, which allowed us to use a genetic RNA interference (RNAi) approach. This systematic analysis not only characterized cellular substrates for 23 acetyltransferases and 8 deacetylases but also allowed us to recognize a "systemic compensation" effect, where the loss of individual KATs not only reduced the abundance of its putative substrate(s) but also led to an increase on other acetylation sites. While most acetyltransferases have a narrow substrate specificity that appears often modulated by modifications close to the putative direct target site (with exception of the CBP/p300 and GCN5/PCAF homologues), deacetylases serve a broader spectrum of lysine sites. These observations are in line with the high connectivity for these enzymes in protein interaction networks [97]. Furthermore, these results provide a starting point for a better understanding of the genesis of combinatorial acetylation motifs, which appear as the prevalent targets for bromodomain recognition modules [98].

4.5 Challenges and New Trends

4.5.1 Challenges and Trends in MS Analysis of Histone PTMs

How many modifications are present on histones, and how many more modifications are yet to be discovered? MS is the chief technology for the identification of new modification types and their precise localization along the polypeptide sequence. During the past few years, eight new modification types have been discovered on histones, raising the number to over 20 modification types placed on over 200 sites forming more than 500 distinct modification motifs [79]. While there is tremendous progress on new algorithms aiding in the identification and site localization of protein modifications, major challenges still remain on correct error estimations, in particular for unrestricted searches where mass shifts are not defined before the search [99, 100]. Modification searches on histones are in particular challenging because of the high sequence homology among histone sequence isoforms, the high number of modification types including many near-isobaric modifications, and the combinations of modifications along the long peptides generated after arginine cleavages (Feller, manuscript in preparation). It is therefore necessary to validate new modifications with synthetic reference peptides, which should yield identical fragmentation patterns and chromatographic retention. In addition to this chemical validation, it is desirable to support findings on new modifications with orthogonal strategies, for example, by describing their genomic distribution with high-quality modification-specific antibodies and characterizing their writers, readers, and erasers. Of note, less than 100 of the over 500 reported histone modifications have been validated by synthetic reference peptides, and biological functions remain to be explored for most of them.

Which modifications co-occur along the same histone molecule? By analyzing the intact protein, top-down MS preserves the complexity of intermolecular modification and sequence variants. In comparison with bottom-up histone PTM workflows, top- and middle-down implementations require substantial amounts of purified histones, are limited to the detection of abundant isoforms, and show more varied identification confidence and quantitative accuracy [10, 101, 102]. Improvements in instrumentation, in particular in separating isobaric isoforms by chromatography or ion mobility, narrow precursor isolation windows to reduce chimeric spectra, efficient isolation and fragmentation at high *m/z*, and automated data analyses workflows, are likely to have a big impact toward transitioning from identifying proteoforms in a few samples to their reliable quantification across many samples [10, 101]. Despite these shortcomings, top- and middle-down studies already identified hundreds of histone proteoforms and revealed fascinating insights into PTM combinatorics. For example, the highly abundant acetylation sites on H3K14 and H3K23 are mostly distributed to different histone H3 proteoforms, while the repressive marks H3K9me2 and H3K27me3 frequently co-occur on the same molecule together with H3K23ac [102, 103]. These observations challenge current views obtained from chromatin immunoprecipitation–DNA sequencing (ChIP-Seq) experiments. Resolving them by characterizing how these combinatorial marks affect reader domain binding and chromatin enzyme activity promises to uncover fundamentally new types of modification cross-talk.

4.5.2 High-Throughput Mass Spectrometry-Based Compound Profiling in Epigenetic Drug Discovery

For various steps of the drug discovery process, MS-based methods have been proposed as more specific and potentially faster replacements for current immune assays that suffer from antibody cross-reactivity and fluorescence-based methods that are prone to generating artifacts (e.g. compound autofluorescence, fluorescence quenching). Compound screening/profiling in ESI-MS/MS (e.g. RapidFire-MS) or MALDI-ToF MS assays that measure *m/z* changes due to PTM changes on modified histone peptides is increasingly feasible and popular in recent years (Table 4.1). Innovative, very fast MALDI-ToF mass spectrometers have even enabled ultrahigh-throughput screening of a million compounds in a matter of days [57].

For selectivity profiling, affinity matrices such as "bromobeads" or "jumonobeads" that feature (combinations of) non-selective modulators of epigenetic targets and that can be used for chemoproteomics profiling against endogenous cellular, posttranslationally modified protein complexes have been complementing panels of recombinant enzyme assays (see Section 4.2). Finally, thermal proteome profiling (TPP), which measures thermal stabilization of protein targets upon ligand binding on a proteome scale, has enabled studies of intracellular drug target engagement [62]. As evidenced for the HDACi panobinostat, these emerging new proteomic workflows can also be used to study cellular targeting of protein complexes [48]. The same compound has been used in MALDI-ToF-based cell assays that could be automated and adapted to new high speed instruments, thus potentially leading to MALDI-ToF cellular compound screening [29].

4.5.3 Mass Spectrometry Imaging of Drug Action

MSI is frequently used for semiquantitative analysis of a compound's tissue distribution in DMPK studies. In principle, MSI being a label-free method allows for simultaneous analysis of parent compound and some of its metabolites, albeit with moderate sensitivity. Because of this limited sensitivity and a mass range that is limited to 30 kDa, most protein substrates of transferase, ligase, or hydrolase enzymes whose mass change could serve as a proximal pharmacodynamics marker are virtually impossible to detect by MSI. Histones, however, may be an exception to this rule, as they are very abundant cellular proteins of

<20 kDa. Consequently, time and concentration dependence of histone PTMs can be assessed in MSI experiments [29].

Acknowledgments

This work was supported by the BMBF grant "NANOKAT II – Netzwerk Biotechnologie" to C.H. C.F. is supported by an EMBO long-term postdoctoral fellowship (ALTF 55-2015). D.W. thanks the Hartmut Hoffmann-Berling International Graduate School of Molecular and Cellular Biology (HBIGS) for support. C.H. and D.W. also thank the CM1406 European COST action "Epigenetic Chemical Biology" for support.

List of Abbreviations

DDA	data-dependent acquisition
DESI	desorption electrospray ionization
DIA	data-independent acquisition
DMPK	drug metabolism and pharmacokinetics
FT-ICR	Fourier transform ion cyclotron resonance
HDACi	histone deacetylase inhibitor
HPLC	high performance liquid chromatography
IAP	immunoaffinity purification
LC–MS/MS	liquid chromatography coupled to tandem mass spectrometry
MALDI-MS	matrix-assisted laser desorption/ionization mass spectrometry
MoA	mechanism of action
PRM	parallel reaction monitoring
PTM	posttranslational modification
SAR	structure–activity relationship
SIMS	secondary ion mass spectrometry
SRM	selected reaction monitoring
SWATH	sequential windowed acquisition of all theoretical fragment ion spectra
ToF	time of flight
TPP	thermal proteome profiling
WC-MALDI-MS	whole-cell MALDI mass spectrometry

References

1 Arrowsmith, C.H., Bountra, C., Fish, P.V. et al. (2012). Epigenetic protein families: a new frontier for drug discovery. *Nat. Rev. Drug Discovery* 11: 384–400.
2 Dawson, M.A. and Kouzarides, T. (2012). Cancer epigenetics: from mechanism to therapy. *Cell* 150: 12–27.

3 Tough, D.F., Tak, P.P., Tarakhovsky, A., and Prinjha, R.K. (2016). Epigenetic drug discovery: breaking through the immune barrier. *Nat. Rev. Drug Discovery* 15: 835–853.
4 Rothbart, S.B., Dickson, B.M., Raab, J.R. et al. (2015). An interactive database for the assessment of histone antibody specificity. *Mol. Cell* 59: 502–511.
5 Blasi, T., Feller, C., Feigelman, J. et al. (2016). Combinatorial histone acetylation patterns are generated by motif-specific reactions. *Cell Syst.* 2: 49–58.
6 Su, Z. and Denu, J.M. (2016). Reading the combinatorial histone language. *ACS Chem. Biol.* 11: 564–574.
7 Marcon, E., Jain, H., Bhattacharya, A. et al. (2015). Assessment of a method to characterize antibody selectivity and specificity for use in immunoprecipitation. *Nat. Methods* 12: 725–731.
8 Weigt, D., Hopf, C., and Medard, G. (2016). Studying epigenetic complexes and their inhibitors with the proteomics toolbox. *Clin. Epigenetics* 8: 76.
9 Noberini, R., Sigismondo, G., and Bonaldi, T. (2016). The contribution of mass spectrometry-based proteomics to understanding epigenetics. *Epigenomics* 8: 429–445.
10 Toby, T.K., Fornelli, L., and Kelleher, N.L. (2016). Progress in top-down proteomics and the analysis of Proteoforms. *Annu. Rev. Anal. Chem.* 9: 499–519.
11 Bantscheff, M., Lemeer, S., Savitski, M.M., and Kuster, B. (2012). Quantitative mass spectrometry in proteomics: critical review update from 2007 to the present. *Anal. Bioanal. Chem.* 404: 939–965.
12 Michalski, A., Cox, J., and Mann, M. (2011). More than 100,000 detectable peptide species elute in single shotgun proteomics runs but the majority is inaccessible to data-dependent LC-MS/MS. *J. Proteome Res.* 10: 1785–1793.
13 Picotti, P. and Aebersold, R. (2012). Selected reaction monitoring-based proteomics: workflows, potential, pitfalls and future directions. *Nat. Methods* 9: 555–566.
14 Gallien, S., Duriez, E., Crone, C. et al. (2012). Targeted proteomic quantification on quadrupole-orbitrap mass spectrometer. *Mol. Cell. Proteomics* 11: 1709–1723.
15 Gillet, L.C., Navarro, P., Tate, S. et al. (2012). Targeted data extraction of the MS/MS spectra generated by data-independent acquisition: a new concept for consistent and accurate proteome analysis. *Mol. Cell. Proteomics* 11, O111 016717: 1–17.
16 Gillet, L.C., Leitner, A., and Aebersold, R. (2016). Mass spectrometry applied to bottom-up proteomics: entering the high-throughput era for hypothesis testing. *Annu. Rev. Anal. Chem.* 9: 449–472.
17 Schubert, O.T., Gillet, L.C., Collins, B.C. et al. (2015). Building high-quality assay libraries for targeted analysis of SWATH MS data. *Nat. Protoc.* 10: 426–441.
18 Ong, S.E., Blagoev, B., Kratchmarova, I. et al. (2002). Stable isotope labeling by amino acids in cell culture, SILAC, as a simple and accurate approach to expression proteomics. *Mol. Cell. Proteomics* 1: 376–386.
19 Carr, S.A., Abbatiello, S.E., Ackermann, B.L. et al. (2014). Targeted peptide measurements in biology and medicine: best practices for mass

spectrometry-based assay development using a fit-for-purpose approach. *Mol. Cell. Proteomics* 13: 907–917.
20 Spengler, B. (2015). Mass spectrometry imaging of biomolecular information. *Anal. Chem.* 87: 64–82.
21 Ellis, S.R., Bruinen, A.L., and Heeren, R.M. (2014). A critical evaluation of the current state-of-the-art in quantitative imaging mass spectrometry. *Anal. Bioanal. Chem.* 406: 1275–1289.
22 Rzagalinski, I. and Volmer, D.A. (2016). Quantification of low molecular weight compounds by MALDI imaging mass spectrometry – a tutorial review. *Biochim. Biophys. Acta* 7: 726–739.
23 Bodzon-Kulakowska, A. and Suder, P. (2016). Imaging mass spectrometry: instrumentation, applications, and combination with other visualization techniques. *Mass Spectrom. Rev.* 35: 147–169.
24 Schulz, S., Gerhardt, D., Meyer, B. et al. (2013). DMSO-enhanced MALDI MS imaging with normalization against a deuterated standard for relative quantification of dasatinib in serial mouse pharmacology studies. *Anal. Bioanal. Chem.* 405: 9467–9476.
25 Zhao, Y.S. and Li, C. (2015). Mass spectrometry imaging: applications in drug distribution studies. *Curr. Drug Metab.* 16: 807–815.
26 Liu, X. and Hummon, A.B. (2015). Mass spectrometry imaging of therapeutics from animal models to three-dimensional cell cultures. *Anal. Chem.* 87: 9508–9519.
27 Baker, T.C., Han, J., and Borchers, C.H. (2017). Recent advancements in matrix-assisted laser desorption/ionization mass spectrometry imaging. *Curr. Opin. Biotechnol.* 43: 62–69.
28 Pote, N., Alexandrov, T., Le Faouder, J. et al. (2013). Imaging mass spectrometry reveals modified forms of histone H4 as new biomarkers of microvascular invasion in hepatocellular carcinomas. *Hepatology* 58: 983–994.
29 Munteanu, B., Meyer, B., von Reitzenstein, C. et al. (2014). Label-free in situ monitoring of histone deacetylase drug target engagement by matrix-assisted laser desorption ionization-mass spectrometry biotyping and imaging. *Anal. Chem.* 86: 4642–4647.
30 Bantscheff, M. and Drewes, G. (2012). Chemoproteomic approaches to drug target identification and drug profiling. *Bioorg. Med. Chem.* 20: 1973–1978.
31 Moellering, R.E. and Cravatt, B.F. (2012). How chemoproteomics can enable drug discovery and development. *Chem. Biol.* 19: 11–22.
32 Jones, L.H. and Neubert, H. (2017). Clinical chemoproteomics-opportunities and obstacles. *Sci. Transl. Med.* 9: 1–6.
33 Ziegler, S., Pries, V., Hedberg, C., and Waldmann, H. (2013). Target identification for small bioactive molecules: finding the needle in the haystack. *Angew. Chem. Int. Ed. Engl.* 52: 2744–2792.
34 Niphakis, M.J. and Cravatt, B.F. (2014). Enzyme inhibitor discovery by activity-based protein profiling. *Annu. Rev. Biochem.* 83: 341–377.
35 Sharma, K., Weber, C., Bairlein, M. et al. (2009). Proteomics strategy for quantitative protein interaction profiling in cell extracts. *Nat. Methods* 6: 741–744.

36 Dittmann, A., Werner, T., Chung, C.W. et al. (2014). The commonly used PI3-kinase probe LY294002 is an inhibitor of BET bromodomains. *ACS Chem. Biol.* 9: 495–502.
37 Chung, C.W., Coste, H., White, J.H. et al. (2011). Discovery and characterization of small molecule inhibitors of the BET family bromodomains. *J. Med. Chem.* 54: 3827–3838.
38 Montgomery, D.C., Sorum, A.W., and Meier, J.L. (2014). Chemoproteomic profiling of lysine acetyltransferases highlights an expanded landscape of catalytic acetylation. *J. Am. Chem. Soc.* 136: 8669–8676.
39 Rye, P.T., Frick, L.E., Ozbal, C.C., and Lamarr, W.A. (2011). Advances in label-free screening approaches for studying histone acetyltransferases. *J. Biomol. Screen.* 16: 1186–1195.
40 Konze, K.D., Pattenden, S.G., Liu, F. et al. (2014). A chemical tool for in vitro and in vivo precipitation of lysine methyltransferase G9a. *ChemMedChem* 9: 549–553.
41 Liu, C., Yu, Y., Liu, F. et al. (2014). A chromatin activity-based chemoproteomic approach reveals a transcriptional repressome for gene-specific silencing. *Nat. Commun.* 5: 5733.
42 Kubicek, S., O'Sullivan, R.J., August, E.M. et al. (2007). Reversal of H3K9me2 by a small-molecule inhibitor for the G9a histone Methyltransferase. *Mol. Cell* 25: 473–481.
43 Guitot, K., Drujon, T., Burlina, F. et al. (2017). A direct label-free MALDI-TOF mass spectrometry based assay for the characterization of inhibitors of protein lysine methyltransferases. *Anal. Bioanal. Chem.* 409: 3767–3777.
44 Bantscheff, M., Hopf, C., Savitski, M.M. et al. (2011). Chemoproteomics profiling of HDAC inhibitors reveals selective targeting of HDAC complexes. *Nat. Biotechnol.* 29: 255–265.
45 Becher, I., Dittmann, A., Savitski, M.M. et al. (2014). Chemoproteomics reveals time-dependent binding of histone Deacetylase inhibitors to endogenous repressor complexes. *ACS Chem. Biol.* 9: 1736–1746.
46 Shan, B., Xu, C., Zhang, Y. et al. (2014). Quantitative proteomic analysis identifies targets and pathways of a 2-aminobenzamide HDAC inhibitor in Friedreich's ataxia patient iPSC derived neural stem cells quantitative proteomic analysis identifies targets and pathways of a 2-aminobenzamide HDAC. *J. Proteome* 13: 4558–4566.
47 Salisbury, C.M. and Cravatt, B.F. (2007). Activity-based probes for proteomic profiling of histone deacetylase complexes. *Proc. Natl. Acad. Sci. U.S.A.* 104: 1171–1176.
48 Franken, H., Mathieson, T., Childs, D. et al. (2015). Thermal proteome profiling for unbiased identification of direct and indirect drug targets using multiplexed quantitative mass spectrometry. *Nat. Protoc.* 10: 1567–1593.
49 Patel, K., Sherrill, J., Mrksich, M., and Scholle, M.D. (2015). Discovery of SIRT3 inhibitors using SAMDI mass spectrometry. *J. Biomol. Screen.* 20: 842–848.

50 Kuo, H.Y., DeLuca, T.A., Miller, W.M., and Mrksich, M. (2013). Profiling deacetylase activities in cell lysates with peptide arrays and SAMDI mass spectrometry. *Anal. Chem.* 85: 10635–10642.

51 Rye, P.T., Frick, L.E., Ozbal, C.C., and Lamarr, W.A. (2011). Advances in label-free screening approaches for studying sirtuin-mediated deacetylation. *J. Biomol. Screen.* 16: 1217–1226.

52 Kruidenier, L., Chung, C.W., Cheng, Z. et al. (2012). A selective jumonji H3K27 demethylase inhibitor modulates the proinflammatory macrophage response. *Nature* 488: 404–408.

53 Joberty, G., Boesche, M., Brown, J.A. et al. (2016). Interrogating the druggability of the 2-oxoglutarate-dependent Dioxygenase target class by chemical proteomics. *ACS Chem. Biol.* 11: 2002–2010.

54 Hutchinson, S.E., Leveridge, M.V., Heathcote, M.L. et al. (2012). Enabling lead discovery for histone lysine demethylases by high-throughput RapidFire mass spectrometry. *J. Biomol. Screen.* 17: 39–48.

55 Leveridge, M., Buxton, R., Argyrou, A. et al. (2014). Demonstrating enhanced throughput of RapidFire mass spectrometry through multiplexing using the JmjD2d demethylase as a model system. *J. Biomol. Screen.* 19: 278–286.

56 Wu, F., Zhou, C., Yao, Y. et al. (2016). 3-(Piperidin-4-ylmethoxy)pyridine containing compounds are potent inhibitors of lysine specific demethylase 1. *J. Med. Chem.* 59: 253–263.

57 Haslam, C., Hellicar, J., Dunn, A. et al. (2016). The evolution of MALDI-TOF mass spectrometry toward ultra-high-throughput screening: 1536-well format and beyond. *J. Biomol. Screen.* 21: 176–186.

58 Wigle, T.J., Swinger, K.K., Campbell, J.E. et al. (2015). A high-throughput mass spectrometry assay coupled with redox activity testing reduces Artifacts and false positives in lysine Demethylase screening. *J. Biomol. Screen.* 20: 810–820.

59 Dawson, M.A., Prinjha, R.K., Dittmann, A. et al. (2011). Effective treatment for MLL-fusion leukaemia. *Nature* 478: 529–533.

60 Chen, P., Chaikuad, A., Bamborough, P. et al. (2016). Discovery and characterization of GSK2801, a selective chemical probe for the Bromodomains BAZ2A and BAZ2B. *J. Med. Chem.* 59: 1410–1424.

61 Hofener, M., Pachl, F., Kuster, B., and Sewald, N. (2015). Inhibitor-based affinity probes for the investigation of JAK signaling pathways. *Proteomics* 15: 3066–3074.

62 Savitski, M.M., Reinhard, F.B., Franken, H. et al. (2014). Tracking cancer drugs in living cells by thermal profiling of the proteome. *Science* 346: 1255784.

63 Salisbury, C.M. and Cravatt, B.F. (2008). Optimization of activity-based probes for proteomic profiling of histone deacetylase complexes. *J. Am. Chem. Soc.* 130: 2184–2194.

64 Taunton, J., Hassig, C.A., and Schreiber, S.L. (1996). A mammalian histone deacetylase related to the yeast transcriptional regulator Rpd3p. *Science* 272: 408–411.

65 Ohana, R.F., Kirkland, T.A., Woodroofe, C.C. et al. (2015). Deciphering the cellular targets of bioactive compounds using a chloroalkane capture tag. *ACS Chem. Biol.* 10: 2316–2324.

66 Schiedel, M., Rumpf, T., Karaman, B. et al. (2016). Structure-based development of an affinity probe for Sirtuin 2. *Angew. Chem. Int. Ed. Engl.* 55: 2252–2256.

67 Schiedel, M., Rumpf, T., Karaman, B. et al. (2016). Aminothiazoles as potent and selective Sirt2 inhibitors: a structure-activity relationship study. *J. Med. Chem.* 59: 1599–1612.

68 Ciceri, P., Muller, S., O'Mahony, A. et al. (2014). Dual kinase-bromodomain inhibitors for rationally designed polypharmacology. *Nat. Chem. Biol.* 10: 305–312.

69 Gavin, A.C. and Hopf, C. (2006). Protein co-membership and biochemical affinity purifications. *Drug Discovery Today Technol.* 3: 325–330.

70 Dunham, W.H., Mullin, M., and Gingras, A.C. (2012). Affinity-purification coupled to mass spectrometry: basic principles and strategies. *Proteomics* 12: 1576–1590.

71 Dittmann, A., Ghidelli-Disse, S., Hopf, C., and Bantscheff, M. (2014). Mapping protein complexes using covalently linked antibodies and isobaric mass tags. *Methods Mol. Biol.* 1156: 279–291.

72 Malovannaya, A., Lanz, R.B., Jung, S.Y. et al. (2011). Analysis of the human endogenous coregulator complexome. *Cell* 145: 787–799.

73 Edfors, F., Bostrom, T., Forsstrom, B. et al. (2014). Immunoproteomics using polyclonal antibodies and stable isotope-labeled affinity-purified recombinant proteins. *Mol. Cell. Proteomics* 13: 1611–1624.

74 Vermeulen, M., Eberl, H.C., Matarese, F. et al. (2010). Quantitative interaction proteomics and genome-wide profiling of epigenetic histone marks and their readers. *Cell* 142: 967–980.

75 Lalonde, M.E., Avvakumov, N., Glass, K.C. et al. (2013). Exchange of associated factors directs a switch in HBO1 acetyltransferase histone tail specificity. *Genes Dev.* 27: 2009–2024.

76 Shen, X., Liu, Y., Hsu, Y.J. et al. (2008). EZH1 mediates methylation on histone H3 lysine 27 and complements EZH2 in maintaining stem cell identity and executing pluripotency. *Mol. Cell* 32: 491–502.

77 Bartke, T., Vermeulen, M., Xhemalce, B. et al. (2010). Nucleosome-interacting proteins regulated by DNA and histone methylation. *Cell* 143: 470–484.

78 Strahl, B.D. and Allis, C.D. (2000). The language of covalent histone modifications. *Nature* 403: 41–45.

79 Huang, H., Lin, S., Garcia, B.A., and Zhao, Y. (2015). Quantitative proteomic analysis of histone modifications. *Chem. Rev.* 115: 2376–2418.

80 Jaffe, J.D., Wang, Y., Chan, H.M. et al. (2013). Global chromatin profiling reveals NSD2 mutations in pediatric acute lymphoblastic leukemia. *Nat. Genet.* 45: 1386–1391.

81 Leroy, G., Dimaggio, P.A., Chan, E.Y. et al. (2013). A quantitative atlas of histone modification signatures from human cancer cells. *Epigenetics Chromatin* 6: 20.

82 Krautkramer, K.A., Kreznar, J.H., Romano, K.A. et al. (2016). Diet-microbiota interactions mediate global epigenetic programming in multiple host tissues. *Mol. Cell* 64: 982–992.
83 Peleg, S., Feller, C., Forne, I. et al. (2016). Life span extension by targeting a link between metabolism and histone acetylation in *Drosophila. EMBO Rep.* 17: 455–469.
84 Schworer, S., Becker, F., Feller, C. et al. (2016). Epigenetic stress responses induce muscle stem-cell ageing by Hoxa9 developmental signals. *Nature* 540: 428–432.
85 Peach, S.E., Rudomin, E.L., Udeshi, N.D. et al. (2012). Quantitative assessment of chromatin immunoprecipitation grade antibodies directed against histone modifications reveals patterns of co-occurring marks on histone protein molecules. *Mol. Cell. Proteomics* 11: 128–137.
86 Feller, C., Forne, I., Imhof, A., and Becker, P.B. (2015). Global and specific responses of the histone acetylome to systematic perturbation. *Mol. Cell* 57: 559–571.
87 Garcia, B.A., Mollah, S., Ueberheide, B.M. et al. (2007). Chemical derivatization of histones for facilitated analysis by mass spectrometry. *Nat. Protoc.* 2: 933–938.
88 Smith, C.M., Gafken, P.R., Zhang, Z. et al. (2003). Mass spectrometric quantification of acetylation at specific lysines within the amino-terminal tail of histone H4. *Anal. Biochem.* 316: 23–33.
89 Tang, H., Fang, H., Yin, E. et al. (2014). Multiplexed parallel reaction monitoring targeting histone modifications on the QExactive mass spectrometer. *Anal. Chem.* 86: 5526–5534.
90 Abshiru, N., Caron-Lizotte, O., Rajan, R.E. et al. (2015). Discovery of protein acetylation patterns by deconvolution of peptide isomer mass spectra. *Nat. Commun.* 6: 8648.
91 Lin, S., Wein, S., Gonzales-Cope, M. et al. (2014). Stable-isotope-labeled histone peptide library for histone post-translational modification and variant quantification by mass spectrometry. *Mol. Cell. Proteomics* 13: 2450–2466.
92 Vedadi, M., Barsyte-Lovejoy, D., Liu, F. et al. (2011). A chemical probe selectively inhibits G9a and GLP methyltransferase activity in cells. *Nat. Chem. Biol.* 7: 566–574.
93 Creech, A.L., Taylor, J.E., Maier, V.K. et al. (2015). Building the connectivity map of epigenetics: chromatin profiling by quantitative targeted mass spectrometry. *Methods* 72: 57–64.
94 Xu, B., On, D.M., Ma, A. et al. (2015). Selective inhibition of EZH2 and EZH1 enzymatic activity by a small molecule suppresses MLL-rearranged leukemia. *Blood* 125: 346–357.
95 Zheng, Y., Sweet, S.M., Popovic, R. et al. (2012). Total kinetic analysis reveals how combinatorial methylation patterns are established on lysines 27 and 36 of histone H3. *Proc. Natl. Acad. Sci. U.S.A.* 109: 13549–13554.
96 Lalonde, M.E., Cheng, X., and Cote, J. (2014). Histone target selection within chromatin: an exemplary case of teamwork. *Genes Dev.* 28: 1029–1041.
97 Joshi, P., Greco, T.M., Guise, A.J. et al. (2013). The functional interactome landscape of the human histone deacetylase family. *Mol. Syst. Biol.* 9: 672.

98 Filippakopoulos, P. and Knapp, S. (2014). Targeting bromodomains: epigenetic readers of lysine acetylation. *Nat. Rev. Drug Discovery* 13: 337–356.

99 Chick, J.M., Kolippakkam, D., Nusinow, D.P. et al. (2015). A mass-tolerant database search identifies a large proportion of unassigned spectra in shotgun proteomics as modified peptides. *Nat. Biotechnol.* 33: 743–749.

100 Kong, A.T., Leprevost, F.V., Avtonomov, D.M. et al. (2017). MSFragger: ultrafast and comprehensive peptide identification in mass spectrometry-based proteomics. *Nat. Methods* 14: 513–520.

101 Dang, X., Scotcher, J., Wu, S. et al. (2014). The first pilot project of the consortium for top-down proteomics: a status report. *Proteomics* 14: 1130–1140.

102 Zheng, Y., Fornelli, L., Compton, P.D. et al. (2016). Unabridged analysis of human histone H3 by differential top-down mass spectrometry reveals hypermethylated proteoforms from MMSET/NSD2 overexpression. *Mol. Cell. Proteomics* 15: 776–790.

103 Schwammle, V., Sidoli, S., Ruminowicz, C. et al. (2016). Systems level analysis of histone H3 post-translational modifications (PTMs) reveals features of PTM crosstalk in chromatin regulation. *Mol. Cell. Proteomics* 15: 2715–2729.

5

Peptide Microarrays for Epigenetic Targets

Alexandra Schutkowski[1], Diana Kalbas[2], Ulf Reimer[3], and Mike Schutkowski[2]

[1] *Martin-Luther-University Halle-Wittenberg, Department of Human Nutrition, Institute of Agricultural and Nutritional Sciences, Karl-Freiherr-von-Fritsch-Straße 4, 06120, Halle/Saale, Germany*
[2] *Martin-Luther-University Halle-Wittenberg, Department of Enzymology, Institute of Biochemistry and Biotechnology, Kurt-Mothes-Straße 3a, 06120 Halle, Germany*
[3] *JPT Peptide Technologies GmbH, Vollmerstraße 5, 12489, Berlin, Germany*

5.1 Introduction

Epigenetic modulations are represented by changes of the function of a genome without changing the sequence of nucleotides. Besides methylation of DNA, epigenetic modulation of DNA function takes place at the nucleosome, which is composed of four pairs of histone molecules (H2A, H2B, H3, and H4) forming an octamer. Double-stranded DNA is bound around each nucleosome mainly by electrostatic interaction between the negatively charged DNA backbone and the positively charged part of the folded histones. Histone H1 stabilizes the chromatin structure by bridging two nucleosomes. Posttranslational modification (PTM) of histone proteins could either lead to increased accessibility of the DNA for interaction with other proteins/protein complexes or to the condensation into the chromosome [1–5]. The unstructured N-terminal and to some extent C-terminal tails of the histone proteins are reaching out of the histone octamer and are well accessible for proteins and protein complexes. Nearly all of the identified PTMs are located at these tails. Enzymes generating or removing PTMs from the side chains of histones are introduced as writers or erasers of epigenetic marks, respectively, (see Figure 5.1). The combination of modifications of amino acid residues results in a histone code inducing subsequent actions by recruitment of protein domains (readers) specifically recognizing different PTMs or combinations of PTMs. There is another layer of complexity because both binding to readers and substrate recognition by writers/erasers can be modulated by PTMs in the surrounding of the recognition site, a phenomenon known as cross-talk [6–9]. Some of the histone modifications are relatively stable like histone H3 methylation at lysine residue 9 or 27 (H3K9Me or H3K27Me) [10], but modifications like histone ac(et)ylation and phosphorylation are transient and more dynamic [11, 12].

Besides well-known phosphorylations of serine, threonine, tyrosine, and histidine residues; methylations of arginines/lysines; and acetylations of lysines, in

Epigenetic Drug Discovery, First Edition. Edited by Wolfgang Sippl and Manfred Jung.

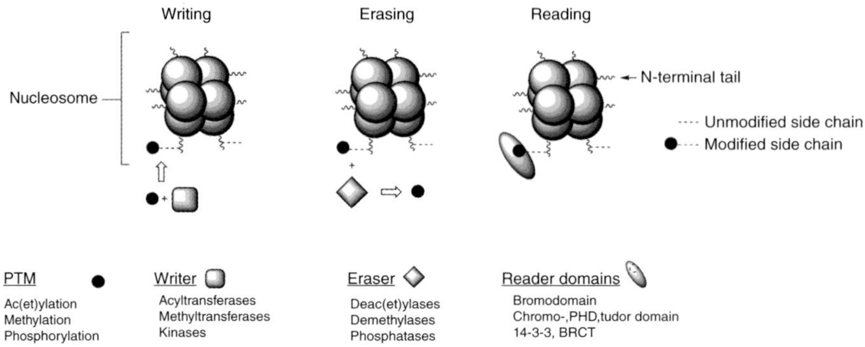

Figure 5.1 Schematic presentation of histone octamers and epigenetic targets modulating histone function.

the last years novel acylations of lysines are identified in histones, like propionylation [13], butyrylation [13], malonylation [14], succinylation [14], glutarylation [15], crotonylation [16], 3-hydroxybutyrylation [17], 4-oxo-nonaylation [18, 19], or hydroxyisobutyrylation [20]. Histone acylations and their biological consequences were recently reviewed [21]. Moreover, transformation of methylated lysines into aldehyde [22] residues, attachment of sugars to serines and threonines [23, 24], ADP-ribose to lysines [25], deamidation of arginines resulting in citrulline residues [26, 27], ubiqutinylation [24, 28], and SUMOylation [29] were described. Therefore, the number of theoretically possible combinations of PTMs in histones is exploding. This makes a complete and systematic analysis of the histone code more difficult. Research is mostly focusing on the N-terminal tails of the histones H3 and H4. Different types of resin-bound histone-derived peptide derivatives were used for uncovering the histone code [30]. A combinatorial peptide library including all possible combinations of known PTMs in the histone H3 N-terminus (10meric peptides) was used to analyze six readers of the methylation status of H3K4 [31]. Arrays of 384 microspheres coated with peptide–cellulose conjugates were used to profile the specificity of 40 different antibodies directed against modified histones [32].

It is assumed that the histone core region is not well accessible for other proteins and that nucleosome structure does not allow efficient enzymatic transformations or reading of PTMs. Nevertheless, it was shown that phosphorylations within the core region of histones are changed dramatically between interphase and mitosis [33]. Moreover, phosphorylations of histone H2B at Tyr 37 by protein tyrosine kinase Wee1 at the border of the histone core [34], methylations of histone H3 on Lys 79 (H3K79) within the histone core by lysine methyltransferase Dot1 [35], and acetylations of H3K122 by acetyltransferases p300/CBP [36] were identified demonstrating that these core regions at least to some extent could be accessible for enzymatic modifications.

We developed a histone code peptide microarray and an acetylome/lysosome peptide microarray for the systematic profiling of binding specificities of readers and substrate specificities of writers/erasers. We used similar technology for comprehensive profiling of human SH2 and SH3 reader domains with microarrays displaying 6202 phosphotyrosine-containing peptides [37] and 9192 proline-rich peptides [38], respectively. Additionally, we successfully analyzed the subsite specificities of erasers like all seven human sirtuin isoforms [39, 40] using an acetylome peptide microarray displaying more than 13 000 peptides derived from human acetylation sites or human protein tyrosine phosphatases using the mentioned phosphotyrosine microarray mentioned for the profiling of the SH2 domains [41, 42]. Moreover, we identified substrates for writers like protein kinases using collections of phosphorylation site peptides immobilized on glass slides [43–49].

In this chapter we would like to introduce a histone code peptide microarray displaying more than 3850 human histone-derived 20meric peptides. The knowledge-based part of the histone microarray displays complete overlapping peptide scans through the four histones H2A, H2B, H3, and H4 including all of their naturally occurring sequence variants (i.e. H3.1, H3.2, H3.3) in form of 20meric peptides. PTMs at all reported sites in the N-terminal tails as well as the known PTM combinations (up to 6 PTMs per peptide) are presented, too. We used acetylated (Kac), propionylated (Kprop), butyrylated (Kbut), methylated (mono-, di-, and trimethylated, Kme1, Kme2, and Kme3, respectively) lysine residues; mono- and dimethylated arginine residues (Rme1, Rme2a, Rme2s for symmetrically and asymmetrically dimethylated arginines, respectively); and arginine–citrulline exchange as well as phosphorylated serine (Sp), threonine (Tp), and tyrosine (Yp) residues. Furthermore, the microarray contains overlapping peptides derived from histone H1 together with all PTMs published by Wisniewski et al. [50]. A scan through histones H2A, H2B, H3, and H4 with single modifications of all PTMs mentioned above plus recently described malonylated (Kmal) and succinylated (Ksucc) lysines at all possible sites was added, too. This histone code peptide microarray should enable the identification of novel histone-based enzyme substrates or binding sites. To address the question if free N-terminus is needed for effective interaction with the target protein, the N-terminal tails are immobilized in addition via their C-termini.

The content of the histone code peptide microarray described in this chapter is more comprehensive as compared with histone peptide arrays published in the past, like histone-derived peptides immobilized on cellulose membranes [51–54], biotinylated histone peptides on streptavidin-coated glass slides [55–60], biotinylated peptide–neutravidin complexes immobilized on activated glass slides [61], or peptide–cellulose conjugates deposited on glass slides [52, 62–65] or nitrocellulose-coated glass slides [53].

Here we want to demonstrate that both the profiling of binding specificities of histone code readers and the reliable determination of substrate specificities of histone code writers and erasers are possible using the same peptide microarray technology. Additionally, we will present an example for the cross-talk between histone H3 phosphorylation and sirtuin-mediated deac(et)ylation. Moreover, we were able to demonstrate that peptide microarrays displaying overlapping scans through epigenetic target enable identification of upstream kinases and phosphorylation sites.

5.2 Applications of Peptide Microarrays for Epigenetic Targets

5.2.1 Profiling of Substrate Specificities of Histone Code Writers

Kinases or methyl-, acetyl-, glycosyl-, and ADP-ribosyltransferases are transferring small moieties from a co-substrate onto the target histone residue, resulting in a modified histone and a coproduct. Strategies used for the readout of enzymatic activities on peptide microarrays could be differentiated by either the nature of the co-substrate or the reagent used for the readout and are comprehensively reviewed in Refs. [66, 67] (Figure 5.2).

Often a tagged co-substrate like radioisotopically labeled ATP, acetyl-CoA, S-adenosylmethionine, or NAD^+ is used for profiling of kinases, acetyltransferases, methyltransferases, or (poly)ADP-ribosyltransferases, respectively, on peptide microarrays. The resulting radioisotopically labeled substrate peptide could be detected by phosphorimaging or exposure to X-ray films. Alternatively, biotinylated co-substrate derivatives were used for the profiling of kinases [68] or ADP-ribosyltransferases followed by treatment with fluorescently labeled streptavidin. Other tagged co-substrates applied for protein kinase profiling are ferrocene-conjugated ATP [69], γ-thio-ATP [70], and dansylated ATP derivatives [71].

A PTM-specific reagent has to be considered for readout of enzymatic activity if the writer is used together with non-modified co-substrates (Figure 5.2a, b, and d). Such a reagent could be either a protein (i.e. an anti-PTM antibody, a lectin, or a reader domain selective for the respective PTM) or a small molecule, which binds specifically to the moiety introduced into the immobilized peptide (i.e. fluorescently labeled [72] or biotinylated phosphate monoester chelators [73, 74]). For example, enzymatic activity of arginine methyltransferase PRMT5-MEP50 and cross-talk between serine phosphorylation and arginine

Figure 5.2 Readout principles for detection of enzymatic activities on peptide microarrays. (a) Binding of PTM-specific antibodies is used for readout. Antibody could be directly fluorescence labeled, or a fluorescence-labeled secondary antibody generates the signal. (b) PTM-specific reader proteins recognize the modified peptide, and bound protein is detected by fluorescence-labeled antibody or is directly labeled. Reader proteins could be PTM-specific reader domains (like chromo- or bromodomains) or so-called trapping mutants of enzymes (like phosphotyrosine phosphatases with active site cysteine replaced by alanine or serine). (c) Labeled co-substrates yield modified peptides carrying a label. Label could be a radioisotope (i.e. radioisotopes of phosphorous in γ-phosphate moiety of ATP) that could be visualized by phosphorimaging. Alternatively, it could be a biotinyl residue (i.e. biotinylated NAD^+ that yields biotinylated ADP-ribosylated peptides subsequent to treatment with poly-ADP-ribosyltransferases) that could be visualized by fluorescence-labeled (strept)avidin. (d) Small molecules chelating the PTM (i.e. chelators for phosphate monoesters) enable readout of enzymatic activity if the chelator is connected either to a fluorophore (phospho-specific fluorescence dyes) or to a biotin molecule. Same assay principles (difficult for (c)) could be used for profiling of erasers, too. Nevertheless, the content of the peptide microarray has to be changed to chemically modified peptide derivatives, and the removal of the PTM is monitored by a signal decrease. Source: Adapted from Thiele et al. 2009 [66].

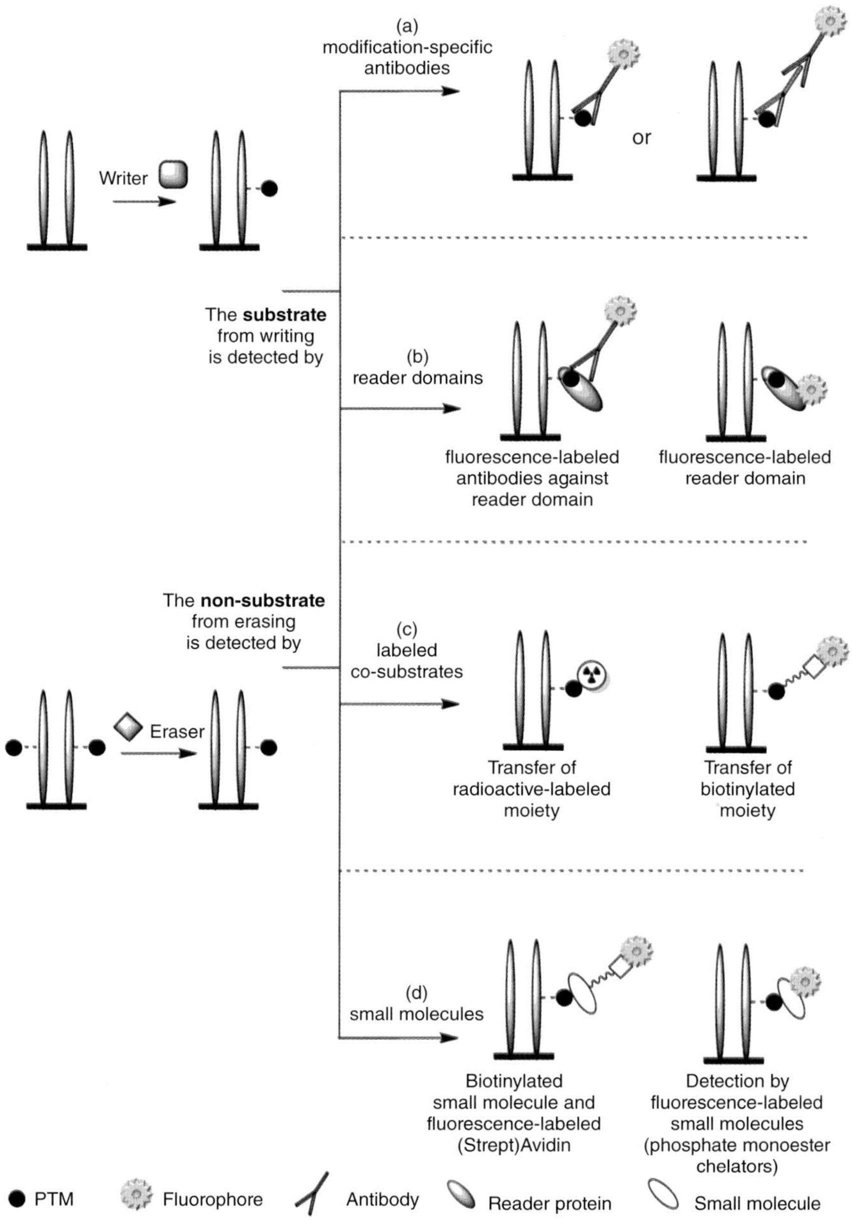

methylation within the histone H4 tail was investigated using a histone peptide microarray in combination with an antibody specific for methylated arginine 3 in histone H4 [75, 76]. A special form of readout for kinase-mediated phosphorylations is the chemical transformation of phosphoserine/phosphothreonine residues into dehydroalanine- or 2-amino-dehydrobutyryl derivatives, respectively, by base-catalyzed ß-elimination [77]. The generated double bond could be targeted by chemoselective reactions of the former kinase substrate with either a fluorophore or a biotin moiety. Shults et al. used chemical formation of a covalent bond between a dye derivative and the phosphorylated amino acid side chain to monitor kinase reaction on peptide microarrays [78].

The activity of histone code writers and, less well studied, of erasers is often modulated by phosphorylations in close proximity. Additionally, such phosphorylations could recruit phospho-specific readers like SH2 domains or 14-3-3 domains (reviewed in Ref. [33]). Activation of p38 MAPK pathways results in activation of kinases like MSK1 and MSK2 known to target serine residues 10 and 28 in histone H3 both preceded by an ARK sequence [79]. These modifications are involved in cancerogenesis, making MSKs attractive targets for drug discovery [80].

We used histone code peptide microarray to analyze MSK1 activity in the presence of [γ-^{33}P]-ATP. Besides the known sites H2AS1 [81], H3S10, and H3S28 [79], phosphorylation of histone H2B (S37) or histone H4 (S48) was detectable. Phosphorylation of H4S48 (catalyzed by PAK2 kinase) promotes formation of nucleosomes composed of histone H3.3 instead of histone H3.1 isoform [82]. The signals for H2BS37 and H4S48 are about five times stronger than for the H3S10 site. Modification of arginine residue in −2 position (methylation or replacement by a citrulline) as well as acylation of lysine residue in −3 position (succinylation, malonylation, and butyrylation) decreases kinase-mediated incorporation of radioisotopically labeled phosphate moieties (Figure 5.3a).

There are some reports of tyrosine phosphorylation in histones [83–87]. Mahajan et al. identified H2BY37 as Wee1 target site, leading to suppression of expression of replication-dependent core histone genes [34]. We profiled Wee1 using the histone code peptide microarray to analyze the cross-talk between different PTMs [88]. Here we present similar data for the Wee1-related, dual-specific kinase Myt1. In the past we were able to show that Myt1 generates robust signals on peptide microarrays using either [γ-^{33}P]-ATP in combination with phosphorimaging or anti-phosphotyrosine antibodies in combination with fluorescence imaging [48].

Principally, Myt1 recognized on the peptide microarray of all substrates described for Wee1 [88] but showed pronounced cross-talk with additional modifications in the substrate peptides. Figure 5.3b shows the relative signal intensities (signal for non-modified histone peptide 100%) for differently modified histone H2B (21–40) peptides containing the Wee/Myt1 target tyrosine residue 37. Both butyrylation of lysine in −9 position and propionylation of lysine in −3 increased substrate efficiency about more than fivefold. On the other hand, succinylation of lysine in −3 position dramatically impedes Myt1-mediated phosphorylation. In addition, methylations of arginine or lysines in −3, −4, and −7 position or phosphorylation of serine in −5 position negatively interferes with Myt1 enzymatic activity. Cross-talk between PTMs over such a long range seems to be unlikely, but (using peptide microarrays displaying systematic substitutions of a NEK6 substrate) we were able to show that kinases could sense changes of the chemical nature of residues up to +8 position specifically [46].

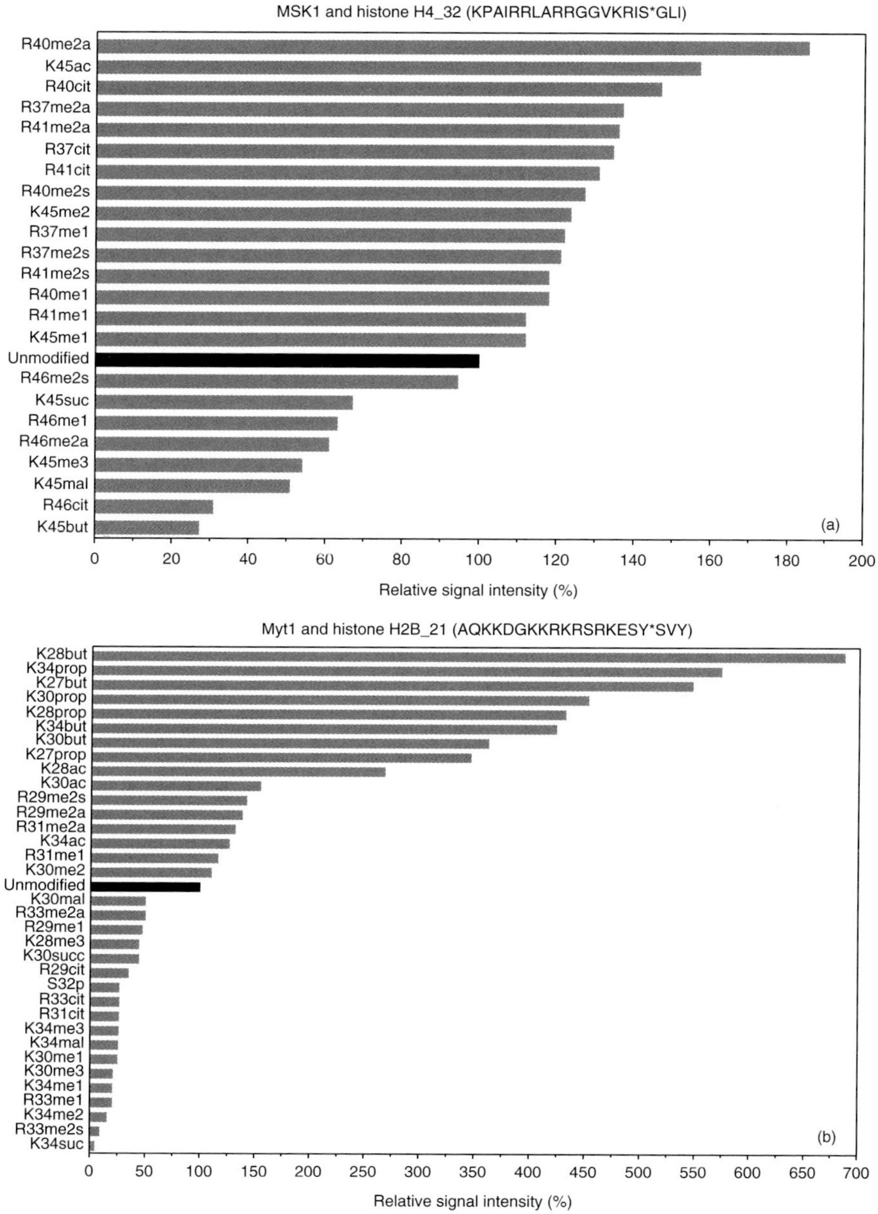

Figure 5.3 Msk1 and Myt1 kinase activity is modulated by cross-talk with other PTMs. Histone peptide microarray was incubated with Msk1 kinase (a) or Myt1 kinase (b) in the presence of [γ-^{33}P]-ATP. Readout was performed by phosphorimaging. (a) Data for several histone H4 peptides covering residues 32–51 are shown. Phosphorylation site is indicated by S*. (b) Data for several histone H2B peptides covering residues 21–40 are shown. Phosphorylation site is indicated by Y*. In both cases signal relative intensities of phosphorimaging are shown for each peptide derivative. The signal intensity for the respective non-modified peptide was set to 100%. Please note that similar data are generated for each histone peptide sequence in one peptide microarray experiment.

5.2.2 Profiling of Substrate Specificities of Histone Code Erasers

Erasers, removing modifications from histone proteins, such as phosphatases, protein arginine deiminases, lysine demethylases, and lysine deac(et)ylases are the counteracting molecules to the appropriate writers. Erasers could either generate the non-modified histone site or introduce a novel modification (i.e. citrulline residues instead of (methylated) arginine introduced by protein arginine deiminases or aldehyde functions instead of methylated lysines introduced by lysyl oxidases [22]). In principle, each readout used for the detection of histone code writer activities could be used for the respective erasers but in opposite direction (Figure 5.2). Nevertheless, the content of the microarray has to be changed starting with the chemically or enzymatically modified peptides as substrates for the eraser enzymes.

Sirtuins are known as NAD^+-dependent lysine deac(et)ylases with additional ADP-ribosylating activity for some isoforms [89, 90]. They regulate activity of metabolic enzymes and stress response, play a major role in expansion of lifespan, and represent target for prevention of age-related diseases. Additionally, sirtuins are considered as tumor suppressors [91–94]. Peptide microarrays displaying more than 6800 acetylated peptides derived from human acetylation sites were used to investigate the subsite specificity of all seven human sirtuin isoforms [40]. We found some specificity for different positions relative to the acetylated lysine but not as pronounced as for kinases or proteases. This fact is not surprising because sirtuins are known to be relatively unspecific regarding the amino acid sequence [95] but more specific with respect to the chemical nature of the acyl moiety [96]. The acetylation status of lysines in histones is regulated by the activity of acetyltransferases and histone deacetylases including sirtuins [90, 97]. Phosphorylation of serine or threonine residues transforms small and neutral amino acid residues into larger ones with (at least at physiological conditions) a double negative charge. We could demonstrate that human sirtuin 3 activity against histone H3K9Ac is suppressed if S10 or T11 is phosphorylated [88].

Here we profiled other human sirtuin isoforms (Sirt1, Sirt2, Sirt5, Sirt6) on the histone peptide microarray using a mixture of PTM-specific antibodies [40] and fluorescently labeled secondary antibody. As a negative control experiment, we used similar reaction mixture but without co-substrate NAD^+ to get an initial signal value for each peptide. Subsequent to treatment of another peptide microarray with sirtuin in the presence of NAD^+, we received a reduced signal for peptides acting as a substrate for the respective sirtuin.

Figure 5.4a–d displays the activities of sirtuins 1 and 2 against histone H3 peptides 1–19 and 23–42. For sirtuin 1 there is no strong influence of serine and/or threonine phosphorylation on lysine deacetylation visible. Reactivity against phosphorylated peptide derivatives drops down to about 60–80% of the non-phosphorylated sirtuin substrate. In contrast, sirtuin 2 activity is strongly diminished if serine in +1 position or threonine in +2 position or both residues are phosphorylated. Recently, a coexisting acetylation at lysine 27 and phosphorylation at serine 28 in histone H3 was postulated to be a result of positive regulation of acetyltransferase activity by MSK1-mediated S28 phosphorylation [98]. Our results could explain the same effect by negative regulation of sirtuin

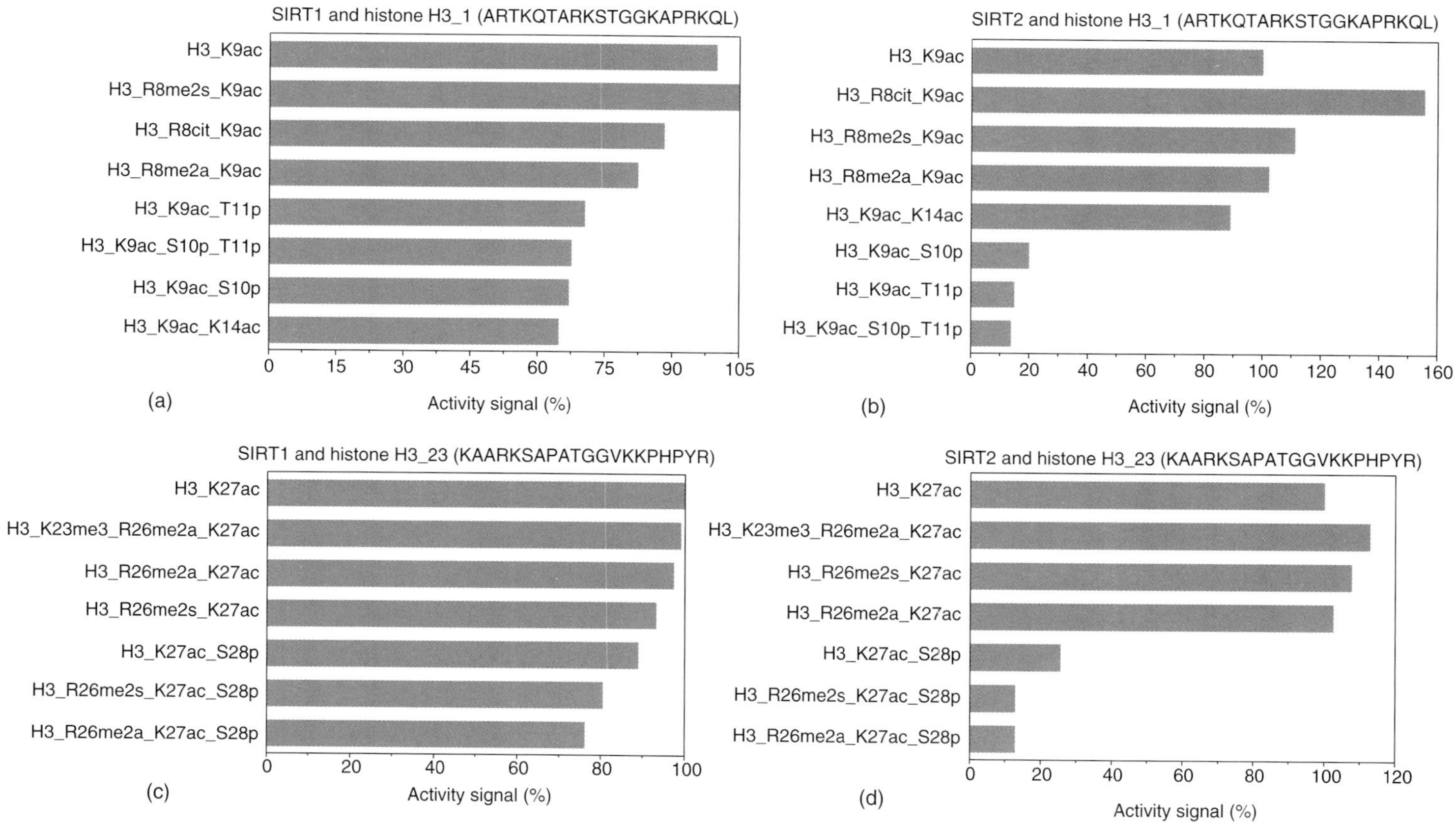

Figure 5.4 Sirtuins 1 and 2 activity on H3K9ac site is differently influenced by phosphorylations. Histone peptide microarray was incubated with sirtuins 1 and 2 in the presence and absence of co-substrate NAD^+. Readout was performed by looking for signal decrease subsequent to treatment with an optimized mixture of anti-acetyl-lysine antibodies and fluorescently labeled secondary antibodies. Activity values for H3K9ac and H3K37ac peptides were set to 100%. (a) Sirtuin 1 activity values for several histone H3 peptides covering residues 1–20 are shown. (b) Sirtuin 2 activity values for several histone H3 peptides covering residues 1–20 are shown. (c) Sirtuin 1 activity values for several histone H3 peptides covering residues 23–42 are shown. (d) Sirtuin 2 activity values for several histone H3 peptides covering residues 23–42 are shown. (e) Kinetic constants for 15meric H3K9ac peptide derivatives were determined using an HPLC-based activity assay for sirtuin 1. (f) Kinetic constants for 15meric H3K9ac peptide derivatives were determined using an HPLC-based activity assay for sirtuin 2.

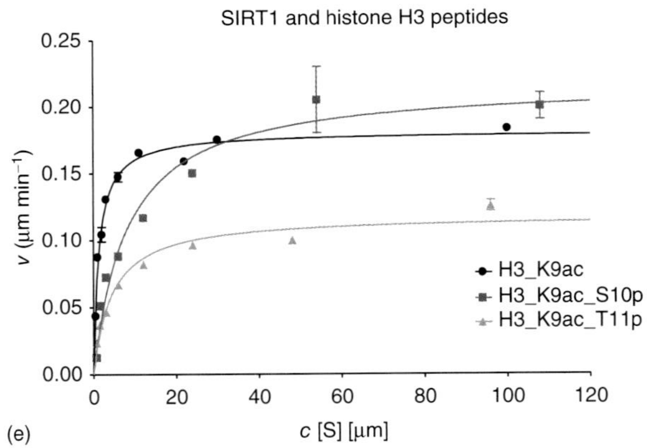

Michaelis–Menten	H3K9ac	H3K9ac_S10p	H3K9ac_T11p
K_M (μm)	1.1	7.9	4.2
k_{cat} (s^{-1})	0.3	0.37	0.2
k_{cat}/K_M ($m^{-1}s^{-1}$)	280 000	46 000	49 000

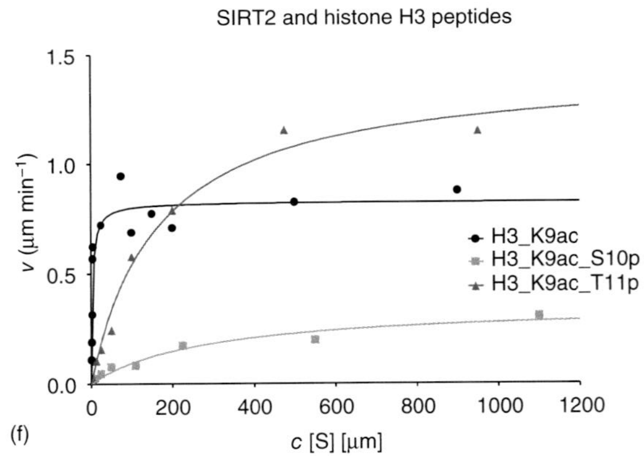

Michaelis–Menten	H3K9ac	H3K9ac_S10p	H3K9ac_T11p
K_M (μm)	3.0	280	160
k_{cat} (s^{-1})	0.14	0.06	0.24
k_{cat}/K_M ($m^{-1}s^{-1}$)	46 000	200	1500

Figure 5.4 *(Continued)*

2-mediated deacetylation activity. We synthesized differently modified histone H3 peptides to determine K_M and k_{cat} values using an high performance liquid chromatography (HPLC)-based activity assay [99]. We found pronounced effect on the K_M value for the sirtuin 2 enzymatic activity, resulting in more than 200-fold decreased specificity constant for the H3K9AcS10p peptide substrate (Figure 5.4f). For sirtuin 1 this constant is decreased only about threefold (Figure 5.4e). Analysis of sirtuins 5 and 6 on the histone code peptide microarray showed no big difference between phosphorylated and non-phosphorylated H3K9 substrates (data not shown). Measurements with the HPLC-based assay [99] and the appropriate succinylated and myristoylated substrates confirmed the peptide microarray results.

5.2.3 Profiling of Binding Specificities of PTM-specific Antibodies and Histone Code Readers

Different readout methods could be used to detect binding events on peptide microarrays (Figure 5.5). First, the use of antibodies directed against the reader

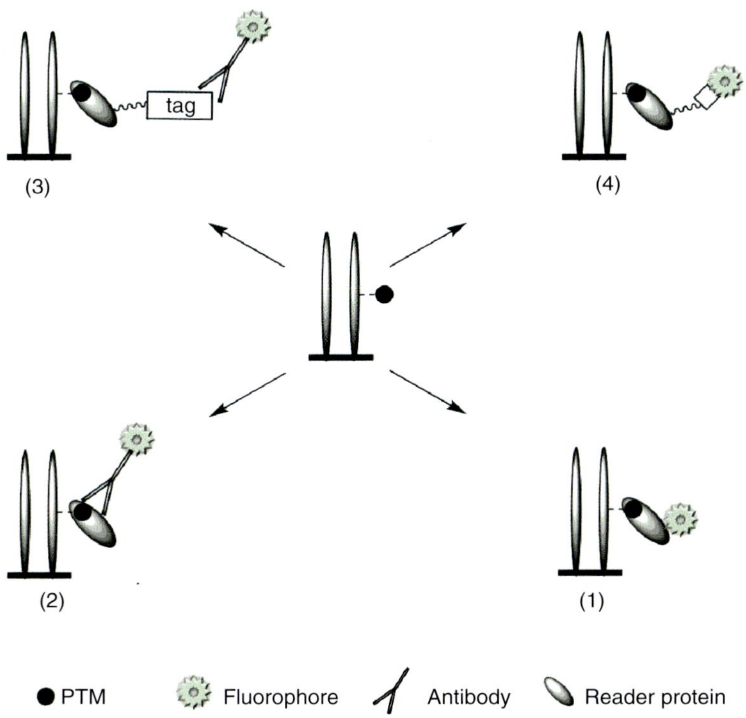

Figure 5.5 Readout principles for detection of binding events on peptide microarrays. Binding of a reader protein to microarray-bound peptides could be detected by (1) direct fluorescence labeling of the protein, (2) a fluorescence-labeled antibody directed against the reader protein, (3) the use of a tagged version of the reader protein followed by a fluorescence-labeled antibody directed against the tag, and (4) the use of a biotinylated version of the reader protein followed by fluorescence-labeled (strept)avidin.

protein will result in a ternary antibody/reader protein/peptide complex, which could be visualized subsequent to treatment of the peptide microarray with a fluorescently labeled secondary antibody. Nevertheless, the anti-reader protein antibody should not recognize an epitope, which is involved in binding to the immobilized peptide, because this would generate false negative results. Alternatively, direct fluorescence labeling of the anti-reader protein antibody was used for the detection of anti-histone peptide antibodies binding or specificity profiling of pan-specific anti-PTM antibodies [40, 100] on peptide microarrays (Figure 5.5).

Second, direct chemical labeling of the target reader protein by fluorophores allows direct readout of binding events on immobilized peptides by fluorescence scanning [101]. Biotin labeling also enables readout subsequent to treatment with fluorescence-labeled streptavidin or avidin. Disadvantages of the chemical labeling using lysine or cysteine side chains are the resulting complex mixture of differently labeled reader protein molecules. This problem could be circumvented by regioselective labeling using more specific chemical methods [102, 103], enzymes (reviewed in Refs. [104, 105]), or genetic methods like the introduction of GST fusions [37, 38], FLAG-, Myc-, His-, and HA-tags, or biotinylation sites. Bound reader protein could be detected using specific anti-tag antibodies or fluorescence-labeled chelators recognizing the His-tag or fluorescence-labeled (strept)avidin.

5.2.3.1 Profiling of Specificities of PTM-specific Antibodies

Antibodies used for chromatin immunoprecipitation experiments should have an exquisite specificity for both the amino acid sequence of the histone peptide recognized and the chemical nature of the PTM. Thus, antibodies directed against dimethylated lysine residues should not see monomethylated or trimethylated lysines, and antibodies directed against acetylated lysines should not see alternative acylations like propionylation, butyrylation, or succinylation. Moreover, the antibodies should be able to discriminate between the modified and non-modified histones. We used histone code peptide microarrays for quality control of several commercially available antibodies directed against PTMs in histones (Abcam: anti-H3K9me1, anti-H3K9me2; Diagenode: anti-H3K9me3; Millipore: anti-H4K16ac, anti-H3S28ph, anti-H4S1ph; Santa Cruz: anti-H4K16ac; Active Motif: anti-H3K4me1, anti-H3K4me2, anti-H3K27me3, anti-H3K36me1, anti-H3K36me2, anti-H3K36me3). Some antibodies showed suboptimal sequence specificity as, for example, anti-H4K16ac-Ab. This antibody generated strong fluorescence signals for H2BK28ac, H2BK30ac, H2BK85ac, H3K115ac, H4K12ac, H4K20ac, H4K44ac, H4K77ac, and H4K91ac peptides with similar intensities as compared with the target site H4K16ac (Figure 5.6). Moreover, at least on our peptide microarray, the binding to the off-targets H4K91ac and H4K12ac seems to be stronger than to H4K16ac target peptide. Detection of binding events using peptide microarrays is very sensitive, and it could be possible that this cross-reactivity to off-targets is an assay artifact caused by the used antibody concentration (recommended 1 : 1000 dilution of 1 mg ml^{-1} stock solution). Therefore, we repeated the experiments using 1 : 5000, 1 : 10 000, and

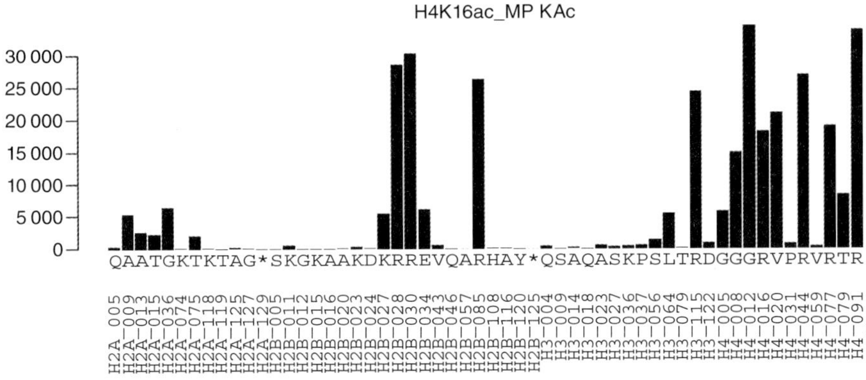

Figure 5.6 Cross-reactivity of anti-H4K16ac antibody. Histone peptide microarray was treated with anti-H4K16ac antibody followed by fluorescence-labeled secondary antibody. Fluorescence signal intensities are plotted for each acetyl-lysine-containing peptide. Amino acid residue following the acetylated lysine (+1 position) is shown directly under the bar. Acetylated lysines representing the C-terminus of the histone are denoted as *.

1 : 50 000 dilutions. There was no change in the cross-reactivity visible. Only the total signal intensities were lower in the 1 : 10 000 and 1 : 50 000 dilution experiments. Obviously, cross-reactivity is mainly determined by the residue in the +1 position relative to the acetylated lysine. If this residue is an arginine, strong signals could be found. But there seems to be more determinants for cross-reactivity because a glycine in +1 position can cause strong antibody binding like for H4K12ac site but is not sufficient for binding (i.e. H2AK127ac or H2BK12ac). Moreover, we were able to demonstrate that this antibody recognizes propionylated and butyrylated lysines with similar specificities. No binding to non-acetylated, succinylated, or malonylated histone peptides was detectable.

5.2.3.2 Profiling of Binding Specificities of Histone Code Readers

Bromodomains represent reader domains with some specificity for acetylated lysine residues within defined peptide sequences. These domains are evolutionary conserved proteins involved in lysine acetylation-dependent assembly of protein complexes regulating transcription of genes. Therefore, several bromodomains represent interesting pharmaceutical targets for the treatment of inflammatory diseases, viral infections, and cancer [106]. In humans a total of 61 bromodomains was found located in 46 different proteins.

Bromodomain-containing protein 4 (BRD4) contains two N-terminal bromodomains (BRD4_1 and BRD4_2) and acts as transcriptional regulator [107]. BRD4 has been suggested as a therapeutic target in many cancers, including acute myeloid leukemia, multiple myeloma, colon cancer, and breast cancer [108, 109].

We profiled the subsite and acyl specificity of these two bromodomains onto the histone peptide microarray using commercially available recombinant BRD4 proteins (Active Motif, Carlsbad, USA). Both proteins contain an N-terminal

His-tag and a C-terminal Flag-tag. We used anti-FLAG-tagged antibody for read-out of binding events (Figure 5.5).

We found a preference in binding of BRD4_1 to H4K91ac, followed by H4K44ac and H4K59ac. Same top binders and similar signal intensities were observed for BRD4_2. Similar results were obtained using histone peptides immobilized on cellulose membranes [54]. Fillipakopoulos and coworkers detected also H4K91ac and H4K44ac peptides as binders for both BRD4 bromodomains. Figure 5.7a illustrates the specificity of BRD4_2 for the amino acid sequence around the 10 lysines found in histone H3. H3K56ac and H3K64ac generated higher signals on the peptide microarray as compared with other H3 acetylation sites. Figure 5.7b shows the acetyl selectivity of BRD_2 recognition. We compared the signal intensities of the histone H3-derived peptides in the acetylated and non-modified state. Signal ratios (Kac/K) of up to 6 for the H3K18 site uncover some acetyl specificity (Figure 5.7b). This is in contrast to results by Fillipakopoulos et al. using histone peptide macroarrays on cellulose membranes. The authors claimed that the Kac-mark only weakly contributes to the binding affinity of BRD proteins to their target sites [54].

Using histone peptide microarrays it is difficult to understand the subsite specificity of bromodomain in a more comprehensive manner because the histone sequences are relatively similar with respect to amino acid composition. Therefore, we used the described acetylome peptide microarray [40] displaying more than 6500 peptides derived from human acetylation sites in acetylated and non-acetylated forms. Figure 5.8 shows exemplarily the sequence logos for four different bromodomains. As expected, each bromodomain has a unique subsite specificity. In contrast to reported subsite specificities for bromodomain binders, we found some preference for amino acid residues relatively far away from the acetylated lysine residue like arginine in −6 position for bromodomains of BRD2 and GCN5L2.

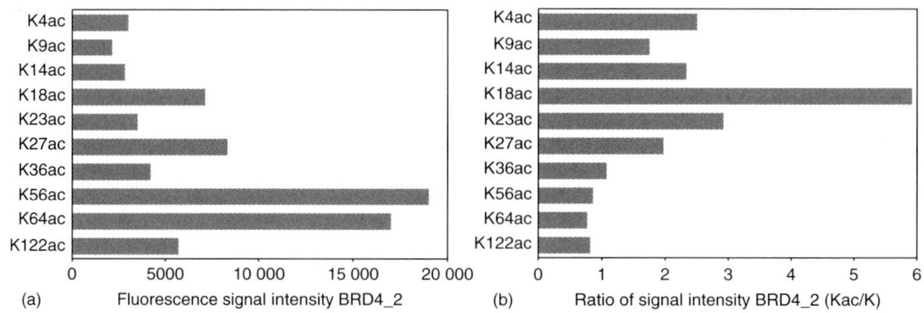

Figure 5.7 Subsite and acetyl specificity of bromodomain BRD4_2. Histone peptide microarray was treated with FLAG-tagged bromodomain followed by anti-Flag-tagged antibody and fluorescence-labeled secondary antibody. (a) Fluorescence signal intensities are shown for all histone H3-derived peptides containing one acetylated lysine residue. (b) Ratio of signal intensities is shown for histone H3-derived peptide pairs (acetylated lysine/non-modified lysine) for BRD4_2. A ratio larger than one means preferred binding to the acetylated version of the peptide.

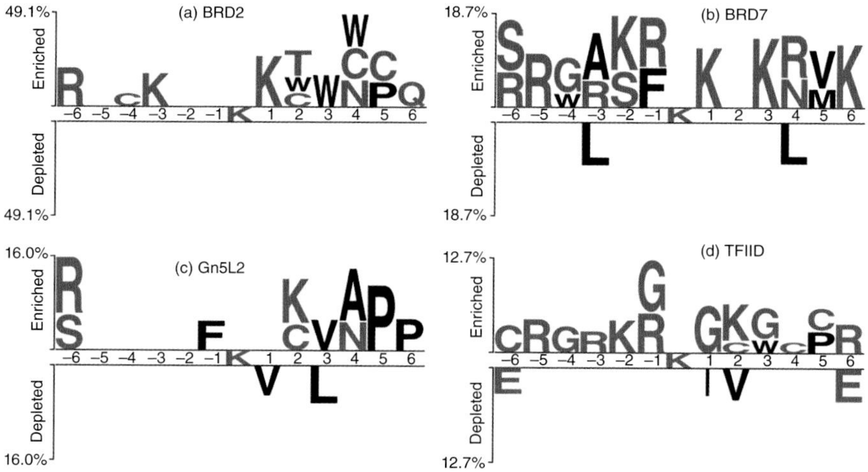

Figure 5.8 Subsite specificity of different bromodomains. Acetylome peptide microarray was treated with bromodomain–GST fusion proteins followed by fluorescence-labeled anti-GST antibody. Shown are sequence logos. The probability of a given amino acid at a defined position relative to the acetylated lysine (in the middle position of the peptides) in the identified binders is compared with the complete library presented on the presented microarray. Enrichment is an indicator for the preference of that amino acid in the position. The size of a residue is proportional to the preference. In a similar manner depleted residues are presented (bottom of the logo). (a) GST-BRD2 domain, (b) GST-BRD7 domain, (c) GST-Gcn5L2 domain, and (d) GST-TFIID domain.

5.2.4 Peptide Microarray-based Identification of Upstream Kinases and Phosphorylation Sites for Epigenetic Targets

Overlapping peptide scans through substrate proteins represent a powerful tool for the identification of phosphorylation sites. We used that technology for the identification of phosphorylation sites in the cytomegalovirus proteome [110], for the detection of autophosphorylation events in the tyrosine kinase Tie2 [47], and for the identification of phosphorylation sites in plakophilin [111]. Sirtuins have relatively low turnover numbers (between 0.1 and 0.001 s^{-1}) *in vitro*, and there are some doubts if the catalytic efficiency is sufficient for regulation of biological processes. On the other hand sirtuins 1 and 2 were described as universal histone deacetylases [112]. Phosphorylations within the catalytic domain could maybe improve the catalytic constants. We created a peptide microarray displaying overlapping peptide scans through all human sirtuin isoforms to search systematically for kinases that are able to phosphorylate side chain within the active site of these enzymes. Peptidyl-prolyl *cis*/*trans* isomerases (PPIases) are able to catalyze the relatively slow *cis*/*trans* isomerization of peptide bonds preceding proline (prolyl bonds). The conformation of the prolyl bonds in histones is critical for methylation events in the surrounding regions [113–116]. FK506 binding protein (FKBP) Fpr4 is able to catalyze the isomerization of all three prolyl bonds in histone H3 [117]. Additionally, parvulin-type PPIase Pin1 interacts with phosphorylated histone H1, catalyzes the slow *cis*/*trans* isomerization of phosphoserine–proline bonds, and influences the subsequent dephosphorylation of histone H1 by isomer-specific phosphatases [118]. PPIases located in the nucleus represent therefore another type of histone code writers influencing the overall conformation of mainly disordered histone tails, thereby fine-tuning the enzymatic activity of other writers and erasers. Enzymatic activity of PPIases is known to affect DNA repair and replication and is involved in regulation of transcription factors and modifications of chromatin structure [119]. Therefore, we generated overlapping peptides covering all human PPIases (16 FKBPs, 16 cyclophilins, and 3 parvulins) and printed it together with the sirtuin-derived peptides onto chip. These PPIase_Sirtuin_Scan microarrays were treated with a panel of recombinant human kinases in the presence of radioisotopically labeled γ-^{32}P-ATP [110]. The use of overlapping peptides allows a fast and efficient quality control of the results. Kinases have often a very simple consensus sequence like RRXS for protein kinase A (PKA) or SDXD for casein kinase II (CKII). Such short motifs are represented in several overlapping peptides, and therefore on the microarray all of these peptides have to be detected as substrates for the kinase (Figure 5.9). PKA phosphorylated only peptides derived from sirtuin 3

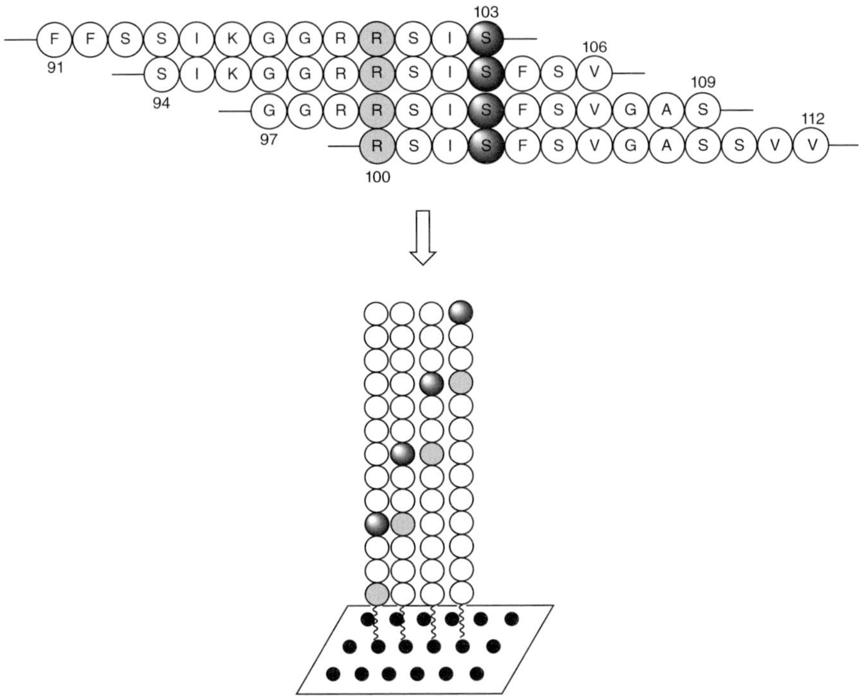

Figure 5.9 General principle of overlapping peptide scans on microarrays. Overlapping peptides covering complete protein sequence were synthesized and printed onto glass slides. Shown are overlapping peptides derived from sirtuin 3 containing a motif recognized by protein kinase A (serine residue 103 of sirtuin 3 is the phosphate acceptor). Subsequent to immobilization and treatment with the protein kinase A in the presence of radioisotopically labeled ATP, we detected strong signals after phosphorimaging for all four peptides containing the RSIS motif.

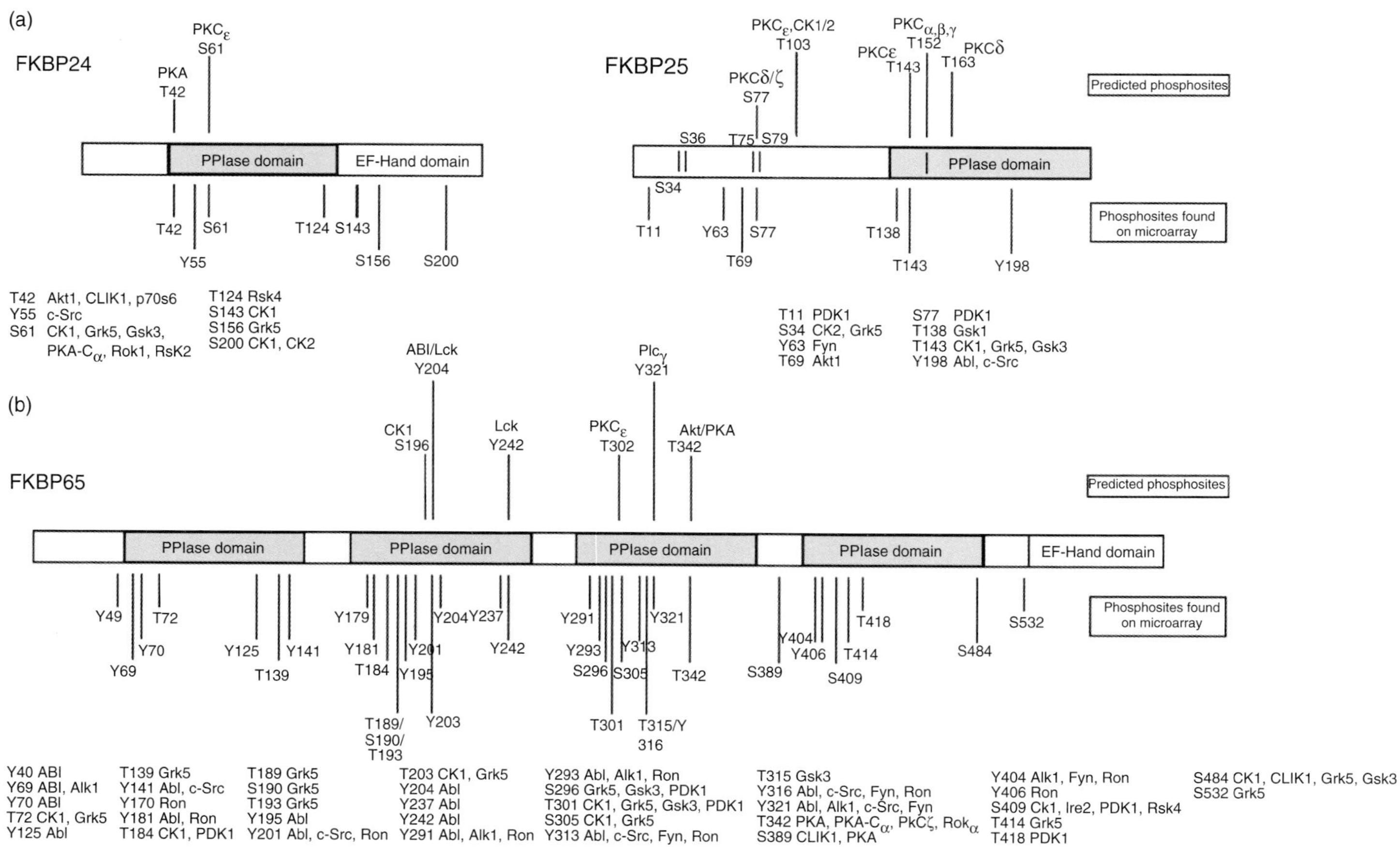

Figure 5.10 Schematic presentation of phosphosite detection for human FKBPs. Predicted phosphorylation sites (a) are shown together with phosphorylation sites detected using the PPIase_Sirtuin_Scan_Microarray (b). Prediction was made using http://www.scansite.mit.edu using medium stringency settings. Peptide microarrays were treated with the respective kinase in the presence of [γ-^{33}P]-ATP, and readout was performed by phosphorimaging.

(serine 103 and serine 337) and sirtuin 4 (threonine 158). Serine 103 is located in the catalytic domain of sirtuin 3 near the NAD^+ binding pocket. For CKII we found strong phosphorylation of all sirtuin 1 peptides containing serine 333. Again this phosphorylation is located in the catalytic domain of the enzyme. Additionally, CKII phosphorylated serine 129 of sirtuin 4. Generally, we detected more phosphorylation sites in the peptides derived from the human PPIases. Figure 5.10 presents the results for three different FKBPs including the human Fpr4 homolog FKBP25. We identified several kinases potentially regulating the enzymatic activity of PPIases including PKA, which phosphorylates serine 71 within the active site of parvulin-type PPIase Pin1. Similar phosphorylation of serine 71 by death-associated protein kinase 1 was shown to completely suppress Pin1 enzymatic activity and prevented its nuclear location [120].

5.3 Conclusion and Outlook

Peptide microarrays enable both identification of substrates for enzymes acting on histones and efficient profiling of binding specificities of histone code readers. The display of all possible histone-derived peptides within one experiment allows discovery of novel substrates or binding sites. The simultaneous presentation of modified peptides (chemically introduced PTMs including recently discovered succinylated and malonylated lysines) allows identification of cross-talk between different PTMs. The number of presented combinations of modifications is one limitation of the histone peptide microarray, but emerging peptide microarray technologies result in about 2 million peptide derivatives per glass slide. If such technologies are extended regarding the number of building blocks that can be used (from 24 so far to at least 50–100 to include all known phosphorylations, methylations, and acylations), a much better coverage of PTM combinations could be displayed dramatically improving our understanding of the histone code and epigenetics.

Acknowledgment

We thank Dr. Antonia Masch for providing some experimental data related to peptide microarrays. We acknowledge the collaboration with Prof. Clemens Steegborn (University of Bayreuth) who provided human sirtuins and with Prof. Wolfgang Sippl and Dr. Frank Erdmann (University of Halle-Wittenberg) providing Myt1 kinase.

References

1 Chi, P., Allis, C.D., and Wang, G.G. (2010). Covalent histone modifications–miswritten, misinterpreted and mis-erased in human cancers. *Nat. Rev. Cancer* 10: 457–469.

2 Rossetto, D., Avvakumov, N., and Cote, J. (2012). Histone phosphorylation: a chromatin modification involved in diverse nuclear events. *Epigenetics* 7: 1098–1108.

3 Cosgrove, M.S., Boeke, J.D., and Wolberger, C. (2004). Regulated nucleosome mobility and the histone code. *Nat. Struct. Mol. Biol.* 11: 1037–1043.

4 Rothbart, S.B. and Strahl, B.D. (2014). Interpreting the language of histone and DNA modifications. *Biochim. Biophys. Acta* 1839: 627–643.

5 Bannister, A.J. and Kouzarides, T. (2011). Regulation of chromatin by histone modifications. *Cell Res.* 21: 381–395.

6 Musselman, C.A., Lalonde, M.E., Cote, J., and Kutateladze, T.G. (2012). Perceiving the epigenetic landscape through histone readers. *Nat. Struct. Mol. Biol.* 19: 1218–1227.

7 Baek, S.H. (2011). When signaling kinases meet histones and histone modifiers in the nucleus. *Mol. Cell* 42: 274–284.

8 Lohse, B., Helgstrand, C., Kristensen, J.B. et al. (2013). Posttranslational modifications of the histone 3 tail and their impact on the activity of histone lysine demethylases in vitro. *PLoS One* 8: e67653.

9 Liokatis, S., Stutzer, A., Elsasser, S.J. et al. (2012). Phosphorylation of histone H3 Ser10 establishes a hierarchy for subsequent intramolecular modification events. *Nat. Struct. Mol. Biol.* 19: 819–823.

10 Smith, E. and Shilatifard, A. (2010). The chromatin signaling pathway: diverse mechanisms of recruitment of histone-modifying enzymes and varied biological outcomes. *Mol. Cell* 40: 689–701.

11 Jackson, V., Shires, A., Chalkley, R., and Granner, D.K. (1975). Studies on highly metabolically active acetylation and phosphorylation of histones. *J. Biol. Chem.* 250: 4856–4863.

12 Barth, T.K. and Imhof, A. (2010). Fast signals and slow marks: the dynamics of histone modifications. *Trends Biochem. Sci* 35: 618–626.

13 Chen, Y., Sprung, R., Tang, Y. et al. (2007). Lysine propionylation and butyrylation are novel post-translational modifications in histones. *Mol. Cell. Proteomics* 6: 812–819.

14 Xie, Z., Dai, J., Dai, L. et al. (2012). Lysine succinylation and lysine malonylation in histones. *Mol. Cell. Proteomics* 11: 100–107.

15 Tan, M., Peng, C., Anderson, K.A. et al. (2014). Lysine glutarylation is a protein posttranslational modification regulated by SIRT5. *Cell Metab.* 19: 605–617.

16 Tan, M., Luo, H., Lee, S. et al. (2011). Identification of 67 histone marks and histone lysine crotonylation as a new type of histone modification. *Cell* 146: 1016–1028.

17 Xie, Z., Zhang, D., Chung, D. et al. (2016). Metabolic regulation of gene expression by histone lysine beta-hydroxybutyrylation. *Mol. Cell* 62: 194–206.

18 Galligan, J.J., Rose, K.L., Beavers, W.N. et al. (2014). Stable histone adduction by 4-oxo-2-nonenal: a potential link between oxidative stress and epigenetics. *J. Am. Chem. Soc.* 136: 11864–11866.

19 Cui, Y., Li, X., Lin, J. et al. (2017). Histone ketoamide adduction by 4-oxo-2-nonenal is a reversible posttranslational modification regulated by Sirt2. *ACS Chem. Biol.* 12: 47–51.

20 Dai, L., Peng, C., Montellier, E. et al. (2014). Lysine 2-hydroxyisobutyrylation is a widely distributed active histone mark. *Nat. Chem. Biol.* 10: 365–370.

21 Sabari, B.R., Zhang, D., Allis, C.D., and Zhao, Y. (2017). Metabolic regulation of gene expression through histone acylations. *Nat. Rev. Mol. Cell Biol.* 18: 90–101.

22 Herranz, N., Dave, N., Millanes-Romero, A. et al. (2012). Lysyl oxidase-like 2 deaminates lysine 4 in histone H3. *Mol. Cell* 46: 369–376.

23 Dehennaut, V., Leprince, D., and Lefebvre, T. (2014). O-GlcNAcylation, an epigenetic mark. Focus on the histone code, TET family proteins, and Polycomb group proteins. *Front. Endocrinol.* 5: 155.

24 Fujiki, R., Hashiba, W., Sekine, H. et al. (2011). GlcNAcylation of histone H2B facilitates its monoubiquitination. *Nature* 480: 557–560.

25 Messner, S., Altmeyer, M., Zhao, H. et al. (2010). PARP1 ADP-ribosylates lysine residues of the core histone tails. *Nucleic Acids Res.* 38: 6350–6362.

26 Thompson, P.R. and Fast, W. (2006). Histone citrullination by protein arginine deiminase: is arginine methylation a green light or a roadblock? *ACS Chem. Biol.* 1: 433–441.

27 Fuhrmann, J. and Thompson, P.R. (2016). Protein arginine methylation and Citrullination in epigenetic regulation. *ACS Chem. Biol.* 11: 654–668.

28 Meas, R. and Mao, P. (2015). Histone ubiquitylation and its roles in transcription and DNA damage response. *DNA Repair* 36: 36–42.

29 Shiio, Y. and Eisenman, R.N. (2003). Histone sumoylation is associated with transcriptional repression. *Proc. Natl. Acad. Sci. U.S.A.* 100: 13225–13230.

30 Garske, A.L., Craciun, G., and Denu, J.M. (2008). A combinatorial H4 tail library for exploring the histone code. *Biochemistry* 47: 8094–8102.

31 Garske, A.L., Oliver, S.S., Wagner, E.K. et al. (2010). Combinatorial profiling of chromatin binding modules reveals multisite discrimination. *Nat. Chem. Biol.* 6: 283–290.

32 Heubach, Y., Planatscher, H., Sommersdorf, C. et al. (2013). From spots to beads-PTM-peptide bead arrays for the characterization of anti-histone antibodies. *Proteomics* 13: 1010–1015.

33 Sawicka, A. and Seiser, C. (2014). Sensing core histone phosphorylation – a matter of perfect timing. *Biochim. Biophys. Acta* 1839: 711–718.

34 Mahajan, K., Fang, B., Koomen, J.M., and Mahajan, N.P. (2012). H2B Tyr37 phosphorylation suppresses expression of replication-dependent core histone genes. *Nat. Struct. Mol. Biol.* 19: 930–937.

35 Frederiks, F., Stulemeijer, I.J., Ovaa, H., and van Leeuwen, F. (2011). A modified epigenetics toolbox to study histone modifications on the nucleosome core. *ChemBioChem* 12: 308–313.

36 Tropberger, P., Pott, S., Keller, C. et al. (2013). Regulation of transcription through acetylation of H3K122 on the lateral surface of the histone octamer. *Cell* 152: 859–872.

37 Tinti, M., Kiemer, L., Costa, S. et al. (2013). The SH2 domain interaction landscape. *Cell Rep.* 3: 1293–1305.

38 Carducci, M., Perfetto, L., Briganti, L. et al. (2012). The protein interaction network mediated by human SH3 domains. *Biotechnol. Adv.* 30: 4–15.
39 Lakshminarasimhan, M., Rauh, D., Schutkowski, M., and Steegborn, C. (2013). Sirt1 activation by resveratrol is substrate sequence-selective. *Aging (Albany NY)* 5: 151–154.
40 Rauh, D., Fischer, F., Gertz, M. et al. (2013). An acetylome peptide microarray reveals specificities and deacetylation substrates for all human sirtuin isoforms. *Nat. Commun.* 4: 2327.
41 Palma, A., Tinti, M., Paoluzi, S. et al. (2017). Both intrinsic substrate preference and network context contribute to substrate selection of classical tyrosine phosphatases. *J. Biol. Chem.* 292: 4942–4952.
42 Ferrari, E., Tinti, M., Costa, S. et al. (2011). Identification of new substrates of the protein-tyrosine phosphatase PTP1B by Bayesian integration of proteome evidence. *J. Biol. Chem.* 286: 4173–4185.
43 Papadopoulos, C., Arato, K., Lilienthal, E. et al. (2011). Splice variants of the dual specificity tyrosine phosphorylation-regulated kinase 4 (DYRK4) differ in their subcellular localization and catalytic activity. *J. Biol. Chem.* 286: 5494–5505.
44 Mah, A.S., Elia, A.E., Devgan, G. et al. (2005). Substrate specificity analysis of protein kinase complex Dbf2-Mob1 by peptide library and proteome array screening. *BMC Biochem.* 6: 22.
45 Rychlewski, L., Kschischo, M., Dong, L. et al. (2004). Target specificity analysis of the Abl kinase using peptide microarray data. *J. Mol. Biol.* 336: 307–311.
46 Lizcano, J.M., Deak, M., Morrice, N. et al. (2002). Molecular basis for the substrate specificity of NIMA-related kinase-6 (NEK6). Evidence that NEK6 does not phosphorylate the hydrophobic motif of ribosomal S6 protein kinase and serum- and glucocorticoid-induced protein kinase in vivo. *J. Biol. Chem.* 277: 27839–27849.
47 Schutkowski, M., Reimer, U., Panse, S. et al. (2004). High-content peptide microarrays for deciphering kinase specificity and biology. *Angew. Chem. Int. Ed* 43: 2671–2674.
48 Rohe, A., Platzer, C., Masch, A. et al. (2015). Identification of peptidic substrates for the human kinase Myt1 using peptide microarrays. *Bioorg. Med. Chem.* 23: 4936–4942.
49 Schonberg, A., Bergner, E., Helm, S. et al. (2014). The peptide microarray "ChloroPhos1.0" identifies new phosphorylation targets of plastid casein kinase II (pCKII) in *Arabidopsis thaliana*. *PLoS One* 9: e108344.
50 Wisniewski, J.R., Zougman, A., Kruger, S., and Mann, M. (2007). Mass spectrometric mapping of linker histone H1 variants reveals multiple acetylations, methylations, and phosphorylation as well as differences between cell culture and tissue. *Mol. Cell. Proteomics* 6: 72–87.
51 Nady, N., Min, J., Kareta, M.S. et al. (2008). A SPOT on the chromatin landscape? Histone peptide arrays as a tool for epigenetic research. *Trends Biochem. Sci* 33: 305–313.

52 Bock, I., Kudithipudi, S., Tamas, R. et al. (2011). Application of CelluSpots peptide arrays for the analysis of the binding specificity of epigenetic reading domains to modified histone tails. *BMC Biochem.* 12: 48.
53 Su, Z., Boersma, M.D., Lee, J.H. et al. (2014). ChIP-less analysis of chromatin states. *Epigenetics Chromatin* 7: 7.
54 Filippakopoulos, P., Picaud, S., Mangos, M. et al. (2012). Histone recognition and large-scale structural analysis of the human bromodomain family. *Cell* 149: 214–231.
55 Bua, D.J., Kuo, A.J., Cheung, P. et al. (2009). Epigenome microarray platform for proteome-wide dissection of chromatin-signaling networks. *PLoS One* 4: e6789.
56 Fuchs, S.M., Krajewski, K., Baker, R.W. et al. (2011). Influence of combinatorial histone modifications on antibody and effector protein recognition. *Curr. Biol.* 21: 53–58.
57 Matthews, A.G., Kuo, A.J., Ramon-Maiques, S. et al. (2007). RAG2 PHD finger couples histone H3 lysine 4 trimethylation with V(D)J recombination. *Nature* 450: 1106–1110.
58 Levy, D., Kuo, A.J., Chang, Y. et al. (2011). Lysine methylation of the NF-kappaB subunit RelA by SETD6 couples activity of the histone methyltransferase GLP at chromatin to tonic repression of NF-kappaB signaling. *Nat. Immunol.* 12: 29–36.
59 Hung, T., Binda, O., Champagne, K.S. et al. (2009). ING4 mediates crosstalk between histone H3 K4 trimethylation and H3 acetylation to attenuate cellular transformation. *Mol. Cell* 33: 248–256.
60 Kleine-Kohlbrecher, D., Christensen, J., Vandamme, J. et al. (2010). A functional link between the histone demethylase PHF8 and the transcription factor ZNF711 in X-linked mental retardation. *Mol. Cell* 38: 165–178.
61 Liu, H., Galka, M., Iberg, A. et al. (2010). Systematic identification of methyllysine-driven interactions for histone and nonhistone targets. *J. Proteome Res.* 9: 5827–5836.
62 Bock, I., Dhayalan, A., Kudithipudi, S. et al. (2011). Detailed specificity analysis of antibodies binding to modified histone tails with peptide arrays. *Epigenetics* 6: 256–263.
63 Zhang, Y., Jurkowska, R., Soeroes, S. et al. (2010). Chromatin methylation activity of Dnmt3a and Dnmt3a/3L is guided by interaction of the ADD domain with the histone H3 tail. *Nucleic Acids Res.* 38: 4246–4253.
64 Dhayalan, A., Tamas, R., Bock, I. et al. (2011). The ATRX-ADD domain binds to H3 tail peptides and reads the combined methylation state of K4 and K9. *Hum. Mol. Genet.* 20: 2195–2203.
65 Rothbart, S.B., Dickson, B.M., Ong, M.S. et al. (2013). Multivalent histone engagement by the linked tandem Tudor and PHD domains of UHRF1 is required for the epigenetic inheritance of DNA methylation. *Genes Dev.* 27: 1288–1298.
66 Thiele, A., Zerweck, J., and Schutkowski, M. (2009). Peptide arrays for enzyme profiling. *Methods Mol. Biol.* 570: 19–65.
67 Thiele, A., Stangl, G.I., and Schutkowski, M. (2011). Deciphering enzyme function using peptide arrays. *Mol. Biotechnol.* 49: 283–305.

68 Green, K.D. and Pflum, M.K. (2007). Kinase-catalyzed biotinylation for phosphoprotein detection. *J. Am. Chem. Soc.* 129: 10–11.
69 Song, H., Kerman, K., and Kraatz, H.B. (2008). Electrochemical detection of kinase-catalyzed phosphorylation using ferrocene-conjugated ATP. *Chem. Commun. (Camb.)* 502–504.
70 Kerman, K. and Kraatz, H.B. (2007). Electrochemical detection of kinase-catalyzed thiophosphorylation using gold nanoparticles. *Chem. Commun. (Camb.)* 5019–5021.
71 Green, K.D. and Pflum, M.K. (2009). Exploring kinase cosubstrate promiscuity: monitoring kinase activity through dansylation. *ChemBioChem* 10: 234–237.
72 Martin, K., Steinberg, T.H., Cooley, L.A. et al. (2003). Quantitative analysis of protein phosphorylation status and protein kinase activity on microarrays using a novel fluorescent phosphorylation sensor dye. *Proteomics* 3: 1244–1255.
73 Shimomura, T., Han, X., Hata, A. et al. (2011). Optimization of peptide density on microarray surface for quantitative phosphoproteomics. *Anal. Sci.* 27: 13–17.
74 Kinoshita, E., Kinoshita-Kikuta, E., Sugiyama, Y. et al. (2012). Highly sensitive detection of protein phosphorylation by using improved Phos-tag biotin. *Proteomics* 12: 932–937.
75 Ho, M.C., Wilczek, C., Bonanno, J.B. et al. (2013). Structure of the arginine methyltransferase PRMT5-MEP50 reveals a mechanism for substrate specificity. *PLoS One* 8: e57008.
76 Burgos, E.S., Wilczek, C., Onikubo, T. et al. (2015). Histone H2A and H4 N-terminal tails are positioned by the MEP50 WD repeat protein for efficient methylation by the PRMT5 arginine methyltransferase. *J. Biol. Chem.* 290: 9674–9689.
77 Akita, S., Umezawa, N., Kato, N., and Higuchi, T. (2008). Array-based fluorescence assay for serine/threonine kinases using specific chemical reaction. *Bioorg. Med. Chem.* 16: 7788–7794.
78 Shults, M.D., Kozlov, I.A., Nelson, N. et al. (2007). A multiplexed protein kinase assay. *ChemBioChem* 8: 933–942.
79 Soloaga, A., Thomson, S., Wiggin, G.R. et al. (2003). MSK2 and MSK1 mediate the mitogen- and stress-induced phosphorylation of histone H3 and HMG-14. *EMBO J.* 22: 2788–2797.
80 Healy, S., Khan, P., He, S., and Davie, J.R. (2012). Histone H3 phosphorylation, immediate-early gene expression, and the nucleosomal response: a historical perspective. *Biochem. Cell Biol.* 90: 39–54.
81 Zhang, Y., Griffin, K., Mondal, N., and Parvin, J.D. (2004). Phosphorylation of histone H2A inhibits transcription on chromatin templates. *J. Biol. Chem.* 279: 21866–21872.
82 Kang, B., Pu, M., Hu, G. et al. (2011). Phosphorylation of H4 Ser 47 promotes HIRA-mediated nucleosome assembly. *Genes Dev.* 25: 1359–1364.
83 Hossain, M.B., Shifat, R., Johnson, D.G. et al. (2016). TIE2-mediated tyrosine phosphorylation of H4 regulates DNA damage response by recruiting ABL1. *Sci. Adv.* 2: e1501290.

84 Warsito, D., Lin, Y., Gnirck, A.C. et al. (2016). Nuclearly translocated insulin-like growth factor 1 receptor phosphorylates histone H3 at tyrosine 41 and induces SNAI2 expression via Brg1 chromatin remodeling protein. *Oncotarget* 7: 42288–42302.
85 Basnet, H., Su, X.B., Tan, Y. et al. (2014). Tyrosine phosphorylation of histone H2A by CK2 regulates transcriptional elongation. *Nature* 516: 267–271.
86 Dawson, M.A., Foster, S.D., Bannister, A.J. et al. (2012). Three distinct patterns of histone H3Y41 phosphorylation mark active genes. *Cell Rep.* 2: 470–477.
87 Xiao, A., Li, H., Shechter, D. et al. (2009). WSTF regulates the H2A.X DNA damage response via a novel tyrosine kinase activity. *Nature* 457: 57–62.
88 Thiele, A., Reimer, U., Zerweck, J., and Schutkowski, M. (2015). Peptide-microarray for profiling of epigenetic targets. In: *Epigenetic Technological Applications* (ed. Y.G. Zheng), 169–186. Academic Press Elsevier.
89 Sauve, A.A., Wolberger, C., Schramm, V.L., and Boeke, J.D. (2006). The biochemistry of sirtuins. *Annu. Rev. Biochem.* 75: 435–465.
90 Morris, B.J. (2013). Seven sirtuins for seven deadly diseases of aging. *Free Radical Biol. Med.* 56: 133–171.
91 Ming, M., Qiang, L., Zhao, B., and He, Y.Y. (2014). Mammalian SIRT2 inhibits keratin 19 expression and is a tumor suppressor in skin. *Exp. Dermatol.* 23: 207–209.
92 Xiao, K., Jiang, J., Wang, W. et al. (2013). Sirt3 is a tumor suppressor in lung adenocarcinoma cells. *Oncol. Rep.* 30: 1323–1328.
93 Csibi, A., Fendt, S.M., Li, C. et al. (2013). The mTORC1 pathway stimulates glutamine metabolism and cell proliferation by repressing SIRT4. *Cell* 153: 840–854.
94 Xiangyun, Y., Xiaomin, N., Linping, G. et al. (2017). Desuccinylation of pyruvate kinase M2 by SIRT5 contributes to antioxidant response and tumor growth. *Oncotarget* 8: 6984–6993.
95 Blander, G., Olejnik, J., Krzymanska-Olejnik, E. et al. (2005). SIRT1 shows no substrate specificity in vitro. *J. Biol. Chem.* 280: 9780–9785.
96 Feldman, J.L., Dittenhafer-Reed, K.E., Kudo, N. et al. (2015). Kinetic and structural basis for acyl-group selectivity and NAD(+) dependence in sirtuin-catalyzed deacylation. *Biochemistry* 54: 3037–3050.
97 Michishita, E., McCord, R.A., Berber, E. et al. (2008). SIRT6 is a histone H3 lysine 9 deacetylase that modulates telomeric chromatin. *Nature* 452: 492–496.
98 Lau, P.N. and Cheung, P. (2011). Histone code pathway involving H3 S28 phosphorylation and K27 acetylation activates transcription and antagonizes polycomb silencing. *Proc. Natl. Acad. Sci. U.S.A.* 108: 2801–2806.
99 Roessler, C., Nowak, T., Pannek, M. et al. (2014). Chemical probing of the human sirtuin 5 active site reveals its substrate acyl specificity and peptide-based inhibitors. *Angew. Chem. Int. Ed.* 53: 10728–10732.
100 Zerweck, J., Masch, A., and Schutkowski, M. (2009). Peptide microarrays for profiling of modification state-specific antibodies. *Methods Mol. Biol.* 524: 169–180.

101 Funkner, A., Parthier, C., Schutkowski, M. et al. (2013). Peptide binding by catalytic domains of the protein disulfide isomerase-related protein ERp46. *J. Mol. Biol.* 425: 1340–1362.

102 Lin, S., Yang, X., Jia, S. et al. (2017). Redox-based reagents for chemoselective methionine bioconjugation. *Science* 355: 597–602.

103 MacDonald, J.I., Munch, H.K., Moore, T., and Francis, M.B. (2015). One-step site-specific modification of native proteins with 2-pyridinecarboxyaldehydes. *Nat. Chem. Biol.* 11: 326–331.

104 Liebscher, S., Schopfel, M., Aumuller, T. et al. (2014). N-terminal protein modification by substrate-activated reverse proteolysis. *Angew. Chem. Int. Ed.* 53: 3024–3028.

105 Schmidt, M., Toplak, A., Quaedflieg, P.J., and Nuijens, T. (2017). Enzyme-mediated ligation technologies for peptides and proteins. *Curr. Opin. Chem. Biol.* 38: 1–7.

106 Filippakopoulos, P. and Knapp, S. (2014). Targeting bromodomains: epigenetic readers of lysine acetylation. *Nat. Rev. Drug Discovery* 13: 337–356.

107 Yang, Z., He, N., and Zhou, Q. (2008). Brd4 recruits P-TEFb to chromosomes at late mitosis to promote G1 gene expression and cell cycle progression. *Mol. Cell. Biol.* 28: 967–976.

108 Wu, S.Y., Lee, A.Y., Lai, H.T. et al. (2013). Phospho switch triggers Brd4 chromatin binding and activator recruitment for gene-specific targeting. *Mol. Cell* 49: 843–857.

109 Liu, Y., Wang, X., Zhang, J. et al. (2008). Structural basis and binding properties of the second bromodomain of Brd4 with acetylated histone tails. *Biochemistry* 47: 6403–6417.

110 Thiele, A., Zerweck, J., Weiwad, M. et al. (2009). High-density peptide microarrays for reliable identification of phosphorylation sites and upstream kinases. *Methods Mol. Biol.* 570: 203–219.

111 Wolf, A., Rietscher, K., Glass, M. et al. (2013). Insulin signaling via Akt2 switches plakophilin 1 function from stabilizing cell adhesion to promoting cell proliferation. *J. Cell Sci.* 126: 1832–1844.

112 Hsu, W.W., Wu, B., and Liu, W.R. (2016). Sirtuins 1 and 2 are universal histone deacetylases. *ACS Chem. Biol.* 11: 792–799.

113 Nelson, C.J., Santos-Rosa, H., and Kouzarides, T. (2006). Proline isomerization of histone H3 regulates lysine methylation and gene expression. *Cell* 126: 905–916.

114 Youdell, M.L., Kizer, K.O., Kisseleva-Romanova, E. et al. (2008). Roles for Ctk1 and Spt6 in regulating the different methylation states of histone H3 lysine 36. *Mol. Cell. Biol.* 28: 4915–4926.

115 Howe, F.S., Boubriak, I., Sale, M.J. et al. (2014). Lysine acetylation controls local protein conformation by influencing proline isomerization. *Mol. Cell* 55: 733–744.

116 Howe, F.S. and Mellor, J. (2014). Proline *cis–trans* isomerization is influenced by local lysine acetylation–deacetylation. *Microb. Cell* 1: 390–392.

117 Monneau, Y.R., Soufari, H., Nelson, C.J., and Mackereth, C.D. (2013). Structure and activity of the peptidyl-prolyl isomerase domain from the histone chaperone Fpr4 toward histone H3 proline isomerization. *J. Biol. Chem.* 288: 25826–25837.

118 Raghuram, N., Strickfaden, H., McDonald, D. et al. (2013). Pin1 promotes histone H1 dephosphorylation and stabilizes its binding to chromatin. *J. Cell Biol.* 203: 57–71.

119 Yao, Y.L., Liang, Y.C., Huang, H.H., and Yang, W.M. (2011). FKBPs in chromatin modification and cancer. *Curr. Opin. Pharmacol.* 11: 301–307.

120 Lee, T.H., Chen, C.H., Suizu, F. et al. (2011). Death-associated protein kinase 1 phosphorylates Pin1 and inhibits its prolyl isomerase activity and cellular function. *Mol. Cell* 42: 147–159.

6

Chemical Probes

Amy Donner[1], Heather King[1], Paul E. Brennan[2], Moses Moustakim[2], and William J. Zuercher[3]

[1] *UNC Eshelman School of Pharmacy, Chemical Probes Portal, 120 Mason Farm Road, Chapel Hill, NC 27599, USA*
[2] *University of Oxford, Structural Genomics Consortium, Nuffield Department of Medicine, Old Road Campus, OX3 7FZ, Oxford, UK*
[3] *UNC Eshelman School of Pharmacy, Structural Genomics Consortium, 120 Mason Farm Road, Chapel Hill, NC 27599, USA*

6.1 Chemical Probes Are Privileged Reagents for Biological Research

A chemical probe is a small-molecule inhibitor or activator of a protein that allows scientists to study its biological function in cells or animals. The potency and selectivity of a chemical probe for its primary target protein are important contributors to its utility as an experimental reagent. A small molecule is not a key that interacts with only a single lock; its ability to discriminate between proteins is dependent of its potency for target engagement with its primary target and a multitude of off-targets within the cell. Therefore, chemical probes must be extensively characterized, and guidelines for their use clearly communicated before they are considered validated, and even then, experiments using them must be designed and interpreted with care.

Chemical probes are among the most valued reagents for selection of new molecular targets for drug discovery. They provide technical and biological validation of a molecular target as an intervention point for disease therapy. However, chemical probes and drug molecules though borne of the same chemical building blocks often have distinct properties. Drugs do not need to be exquisitely selective for their molecular target; indeed, many medicines manifest their clinical effects through polypharmacology [1]. Often drugs are not the optimal reagents to use as chemical probes. Chemical probes, on the other hand, must be selected based on their potency and selectivity for engagement of their molecular target, but do not need to meet the same requirements in terms of pharmacokinetics (PKs), pharmacodynamics, and bioavailability as a drug. Herein, we discuss the development, characterization, and application of chemical probes for the study of proteins involved in epigenetic signaling.

Epigenetic Drug Discovery, First Edition. Edited by Wolfgang Sippl and Manfred Jung.

Although chemical probes have been used as pharmacological reagents for decades, they were often by-products of drug discovery programs rather than the catalysts of innovation. The rise of chemical biology as a scientific discipline provided the impetus to use chemistry to explore the biological function of a myriad of poorly studied proteins in cells. The discovery of over 300 proteins that modify histone tails has demonstrated the value of chemical probes as research tools to study epigenetic signaling. These efforts helped to define many of best practices described in the coming sections.

6.1.1 Best Practices for the Generation and Selection of Chemical Probes

Several excellent articles have been published that establish best practices for the generation of high-quality probes and their application to discover new biology or to validate therapeutic targets [2–4]. We provide a brief synopsis of these reports here and in Table 6.1.

Chemical entities that are developed as chemical probes need to be well defined, meaning that the chemical composition and three-dimensional structure must be known. Furthermore, a robust synthetic route for the reliable production of the chemical entity should be established. The compounds should also be soluble and stable in solution; if they are sensitive to changes in pH, for example, the impact these changes can have on the compound's activity should be explored; they should not aggregate in solution; nor should they be nonspecifically reactive (e.g. redox reactive). Any characteristics of these compounds that impact their solubility and stability should be assessed in detail for impact on their suitability for application to biological systems. Finally, chemical entities must penetrate biological membranes (e.g. plasma and subcellular compartment membranes in mammalian cells, cell walls, and membranes in bacteria) to merit the designation chemical probe; if a compound cannot be applied in a relevant biological system, it is not a chemical probe.

Potency and selectivity are two key properties of chemical probes. Potency refers to the amount of chemical probe a researcher needs to add to a system (e.g. recombinant purified protein or cultured cells) for it to bind to or impact the activity of the target protein. The more potent the chemical probe, the less of it a researcher needs to use. Using a lower concentration is always desirable: the presence of more molecules in an experimental system means that more interactions with unintended targets are possible. Potency requirements vary from target to target, but in general, we recommend researchers to aim for compounds with *in vitro* biochemical K_D or IC_{50} values in the 100 nM range.

Selectivity, which refers to the amount of chemical probe a researcher can use to perturb the protein of interest without also perturbing other proteins, is also critical. At a minimum, chemical probes should be assessed against a broad panel of proteins related to the intended targets. For example, there are 61 bromodomains in the human proteome that are contained in 46 proteins. Any chemical probe designed to inhibit a bromodomain protein should be assessed against as many of these bromodomains as possible. Like potency, selectivity requirements vary across targets; we recommend that compounds

Table 6.1 Chemical, biochemical, and biological criteria for the selection and validation of chemical probes.

Chemical	
Structure	The chemical structure is known and has been validated by state-of-the-art technologies; a robust and reliable synthetic route is known and available
Stability	The compound's response to changing buffer conditions (i.e. acid–base) are known; the compound lacks nonspecifically reactive moieties
Solubility	The compound is soluble and does not aggregate in aqueous solution
Permeability	The compound crosses cell membranes
Biochemical	
Potency	In assays with isolated recombinant protein, the compound should bind to or inhibit the activity of the target at concentrations <100 nM (i.e. K_D or IC_{50}); the compound should dose-dependently impact the target's activity
Selectivity	In assays with isolated recombinant proteins, the compound should be 10–100-fold selective for the intended protein target over other proteins (especially closely related proteins)
Mechanism of action	There should be a general understanding for how the binding of the compound to the target impacts that activity of that target
Biological	
Potency	In cells, the compound should bind to or inhibit the activity of the target at concentrations at low micromolar concentrations (e.g. IC_{50}); the compound should dose-dependently impact the target's activity
Selectivity	In cellular assays, likely off-target proteins (closely related proteins and/or proteins that could yield comparable cellular phenotypes) should be ruled out as targets
Mechanism of action	The cellular phenotype should be consistent with the compound's mechanism of action
Toxicity	The compound should not be generally (nonspecifically) toxic to cells
Target engagement	There should be direct (ideally) or indirect evidence that the compound binds to its intended target in cells

are 10–100-fold selective for the intended protein target over other related targets to merit the designation chemical probe. Ideally, chemical probes will also be assessed for activity against other accessible proteins that are likely to yield phenotypic outcomes (e.g. kinases, GPCRs). Together, data about the potency and selectivity of a probe help a researcher define the upper and lower concentration limits where a probe can be applied with confidence that it will impact the intended protein target and a limited number of off-target proteins.

It is helpful for scientists to understand the mechanism by which the chemical probe interacts with the target protein to impact its activity. This information can help scientists better understand the relationship between the target and its biology that emerge. For example, a chemical probe that disrupts the enzymatic activity of a protein might yield a phenotype distinct from one that disrupts protein–protein interactions of the same target; and both may yield a phenotype that differs from genetic deletion of that same protein. In the absence of any mechanistic understanding of the activity of these different probes, it could be impossible for scientists to understand or reconcile variable but meaningful biological distinctions. It is equally important to verify that a biochemical mechanism of action (i.e. the mechanism in biochemical experiments with recombinant purified protein) matches the mechanism of action in cells. We explore this important concept more fully in Section 6.1.3.

Finally, it is important to ensure that a chemical probe does not lead to nonspecific toxicities that can interfere with its application in cells. For example, a probe that is highly redox reactive will likely perturb cells in a manner unrelated to any impact on the intended protein target. The emergence of such nonspecific toxicities is sufficient to disqualify a compound as a chemical probe. Specific toxicity (i.e. target-related toxicity), on the other hand, is acceptable and can help build a case for a pursuing a target for therapeutic purposes or rule it out altogether. Ideally, target-specific toxicities are considered when designing experiments with the chemical probe (see Section 6.1.2).

6.1.2 Best Practices for Application of Chemical Probes

Many scientists who apply chemical probes use them in cellular systems that are distinct from the ones where they were initially validated. When applying a chemical probe in a new cellular system, it is important to consider several factors. Is the target expressed in the new system at levels comparable to those in original? If not, the concentration of the probe that will be needed to impact the activity of the target will differ. Similarly, are the likely off-target proteins the same, and are they expressed at the same levels? It is important to balance the amount of the probe used in cells with its selectivity, and the best concentration may need to be determined empirically by assessing on- and off-target activity profiles. Similarly, we recommend validating that the chemical probe engages its intended target when moving to a new cellular system. Proteins can adopt different conformations and participate in distinct complexes in different cells; thus, it is essential to demonstrate that the protein target is accessible by the probe in the new system before proceeding with new experiments.

It is essential for scientists applying chemical probes to remember that data supporting the validation of chemical probes are inherently incomplete. A compound that has been shown to modulate its protein target without affecting a known panel of additional proteins may still modulate proteins against which it has not been evaluated. Researchers can and should address this inherent challenge in two ways.

First, a structurally distinct chemical probe (an orthogonal probe), if available, is likely to have a distinct off-target profile. If two chemically unrelated compounds that act against the same target yield the same or very similar phenotypes in cells, the user gains confidence that the phenotypes are caused by interaction at the target protein. Thus, we recommend scientists to design experiments using two orthogonal chemical probes whenever possible.

Second, we strongly recommend the use of a negative control compound that is structurally related to the chemical probe but lacks activity on the target protein.

For a negative control, the concordance between loss of activity against the target and the loss of a target-related phenotype increases confidence that the activity of the probe is indeed related to activity at the intended protein target. In cases where a negative control compound is not available, the application of a series of related compounds that exhibit variable activity against the target *in vitro* (i.e. a range of potencies) can suffice. In this situation, the compounds' relative *in vitro* potencies should track with their relative ability to induce the phenotype of interest in cells.

The application of orthogonal probes together with their negative controls substantially decreases the probability that a researcher will attribute off-target activity of a probe to the protein of interest. Two probes of distinct structure are unlikely to have the same off-target activity profiles; when used in tandem with their negative controls, the likelihood that undesirable off-target activity will impact only the "active" samples decreases even further.

In addition to incorporating these controls in experiments, it is also important to remember that every probe has inherent strengths and weaknesses. The weaknesses should inform the type of follow-on validation experiments that are performed to confirm the connections between a protein target and a phenotype. For example, to distinguish the target–phenotype relationship resulting from the application of a probe that is active against two or more closely related proteins, it may be necessary to use genetic methods (e.g. experiments with a catalytically dead mutants) to define functional roles of the closely related proteins. Likewise, a probe that has cytotoxic activity can cause toxicity because of its interaction with the intended target or an as yet unidentified target. In cases where the toxicity is target related, this activity needs to be considered when applying the probe. For example, if a probe induces cell cycle arrest due to its activity on the target, it is important to distinguish downstream phenotypes resulting from cell cycle arrest from those directly attributable to the target.

6.1.3 Cellular Target Engagement

Chemical probes derive their utility through direct and specific interaction with their target proteins. Traditionally, these interactions have been established *in vitro* with expressed and purified proteins that may vary in structure from the endogenous protein present in cells. A single domain rather than the full-length protein is frequently employed owing to technical feasibility, and beyond the protein construct itself, *in vitro* biochemical assays are unable to account for the effects of partner proteins and cofactors present in the cellular milieu. Accordingly, methods to confirm and monitor target engagement in living cells are critical for complete characterization of a chemical probe. For chemical probes inhibiting the activity of enzymes that modify histone tails (e.g. histone deacetylases (HDACs), histone acetyltransferases (HATs), and histone methyltransferases (HMTs)), cell-based assays can be developed to measure the increase or decrease of the specific acetyl or methyl mark, though additional scrutiny is required to establish whether the effects are direct or indirect. However, for inhibitors of reader domains (e.g. bromodomains and methyl reader domains) where the histone tail is not modified, other techniques are required to demonstrate target engagement in cells.

Recent years have seen rapid progress in the development of technologies for the evaluation of target engagement within cells [5–10]. The available techniques each have their particular strengths and limitations. Factors to consider in selection of a method to evaluate target engagement include the need for modification of the chemical probe or the target protein, the ability to detect interactions with additional proteins in an unbiased manner, the applicability to the target in question, and the instrumentation and data analysis required. For a more inclusive summary of methods for demonstration of small-molecule target engagement in cells, the reader is directed to an outstanding general review [11]. The focus here will be on cellular target engagement approaches of particular relevance for epigenetic targets.

6.1.3.1 Fluorescence Recovery After Photobleaching (FRAP)

Early attempts to establish cellular engagement of epigenetic targets relied on an indirect method, fluorescence recovery after photobleaching (FRAP). FRAP is a confocal microscopy technique measuring the rate at which nonbleached proteins diffuse into a photobleached region [12, 13]. Cells are transfected with plasmids encoding the target of interest fused to a fluorescent protein, such as green fluorescent protein (GFP), and a bleached region is defined by a focused, high-intensity laser pulse. Diffusion of the fluorescing protein back into the bleached region is hindered by protein binding to chromatin and chromatin-associated complexes and is therefore slower compared with a freely diffusible molecule. Thus, the time taken for recovery is related to protein affinity, and the presence of an inhibitor of protein binding reduces recovery time [14]. FRAP was successfully employed in assessing bromodomain inhibitors [15]. FRAP is attractive as it is a target agnostic method but suffers from the disadvantages of being slow and labor intensive and requiring specialized equipment. Moreover, FRAP is not a direct measurement of target engagement and does not work in a native background due to introduction of fusion proteins.

6.1.3.2 Affinity Bead-Based Proteomics

Affinity bead-based proteomics allow a relatively unbiased view of the members of the proteome interacting with a probe. In the technique, the sub-proteome of a lysate binding to an immobilized bait compound is typically identified by mass spectrometry (MS), rather than by simple competition with soluble competitor molecules such as (i) the active probe itself (to reduce the extent of nonspecific binding proteins identified) and/or (ii) a negative control of the probe. These experiments are conducted in lysate rather than in intact cells, and the use of an immobilized bait necessitates that the binding interactions are heterogeneous and may not be competent in identification of all members of the proteome (e.g. membrane-bound targets). Nonetheless, a large body of evidence supports the powerful utility of affinity bead-based proteomic techniques with epigenetic targets [16].

Affinity bead-based proteomic approaches have helped to advance the generation and characterization of small-molecule epigenetic probes. One prominent example involves "bromospheres," an affinity matrix capable of enriching 19 bromodomain-containing proteins used, for example, in the characterization of ATAD2 inhibitors [17]. In analogy to kinobeads [18] and multiplexed inhibitor bead sets (MIBS) [19], matrices that capture kinases from lysates, bromospheres are a useful tool in addressing bromodomain selectivity, a key challenge in targeting this class of protein [20].

6.1.3.3 Cellular Thermal Shift Assay (CETSA)

Ligand-induced thermal stabilization of proteins is the basis for thermal shift assays that measure the change in melting temperature in the absence and the presence of small molecules, ΔTm [21]. More recently, the cellular thermal shift assay (CETSA) has established itself as a useful approach for cellular target engagement [6]. The technique involves heating different aliquots of cells across a range of discrete temperatures. Upon cooling, the unfolded proteins aggregate and precipitate. The ΔTm is then ascertained by separation and quantitation of the soluble proteins. CETSA was initially described with immunoblotting detection of soluble proteins, and the utility of the technique was extended with MS detection, enabling the parallel detection of multiple proteins [5, 22].

Multiple examples demonstrate the utility of CETSA in assessing cellular target engagement with epigenetic proteins. The method was used to confirm the engagement of newly identified small molecules with the Tudor-domain-containing protein Spindlin1 [23]. The ability of CETSA to survey multiple proteins concurrently enabled the unbiased mapping of protein targets of hydroxamic acid-based HDAC inhibitor panobinostat [24]. Similarly, CETSA was employed to define the overlapping and divergent targets of structurally distinct HDAC inhibitors in cardiac fibroblasts [25].

Despite the successful application of CETSA in multiple contexts, it is not a universally applicable method. Proteins of low abundance have problematic detection in complex mixtures due to instrument limitations in MS. Not all proteins exhibit aggregation properties amenable to CETSA and therefore are not part of the "meltome" [22]. Moreover, shifts for certain proteins may not be detectable even if the protein is part of the meltome. For example, if the epigenetic reader/writer/eraser of interest is but a single domain in a large protein, the stability conferred by small-molecule binding may not lead to an observable change in melting temperature [11].

6.1.3.4 Bioluminescence Resonance Energy Transfer

Resonance energy transfer (RET) involves emission from a donor that is transferred to a fluorescent acceptor moiety that relaxes and emits light at a longer wavelength. The donor may be fluorescent or bioluminescent moiety (fluorescence resonance energy transfer (FRET) and bioluminescence resonance energy transfer (BRET), respectively). Importantly, the ability to transfer between a donor and an acceptor is critically dependent on proximity of the two, as the extent of transfer follows an inverse sixth power law. Detectable FRET or BRET signals are restricted to a few nanometers, conveniently a very useful range for monitoring interactions between protein and ligands. Thus, by appending a donor to a protein and the acceptor to a probe or a tracer that can be displaced

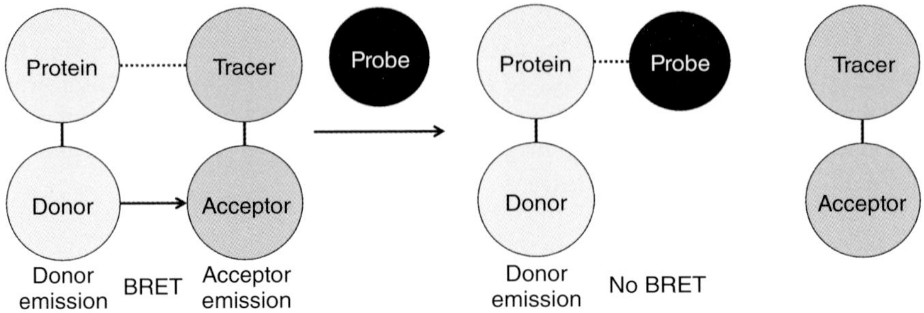

Figure 6.1 Cellular NanoBRET assay. Protein of interest is fused to light producing enzyme nanoluciferase donor and expressed in live cells. A cell-permeable tracer that binds the protein is fused to a fluorophore acceptor. Association of tracer and protein brings the donor and acceptor into proximity necessary for BRET, leading to an observable acceptor emission. Introduction of a probe for the protein induces dissociation of the tracer and protein, leading to dose-dependent loss of BRET signal.

by the probe, RET can applied in cellular target engagement assays (BRET assay; Figure 6.1). FRET requires donor excitation by an external light source. The high energy excitant light required can damage cells or lead to photobleaching, interfering with assay performance. These potential problems do not occur with BRET as the donor light originates from a chemical reaction rather than photostimulation. FRET has been employed to a greater extent than BRET due to technical issues such as a poor signal to noise ratio from the lack of availability of donors of sufficient intensity and limited dynamic range.

The discovery of nanoluciferase (Nluc) by Wood and coworkers has accelerated interest in the use of bioluminescence to monitor target engagement of epigenetic reader proteins. Nluc is an extremely bright luciferase derived from the deep-sea shrimp *Oplophorus gracilirostris* that was generated by a combination of protein engineering and a chemical optimization of the enzyme substrate to be both significantly smaller and 150-fold brighter than firefly or *Renilla* luciferases. When combined with an efficient red-shifted fluorophore acceptor, the combination of greater light intensity and wide spectral resolution gives improved detection sensitivity and dynamic range over conventional BRET technologies [26]. Also owing to the high intensity of Nluc, the protein–Nluc fusion can be expressed at moderate levels that allow for biologically relevant interpretation of data. NanoBRET assays involve real-time monitoring of interactions within the cell and are able to assess binding kinetics. The tracer molecule can be prepared by appropriate functionalization of the chemical probe of interest, if the structure–activity relationship (SAR) is sufficiently understood so that incorporation of the BRET accepting dye molecule will not significantly affect target protein binding. Alternatively, the tracer can be derived from a natural protein partner (e.g. acetylated histone peptide for bromodomains).

NanoBRET has been employed in pioneering work on bromodomains. Wood and colleagues established the applicability of the assay by demonstrating that the interaction between Nluc-BRD4 and an acceptor-tagged histone 3.3 could be competed with the BRD probe I-BET151 [26]. Bradner and coworkers used a BRD4 nanoBRET assay to prioritize compounds based on BRD4 activity and cellular penetrance [27]. Because the compounds had multiple molecular targets, this prioritization would have been challenging with other methods. NanoBRET has also been successfully applied to characterization of HMT inhibitors by providing a means to assay their effects on the interaction between enhancer of zeste homologue 2 (EZH2) and embryonic ectoderm development (EED), components of the polycomb repressive complex 2 (PRC2) [28]. Section 6.2.1.1 contains an additional example of nanoBRET characterization of epigenetic chemical probes.

6.2 Epigenetic Chemical Probes

Epigenetic targets have historically received scant attention from the research community, in large part due to the paucity of suitable reagents for their study. In particular, the increasing availability of small-molecule probes that bind to and affect the function of epigenetic reader, writer, and eraser proteins is enabling basic and biomedical research. In Sections 2.1 and 2.2 we exemplify the application of chemical probes to epigenetic targets by describing the generation and application of chemical probes targeting two different bromodomain families.

6.2.1 Histone Acetylation and Bromodomain Chemical Probes

Of the various epigenetic posttranslational modifications, lysine acetylation is one of the most well studied [29–35]. Addition of this posttranslational modification is carried out by HATs (also lysine acetyltransferases, KATs) and removed by HDACs. This epigenetic mark converts a positively charged lysine (at physiological pH) to a neutral one, significantly altering the local protein environment. The addition or removal of the acetyl group can result in dramatic changes in protein conformation and accordingly affect the protein's ability to interact with binding partners. In one example with implications for gene expression, negatively charged DNA forms nucleosomes with histone proteins that display positively charged lysine residues; upon acetylation, these electrostatic interactions are ablated, relaxing the local chromatin structure and making the DNA more accessible by transcription factors [30, 36].

Bromodomains are epigenetic readers of lysine acetylation on histones and other proteins. Interactions with the acetyl-lysine (KAc) mark occur through the bromodomain KAc recognition site and facilitate the recruitment of transcription-regulating and chromatin modeling proteins and/or complexes. The intricate nature of the bromodomain/KAc-histone interaction and the link to gene expression have expedited significant research efforts in investigating their therapeutic potential [37–39].

Of the 61 identified human bromodomains that derive from 46 bromodomain-containing proteins, 8 subfamilies have been identified (Figure 6.2) [36]. Bromodomain and extra-terminal (BET) subfamily II has received the greatest attention, with seven BET inhibitors being tested in oncology clinical trials as of 2016 [40–42].

The origins for some of these clinical candidates are found in BET bromodomain chemical probes. One of the earliest examples of a chemical probe that

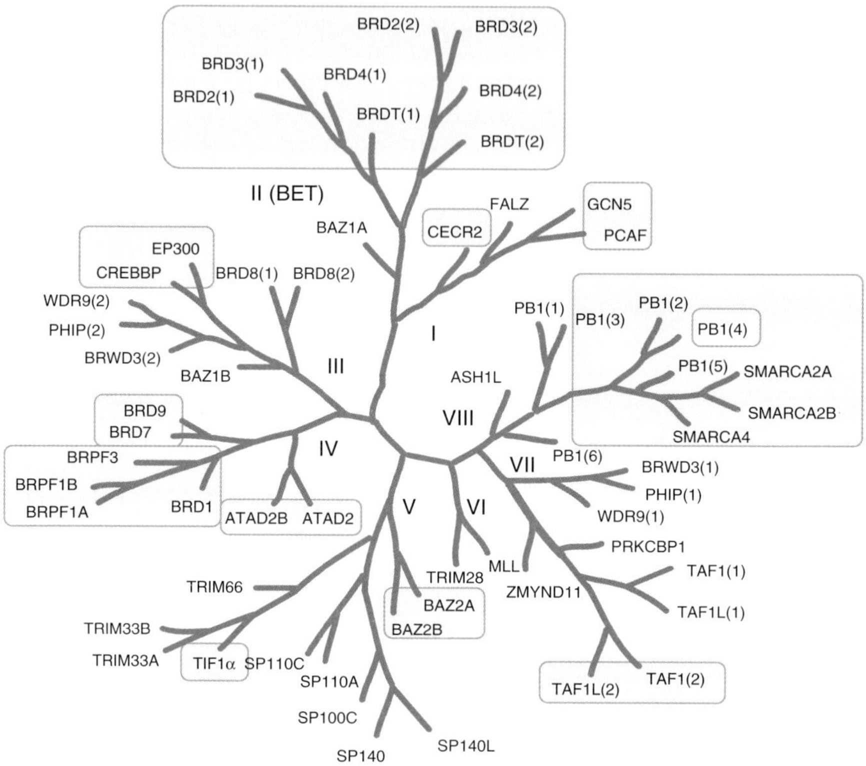

Figure 6.2 Human bromodomain phylogenetic tree. A phylogenetic tree showing the 61 bromodomains encoded by the human genome. Proteins of emphasis in this chapter are shown in branches II and III.

has witnessed significant translational efforts into drug-like compounds is the pan-BET bromodomain chemical probe **(+)-JQ1** [32].

A 2009 patent application from Mitsubishi [43] reported that compound **6** (Figure 6.3) was highly potent in displacement of acetylated histone 4 from BRD4 (IC_{50} = 18.2 nM) in a time-resolved FRET (TR-FRET) assay. The observed interaction with BRD4 corresponded to growth inhibition of chronic myeloid leukemia MV4–11 cells (GI_{50} = 26 nM). One year later, seminal work between the Bradner lab and collaborators at the Structural Genomics Consortium (SGC) described an extensively characterized close derivative of **6**, dubbed **(+)-JQ1** [32, 43]. A combination of biophysical assays (isothermal titration calorimetry (ITC) and differential scanning fluorimetry (DSF)) and co-crystal structures established and rationalized the *in vitro* potency and selectivity of **(+)-JQ1**. On-target cellular activity was demonstrated through the use of a FRAP assay in which **(+)-JQ1** was shown to accelerate fluorescence recovery at <1 μM. **(+)-JQ1** represents one of the first chemical probes [4, 44] that satisfies key criteria such as *in vitro* potency, selectivity, on-target cellular activity, and the availability of a negative control; the enantiomer **(−)-JQ1** displays no activity against BRD4. Since 2010, there have been >300 publications relating to this

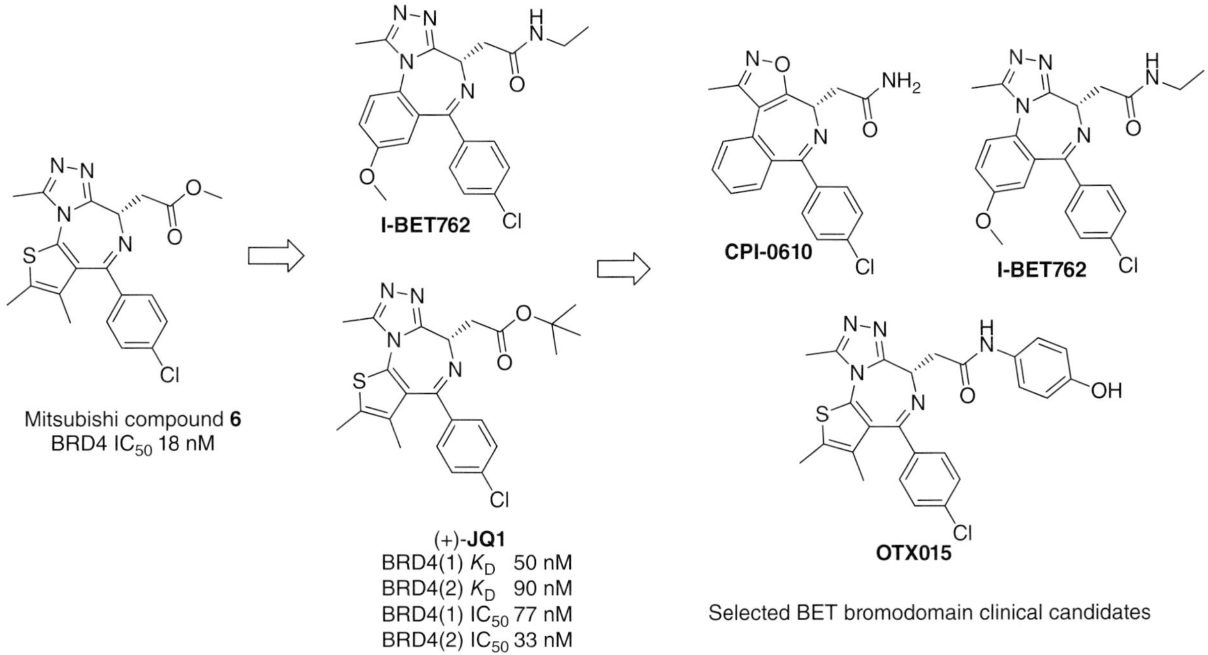

Figure 6.3 BET chemical probes and clinical candidates. The chemical structures of chemical probes and clinical candidates targeting the BET family of bromodomains (branch II from Figure 6.2) are shown in order of their development from left to right.

probe. Concomitant to the **(+)-JQ1** work, GSK published a triazolobenzodiazepine BET bromodomain chemical probe, **I-BET762** (**GSK525762A**) [45]. **I-BET762** has also been used extensively as a BET bromodomain chemical probe, with the original article receiving close to 500 citations. An array of structurally related BET bromodomain inhibitors has since appeared in the patent and academic literature, some of which have progressed toward the clinic (**CPI-0610** from Constellation, **OTX015** from OncoEthix, and **I-BET762** from GSK). More recent extensions inspired by these probes come in the form of bivalent inhibitors that have been used to engage the first and second BET bromodomains [46] and separately as bivalent BRD/E3 ligase ligands that mediate small-molecule-induced protein degradation [47].

6.2.1.1 CBP/p300 Bromodomain Chemical Probes

In the wake of the **(+)-JQ1** and **I-BET762** success stories and as key biological questions about the role BET proteins play in disease begin to be answered, researchers are searching for the next chemical probe success story. Several pathologically relevant bromodomain-containing proteins are garnering attention, including lysine acetyltransferases CBP (KAT3A) and p300 (EP300 or KAT3B). CBP and p300 have been implicated in several human diseases and disorders (e.g. Rubinstein–Taybi syndrome [48], neuropsychiatric disorders [49], inflammation [50], and hematological malignancies) [51]. Thus, there is impetus to discover small-molecule inhibitors of these highly similar bromodomains (96% sequence similarity in their bromodomains) [52].

One of the first reported chemical probes for the CBP and p300 bromodomains was described by Hay and colleagues from the SGC [53]. An early lead in the form of **1** (Figure 6.4) was optimized for potency and selectivity over BRD4, eventually leading to the discovery of **SGC-CBP30**, which shows excellent potency for both CBP and p300 bromodomains (CBP K_D = 21 nM, p300 K_D = 32 nM). **SGC-CBP30** was also shown to be selective over a wide panel of bromodomains in a DSF assay, including BRD4 (40-fold over the first bromodomain of BRD4 (BRD4(1)) [54]. In a FRAP assay [51], **SGC-CBP30** showed accelerated recovery akin to unstimulated levels at 0.1 μM, showing cellular inhibition of and target engagement with CBP. Further activity was shown in a luciferase reporter assay for p21 expression, which is driven by CBP–p53 interaction; p21 expression was decreased by **SGC-CBP30** in a dose-dependent manner (IC_{50} = 1.5 μM) [55].

Since it was first described in 2014, **SGC-CBP30** has been applied in several experiments that highlight the therapeutic potential of targeting CBP and p300. Recently, transcriptional profiling of immune cells treated with **SGC-CBP30** revealed a decrease in the production of pro-inflammatory cytokines, including IL-17A; suppression of IL-17A secretion from Th17 cElls was also observed [52]. Complex cross-talk between the multiple domains of CBP and p300 was revealed in a report describing the opposing effects of **SGC-CBP30** and **I-CBP112**, another CBP and p300 bromodomain inhibitor (Figure 6.5) [56], on HAT activity. Despite this complexity, interest in CBP or p300 inhibition to elicit functional therapeutic effects is growing. Recent reports describe **SGC-CBP30** derivatives as candidates for the treatment of cancer [57, 58].

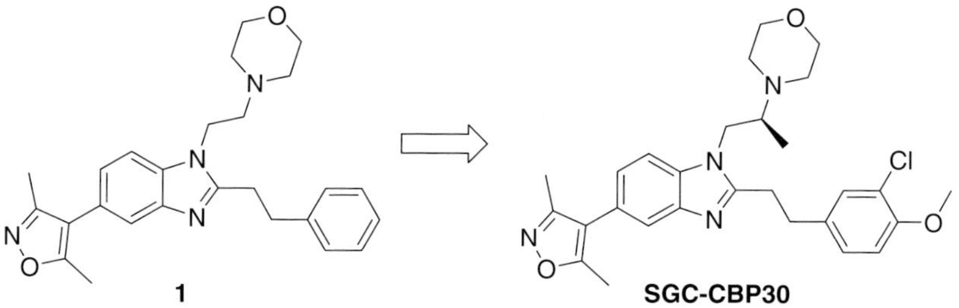

1
CBP K_D 0.32 μM
p300 K_D 0.35 μM

SGC-CBP30
CBP K_D 21 nM
p300 K_D 30 nM

Figure 6.4 **SGC-CBP30.** The chemical structures of compound **1** and **SGC-CBP30** are shown. **SGC-CBP30** is optimized to be selective for CBP and p300 over BRD4.

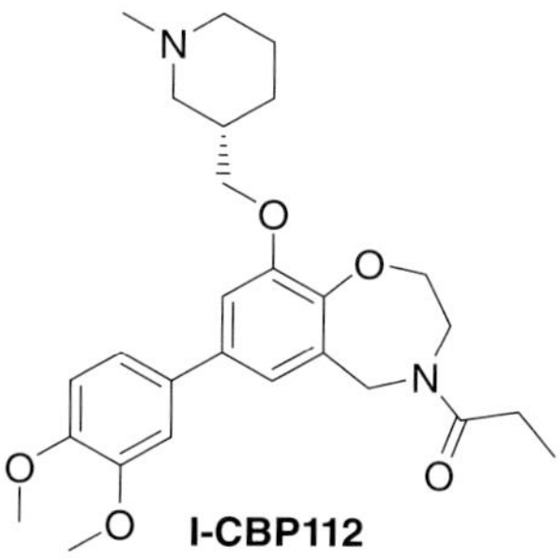

I-CBP112
CBP K_D 0.151 μM
p300 K_D 0.167 μM

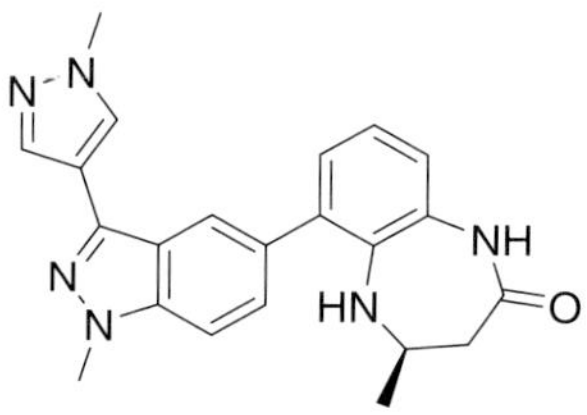

CPI-637
CBP IC_{50} 0.03 μM
p300 IC_{50} 0.051 μM
CBP BRET EC_{50} 0.3 μM

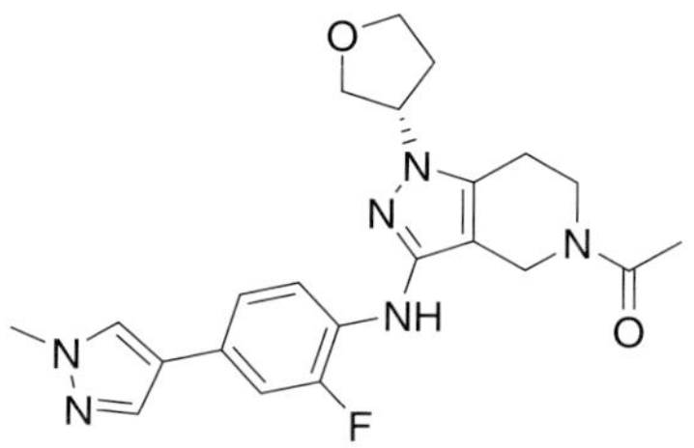

GNE-272
CBP IC_{50} 0.02 μM
p300 IC_{50} 0.03 μM
CBP BRET EC_{50} 0.41 μM

Figure 6.5 CBP, p300 chemical probes **I-CBP112, CPI-637,** and **GNE-272.** The chemical structures of **I-CBP112, CPI-637,** and **GNE-272** are shown. **GNE-272** is optimized for *in vivo* use.

I-CBP112 is an oxazepane chemical probe targeting the CBP and p300 bromodomains. **I-CBP112** shows good potency for both proteins (CBP $K_D = 0.151$ μM, p300 $K_D = 0.167$ μM), is highly selective for CBP and p300 in a panel of 41 bromodomains (by DSF), and demonstrates on-target cellular activity in a FRAP assay [56, 59]. **I-CBP112** perturbs the interaction of CBP-histone 3.3 in a dose-dependent manner ($IC_{50} = 0.6$ μM) using a nanoBRET assay [26].

The application of I-CBP112 in cells is providing additional evidence to support the therapeutic potential of targeting the bromodomains of CBP and p300. **I-CBP112** led to novel phenotypes in a DiscoverX BioMAP Diversity PLUS panel through modulation of IL-10 and VCAM1 anti-inflammatory cytokine expression levels [60]. **I-CBP112** decreased the clonogenic growth of MLL-CBP immortalized murine bone marrow cells without significantly affecting their survival. Leukemic stem cell counts were reduced when MLL-AF9$^+$ leukemic myeloblasts were treated with **I-CBP112**, and **(+)-JQ1** and **I-CBP112** both synergized with doxorubicin to induce cytotoxicity. CBP and p300 have emerged as potential targets in protein misfolding diseases, as both **I-CBP112** and **SGC-CBP30** decrease amyloid-like aggregation [61].

Coworkers at Genentech, WuXi AppTec, and Constellation Pharmaceuticals have described two additional chemical probes targeting bromodomains of CBP and p300: the benzodiazepinone-based **CPI-637** [62] and the pyrazolopiperidine-based **GNE-272** (Figure 6.5) [63].

CPI-637 was developed from fragment screen of a ~2000-membered library using DSF [54]. **CPI-637** shows excellent potency for the CBP (IC_{50} = 0.03 μM) and p300 (IC_{50} = 0.051 μM) bromodomains. Excellent selectivity (>200-fold) for both bromodomains against BRD4 was also observed, as well as selectivity over another representative seven bromodomains. Translation of biochemical potency to cellular target engagement of CBP was established in a CBP-histone 3.3 nanoBRET assay (EC_{50} = 0.3 μM), which concurrently provided evidence for cell permeability. The enantiomer of **CPI-637** shows notably weaker biochemical potency (IC_{50} = 3.4 μM) and cellular activity (BRET IC_{50} > 10 μM) and is therefore suitable for use as a negative control. **CPI-637** also inhibited *MYC* expression (EC_{50} = 0.6 μM), providing orthogonal evidence for target engagement, as CBP and p300 have been reported to regulate *MYC* [64]. Interestingly the comparably potent **SGC-CBP30** demonstrated weaker *MYC* suppression (EC_{50} = 2.7 μM), and **I-CBP112** had no detectable effect (EC_{50} > 20 μM).

These groups also developed a chemical probe suitable for use *in vivo*. **GNE-272** shows excellent potency for the CBP (IC_{50} = 0.02 μM) and p300 (IC_{50} = 0.03 μM) bromodomains [65]. Selectivity was confirmed for CBP and p300 over a representative set of bromodomains from other families with >600-fold selectivity over BRD4(1). Selectivity for CBP and p300 over BET proteins is a critical feature for these chemical probes, as it allows for clear delineation between BET-related and CBP- or p300-related phenotypes. **SGC-CBP30** [53] is also active against BRD4, necessitating that it will be used at a well-defined concentration in cellular assays to deconvolute the resulting phenotypes [64].

In cells, **GNE-272** was active and engaged CBP (CBP IC_{50} = 0.41 μM in nanoBRET assay) [26]. Further cellular activity was shown through dose-dependent modulation of *MYC* expression (EC_{50} = 0.91 μM) commensurate with CBP and p300 inhibition. Throughout this chemical probe discovery program, the PK properties of lead molecule were maintained and optimized. **GNE-272** shows good stability in mouse hepatocytes (Cl_{hep} = 6.2 ml min^{-1} kg^{-1}) and was the best compound in the series based on a balance of cell potency, selectivity, and *in vivo* PK properties. In applications, **GNE-272** showed antiproliferative effects in hematologic cancer cell lines, which held up *in vivo*;

GNE-272 modulated CBP-dependent *MYC* expression, which corresponded with antitumor activity in a *MYC*-dependent AML tumor model. **GNE-272** is the most potent, selective, and cell-active CBP and p300 bromodomain chemical probe so far. Favorable properties for both *in vitro* and *in vivo* applications support the utility of this probe for investigating CBP and p300 phenotypes in a variety of contexts.

6.2.1.2 Future Applications of Bromodomain Chemical Probes

Bromodomain research has intensified over the past decade. Several new chemical probes have been described in the scientific literature for previously untargeted bromodomains [66–68]. Various reviews provide updates on this fast-growing area of research, including chemical probe discovery programs [20, 33–35, 38, 69, 70] and their associated effects [39, 70]. The literature is almost equally populated with reviews and research articles. However, there is high demand for more primary research literature reporting new chemical probe applications. As more bromodomain chemical probes are described and made accessible to the scientific community through sources such as commercial suppliers, the SGC (http://www.thesgc.org/chemical-probes) and the Chemical Probes Portal (http://www.chemicalprobes.org), new phenotypes may be uncovered, such as those seen with BRD7/9 probes [71–74]. In some instances, bromodomain chemical probes for a given target may not recapitulate phenotypes observed after genetic modulation of the same target. In these instances, the hypothesis that bromodomain can be modulated for therapeutic purposes may be invalidated. For example, chemical probes targeting the bromodomain-containing proteins ATAD2A/B [68] and SMARCA2/4 [75] have rendered them unlikely clinical targets, at least for now. These chemical probes are still useful in ascertaining domain-specific contributions to a given pathology and assessing inter-domain "cross-talk." As for any protein target, once multiple literature reports show the potential therapeutic benefit of a small molecule targeting a bromodomain, ideally with small molecules of divergent chemotypes, these chemical probes should provide a translational starting point for drug discovery programs, as seen with **(+)-JQ1**. The success of these translational efforts depends on the drug-like properties of these chemical probes, so desirable PK properties and solubility should be emphasized in second-generation chemical probes [4, 44, 76, 77].

6.3 Summary

Chemical probes are highly valuable reagents for the study of protein targets. When they are freely available, well characterized, selected, and used properly, chemical probes can lead to a step change in our understanding of their cognate targets. The success and impact of chemical probes targeting the BET family and CBP/p300 bromodomains augurs well for current efforts to generate probes for other previously untargeted histone readers, writers, and erasers [78]. These probes are expected to enhance our understanding of these proteins and may reveal opportunities for therapeutic intervention.

References

1 Peters, J.U. (2013). Polypharmacology – foe or friend? *J. Med. Chem.* 56 (22): 8955–8971.

2 Frye, S.V. (2010). The art of the chemical probe. *Nat. Chem. Biol.* 6 (3): 159–161.

3 Workman, P. and Collins, I. (2010). Probing the probes: fitness factors for small molecule tools. *Chem. Biol.* 17 (6): 561–577.

4 Bunnage, M.E., Chekler, E.L., and Jones, L.H. (2013). Target validation using chemical probes. *Nat. Chem. Biol.* 9 (4): 195–199.

5 Huber, K.V., Olek, K.M., Muller, A.C. et al. (2015). Proteome-wide drug and metabolite interaction mapping by thermal-stability profiling. *Nat. Methods* 12 (11): 1055–1057.

6 Martinez Molina, D., Jafari, R., Ignatushchenko, M. et al. (2013). Monitoring drug target engagement in cells and tissues using the cellular thermal shift assay. *Science* 341 (6141): 84–87.

7 Schulze, J., Moosmayer, D., Weiske, J. et al. (2015). Cell-based protein stabilization assays for the detection of interactions between small-molecule inhibitors and BRD4. *J. Biomol. Screening* 20 (2): 180–189.

8 Simon, G.M., Niphakis, M.J., and Cravatt, B.F. (2013). Determining target engagement in living systems. *Nat. Chem. Biol.* 9 (4): 200–205.

9 Sun, Y., Hays, N.M., Periasamy, A. et al. (2012). Monitoring protein interactions in living cells with fluorescence lifetime imaging microscopy. *Methods Enzymol.* 504: 371–391.

10 Tsukiji, S. and Hamachi, I. (2014). Ligand-directed tosyl chemistry for in situ native protein labeling and engineering in living systems: from basic properties to applications. *Curr. Opin. Chem. Biol.* 21: 136–143.

11 Schurmann, M., Janning, P., Ziegler, S., and Waldmann, H. (2016). Small-molecule target engagement in cells. *Cell Chem. Biol.* 23 (4): 435–441.

12 Day, C.A., Kraft, L.J., Kang, M., and Kenworthy, A.K. (2012). Analysis of protein and lipid dynamics using confocal fluorescence recovery after photobleaching (FRAP). *Curr. Protoc. Cytom.* 62: 2.19.1–2.19.29.

13 Deschout, H., Raemdonck, K., Demeester, J. et al. (2014). FRAP in pharmaceutical research: practical guidelines and applications in drug delivery. *Pharm. Res.* 31 (2): 255–270.

14 Sprague, B.L. and McNally, J.G. (2005). FRAP analysis of binding: proper and fitting. *Trends Cell Biol.* 15 (2): 84–91.

15 Philpott, M., Rogers, C.M., Yapp, C. et al. (2014). Assessing cellular efficacy of bromodomain inhibitors using fluorescence recovery after photobleaching. *Epigenetics Chromatin* 7: 14.

16 Weigt, D., Hopf, C., and Medard, G. (2016). Studying epigenetic complexes and their inhibitors with the proteomics toolbox. *Clin. Epigenetics* 8: 76.

17 Bamborough, P., Chung, C.W., Furze, R.C. et al. (2015). Structure-based optimization of naphthyridones into potent ATAD2 bromodomain inhibitors. *J. Med. Chem.* 58 (15): 6151–6178.

18 Bantscheff, M., Eberhard, D., Abraham, Y. et al. (2007). Quantitative chemical proteomics reveals mechanisms of action of clinical ABL kinase inhibitors. *Nat. Biotechnol.* 25 (9): 1035–1044.

19 Duncan, J.S., Whittle, M.C., Nakamura, K. et al. (2012). Dynamic reprogramming of the kinome in response to targeted MEK inhibition in triple-negative breast cancer. *Cell* 149 (2): 307–321.

20 Galdeano, C. and Ciulli, A. (2016). Selectivity on-target of bromodomain chemical probes by structure-guided medicinal chemistry and chemical biology. *Future Med. Chem.* 8 (13): 1655–1680.

21 Kranz, J.K. and Schalk-Hihi, C. (2011). Protein thermal shifts to identify low molecular weight fragments. *Methods Enzymol.* 493: 277–298.

22 Savitski, M.M., Reinhard, F.B., Franken, H. et al. (2014). Tracking cancer drugs in living cells by thermal profiling of the proteome. *Science* 346 (6205): 1255784.

23 Wagner, T., Greschik, H., Burgahn, T. et al. (2016). Identification of a small-molecule ligand of the epigenetic reader protein Spindlin1 via a versatile screening platform. *Nucleic Acids Res.* 44 (9): e88.

24 Franken, H., Mathieson, T., Childs, D. et al. (2015). Thermal proteome profiling for unbiased identification of direct and indirect drug targets using multiplexed quantitative mass spectrometry. *Nat. Protoc.* 10 (10): 1567–1593.

25 Schuetze, K.B., Stratton, M.S., Blakeslee, W.W. et al. (2017). Overlapping and divergent actions of structurally distinct histone deacetylase inhibitors in cardiac fibroblasts. *J Pharmacol. Exp. Ther.* 361 (1): 140–150.

26 Machleidt, T., Woodroofe, C.C., Schwinn, M.K. et al. (2015). NanoBRET – a novel BRET platform for the analysis of protein-protein interactions. *ACS Chem. Biol.* 10 (8): 1797–1804.

27 Koblan, L.W., Buckley, D.L., Ott, C.J. et al. Assessment of bromodomain target engagement by a series of BI2536 analogues with miniaturized BET-BRET. *ChemMedChem* 2016 (23): 11, 2575–2581.

28 Bradley, W.D., Arora, S., Busby, J. et al. (2014). EZH2 inhibitor efficacy in non-Hodgkin's lymphoma does not require suppression of H3K27 monomethylation. *Chem. Biol.* 21 (11): 1463–1475.

29 Dawson, M.A., Kouzarides, T., and Huntly, B.J.P. (2012). Targeting epigenetic readers in cancer. *N. Engl. J. Med.* 367: 647-657.

30 Arrowsmith, C.H., Bountra, C., Fish, P.V. et al. (2012). Epigenetic protein families: a new frontier for drug discovery. *Nat. Rev. Drug Discovery* 11: 384–400.

31 Filippakopoulos, P. and Knapp, S. (2014). Targeting bromodomains: epigenetic readers of lysine acetylation. *Nat. Rev. Drug Discovery* 13 (5): 337–356.

32 Filippakopoulos, P., Qi, J., Picaud, S. et al. (2010). Selective inhibition of BET bromodomains. *Nature* 468: 1067–1073.

33 Romero, F.A., Taylor, A.M., Crawford, T.D. et al. (2016). Disrupting acetyl-lysine recognition: progress in the development of bromodomain inhibitors. *J. Med. Chem.* 59: 1271–1298.

34 Moustakim, M., Clark, P.G.K., Hay, D.A. et al. (2016). Chemical probes and inhibitors of bromodomains outside the BET family. *Med. Chem. Commun.* 7 (12): 2246–2264.

35 Theodoulou, N.H., Tomkinson, N.C.O., Prinjha, R.K., and Humphreys, P.G. (2016). Progress in the development of non-BET bromodomain chemical probes. *ChemMedChem* 11: 477–487.
36 Filippakopoulos, P., Picaud, S., Mangos, M. et al. (2012). Histone recognition and large-scale structural analysis of the human bromodomain family. *Cell* 149 (1): 214–231.
37 Muller, S., Filippakopoulos, P., and Knapp, S. (2011). Bromodomains as therapeutic targets. *Expert Rev. Mol. Med.* 13: e29.
38 Smith, S.G. and Zhou, M.-M. (2016). The bromodomain: a new target in emerging epigenetic medicine. *ACS Chem. Biol.* 11: 598–608.
39 Theodoulou, N.H., Tomkinson, N.C.O., Prinjha, R.K., and Humphreys, P.G. (2016). Clinical progress and pharmacology of small molecule bromodomain inhibitors. *Curr. Opin. Chem. Biol.* 33: 58–66.
40 Garnier, J.M., Sharp, P.P., and Burns, C.J. (2014). BET bromodomain inhibitors: a patent review. *Expert. Opin. Ther. Pat.* 24 (2): 185–199.
41 Owen, D. (2016). Drug discovery: doubling down on BET inhibition. *Nat. Chem. Biol.* 12: 991–992.
42 Andrieu, G., Belkina, A.C., and Denis, G.V. (2016). Clinical trials for BET inhibitors run ahead of the science. *Drug Discovery Today Technol.* 19: 45–50.
43 Miyoshi, S., Ooike, S., Iwata, K. et al. (2009). Antitumor agent. WO2009084693A1.
44 Arrowsmith, C.H., Audia, J.E., Austin, C. et al. (2015). The promise and peril of chemical probes. *Nat. Chem. Biol.* 11 (8): 536–541.
45 Nicodeme, E., Jeffrey, K.L., Schaefer, U. et al. (2010). Suppression of inflammation by a synthetic histone mimic. *Nature* 468 (7327): 1119–1123.
46 Tanaka, M., Roberts, J.M., Seo, H.-S. et al. (2016). Design and characterization of bivalent BET inhibitors. *Nat. Chem. Biol.* 12 (12): 1089–1096.
47 Winter, G.E., Buckley, D.L., Paulk, J. et al. (2015). Phthalimide conjugation as a strategy for in vivo target protein degradation. *Science* 348: 1376–1381.
48 Hennekam, R.C.M. (2006). Rubinstein-Taybi syndrome. *Eur. J. Hum. Genet.* 14: 981–985.
49 Valor, L.M., Viosca, J., Lopez-Atalaya, J.P., and Barco, A. (2013). Lysine Acetyltransferases CBP and p300 as therapeutic targets in cognitive and neurodegenerative disorders. *Curr. Pharm. Des.* 19: 5051–5064.
50 Revilla, Y. and Granja, A.G. (2009). Viral mechanisms involved in the transcriptional CBP/p300 regulation of inflammatory and immune responses. *Crit. Rev. Immunol.* 29: 131–154.
51 Wang, F., Marshall, C.B., and Ikura, M. (2013). Transcriptional/epigenetic regulator CBP/p300 in tumorigenesis: structural and functional versatility in target recognition. *Cell. Mol. Life Sci.* 70: 3989–4008.
52 Hammitzsch, A., Tallant, C., Fedorov, O. et al. (2015). CBP30, a selective CBP/p300 bromodomain inhibitor, suppresses human Th17 responses. *Proc. Natl. Acad. Sci. U.S.A.* 112 (34): 10768–10773.
53 Hay, D.A., Fedorov, O., Martin, S. et al. (2014). Discovery and optimization of small-molecule ligands for the CBP/p300 bromodomains. *J. Am. Chem. Soc.* 136 (26): 9308–9319.

54 Pantoliano, M.W., Petrella, E.C., Kwasnoski, J.D. et al. (2001). High-density miniaturized thermal shift assays as a general strategy for drug discovery. *J. Biomol. Screening* 6 (6): 429–440.

55 Borah, J.C., Mujtaba, S., Karakikes, I. et al. (2011). A small molecule binding to the coactivator CREB-binding protein blocks apoptosis in cardiomyocytes. *Chem. Biol.* 18 (4): 531–541.

56 Picaud, S., Fedorov, O., Thanasopoulou, A. et al. (2015). Generation of a selective small molecule inhibitor of the CBP/p300 bromodomain for leukemia therapy. *Cancer Res.* 75 (23): 5106–5119.

57 Pegg, N.A., Taddei, D.M.A., and Brown, R. (2016). Isoxazolyl-substituted imidazopyridines as p300 and/or CBP modulators and their preparation. WO2016170323A1.

58 Pegg, N.A., Taddei, D.M.A., Onions, S.T. et al. (2016). Isoxazolyl-substituted benzimidazoles as p300 and/or CBP modulators and their preparation. WO2016170324A1.

59 Popp, T.A., Tallant, C., Rogers, C. et al. (2016). Development of selective CBP/P300 benzoxazepine bromodomain inhibitors. *J. Med. Chem.* 59 (19): 8889–8912.

60 Berg, E.L., Yang, J., and Polokoff, M.A. (2013). Building predictive models for mechanism-of-action classification from phenotypic assay data sets. *J. Biomol. Screening* 18: 1260–1269.

61 Olzscha, H., Fedorov, O., Kessler, B.M. et al. (2017). CBP/p300 bromodomains regulate amyloid-like protein aggregation upon aberrant lysine acetylation. *Cell Chem. Biol.* 24: 9–23.

62 Taylor, A.M., Côté, A., Hewitt, M.C. et al. (2016). Fragment-based discovery of a selective and cell-active Benzodiazepinone CBP/EP300 bromodomain inhibitor (CPI-637). *ACS Med. Chem. Lett.* 7 (5): 531–536.

63 Crawford, T.D., Romero, F.A., Lai, K.W. et al. (2016). Discovery of a potent and selective in vivo probe (GNE-272) for the bromodomains of CBP/EP300. *J. Med. Chem.* 59: 10549–10563.

64 Conery, A.R., Centore, R.C., Neiss, A. et al. (2016). Bromodomain inhibition of the transcriptional coactivators CBP/EP300 as a therapeutic strategy to target the IRF4 network in multiple myeloma. *eLife* 5: e10483.

65 Beresini, M.H., Liu, Y., Dawes, T.D. et al. (2014). Small-molecule library subset screening as an aid for accelerating lead identification. *J. Biomol. Screening* 19: 758–770.

66 Moustakim, M., Clark, P.G.K., Trulli, L. et al. (2017). Discovery of a PCAF bromodomain chemical probe. *Angew. Chem. Int. Ed.* 56 (3): 827–831.

67 Humphreys, P.G., Bamborough, P., Chung, C.-w. et al. (2017). Discovery of a potent, cell penetrant, and selective p300/CBP-associated factor (PCAF)/general control nonderepressible 5 (GCN5) bromodomain chemical probe. *J. Med. Chem.* 60 (2): 695–709.

68 Bamborough, P., Chung, C.w., Demont, E.H. et al. (2016). A chemical probe for the ATAD2 bromodomain. *Angew. Chem. Int. Ed.* 55 (38): 11382–11386.

69 Zhang, G., Smith, S.G., and Zhou, M.-M. (2015). Discovery of chemical inhibitors of human bromodomains. *Chem. Rev.* 115: 11625–11668.

70 Ackloo, S., Brown, P.J., and Muller, S. (2017, (Copyright (C) 2017 U.S. National Library of Medicine.)). Chemical probes targeting epigenetic proteins: applications beyond oncology. *Epigenetics* doi: 10.1080/15592294.2017.1279371.

71 Clark, P.G.K., Vieira, L.C.C., Tallant, C. et al. (2015). LP99: discovery and synthesis of the first selective BRD7/9 bromodomain inhibitor. *Angew. Chem. Int. Ed.* 54 (21): 6217–2621.

72 Theodoulou, N.H., Bamborough, P., Bannister, A.J. et al. (2015). Discovery of I-BRD9, a selective cell active chemical probe for bromodomain containing protein 9 inhibition. *J. Med. Chem.* 59 (4): 1425–1439.

73 Martin, L.J., Koegl, M., Bader, G. et al. (2016). Structure-based design of an in vivo active selective BRD9 inhibitor. *J. Med. Chem.* 59 (10): 4462–4475.

74 Karim, R.M. and Schönbrunn, E. (2016). An advanced tool to interrogate BRD9. *J. Med. Chem.* 59 (10): 4459–4461.

75 Vangamudi, B., Paul, T.A., Shah, P.K. et al. (2015). The SMARCA2/4 ATPase domain surpasses the bromodomain as a drug target in SWI/SNF-mutant cancers: insights from cDNA rescue and PFI-3 inhibitor studies. *Cancer Res.* 75 (18): 3865–3878.

76 Vedadi, M., Barsyte-Lovejoy, D., Liu, F. et al. (2011). A chemical probe selectively inhibits G9a and GLP methyltransferase activity in cells. *Nat. Chem. Biol.* 7 (8): 566–574.

77 Liu, F., Barsyte-Lovejoy, D., Li, F. et al. (2013). Discovery of an in vivo chemical probe of the lysine methyltransferases G9a and GLP. *J. Med. Chem.* 56 (21): 8931–8942.

78 Huston, A., Arrowsmith, C.H., Knapp, S., and Schapira, M. (2015). Probing the epigenome. *Nat. Chem. Biol.* 11 (8): 542–545.

Part III

Epigenetic Target Classes

7

Inhibitors of the Zinc-Dependent Histone Deacetylases

Helle M. E. Kristensen, Andreas S. Madsen, and Christian A. Olsen

University of Copenhagen, Center for Biopharmaceuticals & Department of Drug Design and Pharmacology, Faculty of Health and Medical Sciences, Universitetsparken 2, 2100, Copenhagen, Denmark

7.1 Introduction: Histone Deacetylases

Posttranslational acetylation of lysine residues on histone proteins was first described half a century ago, in 1964, by Allfrey et al. [1]. Interestingly, discoveries in previous years had highlighted the inhibitory effect of histone proteins on the transcription of DNA, and Allfrey et al. suggested in their communication that histone acetylation could be an underlying control mechanism to this process [1]. Later, this prediction was confirmed, and it is now acknowledged that transcription of genes is partly regulated by posttranslational modifications on histones with acetylation being a highly abundant example of these modifications [2]. Acetylation patterns on histones are regulated by synergy between a group of epigenetic writers (histone acetyltransferase (HAT) enzymes that introduce the modification) and a group of epigenetic erasers (histone deacetylases (HDACs) that remove the modification) [3]. Although acetylation of lysine residues was first discovered on histone proteins, it is now clear that it is a posttranslational modification for thousands of proteins [4, 5]. It has furthermore become evident that HDACs also play a role in the regulation of the acetylation state of a yet unknown range of non-histone proteins [5]. These discoveries have opened up for association of HDACs with numerous physiological and pathological pathways, with aberrant gene expression in cancer as the best studied example [6]. To date, six different HDAC-targeting drugs have been approved, and though many questions about the biological functions of HDACs are still to be fully illuminated, this family of enzymes has been established as a validated drug target.

The human HDAC family consists of 18 different enzymes, 11 zinc-dependent HDACs, and seven nicotinamide adenine dinucleotide (NAD^+)-dependent enzymes named sirtuins (Table 7.1) [3]. The sirtuins differ from the zinc-dependent enzymes in the catalytic mechanism and will be covered elsewhere in this book.

The zinc-dependent enzymes are divided into four classes based on phylogeny and sequence homology to the yeast HDACs Hda1 and Rpd3 [3]. Class I

Epigenetic Drug Discovery, First Edition. Edited by Wolfgang Sippl and Manfred Jung.

Table 7.1 HDAC subclasses.

Classification	Enzyme	Cofactor
Class I	HDAC1, HDAC2, HDAC3, HDAC8	Zn^{2+}
Class IIa	HDAC4, HDAC5, HDAC7, HDAC9	Zn^{2+}
Class IIb	HDAC6, HDAC10	Zn^{2+}
Class III	SIRT1–7	NAD^{+}
Class IV	HDAC11	Zn^{2+}

constitutes HDAC1, HDAC2, HDAC3, and HDAC8; class IIa includes HDAC4, HDAC5, HDAC7, and HDAC9; and class IIb consists of HDAC6 and HDAC10. Finally, class IV includes HDAC11 as the sole member. The overall structure of the binding domain is conserved between the isozymes and consists of a hydrophobic channel that leads to the active site, where two aspartate residues and a histidine residue coordinate the catalytic Zn^{2+} ion. The enzymatic mechanism has been proposed based on X-ray diffraction studies of HDAC8 and is presented in Figure 7.1 [7].

Despite the overall similarity of the binding domains of HDACs, isozymes of different classes exhibit differences in structure, substrate specificity, and subcellular localization. Class I and IV enzymes are mainly located in the nucleus, class IIb enzymes are mainly located in the cytoplasm, whereas class IIa enzymes are able to shuttle between the nucleus and cytoplasm [3]. Most HDACs are furthermore found in multiprotein complexes in the cell, the exception being HDACs 8 and 10, where no complex associations have been observed. The class I HDACs 1–3 exert their catalytic effect in complexes, such as NuRD, Sin3, MiDAC, and CoREST, where they bind to transcriptional corepressors, other HDACs, HATs, and cofactors [8]. Class IIa enzymes are characterized by extended N-terminal domains with distinct sites for the regulation of nucleocytoplasmic shuttling, recruitment of transcription factors, and binding to other HDACs and corepressors to form multiprotein complexes [9]. The specific function of class IIa HDACs in the multiprotein complexes is still debated because these enzymes show very weak deacetylase activity *in vitro*, due to a loss-of-function tyrosine-to-histidine mutation in the active site [10]. Their specific function in the multiprotein complexes is therefore unknown, though they have been suggested to act as a substrate recognition sites similar to bromodomains [11]. The class IIb enzyme HDAC6 is unique since it contains two distinct binding domains, which are both able to bind inhibitors [12, 13]. It deacetylates several non-histone proteins counting α-tubulin, heat shock protein 90 (HSP90), and Tau, the being latter involved in the pathology of Parkinson's and Alzheimer's diseases [13–15]. In contrast to HDAC6, little is known about the natural substrates of the other class IIb enzyme, HDAC10, and the sole member of class IV, HDAC11.[1]

1 This has changed since submission of the chapter, and indeed reports now exist regarding both HDAC10 and 11.

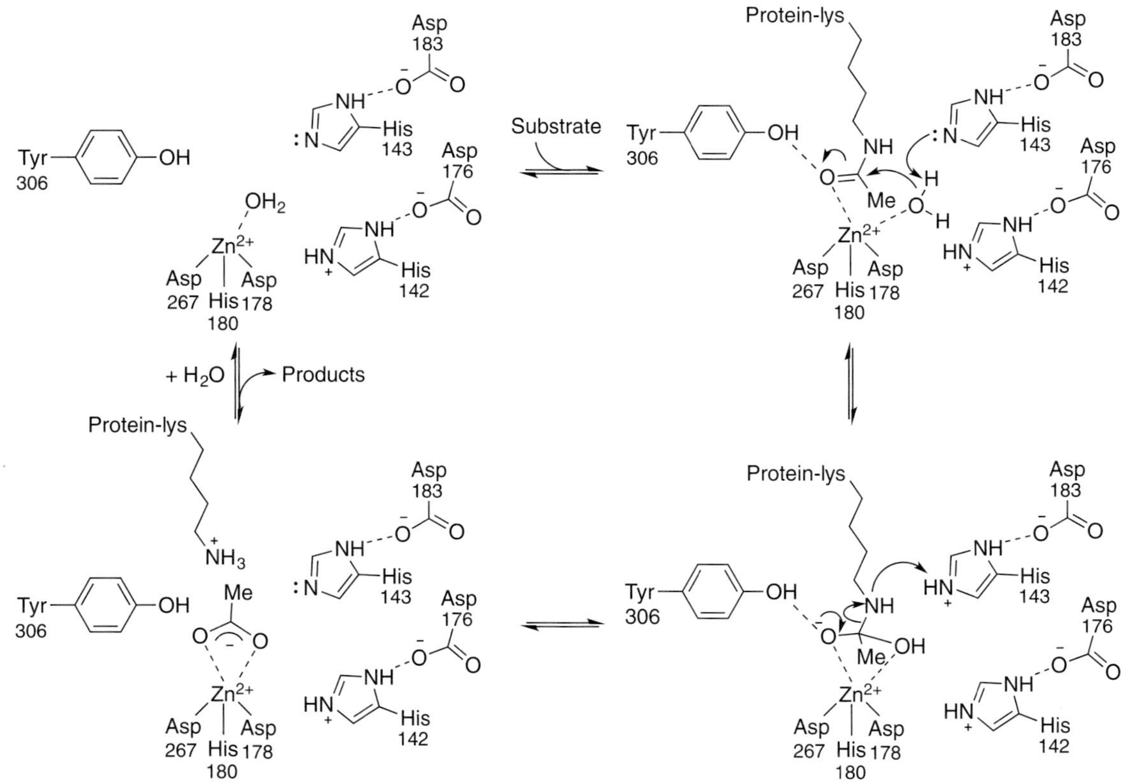

Figure 7.1 Catalytic mechanism of HDAC8, as proposed by Gantt et al. [7]. During the event of HDAC-mediated deacetylation, the acetylated lysine substrate spans the 11 Å hydrophobic channel, and the acetyl carbonyl coordinates to a tyrosine and a Zn^{2+} ion at the bottom of the catalytic site. The Zn^{2+} ion coordinates to a water molecule that is then deprotonated by a His143–Asp183 dyad and activated for a nucleophile attack on the acetyl carbonyl group. The substrate then forms a hydrated intermediate, stabilized by coordination to Tyr306 and bidentate coordination to the Zn^{2+} ion. Subsequently, the His143–Asp183 dyad functions as a general acid, protonating the amine during collapse of the hydrated intermediate, resulting in the deacetylated amine product and an acetate molecule.

7.2 Histone Deacetylase Inhibitors

HDAC inhibitors played an important role in the discovery and characterization of the first mammalian HDAC enzymes. Prior to structural knowledge of the enzymes, HDAC inhibitors indicated their presence and physiological role by inducing cell cycle arrest and differentiation of cancer cells in correlation with hyperacetylation of histones [16]. In 1996, trapoxin B (**2**; Figure 7.2) was furthermore modified and utilized in an affinity matrix to isolate HDAC1 and HDAC2 for the first time [17].

Before achieving structural insight regarding the HDAC proteins, Jung et al. suggested a general pharmacophore model [18], which is still valid for the vast majority of developed inhibitors. Based on the structural features of trichostatin A (TSA) (**1**) and trapoxin B (**2**), they reasoned that an inhibitor should include a "binding region" or "cap," a spacer/linker, and an "enzyme-inhibiting" group [18]. The enzyme-inhibiting group was later renamed to a zinc-binding group, when it was found that the catalytic domain of HDACs contains a Zn^{2+} ion [19]. The classical pharmacophore model is confirmed by X-ray structures of inhibitors bound to their target in a substrate-mimicking manner (Figure 7.2) [19]. The capping group of the inhibitor interacts with the rim of the enzyme, and the linker projects a zinc-binding group through the channel to the Zn^{2+} ion in the catalytic site.

7.2.1 Types of Inhibitors

Based on the pharmacophore model, inhibitors are often characterized on a structural basis with emphasis on the type of zinc-binding group. The majority can thus be divided into six main groups: (i) hydroxamic acids, (ii) macrocyclic

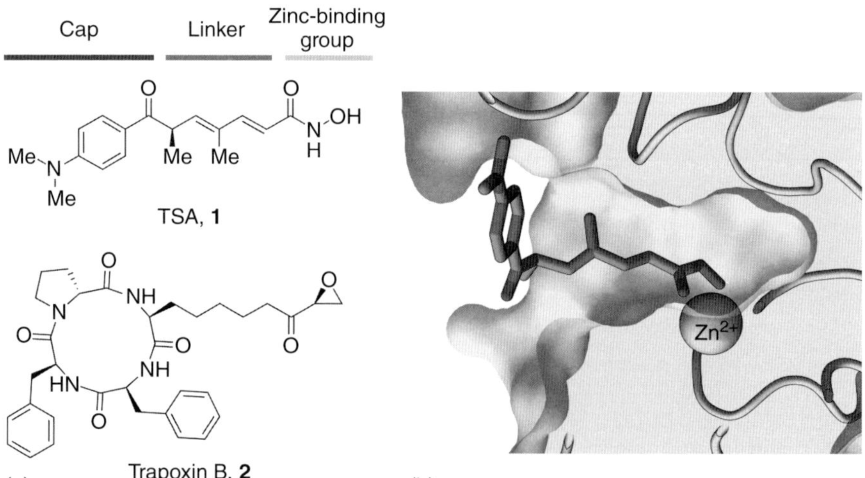

(a) (b)

Figure 7.2 Common pharmacophore model. (a) HDAC inhibitors exemplified by trichostatin A (TSA) (**1**) and trapoxin B (**2**). (b) Crystal structure of TSA–HDAC8 complex showing binding in the active site pocket and bidentate zinc chelation (PDB ID: 1T64).

compounds, (iii) sulfur-containing compounds, (iv) short-chain fatty acids, (v) *ortho*-aminoanilides, and (vi) electrophilic ketones.

Inhibitors containing a hydroxamic acid moiety, such as TSA (**1**) (Figure 7.3), bind to the active site Zn^{2+} ion by forming a favorable bidentate chelate with the hydroxamate group; therefore they often exhibit potent nanomolar inhibition and are well represented among the medically relevant HDAC inhibitors. The inhibitors generally show either nonselective or pan-inhibition across HDAC isozymes, but more selective inhibitors have also been obtained as result of careful design of the linker and surface recognizing capping group [20–22]. The second group is constituted of inhibitors with macrocyclic capping groups that either consists of non-ribosomal cyclic tetrapeptides, like trapoxin B (**2**; Figure 7.3), disulfide-containing bicyclic depsipeptides, or the thioester-containing compound largazole (**3**; Figure 7.3). The thioester is cleaved in the cytoplasm of cells *in vitro* as well as under physiological conditions,

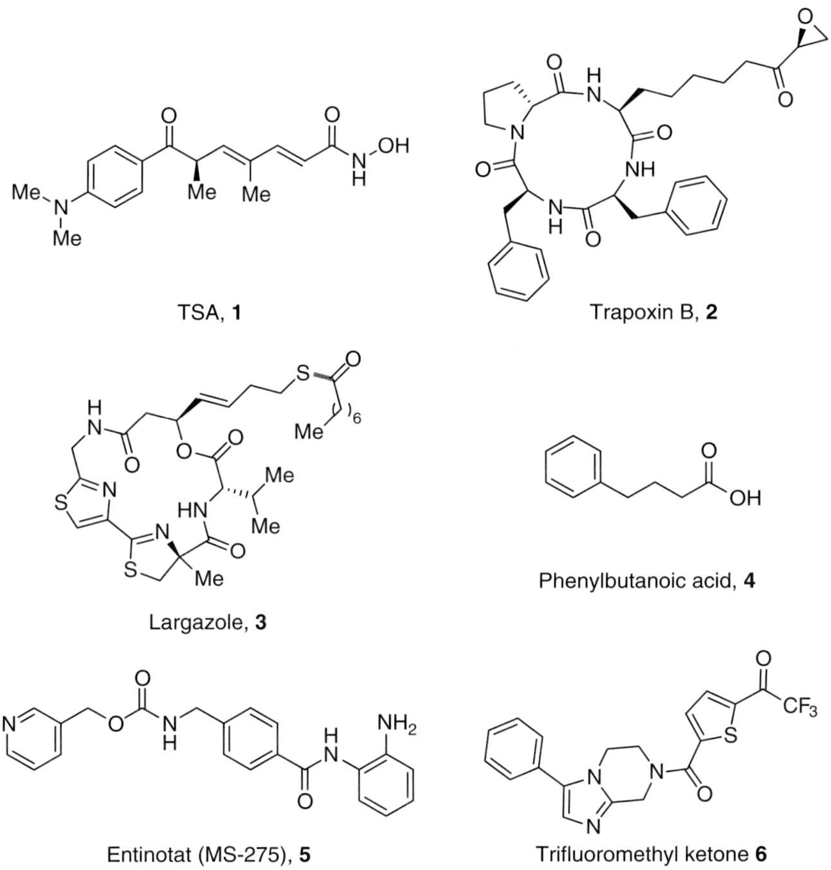

Figure 7.3 Inhibitors representative for different groups of HDAC inhibitors. Hydroxamic acids, macrocyclic compounds, sulfur-containing compounds, short-chain fatty acids, *ortho*-aminoanilides, and electrophilic ketones are all represented. Source: Frey et al. 2002 [27]. Reproduced with permission of Elsevier.

revealing the potent thiol zinc-binding group, and largazole is therefore also an example of a sulfur-containing inhibitor [23]. Many naturally occurring macrocyclic HDAC inhibitors are potent class I inhibitors with IC_{50} values in the nanomolar range and have been used as starting points for the development of both cyclic and linear synthetic inhibitors [23]. The group of short-chain fatty acids is structurally the simplest, exemplified by phenylbutanoic acid (**4**; Figure 7.3). They are weak inhibitors with IC_{50} values in the micromolar range [24]. The fifth group of inhibitors contains *ortho*-aminoanilides as zinc-binding group, represented by entinostat (**5**; Figure 7.3). Inhibition of HDAC1–3 in the nanomolar range is often observed within this group, while only weak inhibition of the remaining isozymes is obtained [25, 26]. The final group of electrophilic ketones contains electron-withdrawing groups in the α-position, such as the trifluoromethyl ketone **6** (Figure 7.3) [27].

7.2.2 HDAC Inhibitors in Clinical Use and Development

HDAC inhibitors have proven to induce apoptosis, cell cycle arrest, and tumor cell differentiation in cancers [6]. In addition to this, they have shown promise for the treatment of neurodegenerative diseases and immune disorders *in vitro* and *in vivo* [6].

During the last decade, focused studies have resulted in five approved drugs for the treatment of cancer (Figure 7.4). The US Food and Drug Administration (FDA) approved the archetypical HDAC inhibitor vorinostat (**7**) as the first for the treatment of cutaneous T-cell lymphoma. It was followed by FDA approval of romidepsin (**8**), panobinostat (**9**), and belinostat (**10**) for the treatment of hematological malignancies. More recently, chidamide (**11**) was approved by Chinese authorities, also for the treatment of peripheral T-cell lymphoma.

Valproate (**12**) is a sixth approved HDAC inhibitor. It has been utilized in the clinic since the 1960s for the treatment of epilepsy and bipolar disorder, decades before the discovery of its HDAC inhibitory effect. Valproate is a multifaceted drug. Its characteristic as an HDAC inhibitor have not been unambiguously linked to the approved indications, and it remains unknown to which degree the HDAC inhibitory effect contributes to its pharmacological effect in epilepsy and bipolar disorder [28].

HDAC inhibitors also show potential in combination therapy of HIV by reactivating latent HIV from resting $CD4^+$ cells that are usually untouched by traditional antiretroviral therapies [29]. Vorinostat (**7**), romidepsin (**8**), and panobinostat (**9**) [30] have entered clinical trials for the treatment of HIV in combination with antiviral therapy, due to their ability to induce activity in latently infected $CD4^+$ T cells in HIV [29].

In addition to the HDAC inhibitors approved for medical treatment of hematological cancers, several are in clinical trials for the treatment of a variety of conditions and often in combination therapy (Table 7.2). These inhibitors are generally nonselective or selective toward HDAC1–3, with the exception of the HDAC6 selective inhibitor, rocilinostat (**20**). Cancers, especially hematologic cancers, dominate the indications tested in clinical trials. There are however noncancerous diseases represented in the pipeline of therapeutic candidates.

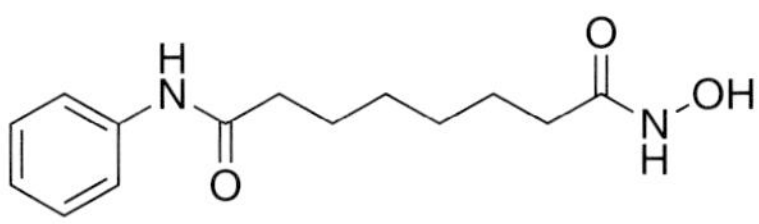

7
Trade name: Zolinza®
Cutaneous T-cell lymphoma
(a)

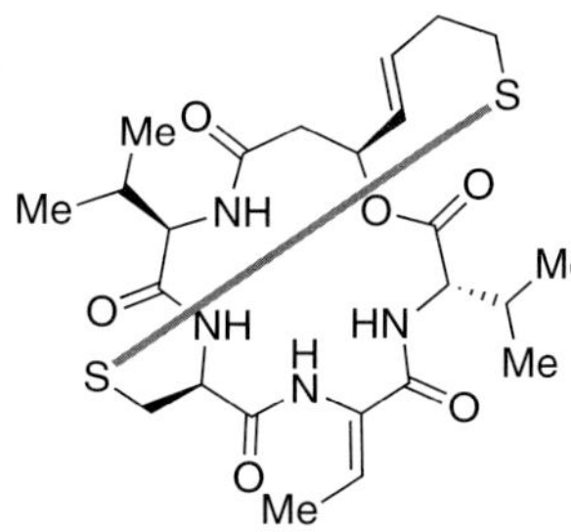

8
Trade name: Istodax®
Periphereal and cutaneous T-cell lymphoma
(b)

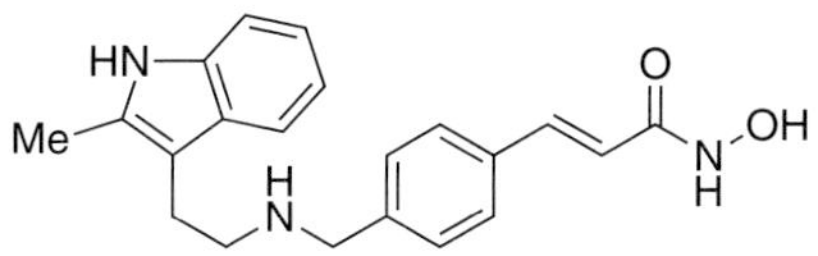

9
Trade name: Farydak®
Multiple myeloma
(c)

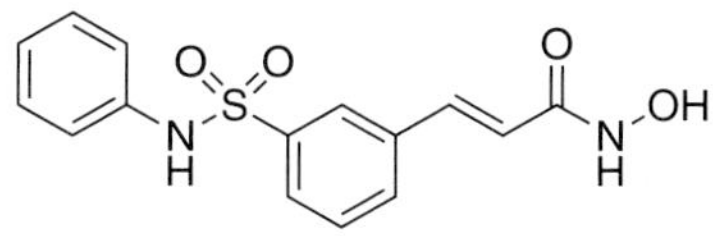

10
Trade name: Beleodaq®
Peripheral T-cell lymphoma
(d)

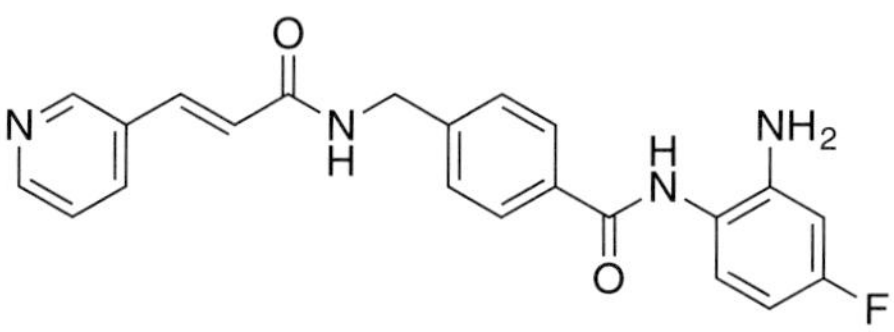

11
Trade name: Epidaza®
Peripheral T-cell lymphoma
(e)

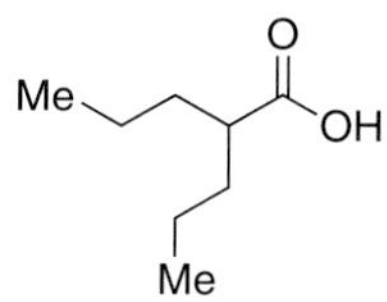

12
Bipolar disorder and epilepsy
(Not approved based on its HDAC inhibition)
(f)

Figure 7.4 Approved HDAC inhibitors. (a) Vorinostat (**7**), (b) romidepsin (**8**), (c) panobinostat (**9**), (d) belinostat (**10**), (e) chidamide (**11**), and (f) valproic acid (**12**) are all approved for medical treatment of various diseases, with strong representation of hematological cancers.

This is exemplified by the hydroxamic acid givinostat (**13**), which is being tested for the treatment of juvenile idiopathic arthritis in phase II and Duchenne muscular dystrophy in phase III. Despite the focus on the treatment of neurodegenerative diseases, only one HDAC inhibitor is currently in clinical trials within this area. FRM-0334 has proceeded to phase II in the treatment of a specific form of dementia with granulin mutation. Finally, remetinostat (**24**) is in clinical development for topical treatment of both cutaneous T-cell lymphoma and alopecia areata, which is often referred to as spot baldness. Remetinostat is a soft drug for topical use and is inactivated if transferred to the bloodstream by esterases to yield an inactive metabolite [31] (Table 7.2).

Two other inhibitors, CUDC-907 (**21**) and CUDC-101 (**28**), show dual- and multifunctionality, respectively, to target different pathological pathways in cancers simultaneously. They represent a new approach in chemotherapy that shows promise of more efficient treatments of cancer with lower degrees of adverse

Table 7.2 Ongoing and completed clinical trials of non-approved HDAC inhibitors registered in the clinicaltrials.gov database.

HDAC inhibitor	Clinical trial	Indications (clinicaltrials.gov identification number, status of trial)
Givinostat (ITF2357), **13** 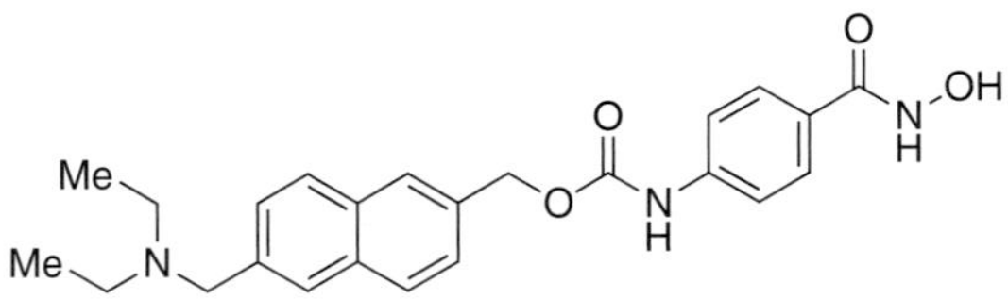	Phase III	Duchenne muscular dystrophy (NCT02851797, ongoing)
	Phase II	Polycythemia vera, mono- or polytherapy (NCT01901432, ongoing and NCT00928707, completed)
		Chronic myeloproliferative neoplasms (NCT01761968, ongoing)
		Juvenile idiopathic arthritis (NCT00570661, completed)
		Hodgkin lymphoma (NCT00792467, completed)
Entinostat (MS275), **5** 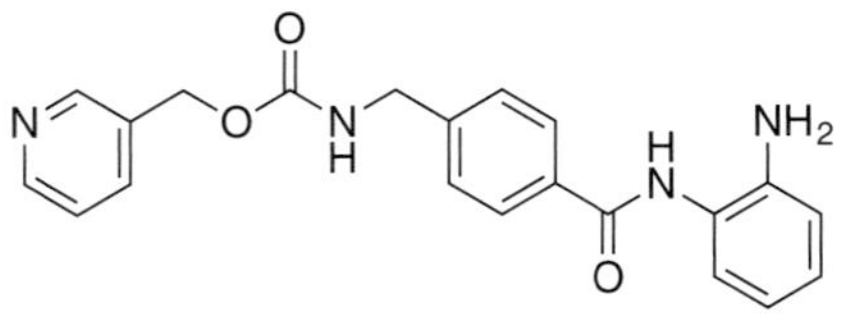	Phase III	Breast cancer, polytherapy (NCT01935947, ongoing)
	Phase II	Ovarian cancer, peritoneal cancer, fallopian tube cancer (NCT02915523, ongoing)
		Hodgkin lymphoma (NCT00866333, completed)
		Colon and rectal cancer, polytherapy (NCT01105377, completed)
		Acute myeloid leukemia, polytherapy (NCT01305499, ongoing)
		Myelodysplastic syndromes, leukemia, mono- and polytherapy (NCT00313586, ongoing and NCT 00462605, completed)
		Metastatic melanoma, mono- and polytherapy (NCT00185302, completed and NCT02697630, ongoing)
		Lung cancer, polytherapy (NCT00750698, completed, NCT01935947 and NCT01928576, ongoing)
	Phase I/II	Renal cancer, polytherapy (NCT01038778, ongoing)
	Phase I	Solid tumors, mono- and polytherapy (NCT02780804 and NCT02909452, ongoing)

Tacedinaline (CI-994), **14** 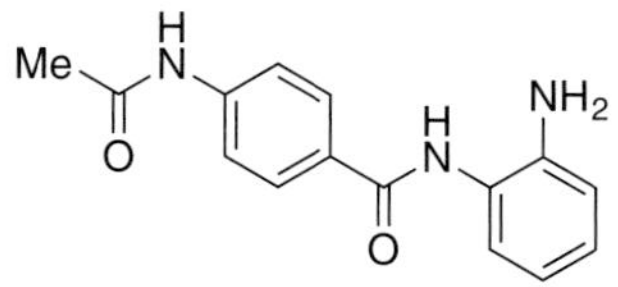	Phase III	Lung cancer, polytherapy (NCT00005093, completed in 2001)
	Phase II	Multiple myeloma (NCT00005624, completed)
Quisinostat (JNJ-26481585), **15** 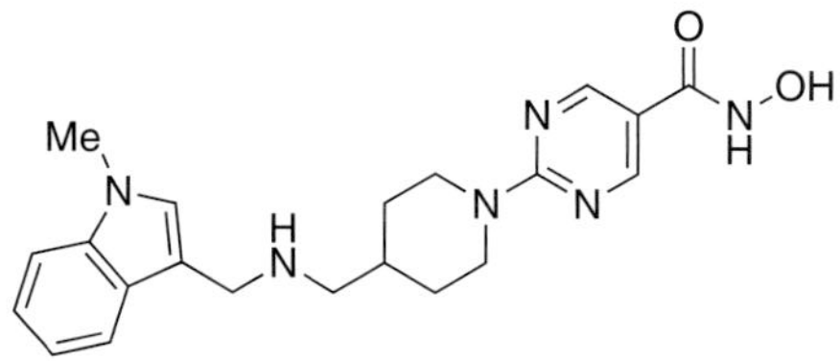	Phase II	Cutaneous T-cell lymphoma (NCT01486277, completed) Ovarian cancer (NCT02948075, ongoing)
	Phase I	Lung cancer, ovarian cancer (NCT02728492, completed) Multiple myeloma (NCT01464112, completed)
Pracinostat (SB939), **16** 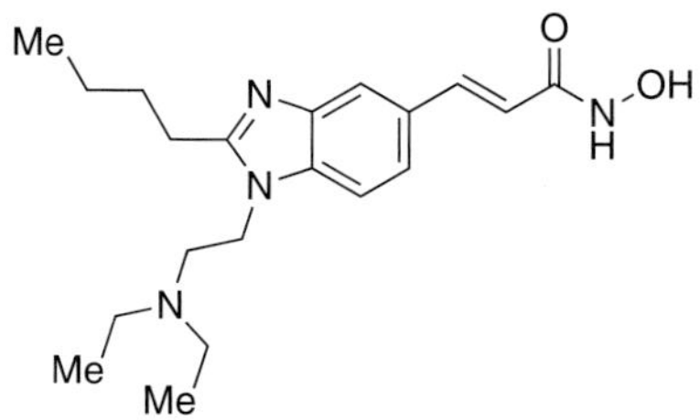	Phase II	Prostate cancer (NCT01075308, completed) Leukemia, polytherapy (NCT01912274, ongoing) Myelodysplastic syndrome, polytherapy (NCT01873703, completed and NCT01993641, completed) Myelofibrosis (Bone marrow cancer), polytherapy (NCT02267278, ongoing) Metastatic sarcoma (NCT01112384, completed) Myeloproliferative disorders (NCT01200498, completed)
	Phase I	Solid tumors (NCT00504296, completed)

(Continued)

Table 7.2 (Continued)

HDAC inhibitor	Clinical trial	Indications (clinicaltrials.gov identification number, status of trial)
Mocetinostat (MGCD0103), **17** 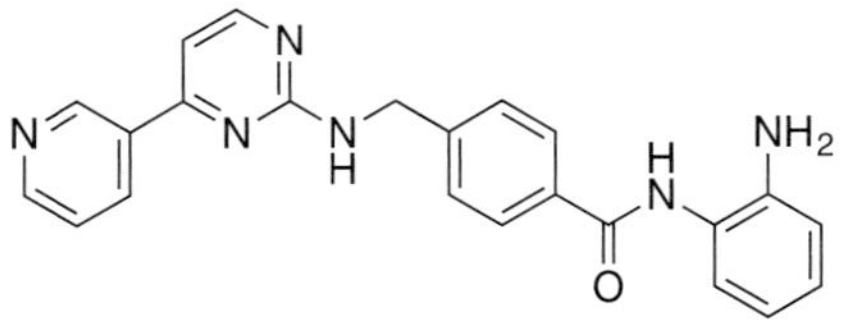	Phase II	Metastatic leiomyosarcoma, polytherapy (NCT02303258, ongoing) Lung cancer, polytherapy (NCT02954991, ongoing) Leukemia (NCT00431873, completed) Urothelial carcinoma (NCT02236195, completed) Lymphoma (NCT00359086, completed)
	Phase I/II	Myelodysplastic syndrome, polytherapy (NCT02018926, completed and NCT00324220, completed) Hodgkin lymphoma, polytherapy (NCT02429375, ongoing) Solid tumors, mono- and polytherapy (NCT00372437, completed and NCT02805660, ongoing)
	Phase I	Squamous cell carcinoma (neck, head, oral cavity, mouth) (NCT02993991, ongoing)
Abexinostat (PCI-24781, CRA24781), **18** 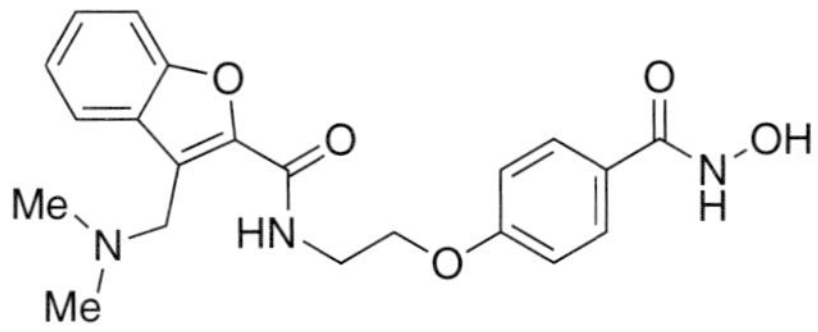	Phase I/II	Solid tumors Sarcoma, polytherapy (NCT01027910, completed) Lymphoma, Hodgkin lymphoma, non-Hodgkin lymphoma (NCT00724984, completed)
	Phase I	Lymphoma, non-Hodgkin lymphoma, multiple myeloma, leukemia (NCT00562224, completed) Solid and hematologic neoplasms (NCT00473577, completed)

Tefinostat (CHR-2845), **19** 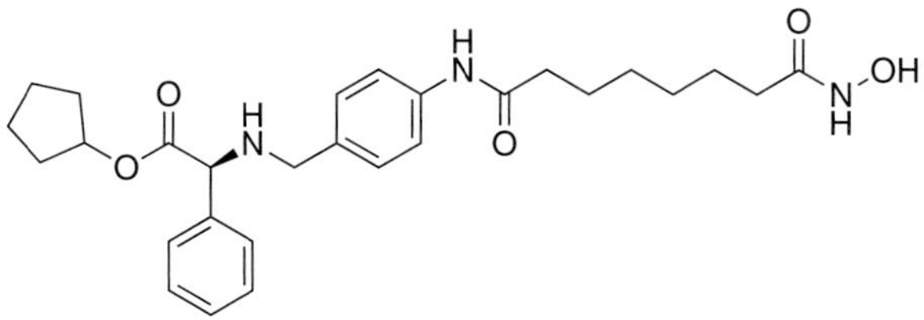	Phase I/II	Liver cancer (NCT02759601, ongoing)
	Phase I	Hematological malignancies, lymphoid malignancies (NCT00820508, completed)
Rocilinostat (ACY-1215), **20** 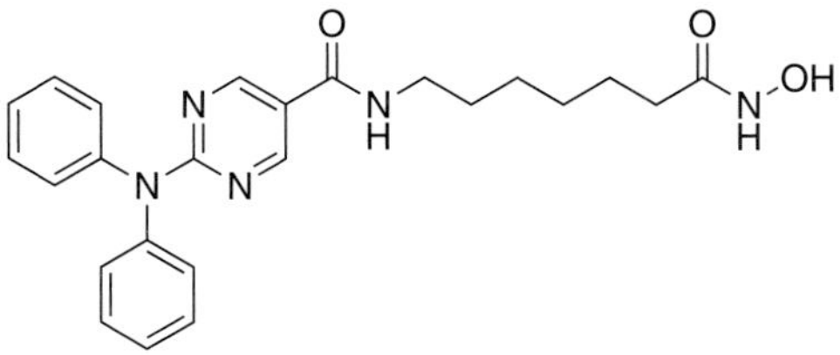	Phase I/II	Multiple myeloma, mono- and polytherapy (NCT01323751, ongoing and NCT01997840, ongoing) Lymphoma, lymphoid malignancies (NCT02091063, ongoing)
	Phase I	Breast cancer, polytherapy (NCT02091063, ongoing) Lymphoid leukemia, polytherapy (NCT02787369, ongoing) Bile duct cancer, polytherapy (NCT02856568, ongoing) Ovarian cancer, fallopian tube cancer, primary peritoneal carcinoma (NCT02661815, ongoing)

(Continued)

Table 7.2 (Continued)

HDAC inhibitor	Clinical trial	Indications (clinicaltrials.gov identification number, status of trial)
CUDC-907, **21** 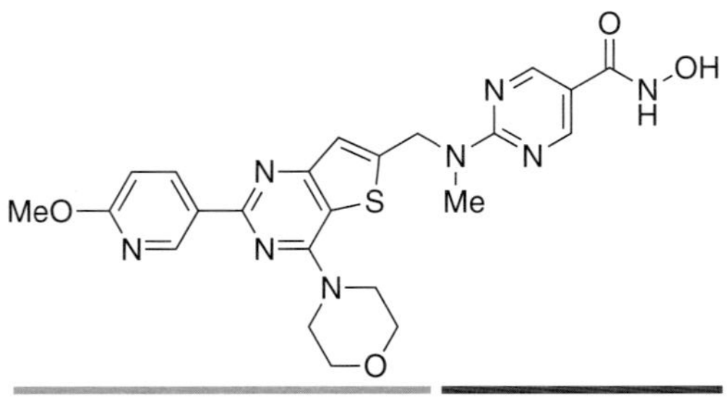	Phase II	B-cell lymphoma, polytherapy (NCT02909777, ongoing) Thyroid cancer, thyroid neoplasms (NCT03002623, ongoing)
	Phase I	Multiple myeloma, lymphoma (NCT01742988, ongoing) NUT midline carcinoma, breast cancer, and solid tumors (NCT02307240, ongoing) Lymphoma, brain tumor, neuroblastoma, solid tumor (NCT02909777, ongoing)
Resminostat (4SC-201, RAS 2410), **22** 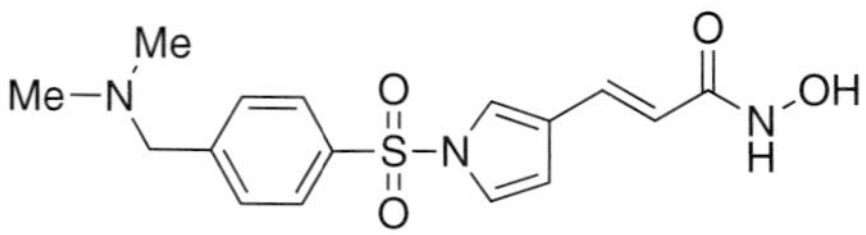	Phase II	Hodgkin lymphoma (NCT01037478, completed) Mycosis fungoides, Sézary syndrome (NCT02953301, ongoing) Hepatocellular carcinoma, polytherapy (NCT00943449, complete)
	Phase I/II	Colorectal carcinoma, polytherapy (NCT01277406, complete) Hepatocellular carcinoma, polytherapy (NCT02400788, ongoing)
Pivanex (AN-9), **23** 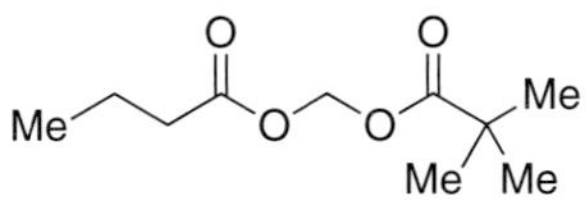	Phase II	Lung carcinoma, polytherapy (NCT00073385, completed 2005)
FRM-0334 (EVP-0334) Structure not available	Phase II	Frontotemporal dementia with granulin mutation (NCT02149160, ongoing)

Remetinostat (SHAPE), **24** 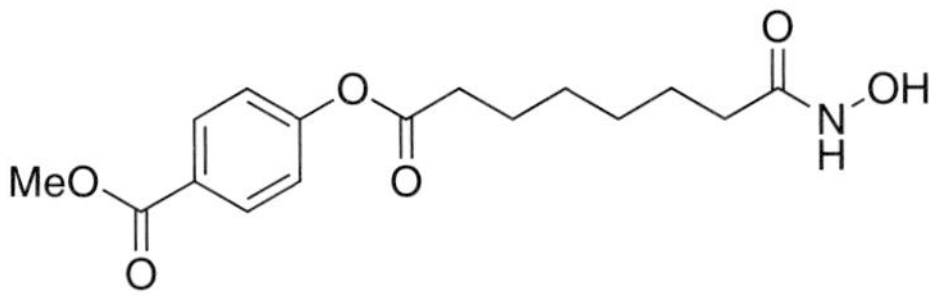	Phase II	Cutaneous T-cell lymphoma (NCT02213861, ongoing) Alopecia areata (spot baldness) (NCT02636244, ongoing)
CKD-581 Structure not available	Phase I	Lymphoma, multiple myeloma (NCT01580371, completed)
CHR-3996, **25** 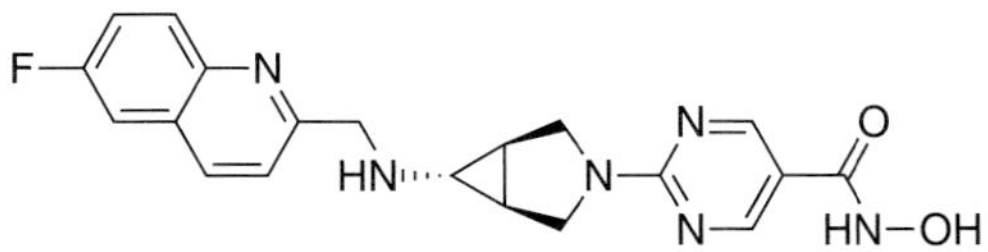	Phase I	Solid tumors (NCT00697879, completed)
MTD KA2507 Structure not available	Phase I	Solid tumor (NCT03008018, ongoing)
R306465, **26** 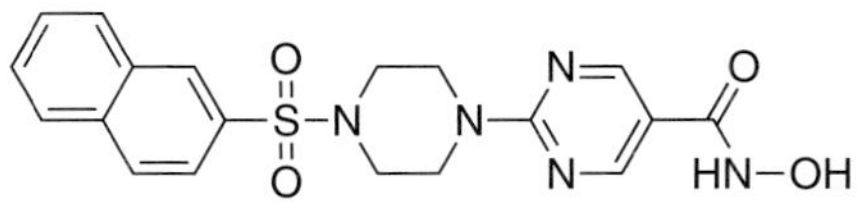	Phase I	Neoplasms (NCT00677001, completed)

(Continued)

Table 7.2 (Continued)

HDAC inhibitor	Clinical trial	Indications (clinicaltrials.gov identification number, status of trial)
4SC-202, **27** 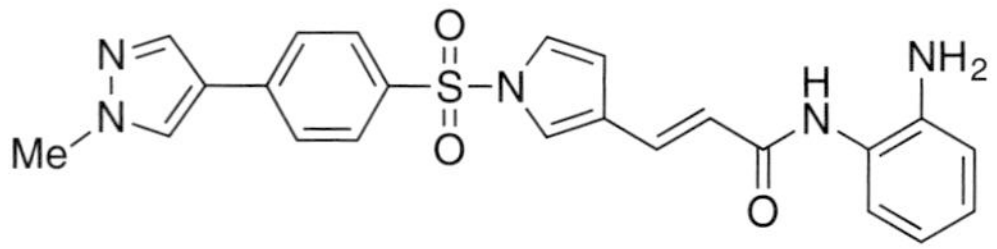	Phase I	Hematologic malignancies (NCT01344707, completed) This compound has also been shown to inhibit lysine-specific demethylase (LSD-1)
CUDC-101, **28** 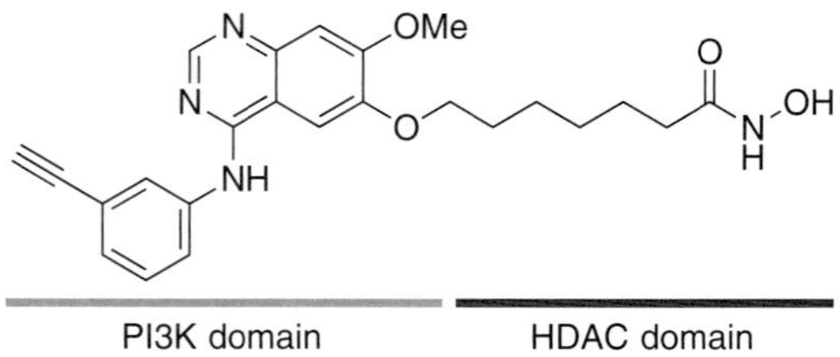	Phase I	Head and neck cancer, polytherapy (NCT01384799, completed) Head and neck cancer, liver cancer, breast cancer, gastric cancer, and lung cancer (NCT01171924, completed)
AR-42 (NSC 731438), **29** 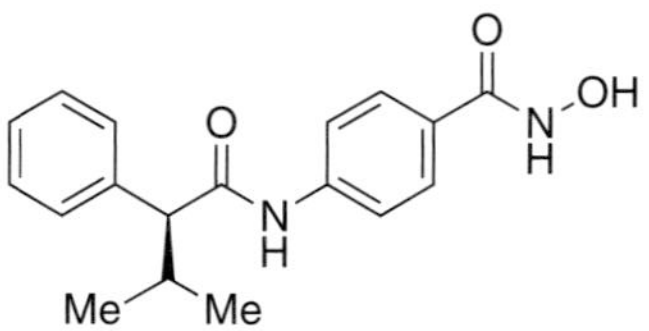	Phase I	Plasma cell myeloma (NCT02569320, ongoing) Leukemia (NCT01798901, ongoing) Multiple myeloma, lymphocytic leukemia, lymphoma (NCT01129193, ongoing)
	Phase 0	Meningioma, acoustic neuroma neurofibromatosis (NCT02282917, ongoing)

effects and limitation of occurrence of drug resistance [32]. CUDC-907 (**21**) is a hybrid drug and consists of two structural units with separate pharmacologic effects. One substructure functions as a phosphatidylinositol 3-kinase (PI3K) inhibitor, whereas the other part is based on the structural units of the common HDAC pharmacophore model and exhibits HDAC inhibition. Promising results show that this combination is 300–500-fold more potent *in vitro* against cancer cell lines than vorinostat alone [32]. Similarly, CUDC-101 (**28**) is a hybrid of vorinostat, and the US FDA approved cancer drug erlotinib, which targets both epidermal growth factor receptor (EGFR) and human epidermal growth factor receptor 2 (HER2). *In vitro* studies on breast, lung, and prostate cancer cell lines confirm that this multi-target drug is two to nine times more potent than a combination of the two parent drugs. CUDC-101 has completed phase I and CUDC-907 is proceeding in phase II studies (Table 7.2).

Nonselective inhibitors dominate the pipeline of HDAC inhibitors with strong hydroxamic acid zinc-binding groups. They are almost exclusively tested against cancers, though *in vivo* studies also show effect in neurological and immune disorders where aberrant activity of HDACs is observed [6]. Hydroxamic acids are generally suspected to be able to bind to other metalloproteins and therefore give rise to off-target effects [33]. Currently approved drugs and HDAC inhibitors in clinical trials show dose-limiting serious adverse effects such as atrial fibrillation, thrombocytopenia, and anemia [34], and clinically used hydroxamates are furthermore suspected to cause mutagenicity [33]. It is debated whether more selective HDAC inhibitors with other zinc-binding groups than hydroxamic acid could limit the toxicity of HDAC inhibitor drugs and expand the therapeutic area.

7.3 Targeting of HDAC Subclasses

Most HDAC inhibitors are nonselective due to the highly conserved binding site among the classes. Between subclasses, minor differences in the active sites and the surrounding enzyme surfaces can however be targeted to obtain class-selective inhibitors. In some cases, subclass selectivity has even been obtained. In the following section we will describe examples of class I, class IIa, and class IIb inhibitors, respectively. No class IV selective inhibitors are known to date.

7.3.1 Class I Inhibitors

Class I enzymes are structurally characterized by an additional cavity adjacent to the Zn^{2+} ion, termed the foot pocket [35–37]. It was first observed in crystal structures of histone deacetylase-like protein (HDLP) from *Aquifex aeolicus* [19]. The unique presence in class I enzymes was later confirmed by crystal structures of HDAC1–3 [35–37]. Some class I hydroxamic acids with well-designed linkers and capping groups, such as the clinical drug candidate CHR-3996, show class I selective inhibition [20]. However, most focus on the development of HDAC1–3 inhibitors has been on exploring the additional space and interaction possibilities originating from the foot pocket [25, 26, 38–42]. HDAC8 is phylogenetically less similar than the remaining class I isozymes [43]. It contains a foot pocket adjacent to the active site like the remaining class I isozymes, but a leucine is replaced with a tryptophan in the opening of this cavity [44]. This larger substitution of amino acids results in restricted space around the foot pocket and changes the preference for ligands [44]. HDAC8 shows a large degree of plasticity, while a widening of the active site tunnel has been observed in some X-ray structures of HDAC8 co-crystalized with inhibitors [45].

7.3.1.1 HDAC1–3 Inhibitors

A major group of class I selective inhibitors contains *ortho*-aminoanilide zinc-binding moieties, including chidamide (**11**) and the clinical candidates entinostat (**5**) and mocetinostat (**17**) (Figure 7.5). Entinostat, which was among the first synthesized *ortho*-aminoanilides, shows both *in vitro* antiproliferative activity and *in vivo* antitumor effects [46]. It contains a benzyl linker adjacent to the *ortho*-aminoanilide, and this structural moiety is commonly used in class I inhibitors. Inhibitors incorporating the aminophenyl benzamide constitute

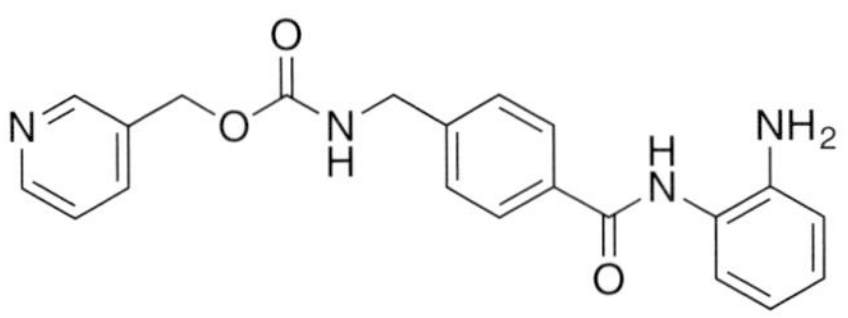

5

IC_{50} (HDAC1)	**0.125 μM**
IC_{50} (HDAC2)	**0.340 μM**
IC_{50} (HDAC3)	**0.370 μM**
IC_{50} (HDAC8)	14 μM
IC_{50} (HDAC4–7)	>50 μM

(a)

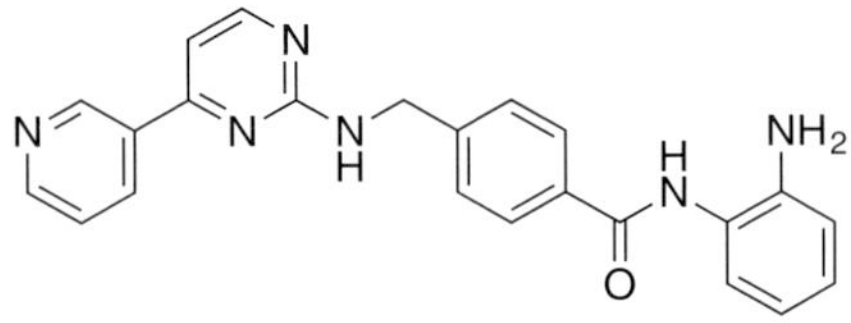

17

IC_{50} (HDAC1)	**0.15 μM**
IC_{50} (HDAC2)	**0.29 μM**
IC_{50} (HDAC3)	**1.66 μM**
IC_{50} (HDAC11)	0.59 μM
IC_{50} (HDAC4–8)	>10 μM

(b)

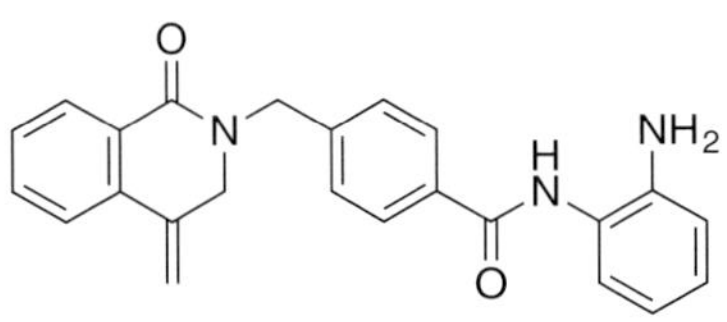

30

IC_{50} (HDAC1)	4.8 μM
IC_{50} (HDAC2)	**0.003 μM**
IC_{50} (HDAC3)	**0.016 μM**
IC_{50} (HDAC6,8)	>10 μM
IC_{50} (HDAC4)	5.0 μM
IC_{50} (HDAC7)	4.1 μM

(c)

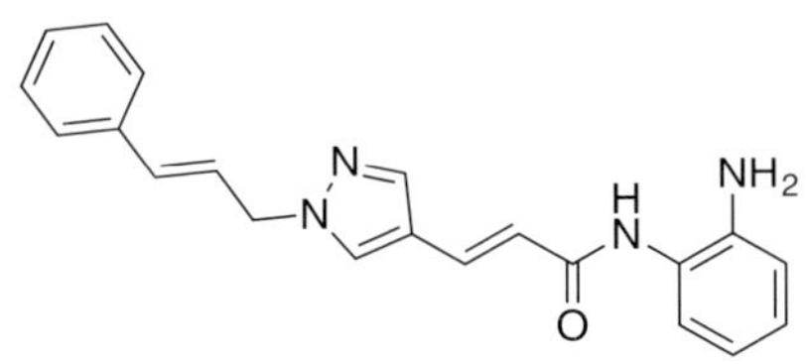

31

IC_{50} (HDAC3)	**0.08 μM**
IC_{50} (HDAC1,2,4–11)	>15 μM

(d)

Figure 7.5 *ortho*-Aminoanilide inhibitors of HDAC1–3. IC_{50} values reported for (a) Entinostat (**5**). Source: Witter et al. 2008 [25]. Reproduced with permission of Elsevier. (b) Mocetinostat (**17**) [26]. (c) MI-192 (**30**). Source: Boissinot et al. 2012 [39]. Reproduced with permission of Elsevier. (d) RGFP966 (**31**) [40].

a significant structural subgroup within the group of *ortho*-aminoanilides, represented here by entinostat, mocetinostat, and MI-192 (**30**), and are often just referred to as benzamides. Mocetinostat is a benzamide developed by MethylGene Inc. as a result of extensive SAR studies [26]. These studies favored a capping group with hydrophobic or basic substituents on an aromatic or heteroaromatic ring system for interaction with the enzyme surface, which resulted in the 4-(pyridin-3-yl)pyrimidine capping group of mocetinostat. Both entinostat and mocetinostat inhibit HDAC1–3 with IC_{50} values in the submicromolar range, and mocetinostat has also been demonstrated to have inhibitory activity against HDAC11, while the remaining HDACs are inhibited to a much lesser extent and often with IC_{50} values greater than 10 μM [25, 26].

The bulky *ortho*-aminoanilides benefit from the space in the foot pocket of HDAC1–3, which enables the accommodation of this zinc-binding group. Interactions in the binding site have been studied via the co-crystal structure of HDAC2 and compound **35** (Figure 7.6), indicating bidentate coordination via the

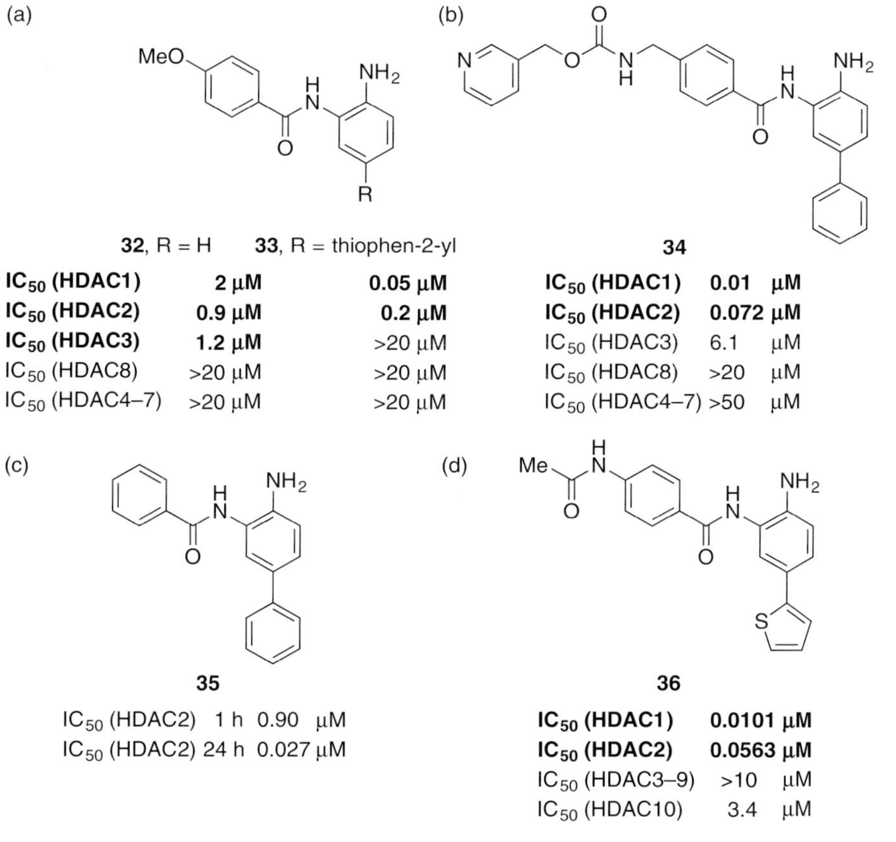

Figure 7.6 Foot pocket binders. 5-Substituted *ortho*-aminoanilide HDAC1 and HDAC2 inhibitors. IC_{50} values reported for (a) Compounds **32** and **33** [41]. (b) Compound **34**. Source: Witter et al. 2008 [25]. Reproduced with permission of Elsevier. (c) Compound **35**. Source: Bressi et al. 2010 [35]. Reproduced with permission of Elsevier. (d) Compound **36** [49].

amide oxygen and the aniline nitrogen [47]. Even though HDAC8 is a member of class I, it has increased resistance toward inhibition by *ortho*-aminoanilides, which has been attributed to the leucine-to-tryptophan mutation in the opening of the HDAC8 foot pocket. This hinders the accommodation of the *ortho*-aminoanilide and impedes proper geometry for zinc binding [44].

Careful design of the linker and capping group has resulted in *ortho*-aminoanilides with increased inhibitory activity against HDAC2 and HDAC3 (Figure 7.5) [39, 40]. This includes MI-192, which shows enhanced activity against HDAC2 and HDAC3 with IC_{50} values 300–1000-fold lower than the remaining isozymes, including HDAC1, and RGFP966, which exhibits inhibition of HDAC3 with nanomolar range IC_{50} values, while the rest of the isozymes are inhibited in the micromolar range [39, 40].

ortho-Aminoanilides are slow-binding HDAC inhibitors as can be observed from kinetic studies. This is exemplified by the change in recorded IC_{50} values after different preincubation times of compound **35** with the enzyme (Figure 7.6) [35, 48]. The slow binding is speculated to arise from either the rearrangement of the active site [49] or the need to break an internal hydrogen bond between one of the hydrogen atoms of the 2-amine and the carbonyl oxygen in the unbound inhibitor [35].

Based on docking studies of entinostat (**5**) into a homology model of HDAC1, it was suggested that the foot pocket could be explored by a thienyl substituent in the 5 position of the *ortho*-aminoanilide, which projects into the hydrophobic pocket and contribute with hydrophobic interactions [41]. By substituting the 5 position of the *ortho*-aminoanilide compound **32** with a thienyl moiety (compound **33**), the potency and selectivity profile drastically shifted toward HDAC1 and HDAC2 over HDAC3 (Figure 7.6). Substitution of the 5-phenyl group in **34** with 5-(2-furyl) or 5-(3-thienyl) (not shown) was also tolerated and was shown by Witter et al. to exhibit similar potencies and selectivity profiles as recorded for **34** [25]. This tolerance is ascribed to hydrophobic interactions of the 5-substituent rather than specific interactions with the sulfur atom [41]. A co-crystal structure of HDAC2 and the phenylanilide-based inhibitor **35** confirms that the phenyl group occupies the foot pocket and makes hydrophobic interactions with a leucine and two phenylalanine residues [35, 47]. Similarly, a co-crystal structure of HDAC2 in complex with compound **36** (Figure 7.6) suggests that the preferred activity of the thienylanilide-based inhibitor against HDAC1 and HDAC2 over HDAC3 is due to interactions in an area where a serine residue in HDAC1 and HDAC2 is mutated to a larger tyrosine in HDAC3 [38, 49, 50].

A more recently discovered group of class I inhibitors contains a benzohydrazide core unit, which was identified via high-throughput screening (HTS) (Figure 7.7) [42]. In addition to this scaffold, a short aliphatic chain at the hydrazide, preferably an *n*-butyl group, and a bulky substituent in *para*-position of the benzoyl moiety seem to ensure inhibition of class I enzymes in the low micromolar range. The lead compound in the series, UF010 (**37**), shows increased selectivity toward HDAC1–3. Molecular modeling suggests that this new class of inhibitors also relies on interaction with the characteristic class I hydrophobic foot pocket, which in this case is occupied by the aliphatic chain.

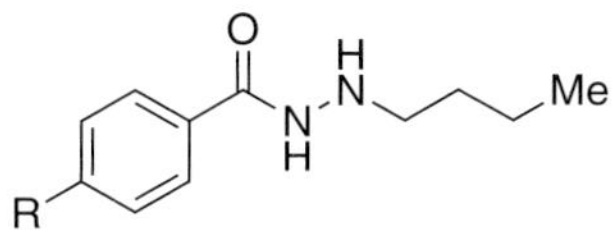

	UF010 (R = Br), **37**	SR-3208 (R = N(Me)$_2$), **38**	SR-3302 (R = C(Me)$_3$, **39**
IC$_{50}$ (HDAC1)	**0.46 μM**	**0.23 μM**	**0.19 μM**
IC$_{50}$ (HDAC2)	**1.33 μM**	**0.88 μM**	**1.04 μM**
IC$_{50}$ (HDAC3)	**0.19 μM**	**0.12 μM**	**0.07 μM**
IC$_{50}$ (HDAC8)	2.83 μM	**0.72 μM**	**0.49 μM**
IC$_{50}$ (HDAC6)	9.09 μM	9.57 μM	6.83 μM

Figure 7.7 Benzoylhydrazide inhibitors. IC$_{50}$ values reported for UF010 (**37**), SR-3208 (**38**), and SR-3302 (**39**). Source: Wang et al. 2015 [42]. Reproduced with permission of Elsevier.

The compound series lacks a traditional zinc-binding group, and molecular modeling studies did not show interaction with the Zn^{2+} ion [42]. The compound series are therefore suggested to rely upon other interactions, rendering them quite unique among small non-macrocyclic HDAC inhibitors [42].

7.3.1.2 HDAC Inhibitors Targeting HDAC8

HDAC8 inhibitors have been obtained by targeting unique structural features of this isozyme. An X-ray structure of HDAC8 co-crystallized with the biaryl hydroxamic acid CRA-A (**40**) (Figure 7.8) revealed the formation of a large HDAC8-specific pocket of the active site that did not occur with the sterically less hindered aliphatic vorinostat [45]. After this discovery, linkerless aryl hydroxamic acids were developed to target the HDAC8-specific pocket, as exemplified by compound **41** (Figure 7.8) [51]. The linkerless and sterically demanding compound **41** shows HDAC8 inhibition with IC$_{50}$ values in the submicromolar range and more than 100-fold decrease in inhibitory activity toward HDAC1 and HDAC6. The indole-based compound PCI-34051 (**42**) and a series of *ortho*-aryl *N*-hydroxycinnamides represent other inhibitors binding in the HDAC8-specific pocket (Figure 7.8) [52, 53]. PCI-34051 targets HDAC8 with IC$_{50}$ values >200-fold lower than HDAC1 and HDAC6 and >1000-fold lower than HDAC2, HDAC3, and HDAC10. In accordance with the hope for selective inhibitors, PCI-34051 appeared to have a selective cytotoxic profile against cell lines derived from T-cell malignancies compared with the nonselective clinical drug candidate abexinostat (**18**), which exhibited cytotoxic effect on a broader range of tumors and other malignant cell lines [52].

Other HDAC8 inhibitors target the "gatekeeper" tryptophan found at the opening of the HDAC8 foot pocket. The two non-hydroxamic acid HDAC8 inhibitors, **46** and **47** (Figure 7.9), are mono- or dichlorosubstituted phenylalanine derivatives where the carbonyl oxygen and amino nitrogen coordinate to the Zn^{2+} ion [54]. These two compounds were discovered by an effort of HTS, and compound **46** targets HDAC8 with more than 150-fold selectivity over the remaining class I enzymes. Based on X-ray crystallography of **46** or **47** bound to HDAC8, the isozyme selectivity was suggested to arise from favorable π-stacking interactions

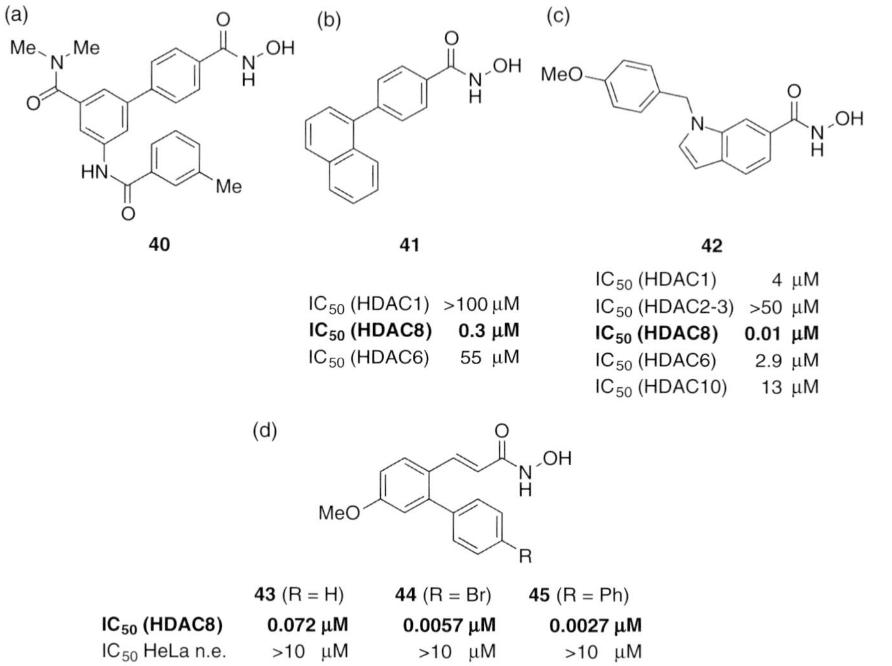

Figure 7.8 Hydroxamate inhibitors targeting the HDAC8-specific pocket, shown with available IC_{50} values (a) CRA-A (**40**). (b) Compound **41**. Source: KrennHrubec et al. 2007 [51]. Reproduced with permission of Elsevier. (c) PCI-34051 (**42**) [52]. (d) Compounds **43**, **44**, and **45** [53].

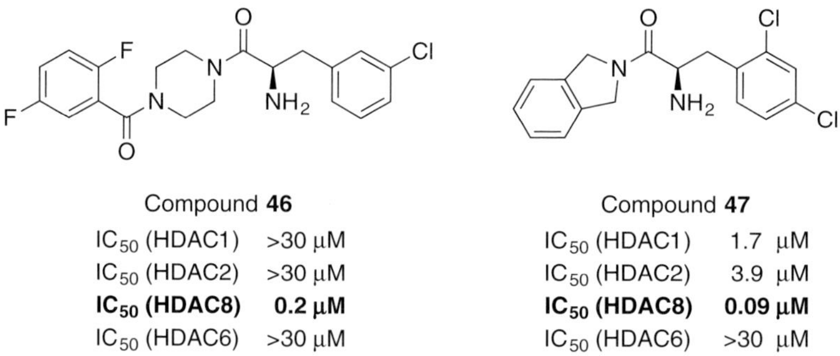

Figure 7.9 Chlorinated phenylalanine derivatives inhibiting HDAC8. IC_{50} values reported for compounds **46** and **47**. Source: Whitehead et al. 2011 [54]. Reproduced with permission of Elsevier.

observed between the benzyl groups and the isozyme-specific tryptophan residue at the opening of the class I foot pocket [54].

7.3.2 Class IIa Inhibitors

All class IIa isozymes contain a tyrosine-to-histidine mutation in the active site, which results in reduced deacetylase activity [10, 55]. However, while inactive against acetylated substrates, class IIa enzymes are able to convert trifluoroacetylated substrates [10, 55]. As a consequence of this mutation, HDAC4 and HDAC7 form a "lower pocket" due to the rotation of the histidine away from the binding sites, as observed in most of the reported crystal structures of HDAC4 and HDAC7 [56, 57]. Except for the Y-to-H mutation, the active site is highly conserved, and an additional pocket appears to be specific for class IIa. Inhibitors targeting this class IIa-specific pocket have been developed [21, 57, 58].

Class IIa inhibitors containing trifluoromethyl ketones were reported after the discovery of class IIa catalytic activity against trifluoroacetylated substrates (Figure 7.10) [55, 59–63]. Compound **48** (Figure 7.10) was one of the first HDAC inhibitors containing a trifluoromethyl ketone group [27]. It exhibits pan-inhibition, including the class IIa isozymes, whereas the corresponding hydroxamate vorinostat (**7**) is a class I, IIb, and IV inhibitor [11, 64]. The trifluoromethyl ketone group was therefore a good starting point for the development of class IIa inhibitors. 2-Trifluoroacetylthiophenes, such as compounds **49** (Figure 7.10) and **6**, were suggested as class II inhibitors with 10-fold higher potencies against HDAC4 and HDAC6 compared with HDAC1 and

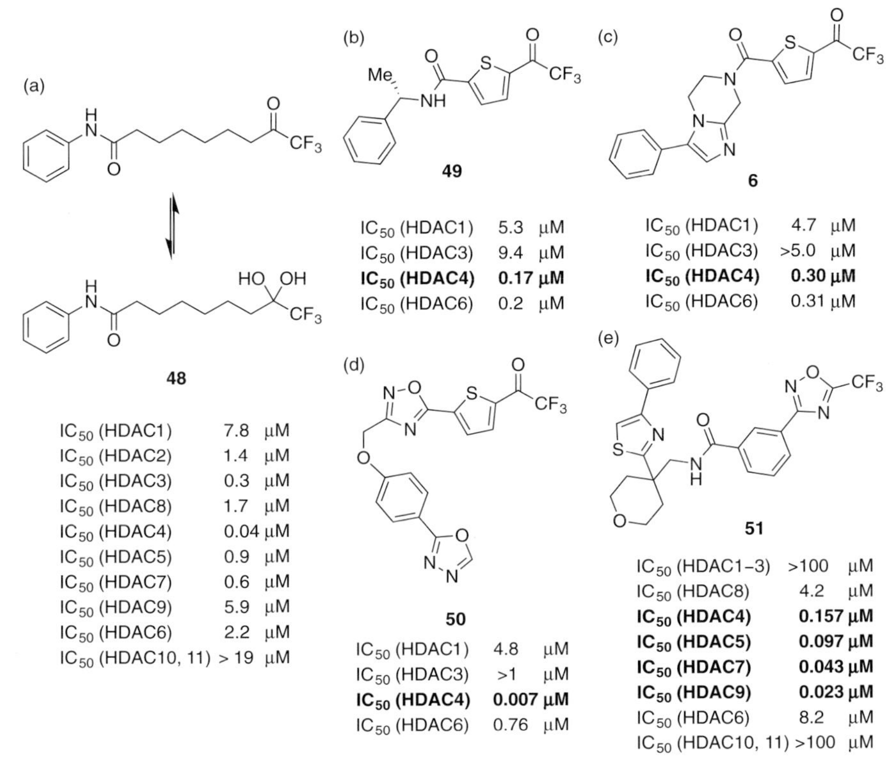

Figure 7.10 Trifluoromethyl ketone inhibitors of class II/IIa. IC_{50} values reported for (a) Compound **48** [64]. (b) Compound **49** [59]. (c) Compound **6**. Source: Scarpelli et al. 2008 [60]. Reproduced with permission of Elsevier. (d) Compound **50**. Source: Muraglia et al. 2008 [61]. Reproduced with permission of Elsevier. (e) TMP969 (**51**) [57].

HDAC2 [59, 60]. SAR studies resulted in a series of 2-trifluoroacetylthiophene oxadiazoles, exemplified by compound **50** (Figure 7.10), with improved metabolic stability and selectivity against class IIa [61].

Trifluoromethyl ketones are electron deficient and undergo hydration in aqueous media, and the functioning zinc-binding group is therefore considered to be the *gem*-diol (Figure 7.10; compound **48**) [27]. X-ray structures of the HDAC4 catalytic domain co-crystallized with the trifluoroacetylthiophene **6** or the hydroxamate analogue (not shown) confirm the trifluoromethyl *gem*-diol binding mode and suggest that the preference (0.3 μM against HDAC4 versus >5 μM against HDAC3) results from a bidentate coordination of the *gem*-diol to the active site Zn^{2+} ion whereas the hydroxamate makes a monodentate interaction [56]. Preference for the trifluoromethylacetylthiophene could further be explained by how the trifluoromethyl moiety occupies the class IIa-specific pocket and interacts with a proline residue [47, 56].

A more selective series of trifluoromethyloxadiazole class IIa inhibitors, exemplified by compound **51** (Figure 7.10), originated from an HTS study [57]. An X-ray structure of HDAC7 in complex with TMP969 suggests that the trifluoromethyloxadiazole is a weak zinc-binding group by coordination to the zinc ion with one fluorine atom and the oxadiazole oxygen. The selectivity was ascribed to the bulky nature of the trifluoromethyloxadiazole, which obtains a U-like conformation that allows the class IIa-specific pocket to be occupied by the inhibitor [57]. This combination of a weak zinc-binding group and class IIa-specific interactions showed promise of lowering the degree of off-target effects compared to the hydroxamic acid analogues, when tested in an *in vitro* metalloproteinase assay [57]. Compounds from this series show the ability to specifically alter gene expression related to immune responses and could be used for further drug development [57].

Despite the general specificity challenge associated with hydroxamic acids, class IIa selective inhibitors containing this moiety have been reported [21, 58]. In the search for a lead compound for targeting class IIa in Huntington's disease, extensive studies on tri- and tetra-substituted cyclopropanes were conducted, with focus on pharmacokinetics and blood–brain barrier penetration [21, 58]. The chiral scaffold (Figure 7.11; compounds **52**, **53**, and **54**) provided steric bulk adjacent to the zinc-binding group in the form of the *cis*-configured phenyl moiety [21, 58]. A co-crystal structure of HDAC4 with compound **52** confirmed that the additional bulk is accommodated by the class IIa-specific pocket [21, 50]. A fluorine atom was added in the α-position to the hydroxamic acid to increase acidity, which resulted in increased potency and improved pharmacokinetics to obtain promising compounds suitable for preclinical testing [58].

7.3.3 Class IIb

Various HDAC6 inhibitors have been developed, and most of them contain hydroxamic acid zinc-binding groups combined with aromatic capping groups designed to target HDAC6, such as tubacin (**55**) [22], tubastatin A (**56**) [65], rocilinostat (**20**) [66], and HPOB (**57**) [67] (Figure 7.12). Other HDAC6 inhibitors do not comply with the contemporary pharmacophore model, such as

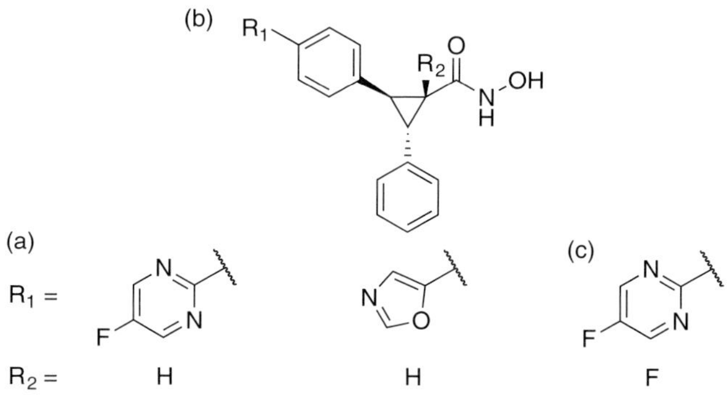

	52	53	54
IC_{50} (HDAC1)	21 µM		14 µM
IC_{50} (HDAC2)		>32 µM	>50 µM
IC_{50} (HDAC3)	9.4 µM	>32 µM	7.4 µM
IC_{50} (HDAC8)	1.4 µM	12 µM	0.28 µM
IC_{50} (HDAC4)	**<0.05 µM**	0.36 µM	**0.01 µM**
IC_{50} (HDAC5)	**<0.05 µM**	**<0.04 µM**	**0.01 µM**
IC_{50} (HDAC7)	**<0.1 µM**	**0.004 µM**	**0.03 µM**
IC_{50} (HDAC9)	**<0.1 µM**	**<0.04 µM**	**0.06 µM**
IC_{50} (HDAC6)	1.9 µM	**<0.04 µM**	3.1 µM

Figure 7.11 Cyclopropane inhibitors binding in the class IIa-specific pocket. IC_{50} values reported for (a, b) compounds **52** and **53** [21] and (c) compound **54** [58].

the naphthoquinone NQN-1 (**58**) [68] or display alternative zinc-binding groups such as mercaptoacetamide **59** [69] (Figure 7.12).

Tubacin (**55**) was first identified by Haggarty et al. in a large library screen for inhibition of α-tubulin deacetylation in mammalian cells, where it induced hyperacetylation of α-tubulin but had no effect on the acetylation state of histones [22]. Docking studies with a homology model of HDAC6 suggest that the selectivity is a result of favorable interactions with the residues on the rim and surface surrounding the active site. These interactions are not present in HDAC1 and HDAC8 [70].

A recent study investigated several crystal structures of the zebrafish HDAC6 catalytic domain co-crystalized with HDAC inhibitors including the HDAC6 selective compound HPOB (**57**) [13]. Interestingly, HPOB only binds to the zinc ion via the hydroxyl group and not via bidentate chelation. Instead the carbonyl oxygen makes a hydrogen bond to a water molecule in the active site, which surprisingly has not been displaced by the inhibitor [13]. This new structural information could direct the design of novel HDAC6 inhibitors.

7.4 Perspectives

HDACs have been targeted for the treatment of diseases such as cancers, neurodegenerative diseases, and immune disorders [6]. The efforts have so far

(a)

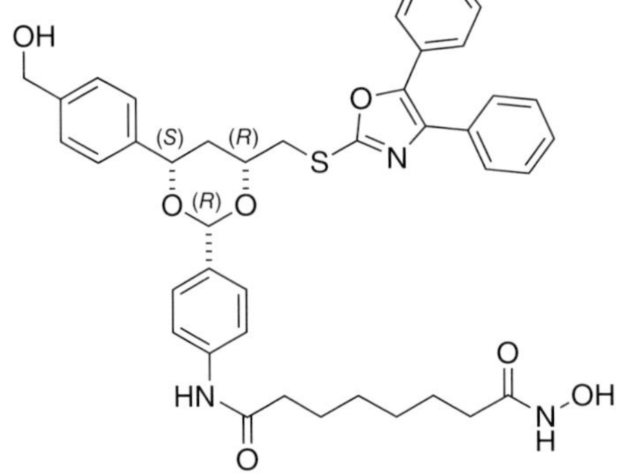

55

IC_{50} (HDAC6)	**0.004 μM**
IC_{50} (HDAC1–5, 7–11)	>1.27 μM

(b)

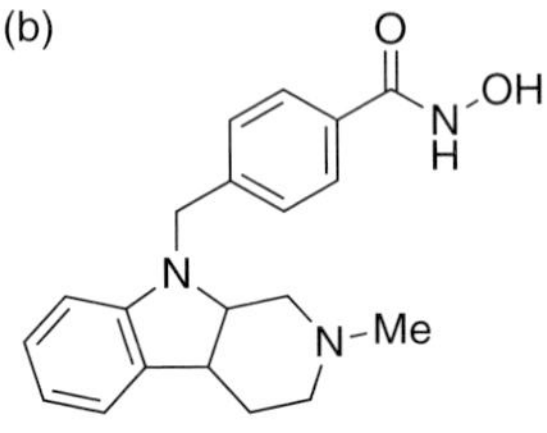

56

IC_{50} (HDAC1)	16.4 μM
IC_{50} (HDAC8)	0.854 μM
IC_{50} (HDAC2-5, 7, 9)	>30 μM
IC_{50} (HDAC6)	**0.015 μM**

(c)

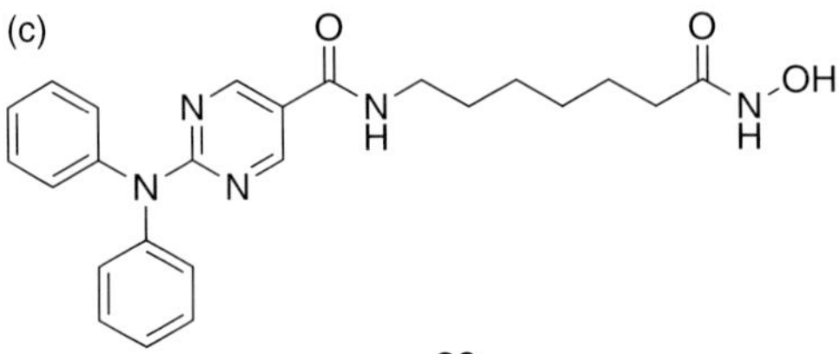

20

IC_{50} (HDAC1)	0.058 μM
IC_{50} (HDAC2)	0.048 μM
IC_{50} (HDAC3)	0.051 μM
IC_{50} (HDAC8)	0.100 μM
IC_{50} (HDAC7)	1.4 μM
IC_{50} (HDAC6)	**0.005 μM**
IC_{50} (HDAC4–5, 9–11)	>5.0 μM

(d)

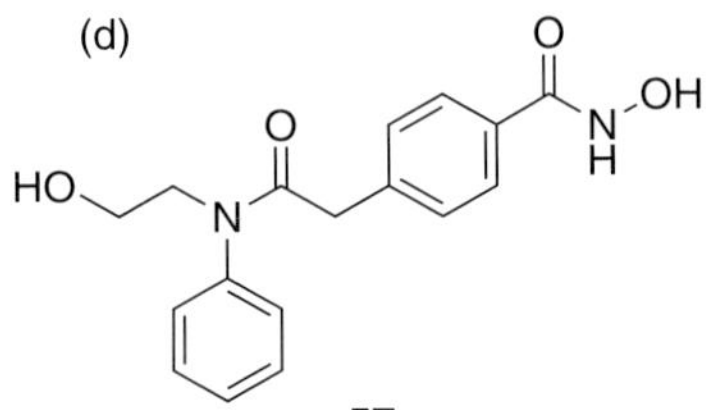

57

IC_{50} (HDAC1)	2.9 μM
IC_{50} (HDAC2)	4.4 μM
IC_{50} (HDAC3)	1.7 μM
IC_{50} (HDAC8)	2.8 μM
IC_{50} (HDAC4, 5, 7, 9)	>10 μM
IC_{50} (HDAC6)	**0.056 μM**
IC_{50} (HDAC10)	3.0 μM
IC_{50} (HDAC11)	>10 μM

(e)

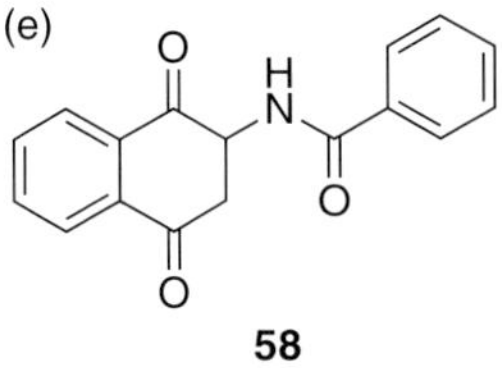

58

IC_{50} (HDAC1–5, 7–11)	>100 μM
IC_{50} (HDAC6)	**5.54 μM**

(f)

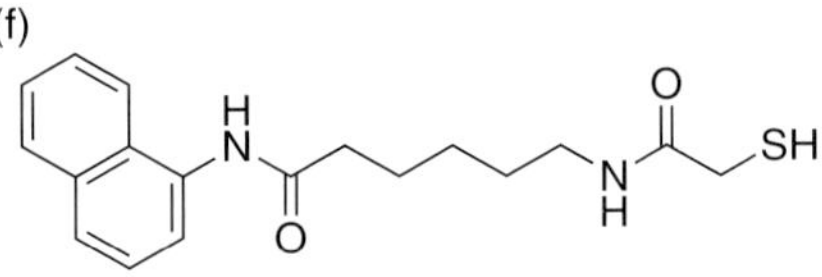

59

IC_{50} (HDAC1)	3.2 μM
IC_{50} (HDAC2)	7.4 μM
IC_{50} (HDAC8)	6.1 μM
IC_{50} (HDAC6)	**0.095 μM**
IC_{50} (HDAC10)	10.7 μM

Figure 7.12 HDAC6 inhibitors. IC_{50} values reported for (a) tubacin (**55**) [65], (b) tubastatin A (**56**) [65], (c) rocilinostat (**20**) [66], (d) HPOB (**57**) [67], (e) NQN-1 (**58**) [68], and (f) compound **59** [69].

resulted in five different drugs for the treatment of hematological malignancies, and numerous HDAC inhibitors are being tested in clinical trials. The pipeline of clinically relevant HDAC inhibitors are dominated by nonselective inhibitors and strong metal chelators such as hydroxamates, which are suspected to result in off-target effects [33]. It is debated whether inhibitors with improved selectivity profiles could limit off-target effects, thereby improving safety and enabling applications in a wider range of diseases. Extensive studies have resulted in the development of class-selective inhibitors, and inhibitors with improved isozyme selectivity have also been discovered [20–22, 25, 26, 38–40, 42, 51–54, 57–61, 65–69, 71]. However, clinical data providing proof of the potential advantages of selective inhibition of HDACs has not been obtained to date. Targeting of HDACs in clinical conditions is further complicated by the fact that HDAC1–3 and class IIa isozymes are recruited into protein complexes with different biological functions [8]. As a result, HDAC inhibitors can target several complexes containing the same isozyme [72, 73]. Interestingly, the two HDAC inhibitors, entinostat (**5**) and tacedinaline (**14**), exhibit some degree of corepressor complex selectivity, which advocate for HDAC complex targeting as a possible future approach within epigenetics, e.g. via the protein–protein interface [8, 72, 73]. Finally, a recently described activator of HDAC8 [74] adds yet another dimension to the possibilities of future manipulation of HDAC activity, which has so far been limited to class III, sirtuins [75].

References

1 Allfrey, V.G., Faulkner, R., and Mirsky, A.E. (1964). Acetylation and methylation of histones and their possible role in the regulation of RNA synthesis. *Proc. Natl. Acad. Sci. U.S.A.* 51: 786–794.

2 Clayton, A.L., Hazzalin, C.A., and Mahadevan, L.C. (2006). Enhanced histone acetylation and transcription: a dynamic perspective. *Mol. Cell* 23: 289–296.

3 Haberland, M., Montgomery, R.L., and Olson, E.N. (2009). The many roles of histone deacetylases in development and physiology: implications for disease and therapy. *Nat. Rev. Genet.* 10: 32–42.

4 Kim, S.C., Sprung, R., Chen, Y. et al. (2006). Substrate and functional diversity of lysine acetylation revealed by a proteomics survey. *Mol. Cell* 23: 607–618.

5 Choudhary, C., Kumar, C., Gnad, F. et al. (2009). Lysine acetylation targets protein complexes and co-regulates major cellular functions. *Science* 325: 834–840.

6 Falkenberg, K.J. and Johnstone, R.W. (2014). Histone deacetylases and their inhibitors in cancer, neurological diseases and immune disorders. *Nat. Rev. Drug Discovery* 13: 673–691.

7 Gantt, S.M.L., Decroos, C., Lee, M.S. et al. (2016). General base–general acid catalysis in human histone deacetylase 8. *Biochemistry* 55: 820–832.

8 Millard, C.J., Watson, P.J., Fairall, L., and Schwabe, J.W. (2017). Targeting class I histone deacetylases in a “complex” environment. *Trends Pharmacol. Sci.* 38 (4): 363–377.

9 Di Giorgio, E. and Brancolini, C. (2016). Regulation of class IIa HDAC activities: it is not only a matter of subcellular location. *Epigenomics* 8: 251–269.
10 Lahm, A., Paolini, C., Pallaoro, M. et al. (2007). Unraveling the hidden catalytic activity of vertebrate class IIa histone deacetylases. *Proc. Natl. Acad. Sci. U.S.A.* 104: 17335–17340.
11 Bradner, J.E., West, N., Grachan, M.L. et al. (2010). Chemical phylogenetics of histone deacetylases. *Nat. Chem. Biol.* 6: 238–243.
12 Miyake, Y., Keusch, J.J., Wang, L. et al. (2016). Structural insights into HDAC6 tubulin deacetylation and its selective inhibition. *Nat. Chem. Biol.* 12: 748–754.
13 Hai, Y. and Christianson, D.W. (2016). Histone deacetylase 6 structure and molecular basis of catalysis and inhibition. *Nat. Chem. Biol.* 12: 741–747.
14 Cohen, T.J., Guo, J.L., Hurtado, D.E. et al. (2011). The acetylation of tau inhibits its function and promotes pathological tau aggregation. *Nat. Commun.* 2: 252.
15 Kovacs, J.J., Murphy, P.J.M., Gaillard, S. et al. (2005). HDAC6 regulates Hsp90 acetylation and chaperone-dependent activation of glucocorticoid receptor. *Mol. Cell* 18: 601–607.
16 Yoshida, M., Horinouchi, S., and Beppu, T. (1995). Trichostatin A and trapoxin: novel chemical probes for the role of histone acetylation in chromatin structure and function. *BioEssays* 17: 423–430.
17 Taunton, J., Hassig, C.A., and Schreiber, S.L. (1996). A mammalian histone deacetylase related to the yeast transcriptional regulator Rpd3p. *Science* 272: 408–411.
18 Jung, M., Hoffmann, K., Brosch, G., and Loidl, P. (1997). Analogues of trichostatin A and trapoxin B as histone deacetylase inhibitors. *Bioorg. Med. Chem. Lett.* 7: 1655–1658.
19 Finnin, M.S., Donigian, J.R., Cohen, A. et al. (1999). Structures of a histone deacetylase homologue bound to the TSA and SAHA inhibitors. *Nature* 401: 188–193.
20 Moffat, D., Patel, S., Day, F. et al. (2010). Discovery of 2-(6-{[(6-fluoroquinolin-2-yl)methyl]amino}bicyclo[3.1.0]hex-3-yl)-*N*-hydroxypyrimidine-5-carboxamide (CHR-3996), a class I selective orally active histone deacetylase inhibitor. *J. Med. Chem.* 53: 8663–8678.
21 Bürli, R.W., Luckhurst, C.A., Aziz, O. et al. (2013). Design, synthesis, and biological evaluation of potent and selective class IIa histone deacetylase (HDAC) inhibitors as a potential therapy for Huntington's disease. *J. Med. Chem.* 56: 9934–9954.
22 Haggarty, S.J., Koeller, K.M., Wong, J.C. et al. (2003). Domain-selective small-molecule inhibitor of histone deacetylase 6 (HDAC6)-mediated tubulin deacetylation. *Proc. Natl. Acad. Sci. U.S.A.* 100: 4389–4394.
23 Maolanon, A.R., Kristensen, H.M.E., Leman, L.J. et al. (2017). Natural and synthetic macrocyclic inhibitors of the histone deacetylase enzymes. *ChemBioChem* 18: 5–49.
24 Fass, D.M., Shah, R., Ghosh, B. et al. (2011). Short-chain HDAC inhibitors differentially affect vertebrate development and neuronal chromatin. *ACS Med. Chem. Lett.* 2: 39–42.

25 Witter, D.J., Harrington, P., Wilson, K.J. et al. (2008). Optimization of biaryl selective HDAC1&2 inhibitors (SHI-1:2). *Bioorg. Med. Chem. Lett.* 18: 726–731.

26 Zhou, N., Moradei, O., Raeppel, S. et al. (2008). Discovery of *N*-(2-aminophenyl)-4-[(4-pyridin-3-ylpyrimidin-2-ylamino)methyl]benzamide (MGCD0103), an orally active histone deacetylase inhibitor. *J. Med. Chem.* 51: 4072–4075.

27 Frey, R.R., Wada, C.K., Garland, R.B. et al. (2002). Trifluoromethyl ketones as inhibitors of histone deacetylase. *Bioorg. Med. Chem. Lett.* 12: 3443–3447.

28 Nalivaeva, N.N., Belyaev, N.D., and Turner, A.J. (2009). Sodium valproate: an old drug with new roles. *Trends Pharmacol. Sci.* 30: 509–514.

29 Wightman, F., Ellenberg, P., Churchill, M., and Lewin, S.R. (2012). HDAC inhibitors in HIV. *Immunol. Cell Biol.* 90: 47–54.

30 clinicaltrials.gov database study identification: NCT01319383, NCT02092116, and NCT02471430.

31 Deng, Y., Ng DiMarco, C., Vakhilt, T. et al. (2016). Process development of the soft histone deacetylate enzyme inhibitor SHP-141: acylation of methyl paraben and suberyl hydroxamic acid formation. *Org. Process Res. Dev.* 20: 1812–1820.

32 Rana, A., Alex, J.M., Chauhan, M. et al. (2015). A review on pharmacophoric designs of antiproliferative agents. *Med. Chem. Res.* 24: 903–920.

33 Shen, S. and Kozikowski, A.P. (2016). Why hydroxamates may not be the best histone deacetylase inhibitors – what some may have forgotten or would rather forget? *ChemMedChem* 11: 15–21.

34 Subramanian, S., Bates, S.E., Wright, J.J. et al. (2010). Clinical toxicities of histone deacetylase inhibitors. *Pharmaceuticals* 3: 2751.

35 Bressi, J.C., Jennings, A.J., Skene, R. et al. (2010). Exploration of the HDAC2 foot pocket: synthesis and SAR of substituted *N*-(2-aminophenyl)benzamides. *Bioorg. Med. Chem. Lett.* 20: 3142–3145.

36 Watson, P.J., Fairall, L., Santos, G.M., and Schwabe, J.W.R. (2012). Structure of HDAC3 bound to co-repressor and inositol tetraphosphate. *Nature* 481: 335–340.

37 Millard, C.J., Watson, P.J., Celardo, I. et al. (2013). Class I HDACs share a common mechanism of regulation by inositol phosphates. *Mol. Cell* 51: 57–67.

38 Methot, J.L., Chakravarty, P.K., Chenard, M. et al. (2008). Exploration of the internal cavity of histone deacetylase (HDAC) with selective HDAC1/HDAC2 inhibitors (SHI-1:2). *Bioorg. Med. Chem. Lett.* 18: 973–978.

39 Boissinot, M., Inman, M., Hempshall, A. et al. (2012). Induction of differentiation and apoptosis in leukaemic cell lines by the novel benzamide family histone deacetylase 2 and 3 inhibitor MI-192. *Leuk. Res.* 36: 1304–1310.

40 Malvaez, M., McQuown, S.C., Rogge, G.A. et al. (2013). HDAC3-selective inhibitor enhances extinction of cocaine-seeking behavior in a persistent manner. *Proc. Natl. Acad. Sci. U.S.A.* 110: 2647–2652.

41 Moradei, O.M., Mallais, T.C., Frechette, S. et al. (2007). Novel aminophenyl benzamide-type histone deacetylase inhibitors with enhanced potency and selectivity. *J. Med. Chem.* 50: 5543–5546.

42 Wang, Y., Stowe, R.L., Pinello, C.E. et al. (2015). Identification of histone deacetylase inhibitors with benzoylhydrazide scaffold that selectively inhibit class I histone deacetylases. *Chem. Biol.* 22: 273–284.

43 Gregoretti, I.V., Lee, Y.M., and Goodson, H.V. (2004). Molecular evolution of the histone deacetylase family: functional implications of phylogenetic analysis. *J. Mol. Biol.* 338: 17–31.

44 Vannini, A., Volpari, C., Filocamo, G. et al. (2004). Crystal structure of a eukaryotic zinc-dependent histone deacetylase, human HDAC8, complexed with a hydroxamic acid inhibitor. *Proc. Natl. Acad. Sci. U.S.A.* 101: 15064–15069.

45 Somoza, J.R., Skene, R.J., Katz, B.A. et al. (2004). Structural snapshots of human HDAC8 provide insights into the class I histone deacetylases. *Structure* 12: 1325–1334.

46 Saito, A., Yamashita, T., Mariko, Y. et al. (1999). A synthetic inhibitor of histone deacetylase, MS-27-275, with marked in vivo antitumor activity against human tumors. *Proc. Natl. Acad. Sci. U.S.A.* 96: 4592–4597.

47 Micelli, C. and Rastelli, G. (2015). Histone deacetylases: structural determinants of inhibitor selectivity. *Drug Discovery Today* 20: 718–735.

48 Chou, C.J., Herman, D., and Gottesfeld, J.M. (2008). Pimelic diphenylamide 106 is a slow, tight-binding inhibitor of class I histone deacetylases. *J. Biol. Chem.* 283: 35402–35409.

49 Lauffer, B.E.L., Mintzer, R., Fong, R. et al. (2013). Histone deacetylase (HDAC) inhibitor kinetic rate constants correlate with cellular histone acetylation but not transcription and cell viability. *J. Biol. Chem.* 288: 26926–26943.

50 Maolanon, A.R., Madsen, A.S., and Olsen, C.A. (2016). Innovative strategies for selective inhibition of histone deacetylases. *Cell Chem. Biol.* 23: 759–768.

51 KrennHrubec, K., Marshall, B.L., Hedglin, M. et al. (2007). Design and evaluation of 'linkerless' hydroxamic acids as selective HDAC8 inhibitors. *Bioorg. Med. Chem. Lett.* 17: 2874–2878.

52 Balasubramanian, S., Ramos, J., Luo, W. et al. (2008). A novel histone deacetylase 8 (HDAC8)-specific inhibitor PCI-34051 induces apoptosis in T-cell lymphomas. *Leukemia* 22: 1026–1034.

53 Huang, W.-J., Wang, Y.-C., Chao, S.-W. et al. (2012). Synthesis and biological evaluation of *ortho*-aryl *N*-hydroxycinnamides as potent histone deacetylase (HDAC) 8 isoform-selective inhibitors. *ChemMedChem* 7: 1815–1824.

54 Whitehead, L., Dobler, M.R., Radetich, B. et al. (2011). Human HDAC isoform selectivity achieved via exploitation of the acetate release channel with structurally unique small molecule inhibitors. *Bioorg. Med. Chem.* 19: 4626–4634.

55 Jones, P., Altamura, S., De Francesco, R. et al. (2008). Probing the elusive catalytic activity of vertebrate class IIa histone deacetylases. *Bioorg. Med. Chem. Lett.* 18: 1814–1819.

56 Bottomley, M.J., Lo Surdo, P., Di Giovine, P. et al. (2008). Structural and functional analysis of the human HDAC4 catalytic domain reveals a regulatory structural zinc-binding domain. *J. Biol. Chem.* 283: 26694–26704.

57 Lobera, M., Madauss, K.P., Pohlhaus, D.T. et al. (2013). Selective class IIa histone deacetylase inhibition via a nonchelating zinc-binding group. *Nat. Chem. Biol.* 9: 319–325.

58 Luckhurst, C.A., Breccia, P., Stott, A.J. et al. (2016). Potent, selective, and CNS-penetrant tetrasubstituted cyclopropane class IIa histone deacetylase (HDAC) inhibitors. *ACS Med. Chem. Lett.* 7: 34–39.

59 Jones, P., Bottomley, M.J., Carfí, A. et al. (2008). 2-Trifluoroacetylthiophenes, a novel series of potent and selective class II histone deacetylase inhibitors. *Bioorg. Med. Chem. Lett.* 18: 3456–3461.

60 Scarpelli, R., Di Marco, A., Ferrigno, F. et al. (2008). Studies of the metabolic stability in cells of 5-(trifluoroacetyl)thiophene-2-carboxamides and identification of more stable class II histone deacetylase (HDAC) inhibitors. *Bioorg. Med. Chem. Lett.* 18: 6078–6082.

61 Muraglia, E., Altamura, S., Branca, D. et al. (2008). 2-Trifluoroacetylthiophene oxadiazoles as potent and selective class II human histone deacetylase inhibitors. *Bioorg. Med. Chem. Lett.* 18: 6083–6087.

62 Ilies, M., Dowling, D.P., Lombardi, P.M., and Christianson, D.W. (2011). Synthesis of a new trifluoromethylketone analogue of L-arginine and contrasting inhibitory activity against human arginase I and histone deacetylase 8. *Bioorg. Med. Chem. Lett.* 21: 5854–5858.

63 Ontoria, J.M., Altamura, S., Di Marco, A. et al. (2009). Identification of novel, selective, and stable inhibitors of class II histone deacetylases. Validation studies of the inhibition of the enzymatic activity of HDAC4 by small molecules as a novel approach for cancer therapy. *J. Med. Chem.* 52: 6782–6789.

64 Madsen, A.S., Kristensen, H.M.E., Lanz, G., and Olsen, C.A. (2014). The effect of various zinc binding groups on inhibition of histone deacetylases 1–11. *ChemMedChem* 9: 614–626.

65 Butler, K.V., Kalin, J., Brochier, C. et al. (2010). Rational design and simple chemistry yield a superior, neuroprotective HDAC6 inhibitor, tubastatin A. *J. Am. Chem. Soc.* 132: 10842–10846.

66 Santo, L., Hideshima, T., Kung, A.L. et al. (2012). Preclinical activity, pharmacodynamic, and pharmacokinetic properties of a selective HDAC6 inhibitor, ACY-1215, in combination with bortezomib in multiple myeloma. *Blood* 119: 2579–2589.

67 Lee, J.-H., Mahendran, A., Yao, Y. et al. (2013). Development of a histone deacetylase 6 inhibitor and its biological effects. *Proc. Natl. Acad. Sci. U.S.A.* 110: 15704–15709.

68 Inks, E.S., Josey, B.J., Jesinkey, S.R., and Chou, C.J. (2012). A novel class of small molecule inhibitors of HDAC6. *ACS Chem. Biol.* 7: 331–339.

69 Kozikowski, A.P., Chen, Y., Gaysin, A. et al. (2007). Functional differences in epigenetic modulators – superiority of mercaptoacetamide-based histone deacetylase inhibitors relative to hydroxamates in cortical neuron neuroprotection studies. *J. Med. Chem.* 50: 3054–3061.

70 Estiu, G., Greenberg, E., Harrison, C.B. et al. (2008). Structural origin of selectivity in class II-selective histone deacetylase inhibitors. *J. Med. Chem.* 51: 2898–2906.

71 Moradei, O., Leit, S., Zhou, N. et al. (2006). Substituted *N*-(2-aminophenyl)-benzamides, (*E*)-*N*-(2-aminophenyl)-acrylamides and their analogues: novel classes of histone deacetylase inhibitors. *Bioorg. Med. Chem. Lett.* 16: 4048–4052.

72 Bantscheff, M., Hopf, C., Savitski, M.M. et al. (2011). Chemoproteomics profiling of HDAC inhibitors reveals selective targeting of HDAC complexes. *Nat. Biotechnol.* 29: 255–265.

73 Becher, I., Dittmann, A., Savitski, M.M. et al. (2014). Chemoproteomics reveals time-dependent binding of histone deacetylase inhibitors to endogenous repressor complexes. *ACS Chem. Biol.* 9: 1736–1746.

74 Singh, R.K., Cho, K., Padi, S.K.R. et al. (2015). Mechanism of *N*-acylthiourea-mediated activation of human histone deacetylase 8 (HDAC8) at molecular and cellular levels. *J. Biol. Chem.* 290: 6607–6619.

75 Villalba, J.M. and Alcaín, F.J. (2012). Sirtuin activators and inhibitors. *Biofactors* 38: 349–359.

8

Sirtuins as Drug Targets

Clemens Zwergel, Dante Rotili, Sergio Valente, and Antonello Mai

Sapienza Università di Roma, Dipartimento di Chimica e Tecnologie del Farmaco, 00185 Roma, Italy

8.1 Introduction

The mammalian sirtuins (SIRTs) are evolutionally highly conserved proteins and belong to class III histone deacetylases (HDACs). Its seven family members (SIRT1–7) share a NAD^+-dependent catalytic protein lysine deacetylase and/or mono-ADP-ribosylase mechanism and are involved in various biological processes acting on diverse substrates. SIRTs vary in length and sequence at their N- and C-termini. This might explain in part their diverse functions and localizations [1]. To date, their protein lysine deacetylation is the most studied function; however recent studies revealed that several SIRTs also are able to cleave other types of acyl groups, e.g. succinyl, malonyl, glutaryl, and long-chain fatty acyl residues [2–4].

In recent years, there is a growing body of literature highlighting the association of SIRTs with various pathologies: SIRT inhibition might be beneficial in cancer treatment, viral infections, muscular diseases, and neurodegenerative disorders, whereas SIRT activation may have a positive impact on metabolic and age-related disorders [1]. Thus, the discovery of SIRT modulators via screening of chemical libraries and catalytic mechanism-based design approaches, often in combination with structure–activity relationship (SAR) investigations, is nowadays a field of active research. In this book chapter, we would like to illustrate the most important SIRT functions and SIRT modulators discovered so far (Figures 8.1–8.3).

8.2 Biological Functions of Sirtuins in Physiology and Pathology

SIRT1 was the first family member discovered among the SIRTs, and it is still the most investigated. It plays a pivotal role in many neuronal processes [5, 6]; hence further studies regarding neurological disorders such as Alzheimer's disease (AD), Parkinson's disease (PD), and Huntington's disease (HD) have been conducted [7–10]. SIRT1 possesses neuroprotective properties, and it is crucial

Epigenetic Drug Discovery, First Edition. Edited by Wolfgang Sippl and Manfred Jung.

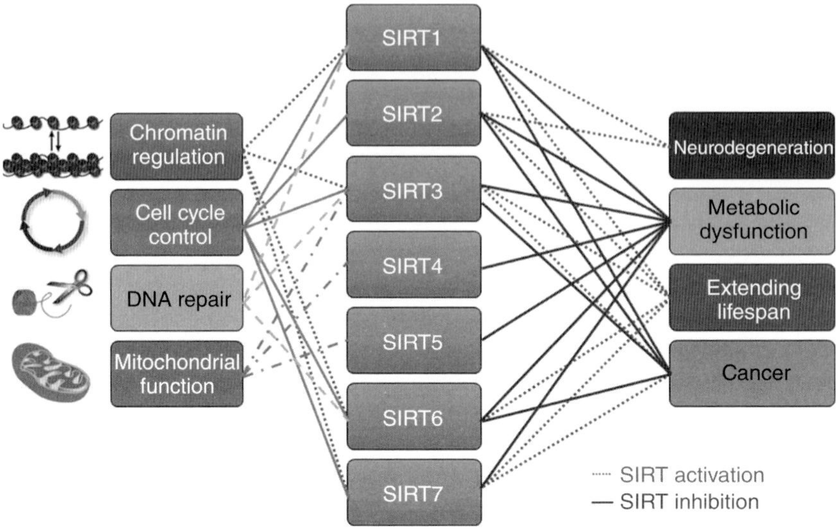

Figure 8.1 Sirtuins: biological functions and influence on various disorders.

for the survival and expression of brain-derived neurotrophic factor (BDMF). SIRT1 knockout in a rodent model of HD led to aggravation of the disease, whereas its re-expression retrieved its neuroprotective properties [11]. Furthermore, SIRT1-mediated deacetylation at K53 of nicotinamide phosphoribosyl transferase (NAMPT), an important therapeutic target against ischemic stroke, resulted in increased activity and secretion [12].

In tumorigenesis, SIRT1 seems to be a double-edged sword, acting as oncogene or tumor suppressor depending on the cell type, status, and context [13, 14]. SIRT1 was found overexpressed in some malignancies (breast, prostate, liver, melanoma, colorectal, and ovarian epithelial cancer) and reduced in others (bladder, ovary, prostate, glioblastoma, and colon cancer). SIRT1 is also able to inactivate at transcriptional and posttranslational level some tumor suppressor proteins (p53, p73, etc.) and to activate the oncoprotein BCL6, to exert antiapoptotic and antidifferentiation activities through deacetylation of transcription factors (Ku70, E2F1, FOXO3a, etc.), and to regulate cell cycle progression, chromosomal stability, and DNA repair [15–20]. On the other hand, Sirt1 knockdown or knockout mice developed spontaneous tumors in multiple organs [21–23].

Wilking et al. investigated the involvement of SIRT1 in melanoma and suggested the use of SIRT inhibitors (SIRTi) such as tenovins, EX-527, and sirtinol, either alone or in combination with other drugs. The aforementioned SIRTi target various SIRTs, supporting the concept that the simultaneous inhibition of several SIRT isoforms could lead to a reduced malignant growth [24].

SIRT2 appears to foster neurodegenerative processes in neurological disorders [25]. Indeed, pharmacological inhibition or genetic knockout of SIRT2 alleviates α-synuclein-mediated toxicity in PD [26]. Recent findings revealed that SIRT2 might also display a regulatory functions being either a tumor promoter or tumor suppressor in tumorigenesis [27, 28]. SIRT2 influences both mitotic and apoptotic processes through p53 and anaphase regulation. SIRT2, as confirmed in several studies, is able to control the cell cycle progression being an important checkpoint for metaphase/anaphase processes and G2/M transition [29, 30]. Additionally, other studies about SIRT2 revealed its participation in metabolic processes, such as adipogenesis [31]. Park et al. showed that the essential enzyme for glucose homeostasis, associated with diabetes mellitus, the acetylated form of the glucokinase regulatory protein (GKRP), can be deacetylated by SIRT2 [32].

SIRT3, together with SIRT4 and SIRT5, is one of the three mitochondrial SIRTs. It has been associated to neurodegenerative diseases, cancer, and metabolic disorders. SIRT3 is able to protect neurons in the cochlea against oxidative degradation under caloric restriction conditions [33, 34] and to regulate under stress the mitochondrial antioxidant manganese superoxide dismutase (MnSOD) in microglia [35, 36]. Additionally, several studies underline the involvement of SIRT3 in cancer progression. SIRT3 mainly blocks ROS production in mitochondria, but also activates many mitochondrial proteins via deacetylation, thus regulating proliferation, differentiation, and survival [37, 38]. Recent findings suggest that SIRT3 might suppress tumors inhibiting the glycolysis metabolism via deacetylation and activation of the pyruvate dehydrogenase. Hence the influence of SIRT3 on cancer needs further and deeper investigation.

The remaining SIRTs need to be studied deeper to discover their precise biological functions. Unlike other isoforms, SIRT4 does not seem to exhibit a NAD^+-dependent deacetylase activity and was initially described as an ADP-ribosyltransferase, which is involved in the regulation of insulin secretion by ADP-ribosylating the glutamate dehydrogenase (GDH). Therefore, SIRT4 could be an attractive therapeutic target to hit for treatment of type 2 diabetes [39].

SIRT4 can be induced by DNA-damaging processes such as chemotherapy and γ-irradiation; furthermore it is leading to cell cycle arrest through the inhibition of the mitochondrial glutamine metabolism. Low expression of SIRT4 led to increased proliferation, invasion, and migration in colorectal cancer cells being associated with a poor prognosis [40].

SIRT5 may interact with carbamoyl phosphate synthetase 1 (CPS1) via deacetylation. The importance of SIRT5 in tumorigenesis needs to be further elucidated, despite the recent study, in which its overexpression in non-small cell lung cancer cells coincided with low survival [41].

Etchegaray et al. described SIRT6 as an epigenetic guard for cellular differentiation via the control of cellular homeostasis, DNA repair mechanisms, maintenance of telomeres, and metabolism; thus SIRT6 can be considered as an oncosuppressor [42].

There is growing evidence that SIRT7 might be an auspicious epigenetic target in cancer therapy. H3K18 is an emerging biomarker in aggressive cancers, influencing various tumor suppressor genes that are selectively deacetylated by SIRT7. In fact, high expression levels of SIRT7 have been correlated with very aggressive forms of cancer and a poor prognosis, whereas its depletion resulted higher survival rates with less aggressive phenotypes [43].

8.3 SIRT Modulators

Nowadays, many SIRT modulators, used alone or in combination with approved drugs or other epigenetic modulators, have been reported as potentially useful in the treatment of cancer and age-related diseases, such as neurodegeneration or metabolic disorders [44].

8.3.1 SIRT Inhibitors

8.3.1.1 Small Molecules

In 2001 splitomicin (**1**) was the first SIRTi found via a yeast-based phenotypic screening. Despite its inactivity against human SIRTs, this compound has been the initial point for the development of several human SIRT1/2 inhibitors [45]. HR-73 (**2**), a single-digit micromolar splitomicin-derived SIRT1i, possessing a phenyl ring in C2 position and a bromine in C8 position, was able to inhibit the HIV transcription via Tat protein deacetylation [46]. Other splitomicin analogues such as **3** and **4**, which possess a phenyl ring in the 3 position and various substituents in C8 position, turned out to exhibit SIRT2 inhibition in the low micromolar range, leading to α-tubulin hyperacetylation and apoptosis in MCF7 cells [47]. EX-527 (selisistat) **5** was identified in 2005 as the first potent, selective (over SIRT2/3), and cell-permeable SIRT1i [48, 49]. Interestingly, despite able to increase p53 acetylation levels, EX-527 administration induced only just negligible effects on viability and proliferation in various cancer cell lines [50]. Presently, **5** is being developed for HD. Indeed, a phase II study has recently confirmed that EX-527 possesses a good safety profile and is well tolerated in HD patients at doses with beneficial effects in preclinical HD models [51]. Another possible application of **5** might be the treatment of human papillomavirus infections through a SIRT1 inhibition-dependent mechanism that blocks the viral amplification [52]. Discovered in 2006, cambinol (**6**) has been shown to lead to hyperacetylated p53, α-tubulin, FOXO3a, and Ku70 in NCI-H460, lymphoma, and HeLa cancer cells, being a moderate SIRT1/2i able to reduce tumor burden in a xenograft mouse model [53]. Furthermore, anticancer properties could be observed both *in vitro* and *in vivo* in hepatocellular carcinoma models [54], as well as in neuroblastoma in N-Myc transgenic mice [55]. These auspicious findings led to various optimized derivatives/analogues with augmented potency and/or selectivity for either SIRT1 or SIRT2 [56–58]. AGK2 (**7**) is a low micromolar SIRT2 inhibitor selective over SIRT1/3 and able to prevent deacetylation of α-tubulin in HeLa cells. AGK2 also protects against PD in various disease models as it salvages dopaminergic neurons from α-synuclein toxicity [26]. It is also assumed that **7** induces caspase-3-dependent cell death in glioma cells via SIRT2 inhibition [59]. The p53 activators tenovin-1 (**8**) and its water-soluble analogue tenovin-6 (**9**) were identified in a cell-based screening as SIRTi. Tenovin-6, a micromolar SIRT1–3i, led to increased levels of p53K382ac, thus displaying cytotoxic effects in melanoma and protracting the growth of ARN8-derived xenograft tumors [60]. Additionally, the induction of apoptosis in gastric cancer [61] as well as in chronic myeloid leukemia cells has been observed. Lastly, a deceleration of cancer progression in rodent models has been demonstrated [62].

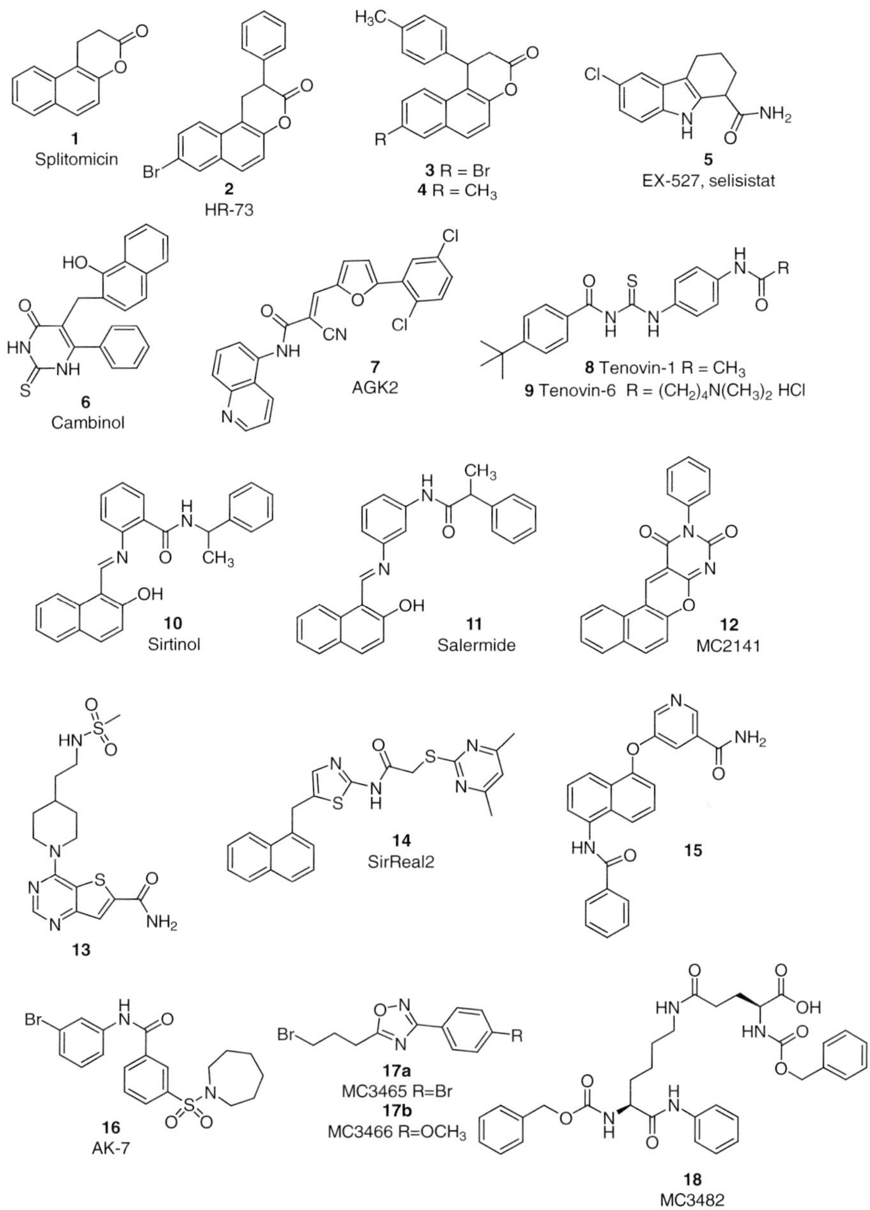

Figure 8.2 Sirtuin inhibitors **1–18**.

In 2001, sirtinol (**10**) was found via a high-throughput cell-based screening, being a micromolar ySir2 and hSIRT1/2 inhibitor endowed with various anticancer effects [63]. Compound **10** inhibits senescence-like growth in human MCF7 and H1299 cells [64], enhances the chemosensitivity of PC3, DU145, and HeLa cells to cisplatin and camptothecin [65, 66], and induces significantly apoptosis in adult T-cell leukemia–lymphoma (ATL) cells [67]. Salermide (**11**), a

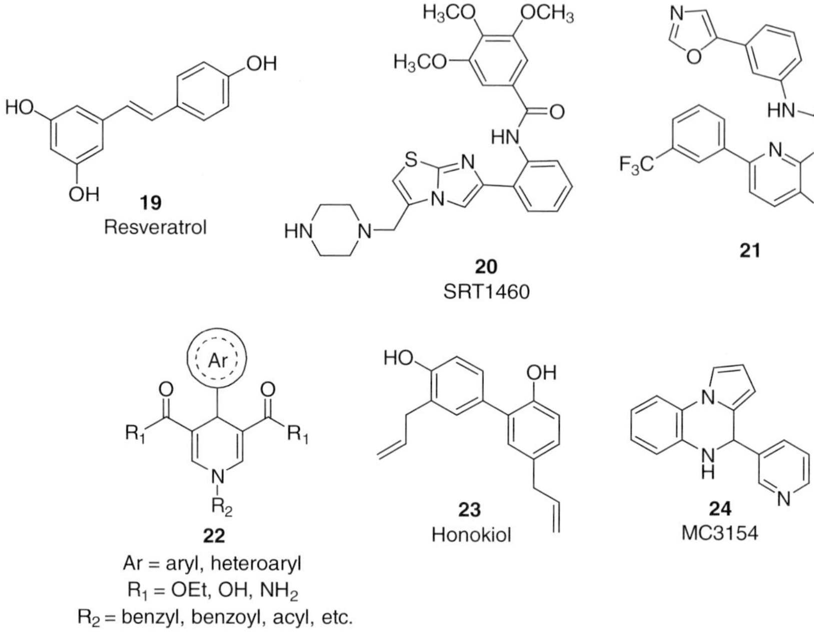

Figure 8.3 Sirtuin activators **18–24**.

moderate SIRT1/2i, derives from a medicinal chemistry optimization of sirtinol (**10**). Compound **11** leads to proapoptotic effects in various cancer cell lines such as MOLT4, KG1A, SW480, or Raji as well as cancer stem cells through the inhibition of SIRT1 via the block of the H4K16ac deacetylation and the subsequent repression of proapoptotic genes such as *CASP8*, *TNF*, *TNFRSF10B*, and *PUMA* [68, 69]. Furthermore it possesses cell protective effects in muscular dystrophy model [70], and **11** exhibits anti-infective properties in *Schistosoma mansoni*, resulting in apoptosis induction, death of schistosomula, and reduction of egg laying [71]. MC2141 (**12**) was described in 2010 as a cambinol derivative prototype of a series of benzodeazaoxaflavins that display SIRT1/2 inhibition at low micromolar levels and exert proapoptotic effects in various cancer cell lines, including cancer stem cells [72, 73]. Some of the most effective SIRTi known so far are the thieno[3,2-*d*]pyrimidine-6-carboxamides. These compounds were discovered through a DNA-encoded small-molecule library screen and have been optimized via a SAR optimization [74]. Despite being nanomolar against SIRT1–3 and the confirmed binding mode by X-ray analysis, for the best compound in the series (**13**), no biological activity has been reported so far.

Recently, SirReal2 (**14**) was described as a highly selective nanomolar SIRT2i with >1000-fold selectivity over SIRT1/3/4/5/6 [75]. X-ray crystallography has shown that this highly potent compound interacts with a previously unexploited SIRT2 binding pocket and is able to induce a structural rearrangement of the enzyme active site. The ability of SirReal2 to inhibit SIRT2 has been demonstrated in HeLa cells, leading to hyperacetylation of α-tubulin and destabilization of the SIRT2 substrate BubR1. Inspired by the SIRTi suramin and nicotinamide, a fragment-based approach led to a nanomolar SIRT2-selective (over SIRT1/3) inhibitor (**15**). This compound possesses cytotoxic effects in various cancer cell lines, leading to hyperacetylation of α-tubulin in MCF-7 cells in a time- and dose-dependent manner [76].

The brain-permeable, selective (over SIRT1/3), and micromolar SIRT2 inhibitor AK-7 (**16**) has been shown in 2011 as able to downregulate the neuronal cholesterol biosynthesis [77]. Interestingly, AK-7 exhibits various other SIRT2 inhibition-dependent neuroprotective properties in diverse models of HD [78] and PD [79], underlining the importance of the development of SIRT2i as a new potential therapeutic approach for neurodegenerative disorders. The potential relevance of SIRT2 inhibition in cancer was recently suggested by Moniot et al. They reported a series of potent SIRT2-selective inhibitors based on the 1,2,4-oxadiazole scaffold. Both compounds **17a** and **17b** are active at single-digit micromolar level by using the Sirt2 substrate α-tubulin-acetylLys40 peptide and inactive up to 100 μM against Sirt1, Sirt3, and Sirt5. Tested in various leukemia cell lines, they showed proapoptotic and antiproliferative effects [80].

8.3.1.2 Peptides and Pseudopeptides

For the first peptide-based SIRTi, the N^{ε}-acetyl-lysine was replaced with a thioacetylated residue, and at the C-terminus, a p53 protein tail was attached [81–83], displaying a strong SIRT1 inhibition ($IC_{50} \sim 2\,\mu M$) being also quite potent against SIRT2/3 ($IC_{50} \sim 2$ and $67\,\mu M$, respectively) [84]. Subsequently other thioacetylated peptidic inhibitors have been described using different SIRT substrates such as human α-tubulin and acetyl-coenzyme A synthetase 2 (AceCS2) inhibiting SIRT1/2/3 at a low micromolar levels [82–90]. The first small pseudopeptidic SIRTi was reported in 2009 [86]. Most of these compounds are SIRT1/2/3 inhibitors; however some of them weakly inhibited SIRT6 as well [88, 90]. Besides the aforementioned thioacetyl group, several other lysine N^{ε}-modifications (such as propionyl, α-hydroxyacetyl, homocitrulline, or homoarginine residues) have been described with improved potency possessing SIRT1/2 IC_{50}s in the (sub)micromolar range [82, 83, 90–92]. Polletta et al. proposed that SIRT5 in non-liver cells controls ammonia production and ammonia-induced autophagy by regulating glutamine metabolism. Their results show that ammonia production increased when using a new specific peptide-based inhibitor of SIRT5 called MC3482 (**18**) or in SIRT5-silenced cells, leading to a higher rate of autophagy and mitophagy [93].

8.3.2 SIRT Activators

The first described SIRT1a in the literature was resveratrol, a natural polyphenolic compound (**19**). This compound has been first demonstrated to prolong lifespan in rather simple forms of life such as yeast, worms, fishes, flies, and bees [94–97]. The encouraging results could be also transferred to mammals: under resveratrol treatment, obese mice displayed an increased health- and lifespan [98], whereas in healthy mice improved mitochondrial functions were found. Furthermore, they were less prone to high-fat diet-induced obesity [99].

A high-throughput screening identified compounds (**20**) as selective SIRT1a keeping most of the beneficial effects of resveratrol *in vivo* with a higher potency [99]. Some of the most potent activators of this type possess very promising properties; thus they are currently evaluated in clinical trials as potential options for the treatment of age-related diseases [100–104].

In the last years, SIRT1 activators have been controversially discussed due to the fact that only artificial substrates such as aminomethylcoumarin (AMC)- or carboxytetramethylrhodamine (TAMRA)-labeled ones led to detect SIRT activation, whereas with natural ones no activation could be detected [105–107]. Park et al. proposed that SIRT1a might be rather an indirect regulator of other pathways, such as AMPK activation and/or PDE inhibition, than a direct biochemical activator [108]. Despite these ambiguous findings, direct SIRT1 activation has recently been confirmed underlining the importance of delicate structural and positional requirements to detect SIRT1 activation in the presence of its natural substrates (e.g. FOXO3a and PGC-1α) [102, 109]. Furthermore, the crystal structure of the SIRT activator **21** bound to an engineered minimally functional hSIRT1 has been published, reassuring the possibility of a direct allosteric activation of SIRT1 by small molecules [110].

1,4-Dihydropyridines (DHPs) with general structure **22** have been discovered by Mai et al. as a new lead structure for SIRT activation. Some of these DHPs led to hypoacetylation of α-tubulin in U937 cells, high NO release in human keratinocyte HaCat cells, and improved wound healing and skin repair in a mouse model. These compounds are able to activate the SIRT1/AMPK pathway in HaCaT keratinocytes and myoblasts improving their mitochondrial functions and density, whereas their combination with either a SIRTi (**5**, EX-527) or an AMPKi (compound C) was capable to hamper these effects. A water-soluble DHP analogue augmented H4K16ac deacetylation and revealed antiproliferative properties in various cancer cell lines at 8–35 μM [111–113].

The natural compound honokiol (**23**) displayed anti-hypertrophic effects in a mouse model of hypertrophy, thus being beneficial for the cardiac system. In this work, Pillai et al. proposed a direct pharmacological activation with increased expression and activation of SIRT3, leading to reduced acetylation of mitochondrial Sirt3 substrates such as MnSOD and oligomycin-sensitivity-conferring protein (OSCP) [36]. Very recently, You et al. disclosed a series of pyrrolo[1,2-*a*]quinoxaline derivatives, yielding the first synthetic SIRT6 activator **24**. Biochemical assays show direct, substrate-independent compound binding to the Sirt6 catalytic core and potent activation of Sirt6-dependent deacetylation of peptide substrates and complete nucleosomes. Furthermore, the crystal structures of Sirt6 in complex with **24** revealed that the compound binds to a Sirt6-specific acyl channel pocket [114].

8.4 Summary and Conclusions

After a critical literature overview regarding SIRT modulation, we can say that a beneficial influence by SIRT small-molecule modulators can be observed in several diseases, such as cancer and neurodegenerative disorders. Obesity, diabetes, and other age-related diseases might benefit from SIRT targeting as well; however further research efforts are needed to confirm the SIRT role in such disorders. However, it is difficult to exactly dissect the role, in physiology as well as in pathology, of SIRTs for a series of reasons. First, SIRT1 is the first human SIRT discovered and thus the most studied in different biological contexts, but not so huge amount of data is available for the other SIRT family members. Second, some SIRTs are expressed ubiquitously in the living organism, making a clear assignment to biological functions rather difficult. Third, in some cases both SIRT activation and inhibition, even of the same isoform (for instance SIRT1, or SIRT2, or SIRT3), have been ascribed to have a positive influence on disorders such as cancer or neurodegeneration. In face of the evident contradiction, this is theoretically a feasible possibility considering as a player not just the catalytic domain of the SIRT, but the various whole SIRT-containing complexes and the several possible natural substrates on which they can act. Another explanation can be that the SIRTs role could change depending on the different states of cells and their physio/pathological contexts (redox status, activation/silencing of specific pathways, etc.)

Considering the recent growing body of literature regarding SIRTa, still in their early stage of development and characterization, and SIRTi, better studied respect to activators, new pathways and alternative strategies are likely to emerge in the years to come.

In closing, an isoform-selective modulation of SIRT might represent a promising way to try to understand the multifaceted biological roles, in the different contexts, of SIRTs; thus identification of new isoform-selective molecules (either inhibitors or activators) is strongly to pursue.

References

1 Schiedel, M., Robaa, D., Rumpf, T. et al. (2017). The current state of NAD^+-dependent histone deacetylases (sirtuins) as novel therapeutic targets. *Med. Res. Rev.* 38 (1): 147–200.

2 Tan, M., Peng, C., Anderson, K.A. et al. (2014). Lysine glutarylation is a protein posttranslational modification regulated by SIRT5. *Cell Metab.* 19: 605–617.

3 Vu, C.B., Dich J.S., Ng P.Y. et al. (2010). Benzimidazoles and related analogs as sirtuin modulators. WO2010/003048 A1.

4 Du, J., Zhou, Y., Su, X. et al. (2011). Sirt5 is a NAD-dependent protein lysine demalonylase and desuccinylase. *Science* 334: 806–809.

5 Li, X.H., Chen, C., Tu, Y. et al. (2013). Sirt1 promotes axonogenesis by deacetylation of Akt and inactivation of GSK3. *Mol. Neurobiol.* 48: 490–499.

6 Araki, T., Sasaki, Y., and Milbrandt, J. (2004). Increased nuclear NAD biosynthesis and SIRT1 activation prevent axonal degeneration. *Science* 305: 1010–1013.

7 Qin, W., Yang, T., Ho, L. et al. (2006). Neuronal SIRT1 activation as a novel mechanism underlying the prevention of Alzheimer disease amyloid neuropathology by calorie restriction. *J. Biol. Chem.* 281: 21745–21754.
8 Donmez, G., Arun, A., Chung, C.Y. et al. (2012). SIRT1 protects against alpha-synuclein aggregation by activating molecular chaperones. *J. Neurosci.* 32: 124–132.
9 Jeong, J.K., Moon, M.H., Lee, Y.J. et al. (2013). Autophagy induced by the class III histone deacetylase Sirt1 prevents prion peptide neurotoxicity. *Neurobiol. Aging* 34: 146–156.
10 Derr, R.S., van Hoesel, A.Q., Benard, A. et al. (2014). High nuclear expression levels of histone-modifying enzymes LSD1, HDAC2 and SIRT1 in tumor cells correlate with decreased survival and increased relapse in breast cancer patients. *BMC Cancer* 14: 604.
11 Jeong, H., Cohen, D.E., Cui, L. et al. (2011). Sirt1 mediates neuroprotection from mutant huntingtin by activation of the TORC1 and CREB transcriptional pathway. *Nat. Med.* 18: 159–165.
12 Yoon, M.J., Yoshida, M., Johnson, S. et al. (2015). SIRT1-mediated eNAMPT secretion from adipose tissue regulates hypothalamic NAD+ and function in mice. *Cell Metab.* 21: 706–717.
13 Deng, C.X. (2009). SIRT1, is it a tumor promoter or tumor suppressor? *Int. J. Biol. Sci.* 5: 147–152.
14 Bosch-Presegue, L. and Vaquero, A. (2011). The dual role of sirtuins in cancer. *Genes Cancer* 2: 648–662.
15 Jung-Hynes, B., Nihal, M., Zhong, W., and Ahmad, N. (2009). Role of sirtuin histone deacetylase SIRT1 in prostate cancer. A target for prostate cancer management via its inhibition. *J. Biol. Chem.* 284: 3823–3832.
16 Chen, J., Zhang, B., Wong, N. et al. (2011). Sirtuin 1 is upregulated in a subset of hepatocellular carcinomas where it is essential for telomere maintenance and tumor cell growth. *Cancer Research* 71: 4138–4149.
17 Jang, K.Y., Kim, K.S., Hwang, S.H. et al. (2009). Expression and prognostic significance of SIRT1 in ovarian epithelial tumours. *Pathology* 41: 366–371.
18 Cha, E.J., Noh, S.J., Kwon, K.S. et al. (2009). Expression of DBC1 and SIRT1 is associated with poor prognosis of gastric carcinoma. *Clin. Cancer Res.* 15: 4453–4459.
19 Chen, X., Sun, K., Jiao, S. et al. (2014). High levels of SIRT1 expression enhance tumorigenesis and associate with a poor prognosis of colorectal carcinoma patients. *Sci. Rep.* 4: 7481.
20 Li, C., Wang, L., Zheng, L. et al. (2015). SIRT1 expression is associated with poor prognosis of lung adenocarcinoma. *Onco Targets Ther.* 8: 977–984.
21 Kabra, N., Li, Z., Chen, L. et al. (2009). SirT1 is an inhibitor of proliferation and tumor formation in colon cancer. *J. Biol. Chem.* 284: 18210–18217.
22 Herranz, D. and Serrano, M. (2010). SIRT1: recent lessons from mouse models. *Nat. Rev. Cancer* 10: 819–823.
23 Wang, R.H., Sengupta, K., Li, C. et al. (2008). Impaired DNA damage response, genome instability, and tumorigenesis in SIRT1 mutant mice. *Cancer Cell* 14: 312–323.

24 Wilking, M.J., Singh, C.K., Nihal, M. et al. (2014). Sirtuin deacetylases: a new target for melanoma management. *Cell Cycle* 13: 2821–2826.

25 Harting, K. and Knoll, B. (2010). SIRT2-mediated protein deacetylation: an emerging key regulator in brain physiology and pathology. *Eur. J. Cell Biol.* 89: 262–269.

26 Outeiro, T.F., Kontopoulos, E., Altmann, S.M. et al. (2007). Sirtuin 2 inhibitors rescue alpha-synuclein-mediated toxicity in models of Parkinson's disease. *Science* 317: 516–519.

27 Li, Z., Xie, Q.R., Chen, Z. et al. (2013). Regulation of SIRT2 levels for human non-small cell lung cancer therapy. *Lung Cancer* 82: 9–15.

28 McGlynn, L.M., Zino, S., MacDonald, A.I. et al. (2014). SIRT2: tumour suppressor or tumour promoter in operable breast cancer? *Eur. J. Cancer* 50: 290–301.

29 Serrano, L., Martinez-Redondo, P., Marazuela-Duque, A. et al. (2013). The tumor suppressor SirT2 regulates cell cycle progression and genome stability by modulating the mitotic deposition of H4K20 methylation. *Genes Dev.* 27: 639–653.

30 Inoue, T., Hiratsuka, M., Osaki, M. et al. (2007). SIRT2, a tubulin deacetylase, acts to block the entry to chromosome condensation in response to mitotic stress. *Oncogene* 26: 945–957.

31 Jing, E., Gesta, S., and Kahn, C.R. (2007). SIRT2 regulates adipocyte differentiation through FoxO1 acetylation/deacetylation. *Cell Metab.* 6: 105–114.

32 Park, J.M., Kim, T.H., Jo, S.H. et al. (2015). Acetylation of glucokinase regulatory protein decreases glucose metabolism by suppressing glucokinase activity. *Sci. Rep.* 5: 17395.

33 Someya, S., Yu, W., Hallows, W.C. et al. (2010). Sirt3 mediates reduction of oxidative damage and prevention of age-related hearing loss under caloric restriction. *Cell* 143: 802–812.

34 Quan, Y., Xia, L., Shao, J. et al. (2015). Adjudin protects rodent cochlear hair cells against gentamicin ototoxicity via the SIRT3-ROS pathway. *Sci. Rep.* 5: 8181.

35 Rangarajan, P., Karthikeyan, A., Lu, J. et al. (2015). Sirtuin 3 regulates Foxo3a-mediated antioxidant pathway in microglia. *Neuroscience* 311: 398–414.

36 Pillai, V.B., Samant, S., Sundaresan, N.R. et al. (2015). Honokiol blocks and reverses cardiac hypertrophy in mice by activating mitochondrial Sirt3. *Nat. Commun.* 6: 6656.

37 Tao, R., Coleman, M.C., Pennington, J.D. et al. (2010). Sirt3-mediated deacetylation of evolutionarily conserved lysine 122 regulates MnSOD activity in response to stress. *Mol. Cell* 40: 893–904.

38 Finley, L.W., Haas, W., Desquiret-Dumas, V. et al. (2011). Succinate dehydrogenase is a direct target of sirtuin 3 deacetylase activity. *PLoS One* 6: e23295.

39 Haigis, M.C., Mostoslavsky, R., Haigis, K.M. et al. (2006). SIRT4 inhibits glutamate dehydrogenase and opposes the effects of calorie restriction in pancreatic beta cells. *Cell* 126: 941–954.

40 Miyo, M., Yamamoto, H., Konno, M. et al. (2015). Tumour-suppressive function of SIRT4 in human colorectal cancer. *Br. J. Cancer* 113: 492–499.
41 Lu, W., Zuo, Y., Feng, Y., and Zhang, M. (2014). SIRT5 facilitates cancer cell growth and drug resistance in non-small cell lung cancer. *Tumour Biol.* 35: 10699–10705.
42 Etchegaray, J.P., Chavez, L., Huang, Y. et al. (2015). The histone deacetylase SIRT6 controls embryonic stem cell fate via TET-mediated production of 5-hydroxymethylcytosine. *Nat. Cell Biol.* 17: 545–557.
43 Barber, M.F., Michishita-Kioi, E., Xi, Y. et al. (2012). SIRT7 links H3K18 deacetylation to maintenance of oncogenic transformation. *Nature* 487: 114–118.
44 Villalba, J.M. and Alcain, F.J. (2012). Sirtuin activators and inhibitors. *Biofactors* 38: 349–359.
45 Bedalov, A., Gatbonton, T., Irvine, W.P. et al. (2001). Identification of a small molecule inhibitor of Sir2p. *Proc. Natl. Acad. Sci. U.S.A.* 98: 15113–15118.
46 Pagans, S., Pedal, A., North, B.J. et al. (2005). SIRT1 regulates HIV transcription via Tat deacetylation. *PLoS Biol.* 3: e41.
47 Neugebauer, R.C., Uchiechowska, U., Meier, R. et al. (2008). Structure-activity studies on splitomicin derivatives as sirtuin inhibitors and computational prediction of binding mode. *J. Med. Chem.* 51: 1203–1213.
48 Napper, A.D., Hixon, J., McDonagh, T. et al. (2005). Discovery of indoles as potent and selective inhibitors of the deacetylase SIRT1. *J. Med. Chem.* 48: 8045–8054.
49 Gertz, M., Fischer, F., Nguyen, G.T. et al. (2013). Ex-527 inhibits sirtuins by exploiting their unique NAD+-dependent deacetylation mechanism. *Proc. Natl. Acad. Sci. U.S.A.* 110: E2772–E2781.
50 Solomon, J.M., Pasupuleti, R., Xu, L. et al. (2006). Inhibition of SIRT1 catalytic activity increases p53 acetylation but does not alter cell survival following DNA damage. *Mol. Cell. Biol.* 26: 28–38.
51 Sussmuth, S.D., Haider, S., Landwehrmeyer, G.B. et al. (2015). An exploratory double-blind, randomized clinical trial with selisistat, a SirT1 inhibitor, in patients with Huntington's disease. *Br. J. Clin. Pharmacol.* 79: 465–476.
52 Langsfeld, E.S., Bodily, J.M., and Laimins, L.A. (2015). The deacetylase sirtuin 1 regulates human papillomavirus replication by modulating histone acetylation and recruitment of DNA damage factors NBS1 and Rad51 to viral genomes. *PLoS Pathog.* 11: e1005181.
53 Heltweg, B., Gatbonton, T., Schuler, A.D. et al. (2006). Antitumor activity of a small-molecule inhibitor of human silent information regulator 2 enzymes. *Cancer Res.* 66: 4368–4377.
54 Portmann, S., Fahrner, R., Lechleiter, A. et al. (2013). Antitumor effect of SIRT1 inhibition in human HCC tumor models in vitro and in vivo. *Mol. Cancer Ther.* 12: 499–508.
55 Marshall, G.M., Liu, P.Y., Gherardi, S. et al. (2011). SIRT1 promotes N-Myc oncogenesis through a positive feedback loop involving the effects of MKP3 and ERK on N-Myc protein stability. *PLos Genet.* 7: e1002135.

56 Medda, F., Russell, R.J., Higgins, M. et al. (2009). Novel cambinol analogs as sirtuin inhibitors: synthesis, biological evaluation, and rationalization of activity. *J. Med. Chem.* 52: 2673–2682.
57 Rotili, D., Carafa, V., Tarantino, D. et al. (2011). Simplification of the tetracyclic SIRT1-selective inhibitor MC2141: coumarin- and pyrimidine-based SIRT1/2 inhibitors with different selectivity profile. *Bioorg. Med. Chem.* 19: 3659–3668.
58 Mahajan, S.S., Scian, M., Sripathy, S. et al. (2014). Development of pyrazolone and isoxazol-5-one cambinol analogues as sirtuin inhibitors. *J. Med. Chem.* 57: 3283–3294.
59 He, X., Nie, H., Hong, Y. et al. (2012). SIRT2 activity is required for the survival of C6 glioma cells. *Biochem. Biophys. Res. Commun.* 417: 468–472.
60 Lain, S., Hollick, J.J., Campbell, J. et al. (2008). Discovery, in vivo activity, and mechanism of action of a small-molecule p53 activator. *Cancer Cell* 13: 454–463.
61 Hirai, S., Endo, S., Saito, R. et al. (2014). Antitumor effects of a sirtuin inhibitor, tenovin-6, against gastric cancer cells via death receptor 5 up-regulation. *PLoS One* 9: e102831.
62 Yuan, H., Wang, Z., Li, L. et al. (2012). Activation of stress response gene SIRT1 by BCR-ABL promotes leukemogenesis. *Blood* 119: 1904–1914.
63 Grozinger, C.M., Chao, E.D., Blackwell, H.E. et al. (2001). Identification of a class of small molecule inhibitors of the sirtuin family of NAD-dependent deacetylases by phenotypic screening. *J. Biol. Chem.* 276: 38837–38843.
64 Ota, H., Tokunaga, E., Chang, K. et al. (2006). Sirt1 inhibitor, sirtinol, induces senescence-like growth arrest with attenuated Ras-MAPK signaling in human cancer cells. *Oncogene* 25: 176–185.
65 Kojima, K., Ohhashi, R., Fujita, Y. et al. (2008). A role for SIRT1 in cell growth and chemoresistance in prostate cancer PC3 and DU145 cells. *Biochem. Biophys. Res. Commun.* 373: 423–428.
66 Jin, K.L., Park, J.Y., Noh, E.J. et al. (2010). The effect of combined treatment with cisplatin and histone deacetylase inhibitors on HeLa cells. *J. Gynecol. Oncol.* 21: 262–268.
67 Kozako, T., Aikawa, A., Shoji, T. et al. (2012). High expression of the longevity gene product SIRT1 and apoptosis induction by sirtinol in adult T-cell leukemia cells. *Int. J. Cancer* 131: 2044–2055.
68 Lara, E., Mai, A., Calvanese, V. et al. (2009). Salermide, a sirtuin inhibitor with a strong cancer-specific proapoptotic effect. *Oncogene* 28: 781–791.
69 Rotili, D., Tarantino, D., Nebbioso, A. et al. (2012). Discovery of salermide-related sirtuin inhibitors: binding mode studies and antiproliferative effects in cancer cells including cancer stem cells. *J. Med. Chem.* 55: 10937–10947.
70 Pasco, M.Y., Rotili, D., Altucci, L. et al. (2010). Characterization of sirtuin inhibitors in nematodes expressing a muscular dystrophy protein reveals muscle cell and behavioral protection by specific sirtinol analogues. *J. Med. Chem.* 53: 1407–1411.

71 Lancelot, J., Caby, S., Dubois-Abdesselem, F. et al. (2013). *Schistosoma mansoni* sirtuins: characterization and potential as chemotherapeutic targets. *PLoS Negl.Trop. Dis.* 7: e2428.
72 Rotili, D., Tarantino, D., Carafa, V. et al. (2010). Identification of tri- and tetracyclic pyrimidinediones as sirtuin inhibitors. *ChemMedChem* 5: 674–677.
73 Rotili, D., Tarantino, D., Carafa, V. et al. (2012). Benzodeazaoxaflavins as sirtuin inhibitors with antiproliferative properties in cancer stem cells. *J. Med. Chem.* 55: 8193–8197.
74 Disch, J.S., Evindar, G., Chiu, C.H. et al. (2013). Discovery of thieno[3,2-d]pyrimidine-6-carboxamides as potent inhibitors of SIRT1, SIRT2, and SIRT3. *J. Med. Chem.* 56: 3666–3679.
75 Rumpf, T., Schiedel, M., Karaman, B. et al. (2015). Selective Sirt2 inhibition by ligand-induced rearrangement of the active site. *Nat. Commun.* 6: 6263.
76 Cui, H., Kamal, Z., Ai, T. et al. (2014). Discovery of potent and selective sirtuin 2 (SIRT2) inhibitors using a fragment-based approach. *J. Med. Chem.* 57: 8340–8357.
77 Taylor, D.M., Balabadra, U., Xiang, Z. et al. (2011). A brain-permeable small molecule reduces neuronal cholesterol by inhibiting activity of sirtuin 2 deacetylase. *ACS Chem. Biol.* 6: 540–546.
78 Chopra, V., Quinti, L., Kim, J. et al. (2012). The sirtuin 2 inhibitor AK-7 is neuroprotective in Huntington's disease mouse models. *Cell Rep.* 2: 1492–1497.
79 Chen, X., Wales, P., Quinti, L. et al. (2015). The sirtuin-2 inhibitor AK7 is neuroprotective in models of Parkinson's disease but not amyotrophic lateral sclerosis and cerebral ischemia. *PLoS One* 10: e0116919.
80 Moniot, S., Forgione, M., Lucidi, A. et al. (2017). Development of 1,2,4-oxadiazoles as potent and selective inhibitors of the human deacetylase sirtuin 2: structure-activity relationship, X-ray crystal structure, and anticancer activity. *J. Med. Chem.* 60: 2344–2360.
81 Fatkins, D.G., Monnot, A.D., and Zheng, W. (2006). Nepsilon-thioacetyl-lysine: a multi-facet functional probe for enzymatic protein lysine Nepsilon-deacetylation. *Bioorg. Med. Chem. Lett.* 16: 3651–3656.
82 Smith, B.C. and Denu, J.M. (2007). Acetyl-lysine analog peptides as mechanistic probes of protein deacetylases. *J. Biol. Chem.* 282: 37256–37265.
83 Smith, B.C. and Denu, J.M. (2007). Mechanism-based inhibition of Sir2 deacetylases by thioacetyl-lysine peptide. *Biochemistry* 46: 14478–14486.
84 Fatkins, D.G. and Zheng, W. (2008). Substituting N(epsilon)-thioacetyl-lysine for N(epsilon)-acetyl-lysine in peptide substrates as a general approach to inhibiting human NAD(+)-dependent protein deacetylases. *Int. J. Mol. Sci.* 9: 1–11.
85 Kiviranta, P.H., Suuronen, T., Wallen, E.A. et al. (2009). N(epsilon)-thioacetyl-lysine-containing tri-, tetra-, and pentapeptides as SIRT1 and SIRT2 inhibitors. *J. Med. Chem.* 52: 2153–2156.
86 Asaba, T., Suzuki, T., Ueda, R. et al. (2009). Inhibition of human sirtuins by in situ generation of an acetylated lysine-ADP-ribose conjugate. *J. Am. Chem. Soc.* 131: 6989–6996.

87 Suzuki, T., Asaba, T., Imai, E. et al. (2009). Identification of a cell-active non-peptide sirtuin inhibitor containing *N*-thioacetyl lysine. *Bioorg. Med. Chem. Lett.* 19: 5670–5672.

88 Huhtiniemi, T., Salo, H.S., Suuronen, T. et al. (2011). Structure-based design of pseudopeptidic inhibitors for SIRT1 and SIRT2. *J. Med. Chem.* 54: 6456–6468.

89 Mellini, P., Kokkola, T., Suuronen, T. et al. (2013). Screen of pseudopeptidic inhibitors of human sirtuins 1–3: two lead compounds with antiproliferative effects in cancer cells. *J. Med. Chem.* 56: 6681–6695.

90 Kokkonen, P., Rahnasto-Rilla, M., Kiviranta, P.H. et al. (2012). Peptides and pseudopeptides as SIRT6 deacetylation inhibitors. *ACS Med. Chem. Lett.* 3: 969–974.

91 Jamonnak, N., Fatkins, D.G., Wei, L., and Zheng, W. (2007). N(epsilon)-methanesulfonyl-lysine as a non-hydrolyzable functional surrogate for N(epsilon)-acetyl-lysine. *Org. Biomol. Chem.* 5: 892–896.

92 Huhtiniemi, T., Suuronen, T., Lahtela-Kakkonen, M. et al. (2010). N(epsilon)-modified lysine containing inhibitors for SIRT1 and SIRT2. *Bioorg. Med. Chem.* 18: 5616–5625.

93 Polletta, L., Vernucci, E., Carnevale, I. et al. (2015). SIRT5 regulation of ammonia-induced autophagy and mitophagy. *Autophagy* 11: 253–270.

94 Howitz, K.T., Bitterman, K.J., Cohen, H.Y. et al. (2003). Small molecule activators of sirtuins extend *Saccharomyces cerevisiae* lifespan. *Nature* 425: 191–196.

95 Wood, J.G., Rogina, B., Lavu, S. et al. (2004). Sirtuin activators mimic caloric restriction and delay ageing in metazoans. *Nature* 430: 686–689.

96 Valenzano, D.R., Terzibasi, E., Genade, T. et al. (2006). Resveratrol prolongs lifespan and retards the onset of age-related markers in a short-lived vertebrate. *Current Biol.* 16: 296–300.

97 Rascon, B., Hubbard, B.P., Sinclair, D.A., and Amdam, G.V. (2012). The lifespan extension effects of resveratrol are conserved in the honey bee and may be driven by a mechanism related to caloric restriction. *Aging (Albany NY)* 4: 499–508.

98 Baur, J.A., Pearson, K.J., Price, N.L. et al. (2006). Resveratrol improves health and survival of mice on a high-calorie diet. *Nature* 444: 337–342.

99 Milne, J.C., Lambert, P.D., Schenk, S. et al. (2007). Small molecule activators of SIRT1 as therapeutics for the treatment of type 2 diabetes. *Nature* 450: 712–716.

100 Minor, R.K., Baur, J.A., Gomes, A.P. et al. (2011). SRT1720 improves survival and healthspan of obese mice. *Sci. Rep.* 1: 70.

101 Dai, H., Kustigian, L., Carney, D. et al. (2010). SIRT1 activation by small molecules: kinetic and biophysical evidence for direct interaction of enzyme and activator. *J. Biol. Chem.* 285: 32695–32703.

102 Hubbard, B.P., Gomes, A.P., Dai, H. et al. (2013). Evidence for a common mechanism of SIRT1 regulation by allosteric activators. *Science* 339: 1216–1219.

103 Sinclair, D.A. and Guarente, L. (2014). Small-molecule allosteric activators of sirtuins. *Annu. Rev. Pharmacol. Toxicol.* 54: 363–380.

104 Hubbard, B.P. and Sinclair, D.A. (2014). Small molecule SIRT1 activators for the treatment of aging and age-related diseases. *Trends Pharmacol. Sci.* 35: 146–154.
105 Borra, M.T., Langer, M.R., Slama, J.T., and Denu, J.M. (2004). Substrate specificity and kinetic mechanism of the Sir2 family of NAD+-dependent histone/protein deacetylases. *Biochemistry* 43: 9877–9887.
106 Kaeberlein, M., McDonagh, T., Heltweg, B. et al. (2005). Substrate-specific activation of sirtuins by resveratrol. *J. Biol. Chem.* 280: 17038–17045.
107 Pacholec, M., Bleasdale, J.E., Chrunyk, B. et al. (2010). SRT1720, SRT2183, SRT1460, and resveratrol are not direct activators of SIRT1. *J. Biol. Chem.* 285: 8340–8351.
108 Park, S.J., Ahmad, F., Philp, A. et al. (2012). Resveratrol ameliorates aging-related metabolic phenotypes by inhibiting cAMP phosphodiesterases. *Cell* 148: 421–433.
109 Lakshminarasimhan, M., Rauh, D., Schutkowski, M., and Steegborn, C. (2013). Sirt1 activation by resveratrol is substrate sequence-selective. *Aging (Albany NY)* 5: 151–154.
110 Dai, H., Case, A.W., Riera, T.V. et al. (2015). Crystallographic structure of a small molecule SIRT1 activator-enzyme complex. *Nat. Commun.* 6: 7645.
111 Mai, A., Valente, S., Meade, S. et al. (2009). Study of 1,4-dihydropyridine structural scaffold: discovery of novel sirtuin activators and inhibitors. *J. Med. Chem.* 52: 5496–5504.
112 Spallotta, F., Cencioni, C., Straino, S. et al. (2013). A nitric oxide-dependent cross-talk between class I and III histone deacetylases accelerates skin repair. *J. Biol. Chem.* 288: 11004–11012.
113 Valente, S., Mellini, P., Spallotta, F. et al. (2016). 1,4-Dihydropyridines active on the SIRT1/AMPK pathway ameliorate skin repair and mitochondrial function and exhibit inhibition of proliferation in cancer cells. *J. Med. Chem.* 59: 1471–1491.
114 You, W., Rotili, D., Li, T.M. et al. (2017). Structural basis of sirtuin 6 activation by synthetic small molecules. *Angew. Chem. Int. Ed.* 56: 1007–1011.

9

Selective Small-Molecule Inhibitors of Protein Methyltransferases

H. Ümit Kaniskan and Jian Jin

Icahn School of Medicine at Mount Sinai, Departments of Pharmacological Sciences and Oncological Sciences, 1425 Madison Avenue, New York, NY 10029, USA

9.1 Introduction

Epigenetic regulation of gene expression depends on the state of chromatin, which can be modified in a variety of ways, including DNA methylation, nucleosome remodeling histone variants, and posttranslational modifications (PTMs) of histones. Given the vital role of epigenetic regulation in important cellular processes, such as cell differentiation, proliferation, development, and maintaining cell identity, epigenetic modifying enzymes have been increasingly recognized as potential therapeutic targets [1]. Histone methylation, along with non-histone protein methylation, is a key process in the regulation of gene expression and transcription and affects wide variety of cellular functions [2, 3]. Therefore, protein methyltransferases (PMTs) are implicated in various cancers and numerous other diseases, and the discovery of selective small-molecule inhibitors of PMTs has become a very active research area [1, 4]. This chapter is not intended to be a comprehensive review, but rather aims to highlight the progress made in the discovery of PMT inhibitors in the last 15 years [4].

9.2 Protein Methylation

PMTs catalyze the transfer of the methyl group from the cofactor *S*-5′-adenosyl-*L*-methionine (SAM) to either lysine or arginine residues of histone and non-histone substrates. They are classified based on the residues they modify: protein lysine methyltransferases (PKMTs) and protein arginine methyltransferases (PRMTs). The lysine residues can be mono-, di-, and/or trimethylated by PKMTs, and the arginine residues can only be mono- and/or dimethylated by PRMTs. Dimethylation of terminal guanidino nitrogens of arginine succeeding monomethylation of arginine (MMA) can occur on the same nitrogen to give asymmetrically dimethylated arginine (aDMA) or on two different guanidino nitrogens, resulting in symmetrically dimethylated arginine (sDMA). Although methylation of lysine or arginine residues does not change the

Epigenetic Drug Discovery, First Edition. Edited by Wolfgang Sippl and Manfred Jung.

charge of these residues, it alters the bulkiness and hydrophobicity of the protein, thus affecting recognition of the methylated protein by methyl-lysine/arginine readers via protein–protein interactions. PKMTs and PRMTs bind lysine or arginine residues of substrate proteins via the substrate binding groove and SAM via the cofactor binding site. A narrow hydrophobic channel that brings the substrate and cofactor in close proximity links these two binding sites. The successive methylation process yields higher states of lysine or arginine methylation.

9.3 Lysine Methyltransferases (PKMTs)

All known PKMTs, with the exception of DOT1L, contain a conserved SET domain, which folds into several small β-sheets that surround a knot-like structure and forms an active site next to the SAM binding pocket [5]. Lysine methylation catalyzed by PKMTs plays a major role in regulating gene expression and transcription and can lead to either transcription activation or repression [2]. While histone H3 lysine 4 (H3K4), H3K36, and H3K79 methylation are generally associated with transcription activation, H3K9 di- and trimethylation (H3K9me2 and H3K9me3) and H3K27 trimethylation (H3K27me3) are typically associated with repression [2]. Mounting evidence suggests clear association of PKMTs with cancer and other diseases [1].

9.4 Inhibitors of PKMTs

9.4.1 Inhibitors of H3K9 Methyltransferases

H3K9 methylation in humans is controlled by SUV39H1, SUV39H2, G9a, GLP, SETDB1, SETDB2, PRDM2, PRDM3, and PRDM16 [6]. The discovery of chaetocin, a fungal mycotoxin, as the first SUV39 methyltransferase inhibitor in 2005, essentially propelled the discovery of selective PKMT inhibitors [7]. However, follow-up studies on chaetocin have elucidated that it is not a selective inhibitor [8].

BIX-01294 (Figure 9.1) is identified as the first potent and selective G9a and GLP inhibitor that is competitive with the peptide substrate [9]. This mechanism of action (MOA) was confirmed by the crystal structure of the GLP SET domain in complex with BIX-01294, which clearly shows that the inhibitor binds to the substrate binding groove of GLP [10]. Based on the 2,4-diaminoquinazoline core of BIX-01294, UNC0224 (Figure 9.1) was designed to have a 7-dimethylaminopropoxy group to occupy the lysine binding channel of G9a and GLP [11]. UNC0224 displays significantly increased potency and selectivity, and the occupation of the G9a lysine binding channel by 7-dimethylaminopropoxy group of UNC0224 has been confirmed by co-crystal structure in complex with G9a. Modifications on UNC0224 resulted in the discovery of the cell-active inhibitor UNC0638 (Figure 9.1) [12]. UNC0638 is more than 200-fold selective for G9a and GLP over 16 other methyltransferases

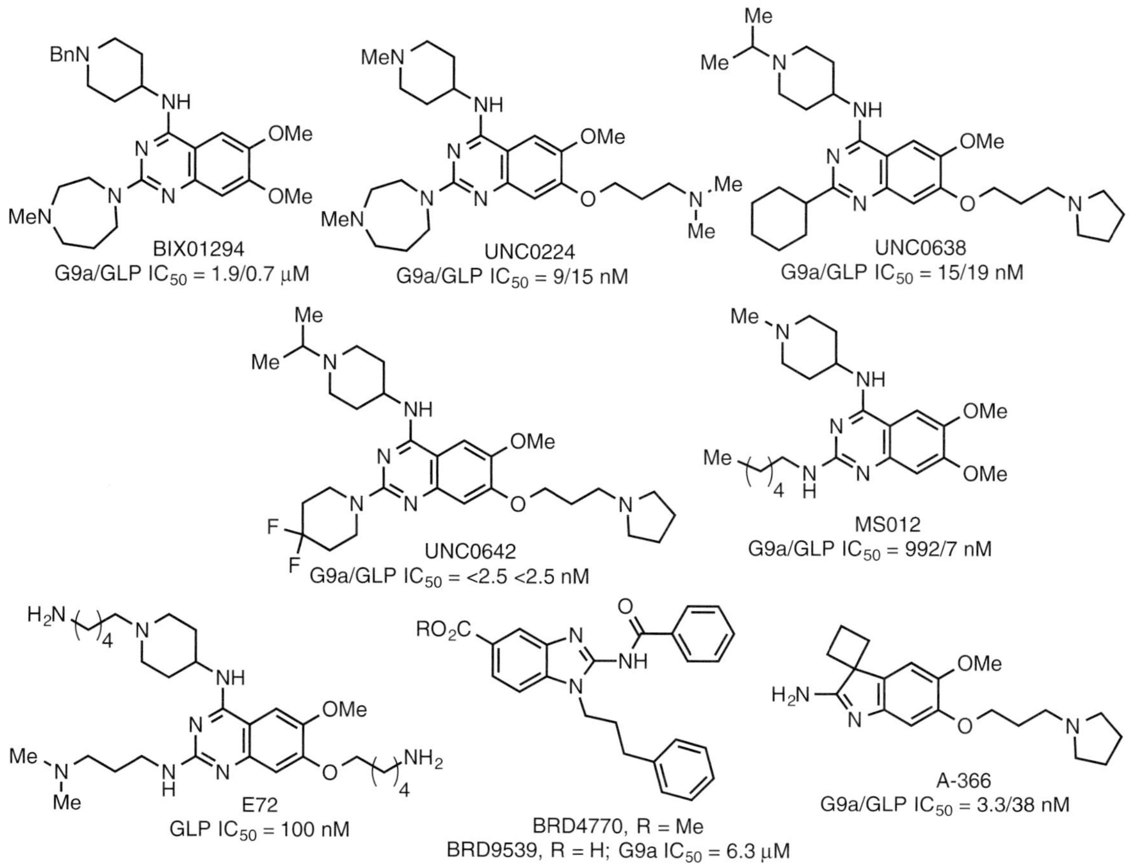

Figure 9.1 Selective small-molecule inhibitors of G9a and GLP. IC_{50} values are highly dependent on the assay conditions used; therefore, it should be interpreted with caution for comparison purposes.

and epigenetic targets. In addition, UNC0638 displays considerably reduced cytotoxicity compared to BIX-01294 in MDA-MB-231 cells (function/toxicity ratio = 138 and 5.6, respectively). Despite its utility as a cell-based chemical probe, UNC0638 displays poor pharmacokinetic (PK) properties in animals. Further optimization has resulted in UNC0642 (Figure 9.1) [13], which retains high *in vitro* potency and selectivity for G9a and GLP, while potently and selectively reducing the H3K9me2 mark and maintaining low cell toxicity in both normal and tumor cell lines. Importantly, UNC0642 exhibits significantly greater exposure in plasma in mouse PK studies, making it suitable for *in vivo* studies.

Very recently, the discovery of a potent GLP inhibitor, MS012, which is 140-fold selective for GLP over G9a, was reported (Figure 9.1) [14]. MS012 is also selective for GLP over a broad range of other PKMTs, PRMTs, DNMTs, and RNMTs. Interestingly, X-ray structures reveal that this substrate-competitive inhibitor binds to GLP and G9a in virtually identical binding modes, highlighting the challenges in structure-based design of selective inhibitors for these highly homologous enzymes.

Another selective G9 and GLP inhibitor that also features a lysine mimic added onto the quinazoline scaffold is E72 (Figure 9.1) [15]. A SAM-competitive inhibitor of G9a (activity against GLP was not reported), BRD9539, and BRD4770 as a prodrug has also been reported (Figure 9.1) [16]. Another potent and selective G9a and GLP inhibitor, A-366, was disclosed in 2014 (Figure 9.1) [17].

9.4.2 Inhibitors of H3K27 Methyltransferases

Methylation of H3K27 is catalyzed by a highly conserved multisubunit protein complex, known as polycomb repressive complex 2 (PRC2) [18]. PRC2 primarily functions to silence its target genes by trimethylating H3K27, resulting in their transcriptional repression. The core PRC2 complex consists of four subunits: (1) EZH1 or EZH2, (2) SUZ12, (3) EED, and (4) RbAp46/48. The EZH1 or EZH2 subunit is the catalytic subunit of PRC2 and contains a SET domain at the C-terminal, which is necessary for the methyltransferase activity. While EZH1 and EZH2 share 96% sequence identity in their SET domains, they possess considerably different tissue distributions [19]. For example, EZH1 exists in both differentiated and dividing cells, while EZH2 is found only in dividing cell populations [18]. Yet another reported distinction between EZH1 and EZH2 is that the PRC2 complex containing EZH2 (PRC2-EZH2) possesses greater methyltransferase activity compared with the PRC2 complex with EZH1 (PRC2-EZH1) [18].

Despite their status as the catalytic subunit of PRC2, EZH2 or EZH1 alone is not catalytically active. Instead, EZH2 or EZH1 requires at least two other subunits, EED and SUZ12, to have the methyltransferase activity [20]. It has also been demonstrated that PRC2 can contain other protein subunits as well, including AEBP2, PCLs, and JARID2 [18]. Point mutations at Y641 in the EZH2 C-terminal SET domain have been observed in 7% of follicular lymphomas and 22% of germinal center B-cell and diffuse large B-cell lymphomas [21]. Another EZH2 point mutation, A677G, has also recently been identified in lymphoma

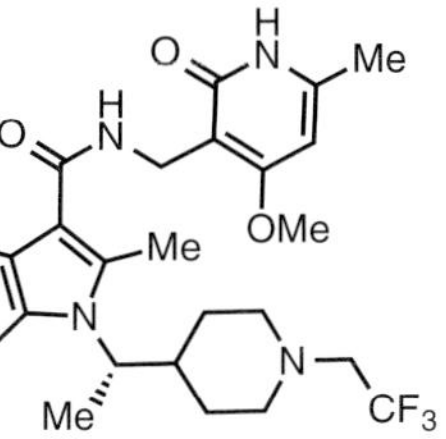

Figure 9.2 Selective small-molecule inhibitors of EZH2 and EZH1.

cell lines [21]. Importantly, overexpression of EZH2 and hypertrimethylation of H3K27 have been associated with multiple human cancers [21, 22].

A number of highly potent and selective SAM-competitive EZH2/EZH1 inhibitors that contain a pyridone core have been reported (Figure 9.2) [4]. The first potent and selective PRC2 inhibitor EPZ005687 targeting the EZH2 subunit was reported in 2012 [23]. EPZ005687 has displayed promising activities in cellular assays. It was also ~50-fold selective for EZH2 over EZH1. The same year, GSK126 another potent inhibitor was disclosed showing >150-fold selectivity for EZH2 over EZH1 [24]. Shortly after the publications of these two inhibitors, the discovery of EI1, which is approximately 90-fold selective for EZH2 over EZH1, was also reported [25]. In 2013, two orally bioavailable, potent, selective, and cell-active inhibitors – UNC1999 and EPZ-6438 – were published [26, 27]. Interestingly, UNC1999 is only ~10-fold selective for EZH2 over EZH1, while EPZ-6438 is about 35-fold selective. Notably, EPZ-6438 displays improved potency and PK profile compared with EPZ005687. EPZ011989 was discovered by modifications of the pyran moiety of EPZ-6438 and showed significantly improved PK and pharmacodynamic properties [28]. Recently the discovery of CPI-1205, displaying high potency, selectivity, and cellular activity, was added to the list of pyridone-containing inhibitors [29]. CPI-1205 exhibited >25-fold less potency for EZH1 than EZH2. Other potent, selective EZH2 inhibitors were also disclosed recently [4]. In general, all these aforementioned EZH2/EZH1 inhibitors potently inhibit both wild-type and mutant forms of EZH2. They are all highly selective for EZH2 over a panel of other methyltransferases and have no appreciable affinity for a broad range of other protein targets including GPCRs, ion channels, transporters, and other chromatin modifiers. It is important to note that EPZ-6438, GSK126, and CPI-1205 are advanced into human clinical trials.

In late 2015, the first crystal structures of an active PRC2 complex from the yeast containing EZH2, EED, and SUZ12-VEFS in complex with inhibiting H3K27M peptide and SAH were reported [30]. In 2016, the structure of the human PRC2 complex was also disclosed [31]. Concurrently, the crystal structure of a small-molecule inhibitor in complex with the wild-type and Y641N-mutated PRC2 complex, consisting of human EED, human SUZ12-VEFS, and engineered American chameleon EZH2 subunits, was also published [32]. Most recently, a co-crystal structure of an analogue of CPI-1205 bound to human PRC2 was also obtained [29]. In these crystal structures, the key interactions with the pyridone motif of the inhibitors elucidate the importance of this core for high affinity binding and the difficulty in finding suitable replacements. These structures represent a landmark in this field and reveal ample information about inhibition of the PRC2 complex. Very recently, several inhibitors that target the EED component of PRC2 have also been reported [4].

9.4.3 Inhibitors of H3K4 and H3K36 Methyltransferases

SETD7, SETD1A, SETD1B, the MLL family proteins (MLL1–5), SETMAR, SMYD1, SMYD2, SMYD3, ASH1L, and PRDM7 and PRDM9 catalyze the methylation of H3K4 in humans [33]. H3K4 trimethylation is associated with transcriptional activation [2].

SETD7 monomethylates H3K4 N-terminal peptides *in vitro* but displays limited activity on nucleosomal substrates. On the other hand, studies have shown that SETD7 targets many non-histone proteins and transcriptional regulators, such as p53, ERα, and pRb, and plays a role in transcriptional regulation and differentiation [33]. In 2014, *(R)*-PFI-2 (Figure 9.3), the first potent, selective, and cell-active inhibitor of SETD7, was reported [34]. It exhibits >1000-fold selectivity for SETD7 over 18 other methyltransferases and is also selective for SETD7 over 134 GPCRs, ion channels, and other enzyme targets. The X-ray crystal structure of SETD7 in complex with *(R)*-PFI-2 shows this compound occupying the substrate binding groove. *(R)*-PFI-2 exhibits a cofactor-dependent and substrate-competitive MOA. In other words, SAM plays a significant role in the binding of the inhibitor to SETD7. The inhibitor exhibits good physicochemical properties and shows no observable cell toxicity in various cell lines tested. There have been several additional studies in recent years toward the discovery of SETD7 inhibitors [4].

The SMYD family of proteins, SMYD1–5, possess a unique SET domain that is divided into two fragments by the zinc ion binding domain MYND and is followed by a cysteine-rich post-SET domain. While SMYD3 was initially described as an H3K4 methyltransferase [35], this activity has not been clearly verified by subsequent studies [36]. On the other hand, SMYD3 has been shown to directly methylate both histone H4K5 and MAP3K2 at K260; the latter seems to be the major target of SMYD3 [36]. A recent report shows that the methylation of MAP3K2 by SMYD3 increases MAP kinase signaling and promotes the formation of Ras-driven carcinomas [36]. EPZ031686 is reported as the first selective small-molecule inhibitor of SMYD3 (Figure 9.3) [37]. It potently inhibits SMYD3 and is highly selective for SMYD3 over 16 other methyltransferases including the highly homologous SMYD2. Co-crystal structure of a potent analogue of EPZ031686 showed that it occupies the lysine binding pocket of SMYD3. In addition, EPZ031686 is orally bioavailable in mouse PK studies, making it suitable for *in vivo* efficacy studies. Recently, SAH analogues with basic amino side chains extending into the lysine binding channel were designed and synthesized, resulting in the discovery of a potent, SAM-competitive inhibitor that displayed poor cell membrane permeability, making it unsuitable for cellular studies [38].

MLL is a large multidomain protein that is specific for H3K4 mono-, di-, and trimethylation [33]. Chromosomal rearrangements associated with MLL have been shown to cause acute myeloid, acute lymphoblastic, or mixed lineage leukemia [39]. While selective small-molecule direct inhibitors of MLLs have

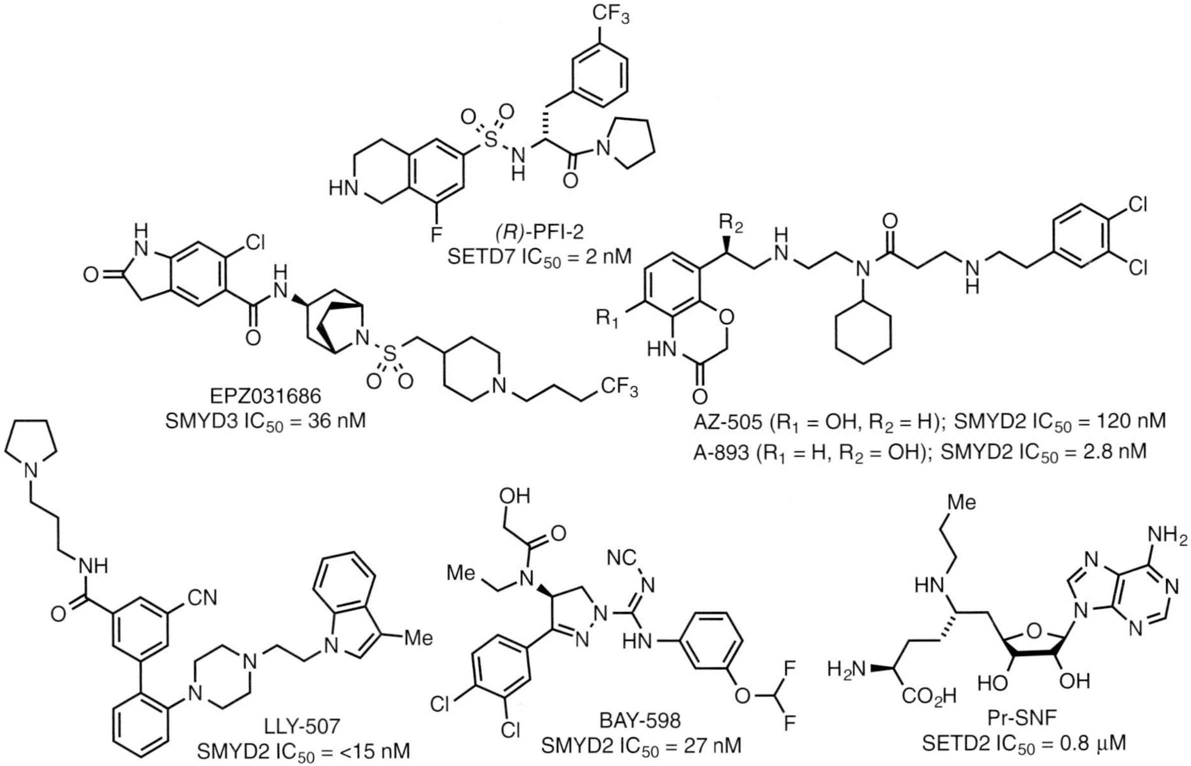

Figure 9.3 Selective small-molecule inhibitors of SETD7, SMYD3, SMYD2, and SETD2.

not yet been reported, small molecules that perturb protein–protein interactions of MLL with its partners, such as WDR5 and menin, have been discovered [4].

Like H3K4 methylation, H3K36 methylation is a hallmark of transcriptional activation. The human genome encodes at least eight SET-domain-containing methyltransferases that are responsible for H3K36 methylation: NSD1, MMSET (NSD2), WHSC1L1 (NSD3), SETD2, SETD3, ASH1L, SETMAR, and SMYD2 [33]. A variety of non-histone substrates of SMYD2, including p53, Rb, and HSP90, have been reported, implicating effects on diverse biological processes [33].

In 2011, the discovery of AZ-505 (Figure 9.3) as a SMYD2 inhibitor was reported [40]. AZ-505 is ~700-fold selective for SMYD2 over 6 other PKMTs, including the closely related SMYD3. It is competitive with the peptide substrate and uncompetitive with SAM. Cellular activities of this inhibitor have not been reported. In 2015, a detailed SAR study of AZ-505 was published and led to the discovery of A-893 (Figure 9.3) with >80-fold improvement in potency over the parent compound AZ-505 [41]. LLY-507 (Figure 9.3) is another potent, selective, and cell-active inhibitor of SMYD2, which is >100-fold selective over 21 other methyltransferases, including SMYD3 [42]. It is inactive against non-epigenetic targets, including kinases, GPCRs, and nuclear hormone receptors, and reduces monomethylation of p53 K370 in several cell systems. Moreover, an HTS campaign and further optimization resulted in the discovery of enantiomerically pure (*S*)-BAY-598 (Figure 9.3) as a cell-active and *in vivo* active inhibitor of SMYD2 [43]. It has >10-fold selectivity for SMYD2 over SMYD3 and >100-fold selectivity over 31 other methyltransferases. It is also highly selective against kinases and other primary molecular targets, including several CNS targets. The inhibitor is competitive with the peptide substrate, but uncompetitive with SAM.

SETD2 is responsible for H3K36 methylation and has been shown to be a tumor suppressor associated with p53-dependent gene regulation. *N*-Propyl sinefungin (Pr-SNF) (Figure 9.3) is a selective inhibitor of SETD2 and is highly selective for SETD2 over 14 other methyltransferases [44]. However, it displays only modest selectivity over SETD7, CARM1, and PRMT1. Structural analysis reveals that the catalytic domain of SETD2 could adopt at least two alternative conformations by flipping its post-SET loop, including an auto-inhibitory closed conformation and a substrate-accessible open conformation. To date, Pr-SNF is the only SETD2 inhibitor; however, no cellular studies have been reported for this inhibitor. Very recently, Pr-SNF was also identified as an MMSET inhibitor [4].

9.4.4 Inhibitors of H4K20 Methyltransferases

In humans, methylation of H4K20 is catalyzed by the methyltransferases SUV420H1, SUV420H2, and SETD8 [33]. SETD8, also known as SET8, PR-SET7, and KMT5A, is the sole methyltransferase that catalyzes monomethylation of H4K20. Monomethylation of H4K20 has been associated with both activation and repression and has been implicated in regulating important biological processes [33]. In addition, SETD8 methylates many non-histone substrates, including the tumor suppressor p53 and proliferating cell nuclear antigen [33]. SETD8 has been shown to be overexpressed in various types of cancers [4].

The first reported inhibitor of SETD8 was a marine natural product, nahuoic acid A [45]. Nahuoic acid A is a competitive inhibitor with respect to the cofactor SAM and is selective for SETD8 over 10 other methyltransferases. In 2014, Ma and coworkers reported the first substrate-competitive, selective inhibitor of SETD8, UNC0379 (Figure 9.4) [46]. UNC0379 displays inhibitory activity against SETD8 with micromolar potency and is selective for SETD8 over 15 other methyltransferases, including G9a and GLP.

Recently, a more potent inhibitor, MS2177 (Figure 9.4), was obtained via installation of an aminoalkyl group to the 7 position of UNC0379 [47]. MS2177 is competitive with the H4 peptide, but noncompetitive with the cofactor SAM. A co-crystal structure of MS2177 complexed with SETD8, which is the first crystal structure of SETD8 with a small-molecule inhibitor, reveals that C311 is near the inhibitor binding site, presenting an opportunity to develop a covalent inhibitor of SETD8. Therefore, MS453 (Figure 9.4) was designed, and it covalently modified C311, but not other cysteine residues of SETD8, as shown by mass spectroscopy (MS)-based analyses. MS453 was incubated with other PMTs, such as PRC2, SMYD2, and SMYD3, and no covalent adduct was observed by MS analysis, suggesting specificity to SETD8. Furthermore, MS453 is selective for SETD8 over 29 other methyltransferases in biochemical assays. The crystal structure of MS453 in complex with SETD8 confirms that C311 is covalently modified by MS453. MS453 has poor cell membrane permeability and a high efflux ratio and is thus not suitable for cellular studies. A few more reports on irreversible small-molecule inhibitors of SETD8, as well as a peptide inhibitor of SETD8, should be noted [4].

SUV420H1 and SUV420H2 are highly homologous methyltransferases that di- and trimethylate H4K20 [33]. Loss of H4K20me3 is a common hallmark of human cancer [48]. A-196 (Figure 9.4) was recently discovered as the first potent, selective, and cell-active inhibitor of SUV420H1 and SUV420H2 [49]. A-196 inhibits SUV420H1 and SUV420H2 potently in a peptide-competitive manner and is >100-fold selective for SUV420H1 and SUV420H2 over other methyltransferases and a broad range of non-epigenetic targets. In cell-based

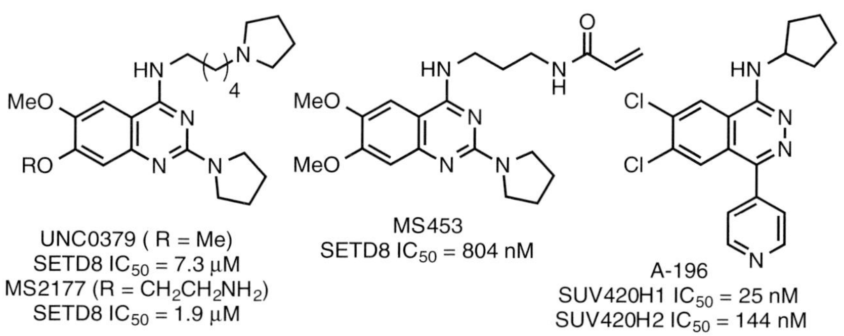

Figure 9.4 Inhibitors of SETD8 and SUV420H1/H2.

assays, A-196 reduced the H4K20me3/me2 marks in multiple cell lines with an $IC_{50} < 1\ \mu M$.

9.4.5 Inhibitors of H3K79 Methyltransferases

It has been shown that DOT1L is responsible for mono-, di-, and trimethylation of H3K79, which is generally correlated with transcriptional activation [50]. H3K79me3 has been associated with transcriptional regulation, DNA repair, embryonic development, cell cycle regulation, hematopoiesis, and cardiac function. It has also been reported that DOT1L interacts with the most commonly seen MLL fusion proteins in MLL-rearranged leukemias [1]. Therefore, DOT1L has been studied increasingly as a potential therapeutic target for the treatment of MLL-rearranged leukemia. DOT1L differs from the other identified human PKMTs by the lack of the SET domain. It contains a non-SET catalytic domain, which adopts a folding topology that is also observed in PRMTs and DNMTs and is therefore more closely related to these families of methyltransferases.

In 2011, the first selective DOT1L inhibitor, EPZ004777, was reported, with picomolar *in vitro* potency (Figure 9.5) [50]. EPZ004777 was designed and synthesized based on the cofactor SAM and the crystal structure of the enzyme active site. It is >1000-fold selective for DOT1L over nine other methyltransferases, despite its structural similarity to the cofactor SAM. Not surprisingly, MOA studies show that it was competitive with SAM. Structural studies reveal the ligand-induced conformational adaptation of the catalytic site [51, 52]. Later on, SGC0946 (Figure 9.5), a DOT1L inhibitor with improved *in vitro* and cellular potencies, was developed [51]. SGC0946 is ~10-fold more potent at reducing H3K79 methylation levels than EPZ004777 in MCF10A cells [51]. Another major

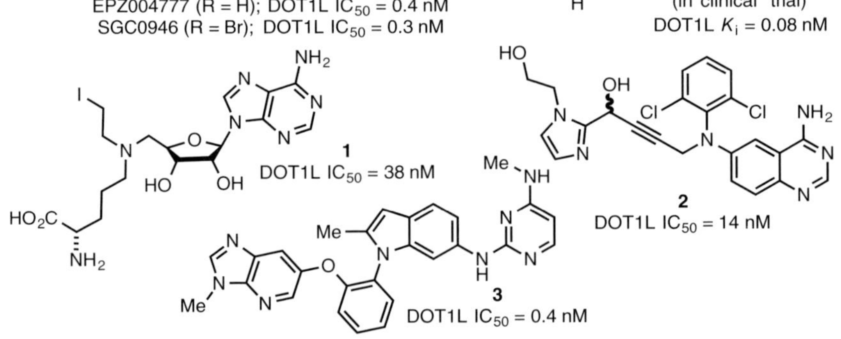

Figure 9.5 Structures of selective small-molecule inhibitors of DOT1L.

advancement in the discovery of DOT1L inhibitors came in 2013 when EPZ-5676 was disclosed (Figure 9.5) [53]. EPZ-5676 (also known as pinometostat) is the first PMT inhibitor advanced to the clinic, the first major breakthrough in the PMT inhibitor field.

In addition, inhibitor **1** that is postulated to covalently modify H3K79 was reported, and it displayed high *in vitro* potency and selectivity for DOT1L over other methyltransferases (Figure 9.5) [54, 55]. In 2016, a new series of DOT1L inhibitors **2** and **3** with chemotypes that differs structurally from previously reported SAM-based inhibitors was published (Figure 9.5) [56, 57]. These inhibitors interact with an induced pocket adjacent to the SAM binding site.

9.5 Protein Arginine Methyltransferases (PRMTs)

Protein arginine methylation is another significant and widely observed PTM in eukaryotic cells. Every methylation of arginine prevents a potential hydrogen bond, creating steric bulkiness and increasing hydrophobicity. Nine PRMTs have been identified to date, and based on their methylation functions, they are divided into three subcategories: type I, type II, and type III PRMTs [58]. Type I PRMTs, which include PRMT1, PRMT2, PRMT3, CARM1 (also known as PRMT4), PRMT6, and PRMT8, catalyze monomethylation and asymmetric dimethylation of arginine residues. PRMT5 and PRMT9 are type II PRMTs, which catalyze monomethylation and symmetrical dimethylation of arginine residues. PRMT7 is the only known type III PRMT, and it catalyzes arginine monomethylation only [59].

All PRMTs contain a conserved core region of about 310 amino acids. The monomeric structure of the PRMT core comprises a methyltransferase domain, a β-barrel that is unique to PRMTs, and a dimerization arm [59]. Type I PRMTs adopt a head-to-tail homodimeric structure. In the homodimer, the dimerization arm that extends out of the β-barrel of one monomeric subunit interacts with the Rossman fold of another subunit. In addition to histones, PRMTs methylate non-histone proteins [59]. Dysregulation of PRMTs and arginine methylation have been implicated in cancer and other diseases [59].

9.5.1 Inhibitors of PRMT1

PRMT1 was the first mammalian PRMT identified and is responsible for most of the type I arginine methyltransferase activity in mammalian cells [60]. PRMT1 catalyzes asymmetric dimethylation of H4R3 (H4R3me2a), which is associated with transcriptional activation. PRMT1 also methylates non-histone substrates. Therefore, PRMT1 has been implicated in numerous cellular processes, including transcription, RNA processing, and signal transduction [61].

In 2016, MS023 (Figure 9.6), a selective inhibitor of type I PRMTs, was developed [62]. MS023 is highly potent for PRMT1, PRMT3, PRMT4, PRMT6, and PRMT8 with IC_{50}s between 4 and 119 nM. Importantly, it is inactive against all type II and III PRMTs, PKMTs, DNMTs, and other epigenetic modifiers. In addition, an X-ray co-crystal structure of MS023 in complex with PRMT6 was obtained and reveals that the inhibitor occupies the substrate binding site and the ethylenediamino group serves as an arginine mimetic.

In 2004, the discovery of arginine methyltransferase inhibitors (AMIs) by HTS was reported [63]. Among the hits identified, only AMI-1, a symmetric sulfonated urea salt, and its analogues showed activity for PRMTs (Figure 9.6). These compounds are selective for PRMT3, CARM1, and PRMT6 over PRMT5 and a series of PKMTs [63]. AMI-1 potently inhibits PRMT1 and is proposed to bind in the substrate binding pocket. Interestingly, it was determined that AMI-1 and related naphthalene sulfonate derivatives directly target the peptide substrates instead of PRMT1 and exert their inhibitory activity by preventing the recognition of substrates by the enzyme [64]. Another derivative, AMI-5 (Figure 9.6), displays better potency, but no selectivity profile has been reported.

A target-based approach was applied to discover stilbamidine and allantodapsone as PRMT1 inhibitors [65]. Other studies to discover PRMT1 inhibitors as well as bisubstrate analogues use strategies that target substrate and cofactor sites [4]. It is important to note, however, that most of these PRMT1 inhibitors reported lack extensive selectivity and characterization data, thereby limiting their use as selective PRMT1 inhibitors.

9.5.2 Inhibitors of PRMT3

PRMT3 is a type I PRMT that contains a zinc finger domain at its N-terminus. It is a cytosolic protein, primarily methylating the 40S ribosomal protein S2 (rpS2). Asymmetric dimethylation of rpS2 by PRMT3 results in stabilization of rpS2 and impacts ribosomal biosynthesis [59].

In 2012, the first selective PRMT3 inhibitor was discovered by screening a diverse library of compounds [66]. Interestingly, this inhibitor was noncompetitive with both the cofactor SAM and the peptide substrate, thereby suggesting an allosteric mechanism of inhibition. The co-crystal structure of this inhibitor with PRMT3 confirms that it occupies a novel allosteric binding site at the interface of the two subunits of the PRMT3 homodimer. This was the first example of an allosteric inhibition of a PMT by a small molecule. Subsequent extensive SAR studies resulted in the discovery of SGC707 (Figure 9.6), the most potent

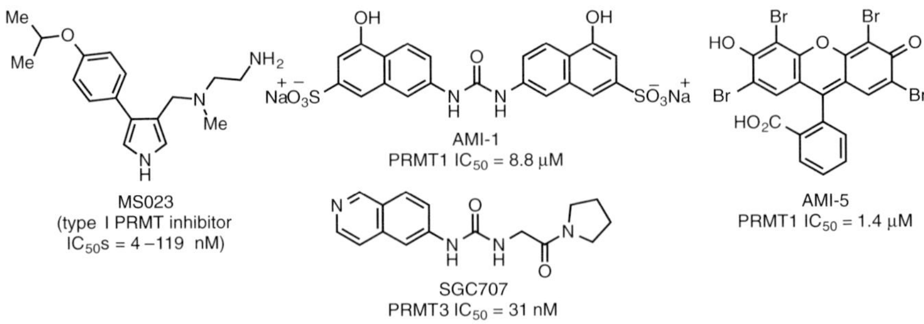

Figure 9.6 Selective small-molecule inhibitors of PRMT1 and PRMT3.

PRMT3 inhibitor to date, binding to the same aforementioned allosteric site [67]. Importantly, SGC707 is selective for PRMT3 over 31 other methyltransferases and a broad range of non-epigenetic targets including kinases, GPCRs, ion channels, and transporters. SGC707, which is a potent, selective, cell-active allosteric inhibitor of PRMT3 is bioavailable and can be used for animal studies.

9.5.3 Inhibitors of CARM1

CARM1 (PRMT4) is a steroid receptor co-activator and the first member of the PRMTs to be associated with transcriptional regulation. It is responsible for the asymmetric dimethylation of H3R17 and H3R26, with preference for H3R17. CARM1 also methylates a variety of non-histone proteins and plays a role in mRNA splicing, RNA processing and stability, cell cycle progression, and the DNA damage response [68].

Several HTS campaigns and SAR studies resulted in the identification of CARM1 inhibitors [69–71]. Further optimization led to the discovery of potent CARM1 inhibitors [72]. These inhibitors display selectivity for CARM1 over PRMT1 and PRMT3, but selectivity for other PRMTs has not been reported. In addition, these inhibitors were either not tested in cells or lack significant cellular activity. Co-crystal structures of the CARM1 catalytic domain in complex with these inhibitors reveal that they are anchored in the PRMT arginine binding channel through a basic alkyl-diamino or alanine-amide tail [72].

In a study aiming to develop PRMT inhibitors via a fragment-based approach, a commercially available diverse fragment library of compounds mimicking basic amino tails was tested against PRMT6, resulting in the discovery of fragment hit **4** that inhibits PRMT6, CARM1, and PRMT8 (Figure 9.7) [73]. Recently, a dual CARM1 and PRMT6 inhibitor, MS049 (Figure 9.7), was discovered via SAR studies based on the aforementioned fragment hit **4**. MS049 is a highly potent, selective, and cell-active dual inhibitor of CARM1 and PRMT6 [74]. It is highly selective for CARM1 and PRMT6 over other type I PRMTs and shows no inhibition against PRMT5, PRMT7, or a broad range of epigenetic modifiers and non-epigenetic targets. The same research group also reported

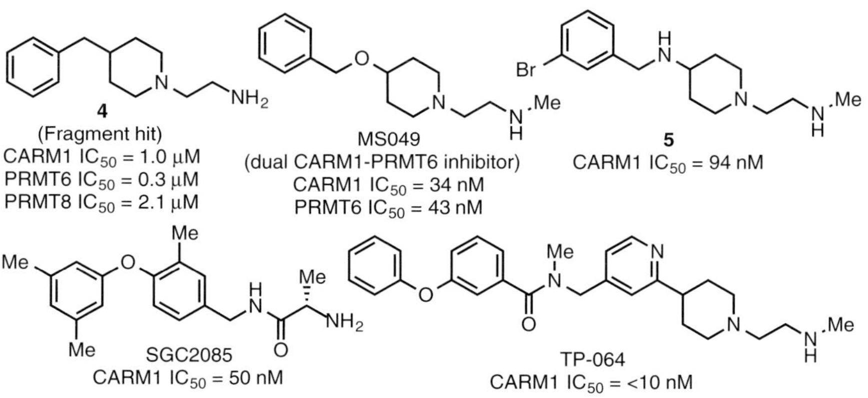

4
(Fragment hit)
CARM1 IC_{50} = 1.0 μM
PRMT6 IC_{50} = 0.3 μM
PRMT8 IC_{50} = 2.1 μM

MS049
(dual CARM1-PRMT6 inhibitor)
CARM1 IC_{50} = 34 nM
PRMT6 IC_{50} = 43 nM

5
CARM1 IC_{50} = 94 nM

SGC2085
CARM1 IC_{50} = 50 nM

TP-064
CARM1 IC_{50} = <10 nM

Figure 9.7 Inhibitors of CARM1.

a potent and selective inhibitor of CARM1, compound **5** (Figure 9.7), based again on the fragment hit **4** [75]. Concurrent with the two studies described above, another potent and selective CARM1 inhibitor was discovered via virtual screening [76]. The hits were then optimized to yield a potent CARM1 inhibitor, SGC2085 (Figure 9.7), which is inactive against all other PRMTs except PRMT6 (~100-fold selective). SGC2085 also shows complete selectivity against a panel of 21 other methyltransferases. However, this inhibitor is inactive in cell-based assays, likely due to its poor cell membrane permeability. Recently, the first potent, selective, and cell-active inhibitor of CARM1, TP-064, was discovered (Figure 9.7) [77]. TP-064 potently inhibits CARM1 and displays a >100-fold selectivity for CARM1 over other methyltransferases and non-epigenetic targets. The negative control of TP-064 exhibits no activity in the same biochemical and cellular assays. Several other CARM1 inhibitors have also been reported [4].

9.5.4 Inhibitors of PRMT5

PRMT5 is the major type II PRMT that is responsible for the symmetric dimethylation of arginine residues [78]. PRMT5 symmetrically dimethylates H2AR3, H4R3, H3R2, and H3R8 *in vivo* [78]. Importantly, these marks are associated with a variety of transcriptional regulatory processes. PRMT5 has also been found to regulate transcription and many downstream events through methylation of transcription factors such as NF-κB, p53, and E2F-1 [59]. PRMT5 interacts with a variety of binding partners, including its most common partner, MEP50. MEP50 is required for PRMT5 enzymatic activity and is likely present in every PRMT5-containing complex *in vivo*. The overexpression of PRMT5 has been reported in several human malignancies [59].

In 2015, EPZ015666 (Figure 9.8), a potent, selective, cell-active, and orally bioavailable inhibitor of PRMT5 with antiproliferative effects in both *in vitro* and *in vivo* models of MCL, was disclosed [79, 80]. It is competitive with the peptide substrate and is inactive against a panel of 20 other PRMTs and PKMTs However, it has not been evaluated against PRMT9. Notably, GSK3326595 (structure is not disclosed), a PRMT5 inhibitor that potently inhibits tumor growth in cellular and animal models, has entered phase I clinical trials. LLY-283, the first potent and selective SAM-competitive chemical probe for PRMT5, was recently discovered (Figure 9.8) [81]. LLY-283 potently inhibits PRMT5, is >100-fold selective for PRMT5 over other methyltransferases and non-epigenetic targets, and shows activity in cellular assays. A few other inhibitors have been reported in recent years [4].

9.5.5 Inhibitors of PRMT6

PRMT6 is a nuclear protein that catalyzes the methylation of H4R3 and H3R2. PRMT6 is the sole methyltransferase modifying the H3R2 mark, which acts as a repressive mark that antagonizes the trimethylation of H3K4 by MLL [59]. PRMT6 has been implicated in a variety of cellular processes, such as regulation of cell cycle, hormone receptor-mediated transcription, maintenance of stem cell pluripotency, and DNA repair [8]. Therefore, PRMT6 is a potential therapeutic

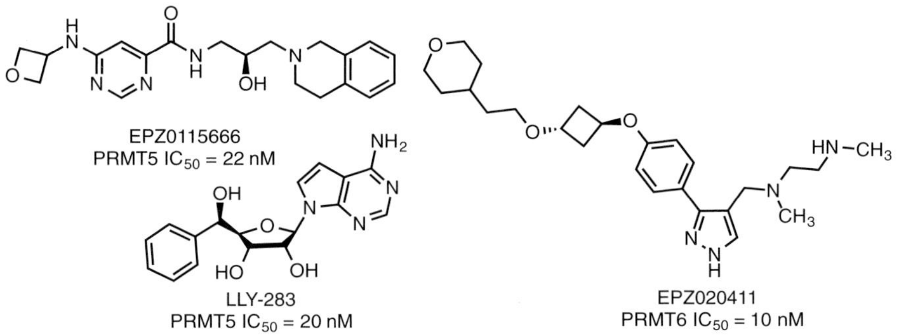

Figure 9.8 Selective small-molecule inhibitors of PRMT5 and PRMT6.

target, and small-molecule inhibitors should be developed for *in vitro* and *in vivo* studies.

The discovery of the first PRMT6 selective inhibitor, EPZ020411 (Figure 9.8), was described in 2015 [82]. This inhibitor is selective for PRMT6 over PRMT1 (12-fold) and PRMT8 (22-fold) and is >100-fold selective for PRMT6, PRMT1, and PRMT8 over other methyltransferases including PRMT3, PRMT4, PRMT5, and PRMT7. A co-crystal structure of EPZ020411 in complex with PRMT6 and SAH was also solved. EPZ020411 demonstrated good bioavailability and is suitable for potential *in vivo* studies.

9.6 Concluding Remarks

This chapter highlighted some of the selective small-molecule PMT inhibitors that have been discovered over the last 15 years [4]. Outstanding progress made in the PMT inhibitor field culminated by DOT1L, EZH2, and PRMT5 inhibitors entering clinical trials in 2012–2016 for diseases such as leukemia and lymphoma. While there has been significant progress, there is still much to be achieved. For example, many individual targets and subgroups of the PMT family such as MLL family and PRDMs lack selective inhibitors. We expect that the progress and successes in inhibitor discovery will continue since potent, selective, and cell-active inhibitors of PTMs are proven to be important tools and novel drug candidates.

References

1 Arrowsmith, C.H., Bountra, C., Fish, P.V. et al. (2012). Epigenetic protein families: a new frontier for drug discovery. *Nat. Rev. Drug Discovery* 11: 384–400.

2 Greer, E.L. and Shi, Y. (2012). Histone methylation: a dynamic mark in health, disease and inheritance. *Nat. Rev. Genet.* 13: 343–357.

3 Biggar, K.K. and Li, S.S. (2015). Non-histone protein methylation as a regulator of cellular signalling and function. *Nat. Rev. Mol. Cell Biol.* 16: 5–17.

4 Kaniskan, H.Ü., Martini, M.L., and Jin, J. (2018). Inhibitors of protein methyltransferases and demethylases. *Chem. Rev.* 118: 989–1068.

5 Jenuwein, T., Laible, G., Dorn, R., and Reuter, G. (1998). SET domain proteins modulate chromatin domains in Eu- and heterochromatin. *Cell. Mol. Life Sci.* 54: 80–93.

6 Krauss, V. (2008). Glimpses of evolution: heterochromatic histone H3K9 methyltransferases left its marks behind. *Genetica* 133: 93–106.

7 Greiner, D., Bonaldi, T., Eskeland, R. et al. (2005). Identification of a specific inhibitor of the histone methyltransferase SU(VAR)3-9. *Nat. Chem. Biol.* 1: 143–145.

8 Kaniskan, H.Ü., Konze, K.D., and Jin, J. (2015). Selective inhibitors of protein methyltransferases. *J. Med. Chem.* 58: 1596–1629.

9 Kubicek, S., O'Sullivan, R.J., August, E.M. et al. (2007). Reversal of H3K9me2 by a small-molecule inhibitor for the G9a histone methyltransferase. *Mol. Cell* 25: 473–481.

10 Chang, Y., Zhang, X., Horton, J.R. et al. (2009). Structural basis for G9a-like protein lysine methyltransferase inhibition by BIX-01294. *Nat. Struct. Mol. Biol.* 16: 312–317.

11 Liu, F., Chen, X., Allali-Hassani, A. et al. (2010). Protein lysine methyltransferase G9a inhibitors: design, synthesis, and structure activity relationships of 2,4-diamino-7-aminoalkoxy-quinazolines. *J. Med. Chem.* 53: 5844–5857.

12 Vedadi, M., Barsyte-Lovejoy, D., Liu, F. et al. (2011). A chemical probe selectively inhibits G9a and GLP methyltransferase activity in cells. *Nat. Chem. Biol.* 7: 566–574.

13 Liu, F., Barsyte-Lovejoy, D., Li, F. et al. (2013). Discovery of an in vivo chemical probe of the lysine methyltransferases G9a and GLP. *J. Med. Chem.* 56: 8931–8942.

14 Xiong, Y., Li, F., Babault, N. et al. (2017). Discovery of potent and selective inhibitors for G9a-like protein (GLP) lysine methyltransferase. *J. Med. Chem.* 60: 1876–1891.

15 Chang, Y., Ganesh, T., Horton, J.R. et al. (2010). Adding a lysine mimic in the design of potent inhibitors of histone lysine methyltransferases. *J. Mol. Biol.* 400: 1–7.

16 Yuan, Y., Wang, Q., Paulk, J. et al. (2012). A small-molecule probe of the histone methyltransferase G9a induces cellular senescence in pancreatic adenocarcinoma. *ACS Chem. Biol.* 7: 1152–1157.

17 Sweis, R.F., Pliushchev, M., Brown, P.J. et al. (2014). Discovery and development of potent and selective inhibitors of histone methyltransferase G9a. *ACS Med. Chem. Lett.* 5: 205–209.

18 Margueron, R. and Reinberg, D. (2011). The polycomb complex PRC2 and its mark in life. *Nature* 469: 343–349.

19 Verma, S.K., Tian, X., LaFrance, L.V. et al. (2012). Identification of potent, selective, cell-active inhibitors of the histone lysine methyltransferase EZH2. *ACS Med. Chem. Lett.* 3: 1091–1096.

20 Cao, R. and Zhang, Y. (2004). SUZ12 is required for both the histone methyltransferase activity and the silencing function of the EED-EZH2 complex. *Mol. Cell* 15: 57–67.

21 Kim, K.H. and Roberts, C.W. (2016). Targeting EZH2 in cancer. *Nat. Med.* 22: 128–134.

22 Xu, B., Konze, K.D., Jin, J., and Wang, G.G. (2015). Targeting EZH2 and PRC2 dependence as novel anticancer therapy. *Exp. Hematol.* 43: 698–712.

23 Knutson, S.K., Wigle, T.J., Warholic, N.M. et al. (2012). A selective inhibitor of EZH2 blocks H3K27 methylation and kills mutant lymphoma cells. *Nat. Chem. Biol.* 8: 890–896.

24 McCabe, M.T., Ott, H.M., Ganji, G. et al. (2012). EZH2 inhibition as a therapeutic strategy for lymphoma with EZH2-activating mutations. *Nature* 492: 108–112.

25 Qi, W., Chan, H., Teng, L. et al. (2012). Selective inhibition of Ezh2 by a small molecule inhibitor blocks tumor cells proliferation. *Proc. Natl. Acad. Sci. U.S.A.* 109: 21360–21365.

26 Konze, K.D., Ma, A., Li, F. et al. (2013). An orally bioavailable chemical probe of the lysine methyltransferases EZH2 and EZH1. *ACS Chem. Biol.* 8: 1324–1334.

27 Knutson, S.K., Warholic, N.M., Wigle, T.J. et al. (2013). Durable tumor regression in genetically altered malignant rhabdoid tumors by inhibition of methyltransferase EZH2. *Proc. Natl. Acad. Sci. U.S.A.* 110: 7922–7927.

28 Campbell, J.E., Kuntz, K.W., Knutson, S.K. et al. (2015). EPZ011989, a potent, orally-available EZH2 inhibitor with robust in vivo activity. *ACS Med. Chem. Lett.* 6: 491–495.

29 Vaswani, R.G., Gehling, V.S., Dakin, L.A. et al. (2016). Identification of (*R*)-*N*-((4-methoxy-6-methyl-2-oxo-1,2-dihydropyridin-3-yl)methyl)-2-methyl-1-(1-(1-(2,2,2-trifluoroethyl)piperidin-4-yl)ethyl)-1*H*-indole-3-carboxamide (CPI-1205), a potent and selective inhibitor of histone methyltransferase EZH2, suitable for phase I clinical trials for B-cell lymphomas. *J. Med. Chem.* 59: 9928–9941.

30 Jiao, L.Y. and Liu, X. (2015). Structural basis of histone H3K27 trimethylation by an active polycomb repressive complex 2. *Science* 350.

31 Justin, N., Zhang, Y., Tarricone, C. et al. (2016). Structural basis of oncogenic histone H3K27M inhibition of human polycomb repressive complex 2. *Nat. Commun.* 7: 11316.

32 Brooun, A., Gajiwala, K.S., Deng, Y.L. et al. (2016). Polycomb repressive complex 2 structure with inhibitor reveals a mechanism of activation and drug resistance. *Nat. Commun.* 7: 11384.

33 Herz, H.M., Garruss, A., and Shilatifard, A. (2013). SET for life: biochemical activities and biological functions of SET domain-containing proteins. *Trends Biochem. Sci.* 38: 621–639.

34 Barsyte-Lovejoy, D., Li, F., Oudhoff, M.J. et al. (2014). (*R*)-PFI-2 is a potent and selective inhibitor of SETD7 methyltransferase activity in cells. *Proc. Natl. Acad. Sci. U.S.A.* 111: 12853–12858.

35 Hamamoto, R., Furukawa, Y., Morita, M. et al. (2004). SMYD3 encodes a histone methyltransferase involved in the proliferation of cancer cells. *Nat. Cell Biol.* 6: 731–740.

36 Mazur, P.K., Reynoird, N., Khatri, P. et al. (2014). SMYD3 links lysine methylation of MAP3K2 to Ras-driven cancer. *Nature* 510: 283–287.

37 Mitchell, L.H., Boriack-Sjodin, P.A., Smith, S. et al. (2016). Novel oxindole sulfonamides and sulfamides: EPZ031686, the first orally bioavailable small molecule SMYD3 inhibitor. *ACS Med. Chem. Lett.* 7: 134–138.
38 Van Aller, G.S., Graves, A.P., Elkins, P.A. et al. (2016). Structure-based design of a novel SMYD3 inhibitor that bridges the SAM-and MEKK2-binding pockets. *Structure* 24: 774–781.
39 Kohlmann, A., Schoch, C., Dugas, M. et al. (2005). New insights into MLL gene rearranged acute leukemias using gene expression profiling: shared pathways, lineage commitment, and partner genes. *Leukemia* 19: 953–964.
40 Ferguson, A.D., Larsen, N.A., Howard, T. et al. (2011). Structural basis of substrate methylation and inhibition of SMYD2. *Structure* 19: 1262–1273.
41 Sweis, R.F., Wang, Z., Algire, M. et al. (2015). Discovery of A-893, a new cell-active benzoxazinone inhibitor of lysine methyltransferase SMYD2. *ACS Med. Chem. Lett.* 6: 695–700.
42 Nguyen, H., Allali-Hassani, A., Antonysamy, S. et al. (2015). LLY-507, a cell-active, potent, and selective inhibitor of protein-lysine methyltransferase SMYD2. *J. Biol. Chem.* 290: 13641–13653.
43 Eggert, E., Hillig, R.C., Koehr, S. et al. (2016). Discovery and characterization of a highly potent and selective aminopyrazoline-based in vivo probe (BAY-598) for the protein lysine methyltransferase SMYD2. *J. Med. Chem.* 59: 4578–4600.
44 Zheng, W., Ibáñez, G., Wu, H. et al. (2012). Sinefungin derivatives as inhibitors and structure probes of protein lysine methyltransferase SETD2. *J. Am. Chem. Soc.* 134: 18004–18014.
45 Williams, D.E., Dalisay, D.S., Li, F. et al. (2013). Nahuoic acid A produced by a *Streptomyces* sp. isolated from a marine sediment is a selective SAM-competitive inhibitor of the histone methyltransferase SETD8. *Org. Lett.* 15: 414–417.
46 Ma, A., Yu, W., Li, F. et al. (2014). Discovery of a selective, substrate-competitive inhibitor of the lysine methyltransferase SETD8. *J. Med. Chem.* 57: 6822–6833.
47 Butler, K.V., Ma, A., Yu, W. et al. (2016). Structure-based design of a covalent Inhibitor of the SET domain-containing protein 8 (SETD8) lysine methyltransferase. *J. Med. Chem.* 59: 9881–9889.
48 Fraga, M.F., Ballestar, E., Villar-Garea, A. et al. (2005). Loss of acetylation at Lys16 and trimethylation at Lys20 of histone H4 is a common hallmark of human cancer. *Nat. Genet.* 37: 391–400.
49 Bromberg, K.D., Mitchell, T.R., Upadhyay, A.K. et al. (2017). The SUV4-20 inhibitor A-196 verifies a role for epigenetics in genomic integrity. *Nat. Chem. Biol.* 13: 317–324.
50 Daigle, S.R., Olhava, E.J., Therkelsen, C.A. et al. (2011). Selective killing of mixed lineage leukemia cells by a potent small-molecule DOT1L inhibitor. *Cancer Cell* 20: 53–65.
51 Yu, W., Chory, E.J., Wernimont, A.K. et al. (2012). Catalytic site remodelling of the DOT1L methyltransferase by selective inhibitors. *Nat. Commun.* 3: 1288.

52 Basavapathruni, A., Jin, L., Daigle, S.R. et al. (2012). Conformational adaptation drives potent, selective and durable inhibition of the human protein methyltransferase DOT1L. *Chem. Biol. Drug Des.* 80: 971–980.
53 Daigle, S.R., Olhava, E.J., Therkelsen, C.A. et al. (2013). Potent inhibition of DOT1L as treatment of MLL-fusion leukemia. *Blood* 122: 1017–1025.
54 Yao, Y., Chen, P., Diao, J. et al. (2011). Selective inhibitors of histone methyltransferase DOT1L: design, synthesis and crystallographic studies. *J. Am. Chem. Soc.* 133: 16746–16749.
55 Anglin, J.L., Deng, L., Yao, Y. et al. (2012). Synthesis and structure-activity relationship investigation of adenosine-containing inhibitors of histone methyltransferase DOT1L. *J. Med. Chem.* 55: 8066–8074.
56 Scheufler, C., Mobitz, H., Gaul, C. et al. (2016). Optimization of a fragment-based screening hit toward potent DOT1L inhibitors interacting in an induced binding pocket. *ACS Med. Chem. Lett.* 7: 730–734.
57 Chen, C., Zhu, H., Stauffer, F. et al. (2016). Discovery of novel DOT1L inhibitors through a structure-based fragmentation approach. *ACS Med. Chem. Lett.* 7: 735–740.
58 Bedford, M.T. and Clarke, S.G. (2009). Protein arginine methylation in mammals: who, what, and why. *Mol. Cell* 33: 1–13.
59 Fuhrmann, J., Clancy, K.W., and Thompson, P.R. (2015). Chemical biology of protein arginine modifications in epigenetic regulation. *Chem. Rev.* 115: 5413–5461.
60 Tang, J., Frankel, A., Cook, R.J. et al. (2000). PRMT1 is the predominant type I protein arginine methyltransferase in mammalian cells. *J. Biol. Chem.* 275: 7723–7730.
61 Yang, Y. and Bedford, M.T. (2013). Protein arginine methyltransferases and cancer. *Nat. Rev. Cancer* 13: 37–50.
62 Eram, M.S., Shen, Y., Szewczyk, M.M. et al. (2016). A potent, selective, and cell-active inhibitor of human type I protein arginine methyltransferases. *ACS Chem. Biol.* 11: 772–781.
63 Cheng, D., Yadav, N., King, R.W. et al. (2004). Small molecule regulators of protein arginine methyltransferases. *J. Biol. Chem.* 279: 23892–23899.
64 Feng, Y., Li, M., Wang, B., and Zheng, Y.G. (2010). Discovery and mechanistic study of a class of protein arginine methylation inhibitors. *J. Med. Chem.* 53: 6028–6039.
65 Spannhoff, A., Heinke, R., Bauer, I. et al. (2007). Target-based approach to inhibitors of histone arginine methyltransferases. *J. Med. Chem.* 50: 2319–2325.
66 Siarheyeva, A., Senisterra, G., Allali-Hassani, A. et al. (2012). An allosteric inhibitor of protein arginine methyltransferase 3. *Structure* 20: 1425–1435.
67 Kaniskan, H.Ü., Szewczyk, M.M., Yu, Z. et al. (2015). A potent, selective and cell-active allosteric inhibitor of protein arginine methyltransferase 3 (PRMT3). *Angew. Chem. Int. Ed.* 54: 5166–5170.
68 Lee, Y.H. and Stallcup, M.R. (2011). Roles of protein arginine methylation in DNA damage signaling pathways is CARM1 a life-or-death decision point? *Cell Cycle* 10: 1343–1344.

69 Huynh, T., Chen, Z., Pang, S.H. et al. (2009). Optimization of pyrazole inhibitors of coactivator associated arginine methyltransferase 1 (CARM1). *Bioorg. Med. Chem. Lett.* 19: 2924–2927.

70 Wan, H.H., Huynh, T., Pang, S.H. et al. (2009). Benzo[d]imidazole inhibitors of coactivator associated arginine methyltransferase 1 (CARM1)-hit to lead studies. *Bioorg. Med. Chem. Lett.* 19: 5063–5066.

71 Allan, M., Manku, S., Therrien, E. et al. (2009). *N*-Benzyl-1-heteroaryl-3-(trifluoromethyl)-1*H*-pyrazole-5-carboxamides as inhibitors of co-activator associated arginine methyltransferase 1 (CARM1). *Bioorg. Med. Chem. Lett.* 19: 1218–1223.

72 Sack, J.S., Thieffine, S., Bandiera, T. et al. (2011). Structural basis for CARM1 inhibition by indole and pyrazole inhibitors. *Biochem. J.* 436: 331–339.

73 Ferreira de Freitas, R., Eram, M.S., Szewczyk, M.M. et al. (2016). Discovery of a potent class I protein arginine methyltransferase fragment inhibitor. *J. Med. Chem.* 59: 1176–1183.

74 Shen, Y., Szewczyk, M.M., Eram, M.S. et al. (2016). Discovery of a potent, selective, and cell-active dual inhibitor of protein arginine methyltransferase 4 and protein arginine methyltransferase 6. *J. Med. Chem.* 59: 9124–9139.

75 Kaniskan, H.Ü., Eram, M.S., Liu, J. et al. (2016). Design and synthesis of selective, small molecule inhibitors of coactivator-associated arginine methyltransferase 1 (CARM1). *MedChemComm* 7: 1793–1796.

76 Ferreira de Freitas, R., Eram, M.S., Smil, D. et al. (2016). Discovery of a potent and selective coactivator associated arginine methyltransferase 1 (CARM1) inhibitor by virtual screening. *J. Med. Chem.* 59: 6838–6847.

77 TP-064: A Chemical Probe For PRMT4. http://www.thesgc.org/chemical-probes/TP-064 (accessed January 2017).

78 Branscombe, T.L., Frankel, A., Lee, J.H. et al. (2001). PRMT5 (Janus kinase-binding protein 1) catalyzes the formation of symmetric dimethylarginine residues in proteins. *J. Biol. Chem.* 276: 32971–32976.

79 Chan-Penebre, E., Kuplast, K.G., Majer, C.R. et al. (2015). A selective inhibitor of PRMT5 with in vivo and in vitro potency in MCL models. *Nat. Chem. Biol.* 11: 432–437.

80 Duncan, K.W., Rioux, N., Boriack-Sjodin, P.A. et al. (2016). Structure and property guided design in the identification of PRMT5 tool compound EPZ015666. *ACS Med. Chem. Lett.* 7: 162–166.

81 LLY-283: A Chemical Probe For PRMT5. http://www.thesgc.org/chemical-probes/LLY-283 (accessed November 2016).

82 Mitchell, L.H., Drew, A.E., Ribich, S.A. et al. (2015). Aryl pyrazoles as potent inhibitors of arginine methyltransferases: identification of the first PRMT6 tool compound. *ACS Med. Chem. Lett.* 6: 655–659.

10

LSD (Lysine-Specific Demethylase): A Decade-Long Trip from Discovery to Clinical Trials

Adam Lee, M. Teresa Borrello, and A. Ganesan

University of East Anglia, School of Pharmacy, Norwich NR4 7TJ, UK

10.1 Introduction

The chemical decoration of the nucleosome through covalent modification lies at the heart of epigenetic regulation. Among the various structural alterations, methylation occupies a special position as it occurs in both DNA and histone proteins (Figure 10.1). Methylation is also distinctive as it is the smallest possible perturbation, adding only a single carbon atom without any change in the net charge of the macromolecule. Nevertheless, this small increase in size is sufficient for selective recognition by protein reader domains that influence gene transcription. Furthermore, lysine or arginine protein sidechains can undergo one, two, or three methylations, and reader domains are able to discriminate between these different methylation states.

As with other epigenetic modifications, the question arises whether methylation is reversible or not. For a long time, the evidence suggested nucleosome methylation was a semi-permanent mark. This is because an active process of demethylation is not obligatory for the removal of methylation patterns. In each round of cell replication, the newly synthesized DNA daughter strands lack methylation. Imperfect fidelity in the reintroduction of methylation would lead to the gradual loss of this mark. Similarly, in histones, the methylation signal should decay over time through the natural turnover of nucleosomes. However, the discovery of specific DNA and lysine demethylase enzymes in the last decade has completely overturned the earlier belief. We now know that these enzymes are capable of actively modifying or removing the methyl group. Cell labeling studies with histone proteins demonstrate a dynamic equilibrium between methyl group introduction and removal on a timescale of hours [1]. By comparison, the interconversion between histone acylation and deacylation is a faster process with a half-life of minutes.

From a mechanistic point of view, there are two distinct possibilities for the demethylation of an oxygen or nitrogen heteroatom, nucleophilic or oxidative (Figure 10.2). In the nucleophilic demethylation, the heteroatom is the leaving group in a S_N2 displacement of the methyl group. While there are many precedents in organic synthesis, it is rare in biology. It occurs in the

Epigenetic Drug Discovery, First Edition. Edited by Wolfgang Sippl and Manfred Jung.

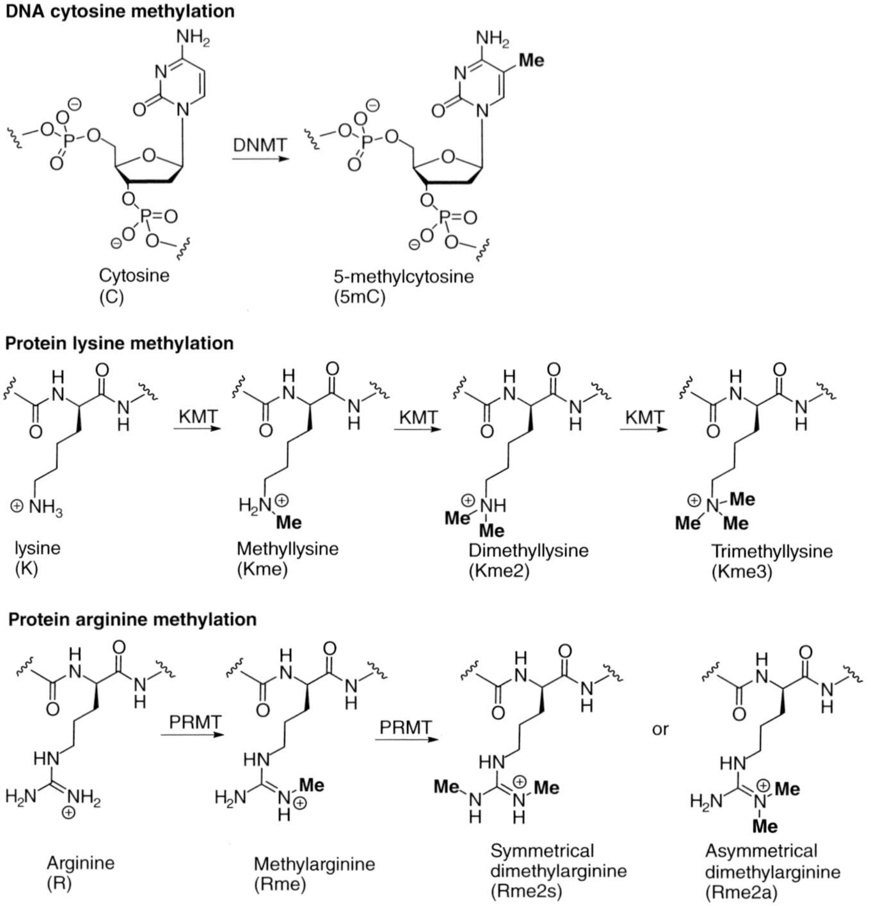

Figure 10.1 Epigenetic DNA and protein methylation.

dealkylation of O^6-methylguanine in DNA, where a cysteine residue in the O^6-alkylguanine-DNA alkyltransferase repair proteins acts as the nucleophile [2]. Since there is no recycling, each molecule of alkyltransferase can only perform a single demethylation. While acceptable for the rapid repair of a highly mutagenic DNA lesion, it is inefficient for a catalytic reaction with turnover. Instead, the demethylating enzymes employ the second oxidative option. The CH_3 group, through homolytic or heterolytic C—H bond fission, is oxidized to a carbinolamine (for C—N oxidation) or hemiacetal (for C—O oxidation) intermediate that spontaneously hydrolyzes to formaldehyde and the demethylated heteroatom. Such oxidations represent the pathway for N- and O-demethylation in drug metabolism by oxidative cytochrome P450 enzymes and for epigenetic N-demethylation in proteins. This chapter summarizes the chemistry and biology of the first discovered protein demethylases, the lysine-specific demethylases (LSDs). The coverage is selective rather than comprehensive with an emphasis on drug discovery and supplements earlier reviews on the subject [3–8].

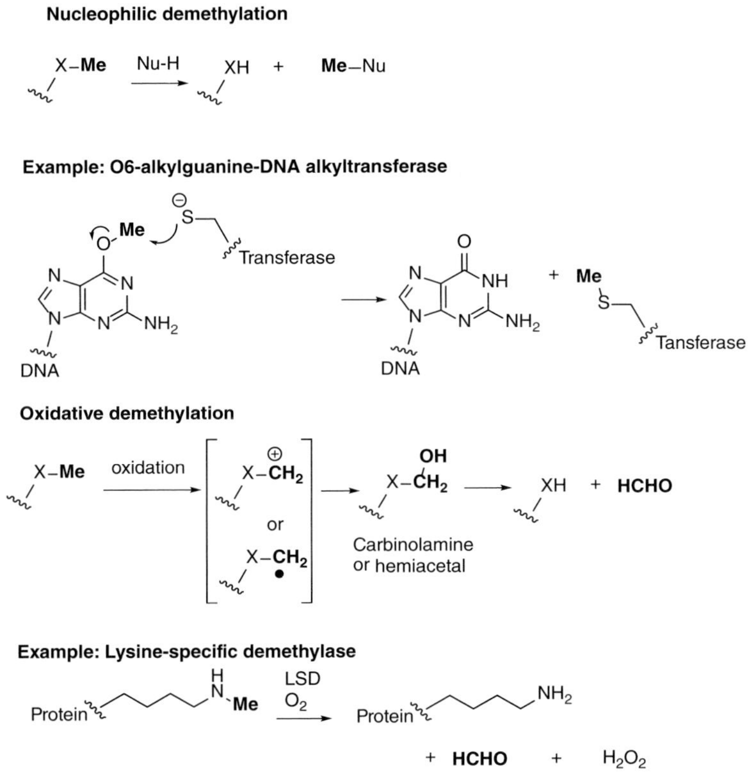

Figure 10.2 Nucleophilic and oxidative mechanisms of biological demethylation.

10.2 LSDs: Discovery and Mechanistic Features

The first report of an enzymatic removal of methyl groups from lysine dates as far back as 1964 [9]. The N^6-methyl-lysine oxidase enzyme accepts mono- and dimethyl derivatives of the amino acid lysine as substrates and carries out an oxidative demethylation with loss of formaldehyde. Forty years later, the Shi group characterized the first enzyme capable of demethylating lysine residues in proteins [10]. This amine oxidase was localized in the nucleus, specifically oxidized the mono- and dimethyl forms of the lysine 4 residue of histone H3 protein (H3K4me and H3K4me2) and was named lysine-specific demethylase 1 (LSD1). Increased levels of H3K4 methylation upon siRNA inhibition of LSD1 in cell-based experiments provided supporting evidence for the enzymatic function of the protein. Sequence homology predicted a second related human amine oxidase, later confirmed to perform an identical H3K4 demethylation and named LSD2 [11]. Concurrently, the Jumonji C domain (JmjC) containing proteins was identified as a second family of histone demethylases with differences in substrate specificity and catalytic mechanism compared to the LSDs [12]. The proposed unified nomenclature for epigenetic proteins classifies the LSD and JmjC enzymes as KDMs (lysine demethylases), with the LSDs subdivided into KDM1A (LSD1) and KDM1B (LSD2). Nevertheless, we adopt the older LSD terminology as it is entrenched in the literature. Much of the following discussion is specific to LSD1 as there is more information available about this isoform compared with LSD2.

The LSDs are part of a subfamily of flavin adenine dinucleotide (FAD)-dependent oxidoreductases (Figure 10.3) that cleave a terminal or internal C—N bond [13, 14]. Typically, strong noncovalent interactions embed the FAD cofactor within the active site as in LSDs, whereas monoamine oxidases (MAOs) feature a covalent linkage to the protein. The catalysis then involves the FAD cofactor acting as an electron acceptor from the substrate, either a low molecular weight amine or a protein in the case of LSDs. The need for electron donation from the substrate to the cofactor explains the LSD specificity for mono- and dimethyllysine residues. These forms of lysine, when bound to the active site in an unprotonated state, can deliver electrons from the nitrogen

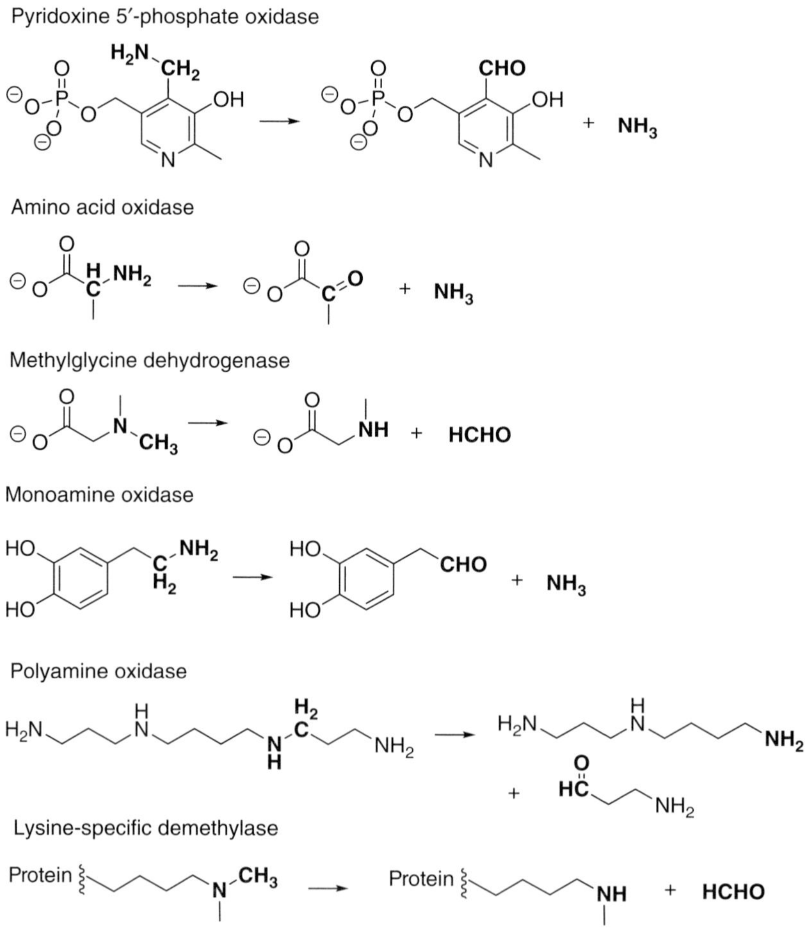

Figure 10.3 Examples of C—N bond cleavage reactions catalyzed by FAD-dependent oxidoreductases.

lone pair, whereas it is impossible with the permanently charged quaternary trimethyllysine residue. Since the FAD cofactor can accept either one or two electrons, there remains the option of single electron transfer or hydrogen atom or hydride transfer pathways for LSD oxidation. Experimental studies with LSD1 suggest a hydride transfer mechanism, which could be operating in LSD2 as well [15]. Nevertheless, the capability for single electron transfer is present, as we shall see in the tranylcypromine class of LSD1 inhibitors (Section 10.5.1). In common with other FAD-dependent enzymes, a separate reduction of molecular oxygen to hydrogen peroxide serves to recycle the reduced FAD cofactor (Figure 10.4). Overall, the chemistry of LSDs is intriguing as each turnover releases two toxic by-products, formaldehyde and hydrogen peroxide. In the other FAD-dependent demethylases, dimethylglycine and sarcosine dehydrogenases, a non-covalently bound tetrahydrofolate takes up the formaldehyde produced. It is possible that LSD1 utilizes the same scavenging pathway, as binding measurements and an X-ray structure indicate that the protein interacts with tetrahydrofolate [16].

Although LSD1 and LSD2 catalyze the same lysine demethylation, the proteins are substantially different (Figure 10.5) in their overall architecture. The two enzymes share a C-terminal catalytic amine oxidase domain (AOD, 33% homology between the isoforms) that accommodates the N-terminal histone tail substrate and the FAD cofactor. The remaining protein domains consist of modules important for the engagement with other macromolecules. Both LSD1 and LSD2 contain a SWIRM (Swi3, Rsc8, and Moira) domain commonly found in proteins associated with chromatin. Since the homology in the SWIRM domain is only 24%, there may be distinct functions for the domain between the isoforms. In addition, LSD1 has a large tower domain that bisects the AOD core and is involved in binding to the REST corepressor 1 (Rcor1, CoREST). This interaction is essential for recruitment of LSD1 to the nucleosome through Rcor1 binding to DNA and has an additional function in stabilizing LSD1 toward protein degradation by the proteasome. LSD2, on the other hand, does not bind Rcor1 and can demethylate nucleosomes on its own. In lieu of the tower domain, LSD2 contains a C4H2C2-type zinc finger and a CW-type zinc finger at its N-terminus. These domains are important for both demethylase activity as well as the independent ability of LSD2 (but not LSD1) to function as an E3 ubiquitin ligase [17].

10.3 LSD Substrates

The most clearly understood substrates of LSDs are histone H3K4me and H3K4me2. *In vitro*, N-terminal H3 peptides of ~20 amino acids containing K4me or K4me2 are sufficient as substrates instead of the full-length protein. LSD enzyme assays routinely use such peptides, while H3K4 peptides without methylation are efficient inhibitors. A number of X-ray structures of LSD1 with stable substrate-like H3K4 peptides are available and are in broad agreement with one another. The general features involve binding of the histone H3 N-terminus in a large anionic pocket in a manner that positions the K4 residue

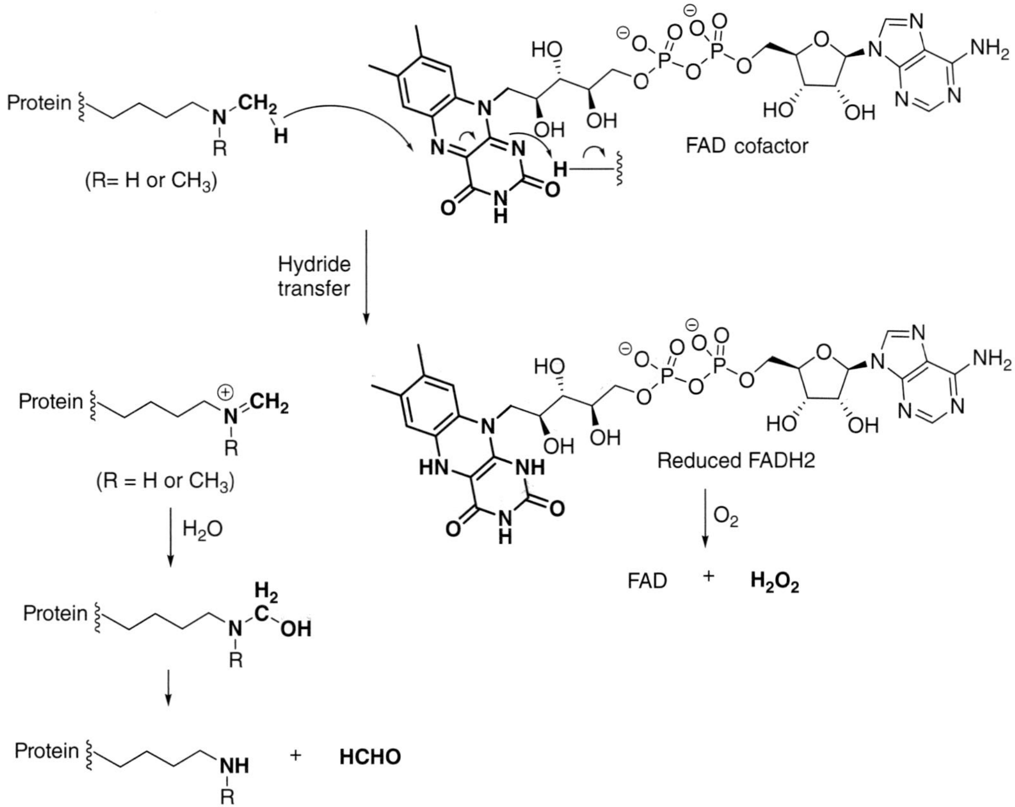

Figure 10.4 Catalytic mechanism of LSD mediated demethylation.

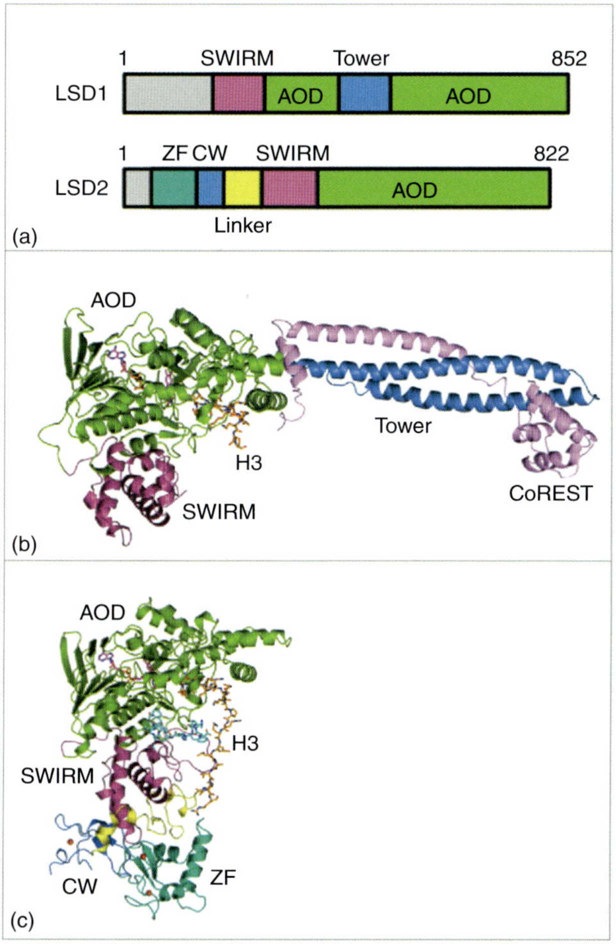

Figure 10.5 (a) Domain architectures of human LSD1 and LSD2. (b) Tertiary structure of LSD1 in complex with CoREST and an H3-like peptide inhibitor (PDB ID: 2V1D). (c) Tertiary structure of LSD2 (PDB ID: 4HSU) in complex with an H3 peptide and the NPAC linker region. Colors in (b) and (c) correspond to domains in panel (a) with peptides in orange and partner proteins CoREST and NPAC in pink and cyan, respectively. Source: Niwa and Umehara 2017 [3]. Reprinted with permission from Taylor and Francis.

adjacent to the FAD cofactor. The X-ray structure obtained with a H3M4 peptide where methionine is a stable methyllysine mimic is illustrative of the active site interactions [18]. The histone tail sits in an extended conformation folded back on itself (Figure 10.6) due to internal hydrogen bonding between Arg2 and Gln5 with distal residues. The Met4 lysine mimetic is facing the FAD cofactor, poised for electron transfer between the two. There are additional hydrogen bonding interactions between LSD1 and the basic N-terminal Ala1 and Arg2, and with Thr6, Arg8, Lys9, and Thr11 residues.

The X-ray structure suggests that LSD1 recognizes an extended region of the histone tail rather than just the K4 site of enzymatic reaction. This hypothesis is

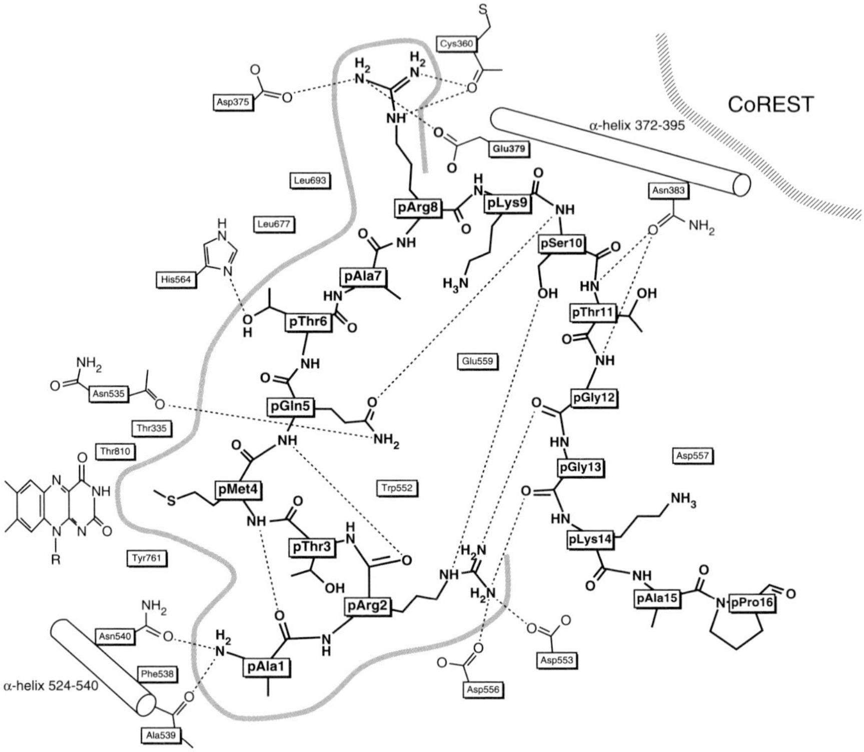

Figure 10.6 Active site interactions in the X-ray structure of LSD1–CoREST bound to a histone H3M4 peptide, shown in bold, acting as a substrate mimic. Source: Reprinted from [18].

supported by peptide affinity measurements: the H3 1–21 peptide with K4me0 (i.e. the product of demethylation) is an LSD1 inhibitor with a K_i of 2 μM, whereas the H3 1–15 peptide has a K_i of 200 μM [19]. Although the two peptides differ only by truncation of six amino acids from the C-terminus that lie in a disordered region outside the binding pocket and remote from the oxidation site, there is a 100-fold loss of binding affinity. Meanwhile, the use of full-length H3 protein substantially improves binding compared to the peptides, and histone H3 is a tight-binding, competitive inhibitor with a K_i of 19 nM [20]. Since the H3 histone tail has multiple sites of posttranslational modification, it is likely that cooperative effects between these structural alterations are important in the modulation of LSD1 binding. Indeed, experiments by the Mattevi group with H3 peptides demonstrate a general decrease in affinity for LSD1 and LSD2 when the sequence contains methylation, acetylation, or phosphorylation [11, 21]. For LSD2, an X-ray structure of the complex with its partner NPAC/GLYR1, a putative glyoxylate reductase, bound to a H3M4 peptide is available [22]. Compared to the LSD1 structure (Figure 10.6), there are hydrogen bonds beyond the 16th amino acid residue involving Lys18, Gln19 and Leu20, and LSD2. Thus, the recognition of H3K4 by LSD2 encompasses an even larger network of noncovalent interactions relative to LSD1.

The Schüle group identified that binding to the androgen receptor (AR) recruits LSD1 to an alternative histone substrate H3K9me and H3K9me2 [23]. Subsequently, the group showed that proline glutamic acid and leucine-rich protein 1 (PELP1) acts as a reader of H3K4me2 and H3K9me2 and forms a functional complex with the estrogen receptor (ER) and LSD1 to demethylate H3K9me2 [24]. This interesting switch from H3K4 to H3K9 is dependent on interactions with the nuclear hormone receptors, and the combination of LSD1 and recombinant ER is sufficient to induce H3K9 demethylation in histones in cell-free experiments [25]. Meanwhile, an alternative splicing variant of LSD1 contains an additional four residues, -Asp-Thr-Val-Lys-, inserted within the catalytic amine oxidase domain. In the X-ray structure of the variant, now known as neuroLSD1 due to its high occurrence in the brain, the tetrapeptide insert forms a miniature loop at the protein surface [26]. While the loop is relatively remote from the histone-binding pocket, it may interact with gatekeeper residues to modulate access by the substrate. The Shi group demonstrated that neuroLSD1 purified from neuronal cells is a H3K9me2 demethylase and suggested that the interacting partner supervilin is necessary for the catalytic activity [27]. Concurrently, the Rosenfeld group showed that neuroLSD1 is a H4K20me demethylase *in vitro* and *in vivo* [28]. Overall, these exciting recent findings expand the histone substrates beyond the well-characterized H3K4 to include H3K9 and H4K20. The X-ray structures of LSD1 with H3K4 peptides (Figure 10.6) show the N-terminus positioned at the entry to the active site, with the K4 substrate close to the FAD cofactor. From this model, it is not obvious how distal lysines like K9 or K20 bind in a conformation suitable for catalysis. Presumably, the histones in these alternative substrates are in a rather different orientation than observed with H3K4, but this has to await confirmation through X-ray structures.

Does the catalytic activity of LSDs extend beyond the nucleosome? This is a controversial question, and numerous non-histone proteins are reportedly LSD1 substrates (Table 10.1). However, the evidence is reliant on antibody-based analytical methods that are susceptible to issues with their specificity. In all non-histone substrates, the proposed sites of demethylation are not located at the N-terminus and there are no examples of peptides based on these internal sites acting as substrates or inhibitors in biochemical cell-free assays. A common feature of the histone substrates is the presence of two or more positively charged residues at the N-terminal end of the lysine that help anchor the protein within the anionic cavity of the catalytic domain. While similar charged residues are there in some of the non-histone sequences, further work is necessary to confirm if these are true LSD substrates.

10.4 LSD Function and Dysfunction

The epigenetic roles of lysine demethylation are complex with each demethylase enzyme having its own preferred binding partners and substrates. Depending on the histone demethylation site and the context, the consequence is either transcriptional activation or repression. The evolutionary conservation of LSDs

Table 10.1 Protein substrates of LSD1 (and LSD2 in the case of H3K4).

Protein	Demethylation sequence	References
Histone H3	ART**K**^{4}QTARK-	[10, 11]
Histone H3	-KQTAR**K**^{9}STGGK-	[27]
Histone H4	-AKRHR**K**^{20}VLRDN-	[28]
p53	-FCQLA**K**^{370}TTPIE-	[29]
DNMT1	-YNAKS**K**^{1095}SFEDP-	[30]
E2F1	-IAKKS**K**^{185}NHIQW-	[31]
MYPT1	-RLGLR**K**^{442}TGSYG-	[32]
STAT3	-AVVTE**K**^{140}QQMLE-	[33]
HSP90α	-KQEEK**K**^{565}TKFEN-	[34]
Tat	-SYGRK**K**^{51}RRQRR-	[35]
MTA1	-KQAVR**K**^{532}PLEAV-	[36]
ERα	-GGRML**K**^{266}HKRQR-	[37]

The bold **K** indicates the demethylation site with the protein residue numbering as a superscript and the sequence of five preceding and five subsequent amino acid residues.

from yeast to man underscores their importance in normal cellular function and indicates they perform key nonredundant roles that cannot be compensated by other demethylases. In mouse models, the knockout of LSD1 is embryonically lethal [30], while that of LSD2 leads to defective DNA methylation of imprinting genes [38]. Through demethylation of H3K4me1/me2 at active promoters and enhancers, LSD1 is a transcriptional repressor and plays a critical role in the maintenance of stem cell pluripotency during embryonic development, hematopoiesis, and neuronal differentiation [39]. In contrast, H3K4 demethylation by LSD2 is associated within the coding regions of target genes and helps maintain Pol II-mediated transcription elongation [40].

Through its binding partners, LSDs are components of many multiprotein transcriptional complexes. These interactions determine the accessibility of the enzymes to specific sites of the nucleosome. LSD1 is part of the CoREST, nucleosome remodeling and histone deacetylase (NuRD), tailless orphan nuclear hormone receptor (TLX), T-cell acute lymphocytic leukemia protein 1 (TAL1), C-terminal-binding protein (CtBP), Pumilio (PUM), and serum response factor (SRF) protein complexes. In addition, the direct binding between LSD1 and the HOX transcript antisense RNA (HOTAIR) long noncoding RNA recruits the demethylase to the polycomb repressor complex 2 (PRC2) complex. All these repressor complexes guide the LSDs toward H3K4 demethylation and typically contain other epigenetic enzymes such as HDACs and KMTs, providing an opportunity for cross-talk and cooperativity between the various histone modifications. In addition to the repressive function, LSD1 in concert with the AR and

ER nuclear hormone receptors is a H3K9 demethylase. In this case, demethylation serves as a transcriptional activator at hormone responsive elements.

Since the histone demethylation effected by LSDs has powerful physiological consequences, the enzyme activity is under tight regulation. Firstly, there are binding partners such as Rcor1 and Rcor2 for LSD1 and NPAC/GLYR1 for LSD2 that recruit the proteins to their substrates and activate catalysis. Among the Rcor family of co-repressors, Rcor1 and Rcor2 promote LSD1 activity, whereas Rcor3 acts as an inhibitor and the ratio between these paralogs is likely to determine cell fate and differentiation [41]. Furthermore, the Snail family of transcription factors act as endogenous inhibitors by competitive binding to the active site. These proteins have a highly conserved N-terminal domain containing a Phe as the fourth amino acid residue (Table 10.2). X-ray crystallography has revealed that the sequence sits in a similar orientation to histone H3 with Phe4 close to the FAD cofactor and acting as a methyllysine substrate mimetic [42]. The Snail proteins are potent reversible inhibitors of LSDs and the effect is reproducible by shorter N-terminal peptides, as discussed below (Section 10.5.3). Finally, as mentioned in Section 10.3, the enzyme products are themselves inhibitors of LSD proteins. Histone H3 tail peptides bearing unmethylated K4 residues are micromolar inhibitors, while the full-length H3 protein is a nanomolar inhibitor. Recently, the enzyme by-product H_2O_2 was reported to inhibit LSD1 with an apparent IC_{50} of 0.6 μM in a mass spectrometric assay by promoting disulfide bond formation between two Cys residues [43]. Overall, such a complex interplay between substrate recognition, product inhibition, and regulation by endogenous inhibitors is unique to LSDs among all the epigenetic enzymes.

The preceding paragraphs suggest that when LSD activity goes awry, it is likely to lead to pathological consequences. Various cancer types including leukemia, breast, prostate, bladder, liver, lung, colon, and neuroblastoma feature

Table 10.2 N-terminal sequences of human members of the Snail family of transcription factors.

Protein	N-Terminal sequence
Snai1	PRS**F**LVRKPS
Snai2	PRS**F**LVKKHF
Snai3	PRS**F**LVKTHS
Gfi1, Gfi1b	PRS**F**LVKSKK
Scrt1	PRS**F**LVKKVK
Scrt2	PRS**F**LVKKIK
Insm1	PRG**F**LVKRSK
Insm2	PRG**F**LVKRTK
Ovol1	PRA**F**LVKKPC
Ovol2	PKV**F**LVKRRS
Ovol3	PRA**F**LVRSRR

The bold **F** highlights the conserved phenylalanine that acts as a substrate mimetic.

an increased expression of LSD1 that is associated with disease progression and poor prognosis [44]. The interaction between LSD1 and the Snail transcription factors is involved in the repression of E-cadherin, an important step in the epithelial–mesenchymal transition necessary for cancer cell migration [45]. There is evidence that LSD1 promotes the Warburg effect in cancer, i.e. the shift from mitochondrial respiration to glycolysis [46]. In leukemia, the normal role of LSD1 in hematopoiesis is subverted to maintain cancer stem cells in a self-renewing state rather than undergo differentiation or apoptosis. There is strong evidence supporting LSD1 as an anticancer target in two particular forms of leukemia. Firstly, in mixed lineage leukemia (MLL) arising from chromosomal translocations that fuse the MLL gene to other proteins, the Somervaille group demonstrated that LSD1 promotes oncogenic transcription [47]. Secondly, in acute promyelocytic leukemia (APL) involving translocations of the retinoic acid receptor (RARα), not all cells respond to therapy with all-*trans*-retinoic acid (ATRA or tretinoin). However, as shown by Zelent's group, the joint inhibition of LSD1 restores sensitivity to tretinoin [48]. Apart from the above effects due to H3K4 demethylation, there is the link between LSD1 and prostate and breast cancer arising through nuclear hormone receptor-driven H3K9 demethylation.

Besides the link with cancer, LSD1 influences infection by the herpes simplex virus. In both *in vitro* and *in vivo* models, LSD1 inhibition reduced viral replication and activation of latent viral reservoirs [49, 50]. In principle, LSD1 could be a target for additional microbial infections such as fungal or parasitic diseases. However, this would require selective inhibitors of the enzyme in these eukaryotes relative to the human ortholog, and such examples are currently unknown. Another potential indication is in metabolic and cardiovascular disorders. LSD1 induces transcription of sterol regulatory element-binding proteins (SREBPs) and the binding of SREBP-1 to the fatty acid synthase promoter, hence suggesting LSD1 inhibition as a target for reducing lipid biosynthesis [51]. Transgenic mouse models have confirmed the importance of LSD1 in adipose tissue metabolism, although there are conflicting reports on whether it increases or reduces oxidative mitochondrial metabolism in white adipose tissue [52, 53]. A recent publication demonstrates that LSD1-mediated regulation of the peroxisome proliferator-activated receptor α (PPARα) leads to the age-related transformation of beige adipocytes to white adipocytes [54]. Meanwhile, the influence of the isoform LSD2 on human disease is less clear, although *in silico* analysis indicates high expression levels in breast cancers [55]. A further complication with LSD2 compared with LSD1 lies in its dual enzymatic ability as a lysine demethylase and E3 ubiquitin ligase. These two functions may have opposing roles in cancer that are tumor promoting or suppressive, respectively.

10.5 LSD Inhibitors

The recognition that LSDs are FAD-dependent amine oxidases greatly facilitated the discovery of the first small-molecule LSD inhibitors. Monoamine oxidases (MAOs) are in the same family and well-validated CNS targets with over half a century of drug approvals. Testing some of the older drugs led to the discovery that two irreversible MAO inhibitors, phenelzine and tranylcypromine, are also capable of inhibiting LSD1 at a micromolar level [56]. Further optimization with second-generation analogues of tranylcypromine has yielded examples with increased potency and selectivity currently undergoing clinical trials. With the passage of time, high-throughput screening and structure-based design have led to additional small molecule scaffolds identified as reversible LSD1 inhibitors. Another starting point for chemical probe design has utilized the histone H3 substrate or Snail transcription factor sequences. Such studies have produced potent inhibitors, albeit of a medium-sized peptidic nature. Before discussing these different approaches in detail, it is helpful to describe briefly how LSD inhibitors are characterized. Cell-free enzymatic assays for LSD inhibition employ recombinant proteins with K4-methylated histone H3 peptides, full-length proteins, or nucleosomes as substrates. The Shi group recommends the non-biotinylated H3K4 1–21 peptide for monitoring LSD1 activity and the assay readouts involve measurement of either the formaldehyde or hydrogen peroxide by-products [57, 58]. Among these, the method involving the oxidation of Amplex Red by H_2O_2 to produce the fluorescent resorufin is perhaps the most popular and suitable for high-throughput screening. Due to differences in assay conditions, the absolute inhibition values from different publications are not comparable unless a reference standard produced similar values. A further complication arises with compounds derived from irreversible MAO inhibitors, which should ideally be reported as $k_{inact}/K_{i(inact)}$ according to the Kitz-Wilson equation but are often given as IC_{50} values.

Although operating at a low throughput, mass spectrometric analysis of substrate peptides to detect demethylation products (*m/z* −14 and *m/z* −28 for H3K4me2 substrates) are less prone to false positives and useful for confirming hits identified from other formats. In cell-based assays, LSD1 inhibitors have antiproliferative effects in cancer cell lines, particularly those derived from leukemia and small cell lung cancer (SCLC). Evidence of target engagement in cells is not straightforward as global increases of H3K4 methylation in the presence of an inhibitor can be difficult to detect. Since LSD1 inhibition often triggers differentiation, the measurement of this process has served as a surrogate biomarker. In the THP-1 leukemia cell line, Somervaille identified the cluster of differentiation protein CD86, a cell surface marker, as increasing in expression levels in a predictable and dose-dependent manner upon LSD1 inhibition [59]. Similarly, others have measured increased levels of other proteins associated with hematopoietic differentiation such as CD11b, CD14, and Gfi1b. For compounds that progress from *in vitro* studies to *in vivo* models, standard mouse xenografts are the method of choice although the observed endpoint of reduction of tumor volume is not associated with a specific mechanism of action.

10.5.1 Irreversible Small Molecule LSD Inhibitors from MAO Inhibitors

The old mechanism-based MAO inhibitors phenelzine and tranylcypromine were the earliest small molecule LSD inhibitors identified. Another MAO

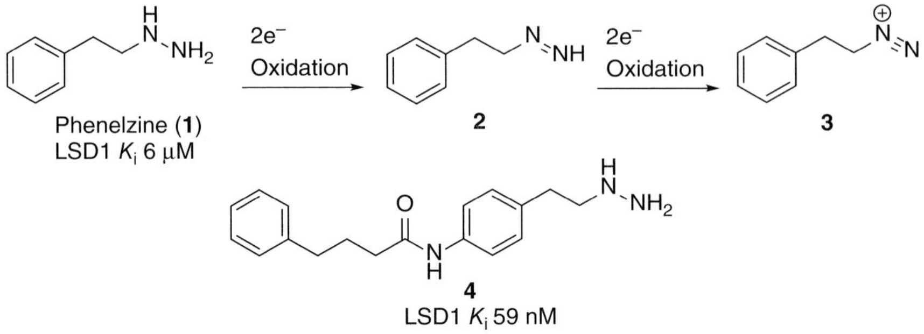

Figure 10.7 Mechanism of action of phenelzine and the structure of a second-generation analogue.

inhibitor, pargyline, weakly inhibits LSD1 but only at high millimolar concentrations [60]. Phenelzine (**1**, Figure 10.7) is believed to undergo two-electron oxidation by the FAD cofactor to give an intermediate diazene (**2**). The diazene itself might be the active compound, or a second two-electron oxidation would generate an alkyldiazonium cation (**3**). Either of these electrophilic species then reacts covalently with the FAD cofactor. Since the tightly bound cofactor does not exchange with external FAD, this effectively results in irreversible LSD inhibition. While phenelzine is nonselective for MAO versus LSD inhibition, SAR studies by Cole et al. led to the second-generation analogue bizine (**4**). Bizine has a selectivity of 23-fold against MAO A, 63-fold against MAO B and >100-fold versus LSD2 [61]. In the prostate cancer LNCaP cell line, 48-hour treatment with bizine caused an increase in H3K4me2 levels with an EC_{50} of 2 μM and growth inhibition with an IC_{50} of 16 μM. Co-treatment with HDAC inhibitors gave contrasting results: trichostatin A or vorinostat together with bizine exhibited moderate antagonism, whereas entinostat or panobinostat was additive. These results suggest that dual inhibition of LSD1 and HDAC inhibition can act synergistically with the right choice of agents. In addition to the repressive effect of both enzymes on gene transcription, a direct link exists in the deacetylation of Lys374 in LSD1 by HDAC1 that increases affinity for the histone H3 substrate [62].

For tranylcypromine (**5**, Figure 10.8), both a single electron transfer (SET) and hydride transfer mechanisms have been proposed for its reaction with FAD. The evidence favors Silverman's SET mechanism for monoamine oxidases, and this is likely to operate for LSD as well. SET to the cofactor gives a radical cation **6** that rapidly undergoes strain-induced cyclopropyl ring opening to give the reactive iminium radical **7** that covalently modifies the FAD cofactor. X-ray structures of tranylcypromine and analogues thereof incubated with LSD1 support the mechanism of action as they reveal FAD adducts such as **8–10** [63, 64]. Tranylcypromine contains two chiral centers, and four stereoisomers are possible. All are active, with the cis diastereomer having an IC_{50} of 29 and 8 μM against MAO A and MAO B, respectively, and the trans having an IC_{50} of 23 and 4 μM against MAO A and MAO B, respectively [65]. Within the *trans*-diastereomer, the (1*R*, 2*S*) or (+) enantiomer was originally reported by

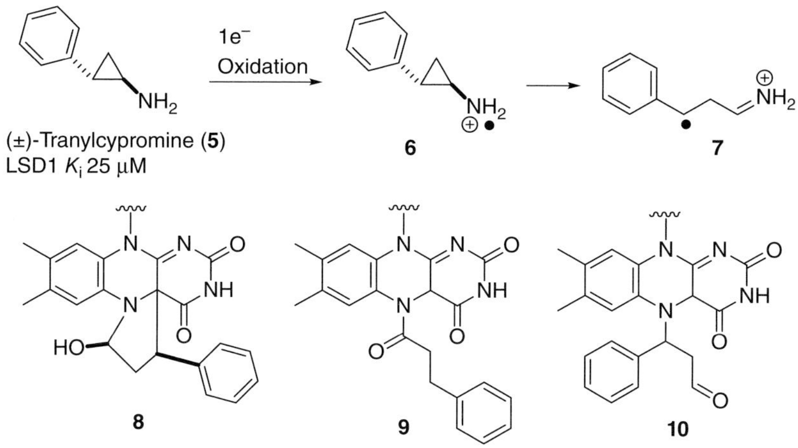

Figure 10.8 Mechanism of action of tranylcypromine and examples of tranylcypromine–FAD covalent adducts observed in X-ray crystallography of drug–LSD1 complexes.

Burger to be approximately fourfold more active than its mirror image [66]. Clinically, the drug is administered as the racemate of the *trans*-diastereomer **5**, thus avoiding the need for expensive separation. Unless stated otherwise, all structures in subsequent schemes refer to racemates. Although both enantiomers converge onto the same achiral reactive intermediate **7**, the initial reversible binding of **5** in the active site can be influenced by stereochemistry and explains the differences between the stereoisomers.

With the identification of tranylcypromine as a LSD1 inhibitor, the importance of stereochemistry for this new target needed to be investigated. Testing the resolved enantiomers of **5** revealed similar activity, with K_i values of ~25 μM [64, 67]. Nevertheless, in tranylcypromine analogues with more elaborate structures, the enantiomers are often no longer equipotent. For practical reasons, many compound series in the literature are racemates and the extent of this difference remains unknown. In a recent example that investigated individual enantiomers, one was 10-fold more potent than the other [68]. The more active enantiomer had the (1*R*, 2*S*) stereochemistry in the cyclopropyl ring, corresponding to the same chirality that is more active for MAO inhibition.

Due to its historic use as an antidepressant drug, there is ample data on tranylcypromine's human pharmacokinetics [69]. The main toxicity associated with the drug is the risk of accumulation of dietary tyramine (normally metabolized by MAOs) leading to cardiovascular issues (the "cheese effect"). Apart from that, the drug's relatively safe profile has rekindled interest in repurposing tranylcypromine as a LSD1 inhibitor, and it is undergoing clinical trials for the treatment of leukemia in combination with tretinoin. Meanwhile, many groups have pursued second-generation analogues with the dual objective of increasing LSD1 potency and improving selectivity versus LSD2 and MAOs to reduce side effects. These efforts have primarily focused on elaboration of tranylcypromine by introducing substituents in either the aromatic ring or the amine.

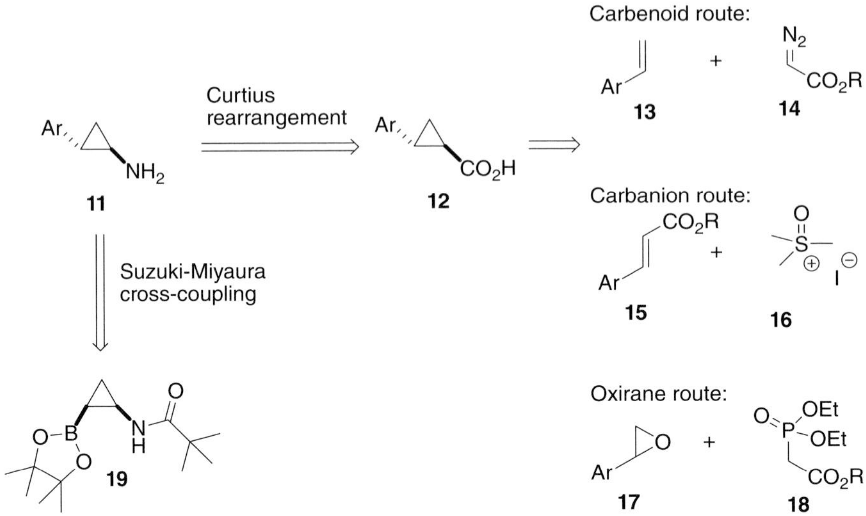

Figure 10.9 Synthetic routes to tranylcypromines.

The most popular synthetic route to tranylcypromine analogues **11** involves Curtius rearrangement of a cyclopropane carboxylic acid intermediate **12** to the amine (Figure 10.9). The carboxylic acid in turn arises from cyclopropanation of styrene **13** by α-diazo esters **14** through generation of a carbenoid, or of cinnamate **15** through generation of an ylide from **16** in the Johnson–Corey–Chaykovsky reaction. A variant involving reaction of a styrene epoxide **17** with **18** in a Wadsworth–Emmons cyclopropanation is suitable for analogues with additional substitution at the C1-position [70]. A recent alternative to these methods proceeds through Suzuki-Miyaura cross-coupling of cyclopropylamine boronate **19** [71]. All these routes are under thermodynamic control and give a preponderance of the *trans*-diastereomer as a racemate. If needed, asymmetric catalysis of the α-diazo ester **14** addition or the use of chiral oxiranes **17** provides an enantioselective synthesis [66, 69].

In addition to modification of the aryl or amine group of tranylcypromine, substitution of the cyclopropane ring is another option for LSD inhibition. While the isomeric 1-phenylcyclopropylamine **20** (Figure 10.10) is a MAO inhibitor, its activity against LSD is undisclosed. Meanwhile, the addition of alkyl or aryl groups to the C-1 position of tranylcypromine generally increases potency. For example, the enantiomeric C-1 phenyl analogues **21** and **22** were submicromolar LSD1 inhibitors (the reference tranylcypromine had an IC_{50} of 11.6 μM), albeit equally active against MAO A [69]. Interestingly, in this case the more active enantiomer **22** has the (1*S*, 2*R*) stereochemistry, the opposite to that of tranylcypromine itself. C1-substitution by fluorine in the diastereomers **23** and **24**, although improving activity against MAOs relative to tranylcypromine, was detrimental for LSD1 inhibition [71]. However, the additional introduction of an aromatic substituent yielded LSD1 inhibitors such as **25** that are more active than tranylcypromine, which had an IC_{50} of 25 μM. A patent from Genentech and Constellation Pharmaceuticals discloses rigidification of the phenyl ring

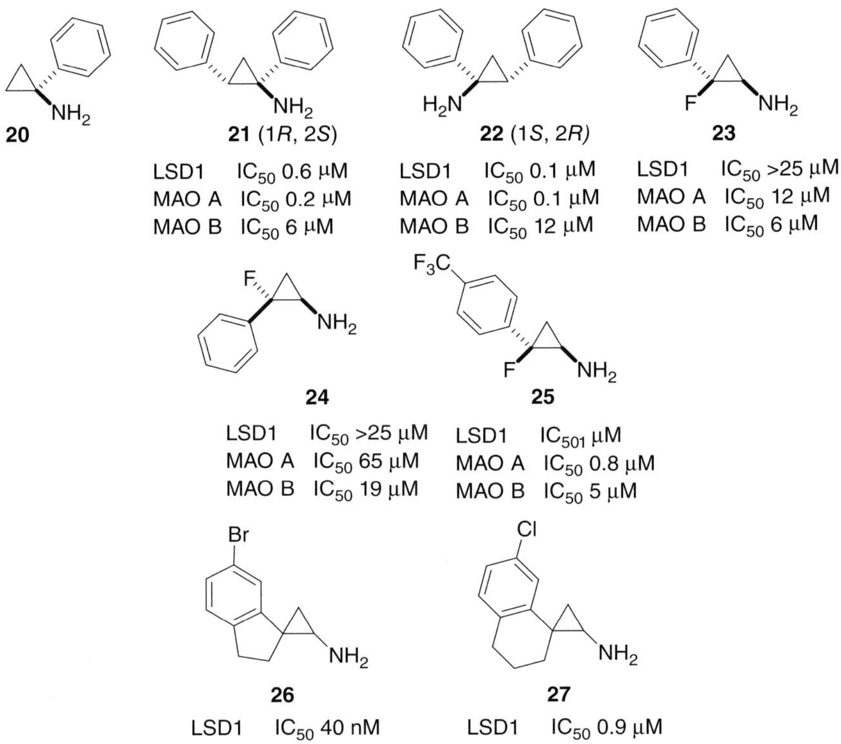

Figure 10.10 Examples of C1 and C2 substituted tranylcypromine analogues.

through C2 cyclization with various ring sizes and examples such as **26** and **27** (stereochemistry undefined) [72].

A large family of tranylcypromine analogues feature expansion of the phenyl ring by the introduction of meta or para substitution. Since X-ray structures of tranylcypromine bound to LSD1 indicate a roomy cavity, the additional substituents should increase compound affinity for the target. The Mai group has extensively published on analogues that contain a *para*-amino substituent further elaborated through amide bond formation (Figure 10.11). The racemate **28** was a submicromolar LSD1 inhibitor and had an IC_{50} of 2.5 μM in the growth inhibition of MV4–11 cells [73]. The single enantiomer **29** (1*S*, 2*R*) was a selective inhibitor of LSD1 and MAO A compared to their isoforms and threefold more active than its enantiomer against LSD1 [74]. The compound increased CD11b and Gfi1b expression in cells and displayed efficacy at an oral dose of 22.5 mg kg^{-1} in a mouse model of promyelocytic leukemia. In another series, attachment of an 8-hydroxyquinoline metal binding motif gave dual inhibitors of LSD1 and members of the Jumonji C demethylase family exemplified by **30** that reduced H3 methylation levels in Western blots [75]. Guided by X-ray structures of LSD1 bound to tranylcypromine and propargyl-lysine H3 peptides, Suzuki designed a *meta*-substituted series that eventually led to **31** [76]. As in Mai's compounds, the (1*S*, 2*R*) enantiomer was more active by approximately threefold in enzyme assays. Both enantiomers were similarly active in HEK293 and SH-SY5Y cells,

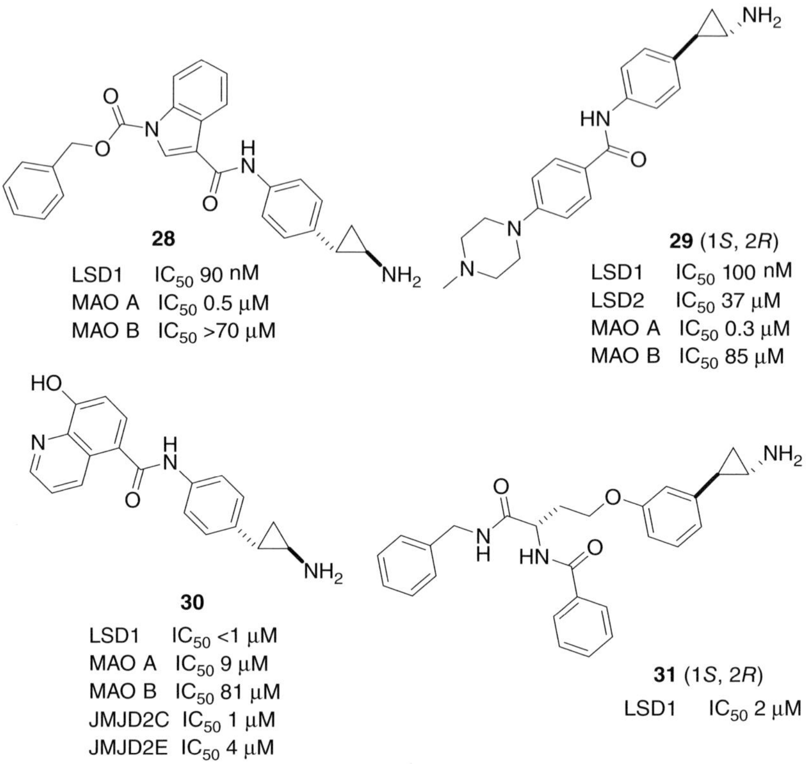

Figure 10.11 Examples of tranylcypromine analogues with *meta*- and *para*-substitution.

while the (1*S*, 2*R*) enantiomer was approximately fourfold more potent in HeLa cells. Compound **31** with racemic stereochemistry at the tranylcypromine ring was effective at an intraperitoneal dose of $1\ mg\ kg^{-1}$ in a mouse model of PC3 prostate cancer [77].

In 2010, a patent from the Spanish company Oryzon disclosed *N*-alkyl tranylcypromine analogues with an improved selectivity for LSD1 over MAOs compared with their nonalkylated counterparts [78]. This is somewhat surprising as hydrolysis of the iminium ion **7** (Figure 10.8) in the tranylcypromine mechanism of action will cleave any *N*-alkyl group that is then released by the enzyme. Indeed, Suzuki has exploited this aspect in the tranylcypromine-tamoxifen conjugate **32** (Figure 10.12) [79]. Upon mechanism-based ring opening by LSD1, iminium ion hydrolysis releases the amine that fragments to produce the AR antagonist 4-hydroxytamoxifen **33**. In the ERα positive MCF7 cell line, **32** increased H3K4me2 levels and inhibited proliferation in the presence of the agonist β-17-estradiol, suggesting the compound acts as a dual LSD1/ER inhibitor.

Further industrial patents clearly indicate that N-alkylation of tranylcypromine not only increases selectivity but also LSD1 target affinity without the need for substitution in the aromatic ring. Although the alkyl group may ultimately exit the enzyme, it presumably facilitates the initial reversible coordination through

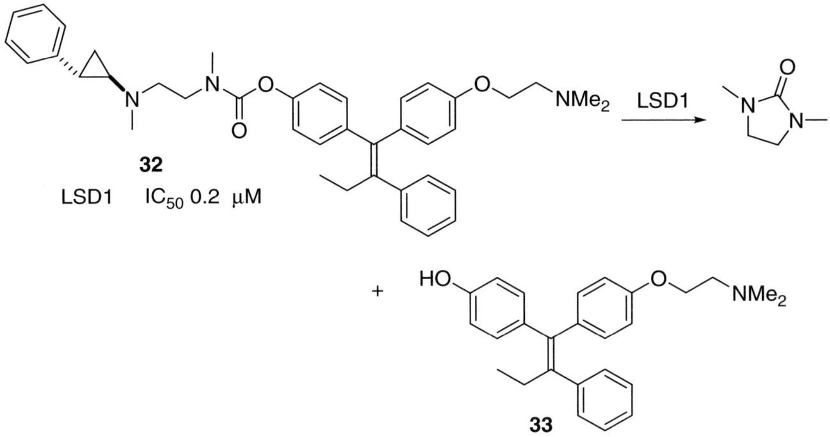

Figure 10.12 Mechanism-based activation of a tranylcypromine-tamoxifen conjugate.

additional bonding interactions, in the same way that the enzyme recognizes an extended region of the histone H3 substrate beyond the K4 demethylation site (Figure 10.6). Four *N*-alkyl tranylcypromines have currently progressed to clinical development as LSD1 inhibitors. Oryzon was the first to announce Phase I clinical trials in 2014 in acute myeloid leukemia (AML) with their candidate ORY-1001. While details of ORY-1001 and its pharmacology are unpublished, a review describes it as a single enantiomer (**34**, Figure 10.13, chirality undisclosed) with >1000-fold selectivity over LSD1B and the MAOs and subnanomolar cellular activity in THP-1 cells [80]. The phase I leukemia trial was successfully completed, and the compound was licensed to Roche who began phase I trials in SCLC in 2017. At the time of writing, Roche has discontinued the partnership based on pipeline priorities rather than any adverse data. Oryzon has a second tranylcypromine candidate, ORY-2001 (structure undisclosed), for the treatment of neurodegenerative disorders through dual LSD1 and MAO B inhibition. Meanwhile, GlaxoSmithKline advanced their LSD1 candidate GSK2879552 (**35**, chirality undisclosed) to phase I clinical trials in SCLC. Apart from LSD1, the compound does not significantly inhibit other FAD-dependent oxidoreductases, with D-amino acid oxidase inhibited only at a 280-fold higher concentration [81]. Compound **35** was highly selective in the growth inhibition of AML and SCLC lines among a panel of 165 cell lines tested. A set of DNA probes was hypomethylated in sensitive cells, suggesting their use as a predictive biomarker. From ChIP-qPCR sequencing after inhibition by **35**, LSD1 was localized at transcription start sites and enhancers rather than redistribution to new regions of the genome, and this phenomenon may be general for other LSD1 inhibitors. Subsequently, Incyte and Imago have announced phase I trials with tranylcypromine analogues for advanced malignancies and myelofibrosis, respectively. The structures of the candidates Imago-7289 and INCB059872 are undisclosed but likely to be similar to **36** and **37**, respectively [82, 83]. A recent patent by Mirati demonstrates further extension of the Oryzon and GlaxoSmithKline clinical candidates with potent single enantiomers such

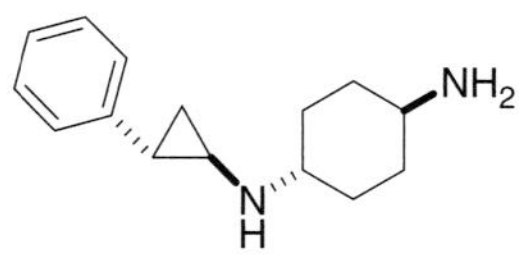

ORY-1001, **34**
LSD1 IC_{50} 18 nM

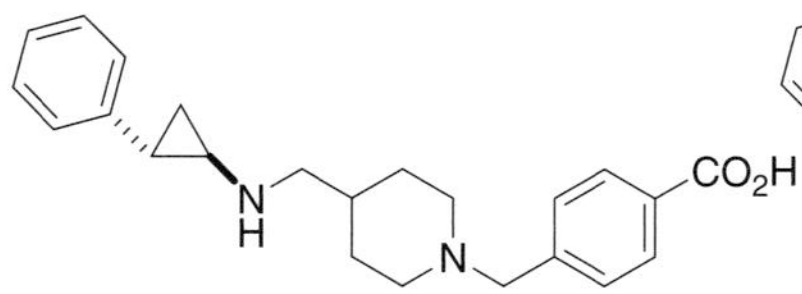

GSK2879552, **35**
LSD1 K_i (app) 2 μM

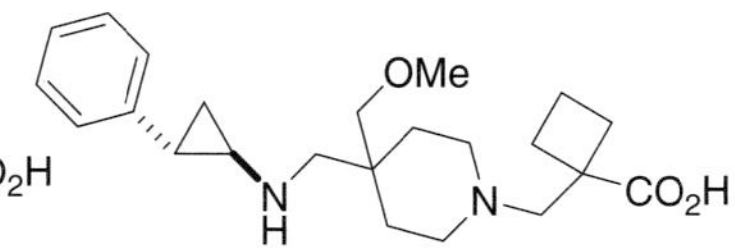

Incyte example, **36** (1*R*, 2*S*)

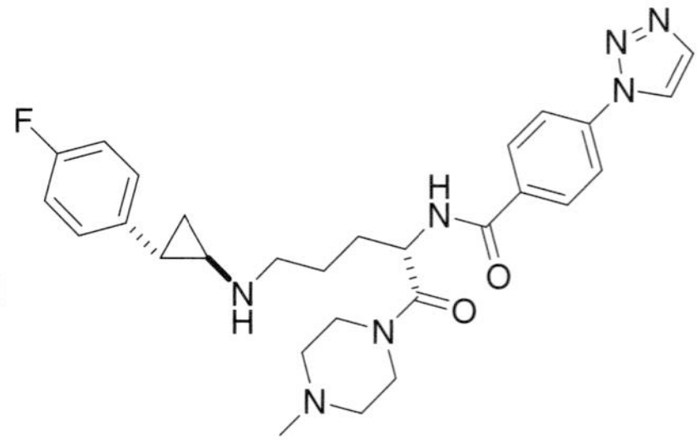

Imago example, **37** (1*R*, 2*S*)

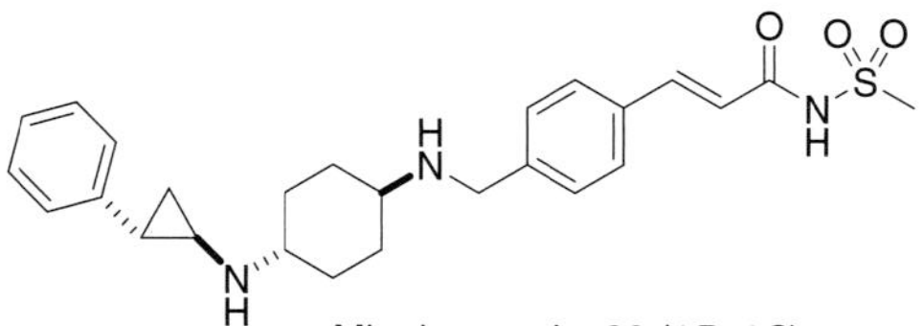

Mirati example, **38** (1*R*, 2*S*)
LSD1 IC_{50} 10 nM

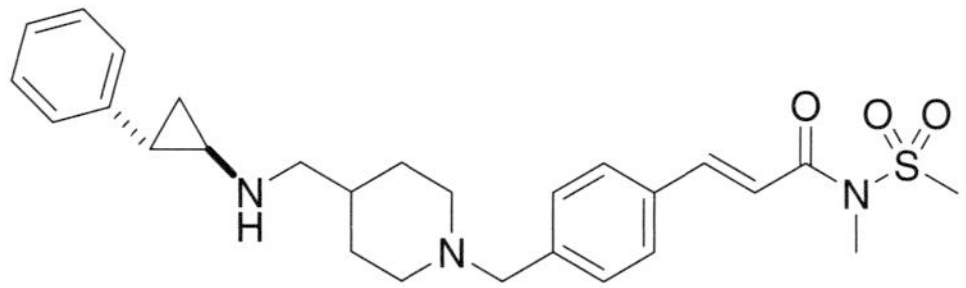

Mirati example, **39** (1*R*, 2*S*)
LSD1 IC_{50} 4 nM

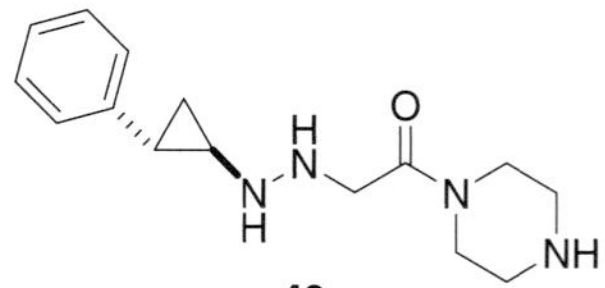

40
LSD1 IC_{50} 0.7 μM

Figure 10.13 Examples of *N*-substituted tranylcypromine LSD1 inhibitors.

as **38** and **39** and suggests **34** and **35** are also (1*R*, 2*S*) stereoisomers [84]. All compounds **34–39** share the structural feature of a nitrogen separated from tranylcypromine by a four carbon linker. A recent publication indicates that hydrazines can replace the amine in tranylcypromines as in **40** [85].

Some examples feature tranylcypromines that contain substitution in both the amine and the phenyl ring. In mouse models, the *para*-benzyl analogue **41** (Figure 10.14) penetrated the blood–brain barrier and impaired long-term memory [86]. In a sickle cell anemia model, the compound induced fetal hemoglobin synthesis [87]. The Takeda compound T-3775440 (**42**) has reportedly >100 000-fold selectivity against MAOs [88, 89]. The compound was particularly potent against acute erythroid leukemia (AEL), acute megakaryoblastic leukemia (AMKL) and SCLC in cell assays and disrupted the LSD1 interactions with Gfi1b and Insm1, two members of the Snail transcription family. In mouse tumor xenografts, the compound was effective at an oral dose of 10 or 20 $mg\,kg^{-1}$, and induction of peptidase inhibitor 16 (PI16) was used as a biomarker of LSD1 inhibition. The mice displayed thrombocytopenia, suggesting that platelet transfusions may be helpful in the clinical use of LSD1 inhibitors. The N-alkylation of racemic **31** to give **43** improved its activity against LSD1, showing how dual modification can be useful [90].

The imine **44** and urea **45** are interesting as the amine is now sp^2-hybridized and expected to be a poorer electron donor, but the compounds are apparently still capable of LSD1 inhibition [85]. A more radical example of dual modification is compound **46** where an ether replaces the phenyl ring altogether [91]. While inactive against LSD1 at 25 μM, **46** inhibited MAO A and MAO B with an IC_{50} of 170 and 5 nM, respectively. As the SAR is limited, it is possible that ethers that more closely resemble tranylcypromine would inhibit LSD1. On the other hand, the extended conjugation of a styrene is a successful replacement for the phenyl ring, as disclosed in a patent by Constellation with examples such as **47** [92]. While heteroaromatic isosteres have also replaced the phenyl ring in patents, such examples have apparently not progressed to candidate compounds so far.

10.5.2 Reversible Small Molecule LSD Inhibitors

The Woster group has published on polyamine containing reversible LSD1 inhibitors that may mimic the polycationic nature of the substrate histone H3 tail. The most potent compound **48** (Figure 10.15) inhibited the proliferation of the Calu-6 human lung adenocarcinoma line with an IC_{50} of 4.2 μM [93]. The polyamine probably binds to the acidic surface at the entry to the LSD1 active site although cellular activity may involve additional targets such as polyamine oxidases. Through computational modeling, Zhang attempted to mimic the histone substrate binding with small molecule guanidines [94]. The bisguanidine **49** increased H3K4me and H3K4me2 levels in mouse F9 embryonic teratocarcinoma cells, increased the expression levels of CHRM4 and SCN3A, and selectively inhibited pluripotent cancer cells. Meanwhile, the Mai and Mattevi groups undertook a screening approach with an amine library to identify quinazoline **50**, originally designed as an inhibitor of the lysine

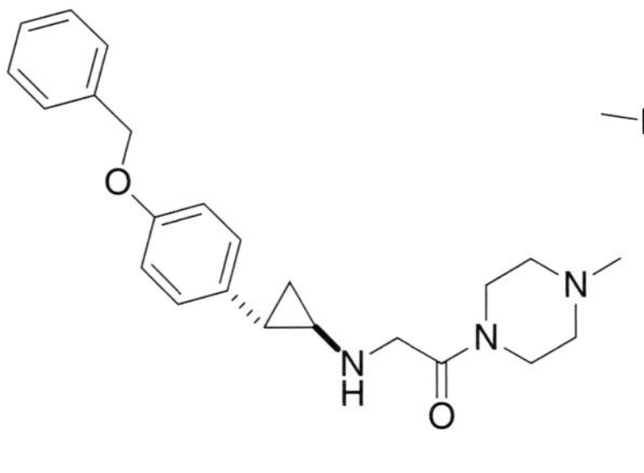

41
LSD1 IC_{50} 70 nM

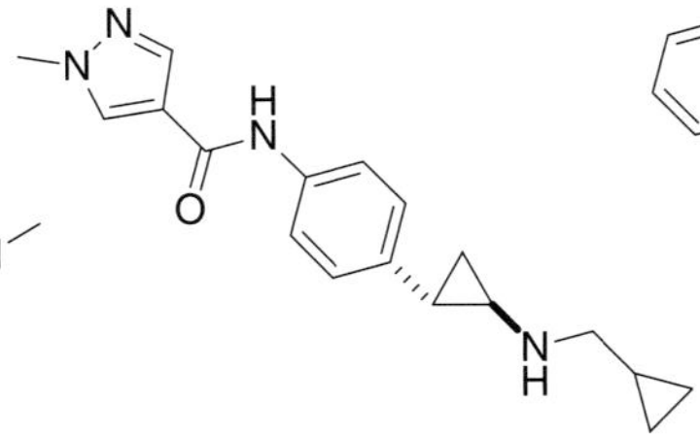

42
LSD1 IC_{50} 2 nM

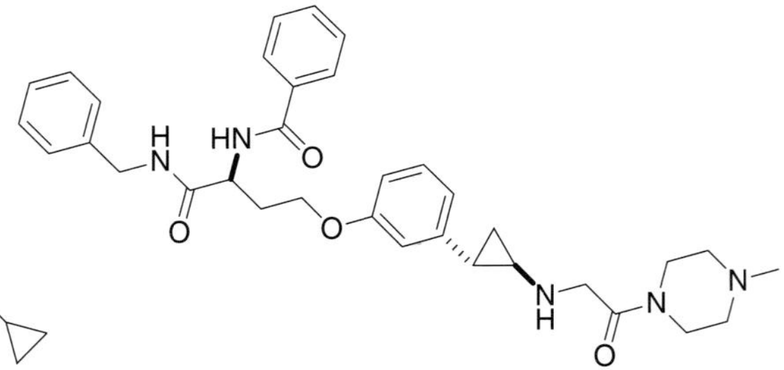

43
LSD1 IC_{50} 0.4 μM

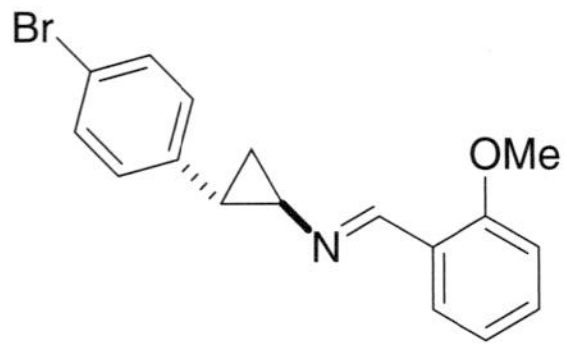

44
LSD1 IC_{50} 0.8 μM

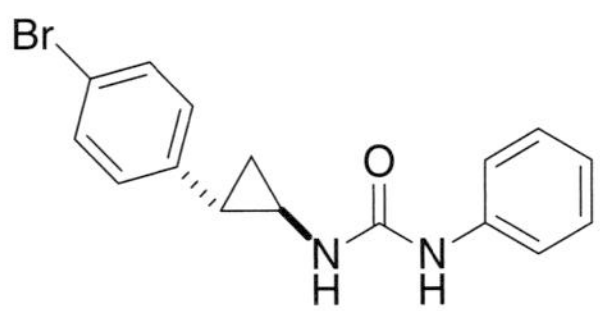

45
LSD1 IC_{50} 7 μM

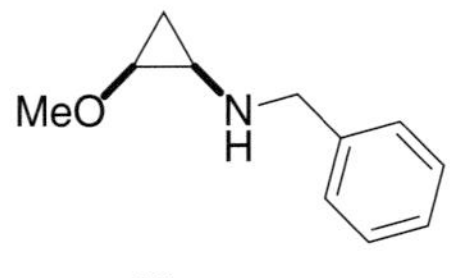

46

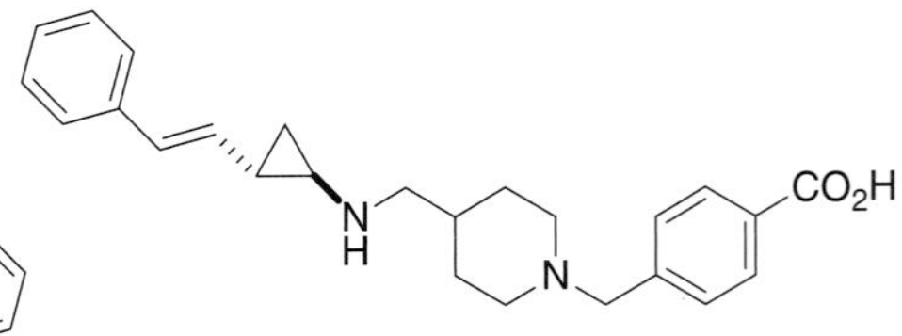

47
LSD1 IC_{50} <100 nM

Figure 10.14 Examples of tranylcypromine analogues featuring dual modification.

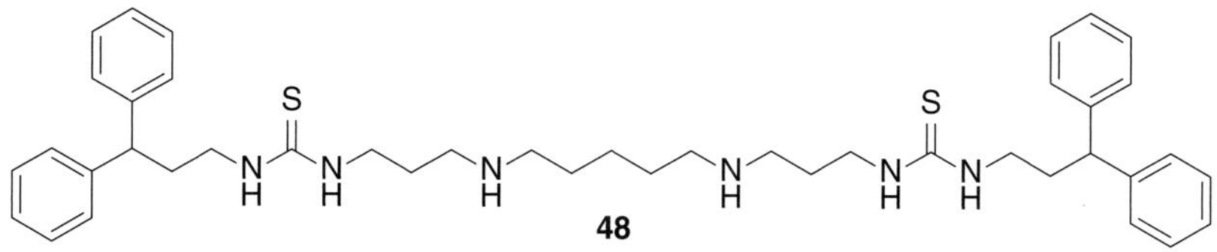

48

LSD1 IC$_{50}$ 5 μM
MAO A IC$_{50}$ >100 μM
MAO B IC$_{50}$ 19 μM

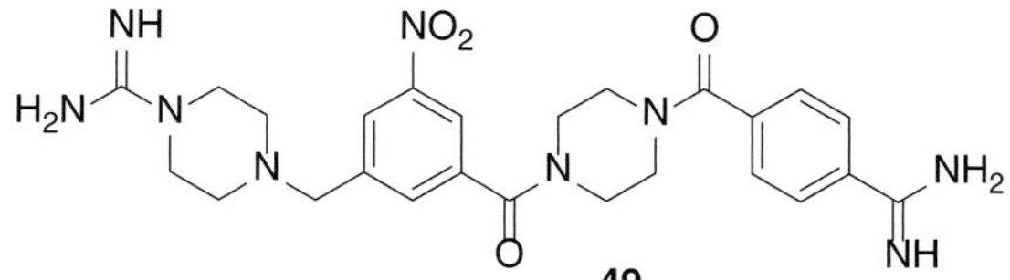

49

LSD1 IC$_{50}$ 11 μM

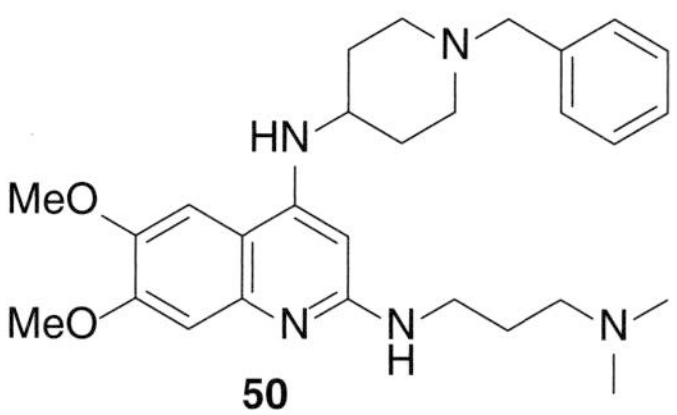

50

LSD1 K_i 0.4 μM

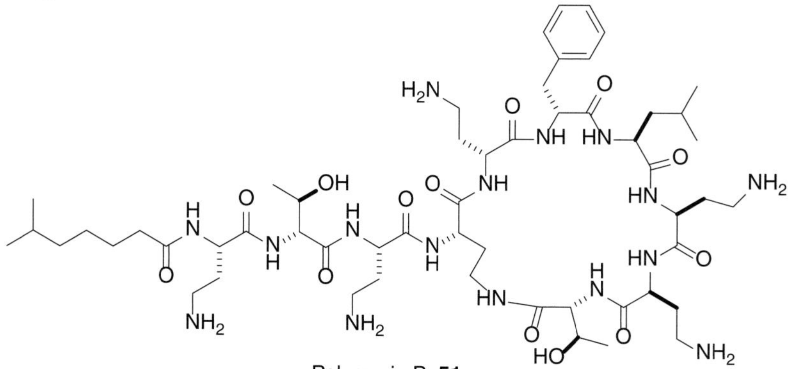

Polymyxin B, **51**

LSD1 K_i 0.2 μM

Figure 10.15 Examples of LSD1 inhibitors with multiple basic functional groups.

methyltransferase G9a, and the natural product antibiotic polymyxin B **51** as submicromolar LSD1 inhibitors [95]. Disordered electron density maps in X-ray crystallography suggest these polycations bind through electrostatic interactions in multiple binding modes at the entry to the active site.

Although FAD-dependent oxidoreductases have been drug targets for decades, there are very few examples of high-affinity natural product leads against these enzymes. Polymyxin B above is the only natural product reported with high affinity for LSD1 although its polycationic nature compromises bioavailability. At a lower micromolar level, the C20 diterpenoid geranylgeranoic acid **52** (Figure 10.16) inhibits LSD1 in a noncompetitive manner and induced TrkB expression in the SH-SY5Y neuroblastoma cell line [96]. Hydrogenation of the terminal alkene improved activity to an IC_{50} of 22 μM, comparable to tranylcypromine. While the literature documents other natural products as LSD1 inhibitors, they are promiscuous compounds that hit multiple targets and unlikely to be leads for further optimization. A natural product-inspired γ-pyrone library led to the discovery of **53** as a micromolar reversible LSD1 inhibitor [97].

Compared to the dearth of natural product leads, de novo design has recently produced a variety of synthetic small molecule scaffolds that potently inhibit LSD1. Virtual screening of the Maybridge Hitfinder compound library by Woster produced 10 potential hits from which arose an amidoxime series including **54** (Figure 10.17) [98]. Elaboration of the aryl group to styrene **55** further increased potency, and the compound increased levels of H3K4me2 and CD86 mRNA in MGC-803 cells [99]. A second hit from the library evolved to aminotriazole **56** [100]. Another virtual screening exercise selected a set of 121 commercial compounds from the docking of 13 million compounds to LSD1. Among these, six acylhydrazides were submicromolar inhibitors in the enzyme assay, and optimization provided the lead **57** [101]. Addition of a ring further increased the potency in **58** [102, 103]. In growth inhibition assays against a panel of cancer cell lines expressing high levels of LSD1, **58** had IC_{50} values of 0.4–1.2 μM and increased levels of both H3K4me2 and H3K9me2 in A2780 cells. Salarius Pharmaceuticals has advanced a compound SP2577 (structure undisclosed) from this series to clinical trials. Potential concerns with the suitability of the acylhydrazide **57** for medicinal chemistry led to investigations of alternative amide isosteres, albeit resulting in less potent compounds such as **59** and **60** [104, 105]. In addition to docking studies, pharmacophore-based modeling has yielded novel LSD1 inhibitors. Starting from the modeling of an oxazole library, a micromolar inhibitor **61** was discovered with an IC_{50} value of 1 nM

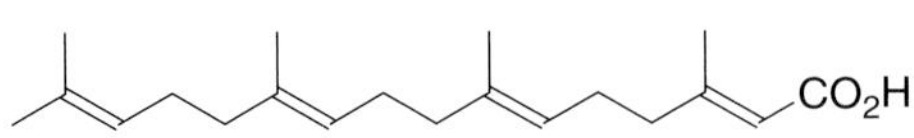

Geranylgeranyl acid, **52**
LSD1 IC_{50} 47 μM

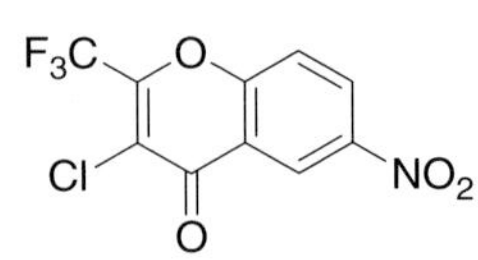

53
LSD1 IC_{50} 51 μM

Figure 10.16 Natural product and natural product-inspired LSD1 inhibitors.

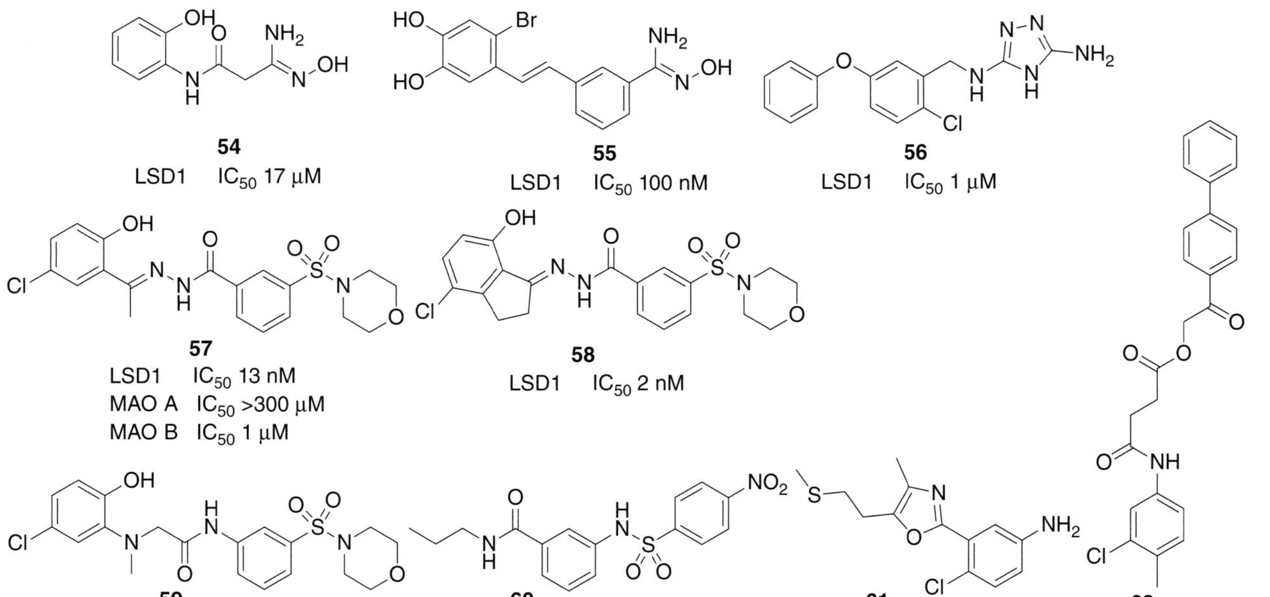

54
LSD1 IC_{50} 17 μM

55
LSD1 IC_{50} 100 nM

56
LSD1 IC_{50} 1 μM

57
LSD1 IC_{50} 13 nM
MAO A IC_{50} >300 μM
MAO B IC_{50} 1 μM

58
LSD1 IC_{50} 2 nM

59
LSD1 IC_{50} 10 μM
MAO A IC_{50} 190 μM
MAO B IC_{50} 66 μM

60
LSD1 IC_{50} 10 μM
MAO A IC_{50} 90 μM
MAO B IC_{50} 230 μM

61
LSD1 IC_{50} 10 μM

62
LSD1 IC_{50} 2 μM
MAO A IC_{50} 700 μM
MAO B IC_{50} 28 μM

Figure 10.17 LSD1 inhibitors originating from virtual screening.

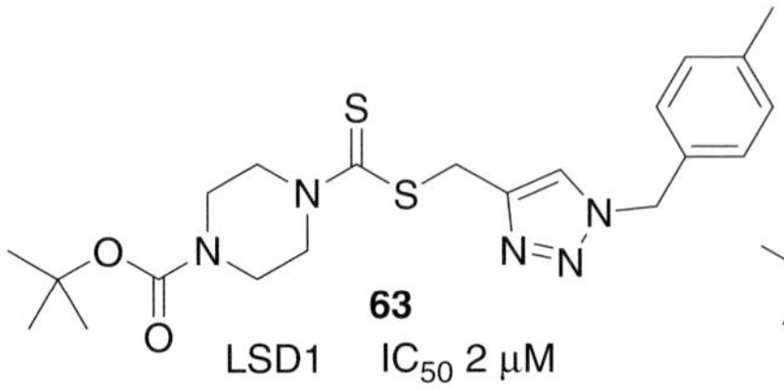

63
LSD1 IC_{50} 2 μM
LSD2 IC_{50} 37 μM
MAO A IC_{50} >1250 μM
MAO B IC_{50} >1250 μM

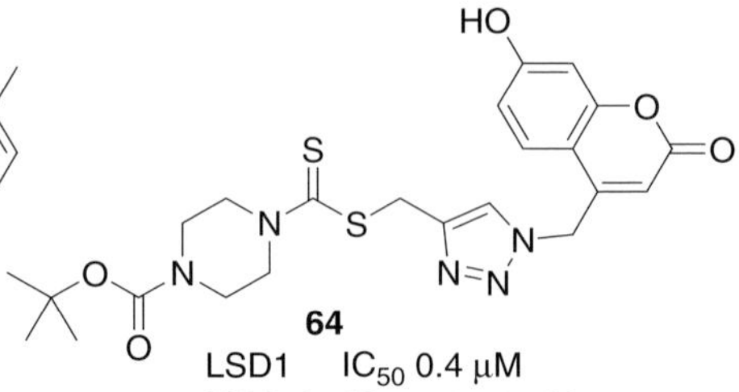

64
LSD1 IC_{50} 0.4 μM
MAO A IC_{50} >1250 μM
MAO B IC_{50} >1250 μM

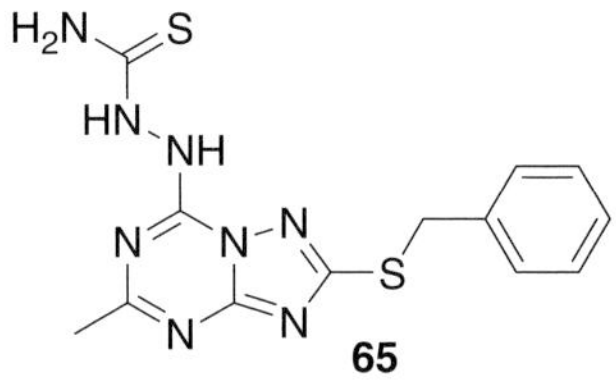

65
LSD1 IC_{50} 0.2 μM

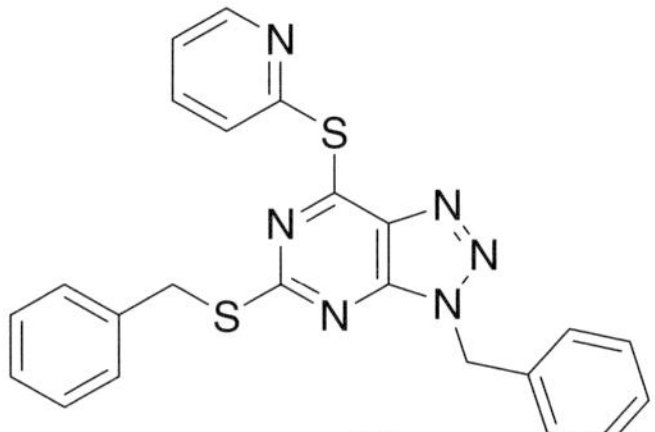

66
LSD1 IC_{50} 0.6 μM

Figure 10.18 Triazole containing LSD1 inhibitors.

in growth inhibition of HeLa and MDA-MB-231 cells [106]. As the cellular activity is superior to enzyme inhibition, additional mechanisms of action may be involved. A 3D QSAR pharmacophore model applied to docking of the SPECS database of 171 000 compounds produced nine hits of which the most potent was **62** with a succinate linker [107].

Liu reported a series of dithiocarbamate triazole containing LSD1 inhibitors such as **63** and **64** (Figure 10.18) prepared by click chemistry [108, 109]. Triazole **63** inhibited the LSD1–Snail association in cells and reduced tumor volume in a mouse gastric cancer xenograft at an intravenous dose of 20 mg kg^{-1}. It was suggested that **63** is a competitive FAD inhibitor, although the cofactor is tightly bound with a K_d of 180 nM. Starting with a thiopyrimidine lead, the group has also reported triazolopyrimidines **65** and **66** [110, 111]. The compounds inhibited migration in MGC-803 cells accompanied by increased levels of the epithelial cell marker E-cadherin and decreased levels of the mesenchymal cell marker N-cadherin.

Although unpublished, GlaxoSmithKline has reported the diarylpyridine GSK690 (**67**, Figure 10.19) as a potent reversible LSD1 inhibitor in conferences. An academic group has subsequently fleshed out the SAR of this series, with the most active compound **68** having an IC_{50} of 0.4 nM in the growth inhibition of MV4-11 cells [112]. While the hERG activity of **67** compromises further development, scaffold hopping at Quanticel, later acquired by Celgene, identified alternative diaryl heterocycles [113]. This has culminated in the candidate CC-90011 (structure undisclosed, likely to be similar to **69**) in phase I clinical trials in solid tumors and non-Hodgkin's lymphomas. SAR around an alternative pyrazole scaffold patented by Quanticel reported the most active compound **70**

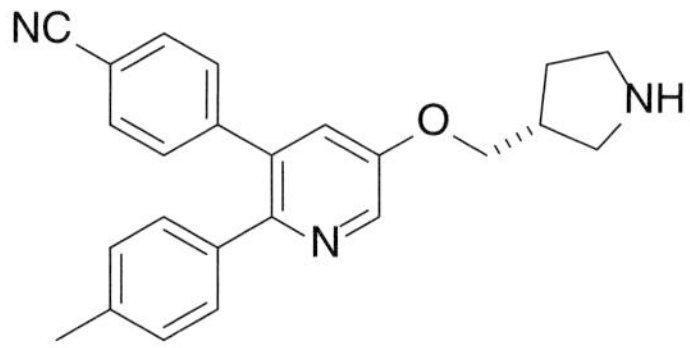

67

LSD1 IC_{50} <100 nM
MAO IC_{50} >200 μM

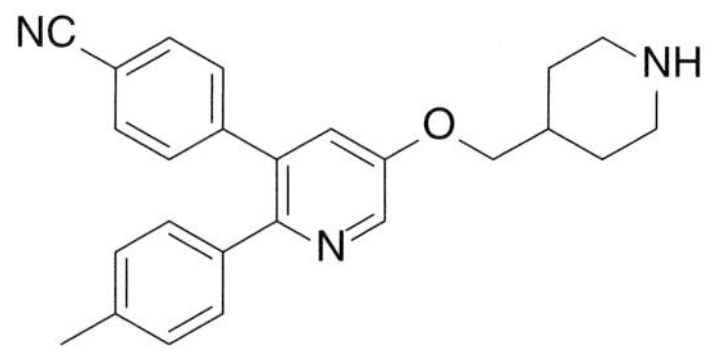

68

LSD1 IC_{50} 29 nM
MAO A IC_{50} >50 μM
MAO B IC_{50} 19 μM

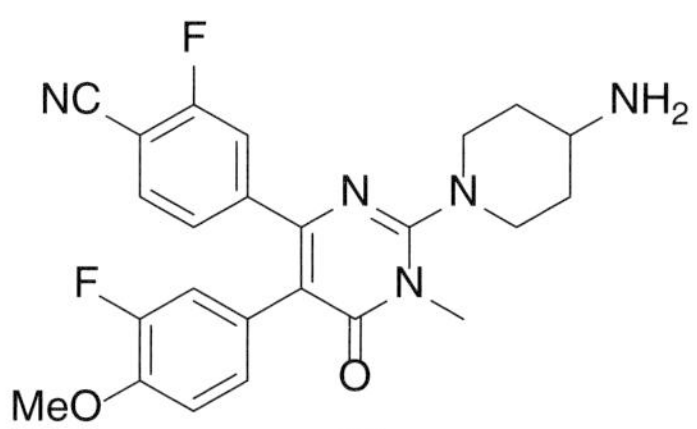

69

LSD1 IC_{50} <100 nM

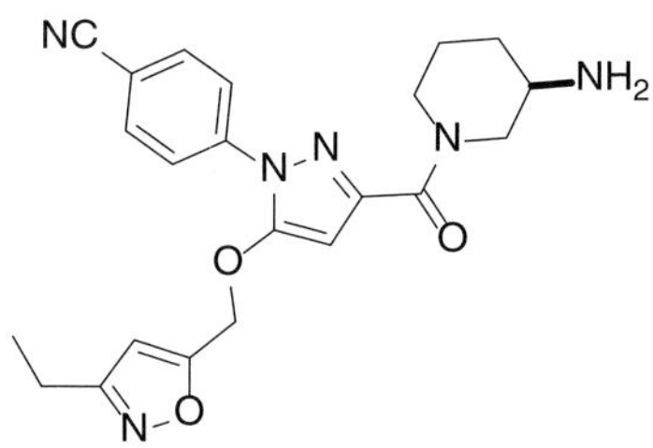

70

LSD1 IC_{50} 79 nM

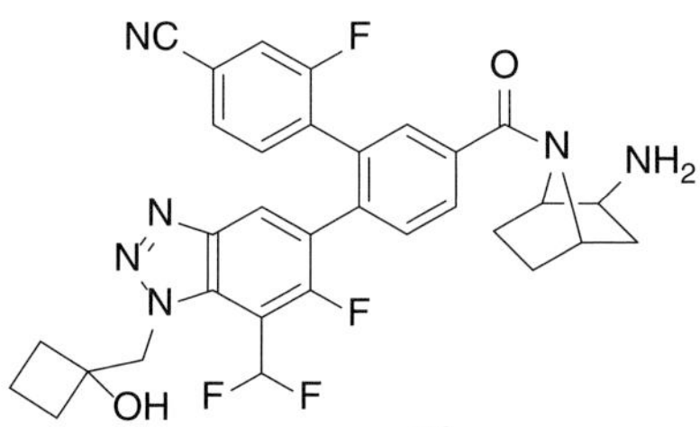

71

LSD1 IC_{50} 0.14 nM

Figure 10.19 Diaryl heterocycles and analogues as LSD1 inhibitors.

to retain hERG inhibition with an IC_{50} of 5 μM [114]. A patent by Taiho has the central heterocycle replaced by a benzoic acid extended by an amide linker with potent inhibitors exemplified by **71**, which had an IC_{50} of 0.13 nM in growth inhibition of the AML HEL cell line [115].

The co-occurrence of LSD1 and histone deacetylases (HDACs) in transcriptional repressor complexes suggests a beneficial effect in the joint inhibition of these two targets in cancer chemotherapy. In rhabdomyosarcoma cells, **67** acted in synergy with HDAC inhibitors to induce apoptosis [116]. The potential for such a dual mechanism of action is currently tested in clinical candidate 4SC-202 (**72**, Figure 10.20), an orally bioavailable inhibitor of nuclear Class I HDACs that additionally inhibits LSD1. However, details of the compound's pharmacology and the extent of dual target engagement *in vivo* are undisclosed. Another unusual LSD1 inhibitor is the rhodium(III) complex **73** that increased H3K4me2 levels in the PC3 prostate cancer cell line [117]. While the preceding paragraphs demonstrate a variety of chemical scaffolds potently inhibit LSD1 in a reversible

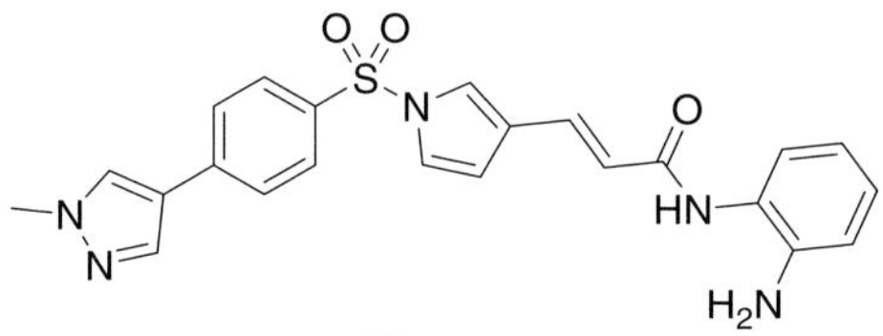

72

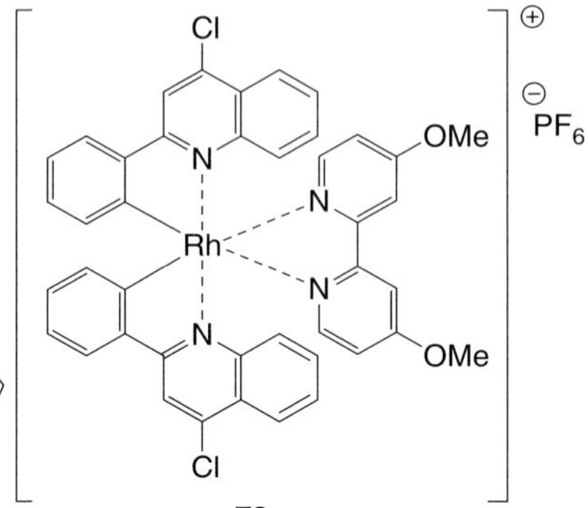

73

LSD1 K_i 0.6 μM

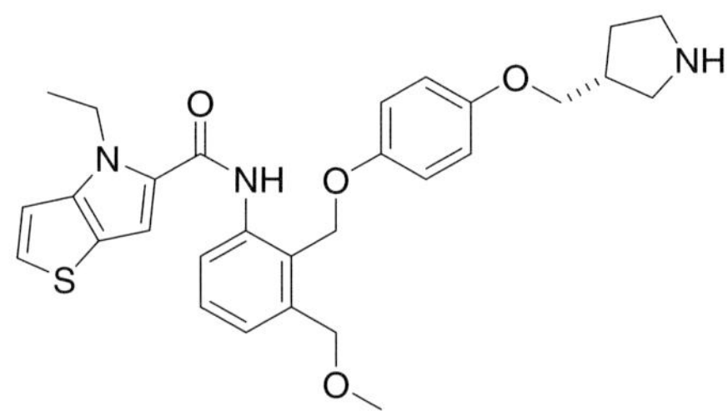

74

LSD1 IC_{50} 7 nM
LSD2 IC_{50} 17 μM
MAO A IC_{50} 47 μM
MAO B IC_{50} >100 μM

Figure 10.20 Additional reversible LSD1 inhibitors.

manner, structural insight into their mode of interaction with the enzyme is currently lacking. A notable exception is the high-throughput screening of a 34 000 compound library by Vianello and coworkers [118]. From 113 hits, optimization of a thieno-pyrrole bicycle generated nanomolar inhibitors such as **74** [119]. The compound induced CD11b, CD14, and CD86 in THP-1 cells and inhibited their growth with an IC_{50} of 1 μM. X-ray structures were successfully obtained for a number of examples bound to the LSD1-Rcor1 complex and showed the thieno-pyrrole moiety facing the FAD cofactor and the pyrrolidine interacting with the charged Asp555 and Asp556 residues.

10.5.3 Synthetic Macromolecular LSD Inhibitors

As described in Section 10.3, the N-terminal sequence of the product histone H3 lacking K4 methylation strongly inhibits LSD1, and our understanding of substrate binding comes from X-ray structures obtained with H3-like peptides. Cole has modified the K4 residue in the H3 1–21 peptide to identify enzyme inhibitors [120]. Peptides **75** and **76** (Figure 10.21) contain 3-chloroallyl and propargyl warheads previously known to react with FAD-dependent enzymes. Both were

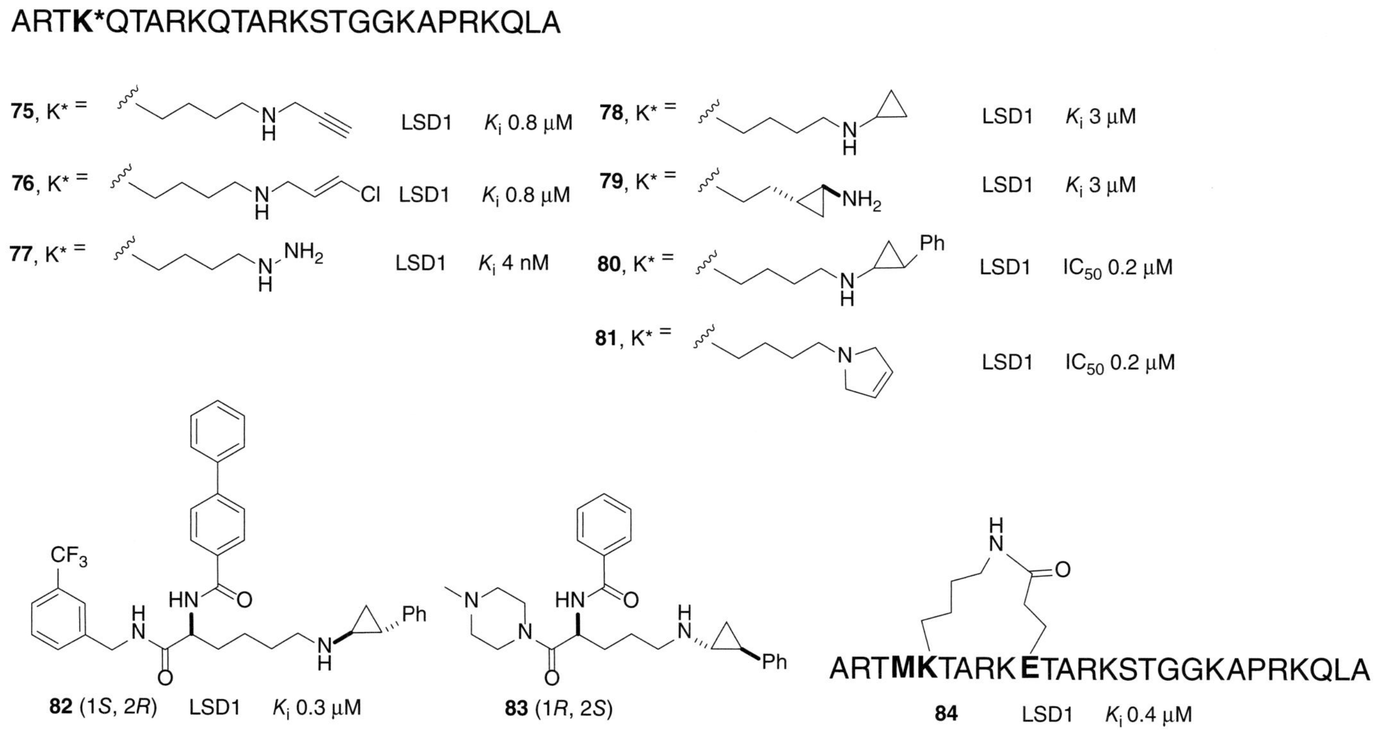

Figure 10.21 LSD1 peptide inhibitors derived from the histone H3K4 substrate. Residues in bold are modified from the native sequence.

irreversible inhibitors, as was the hydrazine **77**. However, tranylcypromine like cyclopropyl peptides **78** and **79** were reversible inhibitors, suggesting ring opening and irreversible inactivation need the additional stabilization from the phenyl ring. A later study by Suzuki supports this conclusion, as peptide **80** with the phenyl ring incorporated and the mechanism-based dihydropyrrolidine inhibitor **81** were more active [121]. C-Terminal truncation studies with **80** gave a respectable IC_{50} of 0.4 μM for the shorter 9-mer ARTKQTARK. With appropriate substitution, even a single modified lysine amino acid can provide potent inhibitors such as **82** and **83** [83, 122]. The (1*S*, 2*R*) enantiomer of **82** was approximately fourfold more potent than the enantiomer, although the order was reversed with other compounds in the series and compounds such as **83** in the Imago patent only contain the (1*R*, 2*S*) stereochemistry. Woster employed an alternative strategy of conformational restriction to histone H3 peptides. The most active peptide **84** with a modified H3 sequence containing Met4 as the K4 mimic and macrocyclization between Lys5 and Glu10 residues had an IC_{50} of 6 μM in MCF-7 cells and was more stable to hydrolysis than the linear peptide [123]. Further gains in potency are likely by altering the H3 sequence, as the single substitution of the N-terminal Ala1 residue by Ser gave a threefold boost in LSD1 affinity for the LSD1 H3M4 1–20 peptide, with the K_i decreasing from 100 to 30 nM [124].

The endogenous inhibition of LSD by the N-terminal domain of the Snail transcription factors offers another opportunity for peptide-based ligands. The initial X-ray crystallography utilized a Snail 1–20 peptide with a K_i of 0.2 μM for LSD1 and 2.0 μM for LSD2 [42]. Later studies showed that reducing the length to the Snail 1–9 peptide **85** (Figure 10.22) did not significantly affect affinity for LSD1 [125]. Further truncation to peptides **86** and **87** demonstrates that a minimum of six residues is required for micromolar inhibition. Alanine scanning of peptide **87** suggests the Pro1, the Arg2, and the Phe4 residues are the major determinants of binding. X-ray crystallography of these short peptides reveals they bind in the same conformation as histone H3 in the active site with the Phe4 residue in close proximity to the FAD cofactor and acting as a H3K4 substrate mimic. Peptides **88**

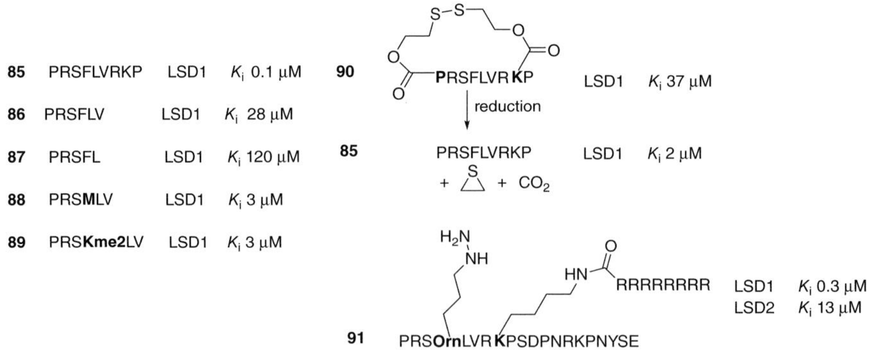

Figure 10.22 LSD1 peptide inhibitors derived from the Snai1 N-terminus. Residues in bold are modified from the native sequence.

and **89** with the Phe4 replaced by methionine and dimethyllysine were superior inhibitors. Interestingly, peptide **89** is an inhibitor of the enzyme rather than a substrate, indicating the importance of additional residues for the correct positioning of the methylated lysine. Umezawa reported the Snail 1–9 peptide **90** with attachment of side chains to Pro1 and Lys9 and formation of an intramolecular disulfide bridge [126]. The cyclic conformation has resulted in loss of activity, whereas reduction of the disulfide causes fragmentation of the linker and regeneration of the active linear peptide **85**. Peptide **90** was inactive in initial cell-based studies suggesting issues with membrane permeability. Suzuki prepared the Snail 1–21 peptide **91** with Phe4 replaced by a hydrazino-ornithine warhead and the attachment of a cell-penetrating Arg8 sequence to Lys9 [127]. In HeLa cells, the peptide increased H3K4me2 levels and decreased viability with an IC_{50} ~50 μM.

Panning a phage library of 7-mers and 12-mers against the catalytic domain of LSD1 led to the identification of the novel sequence SHSEFWDWGPGGG that does not contain the positively charged residues associated with histone H3 or Snail binding [128]. The peptide has an apparent affinity of 2 μM but did not inhibit catalysis. A computational study identified five additional locations outside the active site suitable for ligand binding at the SANT2/Tower interface (refer to Figure 10.5), the SWIRM/AOD interface, the AOD/Tower interface, the back of the AOD domain and a small pocket in the AOD domain [129]. The screening of peptide libraries against full-length protein offers an opportunity to target such sites and potentially modulate LSD through its interactions with binding partners or affect catalysis in an allosteric manner instead of the usual active site occupation. One such example, from the RNA world rather than peptides, is the telomeric repeat containing RNA (TERRA) sequence UUAGGG. The repeat RNA sequence $[UUAGGG]_4U$ inhibited LSD1 enzyme activity with an apparent K_i of 1.0 μM in a noncompetitive manner, while the monomeric pentanucleotide 5′-UUAGG-3′ RNA had a K_i of 75 μM [130]. An X-ray structure of the pentanucleotide with LSD1-Rcor1 shows it complexed in an allosteric binding cleft of the amine oxidase domain and having key interactions between the A and G base pairs with Gly279, Lys280, and Asp619 residues.

10.6 Summary

Our knowledge of the LSD proteins has increased tremendously since their identification as lysine demethylases in 2004. LSD1 is a major transcriptional repressor through the demethylation of histone H3K4 residues, and a transcriptional activator through H3K9 demethylation mediated by nuclear hormone receptors. Nevertheless, the finer details of LSD1 biology including its interacting partners, the downstream pathways, cross-talk with other epigenetic pathways and synergy with non-epigenetic drug targets need further investigation. An added complication is the presence of isoforms with their own distinct roles produced by alternative splicing. In addition to the biological challenges of target validation, LSD1 is a difficult medicinal chemistry target due to the voluminous nature of the active site and its substrate recognition through an extended surface. Nevertheless, multiple chemical scaffolds have produced small molecule inhibitors with

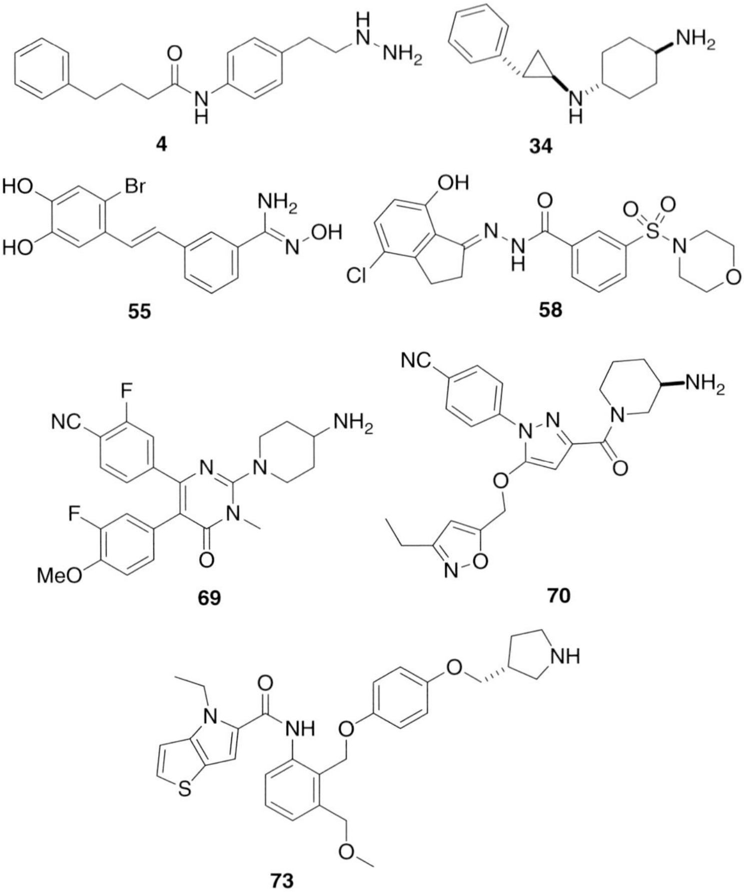

Figure 10.23 Examples of small molecule LSD1 inhibitors with reported IC_{50} or K_i values ≤100 nM.

high potency (Figure 10.23). Furthermore, these compounds are highly selective for LSD1 over the isoform LSD2 or other amine oxidases. The structural and mechanistic diversity available will be invaluable in dissecting on-target effects arising from LSD1 inhibition as opposed to idiosyncratic properties specific to a particular compound. The actual mechanism of action of LSD1 inhibitors *in vivo* is an open question. These compounds not only suppress enzyme turnover but also interfere with the binding of the Snail transcription factors that occupy the active site. Whether inhibition of catalysis or disruption of protein–protein interactions is the more important needs clarification.

Among the epigenetic proteins, LSD1 has advanced from bench to bedside in the remarkably short space of a decade. Nine LSD1 inhibitors are currently in clinical development (Table 10.3). Although the evidence for LSD1 as a target is most compelling in AML and SCLC, other cancer types as well as non-cancer indications are under investigation. There is potential for synergy, either through combination therapy or with compounds having a dual mechanism of action.

Table 10.3 LSD1 inhibitors that are currently in clinical development.

Compound	Mechanism	Organization	Indication
Tranylcypromine (**5**)	Irreversible, in combination with tretinoin	Universities of Freiburg and Halle	Acute myeloid leukemia, myelodysplastic syndrome
ORY-1001 (**34**)	Irreversible	Oryzon	Leukemia, small cell lung carcinoma
ORY-2001	Irreversible, dual LSD1/MAO B inhibitor	Oryzon	Alzheimer's, Parkinson's, multiple sclerosis, Huntington's
GSK2879552 (**35**)	Irreversible	GlaxoSmithKline	Acute myeloid leukemia, small cell lung carcinoma
Imago-7289	Irreversible	Imago	Acute myeloid leukemia, myelodysplastic syndrome
INCB059872	Irreversible	Incyte	Advanced malignancies, sickle cell disease
4SC-202 (**71**)	Reversible, dual LSD1/HDAC inhibitor	4SC	Blood cancer
SP2577	Reversible	Salarius	Ewing sarcoma
CC-90011	Reversible	Celgene	Solid tumors, non-Hodgkin's lymphoma

Time will tell which of these approaches and indications is the most successful for patient benefit. Although the current emphasis is on LSD1, future investigations will undoubtedly shed light on the physiological role of the isoform LSD2 and its potential validity as a target for therapeutic intervention. All in all, the trip began with LSDs will be as fascinating and rewarding in its second decade as in the first.

References

1 Huang, H., Lin, S., Garcia, B.A., and Zhao, Y. (2015). Quantitative proteomic analysis of histone modifications. *Chem. Rev.* 115: 2376–2418.

2 Pegg, A.E. (2011). Multifaceted roles of alkyltransferase and related proteins in DNA repair, DNA damage, resistance to chemotherapy and research tools. *Chem. Res. Toxicol.* 24: 618–639.

3 Niwa, H. and Umehara, T. (2017). Structural insight into inhibitors of flavin adenine dinucleotide-dependent lysine demethylases. *Epigenetics* 12: 340–352.

4 Morera, L., Lübbert, M., and Jung, M. (2016). Targeting histone methyltransferases and demethylases in clinical trials for cancer therapy. *Clin. Epigenet.* 8: 57.

5 McAllister, T.E., England, K.S., Hopkinson, R.J. et al. (2016). Recent progress in histone demethylase inhibitors. *J. Med. Chem.* 59: 1308–1329.
6 Wang, X., Huang, B., Suzuki, T. et al. (2015). Medicinal chemistry insights in the discovery of novel LSD1 inhibitors. *Epigenomics* 7: 1379–1396.
7 Zheng, Y.C., Ma, J., Wang, Z. et al. (2015). A systematic review of histone lysine-specific demethylase 1 and its inhibitors. *Med. Res. Rev.* 35: 1032–1071.
8 Burg, J.M., Link, J.E., Morgan, B.S. et al. (2015). KDM1 class flavin-dependent protein lysine demethylases. *Biopolymers* 104: 213–246.
9 Kim, S., Benoiton, L., and Paik, W.K. (1964). ε-Alkyllysinase. Purification and properties of the enzyme. *J. Biol. Chem.* 239: 3790–3796.
10 Shi, Y., Lan, F., Matson, C. et al. (2004). Histone demethylation mediated by the nuclear amine oxidase homolog LSD1. *Cell* 119: 941–953.
11 Karytinos, A., Forneris, F., Profumo, A. et al. (2009). A novel mammalian flavin-dependent histone demethylase. *J. Biol. Chem.* 284: 17775–17782.
12 Tsukada, Y., Fang, J., Erdjument-Bromage, H. et al. (2006). Histone demethylation by a family of JmjC domain-containing proteins. *Nature* 439: 811–816.
13 Gaweska, H. and Fitzpatrick, P.F. (2011). Structures and mechanism of the monoamine oxidase family. *Biomol. Concepts* 2: 365–377.
14 Lienhart, W.D., Gudipati, V., and Macheroux, P. (2013). The human flavoproteome. *Arch. Biochem. Biophys.* 535: 150–162.
15 Gaweska, H., Pozzi, M.H., Schmidt, D.M.Z. et al. (2009). Use of pH and kinetic isotope effects to establish chemistry as rate-limiting in oxidation of a peptide substrate by LSD1. *Biochemistry* 48: 5440–5445.
16 Luka, Z., Pakhomova, S., Loukachevitch, L.V. et al. (2014). Crystal structure of the histone lysine specific demethylase LSD1 complexed with tetrahydrofolate. *Protein Sci.* 23: 993–998.
17 Yang, Y., Yin, X., Yang, H., and Xu, Y. (2015). Histone demethylase LSD2 acts as an E3 ubiquitin ligase and inhibits cancer cell growth through promoting proteasomal degradation of OGT. *Mol. Cell* 58: 47–59.
18 Forneris, F., Binda, C., Adamo, A. et al. (2007). Structural basis of LSD1-CoREST selectivity in histone H3 recognition. *J. Biol. Chem.* 282: 20070–20074.
19 Forneris, F., Binda, C., Vanoni, M.A. et al. (2005). Human histone demethylase LSD1 reads the histone code. *J. Biol. Chem.* 280: 41360–41365.
20 Burg, J.M., Gonzalez, J.J., Maksimchuk, K.R., and McCafferty, D.G. (2016). Lysine-specific demethylase 1A (KDM1A/LSD1): product recognition and kinetic analysis of full-length histones. *Biochemistry* 55: 1652–1662.
21 Forneris, F., Binda, C., Dall'Aglio, A. et al. (2006). A highly specific mechanism of histone H3-K4 recognition by histone demethylase LSD1. *J. Biol. Chem.* 281: 35289–35295.
22 Fang, R., Chen, F., Dong, Z. et al. (2013). LSD2/KDM1B and its cofactor NPAC/GLYR1 endow a structural and molecular model for regulation of H3K4 demethylation. *Mol. Cell* 49: 558–570.
23 Metzger, E., Wissmann, M., Yin, N. et al. (2005). LSD1 demethylates repressive histone marks to promote androgen-receptor-dependent transcription. *Nature* 437: 436–439.

24 Nair, S.S., Nair, B.C., Cortez, V. et al. (2010). PELP1 is a reader of histone H3 methylation that facilitates oestrogen receptor-α target gene activation by regulating lysine demethylase 1 specificity. *EMBO Rep.* 11: 438–444.
25 Carnesecchi, J., Forcet, C., Zhang, L. et al. (2017). ERRα induces H3K9 demethylation by LSD1 to promote cell invasion. *Proc. Natl. Acad. Sci. U.S.A.* 114: 3909–3914.
26 Zibetti, C., Adamo, A., Binda, C. et al. (2010). Alternative splicing of the histone demethylase LSD1/KDM1 contributes to the modulation of neurite morphogenesis in the mammalian nervous system. *J. Neurosci.* 30: 2521–2532.
27 Laurent, B., Ruitu, L., Murn, J. et al. (2015). A specific LSD1/KDM1A isoform regulates neuronal differentiation through H3K9 demethylation. *Mol. Cell* 57: 957–970.
28 Wang, J., Telese, F., Tan, Y. et al. (2015). LSD1n is an H4K20 demethylase regulating memory formation via transcriptional elongation control. *Nat. Neurosci.* 18: 1256–1264.
29 Huang, J., Sengupta, R., Espejo, A.B. et al. (2007). p53 is regulated by the lysine demethylase LSD1. *Nature* 449: 105–108.
30 Wang, J., Hevi, S., Kurash, J.K. et al. (2008). The lysine demethylase LSD1 (KDM1) is required for maintenance of global DNA methylation. *Nat. Genet.* 41: 125–129.
31 Kontaki, H. and Talianidis, I. (2010). Lysine methylation regulates E2F1-induced cell death. *Mol. Cell* 39: 152–160.
32 Cho, H.S., Suzuki, T., Dohmae, N. et al. (2011). Demethylation of RB regulator MYPT1 by histone demethylase LSD1 promotes cell cycle progression in cancer cells. *Cancer Res.* 71: 655–660.
33 Yang, J., Huang, J., Dasgupta, M. et al. (2010). Reversible methylation of promoter-bound STAT3 by histone-modifying enzymes. *Proc. Natl. Acad. Sci. U.S.A.* 107: 21499–21504.
34 Abu-Farha, M., Lanouette, S., Elisma, F. et al. (2011). Proteomic analyses of the SMYD family interactomes identify HSP90 as a novel target for SMYD2. *J. Mol. Cell Biol.* 3: 301–308.
35 Sakane, N., Kwon, H.S., Pagans, S. et al. (2011). Activation of HIV transcription by the viral Tat protein requires a demethylation step mediated by lysine-specific demethylase 1 (LSD1/KDM1). *PLoS Pathog.* 7: e1002184.
36 Nair, S.S., Li, D.Q., and Kumar, R. (2013). A core chromatin remodeling factor instructs global chromatin signaling through multivalent reading of nucleosome codes. *Mol. Cell* 49: 704–718.
37 Zhang, X., Tanaka, K., Yan, J. et al. (2013). Regulation of estrogen receptor α by histone methyltransferase SMYD2-mediated protein methylation. *Proc. Natl. Acad. Sci. U.S.A.* 110: 17284–17289.
38 Ciccone, D.N., Su, H., Hevi, S. et al. (2009). KDM1B is a histone H3K4 demethylase required to establish maternal genomic imprints. *Nature* 461: 415–418.
39 Nottke, A., Colaiácovo, M.P., and Shi, Y. (2009). Developmental roles of the histone lysine demethylases. *Development* 136: 879–889.

40 Fang, R., Barbera, A.J., Xu, Y. et al. (2010). Human LSD2/KDM1b/AOF1 regulates gene transcription by modulating intragenic H3K4me2 methylation. *Mol. Cell* 39: 222–233.

41 Upadhyay, G., Chowdhury, A.H., Vaidyanathan, B. et al. (2014). Antagonistic actions of Rcor proteins regulate LSD1 activity and cellular differentiation. *Proc. Natl. Acad. Sci. U.S.A.* 111: 8071–8076.

42 Baron, R., Binda, C., Tortorici, M. et al. (2011). Molecular mimicry and ligand recognition in binding and catalysis by the histone demethylase LSD1-CoREST complex. *Structure* 19: 212–220.

43 Ricq, E.L., Hooker, J.M., and Haggarty, S.J. (2016). Activity-dependent regulation of histone lysine demethylase KDM1A by a putative thiol/disulfide switch. *J. Biol. Chem.* 291: 24756–24767.

44 Hino, S., Kohrogi, K., and Nakao, M. (2016). Histone demethylase LSD1 controls the phenotypic plasticity of cancer cells. *Cancer Sci.* 107: 1187–1192.

45 Sun, L. and Fang, J. (2016). Epigenetic regulation of epithelial-mesenchymal transition. *Cell Mol. Life Sci.* 73: 4493–4515.

46 Sakamoto, A., Hino, S., Nagaoka, K. et al. (2015). Lysine demethylase LSD1 coordinates glycolytic and mitochondrial metabolism in hepatocellular carcinoma cells. *Cancer Res.* 75: 1445–1456.

47 Harris, W.J., Huang, X., Lynch, J.T. et al. (2012). The histone demethylase KDM1A sustains the oncogenic potential of MLL-AF9 leukemia stem cells. *Cancer Cell* 21: 473–487.

48 Schenk, T., Chen, W.C., Göllner, S. et al. (2012). Inhibition of the LSD1 (KDM1A) demethylase reactivates the all-*trans*-retinoic acid differentiation pathway in acute myeloid leukemia. *Nat. Med.* 18: 605–611.

49 Liang, Y., Quenelle, D., Vogel, J.L. et al. (2013). A novel selective LSD1/KDM1A inhibitor epigenetically blocks herpes simplex virus lytic replication and reactivation from latency. *MBio* 4: e00558-12.

50 Yao, H.W., Lin, P.H., Shen, F.H. et al. (2014). Tranylcypromine reduces herpes simplex virus 1 infection in mice. *Antimicrob. Agents Chemother.* 58: 2807–2815.

51 Abdulla, A., Zhang, Y., Hsu, F.N. et al. (2014). Regulation of lipogenic gene expression by lysine-specific histone demethylase-1 (LSD1). *J. Biol. Chem.* 289: 29937–29947.

52 Duteil, D., Metzger, E., Willmann, D. et al. (2014). LSD1 promotes oxidative metabolism of white adipose tissue. *Nat. Commun.* 5: 4093.

53 Zeng, X., Jedrychowski, M.P., Chen, Y. et al. (2016). Lysine-specific demethylase 1 promotes brown adipose tissue thermogenesis via repressing glucocorticoid activation. *Genes Dev.* 30: 1822–1836.

54 Duteil, D., Tosic, M., Willmann, D. et al. (2017). Lsd1 prevents age-programed loss of beige adipocytes. *Proc. Natl. Acad. Sci. U.S.A.* 114: 5265–5270.

55 Katz, T.A., Vasilatos, S.N., Harrington, E. et al. (2014). Inhibition of histone demethylase, LSD2 (KDM1B), attenuates DNA methylation and increases sensitivity to DNMT inhibitor-induced apoptosis in breast cancer cells. *Breast Cancer Res. Treat.* 146: 99–108.

56 Lee, M.G., Wynder, C., Schmidt, D.M. et al. (2006). Histone H3 lysine 4 demethylation is a target of nonselective antidepressive medications. *Chem. Biol.* 13: 563–567.

57 Laurent, B. and Shi, Y. (2016). Expression, purification, and biochemical analysis of the LSD1/KDM1A histone demethylase. *Methods Enzymol.* 573: 241–259.

58 Hayward, D. and Cole, P.A. (2016). LSD1 histone demethylase assays and inhibition. *Methods Enzymol.* 573: 261–278.

59 Lynch, J.T., Cockerill, M.J., Hitchin, J.R. et al. (2013). CD86 expression as a surrogate cellular biomarker for pharmacological inhibition of the histone demethylase lysine-specific demethylase 1. *Anal. Biochem.* 442: 104–106.

60 Pollock, J.A., Larrea, M.D., Jasper, J.S. et al. (2012). Lysine-specific histone demethylase 1 inhibitors control breast cancer proliferation in ERα-dependent and -independent manners. *ACS Chem. Biol.* 7: 1221–1231.

61 Prusevich, P., Kalin, J.H., Ming, S.A. et al. (2014). A selective phenelzine analogue inhibitor of histone demethylase LSD1. *ACS Chem. Biol.* 9: 1284–1293.

62 Nalawansha, D.A. and Pflum, M.K. (2017). LSD1 substrate binding and gene expression are affected by HDAC1-mediated deacetylation. *ACS Chem. Biol.* 12: 254–264.

63 Binda, C., Valente, S., Romanenghi, M. et al. (2010). Biochemical, structural, and biological evaluation of tranylcypromine derivatives as inhibitors of histone demethylases LSD1 and LSD2. *J. Am. Chem. Soc.* 132: 6827–6833.

64 Kang, G.I. and Hong, S.K. (1990). Quantitative structure-activity relationships in MAO-inhibitory 2-phenylcyclopropylamines: insights into the topography of MAO-A and MAO-B. *Arch. Pharm. Res.* 13: 82–96.

65 Zirkle, C.L., Kaiser, C., Tedeschi, D.H. et al. (1962). 2-Substituted cyclopropylamines. II. Effect of structure upon monoamine oxidase-inhibitory activity as measured in vivo by potentiation of tryptamine convulsions. *J. Med. Pharm. Chem.* 11: 1243–1265.

66 Benelkebir, H., Hodgkinson, C., Duriez, P.J. et al. (2011). Enantioselective synthesis of tranylcypromine analogues as lysine demethylase (LSD1) inhibitors. *Bioorg. Med. Chem.* 19: 3709–3716.

67 Valente, S., Rodriguez, V., Mercurio, C. et al. (2014). Pure diastereomers of a tranylcypromine-based LSD1 inhibitor: enzyme selectivity and in-cell studies. *ACS Med. Chem. Lett.* 6: 173–177.

68 Ulrich, S., Ricken, R., and Adli, M. (2017). Tranylcypromine in mind (Part I): review of pharmacology. *Eur. Neuropsychopharmacol.* 27: 697–713.

69 Vianello, P., Botrugno, O.A., Cappa, A. et al. (2014). Synthesis, biological activity and mechanistic insights of 1-substituted cyclopropylamine derivatives: a novel class of irreversible inhibitors of histone demethylase KDM1A. *Eur. J. Med. Chem.* 86: 352–363.

70 Miyamura, S., Araki, M., Ota, Y. et al. (2016). C-H activation enables a rapid structure-activity relationship study of arylcyclopropyl amines for potent and selective LSD1 inhibitors. *Org. Biomol. Chem.* 14: 8576–8585.

71 Borrello, M.T., Schinor, B., Bartels, K. et al. (2017). Fluorinated tranylcypromine analogues as inhibitors of lysine-specific demethylase 1 (LSD1, KDM1A). *Bioorg. Med. Chem. Lett.* 27: 2099–2101.

72 Albrecht, B.K., Audia, J.E., Cote, A. et al. (2016). Spirocyclic compounds as LSD1 inhibitors and their preparation. WO patent 2016/123387.

73 Rodriguez, V., Valente, S., Rovida, S. et al. (2015). Pyrrole- and indole-containing tranylcypromine derivatives as novel lysine-specific demethylase 1 inhibitors active on cancer cells. *Med. Chem. Commun.* 6: 665–670.

74 Vianello, P., Botrugno, O.A., Cappa, A. et al. (2016). Discovery of a novel inhibitor of histone lysine-specific demethylase 1A (KDM1A/LSD1) as orally active antitumor agent. *J. Med. Chem.* 59: 1501–1517.

75 Rotili, D., Tomassi, S., Conte, M. et al. (2014). Pan-histone demethylase inhibitors simultaneously targeting Jumonji C and lysine-specific demethylases display high anticancer activities. *J. Med. Chem.* 57: 42–55.

76 Ogasawara, D., Suzuki, T., Mino, K. et al. (2011). Synthesis and biological activity of optically active NCL-1, a lysine-specific demethylase 1 selective inhibitor. *Bioorg. Med. Chem.* 19: 3702–3708.

77 Etani, T., Suzuki, T., Naiki, T. et al. (2015). NCL1, a highly selective lysine-specific demethylase 1 inhibitor, suppresses prostate cancer without adverse effect. *Oncotarget* 6: 2865–2878.

78 Guibourt, N., Ortega, M.A., and Julio Castro-Palomino, L. (2010). Preparation of N-(2-phenylcyclopropyl) amino acid oxidase inhibitory derivatives for treatment of cancer. World patent WO 2010/043721.

79 Ota, Y., Itoh, Y., Kaise, A. et al. (2016). Targeting cancer with PCPA-drug conjugates: LSD1 inhibition-triggered release of 4-hydroxytamoxifen. *Angew. Chem. Int. Ed.* 55: 16115–16118.

80 Maes, T., Carceller, E., Salas, J. et al. (2015). Advances in the development of histone lysine demethylase inhibitors. *Curr. Opin. Pharmacol.* 23: 52–60.

81 Mohammad, H.P., Smitheman, K.N., Kamat, C.D. et al. (2015). A DNA hypomethylation signature predicts antitumor activity of LSD1 inhibitors in SCLC. *Cancer Cell* 28: 57–69.

82 Li, Q., Pan, Y., Han, W. et al. (2017). Preparation of salts of an LSD1 inhibitor. World patent WO 2017/027678.

83 Reinhoff, H.Y. Jr., (2017). Preparation of substituted amides as inhibitors of lysine-specific histone demethylase which is a novel therapeutic target in myeloproliferative neoplasms. World patent WO 2017/079753.

84 Vaisburg, A. and Marx, M.A. (2017). LSD1 inhibitors. World patent WO 2017/079476.

85 Zhou, C., Wu, F., Lu, L. et al. (2017). Structure activity relationship and modeling studies of inhibitors of lysine specific demethylase 1. *PLoS One* 12: e0170301.

86 Neelamegam, R., Ricq, E.L., Malvaez, M. et al. (2012). Brain-penetrant LSD1 inhibitors can block memory consolidation. *ACS Chem. Neurosci.* 3: 120–128.

87 Cui, S., Lim, K.C., Shi, L. et al. (2015). The LSD1 inhibitor RN-1 induces fetal hemoglobin synthesis and reduces disease pathology in sickle cell mice. *Blood* 126: 386–396.

88 Ishikawa, Y., Gamo, K., Yabuki, M. et al. (2017). A novel LSD1 inhibitor T-3775440 disrupts Gfi1b-containing complex leading to transdifferentiation and impaired growth of AML cells. *Mol. Cancer Ther.* 16: 273–284.
89 Takagi, S., Ishikawa, Y., Mizutani, A. et al. (2017). LSD1 inhibitor T-3775440 inhibits SCLC cell proliferation by disrupting LDS1 interactions with SNAG domain proteins INSM1 and GFI1B. *Cancer Res.* doi: 10.1158/0008-5472.
90 Khan, M.N.A., Tsumoto, H., Itoh, Y. et al. (2015). Design, synthesis, and biological activity of *N*-alkylated analogue of NCL1, a selective inhibitor of lysine-specific demethylase 1. *Med. Chem. Commun.* 6: 407–412.
91 Malcomson, T., Yelekci, K., Borrello, M.T. et al. (2015). *cis*-Cyclopropylamines as mechanism-based inhibitors of monoamine oxidases. *FEBS J.* 282: 3190–3198.
92 Albrecht, B.K., Audia, J.E., Cote, A. et al. (2016). LSD1 inhibitors and uses thereof. WO patent 2016/172496.
93 Nowotarski, S.L., Pachaiyappan, B., Holshouser, S.L. et al. (2015). Structure-activity study for (bis)ureidopropyl- and (bis)thioureidopropyldiamine LSD1 inhibitors with 3-5-3 and 3-6-3 carbon backbone architectures. *Bioorg. Med. Chem.* 23: 1601–1612.
94 Wang, J., Lu, F., Ren, Q. et al. (2011). Novel histone demethylase LSD1 inhibitors selectively target cancer cells with pluripotent stem cell properties. *Cancer Res.* 71: 7238–7249.
95 Speranzini, V., Rotili, D., Ciossani, G. et al. (2016). Polymyxins and quinazolines are LSD1/KDM1A inhibitors with unusual structural features. *Sci. Adv.* 2: e1601017.
96 Sakane, C., Okitsu, T., Wada, A. et al. (2014). Inhibition of lysine-specific demethylase 1 by the acyclic diterpenoid geranylgeranoic acid and its derivatives. *Biochem. Biophys. Res. Commun.* 444: 24–29.
97 Willmann, D., Lim, S., Wetzel, S. et al. (2012). Impairment of prostate cancer cell growth by a selective and reversible lysine-specific demethylase 1 inhibitor. *Int. J. Cancer* 131: 2704–2709.
98 Hazeldine, S., Pachaiyappan, B., Steinbergs, N. et al. (2012). Low molecular weight amidoximes that act as potent inhibitors of lysine-specific demethylase 1. *J. Med. Chem.* 55: 7378–7391.
99 Duan, Y.C., Guan, Y.Y., Zhai, X.Y. et al. (2017). Discovery of resveratrol derivatives as novel LSD1 inhibitors: design, synthesis and their biological evaluation. *Eur. J. Med. Chem.* 126: 246–258.
100 Kutz, C.J., Holshouser, S.L., Marrow, E.A., and Woster, P.M. (2014). 3,5-Diamino-1,2,4-triazoles as a novel scaffold for potent, reversible LSD1 (KDM1A) inhibitors. *Med. Chem. Commun.* 5: 1863–1870.
101 Sorna, V., Theisen, E.R., Stephens, B. et al. (2013). High-throughput virtual screening identifies novel *N*′-(1-phenylethylidene)-benzohydrazides as potent, specific, and reversible LSD1 inhibitors. *J. Med. Chem.* 56: 9496–9508.
102 Zhou, Y., Li, Y., Wang, W.J. et al. (2016). Synthesis and biological evaluation of novel (*E*)-*N*′-(2,3-dihydro-1*H*-inden-1-ylidene) benzohydrazides as potent LSD1 inhibitors. *Bioorg. Med. Chem. Lett.* 26: 4552–4557.

103 Vankayalapati, H., Sharma S., Liu, X., and Coburn, C. (2017). Substituted benzohydrazides as histone demethylase inhibitors. WO patent 2017/004519.

104 Xi, J., Xu, S., Wu, L. et al. (2017). Design, synthesis and biological activity of 3-oxoaminobenzenesulfonamides as selective and reversible LSD1 inhibitors. *Bioorg. Chem.* 72: 182–189.

105 Zha, X., Wu, L., Xu, S. et al. (2016). Design, synthesis and biological activity of *N*-(3-substituted-phenyl)benzenesulfonamides as selective and reversible LSD1 inhibitors. *Med. Chem. Res.* 25: 2822–2831.

106 Dulla, B., Kirla, K.T., Rathore, V. et al. (2013). Synthesis and evaluation of 3-amino/guanidine substituted phenyl oxazoles as a novel class of LSD1 inhibitors with anti-proliferative properties. *Org. Biomol. Chem.* 11: 3103–3107.

107 Zhou, C., Kang, D., Xu, Y. et al. (2015). Identification of novel selective lysine-specific demethylase 1 (LSD1) inhibitors using a pharmacophore-based virtual screening combined with docking. *Chem. Biol. Drug Des.* 85: 659–671.

108 Zheng, Y.C., Duan, Y.C., Ma, J.L. et al. (2013). Triazole-dithiocarbamate based selective lysine specific demethylase 1 (LSD1) inactivators inhibit gastric cancer cell growth, invasion, and migration. *J. Med. Chem.* 56: 8543–8560.

109 Ye, X.W., Zheng, Y.C., Duan, Y.C. et al. (2014). Synthesis and biological evaluation of coumarin–1,2,3-triazole–dithiocarbamate hybrids as potent LSD1 inhibitors. *Med. Chem. Commun.* 5: 650–654.

110 Wang, S., Zhao, L.J., Zheng, Y.C. et al. (2017). Design, synthesis and biological evaluation of [1,2,4]triazolo[1,5-a]pyrimidines as potent lysine specific demethylase 1 (LSD1/KDM1A) inhibitors. *Eur. J. Med. Chem.* 125: 940–951.

111 Li, Z.H., Liu, X.Q., Geng, P.F. et al. (2017). Discovery of [1,2,3]triazolo[4,5-*d*]pyrimidine derivatives as novel LSD1 inhibitors. *ACS Med. Chem. Lett.* 8: 384–389.

112 Wu, F., Zhou, C., Yao, Y. et al. (2016). 3-(Piperidin-4-ylmethoxy)pyridine containing compounds are potent inhibitors of lysine specific demethylase 1. *J. Med. Chem.* 59: 253–263.

113 Chen, Y.K., Kanouni, T., Kaldor, S.W., et al. (2015). Preparation of substituted 6-oxo-1,6-dihydropyrimidine compounds as inhibitors of lysine specific demethylase-1. World patent 2015/168466.

114 Mould, D.P., Bremberg, U., Jordan, A.M. et al. (2017). Development of 5-hydroxypyrazole derivatives as reversible inhibitors of lysine specific demethylase 1. *Bioorg. Med. Chem. Lett.* 27: 3190–3195.

115 Yamashita, S., Ogawa, T., and Komatani, H. (2017). Preparation of novel biphenyl compounds or salts thereof as lysine specific demethylase 1 (LSD1) inhibitors. World patent 2017/090756.

116 Haydn, T., Metzger, E., Schuele, R., and Fulda, S. (2017). Concomitant epigenetic targeting of LSD1 and HDAC synergistically induces mitochondrial apoptosis in rhabdomyosarcoma cells. *Cell Death Dis.* 8: e2879.

117 Yang, C., Wang, W., Liang, J.X. et al. (2017). A rhodium(III)-based inhibitor of lysine-specific histone demethylase 1 as an epigenetic modulator in prostate cancer cells. *J. Med. Chem.* 60: 2597–2603.

118 Sartori, L., Mercurio, C., Amigoni, F. et al. (2017). Thieno[3,2-*b*]pyrrole-5-carboxamides as new reversible inhibitors of histone lysine demethylase KDM1A/LSD1. Part 1: high-throughput screening and preliminary exploration. *J. Med. Chem.* 60: 1673–1692.

119 Vianello, P., Sartori, L., Amigoni, F. et al. (2017). Thieno[3,2-*b*]pyrrole-5-carboxamides as new reversible inhibitors of histone lysine demethylase KDM1A/LSD1. Part 2: structure-based drug design and structure-activity relationship. *J. Med. Chem.* 60: 1693–1715.

120 Culhane, J.C., Wang, D., Yen, P.M., and Cole, P.A. (2010). Comparative analysis of small molecules and histone substrate analogues as LSD1 lysine demethylase inhibitors. *J. Am. Chem. Soc.* 132: 3164–3176.

121 Kakizawa, T., Ota, Y., Itoh, Y. et al. (2015). Histone H3 peptide based LSD1-selective inhibitors. *Bioorg. Med. Chem. Lett.* 25: 1925–1928.

122 Itoh, Y., Ogasawara, D., Ota, Y. et al. (2014). Synthesis, LSD1 inhibitory activity, and LSD1 binding model of optically pure lysine-PCPA conjugates. *Comput. Struct. Biotechnol. J.* 9: e201402002.

123 Kumarasinghe, I.R. and Woster, P.M. (2014). Synthesis and evaluation of novel cyclic peptide inhibitors of lysine-specific demethylase 1. *ACS Med. Chem. Lett.* 5: 29–33.

124 Amano, Y., Kikuchi, M., Sato, S. et al. (2017). Development and crystallographic evaluation of histone H3 peptide with N-terminal serine substitution as a potent inhibitor of lysine-specific demethylase 1. *Bioorg. Med. Chem.* 25: 2617–2624.

125 Tortorici, M., Borrello, M.T., Tardugno, M. et al. (2013). Protein recognition by short peptide reversible inhibitors of the chromatin-modifying LSD1/CoREST lysine demethylase. *ACS Chem. Biol.* 8: 1677–1682.

126 Amano, Y., Umezawa, N., Sato, S. et al. (2017). Activation of lysine-specific demethylase 1 inhibitor peptide by redox-controlled cleavage of a traceless linker. *Bioorg. Med. Chem.* 25: 1227–1234.

127 Itoh, Y., Aihara, K., Mellini, P. et al. (2016). Identification of SNAIL1 peptide-based irreversible lysine-specific demethylase 1-selective inactivators. *J. Med. Chem.* 59: 1531–1544.

128 Leurs, U., Lohse, B., Rand, K.D. et al. (2014). Substrate- and cofactor-independent inhibition of histone demethylase KDM4C. *ACS Chem. Biol.* 9: 2131–2138.

129 Robertson, J.C., Hurley, N.C., Tortorici, M. et al. (2013). Expanding the druggable space of the LSD1/CoREST epigenetic target: new potential binding regions for drug-like molecules, peptides, protein partners, and chromatin. *PLoS Comput. Biol.* 9: e1003158.

130 Hirschi, A., Martin, W.J., Luka, Z. et al. (2016). G-quadruplex RNA binding and recognition by the lysine-specific histone demethylase-1 enzyme. *RNA* 22: 1250–1260.

11

JmjC-domain-Containing Histone Demethylases

Christoffer Højrup [1], Oliver D. Coleman [2], John-Paul Bukowski [2], Rasmus P. Clausen [1], and Akane Kawamura [2]

[1] *University of Copenhagen, Department of Medicinal Chemistry, Faculty of Pharmaceutical Sciences 2 Universitetsparken, 2100 Copenhagen, Denmark*
[2] *University of Oxford, Department of Chemistry, Chemistry Research Laboratory, 12 Mansfield Road, Oxford OX1 3TA, UK*

11.1 Introduction

Histone lysine demethylases (KDMs) play a pivotal role in the epigenetic modulation of higher order chromatin structure and transcriptional activity in eukaryotic cells by regulating the methylation states on histones [1]. KDMs are involved in fundamental cellular processes such as proliferation, differentiation, genome integrity, and development [1]. Furthermore, dysregulation of KDMs is associated with various diseases [2]. Thus, histone-modifying enzymes and chromatin-associated proteins are emerging as potential therapeutic targets, with selective modulation of the KDMs being of pharmacological interest [3]. In this chapter, we provide an overview of the Jumonji C (JmjC)-domain-containing KDM family, with a particular focus on their therapeutic potential and the recent progress in the development of JmjC-KDM inhibitors.

11.1.1 The LSD and JmjC Histone Lysine Demethylases

N^{ε}-Methylation of lysine in histones was first detected by analysis of calf thymus histones in the 1960s [4], but it was only in 2000 that the first histone lysine methyltransferase (KMT) was identified [5]. It was assumed that N^{ε}-methylation of lysine in histones was a stable irreversible epigenetic mark [6]. However, the discovery of KDMs confirmed that histone lysine methylation was a dynamic process. In 2004, the first KDM – LSD1/KDM1A – which belongs to the flavin adenine dinucleotide (FAD)-dependent monoamine oxidase superfamily, was identified [7]. Subsequently, in 2006, JHDM1A/KDM2A, a member of the larger JmjC-domain-containing KDMs, was discovered [8]. To date, at least 18 KDMs (including KDM1A/1B) have been identified with substrates that include mono-, di-, and trimethylated H3K4, H3K9, H3K27, and H3K36 [9].

Epigenetic Drug Discovery, First Edition. Edited by Wolfgang Sippl and Manfred Jung.

The demethylation mechanisms of the two KDM families are very distinct. The JmjC-KDMs belong to the 2-oxoglutarate (2OG) oxygenase superfamily and utilize 2OG, ferrous iron (Fe(II)), and molecular oxygen (O_2) to facilitate the demethylation of their histone substrates (Figure 11.1). The catalytic core of the JmjC domain adopts a canonical double-stranded β-helix (DSBH) (also known as jelly-roll, cupin, or JmjC fold) [10], a highly conserved fold made up of eight β-strands in an antiparallel and helical manner [11]. The active site contains two histidine (His) residues that are essential for coordination of the Fe(II) in the catalytic center, while either an aspartic acid (Asp) or a glutamic acid (Glu) residue provides a third Fe(II) coordination site, forming a HXD/E...H motif (Figure 11.2) [11]. Demethylation is proposed to proceed

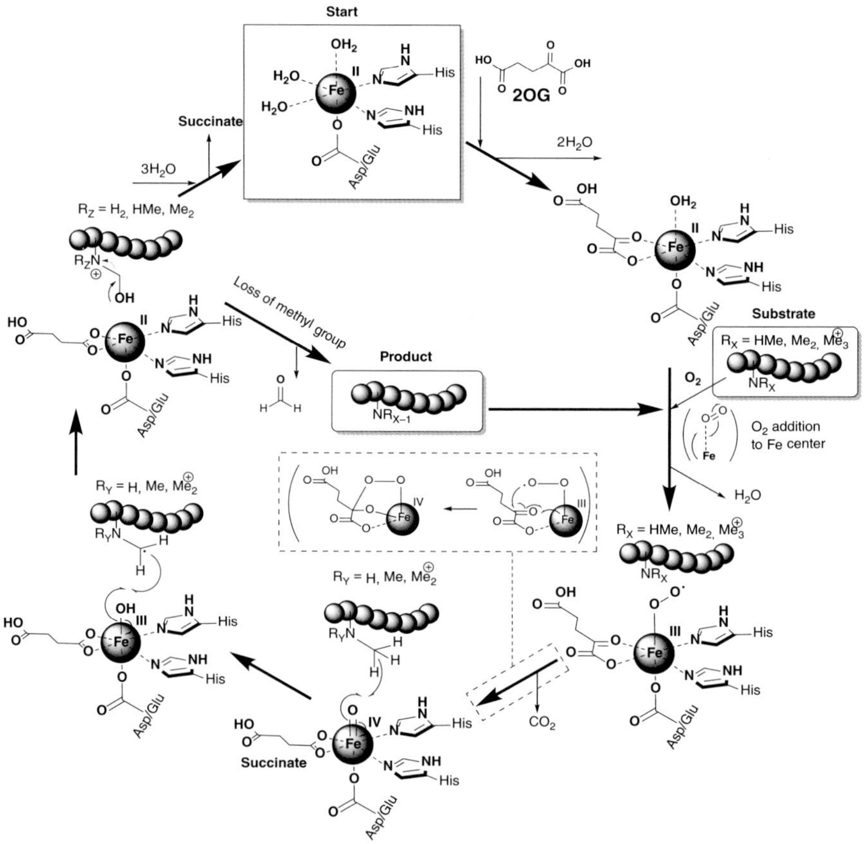

Figure 11.1 Mechanism of JmjC-KDM-catalyzed demethylation. Fe(II) is coordinated by the active site HXD/E...H facial triad. The JmjC is thought to follow a mechanism wherein 2OG binds, followed by substrate and oxygen. Oxidative addition of O_2 to the Fe center initially gives an Fe(III) species followed by further oxidation of 2OG, and loss of CO_2, to give a highly reactive Fe(IV) species. This complex undergoes reductive radical oxygen transfer to the targeted methyl group, enabling its eventual loss as formaldehyde. The starting Fe(II) species is regenerated via release of generated succinate. The substrate may reenter the cycle for further demethylation.

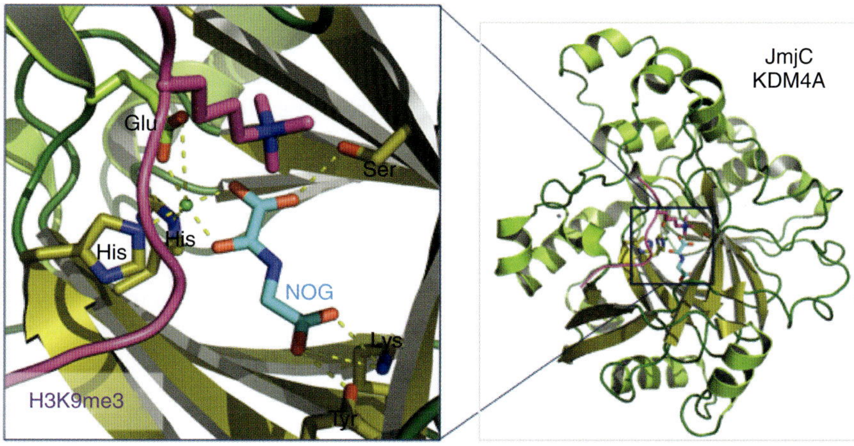

Figure 11.2 Crystal structures of a JmjC-KDM domain in complex with substrate. The JmjC catalytic domain is structurally characterized by a jelly-roll fold of the double-stranded β-helix (DSBH), creating a tunnel in which two crucial His residues and a Asp/Glu residue coordinate Fe(II) (Ni(II) in crystallography), enabling bidentate binding of 2OG (NOG in crystallography) (KDM4A, PDB ID: 2OQ6). Color key: α-helices = pale green, β-sheets = olive yellow, loops = dark green, histone = purple. Source: McDonough et al. 2010 [11]. Reproduced with permission of Elsevier.

via hydroxylation of the N^{ε}-methyl group to give an unstable hemiaminal, which fragments into the demethylated amino group and formaldehyde. This mechanism permits JmjC-KDMs the versatility to catalyze demethylation of mono-, di-, and trimethylated lysine residues [12, 13]. This is in contrast to the FAD-dependent KDM1 family, where the requirement for generation of an iminium ion intermediate during catalysis limits its demethylase activity to mono- and dimethylated lysine residues [12] (see Chapter X). It is also worth noting that, for some substrates, hydroxylation by JmjC-domain-containing oxygenases leads to a stable product, and thus some can function as a hydroxylase [14].

11.1.2 Histone Lysine Methylation and the JmjC-KDMs

In eukaryotic cells histone lysine methylation tightly correlates with transcriptional states and cell identity [15]. Mono-, di-, and trimethylated (me1, me2, me3) lysine residues are found in both the flexible tails and globular domains of the core, linker, and variant histones [16]. The canonical sites for histone lysine methylation are H3K4, H3K9, H3K27, H3K36, H3K79, H4K20, and H1.4K26 [6]. It should be noted that histone arginine residues can be monomethylated (me1), symmetrically dimethylated (me2s), and asymmetrically dimethylated (me2a) by protein arginine methyltransferases (PRMTs) [17]. Genome-wide localization studies, such as chromatin immunoprecipitation followed by sequencing (ChIP-seq), have enabled the mapping of histone lysine methylation distribution in different cell types. These studies demonstrated that the position and degree of lysine methylation are tightly associated with chromatin structure and transcriptional activity, often referred to as "chromatin states" [15]. For example, H3K9me3 is a hallmark of heterochromatin and is enriched surrounding the transcription start site (TSS) of transcriptionally silent genes, whereas H3K9me1 is enriched at the 5′ end of transcriptionally active genes [15].

Of the 31 human JmjC-domain-containing proteins, 16 have been confirmed as bona fide histone demethylases (JmjC-KDMs). The reported substrates of these KDMs and JmjC proteins are summarized in Table 11.1. Based on sequence alignment of the JmjC domain, JmjC-KDMs have been divided into several subfamilies (KDM2–7) that generally cluster with substrate specificity (Table 11.1). For example, KDM5 and KDM6 subfamily members selectively catalyze the demethylation of H3K4 and H3K27, respectively, whereas members of the KDM4 and KDM7 subfamilies catalyze the demethylation of multiple histone lysine residues. KDM4–6 can demethylate trimethylated lysines, whereas KDM2, KDM3, and KDM7 are limited to di- and monodemethylation. Other putative histone and non-histone substrates have been reported but are either disputed or yet to be independently verified. Members of the JmjC family that fall outside of the KDM subfamily assignment (KDM2–7) are either JmjC hydroxylases (e.g. FIH, which hydroxylates Asp803 of hypoxia-inducible factor (HIF1α)), have highly contentious assignments (e.g. both demethylase and hydroxylase activities have been reported for JMJD5, JMJD6, RIOX1, and RIOX2), or have an unknown function (Table 11.1). For example, JMJD6 was initially identified as the first histone arginine demethylase [52] but was later demonstrated to function as a lysyl-5-hydroxylase [53]. Interestingly, KDM4A and KDM5C were recently reported to have *in vitro* histone arginine demethylase activity (Table 11.1), albeit with a lower activity than their canonical methylated histone lysine substrates [54]. HAIRLESS (HR), JARID2, and PHF2 do not contain the HXD/EX...H consensus Fe(II) binding motif and have been predicted to be devoid of catalytic activity, although demethylase activity for HR [33] and PHF2 [39] has been suggested.

Many chromatin-associated proteins contain "reader domains" that specifically bind to epigenetic marks [55]. Reader domains can facilitate the targeting of epigenetic modifiers to particular genomic loci whereby they alter the local chromatin environment to affect transcription. Several members of the JmjC-KDMs contain reader domains such as plant homeodomain (PHD) fingers and double Tudor domains (DTDs) that recognize methylated or non-methylated lysines at specific histone residues (Table 11.1). Methyl-lysine reader domains associated with the KDMs have different putative functions, including the modulation of chromatin recruitment (e.g. DTD in Kdm4c), substrate specificity [56] (e.g. PHD finger in KDM7A/B), and enzymatic activity (e.g. PHD finger in KDM7A/B) [57]. However, the function of many KDM-associated reader domains remains unclear (Table 11.1).

11.1.3 The JmjC-KDMs in Development and Disease

JmjC-KDMs are highly conserved, with orthologues found in model eukaryotic organisms such as *Saccharomyces cerevisiae*, *Caenorhabditis elegans*, *Drosophila* melanogaster, and *Mus musculus* [58]. Knockout (KO) mice phenotypes of JmjC

Table 11.1 The human JmjC family.

Name	Synonyms	Size[a)]	Protein domains and motifs[b)]
KDM6B	JMJD3	1682	
KDM6A	UTX	1429	
UTY	KDM6C	1444	
JARID2	JMJ	1246	
KDM5B	JARID1B, PLU1	1580	
KDM5A	JARID1A, RBP2	1690	
KDM5C	JARID1C, SMCX	1560	
KDM5D	JARID1D, SMCY	1570	
KDM4D	JHDM3D, JMJD2D	523	
KDM4A	JHDM3A, JMJD2A	1064	
KDM4B	JHDM3B, JMJD2B	1130	
KDM4C	GASC1, JHDM3C, JMJD2C	1056	
HR	ALUNC, AU	1189	
JMJD1C	JHDM2C, KDM3C, TRIP8	2540	
KDM3A	JHDM2A, JMJD1A, TSGA	1321	
KDM3B	JHDM2B, JMJD1B, 5QNCA	1761	
KDM2A	CXXC8, JHDM1A, FBXL11	1162	
KDM2B	CXXC2, JHDM1B, FBXL10	1336	
PHF2	JHDM1E, KDM7C	1096	
KDM7A	JHDM1D, KIAA1718	941	
KDM7B	JHDM1F, PHF8	1060	
RIOX2	JMJD10, MINA, MINA53, NO52	465	
RIOX1	JMJD9, MAPJD, NO66	641	
JMJD4		463	
JMJD6	PTDSR	414	
JMJD8		285	
JMJD7		316	
TYW5		315	
HIF1AN	FIH1	349	
HSPBAP1	PASS1	488	
JMJD5	KDM8	454	

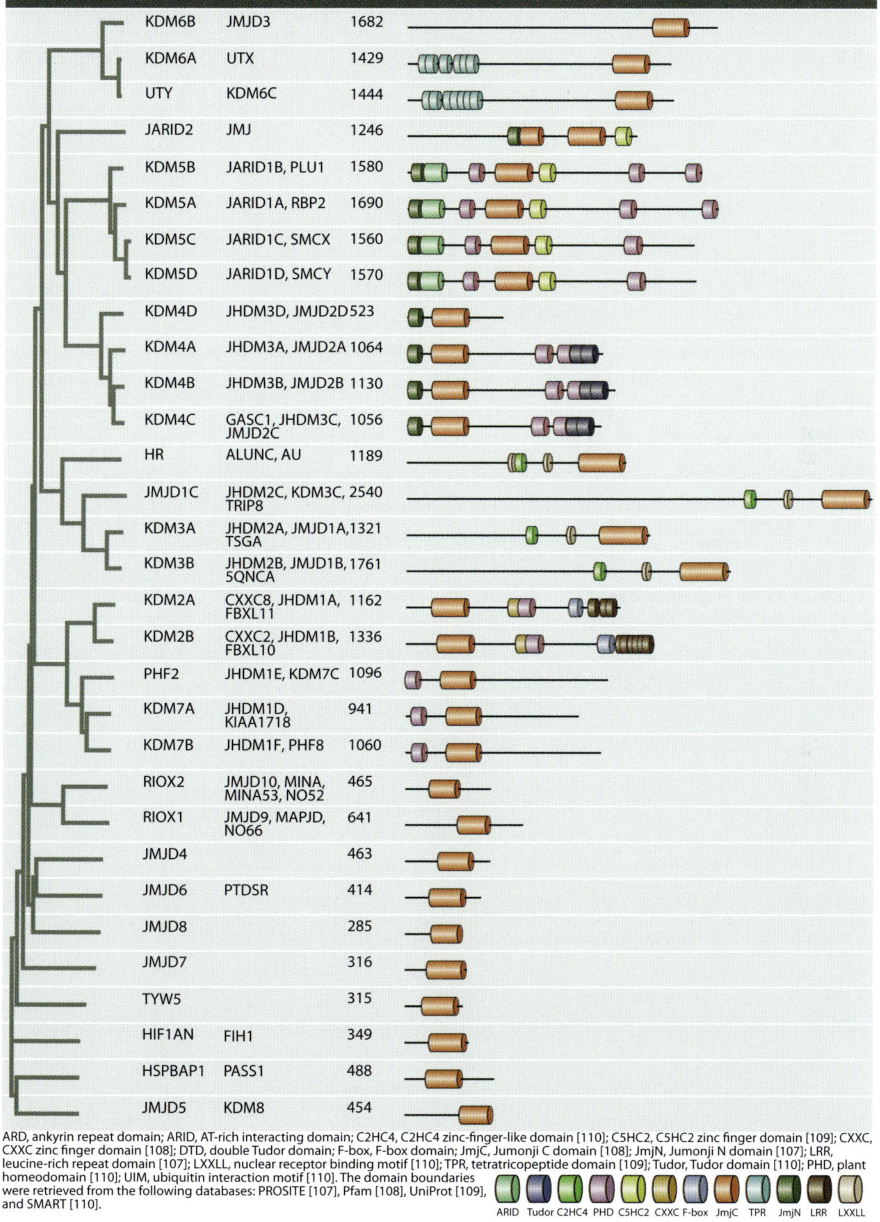

ARD, ankyrin repeat domain; ARID, AT-rich interacting domain; C2HC4, C2HC4 zinc-finger-like domain [110]; C5HC2, C5HC2 zinc finger domain [109]; CXXC, CXXC zinc finger domain [108]; DTD, double Tudor domain; F-box, F-box domain; JmjC, Jumonji C domain [108]; JmjN, Jumonji N domain [107]; LRR, leucine-rich repeat domain [107]; LXXLL, nuclear receptor binding motif [110]; TPR, tetratricopeptide domain [109]; Tudor, Tudor domain [110]; PHD, plant homeodomain [110]; UIM, ubiquitin interaction motif [110]. The domain boundaries were retrieved from the following databases: PROSITE [107], Pfam [108], UniProt [109], and SMART [110].

Name	Substrates[c)]	Putative substrates[d)]	Histone binding[e)]
KDM6B	H3K27me3/me2	H3K27mel [18], RB(K810mel) [19]	
KDM6A	H3K27me3/me2	H3K27mel [20]	
UTY		H3K27me3/me2 [21]	

(Continued)

Table 11.1 (Continued)

Name	Substrates[c)]	Putative substrates[d)]	Histone binding[e)]
JARID2			H2AK119ul (UIM) [22]
KDM5B	H3K4me3/me2	H3K4mel [23]	H3K4me0 (PHD1), H3K4me3 (PHD3) [24]
KDM5A	H3K4me3/me2	H3K4mel [25]	H3K4me0 (PHD1), H3K4me3 (PHD3) [26]
KDM5C	H3K4me3/me2	H3K4mel [18], H3R2me2(s/a)/mel, H3R8me2(s/a), H4R3me2a [27]	H3K9me3 (PHD1) [28]
KDM5D	H3K4me3/me2		
KDM4D	H3K9me3/me2	H1.4K26me3/me2 [29]	
KDM4A	H3K9me3/me2, H3K36me3/me2	H1.4K26me3 [29], H3K27me3/me2 [18] H3R2me2a/mel, H4R3me2a [27]	H3K4me3, H3k23me3, H4K20me3 (DTD) [30]
KDM4B	H3K9me3/me2, H3K36me3/me2	H1.4K26me3 [29], H3K27me3 [18]	H3k23me3 (DTD) [30]
KDM4C	H3K9me3/me2, H3K36me3/me2	H1.4K26me3 [29], H3K27me3/me2 [18], MyoD (K104me2/mel) [31], Pc2 (K191me2) [32]	H3K4me3 (DTD) [30]
HR		H3K9me2/mel [33]	
JMJD1C		H3K9me2/mel [34], MDCl (K45me2) [35]	
KDM3A	H3K9me2/mel	p53 (K372mel) [36]	
KDM3B	H3K9me2/mel		
KDM2A	H3K36me2/mel	p65 (K218mel/K221me2) [37]	
KDM2B	H3K36me2/mel	H3K4me3 [38]	
PHF2		H3K9me2/mel [39, 40], H4K20me3 [41] ARID5B (K336me2) [40]	H3K4me3 (PHD) [39]
KDM7A	H3K9me2/mel, H3K27me2/mel	H3K36me2/mel, H4K20mel [42]	H3K4me3 (PHD) [39]
KDM7B	H3K9me2/mel	H3K36me2, H3K27me2, H4K20mel [42]	H3K4me3 (PHD) [43]
RIOX2		H3K9me3 [44], Rpl27a (H39) [45] [f)]	
RIOX1		H3K4me3/me2/mel, H3K36me3/me2 [46], Rpl8 (H216) [45] [f)]	
JMJD4		eRFl (K63) [47] [f)]	

Table 11.1 (Continued)

Name	Substrates c)	Putative substrates d)	Histone binding e)
JMJD6		H3R2me2(s/a), H4R3me2(s/a)/mel, ERa (R260me2a), HSP70 (R469mel), H3K9/K14 f), H4K5/K8 f), JMJD6 (Kl 11/K167) f), p53 (K382) f), U2AF65 (K15/K276) f), LUC7L2 (K266/K269) f), 7SK snRNA [48] f)	
JMJD8			
JMJD7			
TYW5		$tRNA^{phe}$(yW-72) [49] f)	
HIF1AN	HIFl α (N803) f)	Multiple ARD-containing proteins (N/H/D) f), OTUB1 (N22) [50] f)	
HSPBAP1			
JMJD5		H3K36me2 [51]	

The catalytic domains of human JmjC-domain-containing proteins were aligned and visualized with Clustal Omega (EMBL-EBI) as to generate the phylogenetic tree.
a) Size of largest human isoform in amino acids.
b) The line, the length of the protein, and the cylinders, the domains, are depicted with accurate proportions with respect to position and length.
c) Independently verified consensus *in vitro* and cellular substrate assignment.
d) Not independently verified and/or controversial *in vitro* and/or cellular substrate assignment.
e) Preferential histone binding domain assignment.
f) Substrates are hydroxylated and not demethylated.

proteins vary in severity from no gross abnormalities to mild/severe developmental defects and embryonic lethality (Table 11.2). Mechanistic insights to the biological function of JmjC-KDMs have been gained through KO phenotypes; for example, the male-to-female sex reversal in *Kdm3a* KO mice results from improper transcriptional activation of the sex-determining gene *Sry* on the mammalian Y chromosome due to H3K9me2 promoter hypermethylation [119]. Interestingly, Kdm6b is required for embryonic development but is dispensable for postnatal survival [155], suggesting KDM6B inhibitors could be tolerated in patients. However, a catalytically inactive knock-in (KI) suggests that Kdm6a activity is dispensable for normal mouse development [69].

Functional studies combined with ChIP-seq and genome-wide expression data suggest that KDMs fine-tune the activation or repression of transcription in a context-dependent manner rather than working as essential on/off switches. For example, KDM4A binds to several thousand H3K4me3-positive promoters in KYSE150 cells, but *KDM4A* depletion has only modest effects on target gene expression and distribution of H3K9me3 and H3K36me3 [56]. However, it has been proposed that functional redundancy and/or compensation by subfamily

Table 11.2 The role of JmjC domain proteins in development and disease.

Name	Knockout mouse phenotype	Associated with cancer proliferation
KDM6B	Embryonic lethality before E6.5 [59][a]. Neonatal lethality due to respiratory failure with various developmental defects [60][a] (*Kdm6b* dispensable post E9.5) [60]	BSG (H3.3K27M subtype) [61] [±], cervical [62, 63] [±±], glioblastoma [64] [#], HCC [65] [±], multiple myeloma [66] [#], T-ALL (*NOTCH1* subtype) [67] [#]
KDM6A	Female embryonic lethality E9.5–13.5 with heart and neural tube closure defects. Male partial pre- and postnatal lethality and premature death [68] (X chromosome). *Utx* [K1] (H1146A/E1148A). No overt phenotype [69]	Breast [70, 71] [±,#], cervical [62] [±], glioblastoma [64] [±], T-ALL (*TALI* subtype) [72] [±]
UTY	No overt phenotype [73] (Y chromosome)	–
JARID2	Embryonic lethality E10.5–15.5 with various developmental defects [74]	Rhabdomyosarcoma [75] [±]
KDM5B	Embryonic lethality E4.5–7.5 [76][a]. Partial neonatal lethality due to respiratory failure with various developmental defects and premature death [77][a]. Partial postnatal lethality and premature death, decreased body weight and female fertility, and delayed mammary gland development [78][a]	Breast [23, 79] [±±], melanoma [80] [±]
KDM5A	Mild behavioral and hematological abnormalities [81]	Cervical [82, 83] [±±], gastric [82] [±], lung [83, 84] [±±], NSCLC [85, 86] [±], neuroendocrine [87][a] Osteosarcoma [83, 88] [±±], retinoblastoma [83] [±]
KDM5C	Embryonic lethality, neurulation, and cardiogenesis defects [89] (X chromosome)	—
KDM5D	(Y chromosome)	—
KDM4D	No overt phenotype [90]	—
KDM4A	No overt phenotype. [91] (*Kdm4a/c* KO embryonic lethality before E6.5) [92]	Breast and lung [93, 94] [±±], pancreatic [95] [±], prostate [96, 97] [±±]
KDM4B	No overt phenotype [98]	Breast [99–101] [±±±], colorectal [102] [±], neuroblastoma [103] [±], ovarian [104] [±]
KDM4C	No overt phenotype [56] (*Kdm4a/c* KO embryonic lethality before E6.5) [92]	AML [105] [±], breast [106, 107] [±±], colorectal [108] [±], esophageal [109] [±], Hodgkin lymphoma and PMBL (9p24 amplicon subtype) [110] [±], osteosarcoma [109] [±], prostate [111] [±]
HR	Hair loss and severe skin wrinkling [112]	–
JMJD1C	Age-dependent male infertility [113]	ALL and AML [114–116] [±±,#]
KDM3A	Male infertility [117], adult-onset obesity/metabolic syndrome [118], and male-to-female sex reversal [119]	Bladder [120] [±], breast [121–123] [±±±], DLBCL and multiple myeloma [124] [±], Ewing sarcoma [125] [±], HCC [126] [±], ovarian [127] [±], prostate [128] [±]

KDM3B	Restricted postnatal growth, decreased spermatogenesis [129], impaired male sexual behaviors, and female infertility [130]	—
KDM2A	Embryonic lethality E10.5–12.5 and growth retardation [131]	NSCLC [132] $^{\pm}$
KDM2B	Embryonic lethality E10.5–13.5, neural tube closure defects, and growth retardation (isoform 1 and 2 KO) [133, 134][a]. Partial peri- or postnatal lethality (isoform 1 KO only) [135][a]	ALL and AML [135, 136] $^{\pm\pm}$, AML [137] $^{\pm}$, breast [138] $^{\pm}$, pancreatic [139] $^{\pm}$
PHF2	Partial neonatal lethality, growth retardation, reduced adipose tissue, and adipocytes [140]	—
KDM7A	—	—
KDM7B	Resistance to stress-induced depression and anxiety-like behaviors [141] (X chromosome)	Breast [142, 143] $^{\pm\pm}$, prostate [144] $^{\pm}$, T-ALL (*NOTCH1* subtype) [145] $^{\pm}$
RIOX2	Embryonic lethality [146][a]. Reduced allergic response in the airways [147][a]	—
RIOX1	—	—
JMJD4	No overt phenotype [148]	—
JMJD6	Neo- and perinatal lethality with various developmental defects [149]	Colorectal [150] $^{\pm}$
JMJD8	—	—
JMJD7	—	—
TYW5	—	—
HIF1AN	Reduced body weight, elevated metabolic rate, hyperventilation, improved glucose, and lipid homeostasis [151]	Colorectal and melanoma [152] $^{\pm}$
HSPBAP1	—	—
JMJD5	Embryonic lethality E10.0–11.0, growth retardation, and upregulation of p53 target genes [51, 153]	Breast [154] $^{\pm}$

Reported JmjC family homozygous KO mouse phenotypes. KO, knockout; KI, knock-in.
ALL, acute lymphoblastic leukemia; AML, acute myeloid leukemia; BSG, pediatric brainstem glioma; HCC, hepatocellular carcinoma; NSCLC, non-small cell lung cancer; PMBL, primary mediastinal cell lymphoma; T-ALL, T-cell acute lymphoblastic leukemia.
Selected RNAi $^{\pm}$ or KO $^{\#}$ studies demonstrating decreased proliferation and/or viability of human cancer cells and/or decreased tumor growth in mouse xenograft or mouse KO models. Studies that rescued the growth phenotype with cDNA encoding a wild-type version of the target but not with an enzymatically inactive mutant are underlined.

a) Studies with conflicting KO phenotypes. Genes located on sex chromosomes are highlighted.

paralogues might account for the less severe mouse KO phenotypes. Indeed, the *Kdm4a/Kdm4c* double KO mouse is embryonically lethal, whereas the individual KOs have no overt phenotype [92].

Several KDMs have emerged as interesting new drug targets because they are frequently genetically amplified and/or overexpressed in cancer, e.g. *KDM4C* and *KDM5A* [156, 157], due to their differential requirement in normal and disease states (Table 11.2). Indeed, KO and RNA interference (RNAi) functional studies have demonstrated that some KDMs are required for the proliferation, survival, and drug tolerance of multiple cancers (Table 11.2). Moreover, KDMs have well-defined active site pockets, which can aid in the development of small-molecule inhibitors for transcriptional modulation. However, it should be noted that desirable phenotypes arising from RNAi studies in cancer models could be due to cytotoxic off-target effects. Additionally, KDMs have putative demethylase-independent functions, and rescue experiments with catalytically active and inactive mutants are not often reported. Therefore, data to support small-molecule KDM inhibitors as a therapeutic strategy remains limited (Table 11.2). Moreover, the current paucity of cell-active and selective KDM inhibitors precludes the ability to easily determine if the loss of demethylase activity is the cause of a therapeutically beneficial phenotype. Indeed, the high structural homology of the catalytic domains within JmjC subfamilies has made selective chemical inhibition hard to achieve. However, KDM pan-subfamily selective inhibitors could have therapeutic potential by overcoming paralogue redundancy. For example, MLL-rearranged acute myeloid leukemia (AML) survival is dependent on *Kdm4a/b/c*, yet individually they are dispensable [158]. Clearly, this must be weighed against an increased possibility of toxic side effects to healthy non-transformed cells.

11.2 KDM Inhibitor Development Targeting the JmjC Domain

The JmjC proteins are members of the 2OG oxygenase superfamily (69 in humans) that catalyze the oxidation of a diverse substrate set, presumably via a conserved mechanism [11]. 2OG oxygenases have varied roles in human biology, including collagen biosynthesis, hypoxic response, fatty acid metabolism, DNA repair/modification, and epigenetic regulation [11]. Several 2OG oxygenases are therapeutic targets and chemically tractable, with a number of drugs currently in development or in clinical use. For example, Mildronate is an anti-ischemic drug that inhibits carnitine biosynthesis thought to target γ-butyrobetaine hydroxylase (BBOX1), and multiple drugs targeting the hypoxia-inducible factor (HIF) prolyl hydroxylases (PHD/EGLN) are currently in clinical development for the treatment of anemia in chronic kidney disease. KDM inhibitor strategies targeting the catalytic JmjC domain thus have historic origins in 2OG oxygenase inhibitor scaffolds (see [159] for a review of 2OG oxygenase inhibitors). In this section, we provide an overview of the different inhibitor development strategies employed for targeting the JmjC domain of KDMs and highlight some of the subfamily-targeted inhibitors reported to date.

11.2.1 2-Oxoglutarate Cofactor Mimicking Inhibitors

2OG cofactor mimetics that imitate its iron chelating properties are the most explored class of JmjC-KDM inhibitors. The competitive blockage of the co-substrate binding site is most effective with inhibitors that have a high affinity for iron and contributes toward their potency. However, generic metal-chelating molecules, such as deferoxamine (DFO), may give rise to inhibition of a broad range of metalloenzymes due to their ability to sequester iron in solution. The metal quenching effect thus needs to be differentiated from active site binding. While these 2OG-mimicking scaffolds remain good starting points for broad-spectrum 2OG oxygenase inhibitors, engineering KDM selectivity remains a challenge due to the generic mode of inhibition.

11.2.1.1 Emulation of the Chelating α-Keto Acid Moiety in 2OG

Crystal structures of enzymes from the JmjC family have revealed that the Fe(II) in the catalytic center is coordinated by 2OG in a bidentate manner as seen in other 2OG oxygenases (Figure 11.2) [160]. *N*-Oxalylglycine (NOG) (**2**; Figures 11.2 and 11.3), a 2OG mimic and generic inhibitor of 2OG oxygenases, nonselectively inhibits KDMs with modest potency. Various chelating scaffolds have been used to mimic the α-keto acid moiety in 2OG (Figure 11.3), including 8-hydroxyquinolines (8HQ) (**5**), carboxypicolinic acids (**4**), 2,2′-bipyridines (**4**), hydrazides (**6**), and hydroxamic acids (**7**).

Several studies have assessed the structure–activity relationship (SAR) of KDMs with inhibitors with different chelating properties [161, 162]. While many of these observations have been derived from screening efforts, recent advances in KDM crystallography mean more rational approaches are gaining ground. For example, elaboration of the 2,2′-bipyridine-4,4′-dicarboxylic acid scaffold (**4**) by swapping one of the pyridine moieties with other nitrogen-containing moieties enables fine-tuning of the chelation strength due to the high number of azaheterocycles capable of coordination when connected to a 2-pyridyl moiety (Figure 11.3) [163]. Efforts toward tuning the strength of iron chelation have been made with the 3-aminopyridine-4-carboxylic acid scaffold that was initially disclosed by Quanticel Pharmaceuticals and GlaxoSmithKline in the patent literature (Figure 11.3) [114, 164–166]. This monodentate inhibitor was later elaborated upon in further publications [161, 162]. In another example, CPI-455 (**26**) with monodentate chelation via a cyano group has successfully been utilized for KDM5 selective inhibition. In some instances, the inhibitor potency has been attributed to its ability to translocate the active site metal (e.g. IOX1 (**5**), GSK-J1 (**11**), GSK-J4 (**28**)). Thus fine-tuning of iron chelation balanced with considerations for KDM selectivity is essential in the successful design of KDM inhibitors.

11.2.1.2 Bioisosteres of the Conserved 2OG C5-Carboxylic Acid-Binding Motif

In addition to a metal-chelating motif, the majority of 2OG mimics (**1–7**) carry carboxylic acids that can be superimposed onto the 2OG C5-carboxylic acid as it forms electrostatic and hydrogen bonding interactions with a highly conserved active site basic residue (Lys/Arg) and Tyr (Figure 11.1). SAR analyses of pyridine

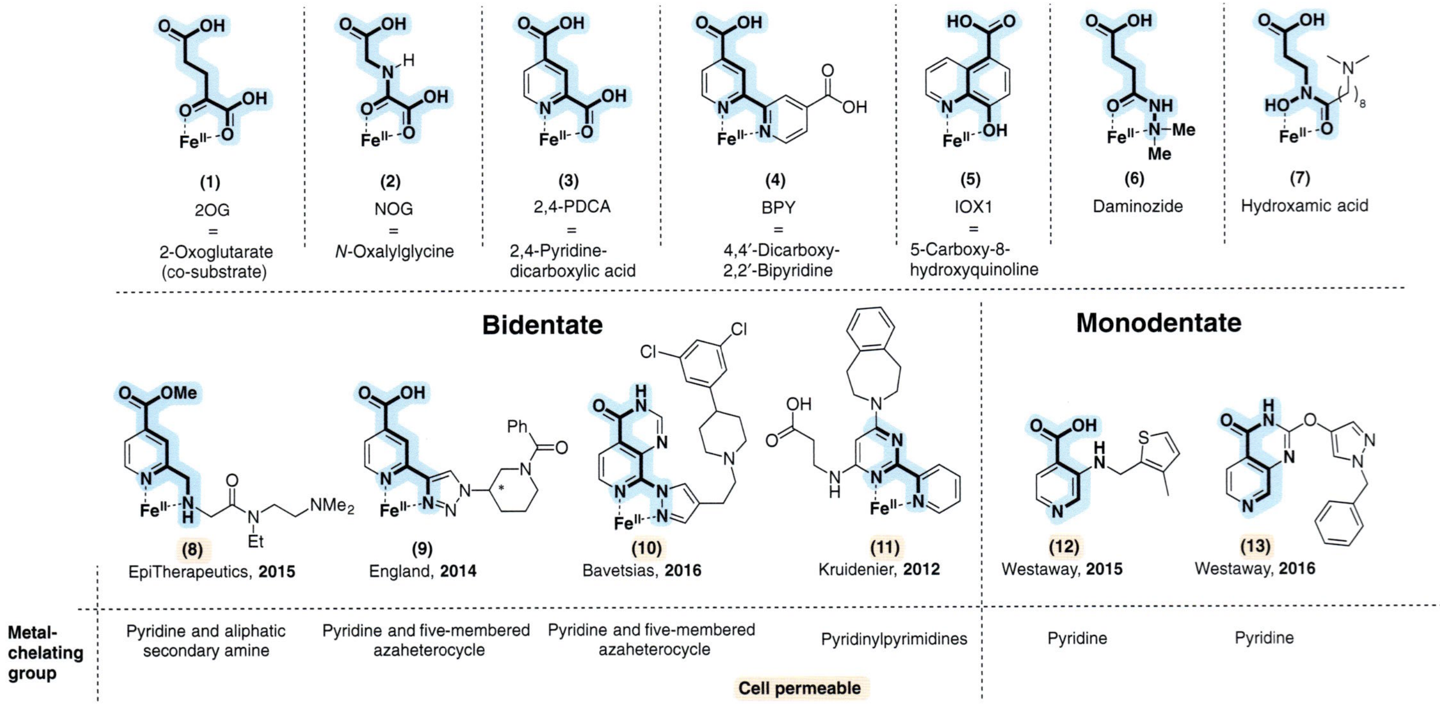

Figure 11.3 Metal-chelating motifs of JmjC-containing KDMs. The backbone of 2OG (**1**) is superimposed onto the inhibitors (**2–7**). Various bidentate Fe(II)-chelating moieties, which emulate the α-keto acid, have been explored. The carboxylic acid that interacts with the JmjC-KDM 2OG-coordinating residues (Lys, Tyr) is generally conserved throughout the mimetics. Metal-chelating strength can be subtly tuned by modifying the bidentate metal-chelating moieties. Examples include 2-pyridine substitutions with aliphatic secondary amines (**8**); five-membered aromatic azaheterocycles, such as triazoles (**9**), imidazoles, and pyrazoles (**10**); or six-membered heterocycles, such as bipyridines (**4**) and pyridinylpyrimidines (**11**). Successful monodentate metal-chelating inhibitors of JmjC-KDM include 3-aminopyridines (**12**), pyrido[3,4-*d*]pyrimidin-4(3*H*)-ones (**13**), and via cyano groups (**27**). The binding modes are supported by X-ray crystallography for the majority of the inhibitors.

[167] and bipyridine [168] dicarboxylates and other scaffolds have revealed the importance of this C5-carboxylate motif positioning for inhibitor potency. However, the development of inhibitors containing carboxylic acids has been challenged by their poor membrane permeability, with the carboxylic acids being ionized at physiological pH, impeding passive diffusion, and a large drop-off between *in vitro* potency and cellular activity. To circumvent this problem, ester prodrug strategies have been employed [168–170]. However, this can add variability and complexity in cellular analysis [171]. 3-Aminopyridine-4-carboxylic acids have shown promise, with intramolecular hydrogen bond formation between the 3-amino group and 4-carboxylic acid improving cellular activity [162]. Recent efforts have focused on bioisosteric replacement of the conserved C5-carboxylic acids (Figure 11.4) [173, 176]. Less acidic bicyclic fragments based on the pyridine core directly mimicking the C5-carboxylic acid in 2OG have been successfully used to enhance cellular activity (Figure 11.4) [173, 176]. Other bioisosteres for the C5-carboxylic acid include tetrazoles, although cellular activity has not reported (Figure 11.4) [175].

11.2.2 Histone Substrate-Competitive Inhibitors

Despite the conserved catalytic mechanism and similarity in the active sites, the substrate binding pockets differ significantly across the KDMs. This is evident from the distinct histone substrate preferences exhibited by KDM subfamilies

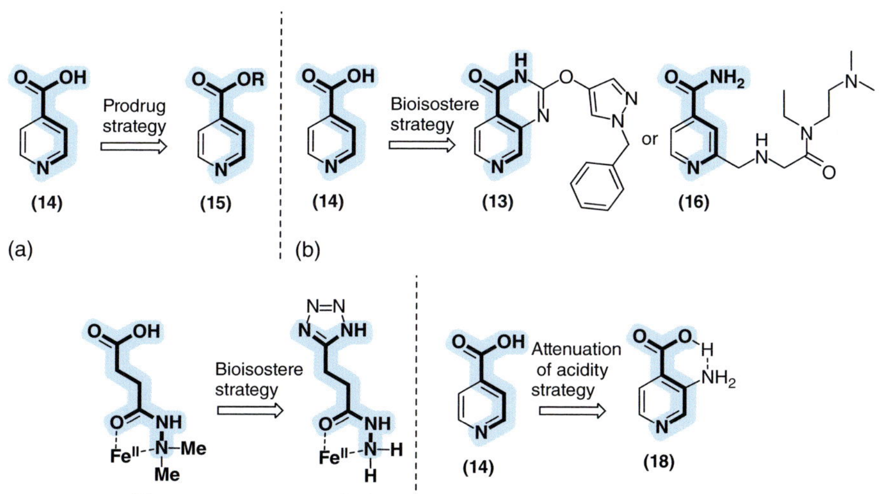

Figure 11.4 Employed strategies for improving the cell permeability of 2OG-mimicking inhibitors. (a) Prodrug ester strategies have been employed to mask carboxylic acids mimicking the 2OG C5-carboxylic acids (e.g. R = methyl, ethyl, octyl) [169, 171, 172]. (b) Bioisosteres of isonicotinic acid, including heterocyclic bioisosteres (**13**) [173] and amides (**16**) [174]. (c) Screening efforts revealed that **17**, a bioisostere of daminozide, was an inhibitor of KDM4A [175]. (d) The intramolecular hydrogen bonding between the amine and carboxylic acid in 3-aminopyridine-4-carboxylic acids stabilizes the acid, thus decreasing its acidity [162].

(Table 11.1). While exploiting the differences in the substrate binding pockets is an attractive strategy, only a few substrate-competitive inhibitors have been reported for the KDMs to date.

11.2.2.1 Small-Molecule Inhibitors

Substrate-competitive scaffolds have been successfully explored for KMTs (see Chapter X). BIX-01294, a quinazoline-based molecule, was initially identified as a potent inhibitor of G9a/KMT1C and G9a-like/KMT1D lysine methyltransferases and mimics the histone substrate [177]. It was later shown that BIX-01294 also inhibited KDM7A in the histone substrate binding pocket [178]. Further crystal structure-based design and SAR analysis of substitutions at the 2, 4, and 7 positions within the quinazoline core yielded E67–2 (**32**), which maintained the potency against KDM7A but enhanced the selectivity over the methyltransferases by >1500-fold and was selective over KDM5C [178]. Further work in "target hopping" of lead histone substrate-competitive scaffolds to other KDMs may prove fruitful.

11.2.2.2 Peptide Inhibitors

Peptides are highly effective in disrupting protein–protein interactions. In this regard, histone peptides provide an ideal starting point for the design of selective inhibitors against the KDMs. Several groups have successfully employed the histone H3 sequence for selectivity, and linked to 2OG/metal-chelator scaffolds for potency, to develop highly selective and potent inhibitors against the KDM4 subfamily [179, 180]. Furthermore, active site-targeted cyclic peptide KDM4A–C inhibitors (CP2 (**31**) and derivatives), which were potent (IC_{50} < 50 nM) and selective, including with respect to subfamily paralogues, have been developed using mRNA-display-based screening [181]. These inhibitors are distinct from histone sequences, revealing that non-histone substrate-competitive and nonmetal-chelating inhibitors can be generated for the KDMs. Despite achieving selectivity and potency, peptides face greater challenges in cell penetration and stability compared with small molecules, and significant work is required to address this.

11.2.3 Allosteric Inhibitors

Leurs et al. used phage display screening of a library of linear peptides and peptides cyclized via a disulfide bridge to identify selective binders of KDM4A and KDM4C [182]. Optimization of cyclic peptide binders with inhibitory activity via SAR analysis on KDM4C yielded a cyclic peptide with IC_{50} of 0.6 μM. Interestingly, inhibition kinetics of the peptides was not competitive toward the histone substrate nor the 2OG co-substrate. Hydrogen/deuterium exchange mass spectrometry (HDX-MS) revealed these cyclic peptides to bind away from the active site of KDM4C and identified potential allosteric inhibition sites for KDM4C. As yet, no small-molecule allosteric inhibitors have been reported.

11.2.4 Inhibitors Targeting KDM Subfamilies

In the following sections, we highlight some inhibitors developed for individual JmjC-KDM subfamilies, focusing on cell-active inhibitors reported in academic literature (Figure 11.5). See [183–185] for comprehensive recent reviews on other KDM inhibitors.

11.2.4.1 KDM4 Subfamily-Targeted Inhibitors

The first reported JmjC-KDM high-throughput screening (HTS) was against the putative pseudogene KDM4E, an H3K9me3/2 demethylase. Several 8HQ were identified as weak inhibitors in a screening of 236 000 library compounds. Substitution at the 2 position of the quinoline core appeared to decrease the inhibitory potency, whereas substitution at the 4, 5, and 7 positions increased the potency [186]. This led to 5-carboxy-8-hydroxyquinoline (IOX1) (**5**), a submicromolar inhibitor of KDM4E (IC_{50} = 200 nM). Co-crystal structure of IOX1 complexed with KDM4A showed a bidentate chelation of IOX1 to the active site metal, with the quinolone nitrogen and the 8-hydroxyl group leading to translocation of the metal approximately 1.5 Å away from the C-terminal His of the HXD/E...H motif [186]. IOX1 (**5**) was later found to be a generic 2OG oxygenase inhibitor against all representative JmjC-KDM subfamilies, HIF hydroxylases (PHD2, FIH), AlkB, and BBOX, albeit with different potencies [187]. IOX1 (**5**) was active in HeLa cells overexpressing Flag-tagged KDM4A, inhibiting the H3K9me3 demethylation activity at EC_{50} 86 μM. Further SAR studies led to the substituted benzamide ML-324 (**21**) without the 5-carboxy group (KDM4E IC_{50} = 920 nM) whereby it exhibited favorable ADME properties (solubility of 308 μM, good Caco-2 cell permeability, and microsomal stability in the presence of both mouse and rat liver microsomes) and demonstrated activity in the herpes simplex virus models for infection [188]. The selectivity of ML-324 (**21**) for KDM4 remains to be confirmed.

In follow-up work, NCGC00244536 (**22**) was developed with submicromolar potency against the KDM4 subfamily (most potent against KDM4B IC_{50} = 10 nM) and little activity against KDM5A and KDM1A [189]. NCGC00244536 (**22**) displayed antiproliferative effects against fast-growing androgen receptor (AR)-negative PC3 cells (IC_{50} = 40 nM) and more than 100-fold selectivity against the immortalized prostate epithelial cell lines PrEC1 and PrEC4. NCGC00244536 (**22**) also decreased proliferation of other AR-positive cancer cell lines (LNCaP and VCaP) in submicromolar ranges and breast cancer cell lines MDA-MB2 and MCF-7 in the micromolar range. NCGC00244536 (**22**) inhibited tumor growth derived from PC3 in a xenograft tumor model and was active in an *ex vivo* human prostate cancer explant model. It was further shown to affect cell cycle arrest, decreasing the number of S-phase cells, phenocopying KDM4B siRNA. KDM4B activates the transcription of Polo-like kinase 1 (PLK1), which can be blocked by NCGC00244536 (**22**), representing a new strategy to treat prostate cancers that are refractive to antiandrogen therapies [189].

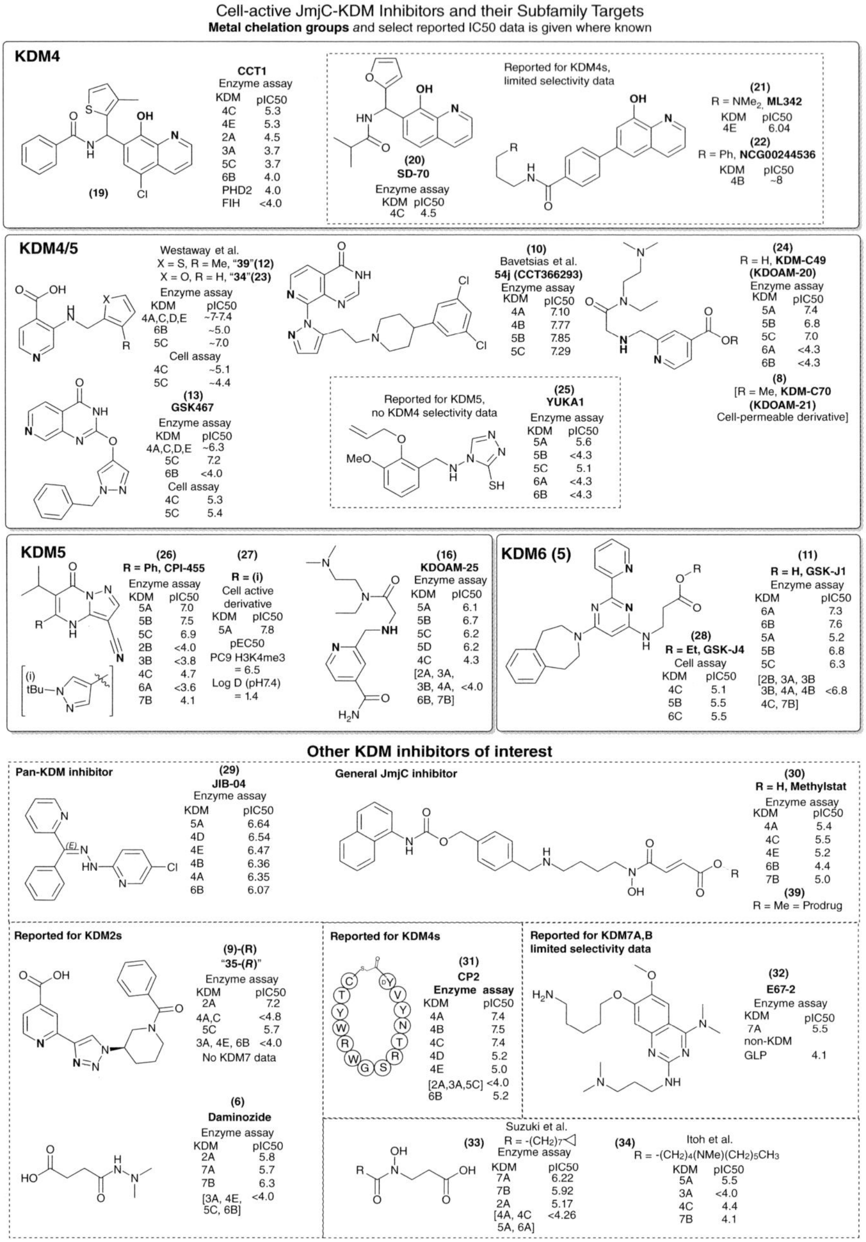

Figure 11.5 Examples of JmjC-KDM inhibitors.

Further exploration of the 8HQ scaffold at the 7 position led to CCT1 (**19**), with 5-Cl substitution, as a relatively selective KDM4 inhibitor *in vitro* ($IC_{50} = 5\ \mu M$), with ≥20-fold selectivity against other KDMs and HIF hydroxylases and sixfold against KDM2A [190]. CCT1 (**19**) also showed better cellular potency than IOX1 in HeLa cells overexpressing KDM4A ($EC_{50} = 9\ \mu M$) and increased H3K9me3 levels in MCF-7 breast cancer cells in a dose-dependent manner (EC_{50} = 12 μM). In patient-matched lung cancer cell lines, an antiproliferative effect was observed in cancerous cells (IC_{50} = 6 μM) but not in non-transformed cells [191].

Independently, phenotypic screening assays for chromosomal translocation events in LNCaP prostate cancer cells under androgen stimulation and genotoxic stress identified SD70 (**20**), a 7-substituted 8HQ, as an inhibitor of a tumor translocation event [192]. Chem-seq assays of LNCaP cells using biotinylated SD70 (**20**) established overlap between SD70 (**20**) and AR-bound enhancers, and further analysis showed SD70 (**20**) to inhibit the AR transcriptional program. SD70 (**20**) showed inhibition of CWR22Rv1 cell growth and a dramatic inhibition of tumor growth in a prostate cancer xenograft mouse model. Modulation of H3K9me3/H3K36me3 was observed upon SD70 (**20**) dosing in HEK293T cells, and SD70 (**20**) was identified to be a KDM4C inhibitor (IC_{50} = 30 μM), although selectivity against other KDMs and 2OG oxygenases remains to be profiled [192]. Overall, as demonstrated for the KDM4s, the 8HQ provide a generic but tunable scaffold for the KDMs [192].

11.2.4.2 KDM4/5 Subfamily-Targeted Inhibitors

Pyridine dicarboxylates were one of the first reported inhibitors of JmjC-KDMs. In particular, the 2,4-dicarboxylates (**3**) showed the best potencies (IC_{50} < 1 μM) against KDM4E [193]. Crystal structures revealed the pyridine nitrogen and 2-carboxylate to act as a bidentate metal chelation motif, with the 4-carboxylate interacting with the basic lysine in the active site pocket.

The first breakthrough in potent cell activity of pyridine carboxylates was the 2-aminomethylpyridine-4-carboxylate derivative KDM5-C49 (**24**) disclosed by EpiTherapeutics [172]. This compound is a bidentate iron chelator and demonstrated a high degree of potency against KDM5A-D (IC_{50} = 4–15 nM) and KDM4C (IC_{50} = 210 nM), with good selectivity against representatives from other JmjC-KDM subfamilies [194–196]. Ethyl-ester prodrug KDM5-C70 (**8**) showed good cellular activity, with KDM5 and KDM4 inhibition in HeLa at EC_{50} = 0.7–4.8 μM and EC_{50} = 10–70 μM, respectively [196]. When KDM5-C70 (**8**) was administered to breast cancer cell lines MCF-7, BT474, and ZR-75-1, their growth was inhibited >70% in colony formation assays; however little effect was observed against KDM5 RNAi-resistant cell lines (MDA-MB-231, PC9, MCF10A) despite seeing upregulation of global H3K4me3 levels [194]. Additionally, differences were seen in KDM5-related gene expression upon either KDM5B knockdown or KDM-C70 (**8**) treatment, which was attributed to as yet uncharacterized regulatory roles of the KDM5s. KDM5-C70 (**8**) also showed antiproliferative effects in myeloma cells [195].

Further work by GlaxoSmithKline on the optimization of a 3-aminopyridine-4-carboxylate series (derived from a HTS hit against KDM6B) led to the identification of 34 (**12**) and 39 (**23**), which are highly potent inhibitors against the KDM4 family *in vitro* (IC_{50} < 100 nM) and in cells (IC_{50} = 6–8 μM). The potency was maintained despite the monodentate chelation, while cellular permeability is improved due to the intramolecular hydrogen bonding between the C4-carboxylate and 3-amino group. While 34 (**12**) and 39 (**23**) were observed to be highly selective for KDM4s over KDM6B (≥50-fold), they also inhibit KDM5C *in vitro* (IC_{50} = 100–125 nM) and in cells (IC_{50} = 25–63 μM). In an effort to improve cell permeability, less acidic bicyclics based on the 3-aminopyridine-4-carboxylates were explored while keeping a metal-chelating sp^2 nitrogen and the interactions with Lys210 and Tyr136 in the KDM4D active site. This led to the generation of pyridopyrimidinones as cell-active alternatives to pyridine carboxylates with substitution at the 2 position – as represented by GSK467 (**13**) – with *in vitro* and cellular potencies against KDM4 of IC_{50} 400–650 nM and IC_{50} = 5 μM, respectively. Remarkably, GSK467 (**13**) also inhibited KDM5C *in vitro* (IC_{50} = 63 nM) and in cells (IC_{50} = 4 μM) but maintained selectivity over KDM6B and HIF hydroxylase PHD3 (EGLN3, IC_{50} > 100 μM) [173, 176].

In parallel, Bavetsias et al. developed 8-substituted pyrido[3,4-*d*]pyrimidin-4(3*H*)-one derivatives as potent cell-active KDM4/5 inhibitors [176]. Structural work confirmed that the pyrimidinones maintain key interactions with the Lys and Tyr of KDM4A (Lys206/Tyr132) and KDM5B (Lys517/Tyr425), where the 2OG C5 carboxylate interacts. The addition of a conformationally constrained 4-phenylpiperidine linker to the main scaffold resulted in 54j (**10**) (and derivatives) that demonstrate equipotent activity against the KDM4 and KDM5 subfamilies (IC_{50} = 12–31 nM) while retaining selectivity over the KDM2, KDM3, and KDM6 subfamilies. The cellular permeability of 54j (**10**) was confirmed in Caco-2 cells, and the cellular activity of the inhibitors was assessed using assays based on cellular H3K9Me3 and H3K4Me3 quantitation.

Despite the potencies achieved for the pyridine series of KDM4 inhibitors, achieving selectivity, in particular over KDM5s, and the discrepancies between the *in vitro* and cellular potencies remain a challenge.

11.2.4.3 KDM5 Subfamily-Targeted Inhibitors

In 2016 Vinogradova et al. published details of the first KDM5 chemical probe, CPI-455 (**26**), an optimized lead compound arising from an HTS of 102 400 commercial compounds in a mass spectrometry-based H3K9me3 demethylation assay against KDM4C [85]. Crystallography was used to confirm the best hits against KDM4C, but follow-up screening revealed a lead fragment that possessed greater affinity for KDM5A. Subsequent chemical modifications arrived at CPI-455 (**26**), which improved both KDM5A/B/C potency (IC_{50} = 3–14 nM) and selectivity over KDM4C. Assays conducted with higher 2OG concentrations negatively affected the observed IC_{50}, indicating competition. Co-crystallization confirmed that CPI-455 (**26**) occupied the 2OG binding site but uniquely showed a monodentate interaction of the nitrile group with the active site metal and the carbonyl group hydrogen bonding to the Asn575 side chain. The aromatic core benefited from π-stacking interactions with nearby Tyr, Phe, and Trp residues, and a KDM5 unique active site cleft was filled by the isopropyl side chain (note: it is suggested as a possible explanation for selectivity over other KDMs, including KDM2B, KDM3B, KDM6A, and KDM7B). Treatment of HeLa cells with CPI-455 (**26**) led to increased global levels of H3K4me3 marks, and a methyl group isotope labeling study demonstrated reduced turnover rate of the K4me3 mark attributable to KDM5 targeted inhibition. Cell activity in PC9 (H3K4me3) was EC_{50} = 5.2 μM. KDM5A inhibition by CPI-455 (**26**) was also shown to reduce the emergence of drug-tolerant persister cells in multiple cancers [85]. CPI-455 (**26**) was optimized for cellular and ADME properties for *in vivo* biological studies, leading to **27** containing substituted pyrazoles, which retained its selectivity for the KDM5 subfamily [197].

In 2017 Tumber et al. reported the discovery of KDOAM-25 (**16**) as a small-molecule inhibitor selective for the KDM5 subfamily [174], inspired by previous work on pyridine-based KDM2A selective inhibitor (**9**) and KDM5-C49 (**24**) [198, 199]. Subsequent chemical derivatization of the common 4-carboxypyridine core led to KDOAM-25 (**16**), which was found to have nanomolar IC_{50} values *in vitro* against KDM5A-D and >100-fold selectivity against other KDM subfamilies, including the closely related KDM4s. It also passed a CEREP express panel of 55 unrelated proteins. Co-crystallization of KDOAM-25 (**16**) with KDM5B (PDB ID: 5A3N) indicated binding in the 2OG/substrate binding site, with the pyridinyl amide forming hydrogen bonds to Tyr425 and Lys517 in the active site. The distortion of Tyr425, as compared to that with an analogous less selective inhibitor, indicates an inherent mobility of Tyr425 that appears unique to the KDM5s and may be responsible for the observed selectivity profile. Dosing of an MM1S multiple myeloma cell line with KDOAM-25 (**16**) gave an IC_{50} of ~30 μM after one week of treatment, alongside global increases in H3K4me3 levels, but activity was not seen in other myeloma cell lines [174].

A biologically active KDM5 inhibitor, YUKA1 (**25**), was published by Gale et al. in 2016 [200]. HTS of a curated ~9000-molecule library in an AlphaScreen KDM5A enzyme inhibition assay gave 34 compounds with $IC_{50} < 5$ μM. YUKA1 demonstrated intra- and inter-subfamily selectivity, >18-fold versus KDM5B, KDM6A, and KDM6B and ~2.5-fold versus KDM5C. Its activity was not affected by 2OG concentration, indicating no competition, but was dependent on Fe(II) concentration, suggesting some form of metal binding-mediated inhibition. However, the mechanism of inhibition of this new scaffold has not yet been firmly established. HeLa cells dosed with YUKA1 exhibited increased levels of H3K4me3 after three days, but MCF-7 did not. This correlated with the effect of CRIPSR/Cas9 deletion of *KDM5A*, curtailing HeLa survival but not affecting the MCF-7 line. Furthermore YUKA1 (**25**) dosing of PC9 and BT474 cancer cell lines demonstrated a decreased propensity for drug resistance to develop versus a small-molecule and antibody therapeutic, respectively, providing further evidence of the KDM5s playing some role in the onset of anticancer drug resistance. YUKA1 (**25**) also increased H3K4me3 levels and decreased proliferation in KDM5A-expressing ZR-75-1 breast cancer cells, with a smaller effect noted in triple-negative MDA-MB-231 breast cancer cells and no effect on immortalized non-transformed MCF10A mammary epithelial cells.

11.2.4.4 KDM6 Subfamily-Targeted Inhibitors

GSK-J1 (**11**) and rapidly hydrolyzed cell-permeable ester derivative GSK-J4 (**28**) were published in 2012 [169] as the first chemical probe for the KDM6 subfamily. Despite its remarkable selectivity over other KDM subfamily member representative enzymes in the study, GSK-J1 (**11**) was later found to also inhibit KDM5B [195, 201], albeit at weaker potency *in vitro*. Nonetheless, GSK-J1 (**11**) remains the most potent and selective KDM6A/B inhibitor to date and is a useful tool compound, providing insight into the biological and pathological roles of the KDM6 subfamily. The crystal structure (PDB ID: 4ASK) of GSK-J1 (**11**) with KDM6B revealed that its propanoic acid side chain occupies the 2OG binding site, with the aromatic core sitting in a cleft normally occupied by histone H3P30 and the pyridyl-pyrimidine motif engaging in bidentate coordination of the active site metal, disturbing its position. It was found to be a competitive inhibitor of both 2OG and Fe(II), but not the substrate [201]. GSK-J1 (**11**) was also screened against 100 kinases and 60 other proteins, including HDACs, without cross-reactivity. Interestingly, GSK-J4 (**28**) decreased the proliferation and/or viability of T-cell acute lymphoblastic leukemia (T-ALL) and pediatric brainstem glioma models and was well tolerated in mice [61, 72, 202]. Importantly, RNAi/KO of KDM6A or KDM6B phenocopied GSK-J4 treatment (Table 11.2). However, it should be noted that GSK-J4 has been reported to inhibit KDM6B, KDM5B, and KDM4C with similar potency in cell-based assays [201].

11.2.4.5 KDM2/7- and KDM3-Targeted Inhibitors

A series of hydroxamates with alkyl chains of different lengths have shown differential inhibitory profiles against KDM subfamilies [203–205]. In particular, cyclopropyl-containing hydroxamate (**7**) demonstrated selectivity toward inhibition of KDM2/KDM7 over other KDMs tested, with antiproliferative activity against HeLa and KYSE150 cells [205].

In 2014, England et al. reported the development of a selective KDM2 inhibitor based on triazolopyridines [199]. Substitution of a pyridinyl ring for a triazole in the classical 2,2′-bipyridine scaffold, readily attached through click chemistry, allowed for facile variation of the side chain to give a selectivity improvement in favor of KDM2A/4A. Co-crystallization of an early intermediate with KDM4A indicated the triazolopyridine occupied the 2OG binding site. Further chemical elaboration to exploit interactions with the histone substrate binding pocket led to a benzoyl 3-piperidine derivative that showed excellent potency and selectivity for KDM2A over KDM4 and other JmjC-KDMs. The (*R*)-enantiomer, **9**, proved most potent. Given the similarity between the active sites of the KDM2/7 subfamilies, **9** is likely to exhibit some inhibition of the KDM7 family. The cellular activities of **9** and other KDM2/7 selective inhibitors, daminozide (**6**) and derivatives [206], are yet to be confirmed. No KDM3 subfamily-targeted inhibitors have been reported to date.

11.2.4.6 Generic JmjC-KDM Inhibitors

A number of reported pan-JmjC-KDM inhibitors have shown promising antiproliferative effects on cancer cells. Methylstat (**30**) is a bivalent molecule, with hydroxamic acid iron-chelating motif linked to a proposed "substrate-mimicking" moiety and an ethyl-ester prodrug. The active acid form inhibits JmjC-KDMs and other 2OG oxygenases at IC_{50} range of 10^{-5}–10^{-6} M in enzyme assays. Methylstat shows antiproliferative activities in a KDM4C-sensitive esophageal carcinoma cell line KYSE150 at $GI_{50} = 5.1$ μM and hypermethylation of H3K4me3 and H3K9me3 in a concentration-dependent manner. JIB-04 (**29**), a pyridine hydrazine, is a JmjC-KDM inhibitor originally identified through phenotypic locus derepression (LDR) cell-based screening using a stably integrated GFP-estrogen receptor transgene, whereby molecules were screened for epigenetic activation of GFP transcription [207]. The E-isomer, but not the Z-isomer, of JIB-04 (**29**) was identified to be an inhibitor of KDM4A–E, KDM6B, and KDM5A, but was inactive against TET1 and PHD2. Kinetic studies established that JIB-04 (**29**) is competitive with respect to iron but not with 2OG. Interestingly, JIB-04 (**29**) showed an antiproliferative effect on a wide range of cancer cell lines but not on HBEC and primary epithelial cells and was selective for cancerous over normal cells in patient-matched lung cells [207]. In *in vivo* models, JIB-04 (**29**) reduced tumor burden and prolonged cancer survival in mice. Recently it was demonstrated that chemoresistant tumors displayed hypersensitivity to JIB-04 (**29**), which prevents the emergence of drug-tolerant non-small cell lung cancer cells [208].

11.2.5 Selectivity and Potency of JmjC-KDM Inhibition in Cells

For many of the JmjC-KDM inhibitors, significant drop-off in potency between biochemical and cellular assays has been observed (Figure 11.5). This has, in part, been attributed to poor cell permeability of 2OG-competitive inhibitors (e.g. the highly polar nature of many 2OG mimics) and the high cellular 2OG concentrations [196]. 2OG concentrations can reach 2 mM levels in cells [209], while IC_{50} values in conventional enzyme inhibition assays are generally evaluated at or near the 2OG K_M, which are in the 10^{-6}–10^{-5} M range for most JmjC-KDMs [187] and thus may overestimate potencies. Increasing the potencies ($pIC_{50} > 7$) and designing inhibitors with different modes of inhibition (e.g. mixed mode/noncompetitive with respect to 2OG) can overcome such limitations, as demonstrated by some of the cell-active KDM inhibitors [85].

It is worth noting that most cellular assays for inhibitor evaluation rely on downstream changes in the global histone methylation states upon compound dosing, either using immunoblotting, immunofluorescence imaging, or mass spectrometry. The on-target effect on KDMs in cells is therefore difficult to ascertain, in particular, as hypermethylation can also be induced by general cytotoxicity [196]. In addition, for metal-chelating scaffolds, off-target effects have been reported, independent of JmjC-KDM inhibition. One example is the upregulation of HIF, despite the lack of direct inhibition of HIF hydroxylases (PHDs/FIH) by JmjC-KDM inhibitors [190]. This has been, in part, attributed to potential disruption of intracellular iron hemostasis, rather than via inhibition of HIF hydroxylases in cells (of which the same cellular effect will be observed).

The expression levels of JmjC-KDMs and other 2OG oxygenases can vary significantly in cells, and compounds dosed at high concentrations may lead to off-target 2OG oxygenase inhibition. Thus, selectivity screening of inhibitors in a cellular context may be more meaningful. Joberty et al. recently described a chemical proteomics approach for profiling ligands on a large set of different 2OG oxygenases [210]. This approach benefits from not depending on isolated and purified enzymes for characterizing the compounds but instead utilizes immobilized ligands to capture the oxygenases from extracts of cell lysates. Tandem mass spectrometry (MS/MS) was used to quantify the relative protein binding with respect to the immobilized ligands. Binding affinity and selectivity profiles of KDM inhibitors across the 2OG oxygenase family were generally in good agreement with the reported biochemical inhibition assays but also identified unexpected off-target binding in some cases. Methylstat (**30**), a generic inhibitor of JmjC-KDMs, showed less potency for KDM4A/C compared with what previous studies had indicated but affected a variety of 2OG oxygenases, with the highest potency for the trimethyl-lysine hydroxylase TMLH, which has been genetically linked to autism. Chemoproteomics profiling of GSK-J1 (**11**), originally reported as a KDM6 chemical probe but subsequently reported to also be weak KDM5 inhibitors, showed that GSK-J1 (**11**) did indeed bind to KDM6A/B together with KDM5C/D. In agreement with biochemical data, compound GSK 34 (**23**) was a mixed inhibitor of KDM4 together with KDM5C/D and showed >1000-fold selectivity against these enzymes compared with other oxygenases. CPI-455 (**26**) was shown to be a potent KDM5C/D inhibitor that had >40-fold selectivity over KDM4C and >300-fold selectivity for other oxygenases.

11.3 KDM Inhibitors Targeting the Reader Domains

Several KDMs possess reader domains, either PHD fingers or Tudor domains, that enable them to bind to specific posttranslationally modified histone marks (Table 11.1). For some, this locates the catalytic JmjC domain closer to a specific target mark for demethylation, and in other cases the reader domain functions as an allosteric regulator of the JmjC catalytic activity. Thus, targeting the reader domains presents an alternative strategy for the inhibition of JmjC-KDMs. Due to the different domain architectures between each subfamily (and in some cases paralogues), as well as greater diversity between the reader domains than between their respective JmjCs, it may offer better opportunities for achieving selectivity.

11.3.1 Plant Homeodomain Fingers (PHD Fingers)

PHD fingers are a major class of reader domains for chromatin modifiers. They are typically 40–80 amino acids in length, consisting of a Cys_4-His-Cys_3 motif that binds two zinc ions, and are structurally characterized by a short double-stranded beta-sheet and alpha helix (Figure 11.6a) [211]. Over 200 human PHD finger-containing proteins have been found to date, with many only partially characterized or uncharacterized.

PHD fingers are found across many of the KDMs (Table 11.1). While the functions of several KDM-associated PHD fingers remain unclear, some have been shown to bind specific lysine methylation states with low micromolar affinity on the tail of histones H3 and H4 (Table 11.1). Often this recognition is for trimethylated lysines [28, 39, 43, 213–215] but can also be for the unmodified histone H3 *N*-terminus, usually requiring H3(1–4) to remain unmodified [24, 26]. The KDM5A/5B PHD1 binds unmodified histone H3 tail and allosterically activates the H3K4me3 demethylase activity [216, 217]. Other

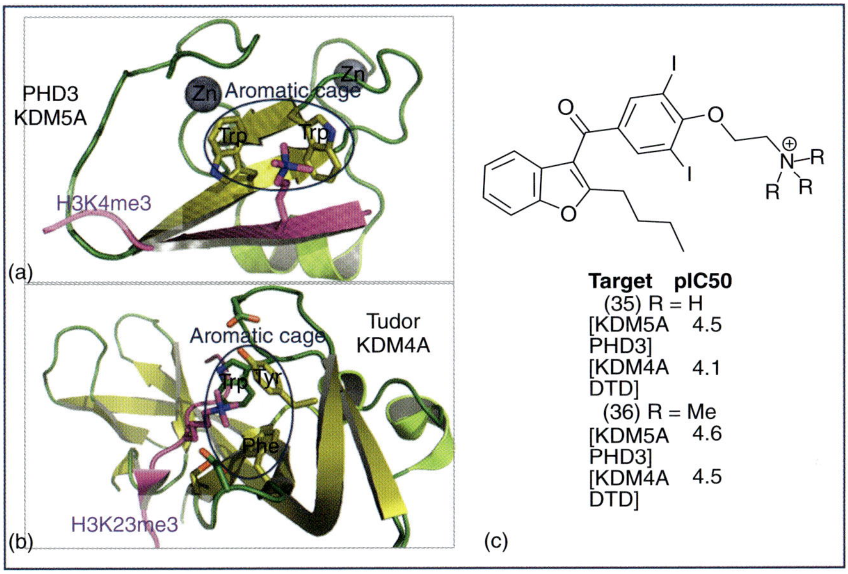

Figure 11.6 Crystal structures of JmjC-KDM reader domains. (a) PHD finger. The PHD finger 3 (PHD3) of KDM5A binds H3K4me3 (PDB ID: 2KGI) in an aromatic cage of Trp residues, and the histone forms a beta-sheet motif. Source: Sanchez and Zhou 2011 [211]. Reproduced with permission of Elsevier. (b) Tudor domain. The KDM4A Tudor domain binds a trimethyl-lysine, H3K23me3, in aromatic cage consisting of Trp, Tyr, and Phe residues (PDB ID: 5D6Y). Color key: α-helices = pale green, β-sheets = olive yellow, loops = dark green, histone = purple. Source: Lu and Wang 2013 [212]. Reproduced with permission of Elsevier. (c) Structure and reported IC_{50s} of the only KDM PHD inhibitor published to date.

PHD fingers have been shown to modulate demethylase activity; KDM7B PHD binding of H3K4me3 significantly enhances demethylation of the H3K9me2 mark [57]. Interestingly, PHD3 of KDM5A forms a fusion oncoprotein with nucleoporin protein 98 (NUP98), a common translocation partner, and drives leukemogenesis [26]. Moreover, a PHD3 mutation that disrupts NUP98-PHD H3K4me3 binding inhibits leukemic transformation [26]. Some KDMs possess multiple PHD fingers (e.g. KDM5 family), bestowing an ability to bind more than one type of histone mark; this is argued to give rise to combinatorial reading of the histone modification network and thus enable more complex gene regulation [24, 216].

The high sequence variability, inherent flexibility, and lack of a well-defined pocket have made chemical targeting of PHD fingers challenging. Wagner et al. reported the screening of a small-molecule library against the KDM5A PHD3 (H3K4me3 reader) via a HaloTag assay. This resulted in the discovery of amiodarone derivatives (Figure 11.6c) with IC_{50} 26–30 μM in a fluorescence polarization assay [218]. Despite the apparent trimethyl-lysine mimicry, the amino group methylation state made little difference to binding and was therefore proposed to unlikely be sitting in the aromatic cage. Cross-screening against other "Kme3" readers revealed similar binding IC_{50} values for KDM4A-DTD and ING2-PHD, but selectivity over other "Kme3" binding PHDs (RAG2 and UHRF1). No inhibition was observed against two "Kme0" reader PHDs (AIRE and BHC8) [218]. While there have been three other reported studies of non-KDM-related PHD finger inhibitor development [219–221], selective potent inhibitors for PHD finger domains are yet to be identified.

11.3.2 Tudor Domains

Tudor domains are similarly sized to the PHD fingers and are also structurally diverse but commonly possess a set of 4–5 antiparallel β-sheets (Figure 11.6b) [212]. Several dozen Tudor domains have been found across the human genome in many classes of proteins, with accordingly diverse associated biology. They bind and direct activity toward specific histone modifications, but with low micromolar affinity for their targets. Some Tudor domains can read histone arginine methylation; however KDM-based Tudor domains are histone methyl-lysine readers. Tudor domains are only present in the KDM4 subfamily, wherein they are found as a DTD [213] and proposed to guide KDM4A–C to their substrates, KDM4A being one of the first reported examples [222] (Table 11.1). The only reported KDM Tudor inhibitors are the nonselective binding amiodarone derivatives (**35** and **36**) that also inhibit the PHD fingers [218].

11.4 Conclusions and Future Perspectives

Since the discovery of JmjC-KDMs in 2006, significant progress has been made on the biochemical, structural, and functional understanding of these enzymes. It is now clear that many of these enzymes have important biological roles in development and in disease (Table 11.2). JmjC-KDMs are emerging as potential therapeutic targets, particularly in cancer, and significant efforts have been made to develop inhibitors. Recent important advances include the development of several potent inhibitor scaffolds for various different JmjC-KDM subfamilies, in particular, the discovery of the first highly selective and potent KDM5 inhibitor CPI-455 and its orally bioavailable derivative, and the emerging target validation results for JmjC-KDMs by small-molecule inhibition. However, substantial challenges still remain in developing selective and potent inhibitors for the JmjC-KDMs for *in vivo* applications. While the focus to date has largely been on targeting the active site metal of JmjC catalytic domains, exploring new mode-of-action inhibitors, such as peptide-competitive inhibitors and allosteric or covalent inhibitors, may help address these challenges. There is also significant scope in targeting the non-catalytic domains within the KDMs.

It is also important to note that the substrates of JmjC-KDMs may not be limited to lysines on histone tails (as summarized in Table 11.1). As new functions of JmjC-KDMs are unraveled, considerations need to be made on how the inhibitors are designed and assessed, including targeting the catalytic or non-catalytic functions of JmjC-KDMs. Also, while potency and selectivity are prerequisite for chemical probes, from a therapeutic perspective, broader JmjC-KDM inhibitors may prove useful in overcoming redundancy, compensation, and resistance.

Acknowledgments

We thank Dr. Richard Hopkinson for critical reading and the Cancer Research UK (C8717/A18245) for funding our work on KDMs. CH is supported by the Danish Research Council and ODC is supported by the EPSRC Centre for Doctoral Training in Synthesis for Biology and Medicine (EP/L015838/1). AK gratefully acknowledges the Royal Society for the Dorothy Hodgkin Fellowship and the European Research Council Starting Grant (EPITOOLS-679479). We apologize for the incomplete citations and for the research that we were not able to cite due to space constraints.

References

1 Kooistra, S.M. and Helin, K. (2012). Molecular mechanisms and potential functions of histone demethylases. *Nat. Rev. Mol. Cell Biol.* 13 (5): 297–311.
2 Højfeldt, J.W., Agger, K., and Helin, K. (2013). Histone lysine demethylases as targets for anticancer therapy. *Nat. Rev. Drug Discovery* 2 (November): 917–930.
3 Helin, K. and Dhanak, D. (2013). Chromatin proteins and modifications as drug targets. *Nature* 502 (7472): 480–488.
4 Murray, K. (1963). The occurrence of epsilon-*N*-methyl lysine in histones. *Biochemistry* 3 (1): 10–14.
5 Rea, S., Eisenhaber, F., O'Carroll, D. et al. (2000). Regulation of chromatin structure by site-specific histone H3 methyltransferases. *Nature* 406 (6796): 593–599.
6 Bannister, A.J., Schneider, R., and Kouzarides, T. (2002). Histone methylation: dynamic or static? *Cell* 109 (7): 801–806.
7 Shi, Y., Lan, F., Matson, C. et al. (2004). Histone demethylation mediated by the nuclear amine oxidase homolog LSD1. *Cell* 119 (7): 941–953.
8 Tsukada, Y., Fang, J., Erdjument-Bromage, H. et al. (2006). Histone demethylation by a family of JmjC domain-containing proteins. *Nature* 439 (7078): 811–816.
9 Hojfeldt, J.W., Agger, K., and Helin, K. (2013). Histone lysine demethylases as targets for anticancer therapy. *Nat. Rev. Drug Discovery* 12 (12): 917–930.
10 Clissold, P.M. and Ponting, C.P. (2001). JmjC: Cupin metalloenzyme-like domains in jumonji, hairless and phospholipase A2β. *Trends Biochem. Sci.* 26 (1): 7–9.
11 McDonough, M.A., Loenarz, C., Chowdhury, R. et al. (2010). Structural studies on human 2-oxoglutarate dependent oxygenases. *Curr. Opin. Struct. Biol.* 20 (6): 659–672.
12 Trewick, S.C., McLaughlin, P.J., and Allshire, R.C. (2005). Methylation: lost in hydroxylation? *EMBO Rep.* 6 (4): 315–320.

13 Klose, R.J. and Zhang, Y. (2007). Regulation of histone methylation by demethylimination and demethylation. *Nat. Rev. Mol. Cell Biol.* 8 (4): 307–318.
14 Markolovic, S., Leissing, T.M., Chowdhury, R. et al. (2016). ScienceDirect Structure – function relationships of human JmjC oxygenases—demethylases versus hydroxylases. *Curr. Opin. Struct. Biol.* 41 (Ii): 62–72.
15 Black, J.C., Van Rechem, C., and Whetstine, J.R. (2012). Histone lysine methylation dynamics: establishment, regulation, and biological impact. *Mol. Cell* 48 (4): 491–507.
16 Huang, H., Sabari, B.R., Garcia, B.A. et al. (2014). SnapShot: histone modifications. *Cell* 159 (2): 458–458.e1.
17 Greer, E.L. and Shi, Y. (2012). Histone methylation: a dynamic mark in health, disease and inheritance. *Nat. Rev. Genet.* 13 (5): 343–357.
18 Williams, S.T. et al. (2014). *Epigenetics* 9: 1596–1603.
19 Zhao, L. et al. (2015). *Cell Death Differ.* 1–11.
20 Agger, K. et al. (2007). *Nature* 449: 731–734.
21 Walport, L.J. et al. (2014). *J. Biol. Chem.* 289: 18302–18313.
22 Cooper, S. et al. (2016). *Nat. Commun.* 7: 13661.
23 Yamane, K. et al. (2007). *Mol. Cell* 25: 801–812.
24 Klein, B.J., Piao, L., Xi, Y. et al. (2014). The histone-H3K4-specific demethylase KDM5B Binds to its substrate and product through distinct PHD fingers. *Cell Rep.* 6 (2): 325–335.
25 Christensen, J. et al. (2007). *Cell* 128: 1063–1076.
26 Wang, G.G., Song, J., Wang, Z. et al. (2009). Haematopoietic malignancies caused by dysregulation of a chromatin-binding PHD finger. *Nature* 459 (7248): 847–851.
27 Walport, L.J. et al. (2016). *Nat Commun.* 7: 11974.
28 Iwase, S., Lan, F., Bayliss, P. et al. (2007). The X-linked mental retardation gene SMCX/JARID1C defines a family of histone H3 lysine 4 demethylases. *Cell* 128 (6): 1077–1088.
29 Trojer, P. et al. (2009). *J. Biol. Chem.* 284: 8395–8405.
30 Su, Z. et al. (2016). *Nat. Commun.* 7: 13387.
31 Jung, E.S. et al. (2015). *Biochim. Biophys. Acta, Gene Regul. Mech.* 1849: 1081–1094.
32 Yang, L. et al. (2011). *Cell* 147: 773–788.
33 Liu, L., Kim, H., Casta, A. et al. (2014). Hairless is a histone H3K9 demethylase. *FASEB J.* 28 (4): 1534–1542.
34 Kim, S.M. et al. (2010). *Nucleic Acids Res.* 38: 6389–6403.
35 Watanabe, S. et al. (2013). *Nat. Struct. Mol. Biol.* 20: 1425–1433.
36 Ramadoss, S. et al. (2017). *Oncogene* 36: 1537–1545.
37 Lu, T. et al. (2010). *Proc. Natl. Acad. Sci. U.S.A* 107: 46–51.
38 Frescas, D., Guardavaccaro, D., Bassermann, F. et al. (2007). *Nature* 450: 309–313.
39 Wen, H., Li, J., Song, T. et al. (2010). Recognition of histone H3K4 trimethylation by the plant homeodomain of PHF2 modulates histone demethylation. *J. Biol. Chem.* 285 (13): 9322–9326.
40 Baba, A. et al. (2011). *Nat. Cell Biol.* 13: 668–675.

41 Stender, J.D. et al. (2012). *Mol. Cell* 48: 28–38.

42 Qi, H.H. et al. (2010). *Nature* 466: 503–507.

43 Kleine-Kohlbrecher, D., Christensen, J., Vandamme, J. et al. (2010). A functional link between the histone demethylase PHF8 and the transcription factor ZNF711 in X-linked mental retardation. *Mol. Cell* 38 (2): 165–178.

44 Lu, Y. et al. (2009). *Cell Cycle* 8: 2101–2109.

45 Ge, W. et al. (2012). *Nat. Chem. Biol.* 8: 960–962.

46 Sinha, K.M., Yasuda, H., Coombes, M.M. et al. (2009). *EMBO J.* 29: 68–79.

47 Feng, T. et al. (2014). *Mol. Cell* 53: 645–654.

48 Kwok, J., O'Shea, M., Hume, D.A., and Lengeling, A. (2017). *Front. Genet.* 8: 1–19.

49 Noma, A. et al. (2010). *J. Biol. Chem.* 285: 34503–34507.

50 Scholz, C.C. et al. (2016). *PLoS Biol.* 14: doi: 10.1371/journal.pbio.l002347.

51 Ishimura, A. et al. (2012). *Development* 139: 749–759.

52 Chang, B., Chen, Y., Zhao, Y., and Bruick, R.K. (2007). JMJD6 is a histone arginine demethylase. *Science* 318 (5849): 444–447.

53 Böttger, A., Islam, M.S., Chowdhury, R. et al. (2015). The oxygenase Jmjd6 – a case study in conflicting assignments. *Biochem. J.* 468 (2): 191–202.

54 Walport, L.J., Hopkinson, R.J., Chowdhury, R. et al. (2016). Arginine demethylation is catalysed by a subset of JmjC histone lysine demethylases. *Nat. Commun.* 7: 11974.

55 Musselman, C.A., Lalonde, M.-E., Côté, J., and Kutateladze, T.G. (2012). Perceiving the epigenetic landscape through histone readers. *Nat. Struct. Mol. Biol.* 19 (12): 1218–1227.

56 Pedersen, M.T., Agger, K., Laugesen, A. et al. (2014). The demethylase JMJD2C localizes to H3K4me3-positive transcription start sites and is dispensable for embryonic Development. *Mol. Cell Biol.* 34 (6): 1031–1045.

57 Horton, J.R., Upadhyay, A.K., Qi, H.H. et al. (2010). Enzymatic and structural insights for substrate specificity of a family of jumonji histone lysine demethylases. *Nat. Struct. Mol. Biol.* 17 (1): 38–43.

58 Klose, R.J., Kallin, E.M., and Zhang, Y. (2006). JmjC-domain-containing proteins and histone demethylation. *Nat. Rev. Genet.* 7 (9): 715–727.

59 Ohtani, K. et al. (2013). *Circ. Res.* 113: 856–862.

60 Li, Q. et al. (2014). *PLoS Genet.* 10: doi: 10.1371/journal.pgen.l004524.

61 Hashizume, R., Andor, N., Ihara, Y. et al. (2014). Pharmacologic inhibition of histone demethylation as a therapy for pediatric brainstem glioma. *Nat. Med.* 20 (12): 1394–1396.

62 McLaughlin-Drubin, M.E., Crum, C.P., and Münger, K. (2011). *Proc. Natl. Acad. Sci. U.S.A.* 108: 2130–2135.

63 McLaughlin-Drubin, M.E., Park, D., and Munger, K. (2013). *Proc. Natl. Acad. Sci. U.S.A.* 110: 16175–16180.

64 Liau, B.B. et al. (2017). *Cell Stern Cell.* 20: 233.e7–246.e7.

65 Tang, B. et al. (2016). *Cancer Res.* 76: 6520–6532.

66 Ohguchi, H. et al. (2017). *Leukemia* 1–9.

67 Ntziachristos, P. et al. (2014). *Nature* doi: 10.1038/naturel3605.

68 Wel-stead, G.G. et al. (2012). *Proc. Natl. Acad. Sci.* 109: 13004–13009.

69 Faralli, H., Wang, C., Nakka, K. et al. (2016). UTX demethylase activity is required for satellite cell-mediated muscle regeneration. *J. Clin. Invest.* 126 (4): 1555–1565.
70 Kim, J.H. et al. (2014). *Cancer Res.* 74: 1705–1717.
71 Xie, G. et al. (2017). *Oncogene* 1–15.
72 Benyoucef, A., Palii, C.G., Wang, C. et al. (2016). UTX inhibition as selective epigenetic therapy against TAL1-driven T-cell acute lymphoblastic leukemia. *Genes Dev.* 30 (5): 508–521.
73 Shpargel, K.B., Sengoku, T., Yo-koyama, S., and Magnuson, T. (2012). *PLoS Genet.* 8: doi: 10.1371/journal.pgen.l002964.
74 Landeira, D. and Fisher, A.G. (2011). Inactive yet indispensable: the tale of Jarid2. *Trends Cell Biol.* 21: 74–80.
75 Walters, Z.S. et al. (2014). *Oncogene* 33: 1148–1157.
76 Catchpole, S. et al. (2011). *Int J. Oncol.* 38: 1267–1277.
77 Albert, M. et al. (2013). *PLoS Genet.* 9: doi: 10.1371/journal. pgen.1003461.
78 Zou, M.R. et al. (2014). *J. Biol. Chem.* 289: 17620–17633.
79 Yamamoto, S. et al. (2014). *Cancer Cell* 25: 762–777.
80 Roesch, A. et al. (2013). *Cancer Cell* 23: 811–825.
81 Klose, R.J. et al. (2007). *Cell* 128: 889–900.
82 Zeng, J. et al. (2010). *Gastroenterology* 138: 981–992.
83 Váraljai, R. et al. (2015). *Genes Dev.* 29: 1817–1834.
84 Teng, Y.C. et al. (2013). *Cancer Res.* 73: 4711–4721.
85 Vinogradova, M., Gehling, V.S., Gustafson, A. et al. (2016). An inhibitor of KDM5 demethylases reduces survival of drug-tolerant cancer cells. *Nat. Chem. Biol.* 12 (7): 531–538.
86 Sharma, S.V. et al. (2010). *Cell* 141: 69–80.
87 Lin, W. et al. (2011). *Proc. Natl. Acad. Sci. U.S.A.* 108: 13379–13386.
88 Benevolenskaya, E.V., Murray, H.L., Branton, P. et al. (2005). *Mol. Cell* 18: 623–635.
89 Cox, B.J. et al. (2010). *Genome Res.* 20: 1154–1164.
90 Iwamori, N., Zhao, M., Meistrich, M.L., and Matzuk, M.M. (2011). *Biol. Reprod.* 84: 1225–1234.
91 Zhang, Q.J. et al. (2011). *J. Clin. Invest.* 121: 2447–2456.
92 Terndrup Pedersen, M., Kooistra, S.M., Radzisheuskaya, A. et al. (2016). Continual removal of H3K9 promoter methylation by Jmjd2 demethylases is vital for ESC self-renewal and early development. *EMBO J.* 35 (14): 1550–1564.
93 Mallette, F.A. et al. (2012). *EMBO J.* 31: 1865–1878.
94 Wang, J. et al. (2016). *Cell Death Differ.* 23: 1886–1896.
95 Neault, M., Mallette, F.A., and Richard, S. (2016). *Cell Rep.* 14: 1966–1978.
96 Kim, T.D. et al. (2016). *J. Clin. Invest.* 126: 706–720.
97 Wang, L.Y. et al. (2016). *Cell Rep.* 16: 3016–3027.
98 Kawazu, M. et al. (2011). *PLoS One* 6: doi: 10.1371/journal.pone.0017830.
99 Yang, J. et al. (2010). *Cancer Res.* 70: 6456–6466.
100 Shi, L. et al. (2011). *Proc. Natl. Acad. Sci. U.S.A.* 108: 7541–7546.
101 Gaughan, L. et al. (2013). *Nucleic Acids Res.* 41: 6892–6904.
102 Castellini, L. et al. (2017). *Nucleic Acids Res.* 45: 3674–3692.

103 Yang, D. et al. (2015). *J. Natl. Cancer Inst.* 107: doi: 10.1093/jnci/djv080.
104 Wilson, C. et al. (2016). *Oncogene* 2565–2576.
105 Cheung, N. et al. (2016). *Cancer Cell* 29: 32–48.
106 Liu, G. et al. (2009). *Oncogene* 28: 4491–4500.
107 Luo, W., Chang, R., Zhong, J. et al. (2012). *Proc. Natl. Acad. Sci. U.S.A.* 109: E3367–E3376.
108 Gregory, B.L. and Cheung, V.G. (2014). *Genome Res.* 24: 52–63.
109 Cloos, P.A.C. et al. (2006). *Nature* 442: 307–311.
110 Rui, L. et al. (2010). *Cancer Cell* 18: 590–605.
111 Wissmann, M. et al. (2007). *Nat. Cell Biol.* 9: 347–353.
112 Zarach, J.M., Beaudoin, G.M.J., Coulombe, P.A., and Thompson, C.C. (2004). *Development* 131: 4189–4200.
113 Kuroki, S. et al. (2013). *Biol. Reprod.* 89: 93.
114 Zhu, N., Chen, M., Eng, R. et al. (2015). MLL-AF9 – and HOXA9-mediated acute myeloid leukemia stem cell self-renewal requires JMJD1C. *J. Clin. Invest.* 126 (35): 1–15.
115 Sroczynska, P. et al. (2014). *Blood* 123: 1870–1882.
116 Chen, M. et al. (2015). *Genes Dev.* 29: 2123–2139.
117 Okada, Y., Scott, G., Ray, M.K. et al. (2007). *Nature* 450: 119–123.
118 Tateishi, K., Okada, Y., Kallin, E.M., and Zhang, Y. (2009). *Nature* 458: 757–761.
119 Kuroki, S., Matoba, S., Akiyoshi, M. et al. (2013). Epigenetic regulation of mouse sex determination by the histone demethylase Jmjd1a. *Science* 341 (6150): 1106–1109.
120 Wan, W. et al. (2017). *Oncogene* 1–10.
121 Wade, M.A. et al. (2015). *Nucleic Acids Res.* 43: 196–207.
122 Kaukonen, R. et al. (2016). *Nat. Commun.* 7: 12237.
123 Ramadoss, S., Guo, G., and Wang, C.-Y. (2016). *Oncogene* doi: 10.1038/onc.2016.174.
124 Ohguchi, H. et al. (2016). *Nat. Commun.* 7: 10258.
125 Parrish, J.K., Sechler, M., Winn, R.A., and Jedlicka, P. (2015). *Oncogene* 34: 257–262.
126 Nakatsuka, T. et al. (2017). *Oncogene* 1–10.
127 Ramadoss, S., Guo, G., and Wang, C.-Y. (2017). *Oncogene* 36: 47–59.
128 Fan, L. et al. (2015). *Oncogene* 1–12.
129 Liu, Z. et al. (2015). *Int. J. Biol. Sci.* 11: 1447–1457.
130 Liu, Z. et al. (2015). *Int. J. Biol. Sci.* 11: 494–507.
131 Kawakami, E., Tokunaga, A., Ozawa, M. et al. (2015). *Mech. Dev.* 135: 31–42.
132 Wagner, K.W. et al. (2013). *J. Clin. Invest.* 123: 5231–5246.
133 Fukuda, T., Tokunaga, A., Sakamoto, R., and Yoshida, N. (2011). *Mol. Cell. Neurosci.* 46: 614–624.
134 Boulard, M., Edwards, J.R., and Bester, T.H. (2015). *Nat. Genet.* 47: 479–485.
135 Andricovich, J., Kai, Y., Peng, W. et al. (2016). *J. Clin. Invest.* 126: 905–920.
136 van den Boom, V. et al. (2016). *Cell Rep.* 14: 332–346.
137 He, J., Nguyen, A.T., and Zhang, Y. (2011). *Blood* 117: 3869–3880.
138 Kottakis, F. et al. (2014). *Cancer Res.* 74: 3935–3946.
139 Tzatsos, A. et al. (2013). *J. Clin. Invest.* 123: 727–739.

140 Okuno, Y. et al. (2013). *Diabetes* 62: 1426–1434.
141 Walsh, R.M. et al. (2017). *Nat. Commun.* 8: 15142.
142 Wang, Q. et al. (2016). *J. Clin. Invest.* 126: 2205–2220.
143 Shao, P. et al. (2017). *Nucleic Acids Res.* 45: 1687–1702.
144 Björkman, M. et al. (2012). *Oncogene* 31: 3444–3456.
145 Yatim, A. et al. (2012). *Mol. Cell* 48: 445–458.
146 Thakur, C. et al. (2015). *Oncotarget* 6: 3722–3736.
147 Mori, T. et al. (2013). *Cell Struct. Funct.* 38: 155–167.
148 Yoo, H., Son, D., Lee, Y.J., and Hong, K. (2016). *Mol. Reprod. Dev.* 83: 588–593.
149 Böse, J. et al. (2004). *J. Biol.* 3: 15.
150 Wang, F. et al. (2014). *PLoS Biol.* 12: doi: 10.1371/journal.pbio.l001819.
151 Zhang, N. et al. (2010). *Cell Metab.* 11: 364–378.
152 Pelletier, J. et al. (2012). *Oncogene* 31: 2989–3001.
153 Oh, S. and Janknecht, R. (2012). *Biochem. Biophys. Res. Commun.* 420: 61–65.
154 Hsia, D.A. et al. (2010). *Proc. Natl. Acad. Sci. U.S.A.* 107: 9671–9676.
155 Li, Q., Wang, H.Y., Chepelev, I. et al. (2014). Stage-dependent and locus-specific role of histone demethylase jumonji D3 (JMJD3) in the embryonic stages of lung development. *PLoS Genet.* 10 (7).
156 Yang, Z.Q., Imoto, I., Fukuda, Y. et al. (2000). Identification of a novel gene, GASC1, within an amplicon at 9p23-24 frequently detected in esophageal cancer cell lines. *Cancer Res.* 60 (17): 4735–4739.
157 Hou, J., Wu, J., Dombkowski, A. et al. (2012). Genomic amplification and a role in drug-resistance for the KDM5A histone demethylase in breast cancer. *Am. J. Transl. Res.* 4 (3): 247–256.
158 Agger, K., Miyagi, S., Pedersen, M.T. et al. (2016). Jmjd2/Kdm4 demethylases are required for expression of Il3ra and survival of acute myeloid leukemia cells. *Genes Dev.* 30 (11): 1278–1288.
159 Rose, N.R., McDonough, M.A., King, O.N.F. et al. (2011). Inhibition of 2-oxoglutarate dependent oxygenases. *Chem. Soc. Rev.* 40: 4364–4397.
160 Ng, S.S., Kavanagh, K.L., McDonough, M.a. et al. (2007). Crystal structures of histone demethylase JMJD2A reveal basis for substrate specificity. *Nature* 448 (7149): 87–91.
161 Bavetsias, V., Lanigan, R.M., Ruda, G.F. et al. (2016). 8-Substituted pyrido[3,4-d]pyrimidin-4(3*H*)-one derivatives as potent, cell permeable, KDM4 (JMJD2) and KDM5 (JARID1) histone lysine demethylase inhibitors. *J. Med. Chem.*
162 Westaway, S.M., Preston, A.G., Barker, M.D. et al. (2016). Cell penetrant inhibitors of the KDM4 and KDM5 families of histone lysine demethylases. 1. 3-Amino-4-pyridine carboxylate derivatives. *J. Med. Chem.* 59 (4): 1357–1369.
163 England, K.S., Tumber, A., Krojer, T. et al. (2014). Optimisation of a triazolopyridine based histone demethylase inhibitor yields a potent and selective. *MedChemComm.*
164 Toufike, K., Stafford, J.A., Marvin, J., and Wallace, M.B. *Histone Demethylase Inhibitors.*

165 Barker, M.D., Campbell, M., Diallo, H. et al. (2012). Histone demethylase inhibitors. WO2013143597, issued 2012.

166 Chen, Y.K. and Wallace, M.B. (2014). Histone dementhylase inhibitors. WO2014164708, issued 2014.

167 Thalhammer, A., Mecinovic, J., Loenarz, C. et al. (2011). Inhibition of the histone demethylase JMJD2E by 3-substituted pyridine 2,4-dicarboxylates. *Org. Biomol. Chem.* 9 (1): 127–135.

168 Chang, K.H., King, O.N., Tumber, A. et al. (2011). Inhibition of histone demethylases by 4-carboxy-2,2′-bipyridyl compounds. *ChemMedChem* 6 (5): 759–764.

169 Kruidenier, L., Chung, C., Cheng, Z. et al. (2012). A selective jumonji H3K27 demethylase inhibitor modulates the proinflammatory macrophage response. *Nature* 488 (7411): 404–408.

170 Luo, X., Liu, Y., Kubicek, S. et al. (2011). A selective inhibitor and probe of the cellular functions of Jumonji C domain-containing histone demethylases. *J. Am. Chem. Soc.* 133 (24): 9451–9456.

171 Schiller, R., Scozzafava, G., Tumber, A. et al. (2014). A cell-permeable ester derivative of the JmjC histone demethylase inhibitor IOX1. *ChemMedChem* 9 (3): 566–571.

172 Labelle, M., Boesen, T., Mehrotra, M. et al. Inhibitors of histone demethylases. WO2014053491.

173 Westaway, S.M., Preston, A.G., Barker, M.D. et al. (2016). Cell penetrant inhibitors of the KDM4 and KDM5 families of histone lysine demethylases. 2. Pyrido[3,4-*d*]pyrimidin-4(3*H*)-one derivatives. *J. Med. Chem.* 59 (4): 1370–1387.

174 Tumber, A., Nuzzi, A., Hookway, E.S. et al. (2017). Potent and selective KDM5 inhibitor stops cellular demethylation of H3K4me3 at transcription start sites and proliferation of MM1S myeloma cells. *Cell Chem. Biol.* 371–380.

175 Ruger, N., Roatsch, M., Emmrich, T. et al. (2015). Tetrazolylhydrazides as selective fragment-like inhibitors of the JumonjiC-domain-containing histone demethylase KDM4A. *ChemMedChem* 10 (11): 1875–1883.

176 Bavetsias, V., Lanigan, R.M., Ruda, G.F. et al. (2016). 8-Substituted pyrido[3,4-*d*]pyrimidin-4(3*H*)-one derivatives as potent, cell permeable, KDM4 (JMJD2) and KDM5 (JARID1) histone lysine demethylase inhibitors. *J. Med. Chem.* 59 (4): 1388–1409.

177 Kubicek, S., O'Sullivan, R.J., August, E.M. et al. (2007). Reversal of H3K9me2 by a small-molecule inhibitor for the G9a histone methyltransferase. *Mol. Cell* 25 (3): 473–481.

178 Upadhyay, A.K., Rotili, D., Han, J.W. et al. (2012). An analog of BIX-01294 selectively inhibits a family of histone H3 lysine 9 Jumonji demethylases. *J. Mol. Biol.* 416 (3): 319–327.

179 Lohse, B., Nielsen, A.L., Kristensen, J.B. et al. (2011). Targeting histone lysine demethylases by truncating the histone 3 tail to obtain selective substrate-based inhibitors. *Angew. Chem. Int. Ed.* 50 (39): 9100–9103.

180 Woon, E.C., Tumber, A., Kawamura, A. et al. (2012). Linking of 2-oxoglutarate and substrate binding sites enables potent and highly selective inhibition of JmjC histone demethylases. *Angew. Chem. Int. Ed.* 51 (7): 1631–1634.
181 Kawamura, A., Münzel, M., Kojima, T. et al. (2017). Highly selective inhibition of histone demethylases by de novo macrocyclic peptides. *Nat. Commun.* 8: 14773.
182 Leurs, U., Lohse, B., Rand, K.D. et al. (2014). Substrate- and cofactor-independent inhibition of histone demethylase KDM4C. *ACS Chem. Biol.* 9 (9): 2131–2138.
183 McAllister, T.E., England, K.S., Hopkinson, R.J. et al. (2016). Recent progress in histone demethylase inhibitors. *J. Med. Chem.* 59 (4): 1308–1329.
184 Kaniskan, H.Ü., Martini, M.L., and Jin, J. (2017). Inhibitors of protein methyltransferases and demethylases. *Chem. Rev.* doi: 10.1021/acs.chemrev.6b00801.
185 Jambhekar, A., Anastas, J.N., and Shi, Y. (2017). Histone lysine demethylase inhibitors. *Cold Spring Harb. Perspect. Med.* 7 (1).
186 King, O.N.F., Li, X.S., Sakurai, M. et al. (2010). Quantitative high-throughput screening identifies 8-hydroxyquinolines as cell-active histone demethylase inhibitors. *PLoS One* 5 (11): e15535.
187 Hopkinson, R.J., Tumber, A., Yapp, C. et al. (2013). 5-Carboxy-8-hydroxyquinoline is a broad spectrum 2-oxoglutarate oxygenase inhibitor which causes iron translocation. *Chem. Sci.* 4 (8): 3110–3117.
188 Rai, G., Kawamura, A., Tumber, A. et al. (2010). Discovery of ML324, a JMJD2 demethylase inhibitor with demonstrated antiviral activity, in Probe Reports from the NIH Molecular Libraries Program, Bethesda (MD).
189 Duan, L., Rai, G., Roggero, C. et al. (2015). KDM4/JMJD2 histone demethylase inhibitors block prostate tumor growth by suppressing the expression of AR and BMYB-regulated genes. *Chem. Biol.* 22 (9): 1185–1196.
190 Thinnes, C.C., Tumber, a., Yapp, C. et al. (2015). Betti reaction enables efficient synthesis of 8-hydroxyquinoline inhibitors of 2-oxoglutarate oxygenases. *Chem. Commun.*
191 Thinnes, C.C., England, K.S., Kawamura, A. et al. (2014). Targeting histone lysine demethylases – progress, challenges, and the future. *Biochim. Biophys. Acta* 27.
192 Jin, C., Yang, L., Xie, M. et al. (2014). Chem-seq permits identification of genomic targets of drugs against androgen receptor regulation selected by functional phenotypic screens. *Proc. Natl. Acad. Sci. U.S.A.* 111 (25): 9235–9240.
193 Thalhammer, A., Mecinovi, J., Loenarz, C. et al. (2011). Inhibition of the histone demethylase JMJD2E by 3-substituted pyridine. *Org. Biomol. Chem.* 9 (1): 127–135.
194 Horton, J.R., Engstrom, A., Zoeller, E.L. et al. (2016). Characterization of a linked Jumonji domain of the KDM5/JARID1 family of histone H3 lysine 4 demethylases. *J. Biol. Chem.* 291 (6): 2631–2646.

195 Johansson, C., Velupillai, S., Tumber, A. et al. (2016). Structural analysis of human KDM5B guides histone demethylase inhibitor development. *Nat. Chem. Biol.* 12 (JUly): 78–80.

196 Hatch, S.B., Yapp, C., Montenegro, R.C. et al. (2017). Assessing histone demethylase inhibitors in cells: lessons learned. *Epigenetics Chromatin* 10: 9.

197 Liang, J., Zhang, B., Labadie, S. et al. (2016). Lead optimization of a pyrazolo[1,5-*a*]pyrimidin-7(4*H*)-one scaffold to identify potent, selective and orally bioavailable KDM5 inhibitors suitable for in vivo biological studies. *Bioorg. Med. Chem. Lett.* 26 (16): 4036–4041.

198 Labelle, M., Boesen, T., Mehrotra, M. et al. (2014). Inhibitors of histone demethylases. WO 2014053491 A1, issued 2014.

199 England, K.S., Tumber, A., Krojer, T. et al. (2014). Optimisation of a triazolopyridine based histone demethylase inhibitor yields a potent and selective KDM2A (FBXL11) inhibitor. *Med. Chem. Commun.* 5 (12): 1879–1886.

200 Gale, M., Sayegh, J., Cao, J. et al. (2016). Screen-identified selective inhibitor of lysine demethylase 5A blocks cancer cell growth and drug resistance. *Oncotarget* 7 (26): 39931–39944.

201 Heinemann, B., Nielsen, J.M., Hudlebusch, H.R. et al. (2014). Inhibition of demethylases by GSK-J1/J4. *Nature* 514 (7520): E1–E2.

202 Ntziachristos, P., Tsirigos, A., Welstead, G.G. et al. (2014). Contrasting roles of histone 3 lysine 27 demethylases in acute lymphoblastic leukaemia. *Nature* 514 (7523): 513–517.

203 Hamada, S., Suzuki, T., Mino, K. et al. (2010). Design, synthesis, enzyme-inhibitory activity, and effect on human cancer cells of a novel series of jumonji domain-containing protein 2 histone demethylase inhibitors. *J. Med. Chem.* 53 (15): 5629–5638.

204 Itoh, Y., Sawada, H., Suzuki, M. et al. (2015). Identification of Jumonji AT-rich interactive domain 1A inhibitors and their effect on cancer cells. *ACS Med. Chem. Lett.* 6: 665–670.

205 Suzuki, T., Ozasa, H., Itoh, Y. et al. (2013). Identification of the KDM2/7 histone lysine demethylase subfamily inhibitor and its antiproliferative activity. *J. Med. Chem.* 56 (18): 7222–7231.

206 Rose, N.R., Woon, E.C.Y., Tumber, A. et al. (2012). Plant growth regulator daminozide is a selective inhibitor of human KDM2/7 histone demethylases. *J. Med. Chem.* 55 (14): 6639–6643.

207 Wang, L., Chang, J., Varghese, D. et al. (2013). A small molecule modulates Jumonji histone demethylase activity and selectively inhibits cancer growth. *Nat. Commun.* 4 (May): 2035.

208 Dalvi, M.P., Wang, L., Zhong, R. et al. (2017). Taxane-platin-resistant lung cancers co-develop hypersensitivity to JumonjiC demethylase inhibitors. *Cell Rep.* 19 (8): 1669–1684.

209 Thirstrup, K., Christensen, S., Møller, H.A. et al. (2011). Endogenous 2-oxoglutarate levels impact potencies of competitive HIF prolyl hydroxylase inhibitors. *Pharmacol. Res.* 64 (3): 268–273.

210 Joberty, G., Boesche, M., Brown, J.A. et al. (2016). Interrogating the druggability of the 2-oxoglutarate-dependent dioxygenase target class by chemical proteomics. *ACS Chem. Biol.* 11 (7): 2002–2010.
211 Sanchez, R. and Zhou, M.M. (2011). The PHD finger: a versatile epigenome reader. *Trends Biochem. Sci.* 36 (7): 364–372.
212 Lu, R. and Wang, G.G. (2013). Tudor: a versatile family of histone methylation "readers". *Trends Biochem. Sci.* 38 (11): 546–555.
213 Su, Z., Wang, F., Lee, J.-H. et al. (2016). Reader domain specificity and lysine demethylase-4 family function. *Nat. Commun.* 7: 13387.
214 Yu, L., Wang, Y., Huang, S. et al. (2010). Structural insights into a novel histone demethylase PHF8. *Cell Res.* 20 (2): 166–173.
215 Horton, J.R., Upadhyay, A.K., Hashimoto, H. et al. (2011). Structural basis for human PHF2 Jumonji domain interaction with metal ions. *J. Mol. Biol.* 406 (1): 1–8.
216 Torres, I.O., Kuchenbecker, K.M., Nnadi, C.I. et al. (2015). Histone demethylase KDM5A is regulated by its reader domain through a positive-feedback mechanism. *Nat. Commun.* 6: 6204.
217 Zhang, Y., Yang, H., Guo, X. et al. (2014). The PHD1 finger of KDM5B recognizes unmodified H3K4 during the demethylation of histone H3K4me2/3 by KDM5B. *Protein Cell* 5 (11): 837–850.
218 Wagner, E.K., Nath, N., Flemming, R. et al. (2012). Identification and characterization of small molecule inhibitors of a plant homeodomain finger. *Biochemistry* 51 (41): 8293–8306.
219 Miller, T.C.R., Rutherford, T.J., Birchall, K. et al. (2014). Competitive binding of a benzimidazole to the histone-binding pocket of the pygo PHD finger. *ACS Chem. Biol.* 9 (12): 2864–2874.
220 Ali, F., Yamaguchi, K., Fukuoka, M. et al. (2016). Logical design of an anti-cancer agent targeting the plant homeodomain in Pygopus2. *Cancer Sci.* 107 (9): 1321–1328.
221 Ali, M., Daze, K.D., Strongin, D.E. et al. (2015). Molecular insights into inhibition of the methylated histone-plant homeodomain complexes by calixarenes. *J. Biol. Chem.* 290 (38): 22919–22930.
222 Huang, Y. (2006). Recognition of histone H3 lysine-4 methylation by the double tudor domain of JMJD2A. *Science* 312 (5774): 748–751.

12

Histone Acetyltransferases: Targets and Inhibitors

Gianluca Sbardella

University of Salerno, Department of Pharmacy, Via Giovanni Paolo II, 132, 84084 Fisciano Salerno, Italy

12.1 Introduction

After 50 years from its discovery [1], lysine acetylation is far from being thoroughly understood and remains an intriguing topic for study. Among the other posttranslational modifications (PTMs), protein acetylation has been identified as one of the key mechanisms in the regulation of biological functions and is involved in several processes that keep a cell working properly, from transcriptional regulation to metabolic functions.

Acetylation is a dynamic process that involves the transfer or removal of acetyl groups to ε-amino group of lysine residues and is mediated by acetyltransferase and deacetylase enzymes, respectively. Because of the broad acceptance of substrates, histone acetyltransferases and deacetylases are hence often referred to as lysine acetyltransferases (KATs) and lysine deacetylases (KDACs), respectively [2]. KATs catalyze the covalent attachment of acetyl groups to lysine residues of histones and other proteins by using acetyl-coenzyme A (Ac-CoA) as a cofactor. Conversely, KDACs catalyze the amide hydrolysis of acetylated lysine. The attachment of acetyl groups to lysine residues neutralizes the positive charge of physiologically protonated ε-amino groups, resulting in altered electrostatic as well as steric properties of the affected protein region. Moreover, acetylation serves as a mark for distinct "reader" domains, which comprise specialized tertiary structures (e.g. bromodomains) in proteins that undergo a selective interaction with acetylated lysines [3]. Lysine acetylation chemically blocks other modifications, such as methylation or ubiquitination, for example, which can in turn increase protein stability, alter subcellular localization, or change the spectrum of interacting proteins. As such, acetylation provides a rich regulatory "switch." Interestingly, other lysine modifications such as propionylation, butyrylation, crotonylation, malonylation, succinylation, and myristoylation recently described as histone marks have been commonly linked to some of the existing lysine acetyltransferases [4]. Consequently, it might be appropriate to expand KATs more generally as "lysine acyltransferases" [5].

The most studied group of proteins that undergo acetylation is histones. All histones, including the four canonical histones, their variants, and linker

Epigenetic Drug Discovery, First Edition. Edited by Wolfgang Sippl and Manfred Jung.

histones, can undergo acetylation, and this was first observed as early as the 1960s [6]. Indeed, histones have served as a model system for the study of acetylation and other PTMs, partially because of their abundance in cells and their high sequence conservation among eukaryotes [7]. Together with other epigenetic PTMs (e.g. phosphorylation, methylation, ubiquitinylation, SUMOylation, ADP-ribosylation), acetylation serves as a regulating tool to control transcriptional activity of specific chromatin loci [8]. Upon acetylation, local affinity of the modified histone protein to negatively charged DNA is decreased, resulting in a less condensed chromatin structure and in exposure of promoter sites. As a consequence of the increased accessibility, the DNA globally becomes more prone to access of the transcriptional machinery [9]. In addition, transcription factors and other regulatory elements are recruited in a modification-specific manner to the relaxed promoter locus by interaction of specialized reader domains with acetyl-lysine moieties [3, 10]. Thus, KATs and histone acetylation are functionally linked with the control of transcription activation, replication, and DNA damage repair [11]. Several mass spectrometry-based studies of proteome-wide lysine acetylation (the "acetylome") have identified thousands of sites in more than 1500 proteins, spanning many important cellular pathways, including chromatin remodeling, cell cycle, splicing, nuclear transport, and actin nucleation [12]. The nucleocytosolic lysine acetylation levels within the cellular proteome are dynamically regulated by the concerted activity of KATs and KDACs, and the acetylation equilibrium is adjusted in response to cellular stimuli like autoacetylation, protein–protein interactions (PPIs), phosphorylation, or the cellular acetyl-CoA level, leading to altered gene transcription and subsequently to phenotype adaption [13]. In addition, lysine acetylation also occurs on a high proportion of mitochondrial proteins [12d]. Mitochondrial protein acetylation is functional and regulates the activity of many metabolic enzymes including those involved in fatty acid oxidation and ketone body production [14]. However, while the enzymatic basis of the deacetylation of N-acetylated lysine residues in mitochondria by SIRT3 is established [15], the mechanisms regulating mitochondrial acetylation are less clear. Although a mitochondrial *N*-acetyltransferase (GCN5L1; see after) has been proposed [16], it is possible that lysine acetylation is largely an uncatalyzed reaction, whereby unprotonated lysine side chains react with the thioester of acetyl-CoA [4c, 17].

Aberrant acetylation levels have been connected with a diversity of disease phenotypes including cancer, neurological disorders, and cardiovascular and metabolic malignancies [18]. Hence, KAT enzymes seem to be deeply involved in the manifestation and progression of such diseases, and therefore the elucidation of their precise mechanism is required to assess their potential as possible drug targets, and small-molecule modulators are of high interest for probing these pathways and as potential drugs.

12.2 Acetyltransferase Enzymes and Families

KAT enzymes could be classified using different classifications on the basis of their subcellular localization or on the basis of their structural homology and/or catalytic mechanism. Traditionally, they are divided into two different classes, type A and type B [19]. Type A KATs comprise a heterogeneous group of proteins that are located in the nucleus and are involved in the regulation of gene expression through acetylation of nucleosomal histones in the context of chromatin [18e]. They contain a bromodomain, which helps them recognize and bind to acetylated lysine residues on histone substrates. Many members of type A KATs cooperate with activators to enhance transcription. On the other hand, type B proteins are located in the cytoplasm and are responsible for acetylating newly synthesized histones prior to their assembly into nucleosomes, to facilitate translocation into the nucleus, where the histone proteins are deacetylated and subsequently incorporated into chromatin fibers. These KATs lack a bromodomain, as their targets are unacetylated. Hat1 (KAT1), the first KAT to be isolated and cloned (from *Saccharomyces cerevisiae* lysates) [20], is the only type B KAT that is highly conserved through evolution in eukaryotes [18e, 21]. It acetylates free H4 protein on lysine residues 5 and 12 (but not in a nucleosome), in humans also on H2A lysine residue 5. Together with the WD40 protein Hat2, Hat1 forms the Hat1/2 complex, which exhibits 10-fold increased activity compared with native Hat1. Interestingly, Hat1 was not solely observed in the cytoplasm but also in the nucleus, thus suggesting a shuttling between different cell compartments [22]. Other members of type B KATs have been uncovered in mitochondria [16, 23], Golgi apparatus [24], and endoplasmic reticulum [25]. Recent data strongly suggest that Nα-acetyltransferase 60 (Naa60) (also known as Hat4 [24]), which is located in the Golgi apparatus where it catalyzes the acetylation of H4 protein on position 79 and 91, is predominantly an N-terminal acetyltransferase (NAT) with the same globular fold and similar catalytic mechanism as all other NAT enzymes characterized thus far [26]. Other type B KATs, like Rtt109 and HatB3.1, have been found in yeast [21c].

Despite this historical classification, some KAT proteins function in multiple complexes or locations and would thus not easily fit into a particular class [11].

A more appropriate classification is based on sequence and structural similarities in the catalytic domain. On this basis, KATs are grouped into five major families (Figure 12.1): general control non-repressible 5 (GCN5)-related *N*-acetyltransferase (GNAT), CREB-binding protein/E1A-associated protein of 300 kDa (CBP/p300), MOZ, Ybf2/SAS3, something about silencing protein 2 (SAS2), Tip60 (MYST), nuclear receptor co-activators, and basal transcription factors [18e, 21a].

12.3 The GNAT Superfamily

The GNAT family is evolutionarily conserved from bacteria to mammals. This family includes both A-type (KAT2A/GCN5 [27], KAT2B/PCAF [28], KAT9/ELP3 [29], KAT10/Hpa2/3 [30], and Nut1 [31]) and B-type KATs (KAT1/Hat1) [20], as well as KATs that reside in other subcellular compartments (GCN5L1, ATAT1, Hat4/Naa60, NAT8, and NAT8B). Altogether, these KATs control a variety of cellular process, such as transcriptional activation [27, 32],

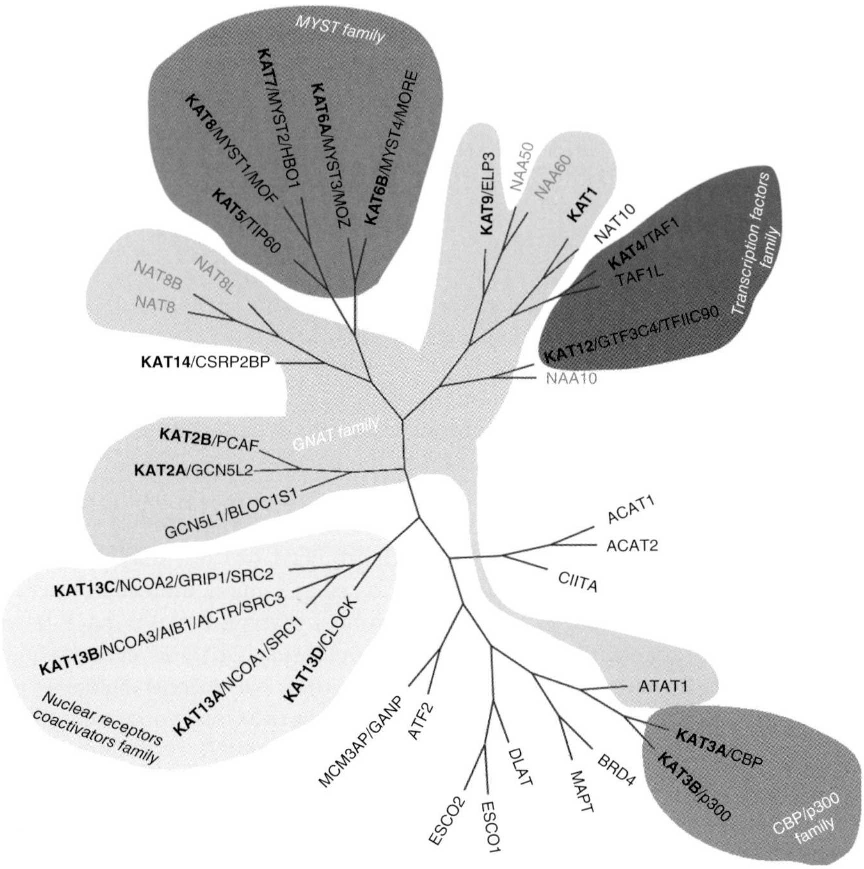

Figure 12.1 Phylogenetic tree of lysine acetyltransferases. Contours in different shades of gray indicate major families. Most common names of the enzymes are reported in bold. N-terminal acetyltransferases are reported in gray.

transcript elongation [29], histone deposition [33], DNA repair [24, 34], and microtubule stability [35]. A characteristic feature of this enzyme family is the presence of up to four sequence motifs (A–D) in the catalytic core [36], with motif C being almost exclusively found in GNAT family enzymes. The A motif is the most highly conserved region, and it contains an acetyl-CoA binding domain that is defined by an Arg/Gln-X-X-Gly-X-Gly/Ala segment [37]. This sequence was also found in other KAT families. While the catalytic domains of KATs show high structural homology within a certain family, enzymes differ in their N- and C-terminal regions, which are responsible for recognition, positioning, and binding of substrates. Structural motifs within these regions, like bromo- and chromodomains, zinc binding moieties, and cysteine/histidine-rich modules (plant homology domain (PHD), TAZ, ZZ), promote the target structure diversity [38]. In addition, KAT2A/GCN5 and KAT2B/PCAF contain a carboxy-terminal bromodomain that associates with acetylated lysine residues [39].

12.3.1 KAT2A/GCN5 and KAT2B/PCAF

Originally discovered in yeast as a transcriptional co-activator [40], the KAT2A/GCN5 orthologue in *Tetrahymena thermophila* was later discovered by Brownell and Allis to have acetyltransferase activity [27, 41]. Therefore, this enzyme was the first transcription-related KAT identified in eukaryotes, and its identification established a molecular link between histone acetylation and the regulation of gene expression, as well as provided a foundation for the discovery of additional KATs, including the closely related family member KAT2B, also referred to as p300/CBP-associated factor (PCAF) [28].

KAT2A is an example for substrate discrimination by the structural motifs that are a typical feature of GNAT family. Human enzyme can be distinguished from its yeast homologue by a 400-residue N-terminal region. Under *in vitro* conditions, recombinant human KAT2A acetylates free histone H3 on lysine 14 and, to a minor degree, histone H4 on lysines 8 and 16. Noteworthy, in contrast to yeast protein, which targets the same substrate residues, the human homologue is capable of acetylating histones in a nucleosomal context, indicating the role of the N-terminal region in substrate recognition [11].

There is evidence that the two KATs catalyze lysine acetylation by a ternary complex mechanism (Figure 12.2). A conserved glutamic acid (KAT2A: Glu-173; KAT2B: Glu-570) in the active site of these enzymes can serve as a general base that facilitates nucleophilic attack on the Ac-CoA thioester by deprotonating the positively charged lysine residue [42]. This mechanism is supported by a kinetic study on KAT2B showing that the enzyme follows a compulsory-order ternary complex mechanism in which Ac-CoA binds first to the enzyme followed by the histone substrate [43]. The current consensus is that all the enzymes of the GNAT family catalyze lysine acetylation through a compulsory-order ternary complex mechanism [44].

KAT2B is a multidomain protein that harbors an HAT and E3 ubiquitin ligase domains as well as a C-terminal bromodomain that may associate with KAT3A and KAT3B [28, 45]. While the roles of the acetyltransferase and the E3 ubiquitin ligase activities have been shown to be required for cell proliferation and apoptosis [28, 45b, 46], the specific function of the PCAF bromodomain in cellular processes is still mostly unknown, yet it probably causes the loss of PCAF acetylation function [47].

Human KAT2A and KAT2B share about 80% sequence similarity, and both exhibit site preference for *in vitro* H3K14 acetylation within free or nucleosomal histones. *In vivo*, they both reside within two major native multisubunit complexes, Spt-Ada-Gcn5 acetyltransferase (SAGA) [48] and ADA-Two-A containing (ATAC) [49]. Incorporation into these complexes enhances KAT activity and has long been suggested to expand lysine target specificity [50] and the functionality of KAT2A and KAT2B. However, recent work indicates that while the SAGA and ATAC increase the catalytic efficiency of human KAT2A, by approximately 10- and 6-fold, respectively, the specificity of KAT2A-mediated acetylation on histone octamers does not change when the enzyme is alone or integrated into these complexes [51]. H3K14 still remains the primary acetylation site *in vitro*, while H3K9, H3K23, H3K27, H3K36, H4K5, and H4K8 residues

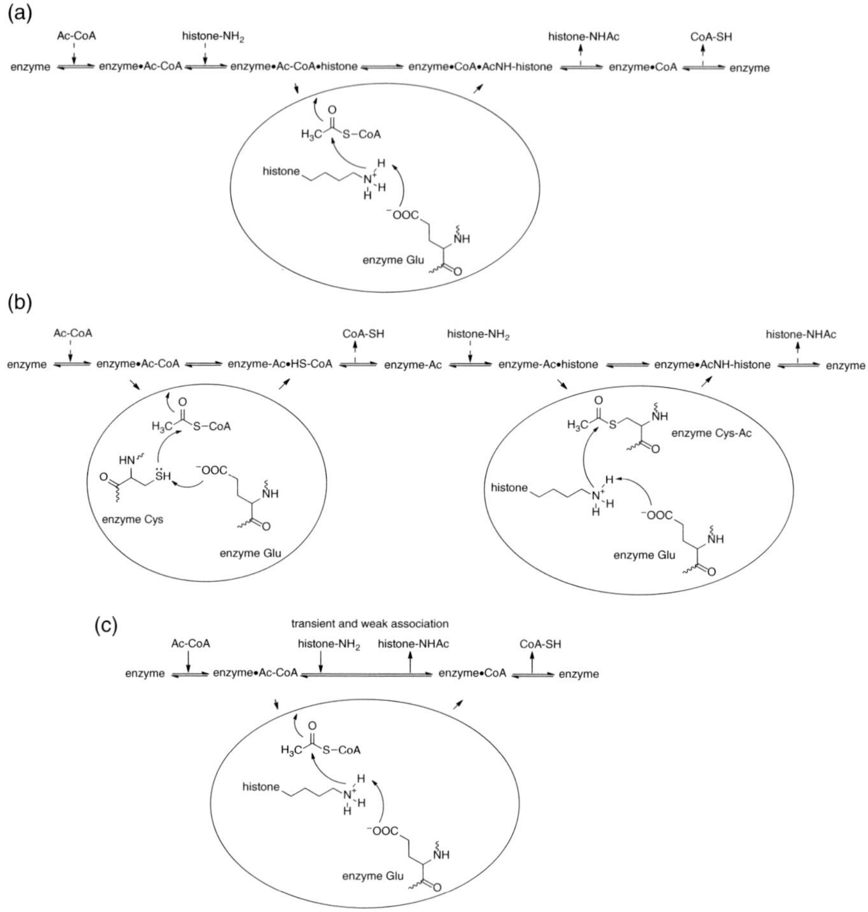

Figure 12.2 Catalytic mechanisms of KATs. In a *compulsory-order ternary complex mechanism* (a), a ternary complex is formed, but acetyl-coenzyme-A (Ac-CoA) has to bind first before the other substrate (histone or other protein) can bind. Catalysis depends on the presence of a general base, such as glutamic acid, which facilitates the nucleophilic attack on the Ac-CoA thioester by deprotonating the lysine residue; in a *ping-pong mechanism* (b), Ac-CoA binds first, and the acetyl group is transferred to an amino acid in the catalytic site of the enzyme. CoA leaves the enzyme and subsequently the substrate binds, to which the acetyl group is transferred. Besides a general base, this mechanism requires also a second amino acid in the catalytic site of the enzyme suitable for accepting the acetyl group, which is commonly a cysteine; in the *Theorell–Chance mechanism* (c), there is no stable ternary complex. Ac-CoA binds first and subsequently, the peptide substrate binds weakly to the enzyme, allowing the lysine to react with the acetyl group. However, kinetically only the interaction with Ac-CoA is important.

are acetylated to a lesser extent by KAT2A [51]. Beyond promoting catalytic activity, the SAGA and ATAC complexes regulate different functions due to their structural modularity. The SAGA complex contains four functional units that sophisticatedly control transcriptional activation, telomere maintenance [52], mRNA export [53], and DNA repair [54]. One module consists of the acetyltransferase unit (Gcn5, Ada2b, Ada3, and Sgf29) that harbors KAT activity and facilitates SAGA recruitment to H3K4me2/3 sites via the tandem Tudor domains of Sgf29 [55] and/or to acetyl-lysine residues through the bromodomain of KAT2A [56]. A second enzymatic module, called the deubiquitination module (Usp22, Atxn7, Atxn7L3, and Eny2), promotes H2BK120 deubiquitination during transcription [57]. The Spt module (Trrap, Spt3, Spt20, Staf42, and Staf65γ) mediates interactions with TATA-binding protein (TBP) [58] and various transcription factors, including Myc and E2f1 [59]. The TBP-associated factor (Taf) module (Taf5l, Taf6l, Taf9, Taf10, and Taf12) further connects SAGA to the general transcription machinery [60].

In comparison with the SAGA complex, ATAC similarly contains a slightly modified KAT module (Gcn5, Ada2a, Ada3, and Sgf29) but lacks many of the general transcription factors present in SAGA. Furthermore, ATAC houses a second KAT activity, which is mediated by Atac2, and contains Yeats2, which is a reader of both acetyl-lysine and crotonyl-lysine modifications [61]. It is worth mentioning that a recent report suggested that KAT2B is a dimer in its functional ATAC complex [62]. The different subunit compositions of SAGA and ATAC reflect their disparate functions, as SAGA primarily associates with gene promoters, while ATAC is found at both promoters and tissue-specific enhancers [63]. The SAGA and ATAC complexes also respond to different stimuli and activate distinct subsets of inducible genes [64]. Additionally, SAGA has been described as a general transcriptional co-activator complex [65], although it is clearly recruited to specific gene promoters in embryonic stem cell (ESCs) [66]. SAGA and ATAC further control distinct biological processes via non-histone acetylation, such as ATAC-specific acetylation of cyclin A, which promotes mitotic progression [67].

In agreement with the wide functionality of both the SAGA and ATAC complexes, KAT2A is required for normal embryonic development in mice [68].

12.3.2 KAT1/Hat1

More commonly referred to as Hat1, KAT1 is a cytoplasmic KAT enzyme that acetylates newly synthesized free histones during chromatin assembly [33]. Together, KAT1 and histone-binding protein RbAp46 form the HAT-B complex [69]. Following new histone synthesis, HAT-B associates with somatic nuclear autoantigenic sperm protein (sNASP)-bound H3–H4 dimers to acetylate H4K5 and H4K12 [70]. These modified histones are then transported to the nucleus for *de novo* chromatin assembly and subsequently deacetylated during chromatin maturation [71]. In addition to functioning within the cytosol, KAT1 is also present in the nuclear compartment [72]. While the nuclear function of KAT1 is still poorly understood, studies performed in yeast suggest that this protein functions as part of the nuclear type B HAT specific for H4 (NuB4) complex, containing KAT1, RbAp46, sNASP, and H3–H4 dimers, to control histone deposition and/or DNA repair-based chromatin reassembly [21b]. Interestingly, the Hat4/Naa60 GNAT family member also regulates many of these processes, but instead acetylates free histone H4 on K20, K79, and K91 residues [24]. KAT1 plays an essential role during embryonic development [73].

12.3.3 GCN5L1

One of the hypotheses to explain the mechanism underlying the acetylation of mitochondrial proteins is the existence of mitochondrial acetyltransferases [4f]. Candidate mitochondrial KATs promoting acetylation of ATPase8, a protein encoded and produced wholly in the mitochondria, were identified by searching a mitochondrial protein database for proteins with homology to KAT2A (one of the first characterized KAT enzymes) through a bioinformatics approach. Then an siRNA screen was performed to assess candidate mitochondrial KATs for their ability to increase oxygen consumption to identify factors, since acetylation of ATPase8 is associated with decreased oxygen consumption, and, therefore, knockdown of a mitochondrial KAT would have been expected to increase oxygen consumption. The mitochondrial-enriched GCN5-like 1 (GCN5L1) protein (also known as BLOC1S1) emerged as the only candidate protein whose knockdown promoted this effect [16]. As can be inferred from the gene name, the weak homology of GCN5L1 to yeast KAT2A (Gcn5) had been noted in an early study [74]. However, the primary biological role attributed to the protein had previously been its involvement in the synthesis of lysosome-related organelles. Interestingly, while isolated GCN5L1 exhibited sluggish KAT activity, acetylation was strongly augmented by the addition of mitochondrial extracts. This indicates GCN5L1 may only manifest its acetylation activity when other members of its endogenous protein complex are present. Since these initial studies, GCN5L1 has also been reported to be involved in the regulation of mitophagy [75]. It has been suggested that mitochondrial GCN5L1 modulates posttranslational control of Forkhead box protein O1 (FOXO1), a transcription factor that plays important roles in regulation of gluconeogenesis and glycogenolysis and is also central to the decision for a pre-adipocyte to commit to adipogenesis. Therefore, GCN5L1 is supposed to regulate gluconeogenesis and controls metabolic pathways via mitochondrial ROS-mediated ERK activation [76]. Exploring mechanisms underpinning GCN5L1-mediated ROS signaling may expand the understanding of the role of mitochondria in gluconeogenesis control. However, due to the challenges of reconstituting its acetyltransferase activity, the biochemical substrates of GCN5L1, as well as the degree to which they overlap with the Sirt3 deacetylation program, remain an open question. Also, given the limited molecular characterization of GCN5L1, the question remains as to whether other mitochondrial KATs may exist.

12.4 KAT3A/CBP and KAT3B/p300 Family

The CBP/p300 family is composed of the two paralogues p300 and CBP (KAT3A and KAT3B, respectively). They were first described as binding partners of the adenovirus early region 1A (E1A) and the cAMP-regulated enhancer (CRE)-binding protein, respectively [77], but it was later demonstrated that these two proteins contain a well-conserved acetyltransferase domain [45] and a number of protein interaction domains that facilitate binding with over 400 proteins and promote many non-histone acetylation events [78]. It has been suggested that the KAT3A/3B-mediated transfer of acetyl groups occurs through a "hit-and-run" (Theorell–Chance) catalytic mechanism (Figure 12.2), which is characterized by stable binding of the cofactor acetyl-CoA followed by transient and rather weak association of the substrate peptide to the enzyme, allowing enough time for the Lys side chain to snake through the enzyme tunnel and react with the acyl group [79]. This catalytic mechanism is distinct from the GNAT and MYST KAT families, which use a ternary complex mechanism [80], and may contribute to the broad substrate acceptance of KAT3A/3B.

In addition to the HAT domain, both proteins contain a CREB binding domain, several zinc finger domains, a PHD, and a bromodomain, one of the most highly conserved domains in KAT3A and KAT3B (96% sequence identity) that bind acetylated Lys residues [47b, 81]. The crystal structure of a section of KAT3B that includes the bromodomain and the HAT domain flanking a middle cysteine/histidine-rich (CH) region 2 (CH2) containing RING and PHD domains revealed that such four domains adopt an assembled configuration with the RING domain positioned over the HAT substrate binding pocket, thus showing the potential for allosteric interactions to modulate KAT3B HAT activity [82]. Structural and biochemical results have showed that the KAT3A bromodomain, CH2, and ZZ domains, which flank the acetyltransferase domain, regulate acetyltransferase activity and also promote SUMOylation of the adjacent CRD1 cell cycle regulatory domain [83].

Given their high sequence and structural similarity, KAT3A/CBP and KAT3B/p300 generally function in an analogous manner yet still modulate distinct processes. Both proteins act as transcriptional co-activators, which are incapable of directly binding to DNA and are hence recruited via interactions with sequence-specific transcription factors [84]. *In vitro*, recombinant KAT3A and KAT3B readily acetylate each of the four core histones in nucleosomes without the assistance of accessory factors [45b, 85] and appear to preferentially acetylate H3K18 and H3K27 residues *in vivo* [86]. KAT3A and KAT3B also contribute to H3K56ac, a modification that is elevated in both ESCs and multiple forms of cancer [87]. Furthermore, the two proteins dually function as crotonyltransferase enzymes that deposit crotonyl moieties on histones to activate transcription [88]. At the genomic level, KAT3A and KAT3B binding is enriched at both promoters and enhancers [89]. However, KAT3A/3B binding and H3K27ac most notably mark active enhancers during early development that drive transcription programs associated with cell and tissue specification [90]. Similarly, KAT3B has also been linked to super-enhancers in mESCs [91], defined as enlarged enhancer regions densely co-bound by the mediator transcriptional co-activator complex and pluripotency regulators Oct4, Sox2, and Nanog [92]. KAT3A and KAT3B are largely co-expressed during mouse embryogenesis and therefore share certain developmental functions. Both are required for normal embryonic development [93]. Interestingly, both KAT2A and KAT2B interact with KAT3A and KAT3B. These two KAT families share both distinct and overlapping functions during development, as illustrated by the finding that about 25% of embryos that carry one null allele of both KAT3A and KAT3B die, even though embryos that are heterozygous for either null allele alone are viable [94].

Interestingly, while being intensively studied in mammals, plant orthologues of KAT3A/3B have also been found and characterized, suggesting a fundamental functional implication of this enzyme family in all metazoan organisms [95].

In addition to diverse substrates, an early study demonstrated that KAT3B is able to utilize propionyl- and butyryl-CoA as histone acylation cofactors *in vitro* [4b]. This property is shared by KAT3A, KAT2A, and KAT2B [4d, 4e]. More recently, it has been reported that KAT3B is also capable of transferring succinyl- and glutaryl-CoA acyl units to a histone H4 peptide substrate [96]. The transfer of these longer acyl chains occurs more slowly than acetylation [4d]. While such studies are in their early stages, an emerging hypothesis is that the cofactor promiscuity exhibited by KAT3B and other KATs may allow them to act as metabolic sensors, relaying signals about cellular acyl-CoA levels to histone or non-histone proteins to mediate their activity [97].

12.5 MYST Family

Named after the initially identified members (MOZ, Ybf2, SAS2, and Tip60) [19, 98], the MYST family actually includes five mammalian members: Tat-interacting protein of 60 kDa (Tip60) (KAT5/HTATIP), males absent on the first (MOF) (KAT8/MYST1), monocytic leukemia zinc finger protein (MOZ) (KAT6A/MYST3), MOZ-related factor (MORF) (KAT6B/MYST4), and histone acetyltransferase binding to ORC1 (HBO1) (KAT7/MYST2). These enzymes show sequence similarities and contain a well-conserved MYST domain that includes a C2HC zinc finger and an acetyl-CoA binding motif homologous to that found in GNAT family members [36, 99]. Furthermore, individual members harbor specialized domains that bind modified histones, including PHD and chromodomains [100]. Similar to the GNAT family, the MYST KATs also function in macromolecular complexes and regulate a wide variety of biological and developmental processes. The *in vivo* activity and specificity of almost all MYST family enzymes is highly determined by the composition of their protein complexes. Different catalytic mechanisms have been described for MYST family proteins (see below) [44].

12.5.1 KAT5/Tip60

KAT5/Tip60 was the first reported human member of the MYST family with preference for H4 acetylation. Transcriptional activation and DNA damage response are mediated by the acetyltransferase activity of this enzyme [101]. In mammals, the enzyme assembles into the multisubunit TIP60 complex. This complex contains at least 16 proteins and has two enzymatic platforms, including Tip60 acetyltransferase activity that drives H2A and H4 acetylation and p400 ATP-dependent chromatin remodeling activity that deposits the H2A.Z histone variant into chromatin [102]. Functionally, the TIP60 complex primarily associates with active promoters, via binding to proximal promoter R-loops and various transcription factors, including Myc, E2f1, and β-catenin [103]. As well, the TIP60 acetyltransferase activity functions in homologous-recombination-based repair of DNA double-strand breaks [102, 104]. In agreement with these observations, multiple members of the TIP60 complex are vital regulators of normal embryogenesis and/or ESC regulation in mice [105].

12.5.2 KAT6A/MOZ, KAT6B/MORF, and KAT7/HBO1

The MYST acetyltransferases KAT6A and KAT6B have similar structural organization and independently form MOZ/MORF tetrameric complexes, containing inhibitor of growth 5 (Ing5), homologue yeast of Esa1-associated factor 6 (Eaf6), and either bromodomain- and PHD finger-containing protein (Brpf) 1/2 or 3 paralogues [106]. Likewise, the KAT7 complex includes the HBO1 acetyltransferase, Ing4/5, Eaf6, and either Brpf1/2/3 or gene for apoptosis and differentiation in epithelia 1/2/3 (Jade1/2/3) [106, 107]. Both KAT6 complexes primarily acetylate H3K14 and function as co-activators for Runx and p53 transcription factors [106], while the Brpf-containing HBO1 complex targets H3K14/K23 acetylation, and the Jade-containing HBO1 complex mediates H4K5/K8/K12 acetylation [108]. As for almost all the members of MYST family, the *in vivo* activity and specificity of these enzymes is highly determined by the composition of their protein complexes. KAT7 interacts with subunits of the origin of replication complex, consequently playing a functional role in DNA replication [109]. The isolated complex acetylates histones H4 at positions 5, 8, and 12 and less effectively H3, whereas recombinant KAT7 shows no significant acetyltransferase activity, implicating PPI and *in vivo* modifications in efficient substrate recognition and turnover [109]. In the case of KAT6B, while recombinant and full-length enzyme catalyzes acetylation of H4 and H3 proteins, under physiological condition solely histone H3 is targeted [110].

12.5.3 KAT8/MOF

KAT8/MOF was originally identified in *Drosophila* as a regulator of dosage compensation, leading to H4K16ac and hyperactivation of the single male X chromosome [111]. However, it is still unclear whether the mammal orthologue manages similar dosage compensation effects. In humans, KAT8 assembles into two primary KAT complexes: the highly conserved male-specific lethal (MSL) multiprotein complex that specifically acetylates H4K16 [112] and the nine subunit nonspecific lethal (NSL) complex that targets H4K5, K8, and K16 acetylation [113]. A kinetic study on the recombinant catalytic domain showed for KAT8 a pattern consistent with a ping-pong mechanism (Figure 12.2), in which Ac-CoA binds first and the acetyl group is transferred onto a residue in the active site of the enzyme. Then CoA leaves the enzyme and subsequently the substrate binds, to which the acetyl group is transferred. Besides a general base, this mechanism requires also a second amino acid in the catalytic site of the enzyme suitable for accepting the acetyl group, which is commonly a cysteine [114]. For KAT8, the subsequent binding of Ac-CoA and the histone peptide was confirmed by calorimetric binding measurements [115]. The conserved glutamate residue Glu-177 can act as a general base, whereas cysteine Cys143 is capable of accepting the acetyl moiety [116]. *In vivo*, KAT8 is required for mouse embryonic development beyond the blastocyst stage [117]. Accordingly, KAT8 is also needed to maintain pluripotency and self-renewal of mESCs [118].

12.5.4 SAS2 and SAS3

Both acetyltransferases SAS2 and SAS3 (also known as Ybf2) are involved in transcriptional silencing processes in *S. cerevisiae* [119]. SAS2 is the acetyltransferase subunit of the SAS complex, a multiprotein complex that acetylates Lys-16 of histone H4 and Lys-14 of histone H3. The SAS complex is however unable to acetylate nucleosomal histones. The complex is involved in transcriptional silencing at telomeres and at HML locus as well as in rDNA silencing and G_0 control [116]. Sas3 is the catalytic component of the histone acetyltransferase NuA3 complex that acetylates Lys-14 of histone H3. Recruitment of NuA3 to nucleosomes requires methylated histone H3. In conjunction with the FACT complex, NuA3 may be involved in transcriptional regulation. *In vitro*, SAS3 acetylates free histones H3 and H4. It is involved in silencing the HMR locus [120].

12.5.5 ESA1

The yeast enzyme ESA1, which shows close homology to human KAT5 and KAT8, is the catalytic subunit of the NuA4 complex. This complex promotes H4 acetylation and subsequently stimulates transcription of genes that are essential for cell cycle progression [121]. In a study with the catalytic domain of ESA1, it was shown that Cys-304 and Glu-338 residues are both essential for enzyme activity, with Glu-338 functioning as a general base, as in GNAT family KATs [122]. A crystal structure of truncated ESA1 co-crystallized with Ac-CoA showed that the acetyl moiety of Ac-CoA had transferred from the cofactor to Cys-304, supporting a ping-pong mechanism (Figure 12.2) [123]. However, this was countered by a study showing that mutation of Cys-304, in contrast to the aforementioned study, did not impair the activity of the enzyme, and kinetic studies showed a pattern indicating catalysis via a ternary complex mechanism [124]. In this study, not the catalytic HAT domain but full-length ESA1 was used, and it was combined with two other proteins forming the piccolo NuA4 complex, which is naturally occurring in yeast. This shows that the catalytic mechanism of ESA1 was influenced by the interaction with other proteins.

12.5.6 Other KATs

In addition to the members of the families described above, there are several other proteins found typically in higher eukaryotes that exhibit acetyltransferase activity.

A few co-activators of nuclear hormone receptors (in particular SRC-1, ACTR, and TIF-2), which exhibit ligand-dependent KAT activity, serve as regulatory elements in hormone-related transcriptional processes and are grouped in a unique KAT family [125]. Human steroid receptor co-activator-1 (SRC-1) is known to interact with KAT3A/B and KAT2B, and its HAT domain is located in its C-terminal region. ACTR (also known as RAC3, AIB1, and TRAM-1 in humans) shares significant sequence homology with SRC-1, in particular in the N-terminal and C-terminal (HAT) regions as well as in the receptor and co-activator interaction domains [11]. ACTR also interacts with KAT3A/B and KAT2B. The former can prevent ACTR from binding to and activating its receptor by acetylating it in its receptor interaction domain. Transcriptional intermediary factor 2 (TIF-2) (also known as GRIP1) is another nuclear receptor co-activator with acetyltransferase activity, and it also interacts with KAT3A/B. The global regulator of circadian gene expression CLOCK shares homology to the HAT domain of ACTR and has been suggested as a member of this family [126].

Another KAT family is formed by protein components of transcription factor complexes like TAF1/TBP and TFIIIC90. These protein complexes affect transcription directly [127].

The fungal-specific lysine acetyltransferase Rtt109 (regulation of Ty1 transposition gene product 109, also known as KAT11) shows very little sequence homology to any of the other KAT family members, but its tertiary fold adopts a structure that is surprisingly similar to KAT3B [128]. However, the catalytic mechanism of Rtt109 is likely different from that of the other KATs, as it involves a ternary intermediate complex [129]. Because of these differences, the enzyme is not assigned to any of the known KAT families. Isolated Rtt109 is not competent to efficiently acetylate lysine substrates, but its activity is stimulated by association with either of two different histone chaperones, Asf1 and Vps75. Upon formation of the enzyme–chaperone complex, the activity of the catalytic domain is dramatically increased, and the substrate specificity is directed toward distinct lysine residues (H3K56 for Asf1 and H3K9,23 for Vps75) (51). Moreover, a general base has not yet been identified for this KAT [130]. Acetyltransferase activity on histone protein H4 was also demonstrated for a member of the Camello protein family, CMLO3, by bioinformatics analysis of the zebrafish genome. Yet, on the basis of their structural divergence and their perinuclear localization, Camello proteins are generally placed outside of any of the other KAT families [131].

It is worth noting that besides histone acetyltransferases that also accept non-histone substrates, specific non-histone acetyltransferases also exist, even if most of them have not been identified yet [12b]. Among the identified ones there are the α-tubulin acetyltransferase αTAT1 [132] and the cohesin acetyltransferase Eco1/ESCO1 [133].

12.6 KATs in Diseases

Being generally linked to activation of transcriptional activity, lysine acetylation of histone and non-histone proteins affects crucial physiological processes within an organism. It appears to have a fundamental role not only in the nucleus but also in the regulation of different cytoplasmic processes, including cytoskeleton dynamics, energy metabolism, endocytosis, autophagy, and signaling from the plasma membrane [134]. Moreover, reversible lysine acetylation may alter enzymatic activity to allow cells to respond to environmental changes in metabolic demands, and it has been proposed to be an evolutionarily conserved mechanism for the regulation of cellular functions [12d, 135]. Hence, aberrations of the acetylation equilibrium induce the manifestation and progression of pathological phenotypes and have been related to a variety of diseases and/or pathological cellular states [136]. Such aberrations could be either due to altered activity of the responsible enzymes or because of changes in their expression levels. The role of distinct KAT subtypes in diseases like cancer, neurodegenerative disorders, viral and parasitic infections, inflammation, and metabolic and cardiovascular malignancies have been extensively investigated [137].

KAT3A and KAT3B are global co-activators of gene transcription and involved in multiple cellular processes. Point mutation and microdeletion of the KAT3A gene (rarely KAT3B) result in Rubinstein–Taybi syndrome (RTS), which is characterized by a short stature, moderate to severe intellectual disability, distinctive facial features, and broad thumbs and first toes [137a, 138]. Both acetyltransferases were demonstrated to be involved in hematopoietic homeostasis, such that mutations in the KAT3A/3B interaction domain of different transcription factors were found in hematologic malignancies [139] and chromosomal translocations involving KAT3A or KAT3B genes are associated with leukemia and lymphomas [98, 140]. Elevated KAT3B expression levels have been associated with several types of cancer, including esophageal squamous cell carcinoma (ESCC) [141], hepatocellular carcinoma (HCC) [142], and melanoma. In the latter case, downregulation of the enzyme catalytic activity retards cell cycle progression in human melanocytes by activating replicative senescence [143]. It is reported that KAT3A/3B interaction with c-Myb facilitates myeloid differentiation block and is required for acute myeloid leukemia (AML) induction [144]. A characteristic feature of this hematologic malignancy is the presence of KAT activity-containing fusion proteins in consequence of chromosomal translocations. Fusion products of KAT3A/3B with KAT6A, KAT6B, or MLL exhibit aberrant KAT activity and substrate specificity and hence lead to abnormal transcription activation [18f]. The AML1/ETO oncogene is acetylated by and co-localized with KAT3B at specific promoter regions, which was found to be essential for leukemogenesis [145]. KAT3A and KAT3B promote prostate cancer progression by activating androgen receptor-regulated transcription [146] and colon cancer progression by microsatellite instability [147] and are involved in the development of drug resistance [148]. KAT3B acetylates the human immunodeficiency virus (HIV) protein integrase after viral infection of the cell, and this acetylation is crucial for incorporation of virus DNA into the cells' genome [149]. Upon integration, KAT3A/3B interaction with the HIV-1-TAT protein promotes transcription of the HI provirus [150]. KAT3B also plays a crucial role in the regulation of human papillomavirus (HPV) genes, and it has been targeted for the inhibition of HPV-associated cervical cancer [151]. Overexpression of either KAT3A or KAT3B significantly enhances transforming growth factor-β1 (TGF-β1)-induced expression of plasminogen activator inhibitor-1 (PAI-1) and p21 in renal mesangial cells, which play a major role in the pathogenesis of diabetic nephropathy [152]. In patients with elevated glucose levels, binding of KAT3B to promoter sites is increased, which leads to upregulation of vasoactive factors and extracellular matrix proteins, suggesting a possible role of KAT3B in chronic diabetes-related complications [153]. KAT3A and KAT3B stimulate cardiac growth, and KAT3B activity is increased in agonist-induced hypertrophy of cardiomyocytes [154]. Recruitment and acetylation of specific transcription factors, such as GATA4 and MEF2, mediate elevated transcription levels of hypertrophy-related effector proteins [18g]. The acetylation of NF-κB by KAT3A/3B is associated with a loss of affinity for IκB, leading to enhanced expression of pro-inflammatory downstream gene products. In neurodegenerative diseases, reduced KAT3A activity is associated with loss of neuronal plasticity and destabilization of long-term memory, implicating KAT3A as an antitarget in this context [18d]. Recent studies showed that in the hippocampus, associated with spatial memory, KAT3A or KAT3B inhibition impairs long-term but not short-term object memory, while inhibition of KAT2B impairs memory at both delays. In perirhinal cortex, associated with object identity processing, KAT inhibition does not impair short-term memory, and only KAT3A and KAT2B inhibition disrupts long-term memory, whereas KAT3B inhibition has no effects. Therefore, distinct KATs can have specific mnemonic roles according to specific brain regions and memory timeframe [155].

Like KAT3 enzymes, the GNAT family KATs serve as regulators of transcriptional activity and are similarly implicated in fundamental physiological processes. Their activity and specificity highly depend on PPI, as these enzymes are usually part of multienzyme complexes. It was shown that KAT2A activity is crucial for cell cycle progression [156]. While KAT2A and KAT2B play important roles in the activation and stabilization of the tumor suppressor p53, KAT2B-mediated acetylation of the cyclin-dependent kinase inhibitor p27 facilitates its degradation and further leads to uncontrolled cell cycle progression [157]. KAT2B knockout hampers the invasive potential and growth rate of urothelial cancer cells [158]. Drug-resistant cancer cells exhibit elevated levels of KAT2B- and KAT2A-mediated H3K9 acetylation in promoter regions of the multidrug-resistant protein 1 (MDR1) gene [159]. Together with p300, KAT2A is competent to acetylate and thus activate HIV integrase and TAT proteins with essential impact on the HIV replication cycle [160]. A *Plasmodium falciparum* homologue of KAT2A (*Pf*Gcn5) was found to play a key role in antigenic switching and expression of plasmodial proteins [161]. Interestingly, knockout or chemical inhibition of KAT2B ameliorates cognitive and behavioral deficits by suppressing NF-κB-mediated neuroinflammation induced by Aβ in a model of Alzheimer's disease, suggesting KAT2B to be a possible target for the treatment of Alzheimer's disease [162]. It has been shown that KAT2B regulates nitroglycerin-dependent arterial relaxation by $N(\varepsilon)$-lysine acetylation of contractile proteins and that KAT2B-mediated acetylation of connexin 43 is implicated in cardiac dystrophy [163]. In type 2 diabetes, KAT2A and KAT2B acetylate the peroxisome proliferator-activated receptor gamma coactivator 1-alpha (PGC-1α), a key co-activator in energy metabolism, thereby regulating its transcriptional activity [164]. KAT2B was found to be downregulated in human cardiac mesenchymal cells of type 2 diabetic patients, and the treatment with a specific KAT2B activator (see after) is able to rescue proliferation and differentiation [165]. Enhancement of KAT2B-mediated lysine acetylation also accelerates wound repair process by a NO-independent mechanism [166]. It was also shown that PCAF is involved in the extinction process of conditioned fear [167].

Domain organization and complex formation in MYST family KATs are more diverse than in KAT3A/3B and GNAT family proteins. Aberrant activity of this KAT family has predominantly been implicated in manifestation and progression of cancer [18b]. The aforementioned fusion proteins of KAT6A and KAT6B with other KAT family members in AML inductions are complemented by the KAT6A–TIF2 fusion protein, which is yielded after chromosomal inversion. In AML, KAT6A–TIF2 interacts with KAT3A and disrupts normal KAT3A-dependent transcriptional activation [168]. In addition, mutation in KAT6A has been associated with esophageal adenocarcinoma [169]. On the other hand, altered activity of KAT6B has been linked to breast cancer, prostate cancer, and leiomyoma [170]. In addition to cancer, these KATs are also involved in developmental processes, and mutation in their encoding genes has been found in several developmental disorders [170b]. The majority of studies investigating the relationship of KAT5 and cancer show that this enzyme functions as a tumor suppressor [18b] and its reduced expression is associated with the colon, lung, mammary, head and neck carcinomas, metastatic prostate cancer, metastatic melanoma, lymphomas, and AML [171]. However, KAT5 is overexpressed in cisplatin-resistant human lung cancer cells, and knockdown of KAT5 expression renders cells sensitive to cisplatin treatment [172]. Resistance to apoptotic signaling cascades in cancer cells after DNA damage was associated with KAT5 mutations [102]. Moreover, the proliferation of prostate cancer has been correlated to increased KAT5 activity [173]. Related to this, KAT5 plays an important role in hormone receptor signaling and DNA damage repair. The androgen receptor is activated in an androgen-independent manner upon KAT5-mediated acetylation. The acetyltransferase KAT7 is a key regulator of DNA replication and proliferation. Its overexpression has been reported in a specific subset of human primary cancers [174]. In breast cancer cell lines, KAT7 increases mammosphere formation when phosphorylated by cyclin E/cyclin-dependent kinase 2 (CDK2) [175], suggesting that its overexpression may lead to a more cancer stem cell-like phenotype.

12.7 KAT Modulators

The increased understanding of the role of major KATs in physiologic pathways as well as in various diseases highlights their distinction as new drug targets and brings the topic of chemical regulation of KAT activity to the forefront of pharmaceutical research. Yet, the identification of such compounds has proven to be challenging, also due to the fact that the design of effective biochemical assays that detect and quantitate KAT activities is highly challenging because the acetyl group is spectroscopically inert and there is no spectral readout from a KAT reaction [176]. Therefore, the development of potent and selective KAT inhibitors lags far behind modulators of other epigenetic enzymes, like KDACs. So far, different approaches have been used to identify KAT modulators, but only few potent substances have been obtained, and none of them has yet entered the clinical stage. With the aid of computational methods and advanced assay technologies [177], progress was made recently toward compounds with improved KAT modulatory properties and *in vivo* characterization. According to their mode of action and origin, the KAT modulators described so far can be assigned to four distinct classes: bisubstrate inhibitors, natural compounds and their analogues and derivatives, synthetic small molecules, and derivatives that inhibit the binding of other proteins targeting the interacting domains, including bromodomain inhibitors.

12.7.1 Bisubstrate Inhibitors

The first published KAT inhibitors were bisubstrate analogues mimicking the ternary complex of cofactor acetyl-CoA and lysine substrate in spatial proximity during the catalytic process (Figure 12.3). Cole and colleagues covalently linked CoA to the lysine residue of a substrate peptide of various chain lengths. In particular, Lys-CoA was yielded by connecting CoA and a single lysine residue via a methylene linker [46c]. The bisubstrates concept was adopted by several other groups to generate specific KAT inhibitors. Lys-CoA is a potent inhibitor of p300 with an IC_{50} value of 0.5 μM and pronounced selectivity toward KAT3B compared with KAT2B. The particular potency of the compound results from the Theorell–Chance kinetic mechanism of KAT3A/KAT2B catalysis (see Figure 12.2), which is characterized by a strong cofactor binding and a weak transient interaction with the histone substrate. The selectivity pattern of bisubstrate inhibitors can efficiently be altered by modifying the length and composition of the peptide chain that comprises the lysine-CoA construct. Derivative H3-CoA-20 mimics the native substrate H3K14 of KAT2B; hence it inhibits the enzyme selectively with an IC_{50} value of 0.3 μM [46c]. Later, to address MYST family enzymes, a similar approach led to the identification of the H4 peptide-containing bisubstrate analogue H4K16-CoA as an acetyl-CoA competitive and potent KAT5 inhibitor with an IC_{50} value in the low micromolar range [178]. Yet, the compound and its analogues are also able to inhibit the KAT5 yeast homologue Esa1 at low micromolar concentrations. Compound *Boc*-C5-CoA (Figure 12.3), bearing a C5 spacing linker that connects the CoA moiety to a *tert*-butyloxycarbonyl (Boc) group, was designed on the basis of the crystal structure of KAT3B, with the aim to occupy also an additional electron-rich pocket (P2) at about 10 Å away from the binding pocket required to accommodate the lysine moiety (P1). The compound is actually capable to occupy both binding pockets and was reported to inhibit KAT3B with an IC_{50} of 0.07 μM [179]. Attributed to the polar phosphate moieties and the partially peptidic structure, bisubstrate inhibitors generally suffer from poor cell permeability and metabolic instability, which limits their use to *in vitro* applications or requires membrane-penetrating techniques like microinjection or lipid permeabilization. Different approaches have been applied to circumvent this limitation.

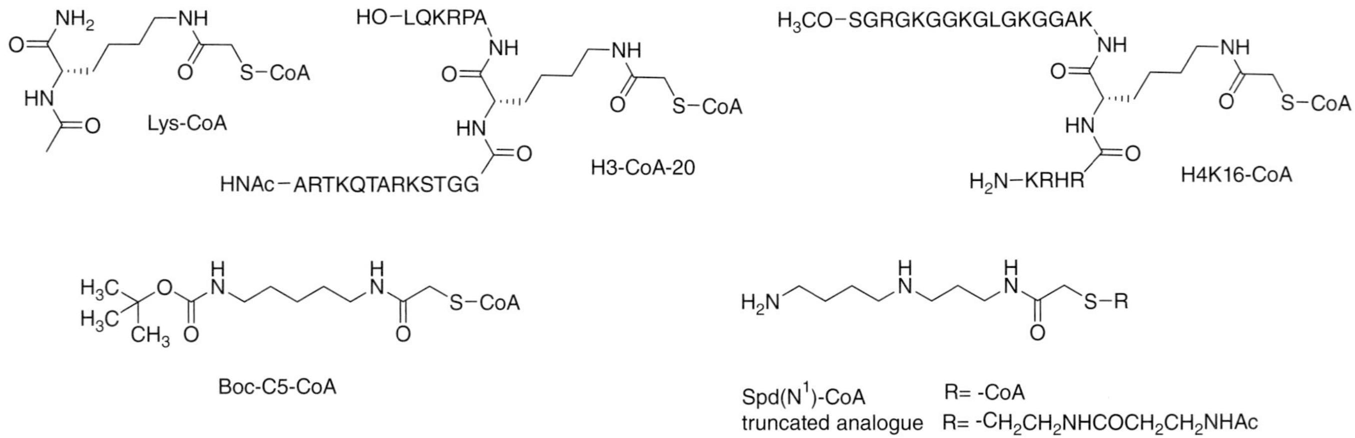

Figure 12.3 Structures of bisubstrate inhibitors.

Truncation of the CoA moiety in 3′ position led to a significant reduction of inhibitory activity [180]. Coupling of the inhibitors' amino acid backbone to the TAT protein transduction domain or to arginine-rich peptides succeeded in facilitating cellular uptake and activity [181]. However, the application of such inhibitor/peptide constructs is rather limited due to their complex handling and elaborate production. An alternative approach was aimed to exploit polyamine transporter uptake for internalized into cells. To this purpose, the polyamine spermidine was fused to CoA (derivative Spd(N^1)-CoA) (Figure 12.3) or to a truncated moiety (cysteamine/β-alanine) [182]. Both polyamine conjugates impeded histone acetylation-dependent repair and synthesis of DNA and consequently led to radio- and chemosensitization.

12.7.2 Natural Products and Synthetic Analogues and Derivatives

One of the most successful strategies in the discovery of KAT modulators derived from the screening of plant or microbial extracts. Several compounds have been identified with activity in the micromolar range (Figure 12.4). Unfortunately, many of them are affected by lack of selectivity and hence often pleiotropic effects in cellular systems, mostly due to common structural scaffolds of natural compounds, such as Michael reaction acceptors and polyphenols. In addition, quite often natural compounds possess unfavorable physicochemical properties, which limit their further development [183]. Synthetic and semisynthetic approaches have been carried out in order to determine structure–activity relationships and to optimize compound properties. Interestingly, the biological evaluation identified also compounds with enzyme-activating capacity and mixed activities, alongside with KAT inhibitors. Thus, even if they are not very drug-like, natural compounds are useful templates for further development into modulators of KAT activity. The first natural product identified as KAT inhibitor was anacardic acid (Figure 12.4), obtained from the liquid of cashew nut (*Anacardium occidentale*) shells together with several other saturated and unsaturated salicylic acid derivatives. A screening of plant extracts with anticancer activity led to the identification of anacardic acid as a nonselective, noncompetitive inhibitor of KAT3A/3B and KAT2B [184].

Following studies performed under similar experimental conditions also reported activity in KAT5 inhibition [185]. The inhibitory potency of anacardic acid varies between different studies, and IC_{50} values between 5 and 1000 μM were obtained for KAT3B inhibition, depending on assay conditions and enzyme source. On the cellular level, repression of the NF-κB signaling pathway by KAT-dependent acetylation of the p65 subunit has been found. However, the application of the compound in numerous pharmacological studies is attributed to its pleiotropic affinity that affects multiple enzyme targets [186]. Moreover, further development of this natural product as a therapeutic tool is also limited by its unfavorable physicochemical properties, especially its high lipophilicity. The 6-alkyl salicylic acid scaffold has been extensively exploited to generate compounds with increased solubility, cell permeability, and inhibitory efficacy. The investigation on a series of 2-(3-(benzyloxy)phenoxy)acetic acid analogues as KAT inhibitors showed that the position and length of the alkyl chain are

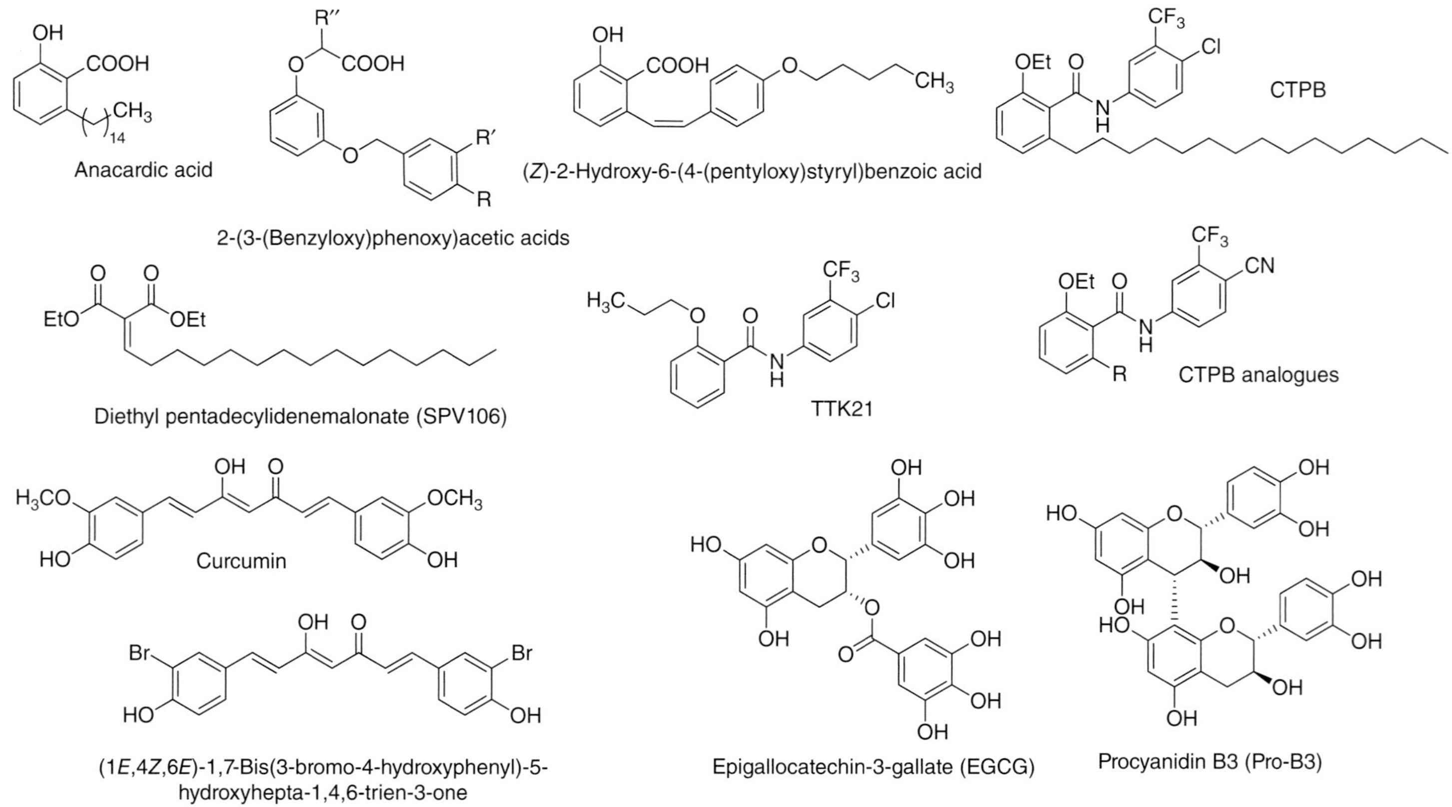

Figure 12.4 Structures of natural products and analogues and derivatives.

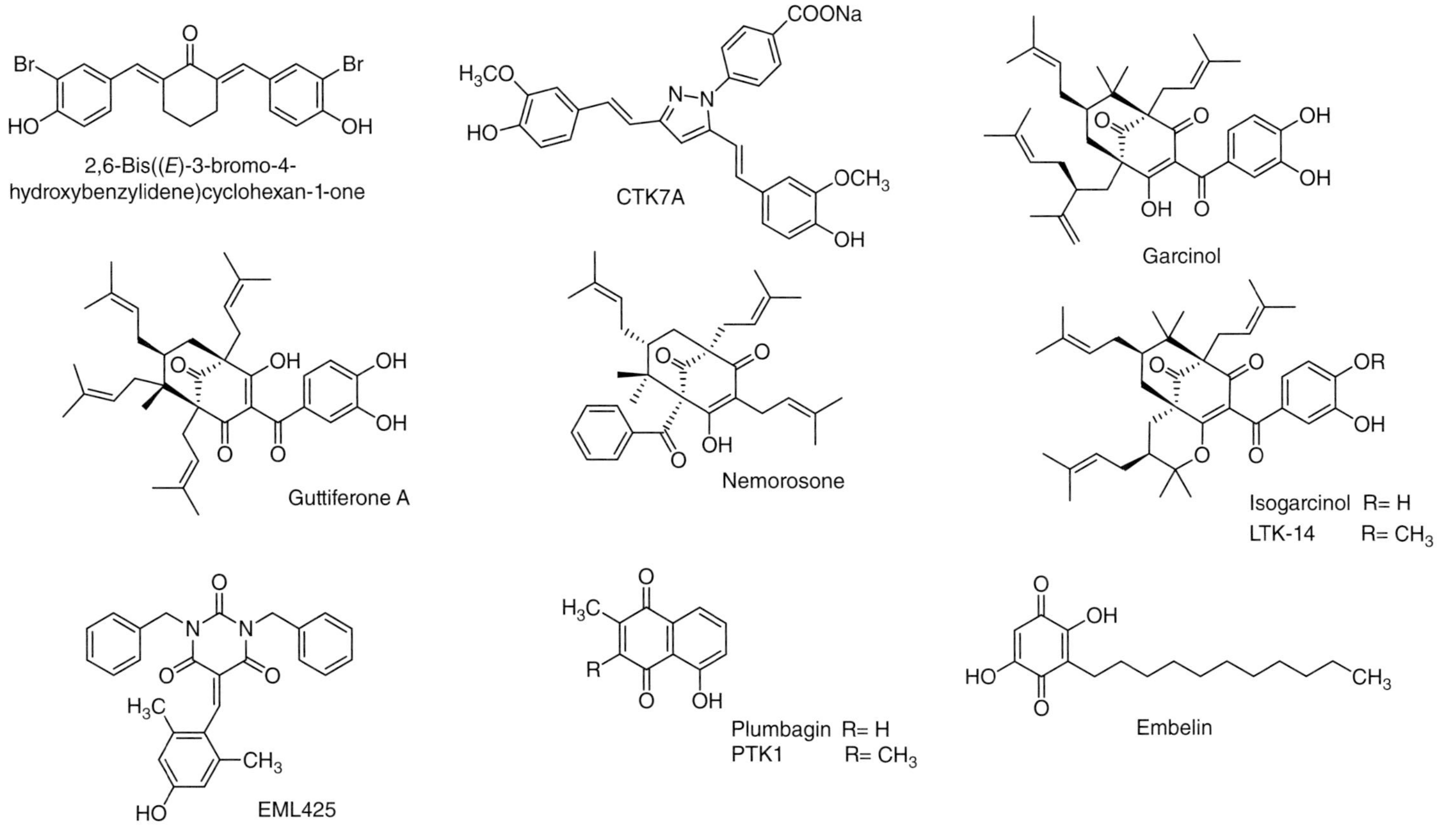

Figure 12.4 (*Continued*)

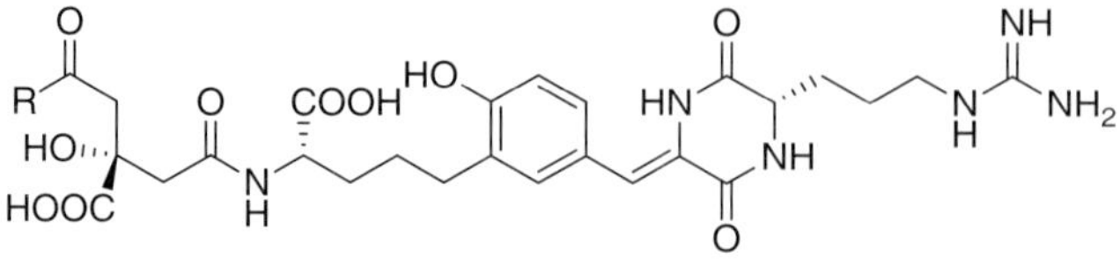

NK13650A R= NH-aspartic acid
NK13650B R= OH

Figure 12.4 (*Continued*)

pivotal for enzyme inhibition and that substitution of the phenolic hydroxyl group is well tolerated [187].

Other compounds with variations in the alkyl chain and the salicylic acid residue were designed and synthesized on the basis of a binding model for anacardic acid built inferred from the co-crystal structure of KAT2B with acetyl-CoA. The structure–activity relationships obtained from these compounds have shown that esterification of the carboxyl group diminished inhibitory activity, whereas modulation of the alkyl chain or hydroxylation in C4 position caused a shift of specificity toward MYST family KATs [188]. Among the compounds, (z)-2-hydroxy-6-(4-(pentyloxy)styryl)benzoic acid showed inhibitory potency superior to anacardic acid in KAT2B inhibition [188]. However, the same compound was later identified as an activator of KAT2B with no effect on KAT3B and inhibitory activity against KAT5 [185]. The shift of specificity toward MYST family KATs has been recently confirmed by a study on alkyl salicylic acid derivatives in KAT8 inhibition [115]. Other small-molecule KAT activators have been also identified. In an inhibitor screen of substituted benzamide analogues of anacardic acid, derivative *N*-[4-chloro-3-(trifluoromethyl)phenyl]-2-ethoxy-6-pentadecylbenzamide (CTPB) (Figure 12.4) promoted activation of KAT3B but not of KAT2B and increased KAT3B-dependent transcription activation [184]. Interestingly, a series of CTPB analogues showed that the selectivity of these compounds toward KAT3B and KAT2B is dependent on the length of the C6 aliphatic chain and the presence of terminal polar groups [189]. A derivative of CTPB, *N*-(4-chloro-3-trifluoromethyl-phenyl)-2-*N*-propoxy-benzamide (TTK21) (Figure 12.4), was found to induce a concentration-dependent activation of both KAT3A and KAT3B, but it was not able to enter into mammalian cells. To overcome the permeability obstacle, the compound was covalently linked to glucose-based carbon nanospheres, and the resulting particles readily penetrated the cell membrane to increase histone acetylation without causing apparent toxicity [190]. Treatment with TTK21-loaded nanospheres was further correlated with beneficial neurologic effects in a mouse xenograft model. The application of a structural simplification approach to anacardic acid led to the identification of diethyl pentadecylidenemalonate (SPV106) (Figure 12.4) as the first mixed activator/inhibitor of KATs [191]. This compound exhibits inhibitory properties against KAT3A/3B with a potency comparable with anacardic acid, and it simultaneously enhances the acetylating activity of KAT2B [191]. As a result of its peculiar activity profile, SPV106 was successfully used as a chemical probe to correlate Duchenne cardiomyopathy to KAT3B-mediated lysine acetylation levels of connexin 43 [163a], to investigate the role of KAT enzymes in the regulation of the extinction of conditioned fear and neuronal plasticity [167c], to study the role of KAT2B acetyltransferase activity in the regulation of nitroglycerin-dependent arterial relaxation [163b], and to investigate their role in wound healing [166]. Treatment with SPV106 has been shown to reverse alterations in human cardiac mesenchymal cells obtained from diabetic patients and restore cellular function [165]. The exploration of the structure–activity relationships of long-chain alkylidenemalonates (LoCAMs) related to SPV106 revealed that the replacement of one or both of the ester moieties with keto or carboxylic acid groups greatly alters the modulation profile of the resulting analogues, ranging from selective activation to unselective inhibition. In terms of the aliphatic chain, it was shown that the introduction of a heteroatom as well as variations in flexibility/rigidity of the core structure is detrimental for binding efficiency and variations in alkyl chain length result in abrogated modulatory activity [136d, 191a]. The precise underlying mechanism of this activator/inhibitor ambivalence is still under investigation. A few polyphenols like epigallocatechin-3-gallate and procyanidin B3 (EGCG and Pro-B3) (Figure 12.4) were reported to disrupt the NF-κB signaling pathway by repression of KAT activity and to inhibit the KAT3B-mediated androgen receptor acetylation [192]. However, these compounds interact with a wide range of other protein targets, and their structural optimization to achieve selectivity seems to be challenging [193].

Another compound with pleiotropic effects is curcumin (Figure 12.4), a major component of turmeric (*Curcuma longa*) rhizome that is commonly used in Indian and Chinese traditional medicine [194]. The compound was reported in 2004 as a KAT3B inhibitor (IC_{50} 25 μM) with no effect on KAT2B. Treatment with curcumin represses KAT3B-dependent H3, H4, and p53 acetylation in different cell lines. Kinetic studies revealed a covalent mode of action at a binding site apart from the substrate and cofactor binding pocket. The double bonds in the cinnamoyl structure serve potentially as Michael reaction acceptors, and their presence is crucial for binding with KAT3B [195]. As mentioned above, although certain selectivity between different KAT enzymes was demonstrated, the inhibitory activity of the compound is rather promiscuous as it also inhibits other epigenetic targets (e.g. KDACs, DNMT1, KDM1A/LSD1) as well as a multitude of non-epigenetic-related proteins [196]. Nevertheless, curcumin is under clinical investigation for several indications. Among "curcuminoids," bis-benzylidene cyclohexanones and their open-chain analogues (Figure 12.4) have been shown to inhibit KAT3B activity with similar potency as the lead structure [197]. The sodium salt of the hydrazinobenzoylcurcumin CTK7A (Figure 12.4) is more water soluble than curcumin while maintaining inhibitory potency [198]. It is important to mention that, like the parent structure, the described curcumin analogues interact with multiple target proteins aside from KAT3B [193]. Such promiscuous effects generally complicate the correlation of *in vitro* with *in vivo* observations. In addition, curcumin is known to be a membrane disruptor, and therefore it is likely that some of its activities could be attributed to nonspecific modes of action [199].

The polyisoprenylated benzophenone garcinol (Figure 12.4) was isolated from the rind of the fruit of kokum (*Garcinia indica*) and identified as a micromolar inhibitor of KAT3B (IC_{50} 7 μM) and KAT2B (IC_{50} 5 μM) acetyltransferase activity [200]. In human cancer cell lines, the compound represses histone acetylation and induces apoptosis, with a mixed-type inhibition kinetics. It has been shown that variations at the C8 side chain in garcinol impinge on its inhibitory effect on KATs [201]. Unfortunately, garcinol (as well as the related isogarcinol) is fairly toxic in cells. A two centered binding mode has been proposed with the catechol hydroxyl groups interacting with the acetyl-CoA binding pocket, while the isoprenoid moieties are placed into the substrate binding domain of the enzyme [202]. Human HeLa cells treated with the intramolecular cyclization product isogarcinol (Figure 12.4) develop concentration-dependent histone hypoacetylation and repression of gene transcription. Monomethylation of isogarcinol in C14 position led to derivative LTK-14, which provides a shifted inhibition pattern toward KAT3B activity, noncompetitive versus both acetyl-CoA and histones, and an attenuated T-cell cytotoxicity [149b]. Similar to isogarcinol, LTK14 induces concentration-dependent histone hypoacetylation and repression of gene transcription in human HeLa cells. Noteworthy, the binding mode proposed for LTK14 is different from that of the parent compound, with a single unique binding side within the enzymes' KAT domain [202]. However, garcinol, isogarcinol, and LTK14 are poorly soluble and unstable, and this has been attempted to be corrected through further investigation of other naturally occurring structural analogues [203]. Interestingly, one of these compounds, nemorosone (Figure 12.4), was found to induce a modest (1.5–2-fold at 10 μM) activation of KAT3B [203b]. The application of a molecular pruning approach to the garcinol core structure, followed by the isosteric replacement of nitrogen for carbon, led to the identification of barbiturate derivatives that were screened for their inhibitory effects against KAT3A/3B and KAT2B using biochemical and biophysical assays as well as in live-cell studies of HAT inhibition using a FRET-based reporter system. One of the compounds, the benzylidene barbituric acid derivative EML425 (Figure 12.4), was shown to be a selective and reversible inhibitor of KAT3A/3B, noncompetitive versus both acetyl-CoA, and a histone H3 peptide, with an inhibitory potency in the low micromolar range (IC_{50} 2.9 μM for KAT3B and 1.1 μM for KAT3A) [204]. The compound is endowed with good cell permeability, and, in human leukemia U937 cells, it induced a marked and time-dependent reduction in the acetylation of lysine H4K5 and H3K9, a marked arrest in the G0/G1 phase, and a significant increase in hypodiploid nuclei percentage. Molecular modeling studies of this derivative proposed high affinity for the same alternative binding pocket as for LTK14. Interestingly, both EML425 and C646 (see below) have been used to demonstrate the role of KAT3A in epigenetic reprogramming of caste-specific foraging and scouting behaviors in carpenter ants (*Camponotus floridanus*) [205].

The hydroxynaphthoquinone plumbagin (Figure 12.4), isolated from *Plumbago rosea*, has been shown to attenuate KAT3B-dependent acetylation of histones H3, H4, and p53 in HepG2 liver cancer cells without causing any effect on KAT2B activity [206]. Structure–activity studies suggested the hydroxyl group in C5 position to be crucial for forming hydrogen bond interactions with

Lys1358 in the active site of the enzyme. Consequently, derivatization of the naphthoquinone in this position diminishes the inhibitory effect. However, plumbagin (as well as related hydroxynaphthoquinone-based inhibitors) is rather cytotoxic, due to free thiol reactivity and redox cycling properties. Methylation in C3 position (compound PTK1) (Figure 12.4) abrogated the thiol reactivity and therefore cytotoxicity of plumbagin while maintaining its function as a p300 inhibitor [207]. Another quinone-containing derivative, the 3-alkyl dihydroxybenzoquinone embelin (Figure 12.4) originally isolated from *Embelia* ribes, has been shown to inhibit KAT activity of recombinant KAT2B and to promote H3K9 hypoacetylation in treated mice [208]. The compound also attenuated KAT2B-mediated MyoD acetylation in HEK239T cells, which was correlated to a block of differentiation in C2C12 cells. Similar to what was reported for anacardic acid analogues (see above), the length of alkyl chain in embelin structure is crucial for the inhibitory capacity as an analogue with one carbon less was reported to be completely inactive.

The screening of a microorganism broth library led to the identification of compounds NK13650A and NK13650B (Figure 12.4), two fungal metabolites of a *Penicillium* strain with a peptidic structure that contains a citric acid moiety [209]. The evaluation of their inhibitory capacity revealed high selectivity for KAT3B (IC_{50} values of 11 and 22 nM, respectively) over KAT5. These compounds are able to repress androgen- and estrogen receptor-dependent activation of gene transcription and are cytotoxic to different cancer cell lines. However, both compounds suffer from poor cell permeability and metabolic instability due to their peptidic nature. Nevertheless, the identification of secondary metabolites endowed with KAT inhibitory activity provides a promising strategy for lead structure discovery.

12.7.3 Synthetic Compounds

Different approaches, such as high-throughput strategies, rational design, and *in silico* screenings have been applied in order to find new potent small-molecule KAT modulators. Structure–activity relationships obtained from identified hits and their derivatives provided new insights into the characteristics of KAT enzymes. With the aid of computational methods, it was possible to identify and develop new compounds with pronounced inhibitory activity (Figure 12.5), even if only few of them are fully characterized *in vitro* and *in vivo* so far.

A series of compounds containing the isothiazolone moiety (Figure 12.5) were identified in 2005 as inhibitors of both KAT2B and KAT3B in biochemical and cell-based assays [210]. An irreversible binding mode was suggested that involves disulfide bond formation between the isothiazolone sulfur and a cysteine residue in the enzyme. Consistent with this, addition of DTT to the reaction abolished inhibitory activity. Since then, several efforts have been dedicated to developing derivatives with reduced general reactivity and promiscuity while maintaining inhibitory potency. A computational screening of public and commercial databases for compounds containing an isothiazolone or isothiazolidinone substructure led to the identification of 51 derivatives that were docked into the substrate binding site observed in the crystal structure of KAT2B. Then,

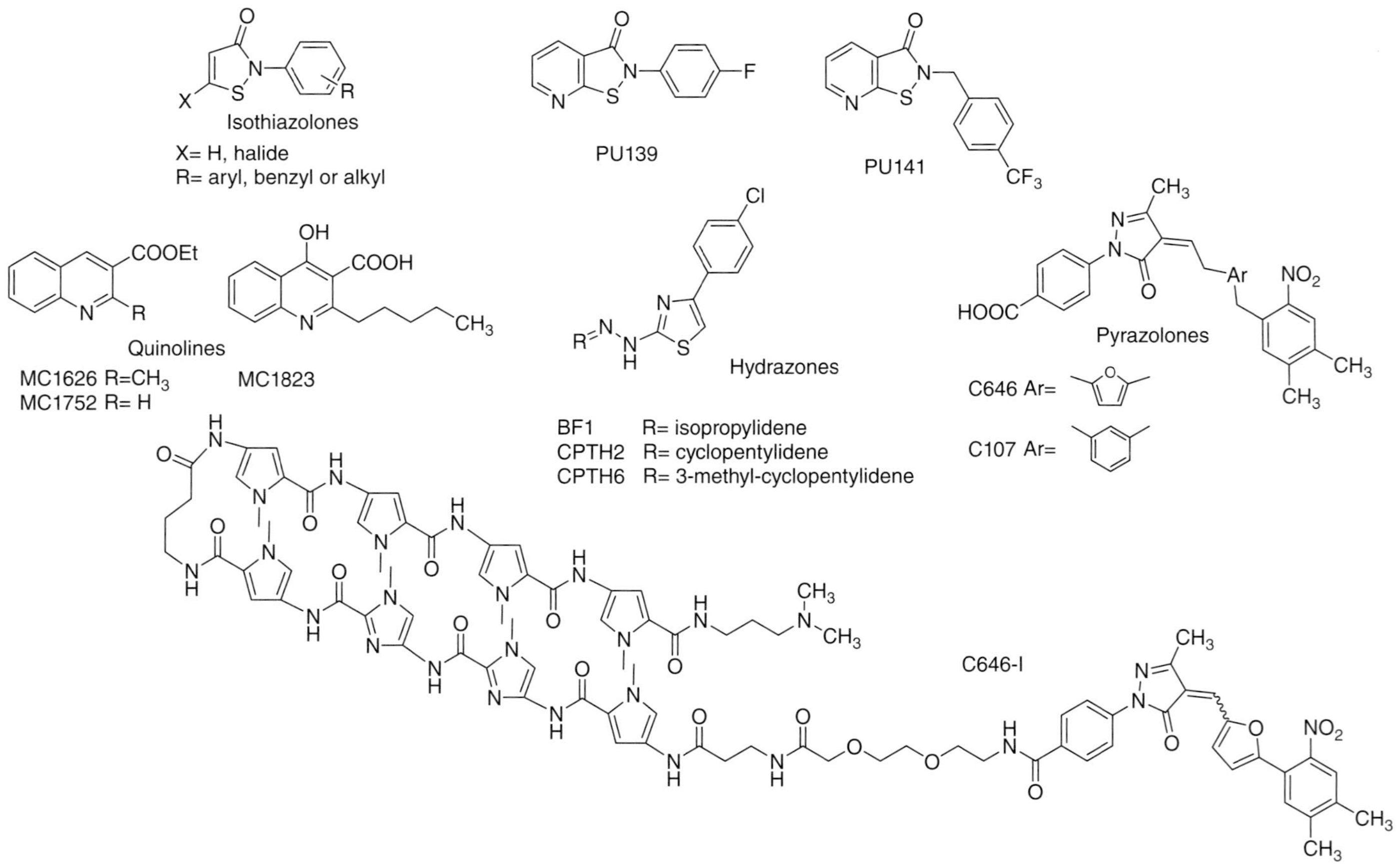

Figure 12.5 Structures of synthetic derivatives.

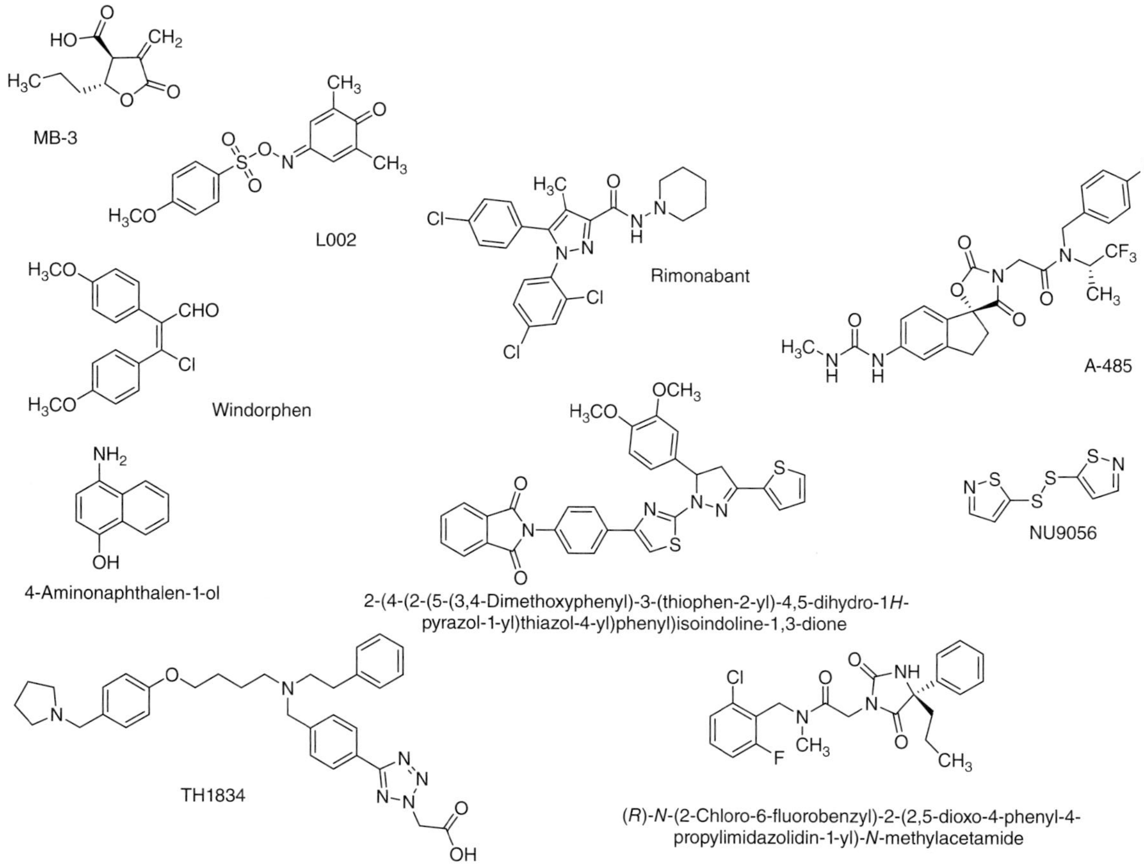

Figure 12.5 (*Continued*)

compounds that showed a reactive S–N moiety in close proximity to the enzyme Cys574 in the docking results were considered for further biological testing. In this way, different pyridoisothiazolones were discovered as novel KAT inhibitors with low micromolar IC_{50} values and reduced general bioreactivity [136c, 211]. Structure–activity relationships revealed a crucial role of the substituent in 2 position for inhibitory activity. *N*-aryl-substituted compounds were shown to cause pan-KAT inhibition on a series of enzymes (KAT2A, KAT2B, KAT3A, KAT3B, and KAT8), while *N*-benzyl or *N*-alkyl substituents led to defined subtype selectivity patterns [212]. The pyridoisothiazolone PU139 (Figure 12.5) was reported to impair promoter activity of the egg shell protein Smp14 in the trematode parasite *Schistosoma mansoni*, probably as a result of diminished activity of the parasite acetyltransferases *Sm*GCN5 and *Sm*CBP1 [213]. The repression of Smp14-regulated genes led to production of abnormal and defective eggs, implying KAT inhibition as novel strategy in the control of schistosomiasis. Compound PU139 and the related derivative PU141 have also been reported to trigger growth inhibition and histone hypoacetylation in multiple cancer cell lines and to block neuroblastoma cell growth in a SK-N-SH xenograft model *in vivo* [214]. Interestingly, no significant effect of the compounds against a cysteine protease panel was observed, indicating that the inhibitors do not have a general reactivity toward all cysteine-dependent enzymes. For compound PU139, the reduction of histone acetylation in healthy mice and the synergistic effects with the DNA-intercalating drug doxorubicin in the xenografts have also been reported.

Yeast phenotypic screenings led to the identification of quinoline and hydrazone derivatives with KAT-inhibiting activity. Structural optimization of the quinolone structure resulted in compound MC1823 (Figure 12.5), which bears structural similarities to anacardic acid [215]. The reduction in cell viability of *S. cerevisiae* was correlated to the hypoacetylation of histone H3 and α-tubulin as a result of KAT2A inhibition. A similar effect has been observed for the hydrazones BF1, CPTH2, and CPTH6 (Figure 12.5) [216]. In addition, compounds BF1 and CPTH6 also inhibit KAT3B and KAT2B, respectively, and induce histone hypoacetylation in different human cancer cell lines [217].

A rational design strategy based on the electrostatic interaction fields within the active site of KAT2A led to the identification of the α-methylene-γ-butyrolactone MB-3 as an inhibitor of KAT2A, selective over KAT3A [218]. Regardless of the apparent Michael reaction acceptor scaffold, the binding mode of MB-3 was reported to follow a noncovalent fashion, with an assessed IC_{50} value of 100 μM comparable with the K_D value of the H3 substrate.

The pyrazolone-containing inhibitor C646 (Figure 12.5) was identified as a potent, selective, and reversible KAT3A/3B inhibitor by means of a virtual screening approach [219]. A database of ~500 000 commercially available compounds was docked into the same binding pocket as occupied by the bisubstrate inhibitor Lys-CoA (see above) in KAT3B and led to the selection of 194 compounds for biological testing. Three among them were found to inhibit KAT3B activity in the micromolar and submicromolar range, where C646 showed the highest inhibitory activity (K_i = 460 nM). The proposed binding mode shows numerous H-bond and salt-bridge interactions and indicates the compound to be a bisubstrate inhibitor of the enzyme, with the side chains of the binding pocket similar to those observed with the CoA moiety of the co-crystallized Lys-CoA. Structure–activity relationships obtained from a number of derivatives indicated that the free carboxylic acid, as well as certain interactions mediated by the nitro group, is essential for inhibitory activity, although replacement of the nitro group with more metabolically favorable functionalities resulted in only minor loss of potency. Also, the conjugated pyrazolone structure is crucial for binding to the KAT3B active site, indicating that a certain level of planarity of the molecule is required for efficient active site targeting. However, this structural entity is potentially prone to nucleophilic attack. Although the binding of C646 to KAT3B was originally described to be non-time dependent and not abrogated by DTT or β-mercaptoethanol, the formation of covalent conjugates with a number of different cellular cysteine-containing proteins has been recently observed [220]. Moreover, the use of the compound in fluorescent-based assays is hampered by the intrinsic fluorescence caused by the large conjugated system in the structure. To elude this hurdle the furan group was replaced with a phenyl ring [219a]. The resultant derivative was absent of intrinsic fluorescence while mostly maintaining its inhibitory capacity against KAT3B (IC_{50} 9 μM). Compound C646 inhibits cell growth and induce histone H3 and H4 hypoacetylation in melanoma and lung cancer cell lines. Cell cycle arrest and induction of apoptosis upon administration were also observed in AML1/ETO-positive leukemia cells [221]. Together with compound EML425 (see above), C646 has been used as a chemical tool to demonstrate the role of KAT3A in epigenetic reprogramming of caste-specific behaviors in carpenter ants [205]. Recently, a sequence-specific epigenetic repressor obtained by conjugating C646 with pyrrole-imidazole polyamide DNA binding domain (C646-I) (Figure 12.5) has been reported to induce antiproliferative activity in p53 wild-type cancer cell lines via the upregulation of p53 target genes and the subsequent initiation of p53-dependent apoptosis [222].

A large high-throughput screening of more than 600 000 substances for their cytotoxic activity against the triple-negative breast cancer cell line MDA-MB-231 resulted in the identification of the small-molecule inhibitor L002 (Figure 12.5) [223]. The compound was found to inhibit both KAT3B (IC_{50} 128 μM) and GNAT family KATs (IC_{50} 34.7 μM for KAT2B and 33.9 μM for KAT2A) in radioactive filter binding assays, whereas no inhibition of MYST family members was observed. *In silico* studies proposed that the sulfonyl-coupled iminoquinone structure engages the acetyl-CoA binding pocket of KAT3B. The compound has been found to promote histone and p53 hypoacetylation and to suppress STAT3 activation in lymphoma, leukemia, and breast cancer cell lines. Reduction of tumor growth and decreased histone acetylation were correlated to *in vivo* administration in a MDA-MB-468 xenograft model.

Another inhibitor of KAT3B was identified through a phenotype-based screening aimed to discover selective small-molecule Wnt pathway modulators. Given the critical roles of Wnt/β-catenin signaling in embryonic dorsoventral patterning, about 30 000 compounds were screened for their capability to perturb dorsoventral patterning in zebrafish (*Danio rerio*) embryos. One of the hits, windorphen (*W*nt *in*hibitor *dor*salizing) (Figure12.5), was found to be a selective Wnt/β-catenin signaling inhibitor acting through selective inhibition of KAT3B. Windorphen has been reported to block Wnt signaling by selectively disrupting the association of KAT3B with the C-terminal transactivation domain of β-catenin-1. No inhibition of KAT3A was observed. It was shown that windorphen induces apoptosis in several Wnt-dependent tumor cell lines including colon adenocarcinoma SW480 and RKO and prostate cancer cell lines DU145 and PC3 [224].

Very recently, it has been reported that the cannabinoid receptor 1 inverse agonist rimonabant (Figure 12.5) induces downregulation of Wnt/β-catenin target genes partially through a direct inhibition of KAT3B [225]. The mechanism of the inhibition has not been characterized, yet.

Another recent report described the identification of A-485 (Figure 12.5), a very potent, selective, and drug-like inhibitor of KAT3A and KAT3B. Approximately 800 000 compounds were virtually screened using an *in silico* method based on compound docking to the binding pocket of the bisubstrate analogue Lys-CoA (see before; Figure 12.3), in a proposed open conformation. Then, around 1300 commercially available compounds were tested in a direct radioactive KAT3A/3B HAT assay to identify two confirmed hits. Iterative rounds of medicinal chemistry optimization finally led to the ureido-substituted indane spirooxazolidinedione A-485 [226]. In a radiometric scintillation proximity assay, the compound was found to inhibit KAT3B catalytic activity with an IC_{50} of 60 nM, thus being about 27-fold more potent than C646 (IC_{50} 1.6 μM). Interestingly, in a time-resolved fluorescence resonance energy transfer (TR-FRET) assay that measured the acetylation of a biotinylated histone H4 peptide, A-485 inhibited the activity of the KAT3B-BHC (bromodomain HAT-C/H3) domain with an IC_{50} of 9.8 nM and also inhibited CBP-BHC with an IC_{50} of 2.6 nM. KAT3B- and KAT3A-BHC are constructs containing the bromodomain, the catalytic acetyltransferase domain, and C/H3 domain and are considerably more efficient as histone acetyltransferases than the single acetyltransferase domains [78, 219a]. Compound C646 was reported to show about four to fivefold reduced potency against BHC acetyltransferase activity versus the isolated acetyltransferase domain [78]. Surface plasmon resonance experiments indicated that A-485 binds to KAT3B with a K_D of 15 nM and high residence time (slow off-rate, $k_{off} = 1.3 \times 10^{-3}\ s^{-1}$). The direct binding was confirmed by a series of other experiments, including the determination of the high resolution (1.95 Å) X-ray co-crystal structure with the fully active human KAT3B catalytic domain, that demonstrated that the A-485 binding site overlaps with that for acetyl-CoA, but not the peptide substrate binding site. The compound did not inhibit the activity of KAT1, KAT2A, KAT2B, KAT5, KAT6A, or KAT6B at 10 μM, and this is probably due to the fact that the methylurea of A-485 makes two equivalent hydrogen bonds to the backbone carbonyl of Gln-1455 of the L1 loop, which is absent in other KATs [226]. The compound was also found to be selective over bromo- and extra-terminal (BET) bromodomain proteins and more than 150 non-epigenetic targets, with the exception of a substantial (greater than 90%) binding to dopamine and serotonin transporters at 10 μM. Yet, the derivative has been reported to not pass the blood–brain barrier, and, therefore, it is unlikely to modulate these transporters *in vivo*. A-485 has been found to selectively inhibit proliferation in lineage-specific tumor types, including several hematological malignancies and androgen receptor-positive prostate cancer. Moreover, it has been shown that the compound inhibits the androgen receptor transcriptional program in both androgen-sensitive and castration-resistant prostate cancer and inhibits tumor growth in a castration-resistant xenograft model [226].

Different scaffolds have been identified and tested in order to find KAT5 inhibitors as a potential therapeutic strategy in malignancies correlated to aberrant activity of the enzyme. A virtual screening campaign based on the KAT5 yeast homologue Esa1 crystal structure resulted in the identification of a phthalimide derivative (Figure 12.5) [178]. The compound displayed an unselective inhibitory activity against different KAT family members (KAT5, Esa1, KAT3B, KAT2B) with IC_{50} values between 100 and 190 μM. Computational docking and kinetic studies suggested that the compound targets the acetyl-CoA binding site of KAT5. One year later, a high-throughput screening for KAT5 inhibitors of ~80 000 substances yielded the identification of the isothiazole-containing derivative NU9056 (Figure 12.5) as a specific inhibitor of KAT5 activity (IC_{50} 2 μM) [227]. It was further shown that the compound blocks acetylation of histone proteins in a concentration-dependent manner and induces apoptosis via caspase activation in prostate cancer cell lines. In 2014, a rational design approach in accordance with the electrostatic surface potential of a KAT5 active site model led to the development of compound TH1834 (Figure 12.5) [228]. The two oppositely charged (under physiological conditions) ends of the structure were predicted to interact with amino acid side chains on different sides of the KAT5 binding pocket. The ethylbenzene side chain was introduced as an isosteric replacement of the adenine residue of acetyl-CoA. The compound promotes induction of apoptosis and radiosensitization in MCF7, DU-145, and PC-3 cancer cell lines, although at high concentration (500 μM).

Another high-throughput screening campaign led to the discovery of the 3,5-imidazolidinedione inhibitor (Figure 12.5) of the fungal KAT Rtt109 [130]. The compound was reported to inhibit enzyme-mediated H3K56 acetylation regardless of which histone chaperone cofactor protein (Asf1 or Vps75) was present. A tight-binding, uncompetitive mechanism was proposed, as inhibitory activity increased over time and was not diminished by dialysis of the inhibitor from the reaction mixture. The compound did not show any apparent effect on KAT3A and KAT3B activity but inhibited Rtt109 potently (IC_{50} 0.56 μM) in a sulfhydryl scavenging assay. The optimization of this compound could lead to promising drug candidates against pathogenic fungal species, such as *Candida albicans*. Unfortunately, the further development of the compound is limited by its inability to induce cellular effects, presumably due to drug efflux or rapid metabolism. Structure–activity studies need to be conducted to assess the overall potential of this scaffold.

A fragment screening approach enabled the identification of 4-amino-1-naphthol (Figure 12.5) as inhibitor of KAT8 [229], even if not selective against other KATs. Kinetic and calorimetric binding studies allowed the calculation of the assay-independent K_i values for both the free enzyme and the acetylated intermediate (K_{i1} = 2.6 μM and K_{i2} = 0.017 μM, respectively), which suggested that the inhibition of KAT8 is mostly due to interaction with the acetylated form of the enzyme. The proposed mechanism of reversible inhibition is intriguing because the scaffold is potentially prone to general bioreactivity.

12.7.4 Compounds Targeting Protein–Protein Interaction Domains

In addition to modulators that directly affect the acetyltransferase activity (by targeting the catalytic domain of KATs or allosteric binding sites), a more recent approach is focused on the identification and development of small-molecule inhibitors of KAT PPI domains.

A secondary metabolite of the filamentous fungi of the *Chaetomium* species, chetomin (Figure 12.6), has been reported to inhibit the interaction between KAT3A/3B and hypoxia-inducible factor 1 (HIF-1) subunit α, although with an unclear mechanism [230]. However, it induces coagulative necrosis, anemia, and leukocytosis in experimental animals. A dimeric epidithiodiketopiperazine (ETP) (Figure 12.6), designed to probe the importance of the bridging disulfide in the ETP moiety for the biological activity, was reported to bind KAT3A/3B with high affinity and disrupt the interaction between the KAT CH1 domain and the HIF-1α C-TAD domain. It has been suggested that the bridging disulfides, after reduction to dithiols within the cell, may bind Zn^{2+} ions, which are required to maintain a properly folded CH1 region. Disrupting the HIF-1α C-TAD-p300/CBP complex results in rapid downregulation of hypoxia-inducible genes critical for cancer progression. Also, ETP is nearly threefold less cytotoxic than chetomin [231].

Other two naturally occurring compounds, the depside sekikaic acid and the depsidone lobaric acid (Figure 12.6), have been found to target PPI of KAT3A/3B. It has been reported that through interaction with a dynamic surface within the protein, these compounds are able to simultaneously modulate two distinct binding sites of the KAT3A/3B GACKIX activator binding domain, thus inhibiting the ability of KAT3A/3B to form a complex with two distinct classes of activators. Cell-based assays have shown sekikaic and lobaric acids causing a dose-dependent downregulation of the c-Jun-driven gene cyclin D1 [232]. Despite the structures suggesting pleiotropic interactions with other proteins, it has been reported that the compounds do not affect other related PPI complexes.

One of the most thoroughly investigated interaction domains is the bromodomain, which recognizes and binds to acetylated lysine residues. The disruption of acetylation-dependent PPI by small-molecule inhibitors abrogates the assembly of transcriptional regulator complexes and subsequently represses the phenotypic consequences of acetylation signaling. Non-BET bromodomains are structural entities of several KATs, and since the successful implementation of the first BET bromodomain ligands, there has been a growing interest in the identification of potent and selective small-molecule inhibitors. As a matter of fact, a series of KAT bromodomain ligands have been identified, mostly through structure-guided approaches (Figure 12.6).

After the BET family inhibitors, KAT3A bromodomain inhibitors are the second most thoroughly studied group. This bromodomain binds the effector

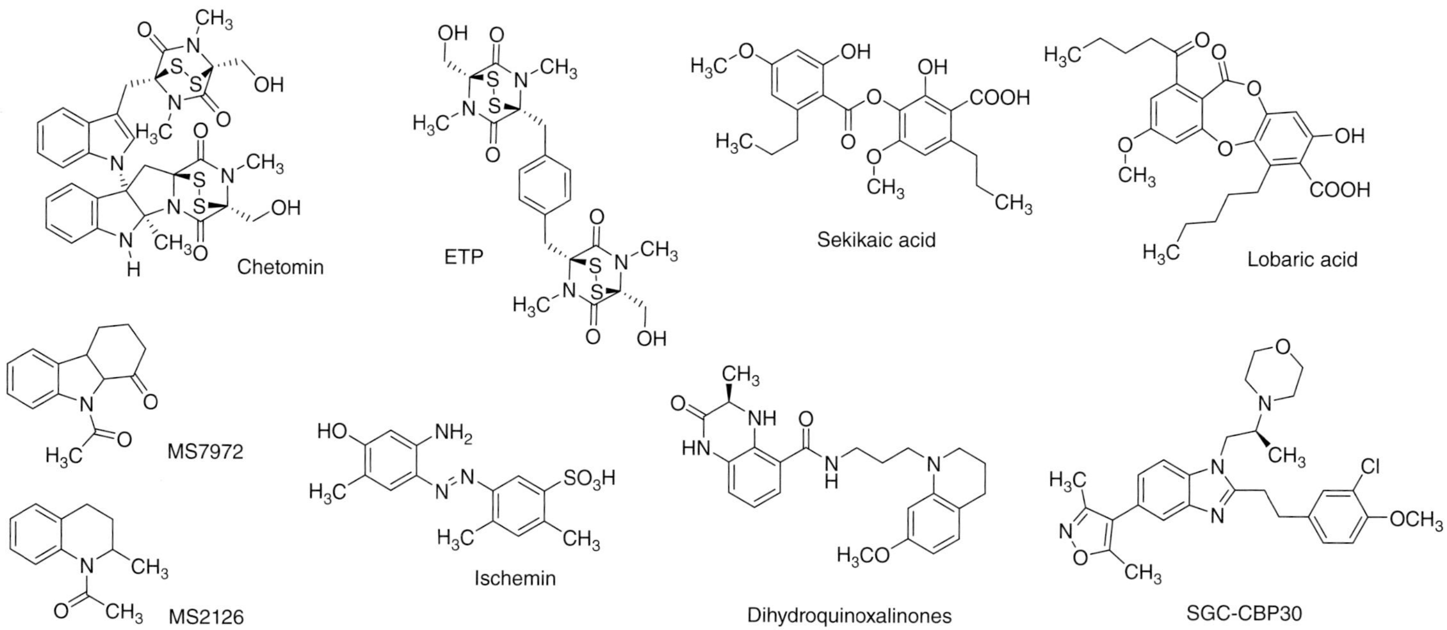

Figure 12.6 Structures of compounds targeting protein–protein interaction domains.

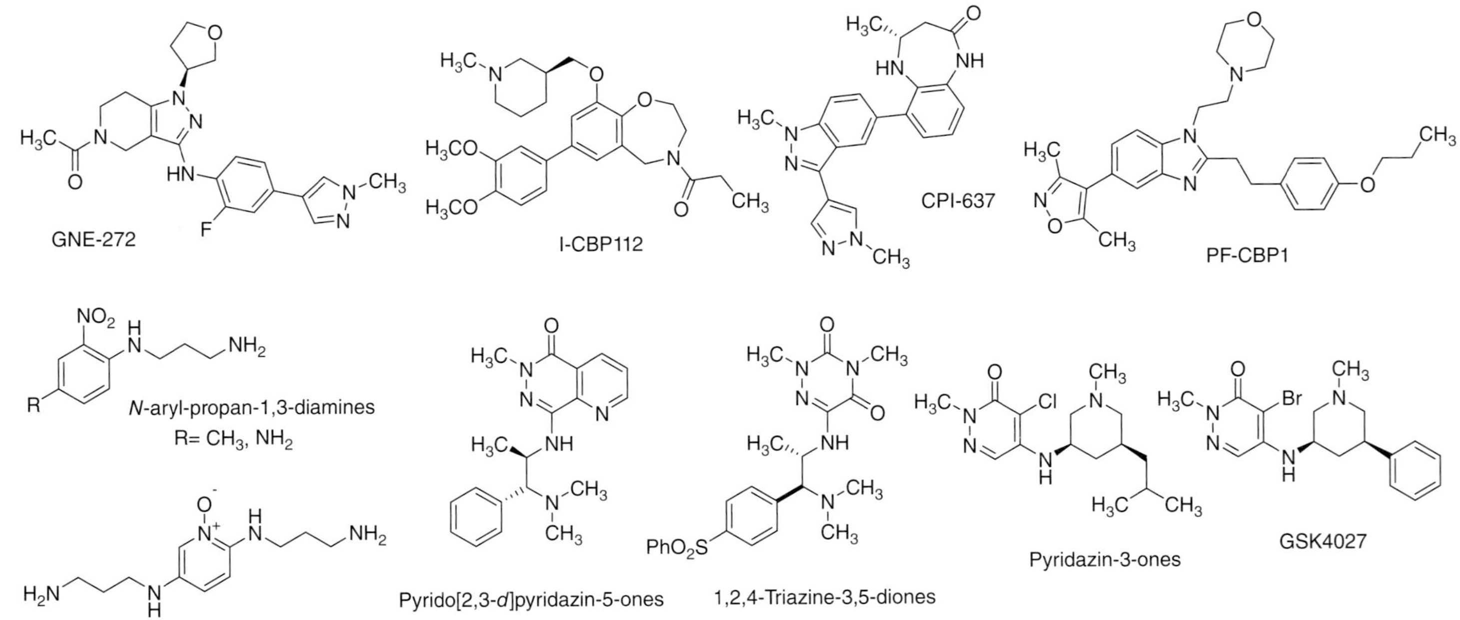

Figure 12.6 (*Continued*)

protein p53 after the acetylation of its lysine residue 382, and upon binding a co-activator complex is formed, which is recruited to p53-controlled promoter sites. The first attempts to obtain KAT3A bromodomain inhibitors were made using a structure-guided nuclear magnetic resonance spectroscopy screening of a focused chemical library, built on the basis on the structural knowledge of KAT3A bromodomain/p53-AcK382 binding. This approach yielded the identification of hits MS2126 and MS7972 (Figure 12.6) that inhibit KAT3A/p53 association by binding to the acetyl-lysine binding site of the bromodomain. Cell-based functional assays demonstrated that the compounds can modulate p53 stability and function in response to doxorubicin-mediated DNA damage [233]. Structural optimization of initial hits yielded the azobenzene derivative ischemin (Figure 12.6), which was demonstrated to inhibit the p53/KAT3A bromodomain interaction with an IC_{50} value of 5 μM and a fivefold selectivity over other bromodomains (KAT2B, BRD4, BAZ2B) [234]. Ischemin was shown to be capable to protect rat cardiomyocytes from p53-induced apoptosis in consequence of doxorubicin treatment [234].

The first submicromolar ligands for the KAT3A bromodomain were later obtained as a result of a fragment-based approach. Chemical expansion of the initial hits yielded a series of dihydroquinoxalinone compounds, the most potent of which (Figure 12.6) was shown to bind to the bromodomain via an induced fit pocket, which is occupied by the tetrahydroquinoline side chain and stabilized by a cation–π interaction of this residue with R1173 of the protein [235]. The K_D of the compound, as assessed in an isothermal titration calorimetry assay, was 390 nM, but the selectivity for the KAT3A bromodomain over BRD4 (K_D 1.4 μM) was only modest.

Another fragment-based approach led to the identification of the 3,5-dimethylisoxazole fragment as an acetyl-lysine mimic in bromodomain inhibitor structures. Chemical optimization of the fragment resulted in the development of a series of 5-isoxazolyl-benzimidazoles as ligands for the bromodomain module of KAT3A/3B. The most potent among these compounds, derivative SGC-CBP30 (Figure 12.6), binds to the KAT3A bromodomain with a K_D value of 21 nM, and the achieved inhibition is highly selective over BRD4. The phenyl ring of the C2 phenethyl substituent of the benzimidazole core forms a cation–π interaction with R1173 in the bromodomain binding site. Moreover, the electron-donating *para*-methoxy group in this moiety is able to enhance KAT3A potency while somewhat improving selectivity over BRD4. In a luciferase-based reporter assay, SGC-CBP30 diminished expression of KAT3A-dependent p53 downstream genes has been observed as a result of SGC-CBP30 incubation [81a]. The compound is selective for KAT3A/3B bromodomains, and it has been reported to suppress the Th17 response that is critical to a variety of human autoimmune diseases. However, it is not suitable for use *in vivo* because it is metabolized so quickly [236]. With the aim to further increase selectivity over BRD4, a structure-based design approach led to the discovery of another 3,5-dimethylisoxazole-containing benzimidazole derivative, PF-CBP1 (Figure 12.6). Noteworthy, the simple replacement of the *para*-methoxy group with a propoxy substituent in the C2 phenethyl moiety increases the steric hindrance with the BRD4 pocket, without negatively affecting CBP binding. As a consequence, the compound is 139-fold selective over BRD4. PF-CBP1 possesses no cytotoxicity in macrophages, and hepatotoxicity in cell-based models was observed only at very high concentrations. The inhibitor downregulates a number of inflammatory genes in macrophages and modulates the mRNA level of the regulator of G-protein signaling 4 (RGS4) gene in neurons, suggesting a potential therapeutic opportunity for CBP inhibitors in the treatment of neurological disorders [237].

Another inhibitor specific for KAT3A and KAT3B, I-CBP112 (Figure 12.6), was developed on the basis of a benzoxazepine scaffold [238]. The acyl group in this structure mimics the acetyl residue of the native recognition motive, thereby forming hydrogen bonds to the binding site, while the aryl substituent interacts with the arginine moiety in position 1173 of the KAT3A bromodomain. The K_D values of I-CBP112 for binding to KAT3A and KAT3B bromodomains were determined to be 151 and 167 nM, respectively, with selectivity over a number of other bromodomains (e.g. BRD2, BRD4, PCAF, TIF1α). The compound proved cellular activity in a FRAP assay on U2OS cells. Further studies revealed that the compound is capable to impair p53 interaction, which results in reduced p21 expression. Treatment of leukemia cell lines with the bromodomain inhibitor led to cell cycle arrest in G1 phase and morphological differentiation. A recent study of I-CBP112 reported that this compound stimulates acetylation activity by KAT3A/KAT3B acting as an activator of these KATs, contributing to its antiproliferative effect in cancer. It is also suggested that activation activity of I-CBP112 could help restore the balance of acetylation levels in tumors that experience KAT3A/KAT3B loss of function due to mutations, although the mechanism and networks involved in tumors affected by this activation are not yet fully understood [81b]. Both I-CBP112 and SGC-CBP30 have been reported to suppress the lymphocyte-specific transcription factor IRF4, which is crucial for the viability of myeloma cells and consequently targets c-MYC [239]. Another fragment-based screening approach identified 4-methyl-1,3,4,5-tetrahydro-2*H*-benzo[*b*][1,4]diazepin-2-one as a promising validated hit, the chemical optimization of which yielded the indazolyl-substituted derivative CPI-637 (Figure 12.6). A co-crystal structure showed that the benzodiazepinone core of the compound occupies the acetyl-lysine binding pocket, with the 4-methyl group positioned in the pocket where the acetyl methyl of the natural substrate binds and the carbonyl group and the lactam NH making key hydrogen bond interactions, while the 3-pyrazolylindazole substituent improves potency by filling space over Pro1110 and the Pro/Arg cleft. CPI-637 is a potent and selective inhibitor of KAT3A/3B (KAT3A IC_{50} = 0.03 μM; BRD4 IC_{50} = 11.0 μM) and is also able to engage the target in a cellular context [240]. However, the compound has been reported to be elusive from further modification. A derivative with similar activity and selectivity profile but more suitable for the development of an *in vivo* tool compound is GNE-272 (KAT3A IC_{50} = 0.02 μM, EC_{50} = 0.41 μM, BRD4(1) IC_{50} = 13 μM) (Figure 12.6), identified through the screening of a lead-like property-restricted diversity set against KAT3A in a thermal shift assay followed by chemical optimization of the ligand-efficient hits. The co-crystal structure with KAT3A shows that the pyrazolopiperidine core of the compound occupies the KAc binding site, the *N*-methylpyrazole packs along the surface defined by the LPF shelf, while the tetrahydrofuran sits under the ZA loop. It has been reported that GNE-72 shows a marked antiproliferative effect in hematologic cancer cell lines and modulates *MYC* expression *in vivo* that corresponds with antitumor activity in an AML tumor model [241].

Different from KAT3A/3B bromodomains, only a small number of inhibitors have been developed for KAT2A/2B bromodomains. Binding of KAT2B bromodomain to the acetylated K50 moiety of the HIV-1-TAT protein is required to induce transcription of the integrated HIV-1 provirus. The first inhibitor identified for KAT2B bromodomain was N^1-(4-methyl-2-nitrophenyl)propane-1,3-diamine (Figure 12.6), identified through an NMR screen [242]. The compound was shown to disrupt the specific interaction between KAT2B and the K50Ac HIV TAT protein. The closely related N^1-(4-amino-2-nitrophenyl)propane-1,3-diamine (Figure 12.6) successfully inhibited the transactivation of the HIV-1 promoter by blocking the KAT2B bromodomain in C1866 cells with an IC_{50} value of 0.6 μM [243]. Subsequent work to optimize this series were done [244], yet it has been reported that higher concentrations of the inhibitors are cytotoxic, which is a drawback for further development of this scaffold. Various fragment-based screening campaigns led to the identification of potential starting points for more potent and selective KAT2B bromodomain inhibitors [47a, 245]. As a matter of fact, three patent applications recently disclosed a number of highly potent dual KAT2A/KAT2B bromodomain inhibitors from three different chemotypes, namely, pyrido[2,3-*d*]pirydazin-5-one, 1,2,4-triazine-3,5-dione, and pyridazin-3-one (Figure 12.6), demonstrating significant improvements in potency from what had been previously reported [246]. Noteworthy, an independent research identified another pyridazine-3-one derivative, GSK4027 (Figure 12.6), as inhibitor of KAT2A/2B bromodomain from the optimization of a weakly potent, nonselective pyridazinone hit. The compound has been reported to be highly potent for the KAT2A/KAT2B bromodomain, highly soluble, capable to engage the target in a cellular context, and ≥18 000-fold selectivity over the BET family, together with ≥70-fold selectivity over the wider bromodomain families [246]. These chemical probes for the KAT2A/KAT2B bromodomains would facilitate further elucidation of the bromodomain function and its role in mammalian cell function.

12.8 Conclusion

After over 20 years since the discovery of the first acetyltransferase, the understanding of the role of KATs in biology has rapidly evolved. This caused a remarkable biomedical impact, with the majority of findings pointing to a pivotal role not only in cancer but also in metabolic diseases, CNS disorders, and infectious pathologies. Yet, many points still need to be addressed, including "orphan" KAT functions that do not involve modification of histones, are not carried out in the nucleus, and/or mediate biological functions distinct from transcription. The role of nonenzymatic versus enzymatic acetylation, in particular in mitochondria, and the effects on KAT catalytic activity of the availability of alternative acyl-CoA cofactors are other relevant open questions. Therefore, for drug discovery but especially chemical epigenetics, there is clearly still a big demand for high-quality KAT modulators.

Surprisingly, despite numerous studies and rich structural data, there is still no drug in clinical trial, and even the development of high-quality chemical probes is hampered by the high conservation of the cofactor binding site, the lack of standardized and reliable assay protocols to achieve comparability of biochemical data, and the drawbacks affecting most inhibitors, such as low *in vivo* potency, metabolic instability, or poor selectivity. Only recently an integrated approach resulting from combined expertise and efforts (synthetic and computational medicinal chemistry, biophysics, biochemistry, molecular, and structural biology) led to the identification of A-485, a highly potent, selective, cell, and *in vivo* active KAT3A/3B inhibitor. Hopefully, similar approaches could also be applied more broadly to improve the chances of developing potent and selective modulators of other KAT enzymes, thus taking important steps toward the validation of these targets for drug development.

References

1 Allfrey, V.G., Faulkner, R., and Mirsky, A.E. (1964). *Proc. Natl. Acad. Sci. U.S.A.* 51 (5): 786–794.

2 Allis, C.D., Berger, S.L., Cote, J. et al. (2007). *Cell* 131 (4): 633–636.

3 Zeng, L. and Zhou, M.-M. (2002). *FEBS Lett.* 513 (1): 124–128.

4 (a) Baumann, K. (2015). *Nat. Rev. Mol. Cell Biol.* 16 (5): 265–265. (b) Chen, Y., Sprung, R., Tang, Y. et al. (2007). *Mol. Cell. Proteom.* 6 (5): 812–819. (c) Wagner, G.R. and Hirschey, M.D. (2014). *Mol. Cell* 54 (1): 5–16. (d) Leemhuis, H., Packman, L.C., Nightingale, K.P., and Hollfelder, F. (2008). *ChemBioChem* 9 (4): 499–503. (e) Montgomery, D.C., Sorum, A.W., and Meier, J.L. (2014). *J. Am. Chem. Soc.* 136 (24): 8669–8676. (f) Montgomery, D.C., Sorum, A.W., and Meier, J.L. (2015). *ACS Chem. Biol.* 10 (1): 85–94.

5 Pande, V. (2016). *J. Med. Chem.* 59 (4): 1299–1307.

6 (a) Johns, E.W., Phillips, D.M.P., Simson, P., and Butler, J.A.V. (1961). *Biochem. J.* 80 (1): 189–193. (b) Phillips, D.M.P. (1963). *Biochem. J.* 87 (2): 258–263.

7 Cohen, I., Poręba, E., Kamieniarz, K., and Schneider, R. (2011). *Genes Cancer* 2 (6): 631–647.

8 Kouzarides, T. (2007). *Cell* 128 (4): 693–705.

9 (a) Bannister, A.J. and Kouzarides, T. (2011). *Cell Res.* 21 (3): 381–395. (b) Görisch, S.M., Wachsmuth, M., Tóth, K.F. et al. (2005). *J. Cell Sci.* 118 (24): 5825–5834. (c) Smith, B.C. and Denu, J.M. (2009). *Biochim. Biophys. Acta, Gene Regul. Mech.* 1789 (1): 45–57.

10 Marmorstein, R. and Zhou, M.-M. (2014). *Cold Spring Harbor Perspect. Biol.* 6(7).

11 Sterner, D.E. and Berger, S.L. (2000). *Microbiol. Mol. Biol. Rev.* 64 (2): 435–459.

12 (a) Chen, Y., Zhao, W., Yang, J.S. et al. (2012). *Mol. Cell. Proteom.* 11 (10): 1048–1062. (b) Choudhary, C., Kumar, C., Gnad, F. et al. (2009). *Science* 325 (5942): 834–840. (c) Kim, S.C., Sprung, R., Chen, Y. et al. (2006). *Mol. Cell* 23 (4): 607–618. (d) Zhao, S., Xu, W., Jiang, W. et al. (2010). *Science* 327 (5968): 1000–1004.

13 Mellert, H.S. and McMahon, S.B. (2009). *Trends Biochem. Sci.* 34 (11): 571–578.

14 (a) Hirschey, M.D., Shimazu, T., Goetzman, E. et al. (2010). *Nature* 464 (7285): 121–125. (b) Shimazu, T., Hirschey, M.D., Hua, L. et al. (2010). *Cell Metab.* 12 (6): 654–661.

15 (a) Lombard, D.B., Alt, F.W., Cheng, H.-L. et al. (2007). *Mol. Cell. Biol.* 27 (24): 8807–8814. (b) Feldman, J.L., Dittenhafer-Reed, K.E., and Denu, J.M. (2012). *J. Biol. Chem.* 287 (51): 42419–42427.

16 Scott, I., Webster, B.R., Li, J.H., and Sack, M.N. (2012). *Biochem. J.* 443 (3): 655–661.

17 Baeza, J., Smallegan, M.J., and Denu, J.M. (2015). *ACS Chem. Biol.* 10 (1): 122–128.

18 (a) Caton, P.W., Nayuni, N.K., Kieswich, J. et al. (2010). *J. Endocrinol.* 205 (1): 97–106. (b) Farria, A., Li, W., and Dent, S.Y.R. (2015). *Oncogene* 34 (38): 4901–4913. (c) Feinberg, A.P. (2007). *Nature* 447 (7143): 433–440. (d) Schneider, A., Chatterjee, S., Bousiges, O. et al. (2013). *Neurotherapeutics* 10 (4): 568–588. (e) Simon, R.P., Robaa, D., Alhalabi, Z. et al. (2016). *J. Med. Chem.* 59 (4): 1249–1270. (f) Sun, X.-J., Man, N., Tan, Y. et al. (2015). *Front. Oncol.* 5 (108). (g) Wang, Y., Miao, X., Liu, Y. et al. (2014). *Oxid. Med. Cell. Longev.* 2014: 11.

19 Lee, K.K. and Workman, J.L. (2007). *Nat. Rev. Mol. Cell Biol.* 8 (4): 284–295.

20 Parthun, M.R., Widom, J., and Gottschling, D.E. (1996). *Cell* 87 (1): 85–94.

21 (a) Roth, S.Y., Denu, J.M., and Allis, C.D. (2001). *Annu. Rev. Biochem.* 70 (1): 81–120. (b) Ge, Z., Wang, H., and Parthun, M.R. (2011). *J. Biol. Chem.* 286 (19): 16790–16799. (c) Zhang, Q., Chen, L., Yu, X. et al. (2016). *J. Basic Microbiol.* 56 (12): 1380–1391.

22 Parthun, M.R. (2007). *Oncogene* 26 (37): 5319–5328.

23 Fan, J., Shan, C., Kang, H.-B. et al. (2014). *Mol. Cell* 53 (4): 534–548.

24 Yang, X., Yu, W., Shi, L. et al. (2011). *Mol. Cell* 44 (1): 39–50.

25 Mak, A.B., Pehar, M., Nixon, A.M.L. et al. (2014). *J. Mol. Biol.* 426 (11): 2175–2182.

26 (a) Støve, S.I., Magin, R.S., Foyn, H. et al. (2016). *Structure* 24 (7): 1044–1056. (b) Aksnes, H., Van Damme, P., Goris, M. et al. (2015). *Cell Rep.* 10 (8): 1362–1374.

27 Brownell, J.E., Zhou, J., Ranalli, T. et al. (1996). *Cell* 84 (6): 843–851.

28 Yang, X.-J., Ogryzko, V.V., Nishikawa, J.-i. et al. (1996). *Nature* 382 (6589): 319–324.

29 Wittschieben, B.Ø., Otero, G., de Bizemont, T. et al. (1999). *Mol. Cell* 4 (1): 123–128.

30 Sampath, V., Liu, B., Tafrov, S. et al. (2013). *J. Biol. Chem.* 288 (30): 21506–21513.

31 Lorch, Y., Beve, J., Gustafsson, C.M. et al. *Mol. Cell* 6 (1): 197–201.

32 Wang, L., Mizzen, C., Ying, C. et al. (1997). *Mol. Cell. Biol.* 17 (1): 519–527.

33 Sobel, R.E., Cook, R.G., and Allis, C.D. (1994). *J. Biol. Chem.* 269 (28): 18576–18582.

34 (a) Guo, R., Chen, J., Mitchell, D.L., and Johnson, D.G. (2011). *Nucleic Acids Res.* 39 (4): 1390–1397. (b) Qin, S. and Parthun, M.R. (2002). *Mol. Cell. Biol.* 22 (23): 8353–8365.

35 Kalebic, N., Sorrentino, S., Perlas, E. et al. (2013). *Nat. Commun.* 4: 1962.

36 Neuwald, A.F. and Landsman, D. (1997). *Trends Biochem. Sci.* 22 (5): 154–155.

37 (a) Dyda, F., Klein, D.C., and Hickman, A.B. (2000). *Ann. Rev. Biophys. Biomol. Struct.* 29 (1): 81–103. (b) Wolf, E., Vassilev, A., Makino, Y. et al. (1998). *Cell* 94 (4): 439–449.

38 Marmorstein, R. (2001). *Nat. Rev. Mol. Cell Biol.* 2 (6): 422–432.

39 Dhalluin, C., Carlson, J.E., Zeng, L. et al. (1999). *Nature* 399 (6735): 491–496.

40 Georgakopoulos, T. and Thireos, G. (1992). *EMBO J.* 11 (11): 4145–4152.

41 Brownell, J.E. and Allis, C.D. (1995). *Proc. Natl. Acad. Sci.* 92 (14): 6364–6368.

42 Tanner, K.G., Trievel, R.C., Kuo, M.-H. et al. (1999). *J. Biol. Chem.* 274 (26): 18157–18160.

43 Tanner, K.G., Langer, M.R., Kim, Y., and Denu, J.M. (2000). *J. Biol. Chem.* 275 (29): 22048–22055.

44 Wapenaar, H. and Dekker, F.J. (2016). *Clin. Epigenet.* 8 (1): 59.

45 (a) Bannister, A.J. and Kouzarides, T. (1996). *Nature* 384 (6610): 641–643. (b) Ogryzko, V.V., Schiltz, R.L., Russanova, V. et al. (1996). *Cell* 87 (5): 953–959.

46 (a) Linares, L.K., Kiernan, R., Triboulet, R. et al. (2007). *Nat. Cell Biol.* 9 (3): 331–338. (b) Lau, O.D., Courtney, A.D., Vassilev, A. et al. (2000). *J. Biol. Chem.* 275 (29): 21953–21959. (c) Lau, O.D., Kundu, T.K., Soccio, R.E. et al. (2000). *Mol. Cell* 5 (3): 589–595.

47 (a) Chaikuad, A., Lang, S., Brennan, P.E. et al. (2016). *J. Med. Chem.* 59 (4): 1648–1653. (b) Pérez-Salvia, M. and Esteller, M. (2017). *Epigenetics* 12 (5): 323–339.

48 (a) Grant, P.A., Duggan, L., Côté, J. et al. (1997). *Genes Dev.* 11 (13): 1640–1650. (b) Ogryzko, V.V., Kotani, T., Zhang, X. et al. (1998). *Cell* 94 (1): 35–44.

49 (a) Guelman, S., Suganuma, T., Florens, L. et al. (2006). *Mol. Cell. Biol.* 26 (3): 871–882. (b) Wang, Y.-L., Faiola, F., Xu, M. et al. (2008). *J. Biol. Chem.* 283 (49): 33808–33815.

50 Grant, P.A., Eberharter, A., John, S. et al. (1999). *J. Biol. Chem.* 274 (9): 5895–5900.

51 Riss, A., Scheer, E., Joint, M. et al. (2015). *J. Biol. Chem.* 290 (48): 28997–29009.

52 Atanassov, B.S., Evrard, Y.A., Multani, A.S. et al. (2009). *Mol. Cell* 35 (3): 352–364.

53 Rodríguez-Navarro, S., Fischer, T., Luo, M.-J. et al. (2004). *Cell* 116 (1): 75–86.

54 Ramachandran, S., Haddad, D., Li, C. et al. (2016). *Cell Rep.* 15 (7): 1554–1565.

55 (a) Bian, C., Xu, C., Ruan, J. et al. (2011). *EMBO J.* 30 (14): 2829–2842. (b) Ringel, A.E., Cieniewicz, A.M., Taverna, S.D., and Wolberger, C. (2015). *Proc. Natl. Acad. Sci.* 112 (40): E5461–E5470.

56 Hassan, A.H., Prochasson, P., Neely, K.E. et al. (2002). *Cell* 111 (3): 369–379.

57 Lang, G., Bonnet, J., Umlauf, D. et al. (2011). *Mol. Cell. Biol.* 31 (18): 3734–3744.

58 Larschan, E. and Winston, F. (2001). *Genes Dev.* 15 (15): 1946–1956.

59 (a) McMahon, S.B., Van Buskirk, H.A., Dugan, K.A. et al. (1998). *Cell* 94 (3): 363–374. (b) McMahon, S.B., Wood, M.A., and Cole, M.D. (2000). *Mol.Cell. Biol.* 20 (2): 556–562.

60 Grant, P.A., Schieltz, D., Pray-Grant, M.G. et al. (1998). *Cell* 94 (1): 45–53.

61 (a) Suganuma, T., Gutierrez, J.L., Li, B. et al. (2008). *Nat. Struct. Mol. Biol.* 15 (4): 364–372. (b) Zhao, D., Guan, H., Zhao, S. et al. (2016). *Cell Res.* 26 (5): 629–632.

62 Shi, S., Lin, J., Cai, Y. et al. (2014). *BMC Struct. Biol.* 14 (1): 2.

63 Arnaud, R., Krebs, K., Karmodiya, M. et al. (2011). *Mol. Cell* 44 (3): 410–423.

64 (a) Nagy, Z., Riss, A., Fujiyama, S. et al. (2010). *Cell. Mol. Life Sci.* 67 (4): 611–628. (b) Nagy, Z., Riss, A., Romier, C. et al. (2009). *Mol. Cell. Biol.* 29 (6): 1649–1660.

65 Bonnet, J., Wang, C.-Y., Baptista, T. et al. (2014). *Genes Dev.* 28 (18): 1999–2012.

66 Hirsch, C.L., Coban Akdemir, Z., Wang, L. et al. (2015). *Genes Dev.* 29 (8): 803–816.

67 Orpinell, M., Fournier, M., Riss, A. et al. (2010). *EMBO J.* 29 (14): 2381–2394.

68 (a) Xu, W., Edmondson, D.G., Evrard, Y.A. et al. (2000). *Nat. Genet.* 26 (2): 229–232. (b) Yamauchi, T., Yamauchi, J., Kuwata, T. et al. (2000). *Proc. Natl. Acad. Sci. U.S.A.* 97 (21): 11303–11306.

69 Verreault, A., Kaufman, P.D., Kobayashi, R., and Stillman, B. (1998). *Curr. Biol.* 8 (2): 96–108.

70 Campos, E.I., Fillingham, J., Li, G. et al. (2010). *Nat. Struct. Mol. Biol.* 17 (11): 1343–1351.

71 Annunziato, A.T. and Seale, R.L. (1983). *J. Biol. Chem.* 258 (20): 12675–12684.

72 Parthun, M.R. (2012). *Biochim. Biophys. Acta, Gene Regul. Mech.* 1819 (3–4): 256–263.

73 Nagarajan, P., Ge, Z., Sirbu, B. et al. (2013). *PLoS Genet.* 9 (6): e1003518.

74 lnoue, M., Isomura, M., lkegawa, S. et al. (1996). *Cytogenet. Genome Res.* 73 (1–2): 134–136.

75 Scott, I., Webster, B.R., Chan, C.K. et al. (2014). *J. Biol. Chem.* 289 (5): 2864–2872.

76 Wang, L., Scott, I., Zhu, L. et al. (2017). *Nat. Commun.* 8 (1): 523.

77 (a) Chrivia, J.C., Kwok, R.P.S., Lamb, N. et al. (1993). *Nature* 365 (6449): 855–859. (b) Harlow, E., Whyte, P., Franza, B.R., and Schley, C. (1986). *Mol. Cell. Biol.* 6 (5): 1579–1589.

78 Dancy, B.M. and Cole, P.A. (2015). *Chem. Rev.* 115 (6): 2419–2452.

79 Liu, X., Wang, L., Zhao, K. et al. (2008). *Nature* 451 (7180): 846–850.

80 Poux, A.N., Cebrat, M., Kim, C.M. et al. (2002). *Proc. Natl. Acad. Sci. U.S.A.* 99 (22): 14065–14070.

81 (a) Hay, D.A., Fedorov, O., Martin, S. et al. (2014). *J. Am. Chem. Soc.* 136 (26): 9308–9319. (b) Zucconi, B.E., Luef, B., Xu, W. et al. (2016). *Biochemistry* 55 (27): 3727–3734.

82 (a) Delvecchio, M., Gaucher, J., Aguilar-Gurrieri, C. et al. (2013). *Nat. Struct. Mol. Biol.* 20 (9): 1040–1046. (b) Rack, J.G.M., Lutter, T., Kjæreng Bjerga, G.E. et al. (2014). *J. Mol. Biol.* 426 (24): 3960–3972.

83 Park, S., Stanfield, R.L., Martinez-Yamout, M.A. et al. (2017). *Proc. Natl. Acad. Sci. U.S.A.* 114 (27): E5335–E5342.

84 (a) Kwok, R.P.S., Lundblad, J.R., Chrivia, J.C. et al. (1994). *Nature* 370 (6486): 223–226. (b) Lundblad, J.R., Kwok, R.P.S., Laurance, M.E. et al. (1995). *Nature* 374 (6517): 85–88. (c) Eckner, R., Ewen, M.E., Newsome, D. et al. (1994). *Genes Dev.* 8 (8): 869–884.

85 Schiltz, R.L., Mizzen, C.A., Vassilev, A. et al. (1999). *J. Biol. Chem.* 274 (3): 1189–1192.

86 Jin, Q., Yu, L.-R., Wang, L. et al. (2011). *EMBO J.* 30 (2): 249–262.

87 (a) Costantini, C., Ko, M.H., Jonas, M.C., and Puglielli, L. (2007). *Biochem. J.* 407 (3): 383–395. (b) Xie, W., Song, C., Young, N.L. et al. (2009). *Mol. Cell* 33 (4): 417–427.

88 Benjamin, R., Sabari, Z., Tang, H. et al. (2015). *Mol. Cell* 58 (2): 203–215.

89 (a) Heintzman, N.D., Stuart, R.K., Hon, G. et al. (2007). *Nat. Genet.* 39 (3): 311–318. (b) Wang, Z., Zang, C., Cui, K. et al. (2009). *Cell* 138 (5): 1019–1031.

90 (a) Visel, A., Blow, M.J., Li, Z. et al. (2009). *Nature* 457 (7231): 854–858. (b) Creyghton, M.P., Cheng, A.W., Welstead, G.G. et al. (2010). *Proc. Natl. Acad. Sci. U.S.A.* 107 (50): 21931–21936. (c) Rada-Iglesias, A., Bajpai, R., Swigut, T. et al. (2011). *Nature* 470 (7333): 279–283.

91 Ding, J., Huang, X., Shao, N. et al. (2015). *Cell Stem Cell* 16 (6): 653–668.

92 Whyte, W.A., Orlando, D.A., Hnisz, D. et al. (2013). *Cell* 153 (2): 307–319.

93 (a) Yao, T.-P., Oh, S.P., Fuchs, M. et al. (1998). *Cell* 93 (3): 361–372. (b) Kung, A.L., Rebel, V.I., Bronson, R.T. et al. (2000). *Genes Dev.* 14 (3): 272–277. (c) Partanen, A., Motoyama, J., and Hui, C.C. (1999). *Int. J. Dev. Biol.* 43 (6): 487–494.

94 Phan, H.M., Xu, A.W., Coco, C. et al. (2005). *Dev. Dyn.* 233 (4): 1337–1347.

95 Bordoli, L., Netsch, M., Lüthi, U. et al. (2001). *Nucleic Acids Res.* 29 (3): 589–597.

96 Tan, M., Peng, C., Anderson, K.A. et al. (2014). *Cell Metab.* 19 (4): 605–617.

97 (a) Cai, L., Sutter, B.M., Li, B., and Tu, B.P. (2011). *Mol. Cell* 42 (4): 426–437. (b) Wellen, K.E., Hatzivassiliou, G., Sachdeva, U.M. et al. (2009). *Science* 324 (5930): 1076–1080.

98 Borrow, J., Stanton, V.P., Andresen, J.M. et al. (1996). *Nat. Genet.* 14 (1): 33–41.

99 Sapountzi, V. and Côté, J. (2011). *Cell. Mol. Life Sci.* 68 (7): 1147–1156.

100 Yang, X.J. (2004). *Nucleic Acids Res.* 32 (3): 959–976.

101 Sapountzi, V., Logan, I.R., and Robson, C.N. (2006). *Int. J. Biochem. Cell Biol.* 38 (9): 1496–1509.

102 Ikura, T., Ogryzko, V.V., Grigoriev, M. et al. (2000). *Cell* 102 (4): 463–473.

103 (a) Fazzio, T.G., Huff, J.T., and Panning, B. (2008). *Cell* 134 (1): 162–174. (b) Frank, S.R., Parisi, T., Taubert, S. et al. (2003). *EMBO Rep.* 4 (6): 575–580. (c) Taubert, S., Gorrini, C., Frank, S.R. et al. (2004). *Mol. Cell. Biol.* 24 (10): 4546–4556. (d) Sierra, J., Yoshida, T., Joazeiro, C.A., and Jones, K.A. (2006). *Genes Dev.* 20 (5): 586–600. (e) Chen, P.B., Chen, H.V., Acharya, D. et al. (2015). *Nat. Struct. Mol. Biol.* 22 (12): 999–1007. (f) Ravens, S., Yu, C., Ye, T. et al. (2015). *Epigenet. Chromatin* 8 (1): 45.

104 Jacquet, K., Fradet-Turcotte, A., Avvakumov, N. et al. (2016). *Mol. Cell* 62 (3): 409–421.

105 (a) Herceg, Z., Hulla, W., Gell, D. et al. (2001). *Nat. Genet.* 29 (2): 206–211. (b) Hu, Y., Fisher, J.B., Koprowski, S. et al. (2009). *Dev. Dyn.* 238 (11): 2912–2921. (c) Mohan, K.N., Ding, F., and Chaillet, J.R. (2011). *Mol. Cell. Biol.* 31 (9): 1861–1869.

106 Doyon, Y., Cayrou, C., Ullah, M. et al. (2006). *Mol. Cell* 21 (1): 51–64.

107 Mishima, Y., Miyagi, S., Saraya, A. et al. (2011). *Blood* 118 (9): 2443–2453.

108 Lalonde, M.-E., Avvakumov, N., Glass, K.C. et al. (2013). *Genes Dev.* 27 (18): 2009–2024.

109 Iizuka, M. and Stillman, B. (1999). *J. Biol. Chem.* 274 (33): 23027–23034.

110 Champagne, N., Bertos, N.R., Pelletier, N. et al. (1999). *J. Biol. Chem.* 274 (40): 28528–28536.

111 Su, J., Wang, F., Cai, Y., and Jin, J. (2016). *Int. J. Mol. Sci.* 17 (1): 99.

112 Smith, E.R., Cayrou, C., Huang, R. et al. (2005). *Mol. Cell. Biol.* 25 (21): 9175–9188.

113 Cai, Y., Jin, J., Swanson, S.K. et al. (2010). *J. Biol. Chem.* 285 (7): 4268–4272.

114 Copeland, R.A. (2002). *Enzymes*, 350–366. Wiley.

115 Wapenaar, H., van der Wouden, P.E., Groves, M.R. et al. (2015). *Eur. J. Med. Chem.* 105: 289–296.

116 Yuan, H., Rossetto, D., Mellert, H. et al. (2012). *EMBO J.* 31 (1): 58–70.

117 Thomas, T., Dixon, M.P., Kueh, A.J., and Voss, A.K. (2008). *Mol. Cell. Biol.* 28 (16): 5093–5105.

118 Li, X., Li, L., Pandey, R. et al. (2012). *Cell Stem Cell* 11 (2): 163–178.

119 Reifsnyder, C., Lowell, J., Clarke, A., and Pillus, L. (1996). *Nat. Genet.* 14 (1): 42–49.

120 (a) John, S., Howe, L., Tafrov, S.T. et al. (2000). *Genes Dev.* 14 (10): 1196–1208. (b) Takechi, S. and Nakayama, T. (1999). *Biochem. Biophys. Res. Commun.* 266 (2): 405–410. (c) Howe, L., Auston, D., Grant, P. et al. (2001). *Genes Dev.* 15 (23): 3144–3154. (d) Martin, D.G.E., Grimes, D.E., Baetz, K., and Howe, L. (2006). *Mol. Cell. Biol.* 26 (8): 3018–3028. (e) Taverna, S.D., Ilin, S., Rogers, R.S. et al. (2006). *Mol. Cell* 24 (5): 785–796.

121 Smith, E.R., Eisen, A., Gu, W. et al. (1998). *Proc. Natl. Acad. Sci. U.S.A.* 95 (7): 3561–3565.
122 Yan, Y., Barlev, N.A., Haley, R.H. et al. (2000). *Mol. Cell* 6 (5): 1195–1205.
123 Yan, Y., Harper, S., Speicher, D.W., and Marmorstein, R. (2002). *Nat. Struct. Mol. Biol.* 9 (11): 862–869.
124 Berndsen, C.E., Albaugh, B.N., Tan, S., and Denu, J.M. (2007). *Biochemistry* 46 (3): 623–629.
125 (a) Spencer, T.E., Jenster, G., Burcin, M.M. et al. (1997). *Nature* 389 (6647): 194–198. (b) York, B. and O'Malley, B.W. (2010). *J. Biol. Chem.* 285 (50): 38743–38750.
126 Doi, M., Hirayama, J., and Sassone-Corsi, P. (2006). *Cell* 125 (3): 497–508.
127 (a) Hsieh, Y.-J., Kundu, T.K., Wang, Z. et al. (1999). *Mol. Cell. Biol.* 19 (11): 7697–7704. (b) Mizzen, C.A., Yang, X.-J., Kokubo, T. et al. (1996). *Cell* 87 (7): 1261–1270. (c) Ogryzko, V.V. (2001). *Cell. Mol. Life Sci.* 58 (5): 683–692.
128 Marmorstein, R. and Trievel, R.C. (2009). *Biochim. Biophys. Acta, Gene Regul. Mech.* 1789 (1): 58–68.
129 Yuan, H. and Marmorstein, R. (2013). *Biopolymers* 99 (2): 98–111.
130 Lopes da Rosa, J., Bajaj, V., Spoonamore, J., and Kaufman, P.D. (2013). *Bioorg. Med. Chem. Lett.* 23 (10): 2853–2859.
131 Karmodiya, K., Anamika, K., Muley, V. et al. (2014). *Sci. Rep.* 4: 6076.
132 Friedmann, D.R., Aguilar, A., Fan, J. et al. (2012). *Proc. Natl. Acad. Sci. U.S.A.* 109 (48): 19655–19660.
133 Hou, F. and Zou, H. (2005). *Mol. Biol. Cell* 16 (8): 3908–3918.
134 (a) Matthias, P., Yoshida, M., and Khochbin, S. (2008). *Cell Cycle* 7 (1): 7–10. (b) Yang, X.-J. and Seto, E. (2008). *Nat. Rev. Mol. Cell Biol.* 9 (3): 206–218. (c) Arif, M., Selvi, B.R., and Kundu, T.K. *ChemBioChem* 2010 (11, 11): 1501–1504. (d) Hosp, F., Lassowskat, I., Santoro, V. et al. (2017). *Mitochondrion* 33: 58–71.
135 (a) von Wantoch Rekowski, M. and Giannis, A. (2010). *Biochim. Biophys. Acta, Gene Regul. Mech.* 1799 (10–12): 760–767. (b) Wang, Q., Zhang, Y., Yang, C. et al. (2010). *Science* 327 (5968): 1004–1007.
136 (a) Dekker, F.J. and Haisma, H.J. (2009). *Drug Discovery Today* 14 (19-20): 942–948. (b) Manzo, F., Tambaro, F.P., Mai, A., and Altucci, L. (2009). *Expert Opin. Ther. Pat.* 19 (6): 761–774. (c) Furdas, S.D., Kannan, S., Sippl, W., and Jung, M. (2012). *Arch. Pharm.* 345 (1): 7–21. (d) Castellano, S., Milite, C., Feoli, A. et al. (2015). *ChemMedChem* 10 (1): 144–157.
137 (a) Portela, A. and Esteller, M. (2010). *Nat. Biotechnol.* 28 (10): 1057–1068. (b) Selvi, B.R., Chatterjee, S., Modak, R. et al. (2013). *Epigenetics: Development and Disease* (ed. T.K. Kundu), 567–596. Dordrecht, Netherlands: Springer.
138 Petrij, F., Dorsman, J.C., Dauwerse, H.G. et al. (2000). *Am. J. Med. Genet.* 92 (1): 47–52.
139 (a) Bayly, R., Chuen, L., Currie, R.A. et al. (2004). *J. Biol. Chem.* 279 (53): 55362–55371. (b) Morin, R.D., Mendez-Lago, M., Mungall, A.J. et al. (2011). *Nature* 476 (7360): 298–303.

140 (a) Chaffanet, M., Gressin, L., Preudhomme, C. et al. (2000). *Genes Chromosomes Cancer* 28 (2): 138–144. (b) Satake, N., Ishida, Y., Otoh, Y. et al. (1997). *Genes Chromosomes Cancer* 20 (1): 60–63.
141 Li, Y., Yang, H.-X., Luo, R.-Z. et al. (2011). *Ann. Thorac. Surg.* 91 (5): 1531–1538.
142 Li, M., Luo, R.-Z., Chen, J.-W. et al. (2011). *J. Transl. Med.* 9 (1): 5.
143 Bandyopadhyay, D., Okan, N.A., Bales, E. et al. (2002). *Cancer Res.* 62 (21): 6231–6239.
144 Pattabiraman, D.R., McGirr, C., Shakhbazov, K. et al. (2014). *Blood* 123 (17): 2682–2690.
145 Sun, X.-J., Wang, Z., Wang, L. et al. (2013). *Nature* 500 (7460): 93–97.
146 Ianculescu, I., Wu, D.-Y., Siegmund, K.D., and Stallcup, M.R. (2012). *J. Biol. Chem.* 287 (6): 4000–4013.
147 Ionov, Y., Matsui, S.-I., and Cowell, J.K. (2004). *Proc. Natl. Acad. Sci. U.S.A.* 101 (5): 1273–1278.
148 Takeuchi, A., Shiota, M., Tatsugami, K. et al. (2012). *Mol. Med. Rep.* 5 (1): 173–176.
149 (a) Cereseto, A., Manganaro, L., Gutierrez, M.I. et al. (2005). *EMBO J.* 24 (17): 3070–3081. (b) Mantelingu, K., Reddy, B.A.A., Swaminathan, V. et al. (2007). *Chem. Biol.* 14 (6): 645–657. (c) Sadoul, K., Wang, J., Diagouraga, B., and Khochbin, S. (2011). *J. Biomed. Biotechnol.* 2011: 970382.
150 Deng, L., de la Fuente, C., Fu, P. et al. (2000). *Virology* 277 (2): 278–295.
151 He, H., Lai, Y., Hao, Y. et al. (2017). *Eur. J. Pharmacol.* .
152 Yuan, H., Reddy, M.A., Sun, G. et al. (2013). *Am. J. Physiol. Renal. Physiol.* 304 (5): F601–F613.
153 Chen, S., Feng, B., George, B. et al. (2010). *Am. J. Physiol.– Endocrinol. Metab.* 298 (1): E127–E137.
154 Gusterson, R., Brar, B., Faulkes, D. et al. (2002). *J. Biol. Chem.* 277 (4): 2517–2524.
155 Mitchnick, K.A., Creighton, S.D., Cloke, J.M. et al. (2016). *Genes Brain Behav.* 15 (6): 542–557.
156 Kikuchi, H., Takami, Y., and Nakayama, T. (2005). *Gene* 347 (1): 83–97.
157 Pérez-Luna, M., Aguasca, M., Perearnau, A. et al. (2012). *Nucleic Acids Res.* 40 (14): 6520–6533.
158 Shiota, M., Yokomizo, A., Tada, Y. et al. (2010). *Cancer Sci.* 101 (8): 1797–1806.
159 Toth, M., Boros, I.M., and Balint, E. (2012). *Cancer Sci.* 103 (4): 659–669.
160 Terreni, M., Valentini, P., Liverani, V. et al. (2010). *Retrovirology* 7 (1): 18.
161 (a) Miao, J., Fan, Q., Cui, L. et al. (2006). *Gene* 369: 53–65. (b) Cui, L., Miao, J., Furuya, T. et al. (2007). *Eukaryotic Cell* 6 (7): 1219–1227.
162 (a) Duclot, F., Meffre, J., Jacquet, C. et al. (2010). *Neuroscience* 167 (3): 850–863. (b) Park, S.Y., Kim, M.J., Kim, Y.J. et al. (2015). *Int. J. Mol. Med.* 35 (4): 1109–1118.
163 (a) Colussi, C., Rosati, J., Straino, S. et al. (2011). *Proc. Natl. Acad. Sci. U.S.A.* 108 (7): 2795–2800. (b) Colussi, C., Scopece, A., Vitale, S. et al. (2012). *Arterioscler. Thromb. Vasc. Biol.* 32 (10): 2435–2443.
164 Sun, C., Wang, M., Liu, X. et al. (2014). *Cell Rep.* 9 (6): 2250–2262.

165 Vecellio, M., Spallotta, F., Nanni, S. et al. (2014). *Diabetes* 63 (6): 2132–2147.

166 Spallotta, F., Cencioni, C., Straino, S. et al. (2013). *Commun. Integr. Biol.* 6 (5): e25466.

167 (a) Blouin, A.M., Sillivan, S.E., Joseph, N.F., and Miller, C.A. (2016). *Learn. Memory* 23 (10): 576–586. (b) Siddiqui, S.A., Singh, S., Ranjan, V. et al. (2017). *Cell. Mol. Neurobiol.* . (c) Wei, W., Coelho, C.M., Li, X. et al. (2012). *J. Neurosci.* 32 (35): 11930–11941.

168 Kindle, K.B., Troke, P.J.F., Collins, H.M. et al. (2005). *Mol. Cell. Biol.* 25 (3): 988–1002.

169 Dulak, A.M., Stojanov, P., Peng, S. et al. (2013). *Nat. Genet.* 45 (5): 478–486.

170 (a) Moore, S.D.P., Herrick, S.R., Ince, T.A. et al. (2004). *Cancer Res.* 64 (16): 5570–5577. (b) Yang, X.-J. (2015). *Biochim. Biophys. Acta, Mol. Cell Res.* 1853 (8): 1818–1826.

171 (a) Zhao, H., Jin, S., and Gewirtz, A.M. (2012). *J. Biol. Chem.* 287 (2): 925–934. (b) Chevillard-Briet, M., Quaranta, M., Grézy, A. et al. (2014). *Hum. Mol. Genet.* 23 (8): 2120–2131. (c) Gorrini, C., Squatrito, M., Luise, C. et al. (2007). *Nature* 448 (7157): 1063–1067. (d) Halkidou, K., Gnanapragasam, V.J., Mehta, P.B. et al. (2003). *Oncogene* 22: 2466–2477. (e) ME, L.L., Vidal, F., Gallardo, D. et al. (2006). *Oncol. Rep.* 16 (3): 603–608. (f) Mattera, L., Escaffit, F., Pillaire, M.J. et al. (2009). *Oncogene* 28 (12): 1506–1517. (g) Chen, G., Cheng, Y., Tang, Y. et al. (2012). *J. Invest. Dermatol.* 132 (11): 2632–2641.

172 Miyamoto, N., Izumi, H., Noguchi, T. et al. (2008). *J. Biol. Chem.* 283 (26): 18218–18226.

173 Shiota, M., Yokomizo, A., Masubuchi, D. et al. (2010). *Prostate* 70 (5): 540–554.

174 Iizuka, M., Takahashi, Y., Mizzen, C.A. et al. (2009). *Gene* 436 (1–2): 108–114.

175 Duong, M.T., Akli, S., Macalou, S. et al. (2013). *Cancer Res.* 73 (17): 5556–5568.

176 (a) Aherne, W.G., Rowlands, M.G., Stimson, L., and Workman, P. (2002). *Methods* 26 (3): 245–253. (b) Benson, L.J. and Annunziato, A.T. (2004). *Methods* 33 (1): 45–52. (c) Berndsen, C.E. and Denu, J.M. (2005). *Methods* 36 (4): 321–331. (d) Li, P., Han, Y., Li, Y. et al. (2016). *Anal. Bioanal. Chem.* 408 (11): 2659–2668. (e) Marmorstein, R. (2003). *Methods Enzymol.* 376: 106–119. (f) Poveda, A. and Sendra, R. (2008). *Anal. Biochem.* 383 (2): 296–300.

177 He, M., Han, Z., Liu, L., and Zheng, Y.G. *Angew. Chem. Int. Ed.* 57 (5): 1162–1184.

178 Wu, J., Wang, J., Li, M. et al. (2011). *Bioorg. Chem.* 39 (1): 53–58.

179 Kwie, F.H.A., Briet, M., Soupaya, D. et al. (2011). *Chem. Biol. Drug Des.* 77 (1): 86–92.

180 Cebrat, M., Kim, C.M., Thompson, P.R. et al. (2003). *Bioorg. Med. Chem.* 11 (15): 3307–3313.

181 (a) Wadia, J.S. and Dowdy, S.F. (2005). *Adv. Drug Delivery Rev.* 57 (4): 579–596. (b) Zheng, Y., Balasubramanyam, K., Cebrat, M. et al. (2005). *J. Am. Chem. Soc.* 127 (49): 17182–17183.

182 Bandyopadhyay, K., Banères, J.-L., Martin, A. et al. (2009). *Cell Cycle* 8 (17): 2779–2788.

183 Dal Piaz, F., Vassallo, A., Rubio, O. et al. (2011). *Mol. Diversity* 15 (2): 401–416.

184 Balasubramanyam, K., Swaminathan, V., Ranganathan, A., and Kundu, T.K. (2003). *J. Biol. Chem.* 278 (21): 19134–19140.

185 Ghizzoni, M., Wu, J., Gao, T. et al. (2012). *Eur. J. Med. Chem.* 47: 337–344.

186 Hemshekhar, M., Sebastin Santhosh, M., Kemparaju, K., and Girish, K.S. (2012). *Basic Clin. Pharmacol. Toxicol.* 110 (2): 122–132.

187 Eliseeva, E.D., Valkov, V., Jung, M., and Jung, M.O. (2007). *Mol. Cancer Ther.* 6 (9): 2391–2398.

188 Ghizzoni, M., Boltjes, A., Graaf, C.D. et al. (2010). *Bioorg. Med. Chem.* 18 (16): 5826–5834.

189 Souto, J.A., Benedetti, R., Otto, K. et al. (2010). *ChemMedChem* 5 (9): 1530–1540.

190 Chatterjee, S., Mizar, P., Cassel, R. et al. (2013). *J. Neurosci.* 33 (26): 10698–10712.

191 (a) Milite, C., Castellano, S., Benedetti, R. et al. (2011). *Bioorg. Med. Chem.* 19 (12): 3690–3701. (b) Sbardella, G., Castellano, S., Vicidomini, C. et al. (2008). *Bioorg. Med. Chem. Lett.* 18 (9): 2788–2792.

192 (a) Choi, K.-C., Jung, M.G., Lee, Y.-H. et al. (2009). *Cancer Res.* 69 (2): 583–592. (b) Choi, K.-C., Park, S., Beom, J. et al. (2011). *Biochem. J.* 433 (1): 235–244.

193 Mai, A., Cheng, D., Bedford, M.T. et al. (2008). *J. Med. Chem.* 51 (7): 2279–2290.

194 Balasubramanyam, K., Varier, R.A., Altaf, M. et al. (2004). *J. Biol. Chem.* 279 (49): 51163–51171.

195 Neckers, L., Trepel, J., Lee, S. et al. (2006). *Med. Chem.* 2 (2): 169–174.

196 Fu, S. and Kurzrock, R. (2010). *Cancer* 116 (20): 4670–4676.

197 (a) Costi, R., Di Santo, R., Artico, M. et al. (2007). *J. Med. Chem.* 50 (8): 1973–1977. (b) Madia, V.N., Benedetti, R., Barreca, M.L. et al. (2017). *ChemMedChem* 12 (16): 1359–1368.

198 Arif, M., Vedamurthy, B.M., Choudhari, R. et al. (2010). *Chem. Biol.* 17 (8): 903–913.

199 (a) Ingólfsson, H.I., Thakur, P., Herold, K.F. et al. (2014). *ACS Chem. Biol.* 9 (8): 1788–1798. (b) Nelson, K.M., Dahlin, J.L., Bisson, J. et al. (2017). *J. Med. Chem.* 60 (5): 1620–1637.

200 Balasubramanyam, K., Altaf, M., Varier, R.A. et al. (2004). *J. Biol. Chem.* 279 (32): 33716–33726.

201 Zhou, X.-Y., Cao, J., Han, C.-M. et al. (2017). *Bioorg. Chem.* 71: 74–80.

202 Arif, M., Pradhan, S.K., Thanuja, G.R. et al. (2009). *J. Med. Chem.* 52 (2): 267–277.

203 (a) Baggett, S., Protiva, P., Mazzola, E.P. et al. (2005). *J. Nat. Prod.* 68 (3): 354–360. (b) Dal Piaz, F., Tosco, A., Eletto, D. et al. (2010). *ChemBioChem* 11 (6): 818–827. (c) Gartner, M., Müller, T., Simon, J.C. et al. (2005). *ChemBioChem* 6 (1): 171–177.

204 Milite, C., Feoli, A., Sasaki, K. et al. (2015). *J. Med. Chem.* 58 (6): 2779–2798.

205 Simola, D.F., Graham, R.J., Brady, C.M. et al. (2016). *Science* 351 (6268).

206 Ravindra, K.C., Selvi, B.R., Arif, M. et al. (2009). *J. Biol. Chem.* 284 (36): 24453–24464.

207 Vasudevarao, M.D., Mizar, P., Kumari, S. et al. (2014). *J. Biol. Chem.* 289 (11): 7702–7717.

208 Modak, R., Basha, J., Bharathy, N. et al. (2013). *ACS Chem. Biol.* .

209 Tohyama, S., Tomura, A., Ikeda, N. et al. (2012). *J. Org. Chem.* 77 (20): 9044–9052.

210 Stimson, L., Rowlands, M.G., Newbatt, Y.M. et al. (2005). *Mol. Cancer Ther.* 4 (10): 1521–1532.

211 Furdas, S.D., Shekfeh, S., Bissinger, E.-M. et al. (2011). *Bioorg. Med. Chem.* 19 (12): 3678–3689.

212 Furdas, S.D., Hoffmann, I., Robaa, D. et al. (2014). *MedChemComm* 5 (12): 1856–1862.

213 Carneiro, V.C., de Abreu da Silva, I.C., Torres, E.J.L. et al. (2014). *PLoS Pathog.* 10 (5): e1004116.

214 Gajer, J.M., Furdas, S.D., Grunder, A. et al. (2015). *Oncogenesis* 4: e137.

215 (a) Mai, A., Rotili, D., Tarantino, D. et al. (2006). *J. Med. Chem.* 49 (23): 6897–6907. (b) Lenoci, A., Tomassi, S., Conte, M. et al. (2014). *ChemMedChem* 9 (3): 542–548.

216 (a) Chimenti, F., Bizzarri, B., Maccioni, E. et al. (2009). *J. Med. Chem.* 52 (2): 530–536. (b) Secci, D., Carradori, S., Bizzarri, B. et al. (2014). *Bioorg. Med. Chem.* 22 (5): 1680–1689. (c) Trisciuoglio, D., Ragazzoni, Y., Pelosi, A. et al. (2012). *Clin. Cancer Res.* 18 (2): 475–486.

217 Di Martile, M., Desideri, M., De Luca, T. et al. (2016). *Oncotarget* 7 (10): 11332–11348.

218 Biel, M., Kretsovali, A., Karatzali, E. et al. (2004). *Angew. Chem. Int. Ed.* 43 (30): 3974–3976.

219 (a) Dancy, B.M., Crump, N.T., Peterson, D.J. et al. (2012). *ChemBioChem* 13 (14): 2113–2121. (b) Bowers, E.M., Yan, G., Mukherjee, C. et al. (2010). *Chem. Biol.* 17 (5): 471–482.

220 Shrimp, J.H., Sorum, A.W., Garlick, J.M. et al. (2016). *ACS Med. Chem. Lett.* 7 (2): 151–155.

221 Gao, X.-N., Lin, J., Ning, Q.-Y. et al. (2013). *PLoS One* 8 (2): e55481.

222 Yu, Z., Taniguchi, J., Wei, Y. et al. (2017). *Eur. J. Med. Chem.* 138: 320–327.

223 Yang, H., Pinello, C.E., Luo, J. et al. (2013). *Mol. Cancer Ther.* 12 (5): 610–620.

224 Hao, J., Ao, A., Zhou, L. et al. (2013). *Cell Rep.* 4 (5): 898–904.

225 Proto, M.C., Fiore, D., Piscopo, C. et al. (2017). *Sci. Rep.* 7 (1): 11678.

226 Lasko, L.M., Jakob, C.G., Edalji, R.P. et al. (2017). *Nature* 550 (7674): 128–132.

227 Coffey, K., Blackburn, T.J., Cook, S. et al. (2012). *PLoS One* 7 (10): e45539.

228 Gao, C., Bourke, E., Scobie, M. et al. (2014). *Sci. Rep.* 4: 5372.

229 Wapenaar, H., van den Bosch, T., Leus, N.G.J. et al. (2017). *Eur. J. Med. Chem.* 136: 480–486.

230 Kung, A.L., Zabludoff, S.D., France, D.S. et al. (2004). *Cancer Cell* 6 (1): 33–43.

231 Block, K.M., Wang, H., Szabó, L.Z. et al. (2009). *J. Am. Chem. Soc.* 131 (50): 18078–18088.

232 Majmudar, C.Y., Højfeldt, J.W., Arevang, C.J. et al. (2012). *Angew. Chem. Int. Ed.* 51 (45): 11258–11262.

233 Sachchidanand, L., R.-S., Yan, S., Mutjaba, S. et al. (2006). *Chem. Biol.* 13 (1): 81–90.

234 Borah, J.C., Mujtaba, S., Karakikes, I. et al. (2011). *Chem. Biol.* 18 (4): 531–541.

235 Rooney, T.P.C., Filippakopoulos, P., Fedorov, O. et al. (2014). *Angew. Chem. Int. Ed.* 53 (24): 6126–6130.

236 Hammitzsch, A., Tallant, C., Fedorov, O. et al. (2015). *Proc. Natl. Acad. Sci. U.S.A.* 112 (34): 10768–10773.

237 Chekler, E.L.P., Pellegrino, J.A., Lanz, T.A. et al. (2015). *Chem. Biol.* 22 (12): 1588–1596.

238 Picaud, S., Fedorov, O., Thanasopoulou, A. et al. (2015). *Cancer Res.* 75 (23): 5106–5119.

239 Conery, A.R., Centore, R.C., Neiss, A. et al. (2016). *eLife* 5: e10483.

240 Taylor, A.M., Côté, A., Hewitt, M.C. et al. (2016). *ACS Med. Chem. Lett.* 7 (5): 531–536.

241 Crawford, T.D., Romero, F.A., Lai, K.W. et al. (2016). *J. Med. Chem.* 59 (23): 10549–10563.

242 Zeng, L., Li, J., Muller, M. et al. (2005). *J. Am. Chem. Soc.* 127 (8): 2376–2377.

243 Wang, Q., Wang, R., Zhang, B. et al. (2013). *MedChemComm* 4 (4): 737–740.

244 Hu, P., Wang, X., Zhang, B. et al. (2014). *ChemMedChem* 9 (5): 928–931.

245 Navratilova, I., Aristotelous, T., Picaud, S. et al. (2016). *ACS Med. Chem. Lett.* 7 (12): 1213–1218.

246 Humphreys, P.G., Bamborough, P., Chung, C.-W. et al. (2017). *J. Med. Chem.* 60 (2): 695–709.

13

Bromodomains: Promising Targets for Drug Discovery

Mehrosh Pervaiz[1], *Pankaj Mishra*[1], *and Stefan Günther*[1,2]

[1] *Albert-Ludwigs-University Freiburg, Institute of Pharmaceutical Sciences, Research Group Pharmaceutical Bioinformatics, Hermann-Herder-Straße 9, 79104 Freiburg, Germany*
[2] *University of Freiburg, Freiburg Institute for Advanced Studies (FRIAS), Albertstraße 19, 79104 Freiburg im Breisgau, Germany*

13.1 Introduction

Bromodomains (BRDs) were first described by Tamkun et al. They studied the drosophila gene *brahma* and detected significant sequence similarity to genes found in other drosophila, yeast, and human regulatory proteins involved in transcriptional activation [1]. The conserved 77 amino acid motif within the gene *brahma* gave the name for the cross-organism occurring BRDs.

BRDs are classified as "reader" domains because they bind noncovalently to acetylated lysine residues that are located on the surface of proteins. These posttranslational modifications exist in thousands of human proteins [2] and play a broad role in diverse signaling networks; for instance, it was shown that virtually every enzyme in the major cellular metabolism pathways such as glycolysis, gluconeogenesis, and fatty acid metabolism is acetylated in human liver tissue [3].

Epigenetic regulation is implemented by the acetylation of the eight N-terminal tails of the histone octamer as well as the core itself, which contain in total more than 100 lysines that can be dynamically deacetylated/acetylated. Since BRD-containing proteins often have more than one BRD and can also work as dimers, the affinity of the proteins is often higher for regions where multiple acetylation sites exist in close proximity [4]. The dynamic lysine acetylation in combination with other histone modifications such as lysine methylation results in a complex pattern ("histone code") that controls the gene regulatory network [5].

The role of histone acetylation for disease progression has been known for more than 20 years. For instance, it was shown in 1997 that the activity of the histone deacetylase HD1 is involved in the regulation of transcriptional repression and tumor suppression [6]. Later, it was also shown that histone acetylation is involved in cancer recurrence in various tumor entities; thus global histone acetylation levels might be a diagnostic marker for universal cancer prognosis [7].

The tremendous opportunities for the development of new anticancer drug candidates that target BRDs were impressively proven in 2010 by Bradner et al.,

Epigenetic Drug Discovery, First Edition. Edited by Wolfgang Sippl and Manfred Jung.

who presented the ground-breaking molecule JQ1. The diazepine binds to the first BRD of BRD4 and is able to displace a BRD4 fusion oncoprotein from chromatin. As a consequence, JQ1 induces immediate and progressive apoptosis in BRD4-dependent human carcinoma cells and reduces tumor growth of NUT midline carcinoma (NMC) in patient-derived xenograft models [8]. With the BRD inhibitor JQ1, a remarkable success story of BRD4 as a novel drug target has been set off that yielded several BRD inhibitors against different diseases that are now in clinical trials. The importance of other BRD-containing proteins as drug targets has been postulated. For some of them it was already validated that they specifically regulate the expression of several oncogenes [9]. However, for many BRDs the exact mechanisms of action and their potential as drug targets have not yet been elucidated.

13.2 The Human Bromodomain Family

The human genome encodes 61 BRDs found together with functionally diverse modules in 46 different proteins. These include acetylases, methyltransferases, ligases, transcription factors and regulators, chromatin adaptors, ATP-dependent chromatin remodelers, and signal transducers.

13.2.1 Structural Features of the Human BRD Family

The human BRDs family has been classified into eight structural families (I–VIII) [10] (Figure 13.1a), all of which share a conserved left-handed α-helical fold, composed of four α-helices (α_Z, α_A, α_B, α_C) connected by loops of different lengths (Figure 13.1b). Outside the two core helices, BRDs show very little sequence similarity. The N- and C-termini of BRDs are highly variable and may constitute additional helices that extend the canonical α-helical bundle (e.g. the C-terminal helix of PB1(6)) or N- and C-terminal extensions (as in TAF1L) that are also present in the form of elongated kinked helices (as in ATAD2).

13.2.1.1 The K_{ac} Binding Site

The K_{ac} binding site of BRDs is lined by the ZA and BC loops and has a largely hydrophobic core composed mainly of hydrophobic and aromatic amino acids (Figure 13.1c). The peripheral region of the binding site is considerably diverse due to the variable length and sequences of the ZA and BC loops. The ZA loop in particular has loop insertions that lead to differences that play an important role in substrate specificity. For example, members of the second BRD family, also known as the bromo- and extra-terminal domain (BET) family (BRD, BRD3, BRD4, BRDT), contain two small helix insertions in the ZA loop. Similarly, all members of the family VIII have a 7–9-residue (23 residues in SMARCA2-1) hairpin insertion between helix α_Z and the ZA loop.

A predominant feature of the acetyl-lysine (K_{ac}) binding site is the presence of a conserved Asn (N140 in BRD4(1)) in the pocket. Forty-eight of the sixty-one

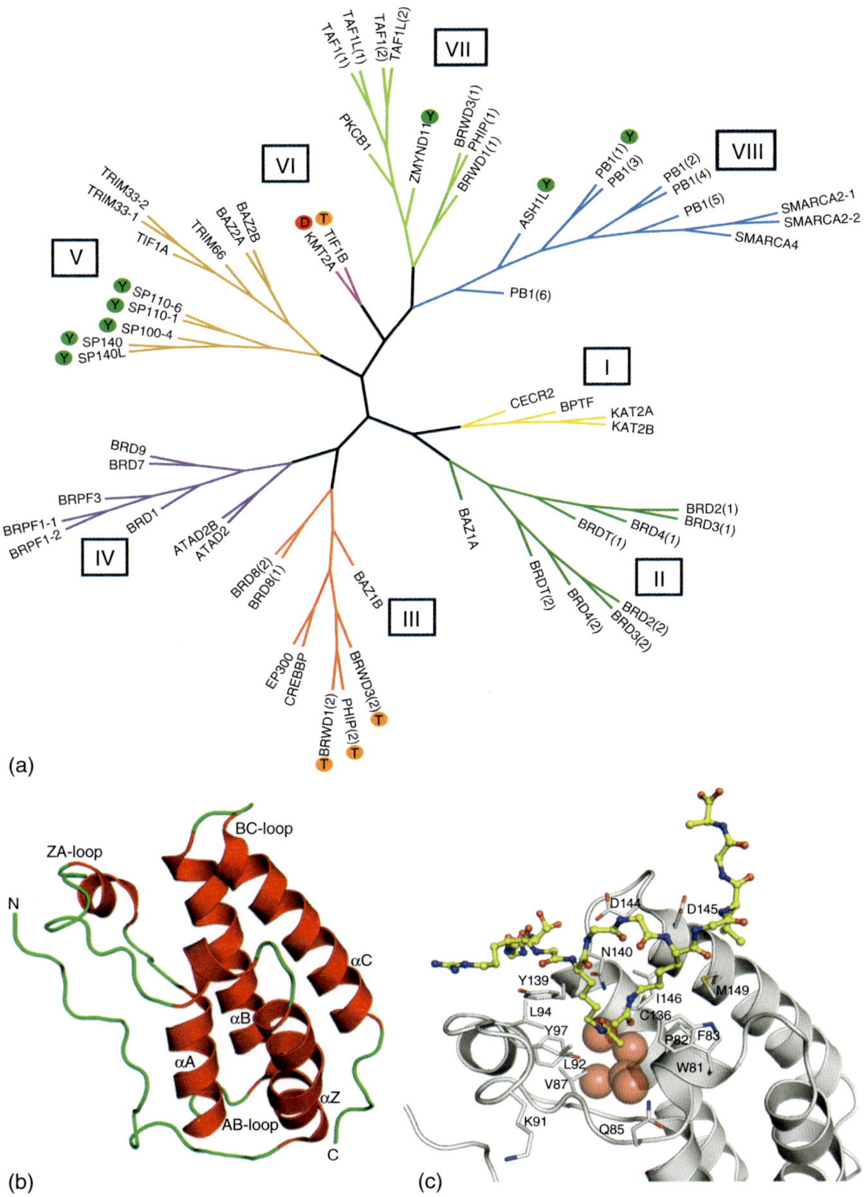

Figure 13.1 (a) Phylogenetic tree of the human bromodomain family. The families are numbered I–VIII. Atypical bromodomains are marked with colored circles showing the residues that substitute the conserved asparagine (aspartate: red, threonine: orange, tyrosine: green). (b) The conserved bromodomain fold (PDB ID: 3UVW). (c) The K_{ac} binding site of BRD4(1) bound by an acetylated lysine peptide (shown in yellow). The water molecules are shown as spheres.

human BRDs contain an asparagine residue in this position and are known as *typical BRDs*. The remaining 13 BRDs are considered *atypical* and contain a Tyr, Thr, or Asp in the same position (Figure 13.1a). The acetylation mark is primarily recognized through a direct hydrogen bonding interaction between $—NH_2$ of the conserved asparagine located in the BC loop and the carbonyl oxygen of the acetyl group.

A second hydrogen bonding interaction to the carbonyl oxygen of acetyl is formed via a structurally conserved water molecule. The presence of well-ordered water molecules at the bottom of the pocket is another important feature of the K_{ac} binding site (Figure 13.1c). These water molecules form a network of interactions inside the binding pocket and can be seen in most crystal structures. In BRD4(1), as well as many other BRDs, the first conserved water molecule bridges the interaction between the carbonyl of an acetylated lysine and the OH of a conserved tyrosine (Y97) located in the ZA loop.

The tyrosine residue is evolutionarily conserved throughout the human BRD family, except for the BRDs of KMT2A, TIF1B, and PB1(6). How the acetylated lysine binds these BRDs is not well understood due to the lack of structures co-crystallized with peptides.

The BRDs of both KMT2A and TIF1B are *atypical* and contain an aspartate and a threonine, respectively, instead of the conserved asparagine. Other *atypical* BRDs that have a threonine residue instead of asparagine include BRWD1(2), BRWD3(2), and PHIP(2), while ASH1L(1), PB1(1), SP100-C, SP110, SP110–6, SP140, SP140-L, and ZMY11 contain a tyrosine in the same position. Both tyrosine and threonine are hydrogen bond donors and can recognize K_{ac} in a way similar to asparagine. The presence of an aspartate residue in KMT2A, however, suggests that it either does not bind acetylated lysines or has a totally different mechanism for substrate recognition.

In the *atypical* binding sites of BRWD1(2) and PHIP(2) (PDB ID: 3Q2E and 3MB3), a water molecule in addition to the conserved five is seen bridging the interaction between —CO (of the acetate ion in BRWD1(2) and —CO of the *N*-methyl pyrrolidone in PHIP(2)) and the *atypical* threonine (Figure 13.2). Because the side chain of threonine is one atom shorter than that of asparagine, there is slightly more space in the threonine-containing binding sites, which is occupied by the additional water molecule in order to mediate the interaction between threonine and the *typical* water network buried deep within the K_{ac} binding site.

Another characteristic feature of some BRDs, including members of the BET family, is the presence of a conserved motif of three amino acids (W, P, and F) in the ZA loop known as the "WPF shelf". In the BET family, this motif together with a flanking KL/AL motif forms the narrow passage that leads to the K_{ac} binding site (Figure 13.1c), a feature that has been utilized to design BET subfamily-specific chemical probes and inhibitors [8, 11].

13.2.1.2 Druggability of the Human BRD Family

The BET family of BRDs as well as many other BRDs have been shown to be druggable (proteins that can bind drug-like molecules). Knowledge of druggability is an important aspect of target validation and assists the screening process by

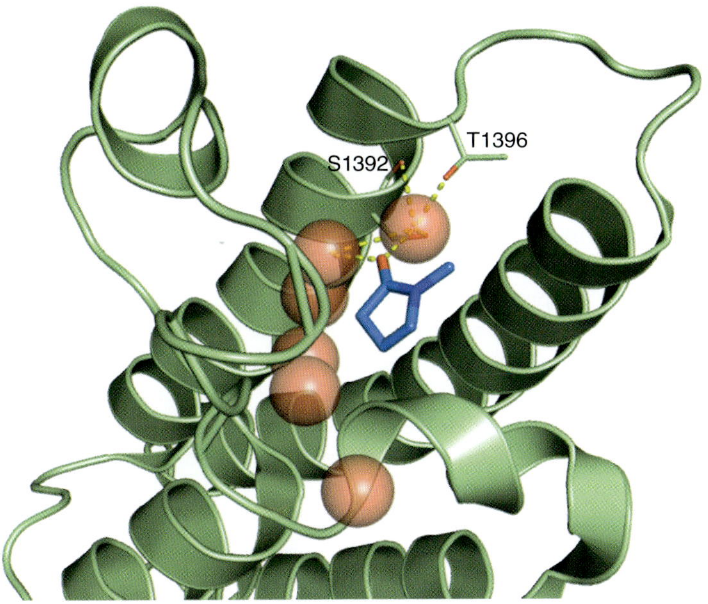

Figure 13.2 The atypical K_{ac} binding site of PHIP(2) (PDB ID: 3MB3). A water molecule in the binding site of PHIP(2) bridges the interaction between the atypical threonine (T1396) and the carbonyl of the *N*-methyl pyrrolidone (shown in blue).

allowing the setting of reasonable expectations regarding the hit rates. Ligands for targets that are considered to be less druggable are expected to have lower potencies and ligand efficiencies compared with the more druggable ones and require alternative approaches (such as higher screening concentrations, the use of more diverse or larger libraries, etc.) to be employed during the screening process. Hoelder and coworkers performed an extensive druggability analysis on BRDs and showed that on the druggability scale, BRDs range from being druggable to being difficult [12]. Using structural features of the binding site, they classified BRDs into eight groups (and outliers), the most druggable of which included members of the BET family as well the BRD subfamily I (Figure 13.1a), BRPF1-B, TAF1(2), and PHIP(2). The classification of BRDs based on the binding site features and druggability does not correlate with the structural classification mentioned in Section 13.2.1.1. This is because the druggability score is affected only by the nature of the binding site and not the complete domain. The BRDs cAMP response element-binding protein (CREBBP), BRD1, PB1(2), PB1(5), and BRD9 were predicted to have intermediate druggability, while ASH1L, ATAD2B, BAZ2B, PB1(1/3/4), EP300, TAF1L(2), and TIF1(2), as well as SMARCA2–2 and SMARCA4, were predicted to be difficult.

The high druggability of the BET family is evident from the large number of potent small-molecule inhibitors known to bind various members of the family and has been attributed to a more enclosed upper part of the binding site, which provides a large surface for interaction with small molecules. The large surface is due to the presence of a tryptophan (W81 in BRD4(1)), the location of which is influenced by a methionine residue (M149 in BRD4(1)) as well as a large ZA loop compared with other members of the human BRD family.

Other proteins such as members of family I predicted to be druggable also possess W81 and exhibit sufficient enclosure as well as significant hydrophobic surface for interaction with small molecules. However, for some druggable BRDs such as bromodomain plant homeodomain transcription factor (BPTF) or PHIP(2), no chemical probes or inhibitors have been reported to date. On the other hand, small-molecule inhibitors with low micromolar affinities are known to bind difficult BRDs including ATAD2B, BAZ2B, PB1(3), and SMARCA4. EP300 was also predicted to be difficult even though it shares significant sequence as well as binding site similarity to CREBBP (intermediate druggability) and is known to bind identical compounds (dual CREBBP and EP300 inhibitors). Both these BRDs show binding site similarity to the BET family through the extended ZA loop. However, the W81 of the WPF shelf of the BET family is substituted by a much smaller leucine in both, and the KL opposite the WPF shelf is replaced by a flexible arginine and a small valine, reducing the enclosure and hydrophobicity and therefore druggability. The difference in druggability of EP300 and CREBBP (median druggability scores [13] 0.72 and 0.75, respectively) may be due to the fact that for EP300 only the apo form of the domain was analyzed, while for CREBBP both apo and holo (ligand- and peptide-bound) forms of the domain were studied. The holo structures are likely to induce a more druggable pocket, due to conformational changes that occur as a result of ligand or peptide binding.

13.2.2 Functions of Bromodomain-containing Proteins

BRDs are often found in functionally diverse, multidomain proteins involved in a wide range of cellular processes including chromatin remodeling, cell cycle and signaling, nuclear transport, and protein stability and metabolism. Some BRD-containing proteins such as the CREBBP and E1A binding protein (EP300) are histone modifiers and have intrinsic histone acetyltransferase (HAT) activity [14], while others such as histone-lysine N-methyltransferase ASH1L and histone-lysine N-methyltransferase 2A (KMT2A also known as myeloid/lymphoid or mixed lineage leukemia (MLL)) protein have intrinsic histone methyltransferase (HMT) activity [15, 16]. Others such as the HATs KAT2A/B (also known as GCN5 and p300/CBP-associated factor (PCAF)), BRD-containing proteins 1 and 8 (BRD1/8), and BRD and plant homeodomain (PHD) finger-containing proteins 1 and 3 (BRPF1/3) serve as transcriptional coactivators by being part of histone-modifying complexes such as Spt-Ada-Gcn5 acetyltransferase (SAGA) [17], Ada-two-A-containing (ATAC) [18], monocytic leukemia zinc finger protein (MOZ)–MOZ-related factor (MORF) [19], and nucleosome acetyltransferase of H4 (NuA4) [20]. BRD-containing proteins are also components of chromatin remodeling complexes involved in gene silencing. For example, the BRD adjacent to zinc finger protein 1A (BAZ1A) is a component of the ATP-utilizing chromatin assembly and remodeling factor (ACF) complex that regulates chromatin structure and generates regularly spaced nucleosomes required for heritable gene silencing [21]; BAZ1B is a part of the nucleolar remodeling complex (NoRC) responsible for silencing of rRNA genes [22]. Other BRD-containing proteins such as BRD and WD repeat protein 1 (BRWD1) and SMARCA2/4 are constituents of the switch/sucrose non-fermentable (SWI/SNF) chromatin remodeling complex, which promotes transcription of certain genes by associating with transcription factors [23]. Similarly, BAZ1B is a component of the Williams syndrome transcription factor–imitation switch (WSTF–ISWI) complex [24], while the proteins CECR2 and BPTF are components of the chromatin remodeling complexes CECR2-containing remodeling factor (CERF) complex [25] and nucleosome remodeling factor (NURF) [26], respectively, all of which are responsible for chromatin decompaction and therefore promote transcription. Another BRD-containing protein that enhances gene expression through chromatin decompaction is the ATPase family AAA domain-containing protein 2 (ATAD2). ATAD2 serves as a positive regulator of transcription and is hypothesized to possess some type of histone chaperone activity [27].

BRD-containing proteins, including members of the BET family as well as the transcription initiation factor TFIID subunit 1 (TAF1) and the zinc finger MYND-type-containing 8 (ZMYND8), serve as histone-recognizing scaffolds for the assembly of large transcriptional complexes, resulting in transcriptional activation (BET, TAF1) [28, 29] or repression (ZMYND8) [30]. They also serve as transcriptional coregulators, where they modulate the activity of transcription factors by acting either as co-activators (e.g. ZMYND8, the tripeptide motif-containing proteins TRIM24 and TRIM33 and proteins from the speckled protein (SP) family) or corepressors (TRIM28). The PHD finger domain of TRIM28 possesses an intrinsic E3 SUMO ligase activity that adds SUMO marks to its BRD, resulting in its interaction with the nucleosome remodeling and deacetylase (NuRD) complex, which stimulates chromatin compaction and transcriptional repression [31]. The HAT proteins CREBBP and EP300 also act as transcriptional coactivators in various signaling pathways, including calcium signaling, Notch signaling, NFκB signaling, and response to hypoxia [32].

The atypical BRD-containing protein, the pleckstrin homology domain-interacting protein (PHIP), acts as a positive regulator of pancreatic β-cell growth and survival [33]. Similar BRD-containing proteins that contain both typical and atypical BRDs (Figure 13.1a) are part of the Janus kinase–signal transducer and activator of transcription (JAK–STAT) signaling pathway (the BRD and WD repeat-containing protein BRWD3) [34] and the SWI/SNF chromatin remodeling complex (BRWD1) and enhance chromatin remodeling at antigen receptor–encoding loci (BRWD1) [34]. The functions of BRD-containing proteins and their role in cellular homeostasis have been comprehensively reviewed by Fujisawa and Filippakopoulos [35].

13.3 Bromodomains and Diseases

State-of-the-art advances in proteomics have shown that BRD-containing proteins exhibit broad and variable expression profiles across different tissues, suggesting their context-dependent role in cellular homeostasis. BRD-containing proteins have been implicated in various developmental, immunological, and neurological disorders, as well as cancer. The role of BRDs in these diseases is briefly discussed below.

13.3.1 The BET Family

The most extensively studied member of the human BRD family, BRD4, is a ubiquitously expressed nuclear protein that acts a cell growth regulator. It recognizes acetylation marks on histones and recruits the positive transcription elongation factor b (P-TEFb), a heterodimer composed of cyclin T1, T2, or K and cyclin-dependent kinase 9 (CDK9), to the transcription start site for phosphorylation of RNA polymerase II (RNA Pol II) at the C-terminal domain (CTD), resulting in transcription elongation [36]. It therefore promotes expression of early G1 and growth-associated genes such as *c-Myc, JunB*, and *cyclin D1*. Consistent with its role in cell growth, BRD4 heterozygous mutant mice ($BRD4^{+/-}$) exhibit a wide range of anatomical abnormalities, while homozygous mutants ($BRD4^{-/-}$) are embryonically lethal [36]. BRD4 is also implicated in transmitting epigenetic memory, as it remains associated with acetylated chromatin throughout mitosis [37].

Chromosomal aberration involving BRD4 due to translocation t(15;19)(q13, p13.1) results in the BRD4-NUT oncogene, which characterizes the rare, aggressive, and lethal NMC affecting young people [37]. Treatment of NMC cell lines with BET inhibitors targeting BRD4 has been shown to displace the fusion oncoprotein from chromatin, causing growth arrest and apoptosis [8]. BET inhibitors have also demonstrated therapeutic potential against MLL-fusion leukemia in murine models as well as human cancer cell lines by selective suppression of BRD3/4 recruitment to chromatin and downregulation of key cell cycle proliferation genes including B-cell lymphoma (*bcl-2*), *cdk6*, and *c-Myc* [38]. Inhibition of BRD4 also reduces the expression of Aurora B kinase, which is required for proper chromosomal segregation and cell division [38]. Overexpression of Aurora B causes aneuploidy and has been implicated in a variety of human cancers including colorectal cancer [39] and invasive ductal carcinoma of the breast [40]. BRD4 is also involved in metastasis and sustenance of triple-negative breast cancer (TNBC) via the BRD4/Jagged1/Notch1 pathway. Downregulation of Jagged1/Notch1 pathway by targeting BRD4 shows potential to prevent TNBC progression, which is otherwise difficult to treat with conventional chemotherapy [41]. BET inhibitors have also demonstrated antiproliferative and cytotoxic effects in various cancer cell lines and models, including those of human lung adenocarcinoma, acute myeloid leukemia (AML), acute lymphoblastic leukemia, hepatocellular carcinoma, lymphoma, glioblastoma, myeloma, medulloblastoma, and neuroblastoma [42–48].

BET proteins also regulate inflammatory gene expression and are crucial for pro-inflammatory cytokine expression [49]. Pro-inflammatory stimuli such as the bacterial lipopolysaccharide (LPS) have the ability to reshape the host epigenome by influencing a variety of epigenetic factors including histone modifications and DNA methylation [50, 51]. Treatment of macrophages with the

BET inhibitor I-BET762 stalled the expression of LPS-inducible genes encoding cytokines, chemokines, and transcription factors involved in inflammatory response. Minimal changes in gene expression were observed in macrophages in the absence of LPS stimulation, implying the selective effects of BET inhibitors on inflammatory genes in macrophages [11]. Similar studies involving BET knockdown in LPS-stimulated macrophages as well as their treatment with the BET inhibitor JQ1 resulted in a dramatic decrease in the levels of inflammatory cytokines including interleukin (IL)-6, monocyte chemotactic protein (MCP)-1, and TNF-α and protected mice from LPS-induced septic shock and death [49]. BRD2 and BRD4 are also important for differentiation of T-helper cells and transcription of various T-helper lineage-associated cytokines [52]. Their inhibition has been shown to be effective in mouse models of autoimmunity, suggesting therapeutic potential of BET inhibitors for autoimmune diseases. Befitting its role in reducing inflammation, BET inhibition also has the potential to treat rheumatoid arthritis (RA), where it has been shown to suppress expression of key molecules responsible for pathology in RA synovial fibroblasts [53].

BRD expression is also induced by viruses such as the respiratory syncytial virus (RSV), which activates BRD4's HAT activity *in vivo* as well as its interaction with CDK9 and NFκB/RelA and recruitment to genes involved in early innate immunity. Inhibition of BRD4 in mouse models abated RSV-induced inflammation, pathogenesis, and airway hyperreactivity, as well as toll-like receptor 3 (TLR3)-induced production of neutrophils, indicating therapeutic significance of BRD4 inhibition in reducing viral-induced mucosal airway inflammation and treatment of acute lung inflammation [54].

A major limitation of current antiretroviral treatments in eliminating HIV infection is that they only target actively replicating viruses. Studies have shown that BRD4 regulates numerous steps of HIV transcription and competes with the viral transcriptional activator Tat (trans-activator of transcription) for interaction with P-TEFb [55], while BRD2 acts as a suppressor of HIV transcription in latent cells [56]. The BET inhibitor JQ1 reactivates latent HIV transcription by making available P-TEFb to associate with Tat, resulting in transcription elongation [55, 57]. Knockdown of BRD2 also activates HIV long terminal repeats (LTRs), suggesting the possibility of Tat-independent mechanism for activation of HIV latency [56]. A combination therapy involving current anti-HIV drugs together with small molecules targeting BET proteins may be the way forward toward completely eliminating the virus in patients with HIV infection.

Apart from their role in cancer, inflammation, and viral infections, BET proteins also affect other organ systems such as the cardiovascular and reproductive systems. BET proteins are involved in heart failure pathogenesis, due to their ability to co-activate multiple master transcription factors such as NFAT, NFκB, and GATA4 that drive heart failure. BET inhibition was shown to suppress cardiomyocyte hypertrophy *in vitro* and pathological ventricular remodeling *in vivo* [58]. The testis-specific BET protein, BRDT, is vital for chromatin remodeling during spermatogenesis, and the BET inhibitor JQ1 possesses reversible contraceptive effects and has been shown to reduce testis mass, sperm count, and motility in rats [59].

Moreover, the second BRD-containing protein BRD2 (also known as RING3) is essential for proper neural development and completion of embryogenesis. Mouse embryos deficient in BRD2 ($BRD2^{-/-}$) did not survive past day 13.5 and showed altered expression of multiple genes, including those known to guide neural development, resulting in defects in neural tube closure [60]. BRD2 was identified as one of the first susceptibility genes for juvenile myoclonic epilepsy (JME) in humans. Single nucleotide polymorphisms (SNPs) in the BRD2 promoter region were observed in probands with JME. It was proposed that these SNPs may affect the levels and timing of BRD2 expression or its tissue specificity, resulting in impaired regulation of brain development, ultimately leading to JME [61]. The exact mechanism, however, remains unknown.

13.3.2 Non-BET Proteins

The success of BET inhibitors due to their compelling antiproliferative properties and rapid advancement to clinical trials has led to the development of chemical probes and small-molecule inhibitors targeting the non-BET proteins. However, the function and therapeutic significance of many of these proteins remain elusive. Like the BET proteins, other members of the human BRD family have been associated with various cancers, as well as developmental and neurological disorders and viral infections.

Because BRD proteins are important chromatin-associated transcriptional regulators, they are often exploited by viruses for their propagation. The PCAF (also known as KAT2B) facilitates the recruitment of P-TEFb to the transactivating responsive RNA (TAR) of HIV via acetylation of the HIV Tat on lysine 28 [62], resulting in transcription elongation by phosphorylation of RNA Pol II. PCAF binds Tat via its BRD, which recognizes the acetylation mark on K50 [63]. It therefore promotes HIV transcription and could serve as a target for anti-HIV therapy.

The K50 of HIV Tat is acetylated by the E1A-binding protein p300 (EP300 or p300), which was first identified as a target for the adenoviral protein E1A [64]. E1A is an oncoprotein, which stimulates cell growth and prevents differentiation by binding to the p300/CBP complex and modulating transcription of important cellular genes [65]. Both the EP300 and the CREBBP (or CBP) are negative regulators of cell growth and act as mediators of interactions between multiple transcription factors and the basal transcriptional machinery. They play a positive role in cellular differentiation, TP53/PCNA-mediated DNA repair, and TP53-mediated apoptosis [66]. The p300/CBP duo is an important target for many viral proteins, which seize the control of transcription by competing with regular transcription factors for binding to p300/CBP, resulting in suppression of cellular genes, and by activating viral transcription. The viruses that maneuver p300/CBP for their proliferation include adenovirus, Epstein–Barr virus, HIV, papillomavirus, polyomavirus, herpesvirus, and the human T-cell leukemia virus type I [62, 63, 65, 67–72]. Development of drugs targeting the function of p300/CBP may provide a cure to viral infections, many of which lead to cancer.

In addition to their involvement in cancer via association with viral oncoproteins, mutations in CBP and p300 have been linked to hematological malignancies as well as solid tumors. The t(8, 16)(p11,p13) translocation fuses the gene encoding MOZ to the N-terminal of CBP [73]. Another translocation t(8, 22)(p11,p13) results in the fusion of MOZ to p300 [74]. Both these translocations are rare and have been implicated in AML [73]. MLL-CBP and MLL-p300 translocations have also been reported in AML [75, 76]. Further mutations have been linked to B-cell non-Hodgkin lymphoma (p300/CBP) [77], ovarian cancer (CBP) [78], and Rubinstein–Taybi syndrome (p300/CBP), a developmental disorder characterized by physical abnormalities, mental retardation, and an increased susceptibility to cancer [79, 80]. Other neurodevelopmental disorders that have been associated with abnormal BRD-containing proteins include Williams syndrome (BAZ1B), neural tube defects (KAT2A or GCN5), X-linked mental retardation (BRWD3), and fragile X syndrome (p300/CBP) [81–84].

The BRD adjacent to zinc finger domain 2A (BAZ2A) is involved in epigenetic modifications in prostate cancer via its cooperation with the enhancer of zeste homologue 2 (EZH2), resulting in abnormal gene silencing. Its overexpression has been associated with the CpG island methylator phenotype (CIMP) in tumors and disease recurrence and may serve as a marker for prostate cancer metastasis [85]. The BRD-containing protein ATAD2 is co-amplified with MYC in a considerable proportion of human tumors, where it regulates MYC-dependent transcription. Overexpression of ATAD2 has been shown to correlate with distant metastasis and grade 3 tumor in breast cancer patients [86]. Its overexpression has also been linked to the pathological development of colorectal cancer, where it may serve as a prognostic marker as well as a target for anticancer therapy [87, 88].

13.4 Methods for the Identification of Bromodomain Inhibitors

A popular approach for the identification of BRD inhibitors is mimicking the acetylated lysine residue following conventional drug design approaches. However, in recent years various other methods were also used to identify potent and selective BRD inhibitors. These methods are broadly categorized as high-throughput screening (HTS), fragment-based lead discovery (FBLD), structure-based drug design (SBDD), and virtual screening (VS).

13.4.1 High-throughput Screening (HTS)

HTS is a technique widely used in drug discovery. It leverages automation to quickly screen large number of compounds in a short time. The most widely adopted HTS for the identification of BRD inhibitors are the fluorescence/luminescence-based assays (such as AlphaScreen and its variants, being the most common peptide displacement assay), differential scanning fluorimetry, and fluorescence polarization assays. In conjunction with modern analytical techniques such as NMR or liquid chromatography–mass spectrometry (LC–MS)/MS and biophysical methods such as isothermal titration calorimetry (ITC) or X-ray crystallography/soaking, HTS offers the analysis of potential biomodulators against defined targets in real time with high sensitivity.

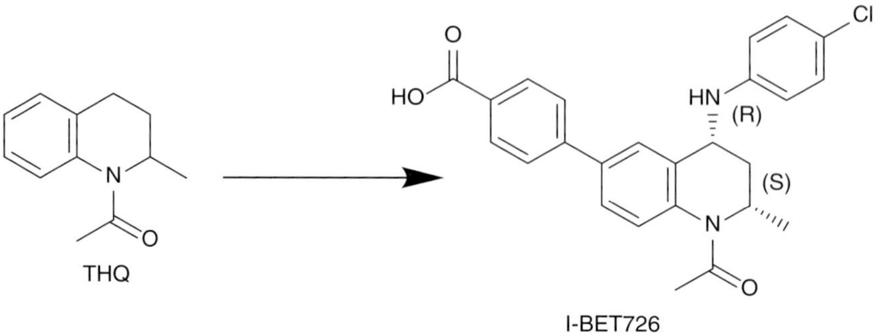

Figure 13.3 An early example of BET bromodomain inhibitor discovered through HTS.

Several of these methods are often used in parallel to cross-validate identified hits.

An early and successful example of the identification of BRD inhibitor hit through a biochemical screening is of the lead fragment tetrahydroquinoline (THQ; Figure 13.3) that has preferably targeted BRD2(1) [89]. Further chemical modifications of this yielded the highly potent and selective BET BRD inhibitor I-BET726 (Figure 13.3), which is currently in clinical trials [90–93].

In another example, Yu et al. reported the identification of four potential hit-to-lead optimization candidates using an NMR-based platform for FBLD. Out of those four candidates, two shared a common quinazoline core structure bound to BRD4(1) in a non-acetylated lysine mimetic mode. These inhibitors also exhibited selectivity to BRD4(1) [94]. Taylor et al. identified the promising hit 4-methyl-1,3,4,5-tetrahydro-2H-benzo[b][1,4]diazepin-2-one fragment **1** (Figure 13.4) against the CBP/EP300 BRDs using a thermal shift assay. The further optimization of the identified fragment through SBDD has yielded a series of cell-potent and selective probes, e.g. **2** (Figure 13.4), of the CBP/EP300 BRDs [95].

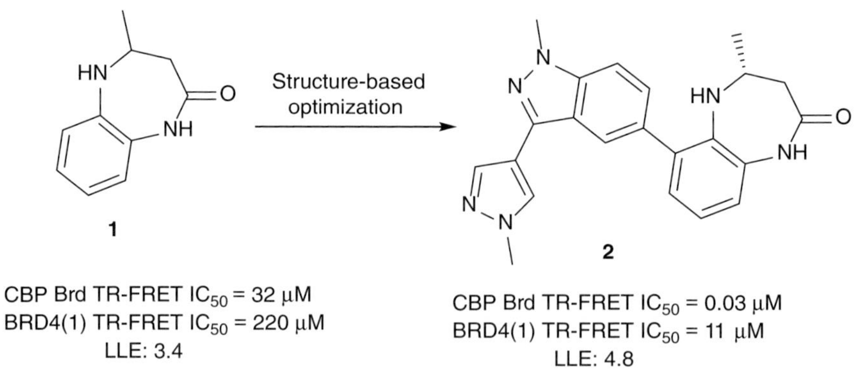

Figure 13.4 Optimization of a CBP/EP300 bromodomain fragment inhibitor identified through HTS.

Chaikuad et al. reported the identification of new classes of millimolar to low micromolar affinity acetyl-lysine mimetic ligands of the PCAF BRD via a thermal shift assay and NMR-based fragment screening as well as downstream optimization by structure-based approaches [96]. Demont et al. identified low micromolar ATAD2 BRD inhibitors by applying a fragment screening of a focused library via time-resolved fluorescence resonance energy transfer (TR-FRET)-labeled peptide displacement assay [97].

13.4.2 Fragment-based Lead Discovery

FBLD is an approach that is increasingly being used in pharmaceutical industry. It starts with the identification of low molecular weight ligands that bind to biologically relevant macromolecules. The identified fragments are then modified into potent drug-like molecules. FBLD is a valuable tool in the early-stage drug discovery for reducing the number of molecules which have to be tested. It is also a valuable tool for obtaining leads for difficult-to-target proteins. FBLD has been utilized extensively in drug discovery campaigns against BRD targets. These fragments can be identified by HTS, SBDD, or VS.

Fragment-based virtual screening has been quite successful in finding novel chemotypes against the BRD of CREBBP. Xu et al. reported the discovery of two chemotypes (4-acylpyrrole **3** and acylbenzene **4**; Figure 13.5) identified through a fragment-based high-throughput docking and molecular dynamics (MD) simulations [98]. The 4-acylpyrrole was earlier identified by Lucas et al. as the novel chemotype inhibiting the BRD4(1) BRD with high affinity [99]. Zhao et al. reported the discovery of two novel chemotypes identified by fragment-based virtual screening (isoxazole based **5** and diazepam **6**; Figure 13.5) that bind to the BRD4(1) BRD [100]. The reported fragments exhibited low micromolar affinity, selectivity over other BRD families, and favorable ligand efficiency. In another study, Zhao et al. reported the discovery of 2-thiazolidinone chemotypes as potential inhibitors of BRD4(1) identified by fragment-based virtual screening [101].

13.4.3 Structure-based Drug Design

In SBDD, the knowledge of the three-dimensional structure of a drug target interacting with small molecules is used as a guide to design drug candidates with high affinity and selectivity. In recent years, the SBDD method for the identification of BRD inhibitors with higher affinity than the primary hits has been particularly successful. For example, Chung and coworkers performed structure-based optimization of the weak binding fragment 4-phenyl 3,5-dimethyl isoxazole to obtain the potent and selective inhibitor **7** (Figure 13.6) of BRD2(1) ($K_d = \sim 500$ nM) [102].

Fish et al. performed a structure-based optimization on a dihydroquinazolinone (DHQ) fragment bound to BRD4(1) [103]. They addressed a hydrophobic pocket in the binding cavity of BRD4(1) by introducing a sulfonamide linker attached to the DHQ scaffold and added an additional aromatic ring. The resulting compound PF-1 (Figure 13.6) was selected as a chemical probe for BET family inhibition with an IC_{50} value of 0.22 μM against BRD4(1).

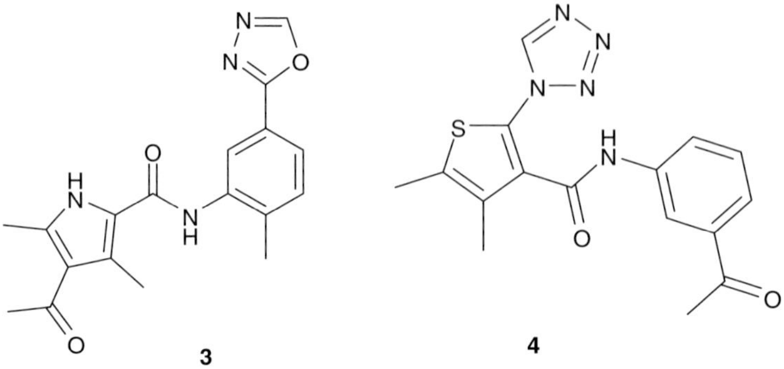

3 4

CREBBP
K_d ; 3 = 13 μM; 4 = 29 μM
Ligand efficiency; 3 = 0.27; 4 = 0.24

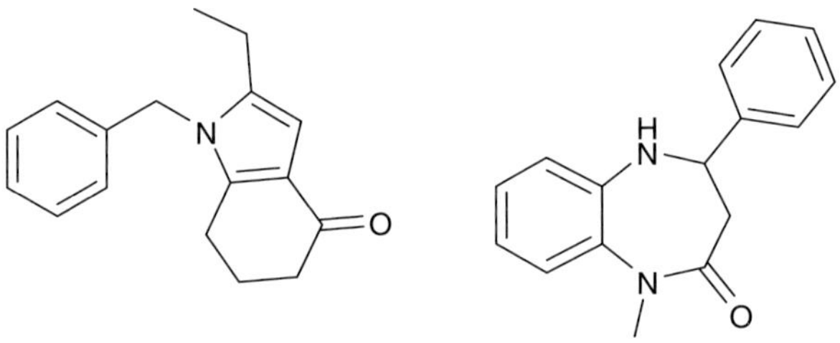

5 6

BRD4(1)
K_d; 5 = ND; 6 = ND
Ligand efficiency; 5 = 0.37; 6 = 0.37

Figure 13.5 Examples of bromodomain inhibitors identified through FBLD approaches.

Palmer et al. reported a selective high-affinity dual inhibitor of the TRIM24-BRPF1 BRDs [104]. The structure-guided design performed with the X-ray co-crystal structures resulted in a potent and selective inhibitor, IACS-9571 (Figure 13.6), of the N,N-dimethylbenzimidazolone structural class. In another study, Martin et al. identified two *in vivo* active, selective, and cell-permeable BRD9 inhibitors [105]. These chemical probes BI-7273 and BI-9564 (Figure 13.6) were derived from the X-ray crystal structure-guided modifications of an initial fragment hit (dimethylpyridinone scaffold) identified through fragment-based HTS screening.

Another potent and selective BRD4(1) BRD inhibitor, MS436 (Figure 13.6), was obtained by using a structure-based design strategy performed on a diazobenzene-based small-molecule inhibitor, MS120 (Figure 13.6), originally discovered as CREBBP BRD inhibitor [106]. Drouin et al. reported the first chemical probes against low druggable BRD targets, viz. BAZ2 BRDs [107]. In this study the authors first identified a weak inhibitor of BAZ2A (IC_{50} = 51 μM) and BAZ2B (IC_{50} = 26 μM) and then performed a structure-guided optimization

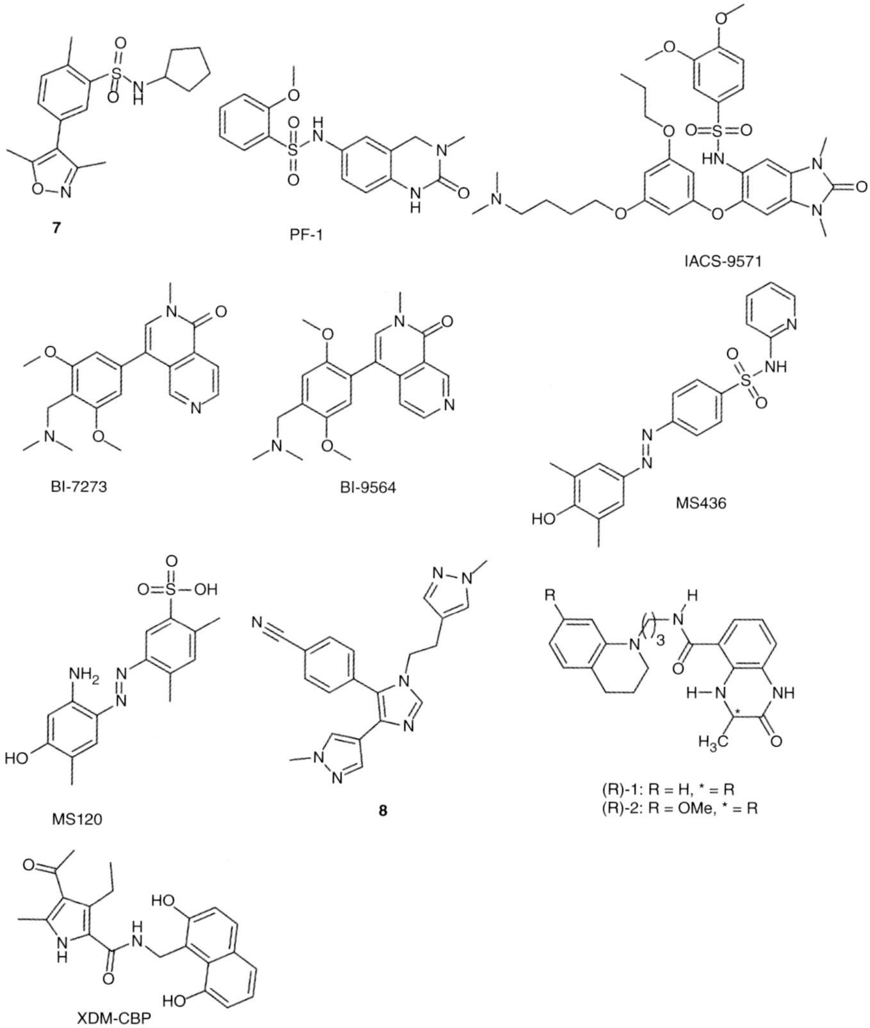

Figure 13.6 Examples of BRD inhibitors identified by applying SBDD approaches.

to obtain the potent and selective chemical probe **8** (Figure 13.6). Recently, Hügle et al. reported new high-affinity derivatives of a 4-acetylpyrrole scaffold originally identified through a structure-based virtual screening (SBVS) protocol performed on BRD4(1) [108]. Another SBDD approach was applied by Rooney et al. to design novel nanomolar inhibitors of the BRD of CREBBP (R-1; $IC_{50} = 0.76\,\mu M$ and R-2; $IC_{50} = 0.32\,\mu M$; Figure 13.6) [109].

In a more recent study, Hügle et al. reported a highly potent and selective inhibitor, XDM-CBP (Figure 13.6), for the CREBBP/p300 BRDs [110]. The authors performed a structure-based optimization on a previously identified pan-selective BET BRD-binding fragment named XD46 [108].

13.4.4 Virtual Screening

Computational HTS, also known as virtual screening, has been defined as "automatically evaluating very large libraries of compounds" using computer programs [111]. VS has presented itself as the inexpensive, fast alternative approach to the *in vitro* HTS methods in early-drug discovery. This field is gaining prominence over the *in vitro* HTS methods because of its effectiveness in handling very large chemical libraries in a cost- and time-effective manner. Another advantage of VS is its success rate to find positive hits in comparison with *in vitro* HTS. Broadly, it is classified as (i) structure-based and (ii) ligand-based virtual screening (LBVS).

In the past decades, the identification of novel BRD inhibitors was mostly focused on the application of *in vitro* HTS methods. However, in recent years, VS has also gained the attention of scientists in academia and industry. Several successful applications of VS to identify BRD inhibitors have been reported recently.

13.4.4.1 Structure-based Virtual Screening

SBVS is a computational approach that involves screening a chemical compound library against the biologically validated drug target to find novel bioactive molecules. In this method, a large collection of chemical compounds is docked into the binding site of the three-dimensional structure of a biological target, obtained either from X-ray, NMR, or computational modeling. After docking, these compounds are ranked by applying a scoring function to estimate their affinity.

An early example of the successful application of SBVS to identify a BRD inhibitor was presented by Vidler et al. in 2013 [112]. They identified six novel hits targeting the binding site of the three-dimensional structure of the BRD4(1) BRD.

In 2014, Muvva et al. identified high-affinity BRD4 BRD inhibitors (ZINC01411240, ZINC19632618, and ZINC04818522) using SBVS methods [113].

Another successful example was reported by Lucas et al. in which they utilized the binding site of the BRD4(1) BRD to analyze a large drug-like molecule library taken from the ChEMBL and ZINC databases [99]. The authors presented a novel promising inhibitor named XD14 (K_d = 237 nM determined via ITC measurements; Figure 13.7).

Xue et al. reported the identification of the benzo[cd]indol-2(1H)-one class of compounds as potent inhibitors of the BRD4(1) BRD using SBVS methods [114]. The proposed lead compound **9** (Figure 13.7) binds with a K_d value of 137 nM determined via ITC measurements.

13.4.4.2 Ligand-based Virtual Screening

LBVS is another popular method in early-stage drug discovery and is particularly helpful in cases where the three-dimensional structures of potential drug targets are not available.

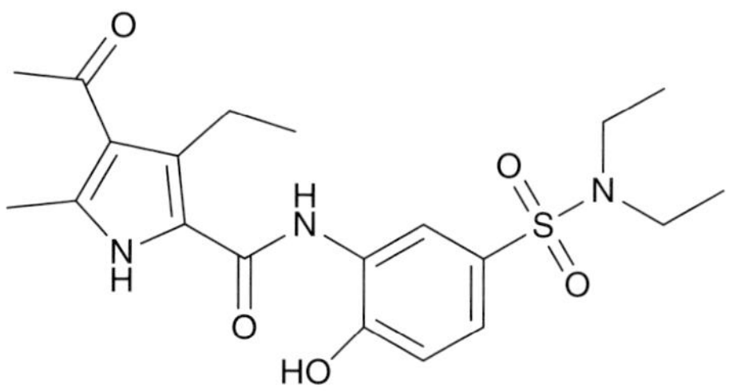

XD14

BRD4(1) K_d = 237 nM

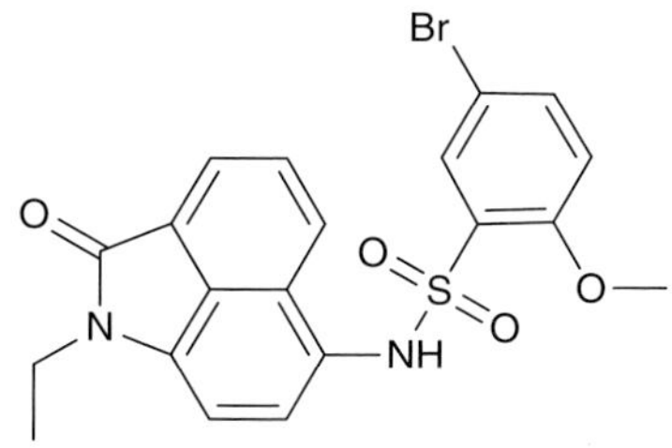

9

BRD4(1) K_d = 137 nM

Figure 13.7 Representative bromodomain inhibitors identified through structure-based virtual screening.

13.4.4.3 Pharmacophore Modeling

A popular approach to LBVS is pharmacophore modeling, which provides predictive models for lead optimization and for VS of molecular libraries. The IUPAC defines a pharmacophore model as "an ensemble of steric and electronic features that is necessary to ensure the optimal supramolecular interactions with a specific biological target and to trigger (or block) its biological response." Recently, Gao et al. established a VS protocol that utilized a structure-based pharmacophore model of a BRD inhibitor JQ1 complexed to the human BRD-containing protein (hBRDT) [115]. The authors reported the novel potent (IC_{50} = 9.02 μM) testis-specific inhibitor **10** (Figure 13.8) of hBRDT.

13.4.4.4 Substructure and Similarity Search

Substructure searches allow the users to search chemical databases using only parts of inhibitors. This method is complemented by similarity searches in which a full molecule is used as reference ligands for the screening of chemical databases. Both these methods have been utilized extensively in early-stage drug development for several drug targets.

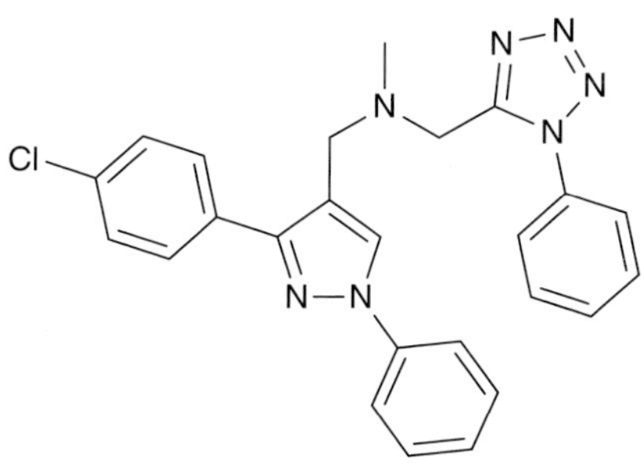

10

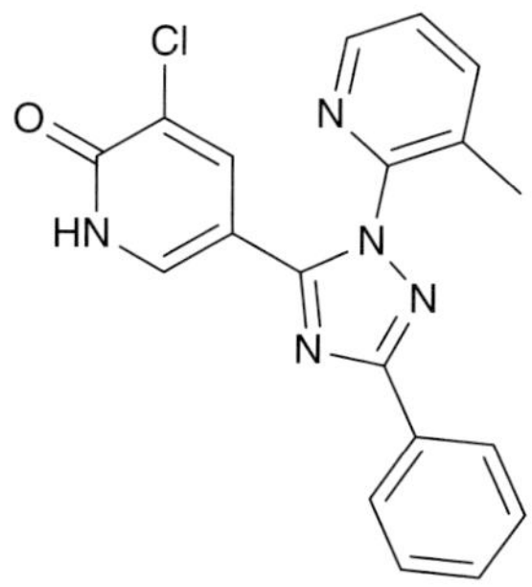

11

Figure 13.8 Representative bromodomain inhibitors identified through ligand-based virtual screening.

Vidler et al. combined these approaches with an SBVS protocol to find novel small-molecule inhibitors of the BRD4(1) BRD [112]. In the first step, they collected K_{ac} mimetic substructures of published BRD inhibitors. Furthermore, they modified the structurally related substructures while maintaining the key pharmacophoric features. For example, 1,2,4-triazole was modified to isoxazole and henceforth used as query. They also utilized similarity searching based on pharmacophore, shape, and 2D fingerprint searches. The published (+)-JQ1 bound to the X-ray crystal structure of BRD4(1) (PDB ID: 3MXF) was used as reference ligand. The substructure and similarity searches were performed against the eMolecules database [116]. Out of the six novel hits identified by the above protocols, **11** (Figure 13.8) showed an IC_{50} value of 4.7 μM determined in an AlphaScreen assay.

13.5 Current Bromodomain Inhibitors

In 2005, a small molecule named NP1 (an N1-aryl-propane-diamine) was reported disrupting the pivotal interaction between the PCAF BRD and acetylated lysine 50 of HIV-1 Tat [117]. It was the first report giving proof of concept that BRDs can be targeted by drug-like molecules. This study has provided a new avenue to the drug discovery community. Scientists from the pharmaceutical industry and academia have started exploring the discovery of small molecules as BRD inhibitors for several therapeutic interventions. Figure 13.9 represents the evolution of BRD inhibitors for drug discovery.

In 2009, Mitsubishi Tanabe Pharma Corporation filed a patent disclosing a clinically relevant and well-known chemotype "thienotriazolodiazepines

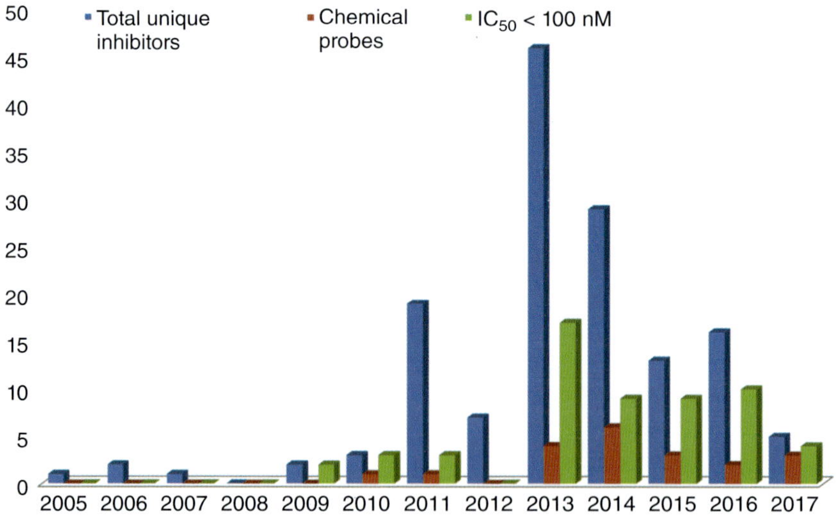

Figure 13.9 Number of published unique bromodomain inhibitors, number of chemical probes derived from them, and inhibitors with the IC_{50} less than 100 nM. Source of the numbers: ChromoHub [118].

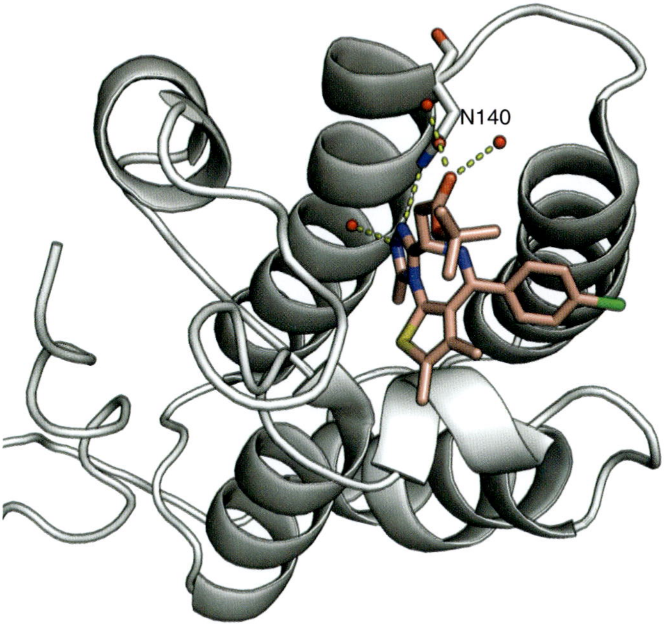

Figure 13.10 Binding mode of JQ1 in the binding pocket of the first BRD of BRD4. JQ1 forms a hydrogen bond with the conserved Asn140 residue.

(TTDs)" that prevents the binding of acetylated lysine into the binding pocket of BRD4 BRDs [119]. The derived low nanomolar BRD4 inhibitors also exhibited remarkable antiproliferative activities against a panel of cancer cell lines.

In 2010, Bradner and his colleagues reported a cell-permeable small molecule (JQ1) that binds selectively to the K_{ac} pocket of BET BRD proteins [8]. JQ1, a close derivative of the TTDs disclosed in the Mitsubishi patent, displayed remarkable potency and selectivity against the BET family.

The X-ray crystal structure of JQ1 bound to the first BRD of BRD4 revealed that its triazole ring acts as the acetyl-lysine mimic forming a hydrogen bond with the conserved asparagine residue Asn140 (Figure 13.10). Examples of BRD inhibitors currently in clinical trials are shown in Figure 13.11. Other chemical scaffolds of BRD inhibitors are discussed in detail by Zhang et al. in a recent review [120].

13.6 Multi-target Inhibitors

13.6.1 Dual Kinase–Bromodomain Inhibitors

The development of inhibitors that can concomitantly target cancer-driving kinases and BRDs is an established strategy now. The rationale for the dual kinase–BRD inhibitors is based on the role of kinases and BET family members in similar disease pathways. In fact, BRD4 was discovered as an atypical kinase [121]. Additionally, FLT3 receptor tyrosine kinase and BRD4 were discovered as important drug targets in AML. JAK kinases and BRD4 demonstrated key roles

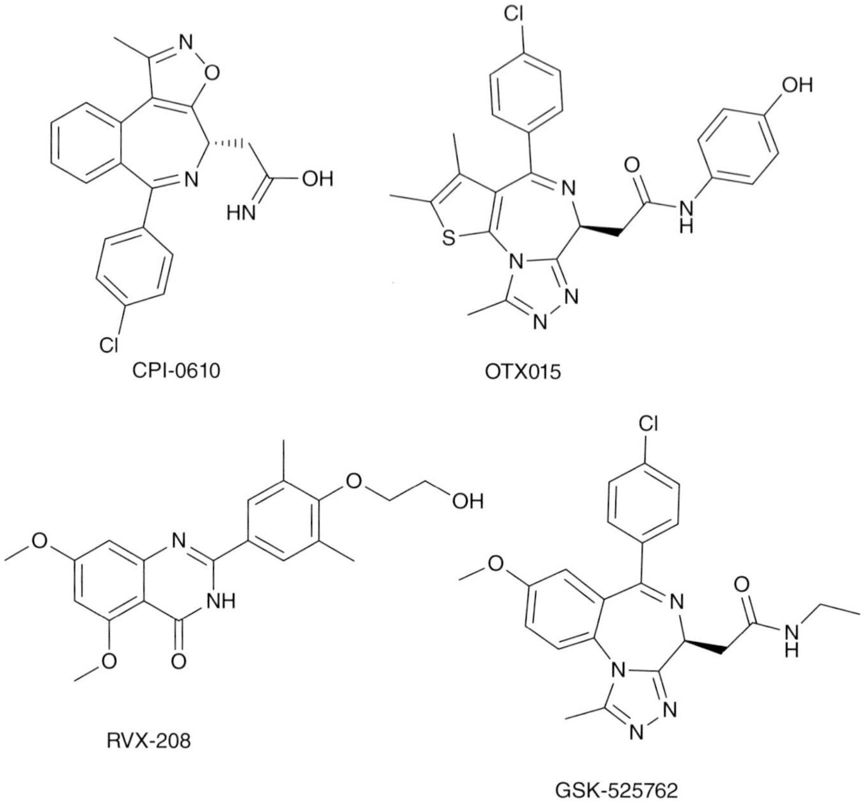

Figure 13.11 Some examples of bromodomain inhibitors in clinical trials (CPI-0610 (Constellation Pharmaceuticals), OTX015 (Oncoethix GmbH), RVX-208 (Resverlogix Corp.), and GSK-525762 (GlaxoSmithKline)).

in multiple myeloma models. Several clinically validated and selective kinase inhibitors are reported to inhibit BRDs with high potency.

In 2013 Martin et al. demonstrated that dinaciclib (SCH727965) (Figure 13.12), a selective CDK inhibitor, also interacts with BRDs [122]. Dinaciclib is a selective inhibitor for CDK1, CDK2, CDK5, and CDK9 with an excellent *in vivo* therapeutic index at low nanomolar concentrations. It has recently moved into phase III clinical trials for refractory chronic lymphocytic leukemia treatment.

An X-ray crystal structure of dinaciclib bound to CDK2 (PDB code 4KD1) revealed that it has a very good shape complementarity with the ATP binding pocket and also exhibits an extensive network of hydrogen bonds and hydrophobic interactions. For example, the pyrazolopyrimidine moiety of dinaciclib formed backbone hydrogen bonds with the hinge region residue Leu83, whereas the hydroxyethyl group interacted with the conserved Leu33 and the nitroxy group interacted with the side chain of Lys89 (Figure 13.13a). Dinaciclib demonstrated activity mainly on the BET family, viz. BRD2, BRD3, BRD4, and BRDT. The ligand shows two different binding modes in the crystal structures

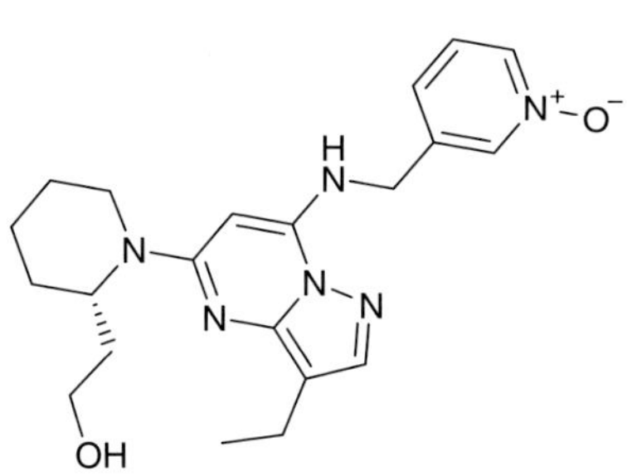

Dinaciclib (SCH727965)

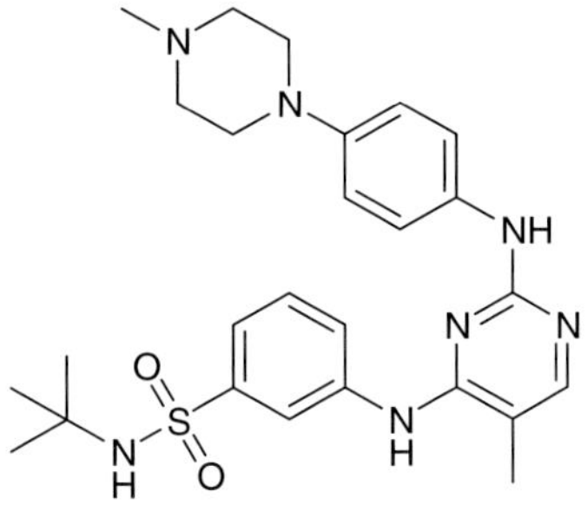

TG101209

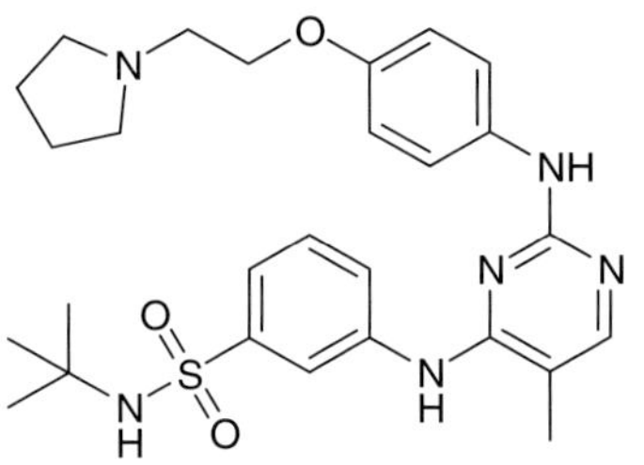

TG101348

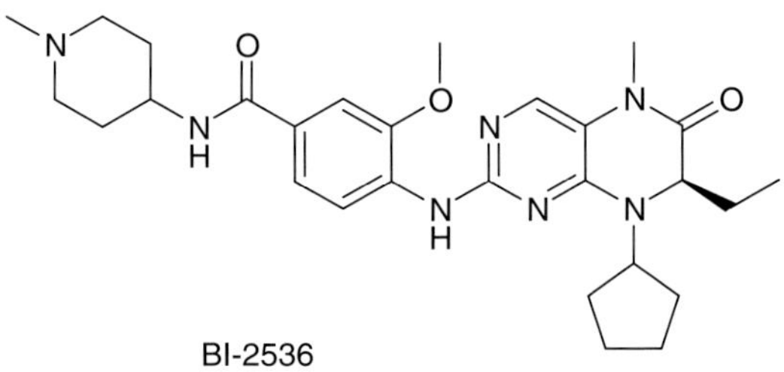

BI-2536

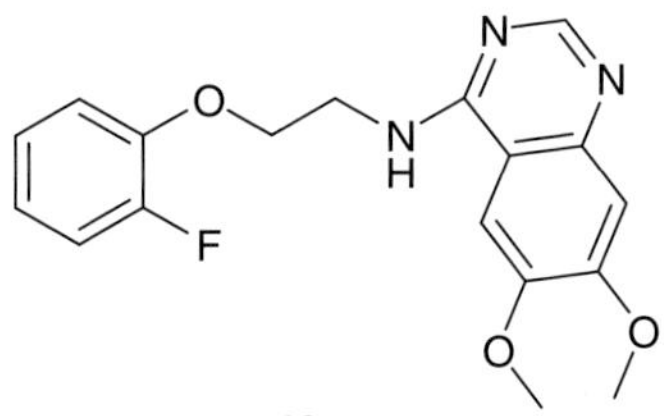

12

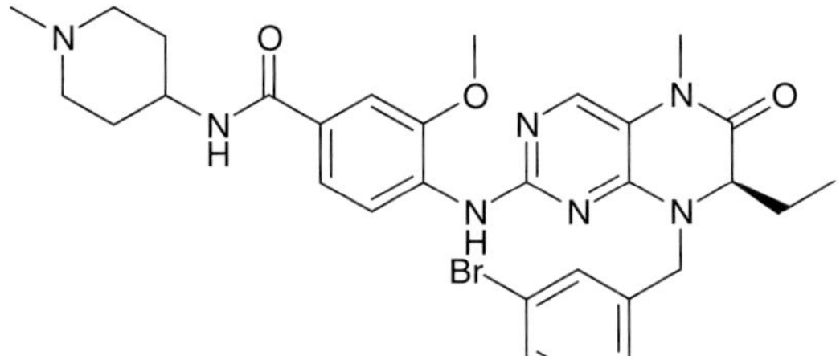

13

Figure 13.12 Representative examples of dual kinase–bromodomain inhibitors.

of different BET BRDs. The structural analysis revealed that the ligand binds to the K_{ac} pocket in both BRDT(1) (PDB code 4KCX) and BRD4(1) (PDB code 4O70). In the BRDT BRD, pyridine-N-oxide acts as acetyl-lysine mimetic moiety in which the nitroxy group of the ligand mediates a hydrogen bond interaction with the side chain of the conserved residue Asn109. The pyrazolopyrimidine moiety located near the WPF shelf is mostly stabilized by the hydrophobic and water-mediated hydrogen bond interactions (Figure 13.13b).

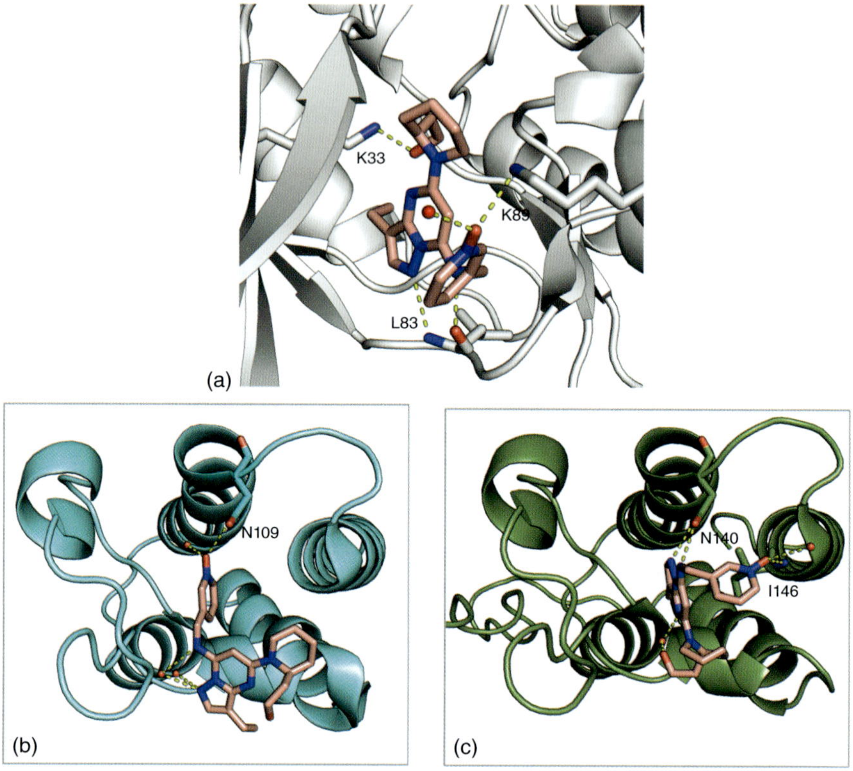

Figure 13.13 Binding modes of the dual kinase–bromodomain inhibitor dinaciclib. (a) Dinaciclib interacts with the ATP binding pocket residues (Lys33, Leu83, and Lys89) of CDK2. (b) The nitroxy group of dinaciclib forms a strong hydrogen bond with the conserved Asn109 of the BRDT(1) bromodomain. (c) The pyrazolopyrimidine moiety of dinaciclib forms two hydrogen bonds with the conserved Asn140 residue of the first bromodomain of BRD4.

In the first BRD of BRD4, the pyrazolopyrimidine moiety of dinaciclib mediates two hydrogen bond interactions with the side chain of Asn140, and the nitroxy group interacts with the backbone of Ile146. The pyrazolopyrimidine moiety also stabilizes the molecule through several hydrophobic as well as water-mediated hydrogen bond interactions (Figure 13.13c).

In subsequent publications, several dual kinase–BRD inhibitors were reported. For example, in 2014 Schönbrunn et al. performed a co-crystallization screening campaign using a total of 581 kinase inhibitors obtained from two distinct chemical libraries (Selleck Chemicals and GlaxoSmithKline published kinase inhibitor set) [123]. They reported that two known JAK2 inhibitors, TG101209 (Figure 13.12) and TG101348 (Figure 13.12), also inhibit BRD4(1) with IC_{50} values in low micromolar range. In an AlphaScreen assay, BI-2536 (Figure 13.12), a well-known polo-like kinase 1 (PLK1) inhibitor, also inhibited the BRD4 protein at nanomolar concentrations ($IC_{50} = 25$ nM).

In an independent study published at the same time, Knapp et al. reported the activity of 10 known kinase inhibitors against BRD proteins [124].

The authors reported that the two most active compounds (TG101348 and BI2536; Figure 13.12) with low micromolar IC_{50} value also suppress *c-Myc* expression in multiple myeloma cells through BET inhibition. Readers are also encouraged to read a recent article on dual kinase–BRD inhibitors by Carlino and Rastelli [125].

Computational approaches were also successfully utilized to obtain dual kinase–BRD inhibitors. In 2015 the Schürer group reported the first successful VS to identify dual kinase–BRD inhibitors [126]. The identified inhibitor **12** (Figure 13.12) exhibited significant affinity against BRD4(1) and epidermal growth factor receptor (EGFR). Chen et al. reported the identification of the more potent dual inhibitor **13** (Figure 13.12) of the PLK1 kinase and the BRD4 BRD identified through the structure–activity relationship study performed on the known inhibitor BI-2536 [127]. Carlino and Rastelli proposed pharmacophore models for the identification of putative dual kinase–BRD putative inhibitors [125]. These pharmacophore models are based on the three different binding modes described by Schönbrunn and coworkers and are available to the scientific community upon request [123].

13.6.2 Dual BET/HDAC Inhibitors

In 2014, GlaxoSmithKline reported a small-molecule dual BET/HDAC (histone deacetylases) inhibitor, DUAL946 (Figure 13.14), obtained through a structure-based design [128]. A BET active THQ core was structurally modified with a hydroxamic acid motif containing the key HDAC pharmacophoric features. The compound displayed submicromolar inhibition activities against both BET and class I and IIb HDAC proteins. Although the compound exhibited significant antiproliferative activities in the cellular assays, no synergy was noted in comparison with the parent compounds.

In 2016, Chen and coworkers reported the design and synthesis of a series of novel 3,5-dimethylisoxazole derivatives as BRD4/HDAC dual inhibitors [129]. The synthesized derivatives showed inhibitory activity against BRD4 and HDAC1. The representative molecules **14** and **15** (Figure 13.14) displayed potent antiproliferative activities in the human leukemia cell line K562 and MV4–11 xenograft mouse models.

13.7 Proteolysis Targeting Chimeras (PROTACs)

The selective degradation of BRD and extra-terminal (BET) proteins via the mediation of a small molecule is a novel and emerging technique. Proteolysis targeting chimeric molecules, or PROTACs, were reported performing the degradation of these proteins via the use of the ubiquitin proteasome system. For example, Raina et al. reported the targeted degradation of BET proteins using the molecule ARV-771 (an active BRD4 PROTAC; Figure 13.15) and its therapeutic potential against castration-resistant prostate cancer [130]. The Ciulli group reported the selective degradation of the BRD4 protein by using a small-molecule PROTAC named MZ1 (Figure 13.15) [131]. This molecule was designed by the rational conjugation of JQ1, a pan-BET inhibitor, and VH032, a potent and specific ligand of

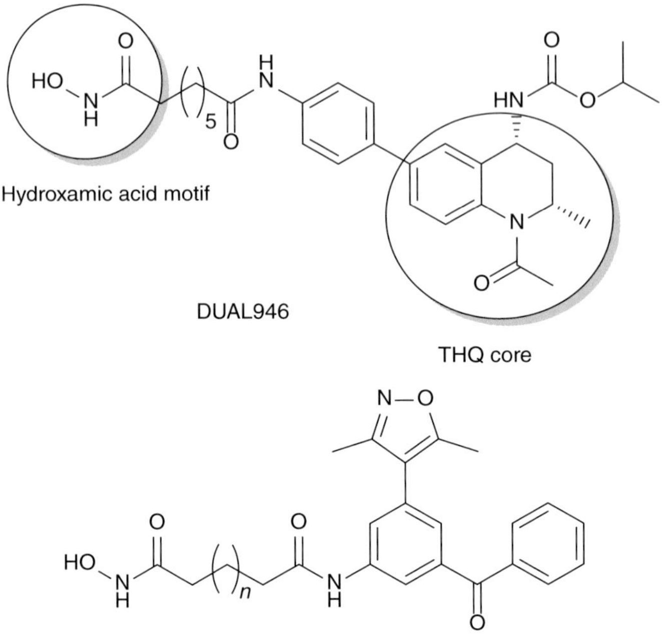

	IC$_{50}$ (nM)		
	n	HDAC1	BRD4
14	3	279	858
15	4	155	673

Figure 13.14 Examples of dual BET/HDAC inhibitors.

the protein von Hippel–Lindau (VHL). In a subsequent report published in 2017, the Ciulli group reported another highly selective BRD4 protein degrader, AT1 (Figure 13.15), a designed molecule based on the crystal structures of MZ1 bound to BRD4 as well as models of VHL proteins [132].

Zhou et al. recently reported a novel PROTAC molecule named BETd-260/ZBC260 (Figure 13.16) that effectively degrades the BRD4 protein at picomolar concentrations (IC$_{50}$ = 51 pM) in the RS4;11 leukemia cell line [133]. This molecule also displayed *in vivo* antiproliferative activity in RS4;11 xenograft tumor models. In 2015 Lu et al. reported the PROTAC molecule, ARV-825 (Figure 13.16), by connecting OTX015 (a BRD4 inhibitor) and pomalidomide (an E3 ligase cereblon binding moiety) using the PROTAC approach [134]. ARV-825 actively degraded BRD4 protein and also exhibited robust antiproliferative activities and apoptosis induction in Burkitt's lymphoma cell lines. In another report published at the same time, Winter et al. developed a new molecule named dBET1 (Figure 13.16) based on the rational combination of JQ1(+) and a thalidomide moiety [134, 135]. The designed molecule, dBET1, showed highly

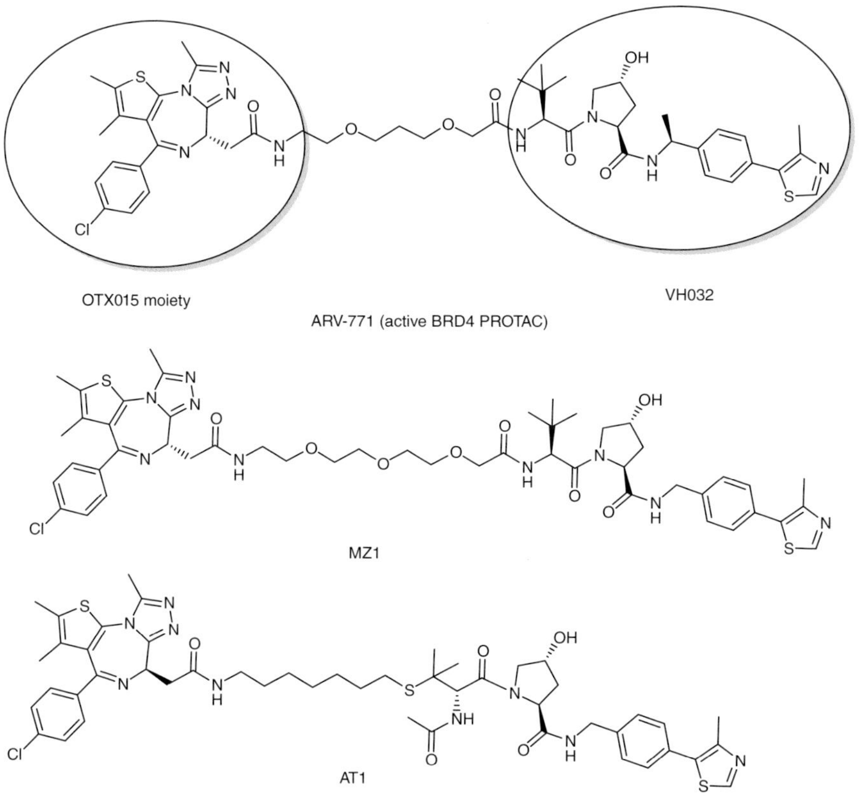

Figure 13.15 Examples of proteolysis targeting chimera molecules.

selective BRD4 degradation in both *in vitro* and *in vivo* studies. dBET1 also delayed leukemia progression in mice.

13.8 Conclusions

It has now been proven that many BRD-containing proteins are promising drug targets with oncogenic potential. Preclinical studies have shown that inhibitors against BRDs are active in a number of hematologic and solid malignancies both *in vivo* and *in vitro*. Several clinical trials are now being conducted to test their efficacy in humans. Most work has been done on the BET inhibitors as therapeutic agents against cancer. However, BRD inhibitors might also be helpful against other diseases, for example as antiparasitic [136] or antiepileptogenic agents [137]. More research on the non-BET BRDs and their role in diseases will certainly reveal novel drug targets. Furthermore, new chemical modulators of BRDs including dual inhibitors and chimeric molecules will facilitate the discovery of novel drugs.

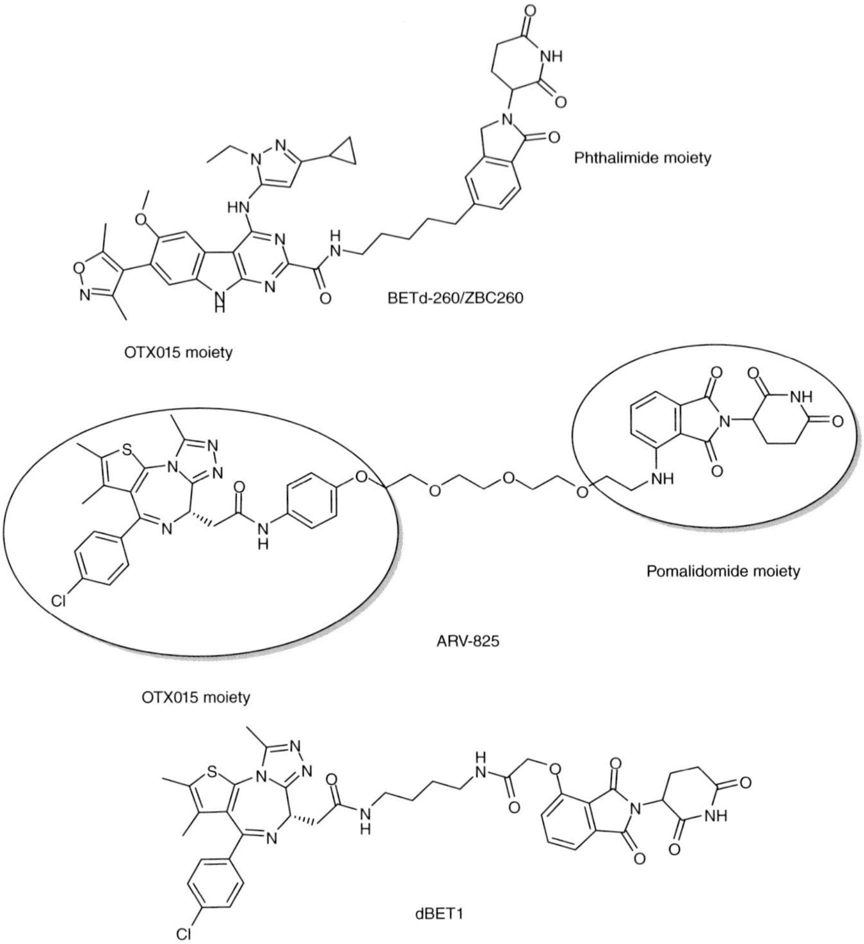

Figure 13.16 Examples of proteolysis targeting chimeras.

Acknowledgments

This work was supported by the German National Research Foundation (DFG: GU1225/3–1, CRC 992, and RTG 1976).

References

1 Tamkun, J.W., Deuring, R., Scott, M.P. et al. (1992). Brahma: a regulator of *Drosophila* homeotic genes structurally related to the yeast transcriptional activator SNF2/SWI2. *Cell* 68: 561–572.

2 Hornbeck, P.V., Zhang, B., Murray, B. et al. (2015). PhosphoSitePlus, 2014: mutations, PTMs and recalibrations. *Nucleic Acids Res.* 43: D512–D520.

3 Zhao, S., Xu, W., Jiang, W. et al. (2010). Regulation of cellular metabolism by protein lysine acetylation. *Science* 327: 1000–1004.

4 Nakamura, Y., Umehara, T., Nakano, K. et al. (2007). Crystal structure of the human BRD2 bromodomain: insights into dimerization and recognition of acetylated histone H4. *J. Biol. Chem.* 282: 4193–4201.

5 Nakayama, J., Rice, J.C., Strahl, B.D. et al. (2001). Role of histone H3 lysine 9 methylation in epigenetic control of heterochromatin assembly. *Science* 292: 110–113.

6 Alland, L., Muhle, R., Hou, H. Jr. et al. (1997). Role for N-CoR and histone deacetylase in Sin3-mediated transcriptional repression. *Nature* 387: 49–55.

7 Mosashvilli, D., Kahl, P., Mertens, C. et al. (2010). Global histone acetylation levels: prognostic relevance in patients with renal cell carcinoma. *Cancer Sci.* 101: 2664–2669.

8 Filippakopoulos, P., Qi, J., Picaud, S. et al. (2010). Selective inhibition of BET bromodomains. *Nature* 468: 1067–1073.

9 Muller, S., Filippakopoulos, P., and Knapp, S. (2011). Bromodomains as therapeutic targets. *Expert Rev. Mol. Med.* 13: e29.

10 Filippakopoulos, P., Picaud, S., Mangos, M. et al. (2012). Histone recognition and large-scale structural analysis of the human bromodomain family. *Cell* 149: 214–231.

11 Nicodeme, E., Jeffrey, K.L., Schaefer, U. et al. (2010). Suppression of inflammation by a synthetic histone mimic. *Nature* 468: 1119–1123.

12 Vidler, L.R., Brown, N., Knapp, S., and Hoelder, S. (2012). Druggability analysis and structural classification of bromodomain acetyl-lysine binding sites. *J. Med. Chem.* 55: 7346–7359.

13 Halgren, T.A. (2009). Identifying and characterizing binding sites and assessing druggability. *J. Chem. Inf. Model.* 49: 377–389.

14 Ogryzko, V.V., Schiltz, R.L., Russanova, V. et al. (1996). The transcriptional coactivators p300 and CBP are histone acetyltransferases. *Cell* 87: 953–959.

15 Nakamura, T., Mori, T., Tada, S. et al. (2002). ALL-1 is a histone methyltransferase that assembles a supercomplex of proteins involved in transcriptional regulation. *Mol. Cell* 10: 1119–1128.

16 Gregory, G.D., Vakoc, C.R., Rozovskaia, T. et al. (2007). Mammalian ASH1L is a histone methyltransferase that occupies the transcribed region of active genes. *Mol. Cell. Biol.* 27: 8466–8479.

17 Koutelou, E., Hirsch, C.L., and Dent, S.Y. (2010). Multiple faces of the SAGA complex. *Curr. Opin. Cell Biol.* 22: 374–382.

18 Suganuma, T., Gutierrez, J.L., Li, B. et al. (2008). ATAC is a double histone acetyltransferase complex that stimulates nucleosome sliding. *Nat. Struct. Mol. Biol.* 15: 364–372.

19 Ullah, M., Pelletier, N., Xiao, L. et al. (2008). Molecular architecture of quartet MOZ/MORF histone acetyltransferase complexes. *Mol. Cell. Biol.* 28: 6828–6843.

20 Doyon, Y. and Cote, J. (2004). The highly conserved and multifunctional NuA4 HAT complex. *Curr. Opin. Genet. Dev.* 14: 147–154.

21 Blosser, T.R., Yang, J.G., Stone, M.D. et al. (2009). Dynamics of nucleosome remodelling by individual ACF complexes. *Nature* 462: 1022–1027.

22 Strohner, R., Nemeth, A., Jansa, P. et al. (2001). NoRC – a novel member of mammalian ISWI-containing chromatin remodeling machines. *EMBO J.* 20: 4892–4900.
23 Kadam, S., McAlpine, G.S., Phelan, M.L. et al. (2000). Functional selectivity of recombinant mammalian SWI/SNF subunits. *Genes Dev.* 14: 2441–2451.
24 Bozhenok, L., Wade, P.A., and Varga-Weisz, P. (2002). WSTF-ISWI chromatin remodeling complex targets heterochromatic replication foci. *EMBO J.* 21: 2231–2241.
25 Banting, G.S., Barak, O., Ames, T.M. et al. (2005). CECR2, a protein involved in neurulation, forms a novel chromatin remodeling complex with SNF2L. *Hum. Mol. Genet.* 14: 513–524.
26 Barak, O., Lazzaro, M.A., Lane, W.S. et al. (2003). Isolation of human NURF: a regulator of engrailed gene expression. *EMBO J.* 22: 6089–6100.
27 Cattaneo, M., Morozumi, Y., Perazza, D. et al. (2014). Lessons from yeast on emerging roles of the ATAD2 protein family in gene regulation and genome organization. *Mol. Cells* 37: 851–856.
28 Jang, M.K., Mochizuki, K., Zhou, M. et al. (2005). The bromodomain protein Brd4 is a positive regulatory component of P-TEFb and stimulates RNA polymerase II-dependent transcription. *Mol. Cell* 19: 523–534.
29 Nogales, E., Fang, J., and Louder, R.K. (2017). Structural dynamics and DNA interaction of human TFIID. *Transcription* 8: 55–60.
30 Gong, F., Chiu, L.Y., Cox, B. et al. (2015). Screen identifies bromodomain protein ZMYND8 in chromatin recognition of transcription-associated DNA damage that promotes homologous recombination. *Genes Dev.* 29: 197–211.
31 Ivanov, A.V., Peng, H., Yurchenko, V. et al. (2007). PHD domain-mediated E3 ligase activity directs intramolecular sumoylation of an adjacent bromodomain required for gene silencing. *Mol. Cell* 28: 823–837.
32 Dancy, B.M. and Cole, P.A. (2016). Correction to protein lysine acetylation by p300/CBP. *Chem. Rev.* 116: 8314.
33 Podcheko, A., Northcott, P., Bikopoulos, G. et al. (2007). Identification of a WD40 repeat-containing isoform of PHIP as a novel regulator of beta-cell growth and survival. *Mol. Cell. Biol.* 27: 6484–6496.
34 Muller, P., Kuttenkeuler, D., Gesellchen, V. et al. (2005). Identification of JAK/STAT signalling components by genome-wide RNA interference. *Nature* 436: 871–875.
35 Fujisawa, T. and Filippakopoulos, P. (2017). Functions of bromodomain-containing proteins and their roles in homeostasis and cancer. *Nat. Rev. Mol. Cell Biol.* 18: 246–262.
36 Yang, Z., Yik, J.H., Chen, R. et al. (2005). Recruitment of P-TEFb for stimulation of transcriptional elongation by the bromodomain protein Brd4. *Mol. Cell* 19: 535–545.
37 Nishiyama, A., Dey, A., Miyazaki, J., and Ozato, K. (2006). Brd4 is required for recovery from antimicrotubule drug-induced mitotic arrest: preservation of acetylated chromatin. *Mol. Biol. Cell* 17: 814–823.
38 Dawson, M.A., Prinjha, R.K., Dittmann, A. et al. (2011). Inhibition of BET recruitment to chromatin as an effective treatment for MLL-fusion leukaemia. *Nature* 478: 529–533.

39 Bischoff, J.R., Anderson, L., Zhu, Y. et al. (1998). A homologue of *Drosophila* aurora kinase is oncogenic and amplified in human colorectal cancers. *Embo J.* 17: 3052–3065.

40 Tanaka, T., Kimura, M., Matsunaga, K. et al. (1999). Centrosomal kinase AIK1 is overexpressed in invasive ductal carcinoma of the breast. *Cancer Res.* 59: 2041–2044.

41 Andrieu, G., Tran, A.H., Strissel, K.J., and Denis, G.V. (2016). BRD4 regulates breast cancer dissemination through Jagged1/Notch1 signaling. *Cancer Res.* 76: 6555–6567.

42 Bandopadhayay, P., Bergthold, G., Nguyen, B. et al. (2014). BET bromodomain inhibition of MYC-amplified medulloblastoma. *Clin. Cancer Res.* 20: 912–925.

43 Delmore, J.E., Issa, G.C., Lemieux, M.E. et al. (2011). BET bromodomain inhibition as a therapeutic strategy to target c-Myc. *Cell* 146: 904–917.

44 Gallagher, S.J., Mijatov, B., Gunatilake, D. et al. (2014). The epigenetic regulator I-BET151 induces BIM-dependent apoptosis and cell cycle arrest of human melanoma cells. *J. Invest. Dermatol.* 134: 2795–2805.

45 Lockwood, W.W., Zejnullahu, K., Bradner, J.E., and Varmus, H. (2012). Sensitivity of human lung adenocarcinoma cell lines to targeted inhibition of BET epigenetic signaling proteins. *Proc. Natl. Acad. Sci. U.S.A.* 109: 19408–19413.

46 Mertz, J.A., Conery, A.R., Bryant, B.M. et al. (2011). Targeting MYC dependence in cancer by inhibiting BET bromodomains. *Proc. Natl. Acad. Sci. U.S.A.* 108: 16669–16674.

47 Pastori, C., Daniel, M., Penas, C. et al. (2014). BET bromodomain proteins are required for glioblastoma cell proliferation. *Epigenetics* 9: 611–620.

48 Puissant, A., Frumm, S.M., Alexe, G. et al. (2013). Targeting MYCN in neuroblastoma by BET bromodomain inhibition. *Cancer Discov.* 3: 308–323.

49 Belkina, A.C., Nikolajczyk, B.S., and Denis, G.V. (2013). BET protein function is required for inflammation: Brd2 genetic disruption and BET inhibitor JQ1 impair mouse macrophage inflammatory responses. *J. Immunol.* 190: 3670–3678.

50 Angrisano, T., Pero, R., Brancaccio, M. et al. (2016). Cyclical DNA methylation and histone changes are induced by LPS to activate COX-2 in human intestinal epithelial cells. *PLoS One* 11: e0156671.

51 Bierne, H., Hamon, M., and Cossart, P. (2012). Epigenetics and bacterial infections. *Cold Spring Harb. Perspect Med.* 2: a010272.

52 Mele, D.A., Salmeron, A., Ghosh, S. et al. (2013). BET bromodomain inhibition suppresses TH17-mediated pathology. *J. Exp. Med.* 210: 2181–2190.

53 Klein, K., Kabala, P.A., Grabiec, A.M. et al. (2016). The bromodomain protein inhibitor I-BET151 suppresses expression of inflammatory genes and matrix degrading enzymes in rheumatoid arthritis synovial fibroblasts. *Ann. Rheum. Dis.* 75: 422–429.

54 Tian, B., Yang, J., Zhao, Y. et al. (2017). BRD4 couples NF-kappaB/RelA with airway inflammation and the IRF-RIG-I amplification loop in respiratory syncytial virus infection. *J. Virol.* 91 (6).

55 Banerjee, C., Archin, N., Michaels, D. et al. (2012). BET bromodomain inhibition as a novel strategy for reactivation of HIV-1. *J. Leukocyte Biol.* 92: 1147–1154.
56 Boehm, D., Calvanese, V., Dar, R.D. et al. (2013). BET bromodomain-targeting compounds reactivate HIV from latency via a Tat-independent mechanism. *Cell Cycle* 12: 452–462.
57 Li, Z., Guo, J., Wu, Y., and Zhou, Q. (2013). The BET bromodomain inhibitor JQ1 activates HIV latency through antagonizing Brd4 inhibition of Tat-transactivation. *Nucleic Acids Res.* 41: 277–287.
58 Spiltoir, J.I., Stratton, M.S., Cavasin, M.A. et al. (2013). BET acetyl-lysine binding proteins control pathological cardiac hypertrophy. *J. Mol. Cell. Cardiol.* 63: 175–179.
59 Matzuk, M.M., McKeown, M.R., Filippakopoulos, P. et al. (2012). Small-molecule inhibition of BRDT for male contraception. *Cell* 150: 673–684.
60 Gyuris, A., Donovan, D.J., Seymour, K.A. et al. (2009). The chromatin-targeting protein Brd2 is required for neural tube closure and embryogenesis. *Biochim. Biophys. Acta* 1789: 413–421.
61 Wallace, R. (2004). Susceptibility gene for juvenile myoclonic epilepsy rings true. *Epilepsy Curr.* 4: 9–10.
62 Kiernan, R.E., Vanhulle, C., Schiltz, L. et al. (1999). HIV-1 Tat transcriptional activity is regulated by acetylation. *Embo J.* 18: 6106–6118.
63 Dorr, A., Kiermer, V., Pedal, A. et al. (2002). Transcriptional synergy between Tat and PCAF is dependent on the binding of acetylated Tat to the PCAF bromodomain. *Embo J.* 21: 2715–2723.
64 Eckner, R., Ewen, M.E., Newsome, D. et al. (1994). Molecular cloning and functional analysis of the adenovirus E1A-associated 300-kD protein (p300) reveals a protein with properties of a transcriptional adaptor. *Genes Dev.* 8: 869–884.
65 Lang, S.E. and Hearing, P. (2003). The adenovirus E1A oncoprotein recruits the cellular TRRAP/GCN5 histone acetyltransferase complex. *Oncogene* 22: 2836–2841.
66 Giles, R.H., Peters, D.J., and Breuning, M.H. (1998). Conjunction dysfunction: CBP/p300 in human disease. *Trends Genet.* 14: 178–183.
67 Avantaggiati, M.L., Carbone, M., Graessmann, A. et al. (1996). The SV40 large T antigen and adenovirus E1a oncoproteins interact with distinct isoforms of the transcriptional co-activator, p300. *Embo J.* 15: 2236–2248.
68 Bernat, A., Avvakumov, N., Mymryk, J.S., and Banks, L. (2003). Interaction between the HPV E7 oncoprotein and the transcriptional coactivator p300. *Oncogene* 22: 7871–7881.
69 Nemethova, M. and Wintersberger, E. (1999). Polyomavirus large T antigen binds the transcriptional coactivator protein p300. *J. Virol.* 73: 1734–1739.
70 Wang, L., Grossman, S.R., and Kieff, E. (2000). Epstein-Barr virus nuclear protein 2 interacts with p300, CBP, and PCAF histone acetyltransferases in activation of the LMP1 promoter. *Proc. Natl. Acad. Sci. U.S.A.* 97: 430–435.

71 Yan, J.P., Garrus, J.E., Giebler, H.A. et al. (1998). Molecular interactions between the coactivator CBP and the human T-cell leukemia virus Tax protein. *J. Mol. Biol.* 281: 395–400.
72 Zimmermann, H., Degenkolbe, R., Bernard, H.U., and O'Connor, M.J. (1999). The human papillomavirus type 16 E6 oncoprotein can down-regulate p53 activity by targeting the transcriptional coactivator CBP/p300. *J. Virol.* 73: 6209–6219.
73 Borrow, J., Stanton, V.P. Jr., Andresen, J.M. et al. (1996). The translocation t(8;16)(p11;p13) of acute myeloid leukaemia fuses a putative acetyltransferase to the CREB-binding protein. *Nat. Genet.* 14: 33–41.
74 Lai, J.L., Jouet, J.P., Bauters, F., and Deminatti, M. (1985). Chronic myelogenous leukemia with translocation (8;22): report of a new case. *Cancer Genet. Cytogenet.* 17: 365–366.
75 Ohnishi, H., Taki, T., Yoshino, H. et al. (2008). A complex t(1;22;11)(q44;q13;q23) translocation causing MLL-p300 fusion gene in therapy-related acute myeloid leukemia. *Eur. J. Haematol.* 81: 475–480.
76 Sobulo, O.M., Borrow, J., Tomek, R. et al. (1997). MLL is fused to CBP, a histone acetyltransferase, in therapy-related acute myeloid leukemia with a t(11;16)(q23;p13.3). *Proc. Natl. Acad. Sci. U.S.A.* 94: 8732–8737.
77 Pasqualucci, L., Dominguez-Sola, D., Chiarenza, A. et al. (2011). Inactivating mutations of acetyltransferase genes in B-cell lymphoma. *Nature* 471: 189–195.
78 Ward, R., Johnson, M., Shridhar, V. et al. (2005). CBP truncating mutations in ovarian cancer. *J. Med. Genet.* 42: 514–518.
79 Petrij, F., Giles, R.H., Dauwerse, H.G. et al. (1995). Rubinstein-Taybi syndrome caused by mutations in the transcriptional co-activator CBP. *Nature* 376: 348–351.
80 Roelfsema, J.H., White, S.J., Ariyurek, Y. et al. (2005). Genetic heterogeneity in Rubinstein–Taybi syndrome: mutations in both the CBP and EP300 genes cause disease. *Am. J. Hum. Genet.* 76: 572–580.
81 Bu, P., Evrard, Y.A., Lozano, G., and Dent, S.Y. (2007). Loss of Gcn5 acetyltransferase activity leads to neural tube closure defects and exencephaly in mouse embryos. *Mol. Cell. Biol.* 27: 3405–3416.
82 Field, M., Tarpey, P.S., Smith, R. et al. (2007). Mutations in the BRWD3 gene cause X-linked mental retardation associated with macrocephaly. *Am. J. Hum. Genet.* 81: 367–374.
83 Lalli, M.A., Jang, J., Park, J.H. et al. (2016). Haploinsufficiency of BAZ1B contributes to Williams syndrome through transcriptional dysregulation of neurodevelopmental pathways. *Hum. Mol. Genet.* 25: 1294–1306.
84 Smith, K.T., Nicholls, R.D., and Reines, D. (2006). The gene encoding the fragile X RNA-binding protein is controlled by nuclear respiratory factor 2 and the CREB family of transcription factors. *Nucleic Acids Res.* 34: 1205–1215.
85 Gu, L., Frommel, S.C., Oakes, C.C. et al. (2015). BAZ2A (TIP5) is involved in epigenetic alterations in prostate cancer and its overexpression predicts disease recurrence. *Nat. Genet.* 47: 22–30.

86 Ciro, M., Prosperini, E., Quarto, M. et al. (2009). ATAD2 is a novel cofactor for MYC, overexpressed and amplified in aggressive tumors. *Cancer Res.* 69: 8491–8498.

87 Hou, M., Huang, R., Song, Y. et al. (2016). ATAD2 overexpression is associated with progression and prognosis in colorectal cancer. *Jpn. J. Clin. Oncol.* 46: 222–227.

88 Luo, Y., Ye, G.Y., Qin, S.L. et al. (2015). ATAD2 overexpression identifies colorectal cancer patients with poor prognosis and drives proliferation of cancer cells. *Gastroenterol. Res. Pract.* 2015: 936564.

89 Chung, C.W., Dean, A.W., Woolven, J.M., and Bamborough, P. (2012). Fragment-based discovery of bromodomain inhibitors part 1: inhibitor binding modes and implications for lead discovery. *J. Med. Chem.* 55: 576–586.

90 Gosmini, R., Nguyen, V.L., Toum, J. et al. (2014). The discovery of I-BET726 (GSK1324726A), a potent tetrahydroquinoline ApoA1 up-regulator and selective BET bromodomain inhibitor. *J. Med. Chem.* 57: 8111–8131.

91 Demont, E.H. and Gosmini, R.L.M. (2011). Thetrahydroquinolines derivatives as bromodomain inhibitors. Google Patents WO2011054848A1.

92 Demont, E.H., Garton, N.S., Gosmini, R.L.M. et al. (2011). Tetrahydroquinoline derivatives and their pharmaceutical use. Google Patents WO2011054841A1.

93 Amans, D., Demont, E.H., Mitchell, D.J., and Watson, R.J. (2012). Tetrahydroquinoline derivatives useful as bromodomain inhibitors. Google Patents WO2012143413A1.

94 Yu, J.-l., Chen, T.-T., Zhou, C. et al. (2016). NMR-based platform for fragment-based lead discovery used in screening BRD4-targeted compounds. *Acta Pharmacol. Sin.* 37: 984–993.

95 Taylor, A.M., Côté, A., Hewitt, M.C. et al. (2016). Fragment-based discovery of a selective and cell-active benzodiazepinone CBP/EP300 bromodomain inhibitor (CPI-637). *ACS Med. Chem. Lett.* 7: 531–536.

96 Chaikuad, A., Lang, S., Brennan, P.E. et al. (2016). Structure-based identification of inhibitory fragments targeting the p300/CBP-associated factor bromodomain. *J. Med. Chem.* 59: 1648–1653.

97 Demont, E.H., Chung, C.-W., Furze, R.C. et al. (2015). Fragment-based discovery of low-micromolar ATAD2 bromodomain inhibitors. *J. Med. Chem.* 58: 5649–5673.

98 Xu, M., Unzue, A., Dong, J. et al. (2016). Discovery of CREBBP bromodomain inhibitors by high-throughput docking and hit optimization guided by molecular dynamics. *J. Med. Chem.* 59: 1340–1349.

99 Lucas, X., Wohlwend, D., Hugle, M. et al. (2013). 4-Acyl pyrroles: mimicking acetylated lysines in histone code reading. *Angew. Chem. Int. Ed.* 52: 14055–14059.

100 Zhao, H., Gartenmann, L., Dong, J. et al. (2014). Discovery of BRD4 bromodomain inhibitors by fragment-based high-throughput docking. *Bioorg. Med. Chem. Lett.* 24: 2493–2496.

101 Zhao, L., Wang, Y., Cao, D. et al. (2015). Fragment-based drug discovery of 2-thiazolidinones as BRD4 inhibitors: 2. Structure-based optimization. *J. Med. Chem.* 58: 1281–1297.

102 Bamborough, P., Diallo, H., Goodacre, J.D. et al. (2012). Fragment-based discovery of bromodomain inhibitors part 2: optimization of phenylisoxazole sulfonamides. *J. Med. Chem.* 55: 587–596.

103 Fish, P.V., Filippakopoulos, P., Bish, G. et al. (2012). Identification of a chemical probe for bromo and extra C-terminal bromodomain inhibition through optimization of a fragment-derived hit. *J. Med. Chem.* 55: 9831–9837.

104 Palmer, W.S., Poncet-Montange, G., Liu, G. et al. (2016). Structure-guided design of IACS-9571, a selective high-affinity dual TRIM24-BRPF1 bromodomain inhibitor. *J. Med. Chem.* 59: 1440–1454.

105 Martin, L.J., Koegl, M., Bader, G. et al. (2016). Structure-based design of an in vivo active selective BRD9 inhibitor. *J. Med. Chem.* 59: 4462–4475.

106 Zhang, G., Plotnikov, A.N., Rusinova, E. et al. (2013). Structure-guided design of potent diazobenzene inhibitors for the BET bromodomains. *J. Med. Chem.* 56: 9251–9264.

107 Drouin, L., McGrath, S., Vidler, L.R. et al. (2015). Structure enabled design of BAZ2-ICR, a chemical probe targeting the bromodomains of BAZ2A and BAZ2B. *J. Med. Chem.* 58: 2553–2559.

108 Hügle, M., Lucas, X., Weitzel, G. et al. (2016). 4-Acyl pyrrole derivatives yield novel vectors for designing inhibitors of the acetyl-lysine recognition site of BRD4(1). *J. Med. Chem.* 59: 1518–1530.

109 Rooney, T.P., Filippakopoulos, P., Fedorov, O. et al. (2014). A series of potent CREBBP bromodomain ligands reveals an induced-fit pocket stabilized by a cation–pi interaction. *Angew. Chem. Int. Ed.* 53: 6126–6130.

110 Hügle, M., Lucas, X., Ostrovskyi, D. et al. (2017). Beyond the BET family: targeting CBP/p300 with 4-acyl pyrroles. *Angew. Chem. Int. Ed.* 56 (41): 12476–12480.

111 Walters, W.P., Stahl, M.T., and Murcko, M.A. (1998). Virtual screening – an overview. *Drug Discovery Today* 3: 160–178.

112 Vidler, L.R., Filippakopoulos, P., Fedorov, O. et al. (2013). Discovery of novel small-molecule inhibitors of BRD4 using structure-based virtual screening. *J. Med. Chem.* 56: 8073–8088.

113 Muvva, C., Singam, E.R.A., Raman, S.S., and Subramanian, V. (2014). Structure-based virtual screening of novel, high-affinity BRD4 inhibitors. *Mol. Biosyst.* 10: 2384–2397.

114 Xue, X., Zhang, Y., Liu, Z. et al. (2016). Discovery of benzo[cd]indol-2(1H)-ones as potent and specific BET bromodomain inhibitors: structure-based virtual screening, optimization, and biological evaluation. *J. Med. Chem.* 59: 1565–1579.

115 Gao, N., Ren, J., Hou, L. et al. (2016). Identification of novel potent human testis-specific and bromodomain-containing protein (BRDT) inhibitors using crystal structure-based virtual screening. *Int. J. Mol. Med.* 38: 39–44.

116 Gozalbes, R. and Pineda-Lucena, A. (2011). Small molecule databases and chemical descriptors useful in chemoinformatics: an overview. *Comb. Chem. High Throughput Screen.* 14: 548–458.

117 Zeng, L., Li, J., Muller, M. et al. (2005). Selective small molecules blocking HIV-1 Tat and coactivator PCAF association. *J. Am. Chem. Soc.* 127: 2376–2377.

118 Liu, L., Zhen, X.T., Denton, E. et al. (2012). ChromoHub: a data hub for navigators of chromatin-mediated signalling. *Bioinformatics* 28: 2205–2206.

119 Miyoshi, S., Ooike, S., Iwata, K. et al. (2009). Antitumor agent. Google Patents WO2009084693A1.

120 Zhang, G., Smith, S.G., and Zhou, M.M. (2015). Discovery of chemical inhibitors of human bromodomains. *Chem. Rev.* 115: 11625–11668.

121 Devaiah, B.N., Lewis, B.A., Cherman, N. et al. (2012). BRD4 is an atypical kinase that phosphorylates serine 2 of the RNA polymerase II carboxy-terminal domain. *Proc. Natl. Acad. Sci. U.S.A.* 109: 6927–6932.

122 Martin, M.P., Olesen, S.H., Georg, G.I., and Schönbrunn, E. (2013). Cyclin-dependent kinase inhibitor dinaciclib interacts with the acetyl-lysine recognition site of bromodomains. *ACS Chem. Biol.* 8: 2360–2365.

123 Ember, S.W.J., Zhu, J.-Y., Olesen, S.H. et al. (2014). Acetyl-lysine binding site of bromodomain-containing protein 4 (BRD4) interacts with diverse kinase inhibitors. *ACS Chem. Biol.* 9: 1160–1171.

124 Ciceri, P., Muller, S., O'Mahony, A. et al. (2014). Dual kinase-bromodomain inhibitors for rationally designed polypharmacology. *Nat. Chem. Biol.* 10: 305–312.

125 Carlino, L. and Rastelli, G. (2016). Dual kinase-bromodomain inhibitors in anticancer drug discovery: a structural and pharmacological perspective. *J. Med. Chem.* 59: 9305–9320.

126 Allen, B.K., Mehta, S., Ember, S.W.J. et al. (2015). Large-scale computational screening identifies first in class multitarget inhibitor of EGFR kinase and BRD4. *Sci. Rep.* 5 (16924).

127 Chen, L., Yap, J.L., Yoshioka, M. et al. (2015). BRD4 structure–activity relationships of dual PLK1 kinase/BRD4 bromodomain inhibitor BI-2536. *ACS Med. Chem. Lett.* 6: 764–769.

128 Atkinson, S.J., Soden, P.E., Angell, D.C. et al. (2014). The structure based design of dual HDAC/BET inhibitors as novel epigenetic probes. *MedChemComm* 5: 342–351.

129 Zhang, Z., Hou, S., Chen, H. et al. (2016). Targeting epigenetic reader and eraser: rational design, synthesis and in vitro evaluation of dimethylisoxazoles derivatives as BRD4/HDAC dual inhibitors. *Bioorg. Med. Chem. Lett.* 26: 2931–2935.

130 Raina, K., Lu, J., Qian, Y. et al. (2016). PROTAC-induced BET protein degradation as a therapy for castration-resistant prostate cancer. *Proc. Natl. Acad. Sci.* 113: 7124–7129.

131 Zengerle, M., Chan, K.-H., and Ciulli, A. (2015). Selective small molecule induced degradation of the BET bromodomain protein BRD4. *ACS Chem. Biol.* 10: 1770–1777.

132 Gadd, M.S., Testa, A., Lucas, X. et al. (2017). Structural basis of PROTAC cooperative recognition for selective protein degradation. *Nat. Chem. Biol.* 13: 514–521.

133 Zhou, B., Hu, J., Xu, F. et al. (2017). Discovery of a small-molecule degrader of bromodomain and extra-terminal (BET) proteins with picomolar cellular potencies and capable of achieving tumor regression. *J. Med. Chem.* 61 (2): 462–481.

134 Lu, J., Qian, Y., Altieri, M. et al. (2015). Hijacking the E3 ubiquitin ligase cereblon to efficiently target BRD4. *Chem. Biol.* 22: 755–763.
135 Winter, G.E., Buckley, D.L., Paulk, J. et al. (2015). Phthalimide conjugation as a strategy for in vivo target protein degradation. *Science* 348: 1376–1381.
136 Jeffers, V., Yang, C., Huang, S., and Sullivan, W.J. Jr. (2017). Bromodomains in protozoan parasites: evolution, function, and opportunities for drug development. *Microbiol. Mol. Biol. Rev.* 81: 1–17.
137 Younus, I. and Reddy, D.S. (2017). Epigenetic interventions for epileptogenesis: a new frontier for curing epilepsy. *Pharmacol. Ther.* 177: 108–122.

14

Lysine Reader Proteins

Johannes Bacher[1], Dina Robaa[2], Chiara Luise[2], Wolfgang Sippl[2], and Manfred Jung[1]

[1] Albert-Ludwigs-University Freiburg, Institute of Pharmaceutical Sciences, Albertstraße 25, 79104, Freiburg im Breisgau, Germany
[2] Martin-Luther-University of Halle-Wittenberg, Institute of Pharmacy, Department of Pharmaceutical Chemistry, Wolfgang-Langenbeck-Straße 4, 06120, Halle (Saale), Germany

14.1 Introduction

Chromatin is composed of negatively charged DNA wrapped around basic and, therefore under physiologic conditions, positively charged histone proteins [1]. In order to regulate the transcription of specific genes, enzymes modify the chromatin structure in highly dynamic processes by adding chemical groups to either the DNA itself or the histones without changing the genetic code itself [2]. These so-called posttranslational modifications (PTMs) lead to differentiation between euchromatin, which is the transcriptional active state, and heterochromatin, the denser, silenced state [3]. The accessibility of the chromatin is highly regulated by epigenetic mechanisms such as direct methylation of cytosine residues in the DNA. This is associated with gene silencing. Another epigenetic mechanism is to modify the histone part of chromatin mostly at the histone tails but also in the core where the actual DNA–histone binding takes place [4]. A broad range of different modifications is known, reaching from addition of a small methyl group to the attachment of large protein structures like ubiquitin. PTMs can have direct effects on the transcription of genes. Acetylation of lysine residues is associated with the formation of euchromatin, as the positive charge of the histone is vanished. Due to the epigenetic modifications being not persistent, a dynamic system of different proteins and enzymes is needed to maintain and rapidly change these epigenetic marks. The proteins responsible for this can be divided into three groups: "writers" that are enzymes capable of adding chemical groups to proteins or histones, "erasers" that in turn can cleave the chemical group off the protein target, and "readers" that are able to detect whether a specific mark is attached to, e.g. a histone [5]. The differentiation between these three groups is sometimes very difficult, as there are examples of proteins containing more than one epigenetically relevant protein domain, like *suppressor of variegation 3–9 homologue 1* (SUV39H1), which includes both a "writer" and a "reader"

Epigenetic Drug Discovery, First Edition. Edited by Wolfgang Sippl and Manfred Jung.

domain for methylated lysine in its sequence [6]. Also some reader proteins are able to bind more than one PTM at once. Altogether, it is assumed that epigenetic reader proteins respectively the requisite reader domains play an important role in interpreting and translating the epigenetic code [7].

In histone tails, lysine and arginine residues can be methylated. Lysine is modified to contain one, two, or three methyl groups attached to the ε-amino group, whereas arginine is methylated once or twice. The two methyl groups can either be attached to the same (asymmetric) or different (symmetric) nitrogen atoms of the guanidine group of the arginine, leading to three different methylation states as well. Methylation is a very small modification and does not affect the charge of the lysine residue, thus keeping it positively charged. There is only a minor change in size and lipophilicity compared with the non-methylated side chain [8]. As the modification does not strongly influence the histone DNA binding globally, most of the biological effect of histone methylation is thought to be due to recognition by highly specific methyl-lysine recognizing proteins. The effects mediated by these methyl-lysine "reader" proteins are diverse, e.g. they are involved in recruiting enzyme complexes or can act as transcription factors or other effector proteins as well [9]. As the biologic effects and the role of reader proteins in different diseases get more and more unveiled, there is a need for therapeutic tools targeting the protein–protein interaction between reader proteins and histones. For acetyl-lysine binding proteins, much effort has been taken, and there is a range of potent and selective inhibitors for these domains; some of them have reached clinical trials for different types of cancer as well as for metabolic diseases [10–12].

Methyl-lysine reader proteins are highly specific proteins capable of differentiation between methylation states. Known methyl-lysine binding domains characteristically contain a so-called aromatic cage, which is formed by two to four aromatic amino acids and serves to anchor the methyl-lysine residue. The binding of the methyl-lysine residue to the aromatic cage is predominantly driven by cation–π interactions between the positively charged methyl-ammonium moiety and the electron-rich residues of the aromatic cage. In addition, favorable van der Waals (vdW) contacts exhibited by the side chain methylene groups within the aromatic cage as well as the expulsion of the structured water molecules occupying the aromatic cage upon substrate binding further contribute to the binding affinity [13]. Lower methylation states (Kme1 and Kme2) are usually recognized via the so-called cavity-insertion recognition mode, where the protonated mono- or dimethyl-ammonium is inserted into a deep and narrow cleft and undergoes additional salt–bridge interactions with an aspartate residue lining the aromatic cage. The narrow nature of the binding cleft is believed to be an essential factor for favoring lower methylation states over Kme3, since the latter would be sterically prevented from binding. Notably, cavity-insertion binding is mainly driven by the interactions between the methyl-lysine residue and the previously described deep protein cleft, with no or only few specific interactions formed by adjacent amino acid residues of Kme. On the other hand, Kme3 residues are commonly recognized via surface-groove recognition, where the binding additionally relies on interactions with peptide residues flanking Kme3 [9, 14]. The group of methyl-lysine recognition proteins is divided into three families according to their binding domain, the plant homeodomain (PHD) zinc finger proteins, the WD40 repeat domain-containing proteins, and the so-called Royal family of reader proteins. The Royal family itself is categorized into four subfamilies: tudor-, chromo-, Pro-Trp-Trp-Pro (PWWP), and malignant brain tumor (MBT) repeat domain-containing proteins.

14.2 The Royal Family of Epigenetic Reader Proteins

The Royal family is the largest family within the group of epigenetic methyl-lysine reader proteins. As mentioned in Section 14.1, the family consists of four subfamilies, which share 3–5 antiparallel β sheets, two of which are connected by a 3_{10} α-helix. This forms a barrel-like structure that can be identified in all reader proteins of the Royal family.

14.2.1 The MBT Domain

The human MBT domain family consists of nine different members [15, 16], mainly recognizing mono- or dimethyllysine residues with binding domains containing about 100 amino acids [17]. Changes in the physiological functions are associated with the progression and development of several diseases [18]. The main function of MBT domain-containing proteins is associated with polycomb complexes [19], mostly leading to the formation of heterochromatin and therefore transcriptional repression [18]. Because of this, some MBT domain-containing proteins are termed histone-methylation-dependent chromatin locks [20]. Like all Royal family members, MBT domains join to contain a barrel-like structure out of five β-strands while β3 and β4 are connected via a 3_{10} helix [21]. The methyl-lysine residue is recognized via the previously described cavity-insertion mode, which only allows the accommodation of lysine residues with lower methylation states such as mono- or dimethyllysine in the binding pocket. As characteristic for Kme1/2 reader domains, the aromatic cage of the MBT domain contains a negatively charged aspartate residue, which is able to form additional salt–bridge interactions with the protonated amino group of Kme. Crystal structures of MBT domain-containing proteins with different histone peptides demonstrated that, apart from interactions of Kme1/2 with the aromatic cage, very little specific interactions with the substrate are formed (Figure 14.1) [22].

L3MBTL1 is a very well-explored member of the MBT domain family. It contains a tandem repeat of three MBT domains [20, 23, 24], of which, interestingly, only the second domain binds methylated lysine, while the function of the remaining two domains is still unknown [25]. Min et al. confirmed via co-crystallization that the second MBT domain of L3MBTL1 is able to bind dimethylated lysine at position 20 of histone H4 (H4K20me2) [26]. Besides this, L3MBTL1 binds rather unselectively to other monomethylated histones, including non-histone substrates [17]. This indicates that only the lysine residue

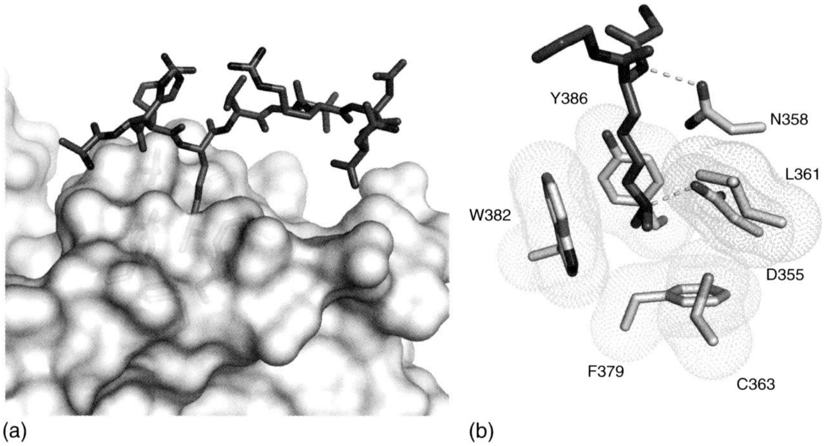

Figure 14.1 L3MBTL1 in complex with H4K20me2 (PDB ID: 2PQW). (a) The surface of the protein is depicted and the H3K20me2 peptide shown as sticks. (b) The aromatic cage of L3MBTL1 is shown as white sticks and the Kme2 residue as gray sticks. H-bond interactions are depicted as dashed lines. Source: Li et al. 2007 [22]. Reproduced with permission of Elsevier.

is recognized very specifically by the aromatic cage, while the surrounding residues only exhibit poor interactions with the MBT domain. The methylated lysine undergoes strong interactions with the aromatic cage (Trp, Tyr, Phe, Cys), including a hydrogen bond to an aspartate that is typical for mono- or dimethyllysine reader domains. The only additional interactions between the bound peptide and the reader protein are two H-bonds, one between the amide NH of Lys20 and Asn358 in addition to a water-mediated H-bond between the backbone carbonyl of His18 and Tyr386/Asn358. In 2012, Sakaguchi et al. confirmed that L3MBTL1 interacts with H4K20me1, a mark established by PR-Set7, resulting in stabilization of the mark. Depletion of the L3MBTL1 protein in S2 cells showed >60% decrease in overall H4K20me1 without great impacts on cell viability itself [27]. By binding to the monomethylated protein p53 (p53K382me1), a mark that is also catalyzed by the methyltransferase PR-Set7/Set8, the activity and consequently the activation of target genes are repressed [25]. Depletion of L3MBTL1 promotes hematopoietic cells toward erythroid differentiation [28]. Moreover, E2F-regulated genes, such as the oncogene *c-myc*, are repressed via L3MBTL1 binding to chromatin [20].

Small-molecule inhibitors of protein–protein interactions are increasingly gaining relevance. In 2010, Kireev et al. reported the finding of inhibitors of the L3MBTL1–histone interaction. For the identification of the scaffolds, they used *in silico* screening methods combined with *in vitro* screening. The primary hits obtained in the virtual screening were purchased and then tested on four different isolated MBT domain-containing proteins using AlphaScreen technology. With this method, they were able to identify a set of small molecule inhibitors with IC_{50} values within the low micromolar range (Kireev-1, Kireev-2, and Kireev-13; Figure 14.2) [29].

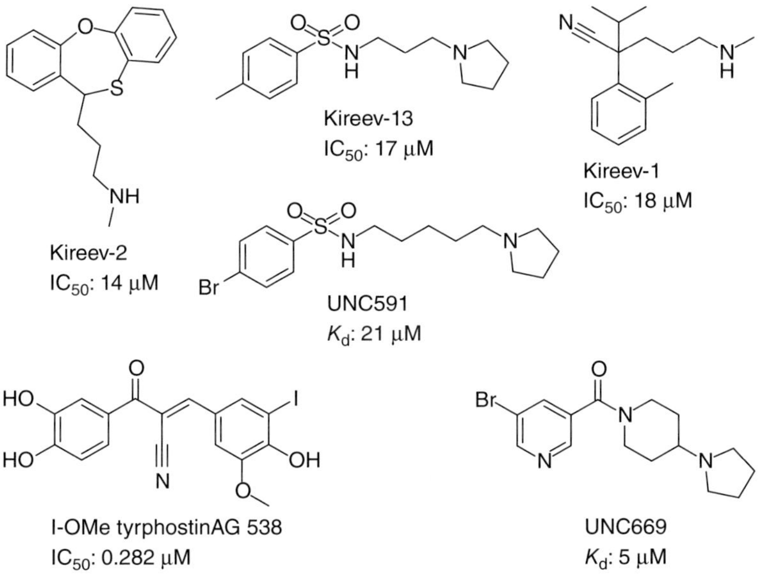

Figure 14.2 Structures of the L3MBTL1 inhibitors.

Another approach in targeting L3MBTL1 was reported in the same year by Wigle et al., where an established AlphaScreen-based assay capable for high-throughput screening (HTS) was used to screen the LOPAC compound library with 1280 compounds. With this screen, they were able to identify two compounds inhibiting the interaction between L3MBTL1 and H3K9me1 within the nanomolar range (cefsulodin IC_{50}: 98 nM, I-OMe tyrphostin AG 538 (Figure 14.2) IC_{50}: 282 nM), while three more hits were identified as micromolar inhibitors [30].

In 2011, Gao et al. reported the finding of UNC591 (Figure 14.2) as an inhibitor of L3MBTL1 [31]. The compound was identified among a set of biophysical probes and determined via use of molecular dynamics (MD) and free energy perturbation (FEP). The computational findings were then validated by isothermal titration calorimetry (ITC), showing a K_d value of 21 µM for UNC591. The assumption was made that the probes mimic the native lysine and therefore interrupt the interaction [31].

In the same year, Herold et al. successfully identified a further micromolar inhibitor for L3MBTL1. UNC669 (Figure 14.2) was reported with a K_d value of 5 µM, an affinity that is fivefold higher than that of the native peptide H4K20me1. Additionally, the authors showed that UNC669 is sixfold more selective for L3MBTL1 over the close homologue L3MBTL3 and 10-fold over L3MBTL4 [32]. The crystal structure of L3MBTL1 in complex with UNC669 reveals that the pyrrolidine moiety acts as a Kme mimetic and exhibits the same cation–π and H-bond interactions in the second MBT domain. Moreover, the bromine substituent in UNC669 forms a halogen bond with Ser360 of the L3MBTL1 protein (Figure 14.3) [32].

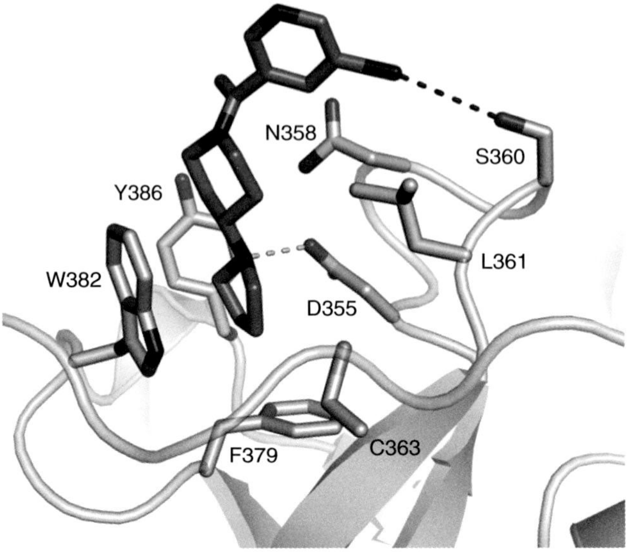

Figure 14.3 Cocrystal structure of L3MBTL1 with UNC669 (PDB ID: 2P8H). H-bond and halogen bond interactions are depicted as dashed lines.

L3MBTL3 is another member of the MBT domain family and a close homologue to L3MBTL1 described in Section 14.2.1. This reader protein prefers dimethyllysine as binding partner while being rather unselective for the sequence of the ligand. Therefore, it shows high affinities toward H4K20me2, H3K4me2, K3K9me2, and H4K9me2 [17]. *L3mbtl3* was shown to be a tumor suppressor gene, which is able to block transcription. In fact, re-expression of *l3mbtl3* after knockout in a medulloblastoma (malignant mostly pediatric brain tumor) cell line attenuated malignancy of the tumor. This finding was supported by single nucleotide polymorphism (SNP) analysis in 212 medulloblastoma patients. The study found out that *l3mbtl2* and *l3mbtl3* show both homozygous and hemizygous deletions in these tumors. This data supports the critical role of epigenetic methyl-lysine reader proteins and lysine methylation itself in the development and progression of cancer, especially in medulloblastoma [33]. Therefore, it is crucial to gain selective and potent probes to further study these protein–protein interactions and to gain insight in the pathogenesis of malignancies.

Similar to L3MBTL1, promising results in addressing the interactions of L3MBLT3 have been achieved. James et al. identified a symmetrical small molecule, UNC1215 (Figure 14.4), which addresses the MBT domain of L3MBTL3. This compound shows about 40-fold increase in affinity toward L3MBLT3 compared with UNC669 toward L3MBTL1, resulting in a K_d value of 120 nM. Besides its strong affinity, UNC1215 is over 50-fold selective against other MBT domain-containing proteins and over 200 different reader proteins [34]. Unlike UNC669, UNC1215 consists of two pyrrolidine substructures linked via an aromatic residue. The crystal structure of UNC1215 in complex with L3MBTL3 revealed a 2 : 2 binding mode. Here, one pyrrolidine residue binds to

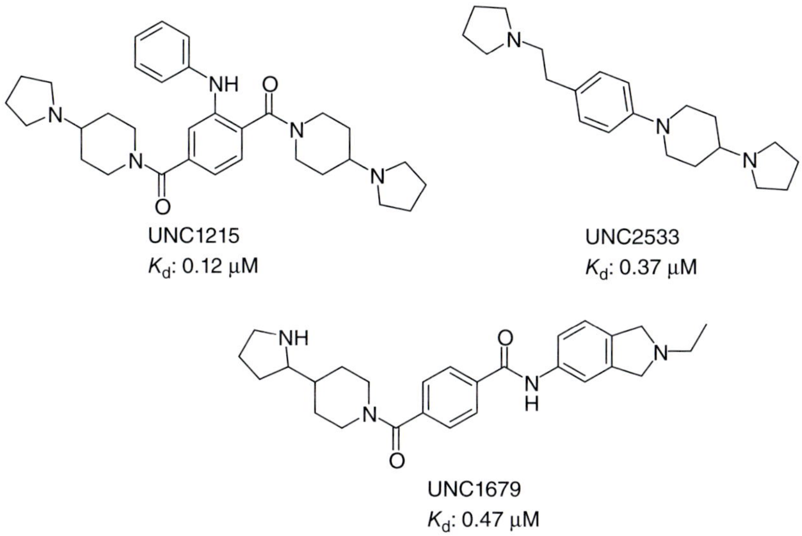
UNC1215
K_d: 0.12 μM

UNC2533
K_d: 0.37 μM

UNC1679
K_d: 0.47 μM

Figure 14.4 Structures of the L3MBTL3 inhibitors.

the second MBT domain of one protein, while the other pyrrolidine binds the first MBT domain of a second L3MBTL3 protein (Figure 14.5) [34]. Guo et al. predicted these findings in 2009 claiming that dimerization would occur in some MBT members including L3MBTL3 [35]. Recently, Baughman et al. showed with an AlphaScreen-based assay that dimerization between domain one and two of L3MBTL3 occurs in the absence of small molecule inhibitors in isolated protein as well as in cells. The dimerization of L3MBTL3 was disrupted upon mutation of the MBT domain 1 and 2 (D274A and D381A mutants). This suggests that L3MBTL3 functions as a dimer to recognize the histone substrate and could

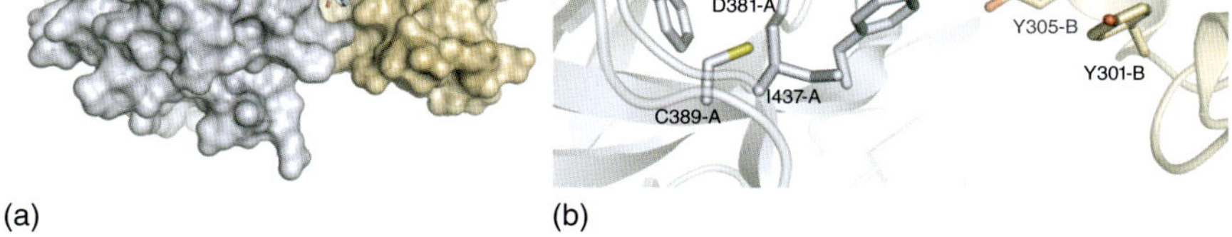

Figure 14.5 (a) The overall structure of L3MBTL3 in complex with UNC1215 showing a 2 : 2 binding mode (PDB ID: 4FL6). The surface of the protein is shown in gray and beige for chains A and B, respectively, and the ligands as cyan sticks. (b) Binding mode of UNC1215 in MBT2 and MBT1 of chains A and B.

serve as an explanation for the replicative MBT domains [36]. In 2013, Camerino et al. revealed UNC2533 (Figure 14.4), a structural descendent of UNC1215, as another high-affinity ligand of L3MBTL3 (K_d = 370 nM) with almost the same selectivity as UNC1215 [37]. In order to investigate the cellular effects of UNC1215, James et al. established an assay based on fluorescence recovery after photobleaching (FRAP). A fusion protein consisting of the three MBT domains of L3MBTL3 fused with green fluorescent protein (GFP) was introduced to human embryonic kidney 293 cells (HEK293). After photobleaching, the recovery time of the UNC1215 treated cells was reduced, compared with control cells. This effect was observed in a dose-dependent manner that indicates that UNC1215 is competing with cellular factors for binding L3MBTL3 [38]. They could also show that in G-401 cells, which contain high levels of L3MBTL3 compared with HEK295, biotinylated UNC1215 could purify L3MBTL3 out of cell lysates. This binding is antagonized with untagged UNC1215 indicating specific binding [38]. Further lead optimization studies led to the discovery of UNC1679 (Figure 14.4) that maintained high potency with improved selectivity [38].

14.2.2 The PWWP Domain

Another member of the Royal family of epigenetic readers is composed of PWWP domain-containing proteins (PDPs). There are 24 proteins known to contain a PWWP domain [39]. Originally, PWWP domains were discovered as DNA binding domains [40], while the interaction with histones was unveiled later [41]. The affinity of the PWWP domain to histone interaction is weak compared with the DNA binding, therefore a dual double-stranded DNA–histone binding mechanism is proposed [41]. For the PDP1, this dual binding mechanism has been shown, supporting the hypothesis described in Section 14.2.2 [42]. Similar to the other members of the Royal family, PWWP domain proteins contain a barrel-like structure, which is built out of five β-strands [41]. They typically possess a positively charged surface, rich in arginine and lysine residues, which is essential for DNA binding. Co-crystal structures of PWWP domains with histone peptides generally reveal that the peptide resides in a wide surface groove. The Kme3 residue is embedded in an aromatic cage formed by three conserved aromatic acid residues, while the surrounding peptide residues undergo several interactions with the protein residues in the outer surface groove [41]. PWWP domains are able to recognize higher methylation states, mainly trimethylated lysine (Kme3). Since the Kme3 moiety is bulkier than Kme1 or Kme2, the binding pocket of the PWWP domain is more open and surface exposed (surface recognition mode) than it is in the previously discussed MBT domains. As the binding site is easier to access, discrimination between different methylation states is not as distinct.

To date, there is no reported success in targeting this family of lysine readers with a small molecule. Only one crystal structure of the PWWP domain of BRPF1 in complex with an isoquinoline fragment is deposited in the Protein Data Bank (PDB), which shows the isoquinoline ring nicely placed in the aromatic cage (Figure 14.6) [43]. Since PDPs are involved in several diseases,

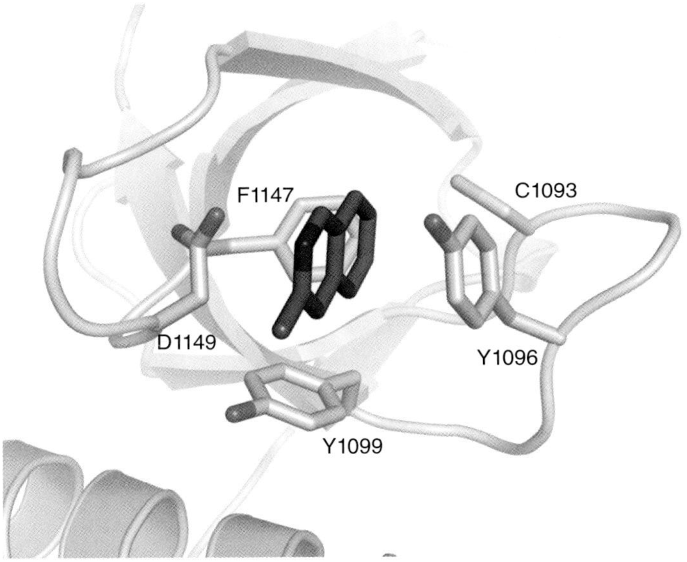

Figure 14.6 PWWP domain of BRPF1 in complex with an isoquinoline fragment (PDB ID: 5C6S).

inhibitors of the interaction of PWWP domains with their targets could be useful tools in further exploring the molecular mechanisms of related diseases. One member of the PWWP domain family is heparin binding hepatoma-derived growth factor (HDGF), which is a cardiovascular mitogen. HDGF is expressed during developmental stages and can be re-expressed in injured vascular smooth muscle cells and atherosclerosis [44]. This suggests that *hdgf* plays an important role in both development of the cardiovascular system and disease. Besides, overexpression of *hdgf* has been shown to appear in cancer such as hepatocellular carcinoma [45, 46], non-small cell lung cancer [47], gastric carcinoma [48], colorectal stromal tumors [49], and pancreatic cancer [50]. HDGF proteins contain an N-terminal PWWP domain that binds very specifically to DNA in promoter regions of their target genes, e.g. on *smyd1* gene that encodes for a methyltransferase essential for cardiogenesis [44]. For the PWWP domain of HDGF2, hints exist that the PWWP domain also binds weakly to histone marks such as H3K79me3 and H4K20me3 [41].

Another interesting member of the PWWP family is, as already mentioned in Section 14.2.2, PDP1. This protein binds to methylated H4K20 on chromatin and is associated with the histone methyltransferase Set9. This writer enzyme catalyzes the mono-, di-, and trimethylation of H4K20. By binding to H4K20me, PDP1 directs methylation to higher states via Set9 [42]. This histone mark is involved in recruiting checkpoint proteins to DNA damages and activation of a DNA lesion checkpoint [51].

As mentioned in Section 14.1, reader domains also occur as a module within a multidomain protein. An example for this complexity is Wolf–Hirschhorn syndrome candidate 1 (WHSC1), which includes a methyltransferase and two PWWP domains and five PHD finger domains within its sequence [52]. WHSC1 is associated with the genesis of multiple myeloma [53], and overexpression via a translocation of the *whsc1* gene leads to increased proliferation of multiple myeloma cells [54], resulting in poor prognosis.

Another interesting protein is PC4 and SFRS1 interacting protein 1 (PSIP1), which exists in two isoforms (p52 and p75). One function of the p75 isoform, namely, the binding of chromatin via its PWWP domain, is essential for human immunodeficiency virus 1 (HIV-1) for recruitment to active transcription units on chromatin [55]. Furthermore, PSIP is shown to be antiapoptotic [56] and to enhance transcriptional activity, leading to resistance to oxidative stress-mediated and chemotherapy-induced apoptosis [57]. The PWWP domain of PSIP1 shows high affinity toward H3K36me3 when it is in complex with nucleosomes, but this affinity decreases 1000-fold for the isolated histone peptide. This supports the hypothesis of a dual dsDNA–histone binding mode of PDPs [58]. H3K36me3 is associated with the formation of euchromatin and, hence, transcriptional activation [55].

14.2.3 The Tudor Domain

The third group in the Royal family is referred to as the Tudor-domain-containing group of proteins. Compared with the two families discussed in Sections 14.2.1 and 14.2.2, the Tudor domain family consists of a larger number of over 40 different proteins [59, 60]. Another unique feature compared with MBT and PDP is the more diverse recognition of methylated residues. The family contains a large variety of proteins recognizing methyl-lysine residues in different methylation states and also proteins that detect arginine methylation [61–63]. Although there are many different proteins with a variety of binding motifs, only little progression is made in finding inhibitors of the interactions. This might also be due to the low druggability of Tudor domains showing [64].

An example of these diverse binding motifs is the Tudor-like domain-containing protein Spindlin1. It consists of three Tudor domain repeats, of which only the second domain binds to H3K4me3 [65] that is the main target of histone recognition. Interestingly, affinity toward H3K4me3 is increased when the arginine residue in position 8 is asymmetrically dimethylated (H3R8me2a). This modification is recognized by the first Tudor domain repeat. Furthermore, Spindlin1 is supposed to activate Wnt/β-catenin, which is mediated by protein arginine methyltransferase 2 (PRMT2) and mixed lineage leukemia (MLL) complex. PRMT2 and MLL complex are responsible for the generation of the H3R8me2a and H3K4me3 mark, resulting in the specific "K4me3–R8me2a" pattern, recognized by Spindlin1. Mutations in the binding pockets of Spindlin1 lead to a decreased activation of Wnt target genes, implicating an important role in this pathway [66]. Spindlin1 derives its name from the first discovery, being associated with the meiotic spindle apparatus in mice [67]. In humans, overexpression of *spindlin1* was discovered to play a role in a variety of cancers like ovarian, non-small cell lung cancer, hepatic tumors [68], and liposarcoma [69]. Additionally, Gao et al. showed that overexpression of *spindlin1* leads to different cell cycle phases, faster cell growth, and morphological changes in NIH3T3 cells [70]. These findings indicate that *spindlin1* influences tumorigenesis, supported by the fact that the cells are delayed in metaphase of the cell cycle and show chromosomal instability [71, 72]. A possible way for the promotion of cell growth is the Wnt/TCF4 signaling pathway mentioned in Section 14.2.3[73]. As these pathways are involved in the genesis and progression of a variety of diseases, inhibitors of Spindlin1 can be useful therapeutic tools.

In 2016, Wagner et al. reported the finding of a potent small-molecule inhibitor for Spindlin1. They established a screening platform using AlphaLISA technology for initial testing and an orthogonal fluorescence polarization (FP) assay for hit conformation. In an exploratory screen, they were able to confirm A366 (Figure 14.7) as potent inhibitor of Spindlin1, showing an IC_{50} value of 182 nM [74]. As A366 is originally a potent inhibitor of the histone methyltransferase G9a [75], off-target effects of A366 are expected. Robaa et al. presented the first set of small-molecule inhibitors bearing another scaffold, with Robaa-1k (Figure 14.7) exhibiting an IC_{50} of 3.5 μM [76]. The compounds did not inhibit the lysine methyltransferases EZH2 and Dot1L.

As previously mentioned, there are proteins that can be classified into more than one category of epigenetic modifiers. SETDB1 is an example for this, as it was first discovered to be a histone methyltransferase capable of H3K9 methylation [77, 78] in an H3K4me3-dependent pathway [79]. In the N-terminal region, SETDB1 also contains a Tudor domain, which function still remains unclear. Nevertheless, Dong et al. recently presented two tetrahydroisoquinoline-carboxamide fragments co-crystallized with this Tudor domain [80, 81]. Interestingly, the phenyl moiety of these fragments is anchored in the aromatic cage undergoing π–π stacking with two tryptophan residues, while the protonated isoquinoline-NH forms salt–bridge interactions with an aspartate residue lining the aromatic cage (Figure 14.8).

Another protein for which there is a known inhibitor is the p53 binding protein (53BP1). This protein binds, as the name suggests, to the tumor suppressor p53 at dimethylated lysine 382 (p53K382me2) and furthermore to H4K20me2 [82]. H4K20me2 typically marks DNA damage sites, which are the target for p53. By the dual binding mode to both H4K20me2 and p53, 53BP1 is able to direct p53 to its targets more directly. Furthermore, 53BP1 is able to bind ubiquitinylated H2AK15 [83], which in combination with H4K20me2 is a specific histone code for DNA double-strand breaks (DNA-DSBs). 53BP1 is known to play a role in the progression of breast [84] and ovarian cancer [85]. In 2015, Perfetti et al. were successful in targeting the Tudor domain of 53BP1. In a cross-screening approach, they were able to identify UNC2170 (Figure 14.9) as an inhibitor of

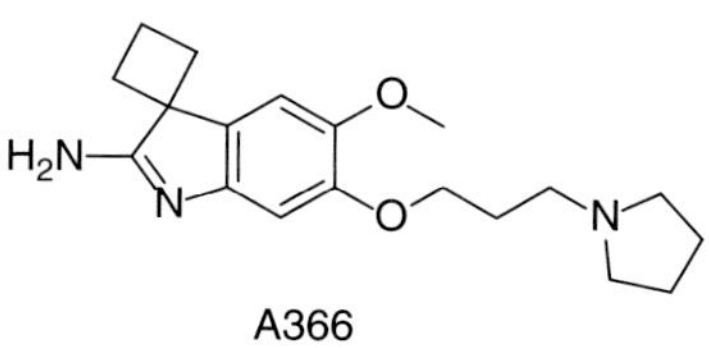

A366
IC_{50}: 0.183 μM

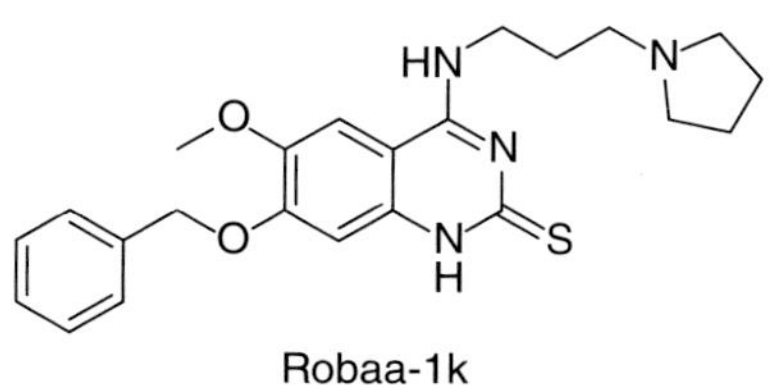

Robaa-1k
IC_{50}: 3.5 μM

Figure 14.7 Structure of the Spindlin1 inhibitors.

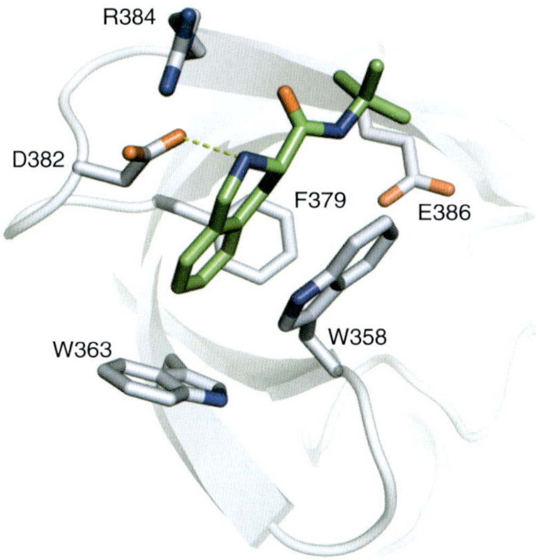

Figure 14.8 SETDB1 co-crystallized with a tetrahydroisoquinoline-carboxamide fragment (PDB ID: 5KE2).

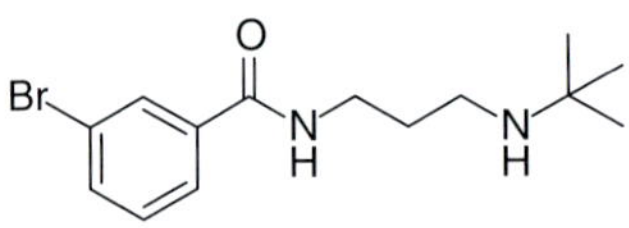

Figure 14.9 Structure of the p53BP inhibitor UNC2170.

53BP1. The micromolar fragment-like molecule is more than 17 times selective against other methyl-lysine binding proteins and shows cellular activity. Crystal structure analysis revealed that UNC2170 binds to a dimeric form of 53PB1, making distinct interactions with both Tudor domains (Figure 14.10). The *tert*-butyl group of UNC2170 is anchored in the aromatic cage of one Tudor domain by vdW interactions, while the protonated secondary amine undergoes salt–bridge interactions with both Asp1521 lining the aromatic cages of both protein monomers. Additionally, the amide NH forms H-bond interactions with Asp1521 of one protein chain. Notably, cation–π interactions seem to play a minor role in ligand binding, since the distance between the protonated amine and the aromatic acid residues of the cage is beyond 5 Å [86].

JMJD2A, which is also an eraser protein that catalyzes the demethylation of H3K9me2/3 and H3K4me2/3, contains two tandem Tudor domains and two PHD finger domains. These domains are shown to bind H3K4me3 and H4K20me3 [87]. The Tudor domains of JMJD2A are associated with different diseases such as gastric [88] and endometrial carcinoma [89]. Despite their potential relevance as therapeutic targets, there are still no reported inhibitors targeting the JMJD2A Tudor domains.

Enzyme complexes like polycomb repressive complex 2 (PRC2) also contain Tudor domain proteins like PHF1, MTF2, or PHF19. Via interaction with H3K36me3, these proteins are crucial for the recruitment of the enzyme complex to its target genes of methylation and, therefore, for the activity of the whole

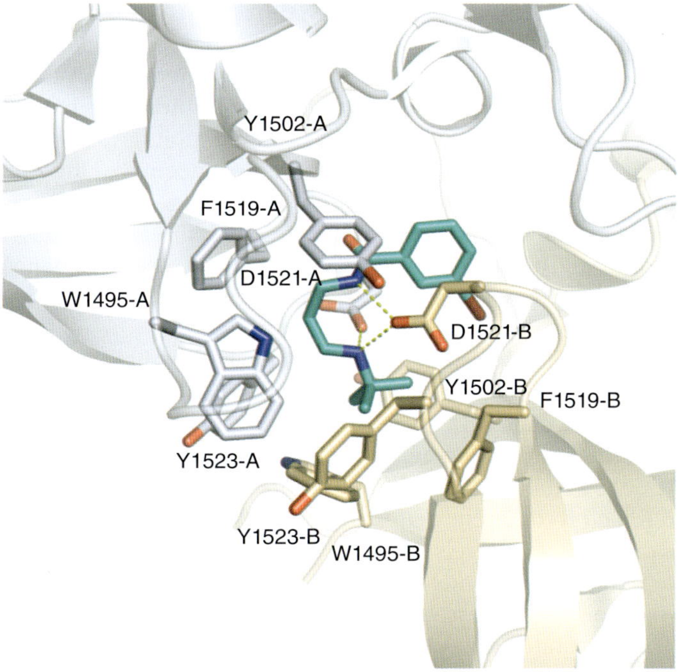

Figure 14.10 UNC2170 bind to a dimeric form of 53PB1 (PDB ID: 4RG2). Chains A and B of the protein are shown in white and beige color, respectively, and the ligand in cyan.

complex [90]. In endometrial sarcoma, fusion proteins of PHF1 with MEAF6, JAZF1, or EPC1 were discovered to play an important role in the disease [91, 92].

14.2.4 The Chromodomain

Another group of the Royal family of reader proteins is called the chromodomain group. This family with about 30 members [93] derives its name from "chromatin organization modifier," which already implicates the function as an epigenetic reader protein. Typically, all chromodomains contain a three-stranded β-barrel structure followed by two α-helices and another α-helix connecting β1 and β2. Like in the Tudor-domain-containing proteins, recognition of the methylated substrate mainly takes place in a surface groove with aromatic cages built up of three aromatic amino acids. Due to this type of methyl-lysine recognition, chromodomain proteins are mainly binding to higher methylation grades with the highest affinity toward trimethylated substrates [94]. One subfamily of the chromodomain group constitutes heterochromatin-associated proteins 1 (HP1). Two members of this subfamily, namely, HP1β and HP1γ, are known to play critical roles in prostate cancer and various other diseases via binding to H3K9me3 [95]. To date, no inhibitors were unveiled for these proteins.

The chromobox homologue (CBX) proteins, which represent the largest subfamily of the chromodomain-containing proteins, consist of five members (CBX 2, 4, 6, 7, 8), which are associated with PRCs and are referred to as the polycomb group of proteins. The name of this complex is derived from the polycomb protein found in *Drosophila* of which CBX proteins are the human paralogue [96]. PRCs in mammalian cells (PRC1 and PRC2) are clusters of different polycomb proteins with various functions. As the name suggests, the complexes are mainly associated with gene repression, and imbalances in their function have impacts on development, cancer progression, and stem cell maintenance [97]. Unifying for the five CBX proteins associated with PRC is the recognition of H3K27me3 as substrate. This methylation mark is catalyzed by EZH2, a lysine methyltransferase also found within the PRC. Notably, CBX proteins share a high sequence similarity, especially in the substrate binding pocket, which renders their selective targeting very challenging.

A representative of the chromobox homologues is chromobox homologue 7 (CBX7), which is also the best investigated protein of this group. Furthermore, it presents a promising drug target as CBX7 is involved in several diseases via gene repression of different tumor repression genes [98]. In different cell lines, overexpression of *cbx7* is linked to proliferation of the tumor cells. Overexpression of this gene was also linked to different regulation processes as, e.g. in prostate cancer cells expression of higher levels of *cbx7* is connected with hormone resistance [98]. The same observations were obtained in ovarian clear cell adenocarcinomas where highly expressed *cbx7* results in poor prognosis [99]. Knockdown of CBX7 in prostate cancer cells results in decreased proliferation rates and increased aging of the tumor cells [100, 101]. Moreover, altering levels of CBX7 are present in leukemia [102] and lymphomas [103]. Contrary to these findings, antiproliferative effects of CBX7 were revealed in other cancer types like lung adenoma, suggesting a more complex involvement in cell regulation [104].

Many efforts in targeting the chromodomain of CBX7 were made in the past, which resulted in the discovery of several peptidomimetic and small-molecule inhibitors. Simhadri et al. tried to identify small-molecule inhibitors of the CBX7–H3K27me3 interaction via a virtual screening-based approach. As the hits generated by *in silico* methods were not active *in vitro*, they focused on peptidomimetic inhibitors. As a starting point, they used a short sequence derived from SETDB1 (Ac-FALKme$_3$S-NH$_2$) that had slightly higher-affinity (IC$_{50}$: 12 μM) toward CBX7 than the original ligand, an H3K27me3-containing protein. Within this initial scaffold, mutations of the amino acid residues, excluding the trimethyllysine, were carried out in order to increase potency. After several SAR analysis steps, a set of compounds merging together all positive variations of the SAR studies was synthesized and shown to be the first set of CBX7 inhibitors and the first chromodomain inhibitors ever reported. Affinity of the optimized inhibitor Simhadri-64 (Figure 14.11) was around 200 nM with a 10- and 400-fold selectivity over CBX8 and 1, respectively [98]. Only recently, Stuckey et al. reported the finding of another potent peptidic inhibitor of CBX7, UNC3866 (Figure 14.11, K_d: 97 nM), which exhibits similar affinities toward CBX4 while being at least 10-fold selective against other chromodomains [105]. The design of UNC3866 was primarily based on optimizing known peptidic CBX substrates, including H3K27me3, with the goal of enhancing potency and increasing cellular permeability. The latter was markedly improved by introducing a leucine residue instead of the arginine preceding Kme3 in H3K27me3 and replacing the Kme3

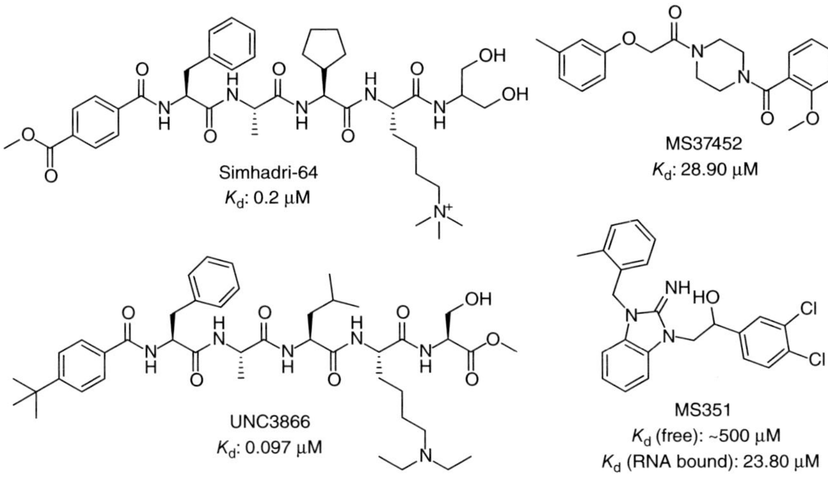

Figure 14.11 Structures of CBX7 inhibitors.

with diethyl-lysine. Guided by MD simulation studies, the interactions of the N-terminal part of the peptide with CBX7 were optimized by introducing the *tert*-butyl-benzoyl cap. In an attempt to explain the selectivity of the compound to CBX4/7 over other chromobox proteins, crystal structures of UNC3866 in complex with CBX2/4/7/8 were resolved. Structural analyses revealed that the compound binds in the same manner to all the aforementioned proteins. Indeed, the binding pocket of the peptide contains highly conserved residues, which renders the explanation of the selectivity rather difficult. However, it was hypothesized that subtle differences in the pocket accommodating the alanine residue of the peptide (at position −2 of KMe3) are responsible for the selectivity of UNC3866. This pocket is composed of an alanine, a leucine, and a valine residue in CBX2/8, while in CBX4/7 the alanine is replaced by a valine (Val13 in CBX7) rendering the pocket more ideal for undergoing vdW interactions with the alanine residue of the peptide [106].

Another approach in targeting CBX7 with peptidomimetic inhibitors was reported by Traoré et al. in 2017, where the amide bonds of the peptide backbone were replaced with *aza*-amino residues, thereby obtaining inhibitors with activities within the micromolar range [107].

In 2016, Ren et al. reported the finding of a small-molecule inhibitor for CBX7. By screening a library of all drugs approved by the FDA and over 100 000 commercially available compounds, they could identify 56 initial hits. To validate the hits, they established a fluorescence anisotropy-based assay, using fluorescein-tagged SETDB1-K1170me3-derived binding peptide. Additionally, 2D ^{1}H–^{15}N HSQC analysis was used in order to confirm the hits as inhibitors. Besides five compounds of the FDA-approved drugs (sennoside A, suramin, aurin tricarboxylic acid, Evans blue, and trypan blue), one commercially available compound (MS37452; Figure 14.11) raised further interest. By using NMR titration, this compound could be confirmed as an inhibitor of CBX7 (K_d: 29 μM) showing a K_i value of 43 μM for the native H3K27me3 binding assessed via FP displacement assay. Selectivity assessment revealed that MS37452 is threefold more selective for CBX7 over CBX4 and more than 10-fold selective over the other chromobox homologues. Interestingly, the compound shows no binding toward other chromodomain-containing proteins. By mutations of the aromatic cage of CBX7, Ren et al. showed that this region is crucial for the interaction of CBX7 with MS37452. Notably, MS37452 shows two different rotameric conformations in the resolved crystal structure (Figure 14.12). The dimethoxybenzene and piperazine moieties adopt the same orientation in the aromatic cage of CBX7, interacting with Phe11, Trp32, and Trp35, while the tolyl moiety is bound in either a *cis* or *trans* conformation with respect to the dimethoxybenzene. In the *trans* conformation, the tolyl group is positioned similar to H3K27me3, forming interactions with His47, Val13, and Glu43. Meanwhile, in the *cis* conformation, the tolyl moiety shifts upward toward Val10. The selectivity of MS37452 for CBX7 over CBX2/4/6/8 was mainly attributed to the presence of residues Val13 and His47 in CBX7, which are able to stabilize the tolyl moiety by vdW contacts and π–π stacking, respectively. In CBX2/6/8, Val13 is replaced by an alanine, whereas His47 is substituted by an asparagine residue in CBX2/4/6/8. Ren et al. also revealed MS351 (Figure 14.11) that shows low affinity (K_d: ~500 μM) toward the CBX7–H3K27me3. Interestingly, the affinity of MS351 is increased by 21-fold if CBX7 is bound to RNA (K_d: 23.8 μM) while being selective over a set of chromo- and bromodomain-containing proteins [108, 109].

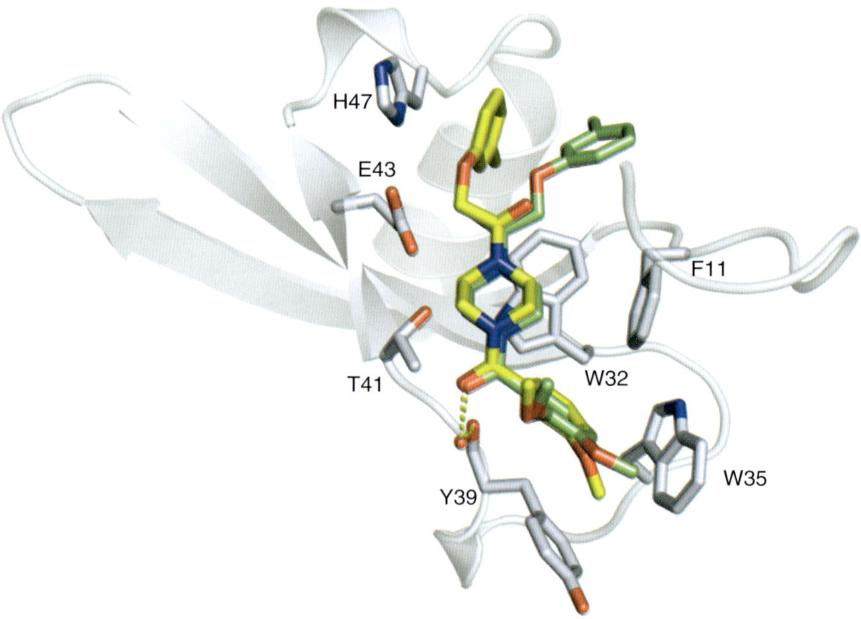

Figure 14.12 MS37452 binds to CBX7 in two rotameric forms (PDB ID: 4X3T). The *trans*-rotamer is depicted as yellow sticks, and the *cis*-rotamer as green sticks. Only side chains of chain A accommodating the *trans*-rotamer are depicted as white sticks.

CBX6 also belongs to the chromobox homologues and is part of the PRC. Like its homologue CBX7, CBX6 is overexpressed in cancer, like hepatocellular carcinoma [110]. Milosevich et al. reported the finding of an inhibitor for CBX6 in 2016, which was developed by introducing structural variations on the previously reported CBX7 peptidic inhibitors [111]. Since the chromobox homologues show high similarities in their chromodomain, Milosevich et al. took the non-selective CBX inhibitor pBr-FAYKme$_3$S-NH$_2$ with K_d values of 300 nM for CBX6 and 100 nM for CBX7, respectively, as a starting point for selectivity optimizations. By altering the alanine residue (position −2) to a valine, the obtained pBr-FVYKme$_3$S-NH$_2$ (Figure 14.13) peptide was found to retain high potency toward CBX6 with K_d: 900 nM, as determined via FP titration assay, while showing a decreased binding affinity to CBX2/4/8/7. pBr-FVYKme$_3$S-NH$_2$ showed about sixfold selectivity for CBX6 over CBX7. Docking and MD studies were performed in an attempt to explain the preference of the bulkier valine group of the peptide for CBX6 over CBX7. As proposed for the selective CBX7 peptidic inhibitor, UNC3866, the Val/Ala substitution at the floor of the (−2) pocket of the different CBX proteins was suggested to play a role in defining CBX protein selectivity. However, it becomes clear that the selectivity of the reported CBX-peptidic inhibitors cannot be solely explained by structural analysis and that there are obviously other contributing factors [111].

The chromodomain Y peptides are involved in spermatogenesis as well as being part of repressing complexes [112]. Since the peptidic inhibitors evaluated for CBX7, despite their selectivity, still showed off-target effects in the nanomolar range against other chromodomain-containing protein, Barnash et al. used UNC3866 as a starting point in order to gain selective chemical probes for CDYL proteins. UNC3866 shows K_d of 850 nM against chromodomain Y like protein 2 (CDYL2). In a combinatory approach, they synthesized a set of peptides to create a peptide library. Afterward, this library was screened via magnetic screening, and initial hits were tested for their CDYL2 selectivity. The most promising hits were then conjugated with a fluorescein label, and selectivity measurements were carried out using FP. UNC4980 and UNC4981 (Figure 14.14) were identified to be at least three- to fivefold selective for CDYL2 with affinities in the nanomolar range (K_d = 430 nM (UNC4980) and K_d = 577 nM (UNC4981)). The untagged analogues of UNC4990 and UNC4991 showed similar potencies against CDYL2. The binding to the chromodomain was confirmed by pulldown assays [113].

Figure 14.13 Structure of a CBX 6 inhibitor.

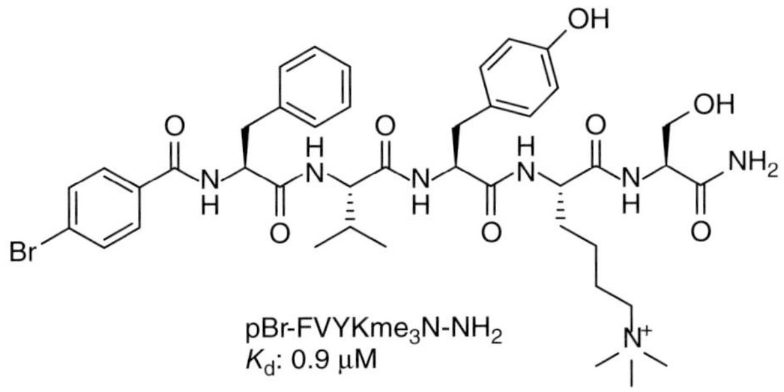

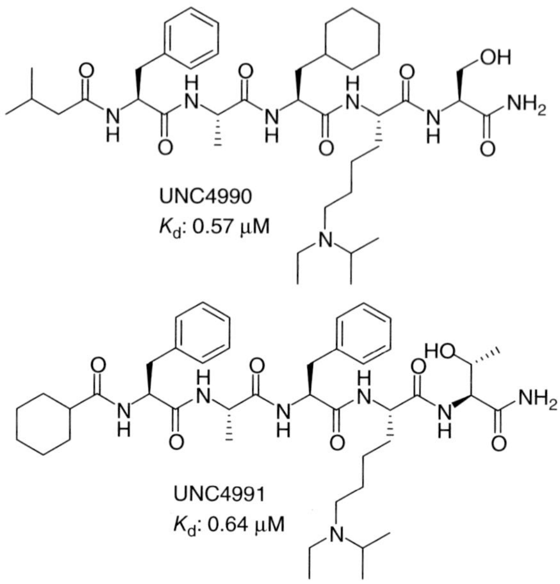

Figure 14.14 Structures of the CDYL2 ligands.

14.3 The PHD Finger Family of Epigenetic Reader Proteins

The second family of epigenetic reader proteins presented herein is the PHD finger domain-containing proteins. This family is separated from the Royal family despite sharing some similarities. The PHD finger proteins are able to differ between diverse states of lysine methylation, ranging from unmodified lysine residues to trimethylated ones mostly located at H3K4. The family is diverse with about 100 different proteins containing a PHD finger domain [114], many of which are implicated in different neurological and immune diseases [115] as well as in cancer like hepatocellular carcinoma [116], glioblastoma [117], non-small cell lung cancer [118], prostate cancer [119], and bone cancer [120]. The PHD finger is a small protein domain built up of 50–80 amino acid residues, which fold into two antiparallel β-sheets and a C-terminal α-helix. The protein fold is stabilized by two zinc ions, which are anchored by the conserved Cys3–His–Cys3 motif. Crystal structures of various PHD domains with H3K4 peptides reveal that the peptide substrate binds in an extended conformation, usually forming an additional antiparallel β-strand which pairs with the double-stranded β-sheet of the PHD domain by undergoing extensive backbone intermolecular hydrogen bonds. In the case of H3K4me3 readers, the trimethyllysine residue is embedded in an aromatic cage formed by 2–4 aromatic acid residues, all containing a conserved tryptophan residue. Conversely, PHD domains responsible for the recognition of the unmodified H3K4 peptide lack the aromatic cage. The unmodified lysine residue is embedded in a subpocket containing polar or acidic amino acid residues and is, hence, stabilized by H-bond and salt–bridge interactions [121, 122].

As mentioned in Section 14.1, most of the epigenetic proteins contain more than one domain within their sequence and can therefore exhibit several different functions. JARID1A (KDM5A or RBBP2) is another protein exhibiting eraser functions as a lysine demethylase as well as containing three PHD finger reader domains. The third PHD finger domain is involved in acute myeloid leukemia as part of a fusion protein with nucleoporin protein 98 (NUP98) [123]. Further studies with a fusion protein of the same PHD finger and PHF23 have shown that the specific interaction of the PHD finger and H3K4me2/3 is essential for the induction of AML [124]. In 2012, Wagner et al. reported the first inhibitors targeting the third PHD finger domain of JARID1A. By using the HaloTag technology, they screened 446 compounds of the NIH clinical collection 1 for displacement of an H3K4me3-derived peptide. All the compounds found within this library have undergone clinical trials up to phase III in various diseases. Out of this screen, three initial hits, namely, disulfiram, amiodarone, and tegaserod (Figure 14.15), which are all approved drugs, were confirmed by FP and pulldown assays. Further dose–response testing using fragments of amiodarone revealed that an aminoalkyl chain is essential for the affinity toward JARID1A. By taking together the findings of the SAR studies and the fragment-based approach, new amiodarone derivatives were synthesized, among which di-*N*-desethylamiodarone (Figure 14.15; IC_{50}: 26 μM) and trimethylamiodarone (WAG-003 IC_{50}: 30 μM; Figure 14.15) were found to exhibit enhanced potencies that were roughly 10-fold higher than the original hit amiodarone [125].

As already mentioned in Section 14.2.3 in connection with the inhibition of Tudor domains, the Wnt/β-catenin signaling pathway plays crucial roles in the genesis and development of several diseases. Not only misdirected upstream signaling but also mutations in β-catenin protein itself are sufficient for dysregulation [126]. β-Catenin interacts with the transcription factor TCF via its armadillo

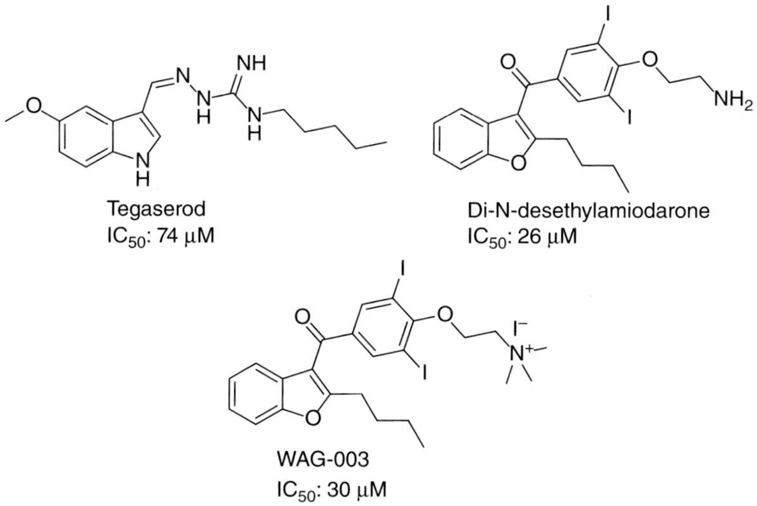

Figure 14.15 Structures of ligands for the PHD domain of JARID1a.

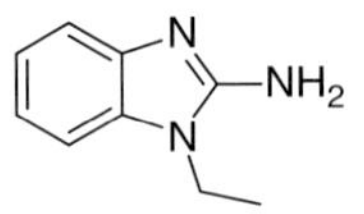

CF16

Figure 14.16 Structure of the fragment Pygo inhibitor CF16.

repeat domain (ARD). As β-catenin is constitutively active, negative regulators like APC and axin mediate its signal transmission rate. Both transcription factors and regulators share ARD as site of interaction [127], which makes ARD a difficult drug target. It comes as little surprise that so far all attempts to target the ARD TCF interaction failed [126]. The N-terminal domain of β-catenin hosts another interaction surface enabling the binding to BCL9 proteins. BCL9 itself consists of two short α-helical domains (HD1/2), of which HD2 interacts with ARD [128]. If bound to β-catenin, the HD1 site is able to bind to Pygo PHD fingers, which leads to a slight conformational change in the PHD finger enabling the binding to H3K4me2 [129]–[131]. As these three interactions are much more confined in space than the ARD-TCF, they show good potential as targets for small-molecule disruptors [132]. In 2014, Miller et al. reported the finding of fragment-based disruptors of the abovementioned interactions. Because of the low stability of the Pygo PHD domain in aqueous solution, the HD1-Pygo PHD complex was used for the screening. An NMR-based fragment screening approach led to the identification of 6-fluoro- and 6-chloro-2-aminobenzothiazoles as potential lead scaffold. Interestingly, these derivatives, however, were found to bind in a narrow cleft at the PHD–HD1 interface, where function is still unclear. Meanwhile, studies on the 2-aminobenzimidazole analogue, CF16 (Figure 14.16), implied that it competitively inhibits histone binding and that it resides in the K4me2 pocket. CF16, hence, represents the first fragment ligand reported to interrupt this specific binding [133].

Another example for the complex structures of epigenetic reader proteins is the multidomain protein UHRF1. It encompasses a PHD domain capable of reading unmodified histone H3 tails, whereupon it recruits histone methyltransferases that convert H3K9 to H3K9me3. This mark is recognized by the tandem Tudor domain of UHRF1 [134]. Moreover, UHRF1 can bind to methylated cytosines on the DNA via an SRA domain, and it has catalytic functionality acting as E3 ubiquitin ligase via a RING finger domain. Overexpression of UHRF1 is associated with DNA repair inhibition, tumor suppressor gene silencing, tumor growth, and metastasis [135]. This example shows that chemical probes selectively inhibiting one of these domains are necessary to fully understand the biological effect of each single domain. Knockout of *Uhrf1* results in the deletion of more than one domain, and, therefore, the assessment of the detailed impact of every domain alone gets exceedingly difficult.

14.4 The WD40 Repeat Domain Family

The last family of epigenetic reader proteins discussed herein is the WD40 repeat domain family. Alongside with the Royal family and the PHD finger family, it represents the third large group of methyl-lysine reader proteins. WD40 repeat domains are formed of a 7/8-bladed β-propeller fold, where each blade is built of a four-stranded antiparallel β-sheet. With over 200 proteins containing a WD40 repeat [136], this family is involved in many cellular pathways. Proteins belonging to this family play various roles participating not only in histone modification readout but also in many other protein–protein interactions, e.g. the formation of protein complexes [137]. Of this large variety of interactions, 18 proteins are involved in chromatin complexes [16], and for 2 of them, inhibitors are known to date.

WD40 repeat protein 5 (WDR5) constitutes one component of a macromolecular complex, which additionally comprises members of the SET1 methyltransferase family (SET1a, SET1b, and MLL). These methyltransferases are capable of writing the H3K4me3 mark, which is usually coupled with gene activation. Binding of WDR5 to H3K4me2 is essential for the activation of the methyltransferase activity of the complex by enabling it to reach its target and set the trimethyl mark [138, 139]. The crystal structure of WDR5 with the histone peptide, however, revealed WDR5 to be an arginine reader rather than a methyl-lysine reader. WDR5 lacks an aromatic cage, the characteristic feature of Kme readers, and the dimethyllysine residue resides on the surface of the protein only interacting with Glu322 [140, 141]. In fact, H3 recognition is mainly based on the binding to H3R2. This is supported by the fact that WDR5 does not discriminate between different states of lysine methylation [141, 142]. As mentioned before, the binding of WDR5 to other proteins is crucial for their activity. The activity of the SET1 family histone methyltransferase mixed lineage leukemia protein 1 (MLL1; KMT2) is dependent of the interaction with WDR5 via the Win motif. This structural feature represents the smallest interaction surface needed to bind WDR5. As the name of MLL1 already suggests, mutations and fusion proteins of MLL1 are associated with the development and progression of acute leukemia in about 80% of infants and up to 10% of adults [143]. As the WDR5–MLL interaction is crucial for the activity of the SET1 complex, disruption of this binding has the potential to provide new therapeutic strategies for leukemia [144, 145].

The first efforts in finding inhibitors of the WDR5–MLL1 binding were reported by Senisterra et al. in 2013. By screening 16 000 compounds using FP assay, they were able to identify WDR5-0101 (Figure 14.17; K_d: 5.5 μM) as the first small-molecule inhibitor of the WDR5–Win binding. In order to increase the potency of this lead structure, 6 000 000 commercially available compounds were screened for their structural similarity. Two hits were confirmed (WDR5-0102; K_d: 4.0 μM and WDR5-0103; K_d: 450 nM; Figure 14.17) of which WDR5-0103 shows the highest affinity toward WDR5. Furthermore, they could confirm that WDR5-0103 also inhibited the methyltransferase activity of MLL by targeting the WDR5–MLL1 interaction (IC_{50}: 39 μM), which proves this binding to be a potential drug target [146]. Later on, Grebien et al. optimized the lead structure based on SAR studies. They obtained OICR-9429 (Figure 14.17; K_d: 93 nM) that is highly selective and shows no binding to a set of over 250 proteins and enzymes [147]. The crystal structure of WDR5 with OICR-9429 clearly reveals that the methylpiperidine moiety occupies the H3R2 binding pocket, where it

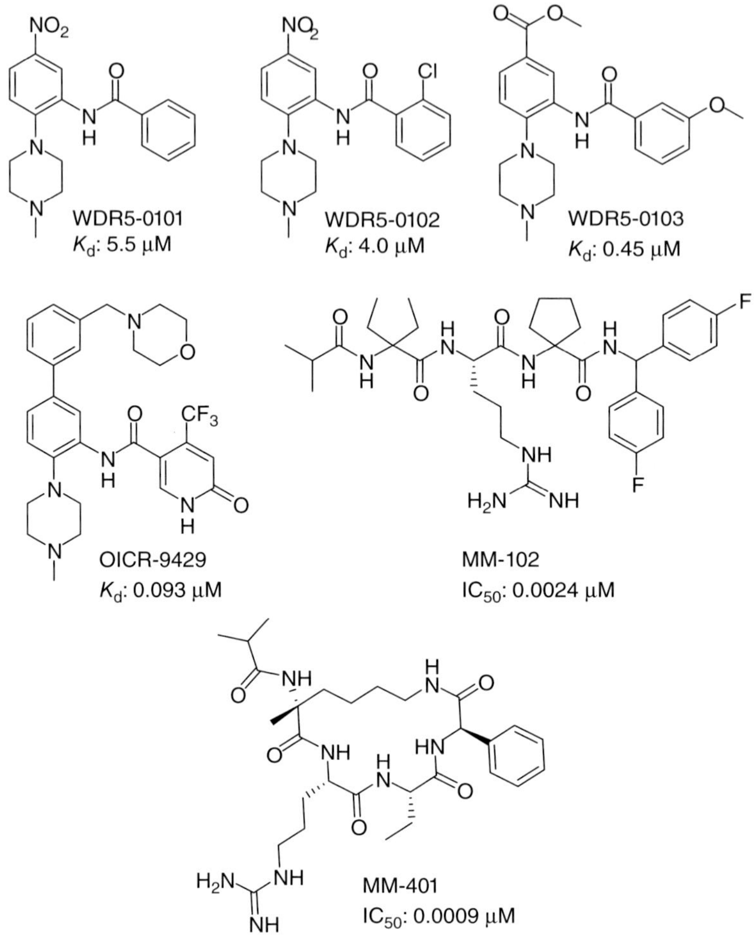

Figure 14.17 Structures of WDR5 inhibitors.

undergoes water-mediated H-bond interactions as well as CH—O interactions. One of the phenyl rings is stabilized by π–π stacking with Phe333, whereas the second one is embedded in a hydrophobic region lined by Phe149, Pro173, and Tyr191. Additionally, the amide linker and the pyridone group are further stabilized by extensive polar interactions (Figure 14.18). Notably, several other analogues of OICR-9429 were successfully crystallized with WDR5, all of them showing a highly similar binding mode.

Another approach in targeting this interaction was carried out by Karatas et al. in 2013. In their work, they used the smallest arginine binding motif (CO-ARA-NH) as a starting point to target the arginine binding site with peptidomimetic inhibitors. They identified three molecules (MM-101, MM-102, and MM-103) that bind to WDR5 with high affinities (K_i: <1 nM) detected by BioLayer Inferometry (BLI). MM-102 (Figure 14.17) was also shown to inhibit cell growth in cell lines exhibiting MLL fusion proteins (IC_{50}: 25 μM) [148]. Based

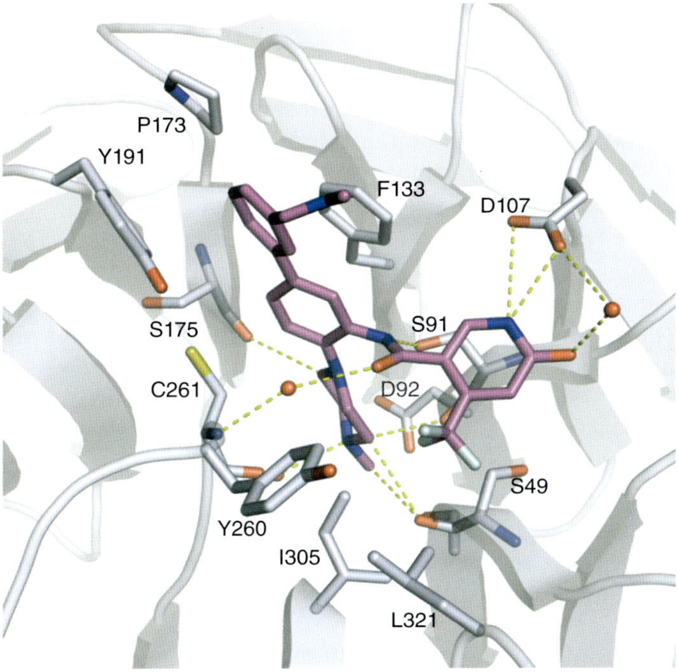

Figure 14.18 Crystal structure of WDR5 in complex with OICR-9429 (PDB ID: 4QL1, the morpholine ring is not resolved in the X-ray structure).

on the discovery of the MM compounds, Cao et al. carried out the cyclization of the peptide. The obtained compound (MM-401; Figure 14.17) maintained high affinity toward WDR5 (K_i: <1 nM). In an FP-based displacement assay, they confirmed a threefold higher potency (IC_{50}: 0.9 nM) compared with MM-101 (IC_{50}: 2.9 nM) and 700-fold selectivity over the Win motif in MLL1 (IC_{50}: 750 nM). Proper stereochemistry seems to be crucial for this compound as the enantiomer of MM-401 is not active [149]. In cell assays, MM-401 showed induction of apoptosis in leukemic cells lines, whereas normal cell lines were unaffected by the compound. This is an important point, making MM-401 an interesting candidate for further drug development.

The second WD40 repeat protein evaluated as a drug target is the embryonic ectoderm development (EED) protein, which is part of the abovementioned PRC2. PRC2 complexes catalyze the mono-, di-, and trimethylation of H3K27 through their SET-domain methyltransferase subunit EZH2. An enzymatically active PRC2 complex requires the association of three subunits, namely, the catalytic methyltransferase subunit EZH2, the WD40 repeat subunit EED, and the zinc finger domain Suz12 [150–152]. Importantly, EZH2 alone is catalytically inactive, and the isolated EZH2 SET domain was found to adopt an auto-inhibited conformation where the cofactor and substrate binding pockets are incomplete [153, 154]. Basal PRC2 activity is, however, achieved by binding to EED and Suz12 [150–152]. The WD40 repeat domain EED was found to be essential not only for the activation of the PRC2 complex but also for its recruitment to preexisting H3K27me3 marks, which results in the propagation of this repressive mark. EED allosterically activates EZH2 by bringing the N-terminal activation loop of EZH2 closer to the catalytic SET-domain pocket [155]. Further catalytic stimulation of EZH2 is achieved upon binding of the H3K27me3 mark to EED, which causes the stimulatory recognition motif (SRM) of EZH2 to adopt an α-helical structure, thus stabilizing the SET-I domain [155]. As characteristic for methyl-lysine readers, the recognition of the trimethylated lysine mark of H3K27me3 is mainly mediated by an aromatic cage in the WD40 domain of EED, which is composed of Phe97, Tyr148, and Tyr365 [152]. The development of small molecules that disrupt the recognition of H3K27me3 by EED is, hence, a promising strategy to modulate the PRC2 activity and can be used as an alternative to direct EZH2 inhibitors. Indeed, recent years have seen the development of several small-molecule inhibitors that allosterically modulate the activity of EZH2 by binding to EED.

First attempts to target EED were reported in 2014. In a virtual screening approach, a library of 1000 compounds was screened *in silico*. The top 50 compounds were then assessed in an FP-based displacement assay, directly targeting the EED–EZH2 binding. Kong et al. identified the FDA-approved anti-allergic drug astemizole (Figure 14.19) as disruptor of the EED–EZH2 complex (IC_{50}: 94 μM) [156].

Lingel et al. were able to discover several small-molecule inhibitors of PRC2 that compete with H3K27me3 binding to EED. In this study, they implemented an HTS approach using recombinant five-membered PRC2 complex as enzyme, H3K27me0 as a peptide substrate, and the homogeneous time-resolved fluorescence (HTRF) method to detect the H3K27me2 product. The primary HTS hit showed one digit micromolar activity against PRC2 and was confirmed to bind to the H3K27me3 binding site of EED. Lead-optimization studies, which involved ligand deconstruction to smaller fragments followed by efficient fragment–regrowth approaches to improve the potency, led to the discovery of a set of potent inhibitors of the PRC2 including Lingel-17 (Figure 14.19; IC_{50}: 0.43 μM) and Lingel-19 (Figure 14.19; IC_{50}: 1.3 μM). The affinities of these compounds were further validated by an AlphaScreen binding assay as well as ITC and surface plasmon resonance (SPR) assays. Several of the described inhibitors were successfully co-crystallized with EED and an EZH2 fragment. Crystal structures reveal that the compounds bind to the Kme3 pocket, which undergoes conformational changes not previously observed in the apo- and peptide-bound structures. The most pronounced changes can be observed for Arg367 and Tyr365 of the aromatic cage; the latter exhibits a strong outward movement, thereby allowing the compounds to deeply reside in the Kme3 pocket. The aromatic moieties of the ligands (methylenedioxyphenyl of Lingel-17 and fluoromethyoxyphenyl of Lingel-19) are placed deep into the Kme3 pocket and undergo π-stacking interactions with Arg368. Meanwhile, the piperidine ring is located within the aromatic cage and is stabilized by cation–π interactions, vdW contacts, and water-mediated H-bonds. Additionally, the amine at the imidazole moiety forms water-mediated H-bond interactions with the side chains of Asn194 and Glu238 [157, 158].

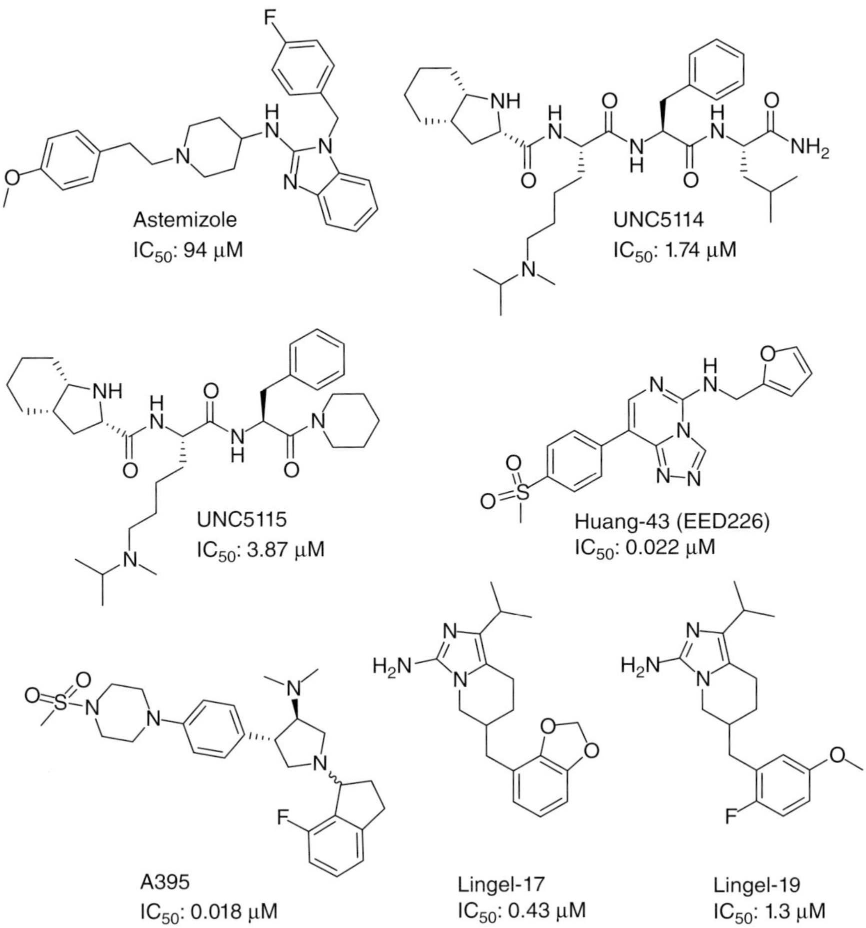

Figure 14.19 Structures of EED ligands.

Figure 14.20a shows the co-crystal structure of EED with Lingel-17 as an example.

Using the same HTS approach described in Section 14.4, Huang et al. were able to discover another promising lead scaffold, which could intercept EED–histone binding. Lead optimization by adopting a fragmentation and regrowth approach finally resulted in the discovery of the highly potent and orally bioavailable compound, Huang-43 (EED226, IC_{50}: 0.022 μM; Figure 14.19), which could efficiently inhibit PRC2 activity. Crystal structures of Huang-43 with EED and an EZH2 peptide show a similar binding mode as observed for Lingel-17 and Lingel-19. The furan moiety is embedded deep in the Kme3 pocket and is stabilized by cation–π interactions with Arg367; the triazolopyrimidine core is stabilized by π–π stacking with Tyr148 and Tyr365 in the aromatic cage as well as several H-bond interactions, while the methylsulfonylphenyl moiety is majorly solvent exposed and forms π–π interactions with Phe97 (Figure 14.20b). Notably, Huang-43 showed

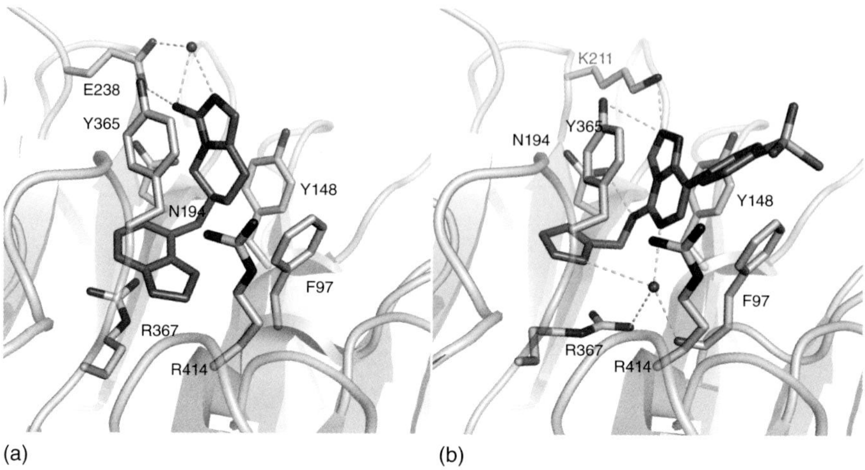

Figure 14.20 Crystal structure of the EED domain in complex with (a) Lingel-17 (PDB ID: 5U62) and (b) Huang-43 (PDB-ID: 5GSA).

high efficacy in a xenograft mouse model while showing a promising pharmacokinetic profile [159]. Furthermore, it shows tumor regression effects *in vivo*, even in cells resistant to S-Adenosyl methionine (SAM)-competitive EZH2 inhibitors [160].

Another HTS approach was carried out using fluorescence thermal shift assay by He et al. in 2017. By assessment of inhibition values using time-resolved fluorescence resonance energy transfer (TR-FRET), the analogues of the primary hits were further optimized. The optimized compound A-395 (Figure 14.19) showed K_d of 1.5 nM toward EED while inhibiting the activity of EZH2 with IC_{50} of 18 nM. In their study, they showed that inhibition of the EED–histone binding motif leads to the same phenotype as inhibitors of the catalytic domain of EZH2. To this date, no resistance in cells is known for this kind of inhibition [161]. The crystal structure reveals that A-395 adopts a similar binding mode as for the previously discussed inhibitors, with the indane ring stacking Arg367 and the aminopyrrolidine moiety being sandwiched between the aromatic acid residues of the cage.

Barnash et al. adopted a different approach, where they reported the development of peptidomimetic ligands that can disrupt the H3K27me3 recognition using a combinatorial chemistry approach. Starting from the high-affinity binding peptide $Jarid2_{114-118}K116me3$, they were able to reduce the active structure to four amino acids. Two compounds, UNC5114 and UNC5115 (Figure 14.19), exhibited IC_{50} values in the low micromolar range in a FP-based displacement assay. Follow-up via ITC measurements confirmed them as high-affinity ligands (UNC5114 K_d: 0.68 μM; UNC5115 K_d: 1.14 μM). Both compounds were shown to inhibit EZH2 methyltransferase activity by allosterically disrupting the EED–EZH2 interaction [162].

14.5 Conclusion and Outlook

Despite significant structural and biological knowledge, the chemical epigenetics of methyl-lysine reader proteins is still very much underdeveloped. Only for very few proteins, potent ligands have been identified, and even for those, selectivity remains a challenge. But initial potent leads and virtual and high-throughput screens should lead to optimized potent and selective ligands to probe the role of these reader domains in health and disease. Especially interesting will be to dissect the different contribution of reader and enzymatic domains in the same protein. Clearly, the disease association of methyl-lysine readers makes them promising candidates for further investigations.

Acknowledgment

J.B. and M.J. thank the Deutsche Forschungsgemeinschaft (DFG, Project A04 within CRC992) for funding.

References

1 Luger, K. and Richmond, T.J. (1998). The histone tails of the nucleosome. *Curr. Opin. Genet. Dev.* 8 (2): 140–146.
2 Berger, S.L., Kouzarides, T., Shiekhattar, R., and Shilatifard, A. (2009). An operational definition of epigenetics. *Genes Dev.* 23 (7): 781–783.
3 Bannister, A.J. and Kouzarides, T. (2011). Regulation of chromatin by histone modifications. *Cell Res.* 21 (3): 381–395.
4 Daujat, S., Weiss, T., Mohn, F. et al. (2009). H3K64 trimethylation marks heterochromatin and is dynamically remodeled during developmental reprogramming. *Nat. Struct. Mol. Biol.* 16 (7): 777–781.
5 Lu, Q., Quinn, A.M., Patel, M.P. et al. (2012). Perspectives on the discovery of small-molecule modulators for epigenetic processes. *J. Biomol. Screening* 17 (5): 555–571.
6 Aagaard, L., Laible, G., Selenko, P. et al. (1999). Functional mammalian homologues of the drosophila PEV-modifier Su(var)3–9 encode centromere-associated proteins which complex with the heterochromatin component M31. *EMBO J.* 18 (7): 1923–1938.
7 Jenuwein, T. and Allis, C.D. (2001). Translating the histone code. *Science* 293 (5532): 1074–1080.
8 Bissinger, E.M., Heinke, R., Sippl, W., and Jung, M. (2010). Targeting epigenetic modifiers: inhibitors of histone methyltransferases. *MedChemComm* 1 (2): 114–124.
9 Taverna, S.D., Li, H., Ruthenburg, A.J. et al. (2007). How chromatin-binding modules interpret histone modifications: lessons from professional pocket pickers. *Nat. Struct. Mol. Biol.* 14 (11): 1025–1040.

10 Zhao, Y., Yang, C.Y., and Wang, S. (2013). The making of I-BET762, a BET bromodomain inhibitor now in clinical development. *J. Med. Chem.* 56 (19): 7498–7500.
11 Matzuk, M.M., McKeown, M.R., Filippakopoulos, P. et al. (2012). Small-molecule inhibition of BRDT for male contraception. *Cell* 150 (4): 673–684.
12 Wyspianska, B.S., Bannister, A.J., Barbieri, I. et al. (2014). BET protein inhibition shows efficacy against JAK2V617F-driven neoplasms. *Leukemia* 28 (1): 88–97.
13 Kamps, J.J., Huang, J., Poater, J. et al. (2015). Chemical basis for the recognition of trimethyllysine by epigenetic reader proteins. *Nat. Commun.* 6: 8911.
14 Herold, J.M., Ingerman, L.A., Gao, C., and Frye, S.V. (2011). Drug discovery toward antagonists of methyl-lysine binding proteins. *Curr. Chem. Genomics* 5: 51–61.
15 SGC (2017). Annotated phylogenetic tree of MBT domain family. http://apps.thesgc.org/resources/phylogenetic_trees/?domain=MBT#options.
16 Liu, L., Zhen, X.T., Denton, E. et al. (2012). ChromoHub: a data hub for navigators of chromatin-mediated signalling. *Bioinformatics* 28 (16): 2205–2206.
17 Nady, N., Krichevsky, L., Zhong, N. et al. (2012). Histone recognition by human malignant brain tumor domains. *J. Mol. Biol.* 423 (5): 702–718.
18 Bonasio, R., Lecona, E., and Reinberg, D. (2010). MBT domain proteins in development and disease. *Semin. Cell Dev. Biol.* 21 (2): 221–230.
19 Lemmens, B.B., Johnson, N.M., and Tijsterman, M. (2013). COM-1 promotes homologous recombination during *Caenorhabditis elegans* meiosis by antagonizing Ku-mediated non-homologous end joining. *PLos Genet.* 9 (2): e1003276.
20 Trojer, P., Li, G., Sims, R.J. III, et al. (2007). L3MBTL1, a histone-methylation-dependent chromatin lock. *Cell* 129 (5): 915–928.
21 Wang, W.K., Tereshko, V., Boccuni, P. et al. (2003). Malignant brain tumor repeats: a three-leaved propeller architecture with ligand/peptide binding pockets. *Structure* 11 (7): 775–789.
22 Li, H., Fischle, W., Wang, W. et al. (2007). Structural basis for lower lysine methylation state-specific readout by MBT repeats of L3MBTL1 and an engineered PHD finger. *Mol. Cell* 28 (4): 677–691.
23 Kalakonda, N., Fischle, W., Boccuni, P. et al. (2008). Histone H4 lysine 20 monomethylation promotes transcriptional repression by L3MBTL1. *Oncogene* 27 (31): 4293–4304.
24 Boccuni, P., MacGrogan, D., Scandura, J.M., and Nimer, S.D. (2003). The human L(3)MBT polycomb group protein is a transcriptional repressor and interacts physically and functionally with TEL (ETV6). *J. Biol. Chem.* 278 (17): 15412–15420.
25 West, L.E., Roy, S., Lachmi-Weiner, K. et al. (2010). The MBT repeats of L3MBTL1 link SET8-mediated p53 methylation at lysine 382 to target gene repression. *J. Biol. Chem.* 285 (48): 37725–37732.
26 Min, J., Allali-Hassani, A., Nady, N. et al. (2007). L3MBTL1 recognition of mono- and dimethylated histones. *Nat. Struct. Mol. Biol.* 14 (12): 1229–1230.

27 Sakaguchi, A., Joyce, E., Aoki, T. et al. (2012). The histone H4 lysine 20 monomethyl mark, set by PR-Set7 and stabilized by L(3)mbt, is necessary for proper interphase chromatin organization. *PLoS One* 7 (9): e45321.

28 Perna, F., Gurvich, N., Hoya-Arias, R. et al. (2010). Depletion of L3MBTL1 promotes the erythroid differentiation of human hematopoietic progenitor cells: possible role in 20q-polycythemia vera. *Blood* 116 (15): 2812–2821.

29 Kireev, D., Wigle, T.J., Norris-Drouin, J. et al. (2010). Identification of non-peptide malignant brain tumor (MBT) repeat antagonists by virtual screening of commercially available compounds. *J. Med. Chem.* 53 (21): 7625–7631.

30 Wigle, T.J., Herold, J.M., Senisterra, G.A. et al. (2010). Screening for inhibitors of low-affinity epigenetic peptide–protein interactions: an AlphaScreen-based assay for antagonists of methyl-lysine binding proteins. *J. Biomol. Screening* 15 (1): 62–71.

31 Gao, C., Herold, J.M., Kireev, D. et al. (2011). Biophysical probes reveal a "compromise" nature of the methyl-lysine binding pocket in L3MBTL1. *J. Am. Chem. Soc.* 133 (14): 5357–5362.

32 Herold, J.M., Wigle, T.J., Norris, J.L. et al. (2011). Small-molecule ligands of methyl-lysine binding proteins. *J. Med. Chem.* 54 (7): 2504–2511.

33 Northcott, P.A., Nakahara, Y., Wu, X. et al. (2009). Multiple recurrent genetic events converge on control of histone lysine methylation in medulloblastoma. *Nat. Genet.* 41 (4): 465–472.

34 James, L.I., Barsyte-Lovejoy, D., Zhong, N. et al. (2013). Discovery of a chemical probe for the L3MBTL3 methyllysine reader domain. *Nat. Chem. Biol.* 9 (3): 184–191.

35 Guo, Y., Nady, N., Qi, C. et al. (2009). Methylation-state-specific recognition of histones by the MBT repeat protein L3MBTL2. *Nucleic Acids Res.* 37 (7): 2204–2210.

36 James, L.I., Frye, S.V. (2016). Chemical probes for methyl lysine reader domains. *Curr. Opin. Chem. Biol.* 33: 135–41.

37 Camerino, M.A., Zhong, N., Dong, A. et al. (2013). The structure–activity relationships of L3MBTL3 inhibitors: flexibility of the dimer interface. *Medchemcomm* 4 (11): 1501–1507.

38 James, L.I., Korboukh, V.K., Krichevsky, L. et al. (2013). Small-molecule ligands of methyl-lysine binding proteins: optimization of selectivity for L3MBTL3. *J. Med. Chem.* 56 (18): 7358–7371.

39 SGC (2017). Annotated phylogenetic tree of PWWP domain family. http://apps.thesgc.org/resources/phylogenetic_trees/index.php?domain=PWWP#options.

40 Qiu, C., Sawada, K., Zhang, X., and Cheng, X. (2002). The PWWP domain of mammalian DNA methyltransferase Dnmt3b defines a new family of DNA-binding folds. *Nat. Struct. Biol.* 9 (3): 217–224.

41 Wu, H., Zeng, H., Lam, R. et al. (2011). Structural and histone binding ability characterizations of human PWWP domains. *PLoS One* 6 (6): e18919.

42 Qiu, Y., Zhang, W., Zhao, C. et al. (2012). Solution structure of the Pdp1 PWWP domain reveals its unique binding sites for methylated H4K20 and DNA. *Biochem. J.* 442 (3): 527–538.

43 Dobrovetsky, E., Dong, A., Mader, P. et al., Structural Genomics Consortium (2015). Human bromodomain and PHD finger containing 1, PWWP domain in complex with XST005904a. PDB ID: 5C6S.

44 Yang, J. and Everett, A.D. (2009). Hepatoma-derived growth factor represses SET and MYND domain containing 1 gene expression through interaction with C-terminal binding protein. *J. Mol. Biol.* 386 (4): 938–950.

45 Han, Y., Zhang, W., and Liu, Y. (2013). Identification of hepatoma-derived growth factor as a potential prognostic and diagnostic marker for extrahepatic cholangiocarcinoma. *World J. Surg.* 37 (10): 2419–2427.

46 Yoshida, K., Tomita, Y., Okuda, Y. et al. (2006). Hepatoma-derived growth factor is a novel prognostic factor for hepatocellular carcinoma. *Ann. Surg. Oncol.* 13 (2): 159–167.

47 Ren, H., Tang, X., Lee, J.J. et al. (2004). Expression of hepatoma-derived growth factor is a strong prognostic predictor for patients with early-stage non-small-cell lung cancer. *J. Clin. Oncol.* 22 (16): 3230–3237.

48 Yamamoto, S., Tomita, Y., Hoshida, Y. et al. (2006). Expression of hepatoma-derived growth factor is correlated with lymph node metastasis and prognosis of gastric carcinoma. *Clin. Cancer Res.* 12 (1): 117–122.

49 Hu, T.H., Lin, J.W., Chen, H.H. et al. (2009). The expression and prognostic role of hepatoma-derived growth factor in colorectal stromal tumors. *Dis. Colon Rectum* 52 (2): 319–326.

50 Uyama, H., Tomita, Y., Nakamura, H. et al. (2006). Hepatoma-derived growth factor is a novel prognostic factor for patients with pancreatic cancer. *Clin. Cancer Res.* 12 (20 Pt 1): 6043–6048.

51 Wang, Y., Reddy, B., Thompson, J. et al. (2009). Regulation of Set9-mediated H4K20 methylation by a PWWP domain protein. *Mol. Cell* 33 (4): 428–437.

52 He, C., Li, F., Zhang, J. et al. (2013). The methyltransferase NSD3 has chromatin-binding motifs, PHD5-C5HCH, that are distinct from other NSD (nuclear receptor SET domain) family members in their histone H3 recognition. *J. Biol. Chem.* 288 (7): 4692–4703.

53 Pei, H., Wu, X., Liu, T. et al. (2013). The histone methyltransferase MMSET regulates class switch recombination. *J. Immunol.* 190 (2): 756–763.

54 Min, D.J., Ezponda, T., Kim, M.K. et al. (2013). MMSET stimulates myeloma cell growth through microRNA-mediated modulation of c-MYC. *Leukemia* 27 (3): 686–694.

55 Pradeepa, M.M., Sutherland, H.G., Ule, J. et al. (2012). Psip1/Ledgf p52 binds methylated histone H3K36 and splicing factors and contributes to the regulation of alternative splicing. *PLos Genet.* 8 (5): e1002717.

56 Shun, M.C., Botbol, Y., Li, X. et al. (2008). Identification and characterization of PWWP domain residues critical for LEDGF/p75 chromatin binding and human immunodeficiency virus type 1 infectivity. *J. Virol.* 82 (23): 11555–11567.

57 Leoh, L.S., van Heertum, B., De Rijck, J. et al. (2012). The stress oncoprotein LEDGF/p75 interacts with the methyl CpG binding protein MeCP2 and influences its transcriptional activity. *Mol. Cancer Res.* 10 (3): 378–391.

58 van Nuland, R., van Schaik, F.M., Simonis, M. et al. (2013). Nucleosomal DNA binding drives the recognition of H3K36-methylated nucleosomes by the PSIP1-PWWP domain. *Epigenetics Chromatin* 6 (1): 12.

59 SGC (2017). Annotated phylogenetic tree of Tudor domain family. http://apps.thesgc.org/resources/phylogenetic_trees/index.php?domain=TUDOR#options.

60 SGC (2017). Annotated phylogenetic tree of Spindlin domain family. http://apps.thesgc.org/resources/phylogenetic_trees/index.php?domain=SPINDLIN#options.

61 Liu, K., Guo, Y., Liu, H. et al. (2012). Crystal structure of TDRD3 and methyl-arginine binding characterization of TDRD3, SMN and SPF30. *PLoS One* 7 (2): e30375.

62 Cote, J. and Richard, S. (2005). Tudor domains bind symmetrical dimethylated arginines. *J. Biol. Chem.* 280 (31): 28476–28483.

63 Saxe, J.P., Chen, M., Zhao, H., and Lin, H. (2013). Tdrkh is essential for spermatogenesis and participates in primary piRNA biogenesis in the germline. *EMBO J.* 32 (13): 1869–1885.

64 Santiago, C., Nguyen, K., and Schapira, M. (2011). Druggability of methyl-lysine binding sites. *J. Comput.-Aided Mol. Des.* 25 (12): 1171–1178.

65 Yang, N., Wang, W., Wang, Y. et al. (2012). Distinct mode of methylated lysine-4 of histone H3 recognition by tandem tudor-like domains of spindlin1. *Proc. Natl. Acad. Sci. U.S.A.* 109 (44): 17954–17959.

66 Su, Y.Q., Sun, F., Handel, M.A. et al. (2012). Meiosis arrest female 1 (MARF1) has nuage-like function in mammalian oocytes. *Proc. Natl. Acad. Sci. U.S.A.* 109 (46): 18653–18660.

67 Oh, B., Hwang, S.Y., Solter, D., and Knowles, B.B. (1997). Spindlin, a major maternal transcript expressed in the mouse during the transition from oocyte to embryo. *Development* 124 (2): 493–503.

68 Yue, W., Sun, L.Y., Li, C.H. et al. (2004). Screening and identification of ovarian carcinomas related genes. *Ai Zheng* 23 (2): 141–145.

69 Franz, H., Greschik, H., Willmann, D. et al. (2015). The histone code reader SPIN1 controls RET signaling in liposarcoma. *Oncotarget* 6 (7): 4773–4789.

70 Gao, Y., Yue, W., Zhang, P. et al. (2005). Spindlin1, a novel nuclear protein with a role in the transformation of NIH3T3 cells. *Biochem. Biophys. Res. Commun.* 335 (2): 343–350.

71 Zhao, Q., Qin, L., Jiang, F. et al. (2007). Structure of human spindlin1. Tandem tudor-like domains for cell cycle regulation. *J. Biol. Chem.* 282 (1): 647–656.

72 Zhang, P., Cong, B., Yuan, H. et al. (2008). Overexpression of spindlin1 induces metaphase arrest and chromosomal instability. *J. Cell. Physiol.* 217 (2): 400–408.

73 Bachmayr-Heyda, A., Reiner, A.T., Auer, K. et al. (2015). Correlation of circular RNA abundance with proliferation – exemplified with colorectal and ovarian cancer, idiopathic lung fibrosis, and normal human tissues. *Sci. Rep.* 5: 8057.

74 Wagner, T., Greschik, H., Burgahn, T. et al. (2016). Identification of a small-molecule ligand of the epigenetic reader protein spindlin1 via a versatile screening platform. *Nucleic Acids Res.* 44 (9): e88.
75 Sweis, R.F., Pliushchev, M., Brown, P.J. et al. (2014). Discovery and development of potent and selective inhibitors of histone methyltransferase g9a. *ACS Med. Chem. Lett.* 5 (2): 205–209.
76 Robaa, D., Wagner, T., Luise, C. et al. (2016). Identification and structure–activity relationship studies of small-molecule inhibitors of the methyllysine reader protein spindlin1. *ChemMedChem* 11 (20): 2327–2338.
77 Schultz, D.C., Ayyanathan, K., Negorev, D. et al. (2002). SETDB1: a novel KAP-1-associated histone H3, lysine 9-specific methyltransferase that contributes to HP1-mediated silencing of euchromatic genes by KRAB zinc-finger proteins. *Genes Dev.* 16 (8): 919–932.
78 Yang, L., Xia, L., Wu, D.Y. et al. (2002). Molecular cloning of ESET, a novel histone H3-specific methyltransferase that interacts with ERG transcription factor. *Oncogene* 21 (1): 148–152.
79 Binda, O., LeRoy, G., Bua, D.J. et al. (2010). Trimethylation of histone H3 lysine 4 impairs methylation of histone H3 lysine 9: regulation of lysine methyltransferases by physical interaction with their substrates. *Epigenetics* 5 (8): 767–775.
80 Dong, A., Iqbal, A., Mader, P. et al., Structural Genomics Consortium (2016). Crystal structure of SETDB1 Tudor domain in complex with fragment MRT0181a. www.rcsb.org/pdb/explore/explore.do?structureId=5KE3.
81 Dong, A., Iqbal, A., Mader, P. et al., Structural Genomics Consortium (2017). Crystal structure of SETDB1 Tudor domain in complex with inhibitor xst06472a. www.rcsb.org/pdb/explore/explore.do?structureId=5KE2.
82 Botuyan, M.V., Lee, J., Ward, I.M. et al. (2006). Structural basis for the methylation state-specific recognition of histone H4-K20 by 53BP1 and Crb2 in DNA repair. *Cell* 127 (7): 1361–1373.
83 Fradet-Turcotte, A., Canny, M.D., Escribano-Diaz, C. et al. (2013). 53BP1 is a reader of the DNA-damage-induced H2A Lys 15 ubiquitin mark. *Nature* 499 (7456): 50–54.
84 Li, X., Kong, X., Wang, Y., and Yang, Q. (2013). 53BP1 is a novel regulator of angiogenesis in breast cancer. *Cancer Sci.* 104 (11): 1420–1426.
85 Pennington, K.P., Wickramanayake, A., Norquist, B.M. et al. (2013). 53BP1 expression in sporadic and inherited ovarian carcinoma: relationship to genetic status and clinical outcomes. *Gynecol. Oncol.* 128 (3): 493–499.
86 Perfetti, M.T., Baughman, B.M., Dickson, B.M. et al. (2015). Identification of a fragment-like small molecule ligand for the methyl-lysine binding protein, 53BP1. *ACS Chem. Biol.* 10 (4): 1072–1081.
87 Lee, J., Thompson, J.R., Botuyan, M.V., and Mer, G. (2008). Distinct binding modes specify the recognition of methylated histones H3K4 and H4K20 by JMJD2A-tudor. *Nat. Struct. Mol. Biol.* 15 (1): 109–111.
88 Hu, C.E., Liu, Y.C., Zhang, H.D., and Huang, G.J. (2014). JMJD2A predicts prognosis and regulates cell growth in human gastric cancer. *Biochem. Biophys. Res. Commun.* 449 (1): 1–7.

89 Wang, H.L., Liu, M.M., Ma, X. et al. (2014). Expression and effects of JMJD2A histone demethylase in endometrial carcinoma. *Asian Pac. J. Cancer Prev.* 15 (7): 3051–3056.
90 Qin, S., Guo, Y., Xu, C. et al. (2013). Tudor domains of the PRC2 components PHF1 and PHF19 selectively bind to histone H3K36me3. *Biochem. Biophys. Res. Commun.* 430 (2): 547–553.
91 Panagopoulos, I., Micci, F., Thorsen, J. et al. (2012). Novel fusion of MYST/Esa1-associated factor 6 and PHF1 in endometrial stromal sarcoma. *PLoS One* 7 (6): e39354.
92 Micci, F., Panagopoulos, I., Bjerkehagen, B., and Heim, S. (2006). Consistent rearrangement of chromosomal band 6p21 with generation of fusion genes JAZF1/PHF1 and EPC1/PHF1 in endometrial stromal sarcoma. *Cancer Res.* 66 (1): 107–112.
93 SGC (2017). Annotated phylogenetic tree of chromodomain family. http://apps.thesgc.org/resources/phylogenetic_trees/index.php?domain=CHROMO#options.
94 Kaustov, L., Ouyang, H., Amaya, M. et al. (2011). Recognition and specificity determinants of the human cbx chromodomains. *J. Biol. Chem.* 286 (1): 521–529.
95 Itsumi, M., Shiota, M., Yokomizo, A. et al. (2013). Human heterochromatin protein 1 isoforms regulate androgen receptor signaling in prostate cancer. *J. Mol. Endocrinol.* 50 (3): 401–409.
96 Whitcomb, S.J., Basu, A., Allis, C.D., and Bernstein, E. (2007). Polycomb group proteins: an evolutionary perspective. *Trends Genet.* 23 (10): 494–502.
97 Gieni, R.S. and Hendzel, M.J. (2009). Polycomb group protein gene silencing, non-coding RNA, stem cells, and cancer. *Biochem. Cell Biol.* 87 (5): 711–746.
98 Simhadri, C., Daze, K.D., Douglas, S.F. et al. (2014). Chromodomain antagonists that target the polycomb-group methyllysine reader protein chromobox homolog 7 (CBX7). *J. Med. Chem.* 57 (7): 2874–2883.
99 Shinjo, K., Yamashita, Y., Yamamoto, E. et al. (2014). Expression of chromobox homolog 7 (CBX7) is associated with poor prognosis in ovarian clear cell adenocarcinoma via TRAIL-induced apoptotic pathway regulation. *Int. J. Cancer* 135 (2): 308–318.
100 Bernard, D., Martinez-Leal, J.F., Rizzo, S. et al. (2005). CBX7 controls the growth of normal and tumor-derived prostate cells by repressing the Ink4a/Arf locus. *Oncogene* 24 (36): 5543–5551.
101 Yap, K.L., Li, S., Munoz-Cabello, A.M. et al. (2010). Molecular interplay of the noncoding RNA ANRIL and methylated histone H3 lysine 27 by polycomb CBX7 in transcriptional silencing of INK4a. *Mol. Cell* 38 (5): 662–674.
102 Klauke, K., Radulovic, V., Broekhuis, M. et al. (2013). Polycomb Cbx family members mediate the balance between haematopoietic stem cell self-renewal and differentiation. *Nat. Cell Biol.* 15 (4): 353–362.
103 Scott, C.L., Gil, J., Hernando, E. et al. (2007). Role of the chromobox protein CBX7 in lymphomagenesis. *Proc. Natl. Acad. Sci. U.S.A.* 104 (13): 5389–5394.

104 Forzati, F., Federico, A., Pallante, P. et al. (2012). CBX7 is a tumor suppressor in mice and humans. *J. Clin. Invest.* 122 (2): 612–623.
105 Stuckey, J.I., Simpson, C., Norris-Drouin, J.L. et al. (2016). Structure–activity relationships and kinetic studies of peptidic antagonists of CBX chromodomains. *J. Med. Chem.* 59 (19): 8913–8923.
106 Stuckey, J.I., Dickson, B.M., Cheng, N. et al. (2016). A cellular chemical probe targeting the chromodomains of Polycomb repressive complex 1. *Nat. Chem. Biol.* 12 (3): 180–187.
107 Traoré, M., Gignac, M., Doan, N.D. et al. (2017). Aza-amino acid scanning of chromobox homolog 7 (CBX7) ligands. *J. Pept. Sci.* 23 (4): 266–271.
108 Ren, C., Smith, S.G., Yap, K. et al. (2016). Structure-guided discovery of selective antagonists for the chromodomain of polycomb repressive protein CBX7. *ACS Med. Chem. Lett.* 7 (6): 601–605.
109 Ren, C., Morohashi, K., Plotnikov, A.N. et al. (2015). Small-molecule modulators of methyl-lysine binding for the CBX7 chromodomain. *Chem. Biol.* 22 (2): 161–168.
110 Zheng, H., Jiang, W.H., Tian, T. et al. (2017). CBX6 overexpression contributes to tumor progression and is predictive of a poor prognosis in hepatocellular carcinoma. *Oncotarget* 8 (12): 18872–18884.
111 Milosevich, N., Gignac, M.C., McFarlane, J. et al. (2016). Selective inhibition of CBX6: a methyllysine reader protein in the polycomb family. *ACS Med. Chem. Lett.* 7 (2): 139–144.
112 Fischle, W., Franz, H., Jacobs, S.A. et al. (2008). Specificity of the chromodomain Y chromosome family of chromodomains for lysine-methylated ARK(S/T) motifs. *J. Biol. Chem.* 283 (28): 19626–19635.
113 Barnash, K.D., Lamb, K.N., Stuckey, J.I. et al. (2016). Chromodomain ligand optimization via target-class directed combinatorial repurposing. *ACS Chem. Biol.* 11 (9): 2475–2483.
114 SGC (2017). Annotated phylogenetic tree of PHD finger family. http://apps.thesgc.org/resources/phylogenetic_trees/index.php?domain=PHD#options.
115 Baker, L.A., Allis, C.D., and Wang, G.G. (2008). PHD fingers in human diseases: disorders arising from misinterpreting epigenetic marks. *Mutat. Res.* 647 (1–2): 3–12.
116 Zhou, P., Wu, L.L., Wu, K.M. et al. (2013). Overexpression of MMSET is correlation with poor prognosis in hepatocellular carcinoma. *Pathol. Oncol. Res.* 19 (2): 303–309.
117 Li, J., Yin, C., Okamoto, H. et al. (2008). Identification of a novel proliferation-related protein, WHSC1 4a, in human gliomas. *Neuro Oncol.* 10 (1): 45–51.
118 Qi, L., Zhu, F., Li, S.H. et al. (2014). Retinoblastoma binding protein 2 (RBP2) promotes HIF-1alpha-VEGF-induced angiogenesis of non-small cell lung cancer via the Akt pathway. *PLoS One* 9 (8): e106032.
119 Bjorkman, M., Ostling, P., Harma, V. et al. (2012). Systematic knockdown of epigenetic enzymes identifies a novel histone demethylase PHF8 overexpressed in prostate cancer with an impact on cell proliferation, migration and invasion. *Oncogene* 31 (29): 3444–3456.

120 Tian, J., He, H., and Lei, G. (2014). Wnt/beta-catenin pathway in bone cancers. *Tumour Biol.* 35 (10): 9439–9445.

121 Sanchez, R. and Zhou, M.M. (2011). The PHD finger: a versatile epigenome reader. *Trends Biochem. Sci.* 36 (7): 364–372.

122 Musselman, C.A. and Kutateladze, T.G. (2011). Handpicking epigenetic marks with PHD fingers. *Nucleic Acids Res.* 39 (21): 9061–9071.

123 van Zutven, L.J., Onen, E., Velthuizen, S.C. et al. (2006). Identification of NUP98 abnormalities in acute leukemia: JARID1A (12p13) as a new partner gene. *Genes Chromosomes Cancer* 45 (5): 437–446.

124 Wang, G.G., Song, J., Wang, Z. et al. (2009). Haematopoietic malignancies caused by dysregulation of a chromatin-binding PHD finger. *Nature* 459 (7248): 847–851.

125 Wagner, E.K., Nath, N., Flemming, R. et al. (2012). Identification and characterization of small molecule inhibitors of a plant homeodomain finger. *Biochemistry* 51 (41): 8293–8306.

126 Polakis, P. (2012). Wnt signaling in cancer. *Cold Spring Harbor Perspect. Biol.* 4 (5): a008052.

127 Daniels, D.L., Eklof Spink, K., and Weis, W.I. (2001). Beta-catenin: molecular plasticity and drug design. *Trends Biochem. Sci.* 26 (11): 672–678.

128 Sampietro, J., Dahlberg, C.L., Cho, U.S. et al. (2006). Crystal structure of a beta-catenin/BCL9/Tcf4 complex. *Mol. Cell* 24 (2): 293–300.

129 Fiedler, M., Sanchez-Barrena, M.J., Nekrasov, M. et al. (2008). Decoding of methylated histone H3 tail by the Pygo-BCL9 Wnt signaling complex. *Mol. Cell* 30 (4): 507–518.

130 Gu, B., Sun, P., Yuan, Y. et al. (2009). Pygo2 expands mammary progenitor cells by facilitating histone H3 K4 methylation. *J. Cell Biol.* 185 (5): 811–826.

131 Miller, T.C., Rutherford, T.J., Johnson, C.M. et al. (2010). Allosteric remodelling of the histone H3 binding pocket in the Pygo2 PHD finger triggered by its binding to the B9L/BCL9 co-factor. *J. Mol. Biol.* 401 (5): 969–984.

132 Edfeldt, F.N., Folmer, R.H., and Breeze, A.L. (2011). Fragment screening to predict druggability (ligandability) and lead discovery success. *Drug Discovery Today* 16 (7-8): 284–287.

133 Miller, T.C., Rutherford, T.J., Birchall, K. et al. (2014). Competitive binding of a benzimidazole to the histone-binding pocket of the Pygo PHD finger. *ACS Chem. Biol.* 9 (12): 2864–2874.

134 Xie, S., Jakoncic, J., and Qian, C. (2012). UHRF1 double tudor domain and the adjacent PHD finger act together to recognize K9me3-containing histone H3 tail. *J. Mol. Biol.* 415 (2): 318–328.

135 Bronner, C., Krifa, M., and Mousli, M. (2013). Increasing role of UHRF1 in the reading and inheritance of the epigenetic code as well as in tumorogenesis. *Biochem. Pharmacol.* 86 (12): 1643–1649.

136 SGC (2017). Annotated phylogenetic tree of WD40 repeat family. http://apps.thesgc.org/resources/phylogenetic_trees/index.php?domain=WDR_300#options.

137 Stirnimann, C.U., Petsalaki, E., Russell, R.B., and Muller, C.W. (2010). WD40 proteins propel cellular networks. *Trends Biochem. Sci.* 35 (10): 565–574.

138 Wysocka, J., Swigut, T., Milne, T.A. et al. (2005). WDR5 associates with histone H3 methylated at K4 and is essential for H3 K4 methylation and vertebrate development. *Cell* 121 (6): 859–872.

139 Couture, J.F., Collazo, E., and Trievel, R.C. (2006). Molecular recognition of histone H3 by the WD40 protein WDR5. *Nat. Struct. Mol. Biol.* 13 (8): 698–703.

140 Ruthenburg, A.J., Wang, W., Graybosch, D.M. et al. (2006). Histone H3 recognition and presentation by the WDR5 module of the MLL1 complex. *Nat. Struct. Mol. Biol.* 13 (8): 704–712.

141 Schuetz, A., Allali-Hassani, A., Martin, F. et al. (2006). Structural basis for molecular recognition and presentation of histone H3 by WDR5. *EMBO J.* 25 (18): 4245–4252.

142 Han, Z., Guo, L., Wang, H. et al. (2006). Structural basis for the specific recognition of methylated histone H3 lysine 4 by the WD-40 protein WDR5. *Mol. Cell* 22 (1): 137–144.

143 Muntean, A.G., Tan, J., Sitwala, K. et al. (2010). The PAF complex synergizes with MLL fusion proteins at HOX loci to promote leukemogenesis. *Cancer Cell* 17 (6): 609–621.

144 Zhang, P., Lee, H., Brunzelle, J.S., and Couture, J.F. (2012). The plasticity of WDR5 peptide-binding cleft enables the binding of the SET1 family of histone methyltransferases. *Nucleic Acids Res.* 40 (9): 4237–4246.

145 Dharmarajan, V., Lee, J.H., Patel, A. et al. (2012). Structural basis for WDR5 interaction (win) motif recognition in human SET1 family histone methyltransferases. *J. Biol. Chem.* 287 (33): 27275–27289.

146 Senisterra, G., Wu, H., Allali-Hassani, A. et al. (2013). Small-molecule inhibition of MLL activity by disruption of its interaction with WDR5. *Biochem. J.* 449 (1): 151–159.

147 Grebien, F., Vedadi, M., Getlik, M. et al. (2015). Pharmacological targeting of the Wdr5-MLL interaction in C/EBPalpha N-terminal leukemia. *Nat. Chem. Biol.* 11 (8): 571–578.

148 Karatas, H., Townsend, E.C., Cao, F. et al. (2013). High-affinity, small-molecule peptidomimetic inhibitors of MLL1/WDR5 protein–protein interaction. *J. Am. Chem. Soc.* 135 (2): 669–682.

149 Cao, F., Townsend, E.C., Karatas, H. et al. (2014). Targeting MLL1 H3K4 methyltransferase activity in mixed-lineage leukemia. *Mol. Cell* 53 (2): 247–261.

150 Cao, R. and Zhang, Y. (2004). SUZ12 is required for both the histone methyltransferase activity and the silencing function of the EED-EZH2 complex. *Mol. Cell* 15 (1): 57–67.

151 Cao, R. and Zhang, Y. (2004). The functions of E(Z)/EZH2-mediated methylation of lysine 27 in histone H3. *Curr. Opin. Genet. Dev.* 14 (2): 155–164.

152 Margueron, R., Justin, N., Ohno, K. et al. (2009). Role of the polycomb protein EED in the propagation of repressive histone marks. *Nature* 461 (7265): 762–767.

153 Antonysamy, S., Condon, B., Druzina, Z. et al. (2013). Structural context of disease-associated mutations and putative mechanism of autoinhibition

revealed by X-ray crystallographic analysis of the EZH2-SET domain. *PLoS One* 8 (12): e84147.
154 Wu, H., Zeng, H., Dong, A. et al. (2013). Structure of the catalytic domain of EZH2 reveals conformational plasticity in cofactor and substrate binding sites and explains oncogenic mutations. *PLoS One* 8 (12): e83737.
155 Jiao, L. and Liu, X. (2015). Structural basis of histone H3K27 trimethylation by an active polycomb repressive complex 2. *Science* 350 (6258): aac4383.
156 Kong, X., Chen, L., Jiao, L. et al. (2014). Astemizole arrests the proliferation of cancer cells by disrupting the EZH2-EED interaction of polycomb repressive complex 2. *J. Med. Chem.* 57 (22): 9512–9521.
157 Lingel, A., Sendzik, M., Huang, Y. et al. (2017). Structure-guided design of EED binders allosterically inhibiting the epigenetic polycomb repressive complex 2 (PRC2) methyltransferase. *J. Med. Chem.* 60 (1): 415–427.
158 Li, L., Zhang, H., Zhang, M. et al. (2017). Discovery and molecular basis of a diverse set of polycomb repressive complex 2 inhibitors recognition by EED. *PLoS One* 12 (1): e0169855.
159 Huang, Y., Zhang, J., Yu, Z. et al. (2017). Discovery of first-in-class, potent, and orally bioavailable embryonic ectoderm development (EED) inhibitor with robust anticancer efficacy. *J. Med. Chem.* 60 (6): 2215–2226.
160 Qi, W., Zhao, K., Gu, J. et al. (2017). An allosteric PRC2 inhibitor targeting the H3K27me3 binding pocket of EED. *Nat. Chem. Biol.* 13 (4): 381–388.
161 He, Y., Selvaraju, S., Curtin, M.L. et al. (2017). The EED protein–protein interaction inhibitor A-395 inactivates the PRC2 complex. *Nat. Chem. Biol.* 13 (4): 389–395.
162 Barnash, K.D., The, J., Norris-Drouin, J.L. et al. (2017). Discovery of peptidomimetic ligands of EED as allosteric inhibitors of PRC2. *ACS Comb. Sci.* 19 (3): 161–172.

15

DNA-modifying Enzymes

Martin Roatsch[1], Dina Robaa[2], Michael Lübbert[3], Wolfgang Sippl[2], and Manfred Jung[1]

[1] *Albert-Ludwigs-University Freiburg, Institute of Pharmaceutical Sciences, Albertstraße 25, 79104, Freiburg im Breisgau, Germany*
[2] *Martin-Luther-University Halle-Wittenberg, Institute of Pharmacy, Wolfgang-Langenbeck-Straße 4, 06120, Halle (Saale), Germany*
[3] *University of Freiburg Medical Centre, Clinic for Internal Medicine I, Division of Haematology, Oncology and Stem Cell Transplantation, Hugstetter Straße 55, 79106, Freiburg, Germany*

15.1 Introduction

The two major biochemical pathways of epigenetic gene regulation are DNA methylation and posttranslational modifications of amino acid side chains in histone proteins, around which the DNA is wrapped. The latter actually comprise a whole set of different biochemical modifications such as reversible acetylation of lysine residues, methylation of lysine and arginine residues, and phosphorylation of serine and threonine residues and modifications with larger biomolecules such as SUMOylation and ubiquitinylation [1, 2]. These histone modifications interact with each other and constitute a pattern of alterations in the chromatin structure, sometimes referred to as the "histone code" [2, 3] or as a complex combinatorial language [4, 5]. Multiple enzymes are involved that deposit these modifications ("writers") and remove them ("erasers") as well as protein domains, which can recognize them and recruit further effector proteins ("readers") [6]. The same modification on different amino acids in the histone proteins may lead to opposing effects on transcription [1, 2]. The various modifications of histones have attracted attention as drug targets due to their prominent involvement in disease, in particular in oncology [6–11], and a large number of drug discovery programs toward such "epi-drugs" have been initiated with some histone deacetylase (HDAC) inhibitors reaching market approval and many more in clinical trials [12–14]. These are outlined in Chapters 7-14 of this book.

DNA methylation refers to the enzymatic addition of a methyl group directly to nucleotide bases. Several types of base methylation are known: adenine can be modified at the N-6 position (in bacteria and in most eukaryotes), guanine has been found to be methylated at the C-7 position, and cytosine can be modified at the N-4 (mostly in thermophilic bacteria) and the C-5 position.

Epigenetic Drug Discovery, First Edition. Edited by Wolfgang Sippl and Manfred Jung.

The latter modification is the most common form of DNA alteration among all organisms [15]. Cytosine is methylated by DNA nucleotide methyltransferases (DNMTs) to 5-methylcytosine. The differences in biological significance of various patterns of methylation result from the stretch of DNA where the methylation takes place. Usually, methylation of the promoter regions of genes will result in repression of transcription. For DNMTs, the first inhibitors, which are structurally nucleoside-mimicking molecules, have already been approved for clinical use. Together with HDAC inhibitors, these are the only approved "epi-drugs." However, there is much less work published relating to novel non-nucleoside inhibitors or structure–activity relationships [16] as compared with other epigenetic modulators, e.g. HDACs.

15.2 DNA Methylation

DNMTs catalyze the methylation of cytosine bases located 5′ of a guanosine base as part of a cytosine–guanine dinucleotide (CpG) dinucleotide in the DNA to 5-methylcytosine using *S*-adenosylmethionine (SAM) as the methyl donor (see Figure 15.1) [2, 15]. Therefore, the DNA contains at least five and not only four different bases, i.e. adenine, thymine, guanine, cytosine, and 5-methylcytosine (including oxidized forms). At first, this seems to be a contradiction to the conserved genetic code and to the definition of epigenetics. In a narrower sense, epigenetics is regarded as "somatically inherited changes in gene expression that occur without a change in the genetic code" [17]. In the case of DNA methylation, the biochemical modification is occurring directly on the DNA, but the base pairing of cytosine to the guanine base on the complementary DNA strand is not altered upon methylation (see Figure 15.2), and, thus, the genetic code is preserved. Frequently, longer hypermethylated stretches of CpG sequences are observed, which are recognized by proteins with methyl-CpG-binding domain proteins (MBD), which mediate epigenetic regulation (see Section 15.6) [18]. Single 5-methylcytosines in the DNA that are not complexed by proteins are prone to hydrolysis to thymines, leading to a 5-methyl-C to A transition (see Figure 15.3) [19]. This may occur spontaneously but also catalyzed by DNA deaminases from the Aid/Apobec family [20]. This has the risk of inducing point mutations but may on the other hand be beneficial. For example, it plays a role in "hypermutations" of immunoglobulin loci that lead to an expansion of the armory of antigen receptors and hence strengthening of the immune system [21].

Altogether, about 70–80% of the CpG sites are methylated [22–24]. CpG dinucleotides are found in short DNA regions with a length of about 0.5–4 kb, which are known as CpG islands and which are located in the proximal promoter regions of approximately half of the genes in our genome. These CpG islands represent about 1–2% of the total genome and contain more than half of the unmethylated CpG sites [22, 23, 25, 26]. High degrees of methylated sequences can be found in satellite DNA, repetitive elements, nonrepetitive intergenic DNA, and exons of genes [27].

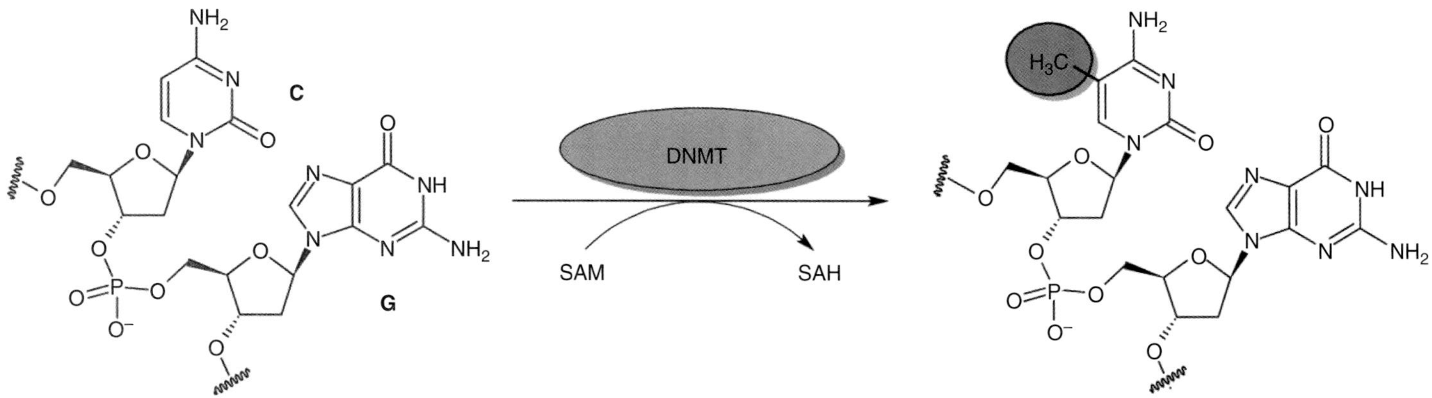

Figure 15.1 Methylation reaction catalyzed by DNA nucleotide methyltransferases (DNMTs). An exemplary cytosine–guanine dinucleotide (CpG) is shown. DNMTs use *S*-adenosylmethionine (SAM) as cofactor, which is converted to *S*-adenosylhomocysteine (SAH).

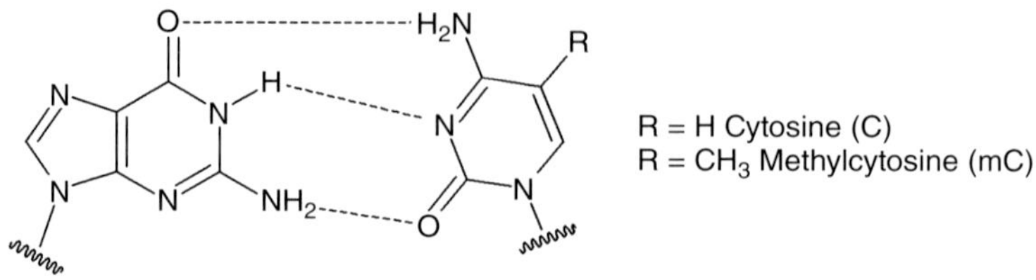

Figure 15.2 Watson–Crick base pairing is unaffected by methylation of cytosine.

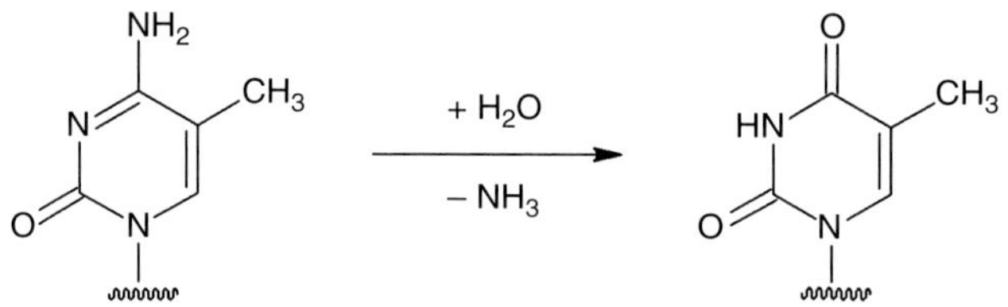

Figure 15.3 Deamination of 5-methylcytosine to thymine can occur both spontaneously and by enzymatic catalysis, introducing a point mutation in the DNA sequence.

15.3 Further Modifications of Cytosine Bases

While most histone modifications exist in a dynamic equilibrium between being "written" and "erased," this has not been rigorously demonstrated for cytosine methylation. To date, no enzyme with true DNA demethylase activity is known, despite lively discussions in the field [28–30]. In the absence of an active demethylation mechanism, loss of 5-methylcytosine can occur only by passive demethylation, i.e. by dilution during replication when newly synthesized DNA strands remain unmethylated, a slow process that takes several rounds of cell division to completely remove a previously established methylation mark [31].

However, a class of enzymes, the so-called ten-eleven translocation (TET) enzymes, has been discovered, which further oxidize methylcytosine (mC) in a stepwise manner as depicted in Figure 15.4 [32, 33]. This class contains three highly similar members (TET1, TET2, TET3), which belong to the much larger cupin superfamily of iron(II)- and 2-oxoglutarate-dependent oxygenases [34]. Given their high degree of similarity, they likely have redundant functions.

The TET enzymes oxidize methylcytosine to hydroxymethylcytosine (hmC) and further to formylcytosine (fC) and carboxylcytosine (caC). The function of these oxidized variants is still under debate, but it is speculated that they can play a role in active reversal of DNA methylation [35]. This could occur either by spontaneous decarboxylation of caC or if the other oxidized species function as triggers that are recognized by the DNA repair machinery leading to removal by base excision repair (BER) or nucleotide excision repair (NER) (Figure 15.4). The missing base or nucleotide would then be replaced by unmethylated cytosine [33, 35–37].

Recently, it has been reported that TET enzymes can also oxidize thymine to hydroxymethyluracil in the DNA of mouse embryonic stem cells [38].

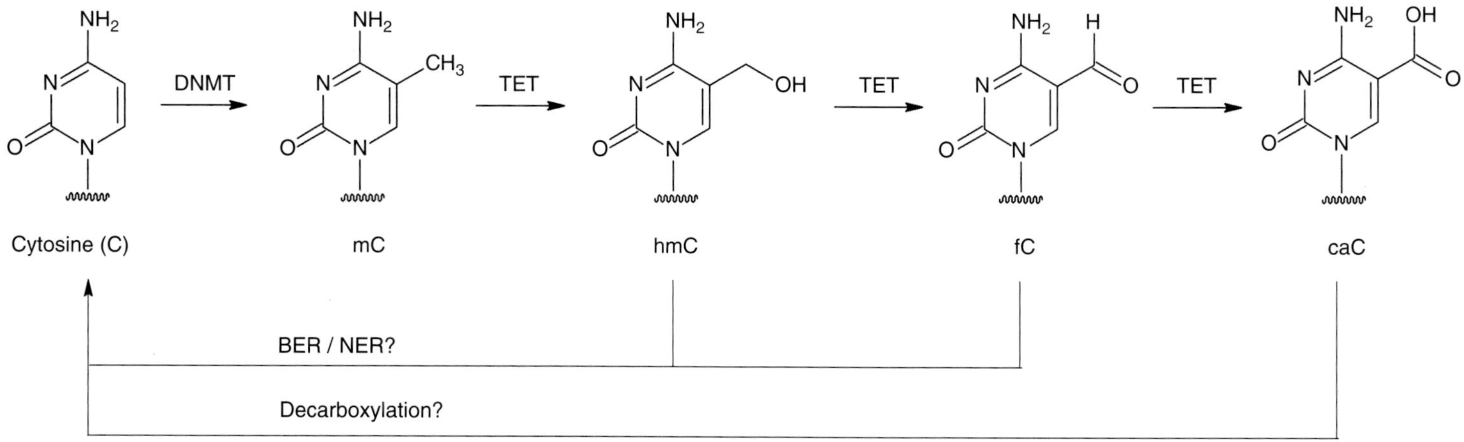

Figure 15.4 Methylcytosine (mC) can be further oxidized by TET enzymes to hydroxymethylcytosine (hmC), formylcytosine (fC), and carboxylcytosine (caC). These oxidized species are thought to play a role in active DNA demethylation either by decarboxylation or by base or nucleotide excision repair mechanisms (BER or NER).

A crystal structure of TET2 in complex with the DNA has recently been published [39]. In spite of this and the close relationship to other Fe(II)-/2-oxoglutarate-dependent enzymes, such as JumonjiC histone demethylases, for which a large number of inhibitors have been reported in recent years [9, 40–42], no specific TET inhibitors are available to date.

15.4 DNA Methyltransferases: Substrates and Structural Aspects

So far, three groups of DNMTs are known in mammals [43–46]: DNMT1, DNMT2, and DNMT3s with members DNMT3A, DNMT3B, the recently discovered DNMT3C, and a protein called DNMT3-like (DNMT3L), which does not possess catalytic activity itself, but which interacts with HDAC1 to function as a transcriptional repressor [47, 48].

Whereas DNA methylation in early embryogenesis is carried out by the *de novo* methyltransferases DNMT3A and DNMT3B, the methylation level in differentiated cells is regulated by DNMT1. Therefore, DNMT1 is also called the maintenance methyltransferase. DNMT1 was the first cytosine methyltransferase identified and is closely associated with the DNA replication process. It transfers the DNA methylation patterns from the parent strand to the newly synthesized strand.

DNMT2 (now known as tRNA aspartic acid methyltransferase 1 or TRDMT1) is a relatively small protein (391 amino acids) and shows all the sequence and structural characteristics of the other DNMTs except for the large N-terminal domain present in DNMT1 and the DNMT3s. Despite the sequence and structural similarity between DNMT2 and other DNMTs, DNMT2 is not necessary for DNA methylation *in vivo*. It has been shown that genomic methylation levels are not measurably altered in DNMT2-deficient mouse embryonic stem cells [43]. It was reported that human DNMT2 methylates cytosine 38 in the anticodon loop of aspartic acid transfer RNA [49].

Very recently, DNMT3C, another *de novo* DNA methyltransferase, has been identified [50]. It could be shown that DNMT3C is the enzyme responsible for methylating the promoters of evolutionarily young retrotransposons in the male germ line. DNMT3C reveals the plasticity of the mammalian DNA methylation system and expands the scope of the mechanisms involved in epigenetic control of retrotransposons.

The protein DNMT3L shows high sequence similarity with DNMT3A, but it lacks a conserved segment in the C-terminal domain required for methyltransferase activity. Therefore, DNMT3L is not an active methyltransferase, but has dual functions of binding the histone tail as well as activating DNMT3A. DNMT3L was found to stabilize the conformation of the active site loop of DNMT3A [47, 48].

Mammalian DNMTs (expect for the tRNA methyltransferase DNMT2) are composed of two domains: the N-terminal regulatory domain and the C-terminal domain bearing the catalytic site. The catalytic methyltransferase domain carries 10 characteristic sequence motifs, 6 of which are conserved in nearly all cytosine methyltransferases from bacteria through plants to mammals [51]. The different motifs have been described according to their function [52, 53]: motifs I to III and X form the cofactor binding pocket; motif IV contains the essential cysteine residue, which forms the thiolate initiating the methyl group transfer; motifs V, VI, and VII compose the substrate binding site; and the nonconserved region between motifs VIII and IX forms the so-called target recognition domain (TRD) (Figure 15.5) [54, 55].

The first methyltransferases that could be crystallized were prokaryotic enzymes (e.g. the bacterial cytosine methyltransferases M.HhaI [56–58] and M.HaeIII [59]), which are simpler in structure and less complex than the mammalian DNMT family members DNMT1 and DNMT3. Later on, crystal structures of mammalian DNMTs were also reported. The structures of the resolved DNMTs can be divided into a C-terminal catalytic domain and an N-terminal regulatory domain. As described above, the C-terminal region includes the SAM binding site, whereas the N-terminal domain plays a regulatory role and is important for distinguishing hemi- from unmethylated DNA. A detailed analysis of all available structures of DNMTs has recently been reviewed [31, 60, 61]. Currently, 3D structures are available for DNMT2 [62], the complex between the C-terminal domains of DNMT3A and DNMT3L [63, 64], and as of recently also for DNMT1 [65–68]. The crystal structure of intact DNMT3L in complex with a histone H3 amino-tail peptide has also been resolved [69]. With regard to the N-terminal domains, 3D structures are also available for the PWWP domain of DNTM3A and DNMT3B [70, 71], the ADD domain of DNMT3A [72], and the replication focus targeting domain (RFS) of DNMT1 [73].

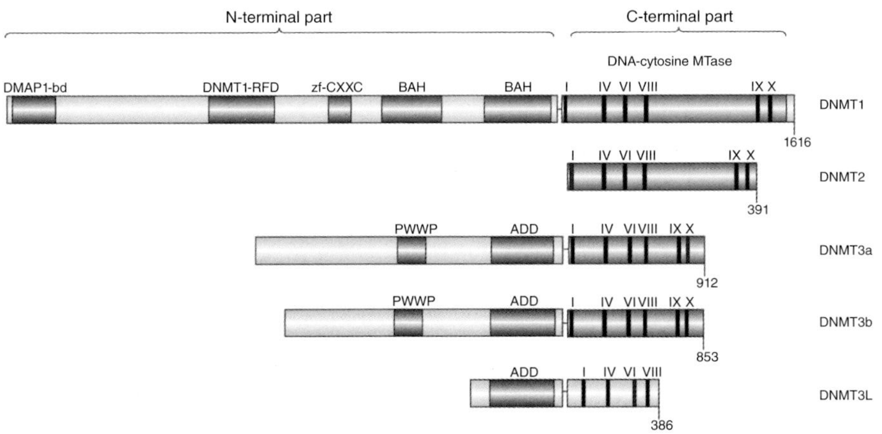

Figure 15.5 Domain architecture of mammalian DNMT1, DNMT2, DNMT3A, DNMT3B, and DNMT3L. Functional domains in the N-terminal part are shown, and conserved methyltransferase motifs in the C-terminal part are labeled. Domain abbreviations: ADD, ATRX–DNMT3–DNMT3L domain; BAH, bromo-adjacent homology domain; DMAP1-bd, DNA methyltransferase associated protein 1 binding domain; DNMT1-RFD, DNA methyltransferase 1 replication foci domain; PWWP, Pro–Trp–Trp–Pro domain; zf-CXXC, zinc finger-CXXC.

Crystal structures of mammalian DNMTs and the bacterial cytosine methyltransferases are essentially superimposable over the C-terminal domain including the catalytic site. The binding pockets show high structural conservation of residues interacting with deoxycytidine and the SAM cofactor.

The crystal structure of the DNMT3A–DNMT3L complex contains two monomers of each protein, resulting in a tetramer and was crystallized without bound DNA. In this complex, DNMT3L stabilizes the conformation of the DNMT3A active site loop. Since the DNMT3A structure is similar to the bacterial methyltransferase M.HhaI, which was co-crystallized with DNA, the DNMT3A–DNMT3L–DNA complex could be modeled from the bacterial complex [63]. DNTM3A has a relatively small DNA binding domain (~50 base pairs) compared with the bacterial enzymes. However, in the tetramer, two active sites come together closely, which leads to the doubling of the DNA binding surface. It was suggested that the DNA is bound to both binding sites, such that the two active sites are located in the major groove ~40 Å apart, which corresponds to one DNA helical turn. In the DNMT3A–DNA complex, the cytosine is located between the nucleophilic Cys706 and the cofactor SAM (Figure 15.6).

More recent structural and biochemical analyses elucidated the role of the ADD domain in the regulation of DNMT3A activity. Crystal structures of DNMT3A, encompassing both the catalytic and the ADD domains, in complex with DNMT3L were obtained in the presence and absence of histone H3 peptide. In absence of the histone peptide, DNMT3A exists in an auto-inhibited form, where the ADD domain blocks the DNA from binding. The crystal structure shows how the ADD domain is situated near the DNA binding site and inserts a loop region (residues 526–533) into the catalytic domain. Binding of H3K4me0 to the ADD domain induces large movement of the ADD domain, thereby opening the DNA binding region (Figure 15.7) [64].

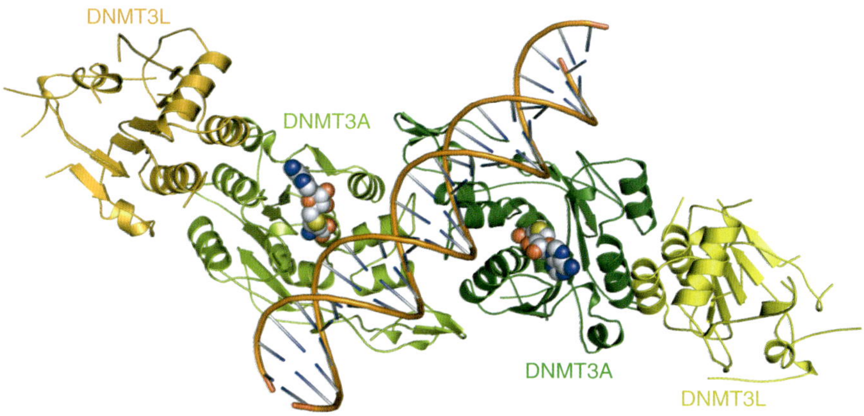

Figure 15.6 Model of DNMT3A–DNMT3L tetramer with DNA. The resolved crystal structure of the tetramer (PDB ID: 2QRV) was used, while the DNA molecule was modeled based on the superimposition with M.HhaI–DNA complex structure. The cofactor SAM is shown as white spheres.

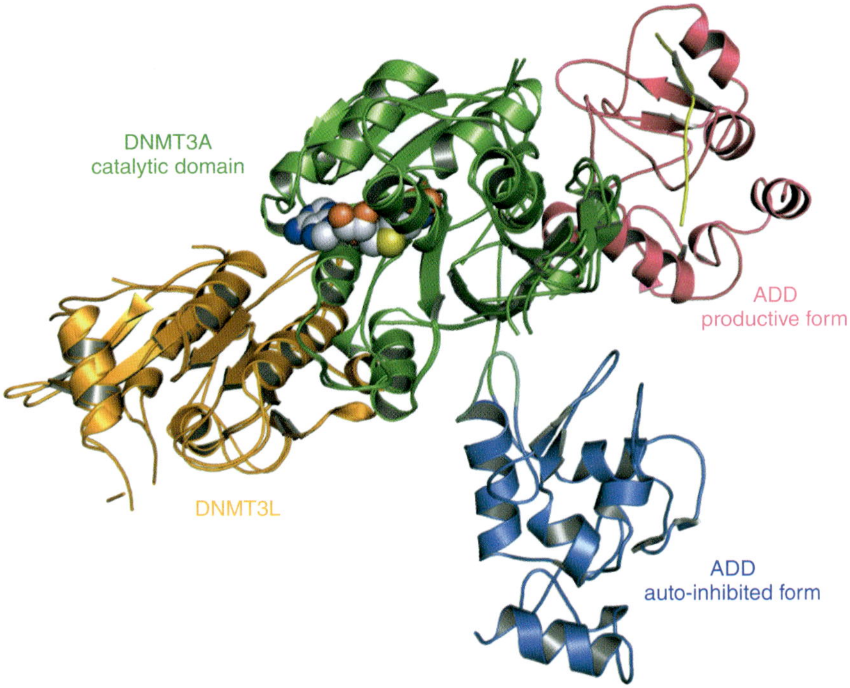

Figure 15.7 Role of the ADD domain in autoregulation of DNMT3A. Comparison between the auto-inhibited and productive forms. Crystal structures of the auto-inhibited and productive forms (PDB ID: 4U7P and 4U7T, respectively) are superimposed, showing the strong movement of the ADD domain.

Since, in differentiated cells, DNMT1 is responsible for the transfer of the methylation patterns, it represents the more interesting target for experimental cancer therapies.

Crystal structures of the auto-inhibited and productive form of truncated mouse DNMT1 (mDNMT1) complexed with DNA have recently been reported [65, 66]. In the auto-inhibited form, two factors seem to play a role in preventing the unmethylated DNA from binding to the catalytic domain, thereby hindering *de novo* DNA methylation. Firstly, the zf-CXXC domain undergoes extensive interactions with the DNA, thereby anchoring it in a position distant from the active site. Secondly, the CXXC–BAH1 linker, which contains highly acidic residues, is located between the DNA and the active site and excludes the catalytic loop (containing the catalytically important cysteine) from the DNA minor groove. It is, however, noteworthy that DNMT1 in absence of both domains still retains considerable preference for hemimethylated CpG sites, indicating that the catalytic domain itself possesses the capability of distinguishing between the methylation statuses of the DNA.

In the productive complex, the bound DNA strand contains a central hemimethylated CpG site, and the methyl group of mC is located within a shallow hydrophobic pocket in the TRD domain, where the cytosine ring undergoes π–π stacking with Trp1512. Meanwhile, the unmodified cytosine of

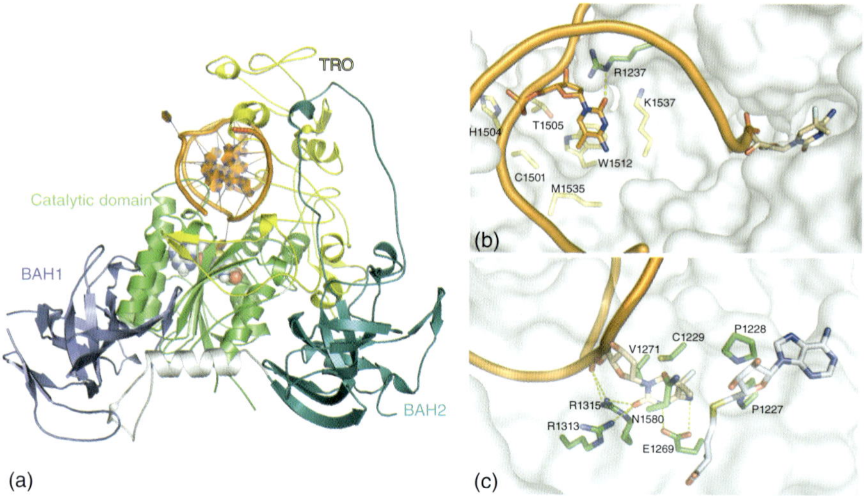

Figure 15.8 (a) Structure of productive form of mouse DNMT1 bound to hemimethylated CpG DNA (PDB ID: 4DA4). SAH is shown as white spheres. (b) Methylcytosine (orange) is anchored in a hydrophobic pocket in the TRD domain (side chains are shown as yellow sticks). The target cytosine residue (beige) is looped out into the catalytic pocket. (c) Interactions of the cytosine base (beige) of the target strand with residues in the catalytic pocket (green sticks). H-bonds are shown as yellow-dashed lines and SAH as white sticks.

the target strand is looped out and fixed within the catalytic pocket by forming extensive H-bond interactions with polar residues (Figure 15.8).

Unlike in the auto-inhibited form, the zf-CXXC domain and the flanking loop were not resolved in the crystal structure of the productive form. The major conformational change is observed for the catalytic loop (residues 1227–1243 in mDNMt1), which inserts into the DNA minor groove, forming extensive interactions with the DNA. Meanwhile, two TRD loops undergo a slight movement toward the DNA major groove and participate in the recognition of mC by hydrophobic contacts, as previously described.

The availability of structural data for DNMTs has enabled virtual screening campaigns, which were the basis for the identification of further DNMT inhibitors [74–76]. For details, the reader is referred to Section 15.8.

15.5 Mechanism of Enzymatic DNA Methylation

The enzymatic mechanism of cytosine methylation has been studied in detail [57, 77] and is depicted in Figure 15.9. DNMTs form a complex with the DNA, and the cytosine base to be methylated is flipped out of the DNA double helix [78]. The methyltransferase then forms a covalent adduct **I** with the DNA by means of a conjugate addition of the thiol group of a cysteine residue in the active site to the double bond of cytosine. This adduct **I** then reacts with the co-substrate SAM **II** present in the active site, giving rise to a

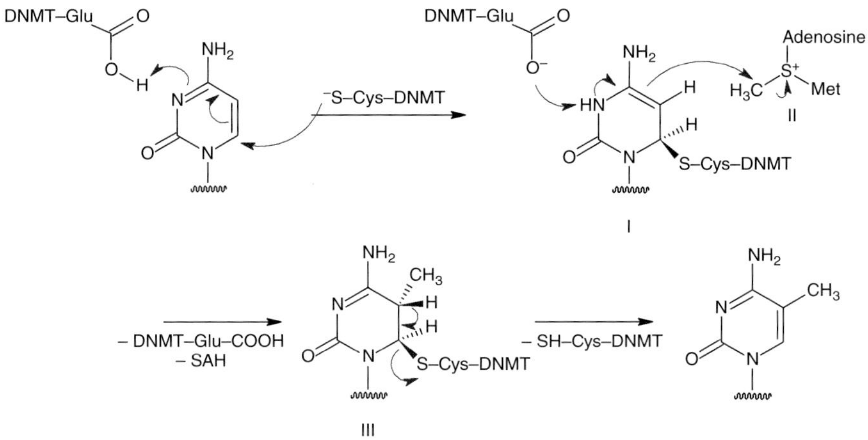

Figure 15.9 Enzymatic mechanism of cytosine methylation.

methylated adduct **III**. A glutamate residue in the active site of the DNMT delivers the carboxylate moiety that participates in the process by protonating the cytosine N-3 atom. Finally, a retro-Michael-type elimination of the enzyme and an adjacent proton liberates the enzyme from the DNA under formation of the 5-methylated cytosine base, which finally flips back into its original position within the DNA double helix [78]. This mechanism is the basis for the inhibitory potency of the two inhibitors approved for human use [79], and this knowledge can be exploited for further inhibitor design (see Section 15.8). It has been proposed that a deamination to thymine can occur from intermediate **I** in the absence of SAM with subsequent BER to cytosine [29]. This would formally be a demethylation reaction but not in a direct biochemical sense.

15.6 Physiological Role of DNA Methylation

DNA methylation is an essential function in normal mammalian cells. It is involved in genomic imprinting (restriction of the expression of a gene to only one of the two parental chromosomes) [80], which is important for development. It is also necessary for X chromosome inactivation [81], the mechanism of dosage compensation in female mammals.

DNA methylation can in some cases directly decrease the binding affinity of certain transcription factors to the DNA [82]. Additionally, methylated CpG sites recruit MBD proteins, which in turn leads to transcriptional repression [83]. siRNA-mediated knockdown of MBD proteins leads to re-expression of silenced tumor suppressor protein candidates [84]. DNA methylation and histone modifications are dependent on each other [85, 86], and because of this crosstalk, a synergy, e.g. of HDAC and DNMT inhibitors, in derepression can be observed [87–91]. In fact, a number of different combinations of HDAC and

DNMT inhibitors, both approved and experimental, are currently used in the clinic or are undergoing clinical trials [92–95] as recently reviewed [11, 14, 96].

The importance of DNA methylation for normal genome function could also be shown by the finding that homozygous mutation of the DNMT genes results in embryonic lethality [97]. Thus, unwanted activation of certain genes by unselective DNMT inhibitors may pose a risk, e.g. by derepression of pro-invasive genes [98], which puts a caveat to these therapeutic approaches.

15.7 DNA Methylation in Disease

Epigenetics and hence also DNA methylation can be seen as biochemical manifestations of environmental stimuli [99]. These epigenetic alterations may lead to an increased susceptibility to certain diseases, e.g. cancer [11, 96, 100, 101] and psychiatric diseases [102–104]. This increases over the lifetime as has been elegantly described in monozygotic twins who have the same gene set but become more and more different in terms of DNA methylation pattern with age [105]. Genome-wide association studies (GWAS) or epigenome-wide association studies (EWAS) of epigenetic patterns have become increasingly important in order to correlate those patterns with human disease [106–108]. This requires the observation of large numbers of participants in population studies over a long period of time (longitudinal study design) as, for example, performed in the Human Epigenome Project (HEP). Ultimately, the goal will be to use those patterns as predictive markers for disease risk, disease progression, and optimized treatment [11, 91, 109].

In cancer cells, widespread hypomethylation together with hypermethylation of certain promoter regions has been recognized as one of the most important epigenetic changes taking place in tumors ("hallmarks of cancer"), leading to transcriptional silencing of tumor suppressor genes [110–112]. This may be due to overexpression of the enzyme [113, 114], aberrant recruitment of the enzyme [115], or expression of truncated mutants [116]. Conversely, it was also found that hypermethylation may already be observed in pre-cancerous tissue and reflect cell lineage rather than disease state [117]. A large variety of genes were discovered to be aberrantly methylated in cancer cells including genes that are involved in regulating DNA repair (e.g. *BRCA1*, *MLH1*), signal transduction (e.g. *RASSF1*), the cell cycle (e.g. $p16^{INK4a}$, $p15^{INK4b}$), metabolism of carcinogenics (e.g. *GSTP1*), cell adherence (e.g. *CDH1*, *CDH13*), apoptosis (e.g. *DAPK*, *TMS1*), and angiogenesis (e.g. *THBS1*) [111, 112].

Besides the promotor regions of protein-coding genes, also non-protein-coding regions of DNA, can be affected, and DNA methylation in the regulatory region also has an impact on microRNA expression [118]. An example is suppression of the formation of miRNA-124a by DNA hypermethylation, which in turn leads to an activation of the oncogenic cyclin D kinase 6 [119].

Moreover, DNA methylation has also been associated with acquired drug resistance as shown in ovarian cancer cells that initially responded well to treatment with standard chemotherapeutics like cisplatin but, after several rounds of treatment, became resistant, consistent with increased CpG island methylation [120].

15.8 DNMT Inhibitors

The DNMT inhibitors published so far can be categorized into two big classes [31, 96]. One group, including the two licensed DNMT inhibitors, consists of nucleoside analogues that are incorporated into the DNA and act as suicide inhibitors by irreversibly trapping the DNMT enzyme via formation of a covalent adduct. The other group contains reversible small-molecule inhibitory molecules as discussed in Section 15.8.2.

15.8.1 Nucleoside-mimicking DNMT Inhibitors

This group comprises the two approved DNMT inhibitors azacitidine **1** (5-azacytidine, AZA, Vidaza™, Celgene Corporation, approved in 2004) and decitabine **2** (5-aza-2′-deoxycytidine, DAC, Dacogen™, Otsuka Pharmaceuticals, approved in 2006) [121, 122] and a number of analogues as depicted in Figure 15.10. These compounds act as prodrugs and show no activity in *in vitro* DNMT assays [123]. The reason for this is based in their mode of action as depicted in Figures 15.11 and 15.12. These compounds require metabolic activation by kinases and, in the case of azacitidine **1**, by a reductase in order to be transformed into the triphosphate nucleotide building blocks that are then incorporated into the DNA. While the deoxyriboside decitabine **2** can be directly transformed to the triphosphate (by deoxycytidine kinase), which then serves as a building block for DNA polymerases, the riboside azacytidine **1** needs to be 2′-deoxygenated in the sugar moiety before it is phosphorylated by cytidine–uridine kinase. This step is performed by a ribonucleotide reductase on the level of the diphosphate. Then, the same triphosphate as from decitabine **2** is formed, and the false base is incorporated into the DNA (see Figure 15.11). Hence, the rate of DNA incorporation of azacitidine **1** is much lower than that of decitabine **2**. Additionally, azacitidine **1** as a riboside can also be incorporated into RNA [124], which can explain some of the additional side effects. The activity of both inhibitors is in the low micromolar region [125].

Once the azacytosine analogues are incorporated into the DNA, they are also subject to covalent addition of the thiol group of a cysteine residue in the DNMT enzyme, and an adduct **Ia** is formed (Figure 15.12, compared with enzymatic mechanism in Figure 15.9). In most cases, this adduct **Ia** also reacts with *S*-adenosyl methionine **II** to a methylated adduct **IIIa**. However, due to the absence of an α-proton, the enzyme cannot be liberated by elimination and remains trapped to the DNA (Figure 15.12) [126–128]. The covalent trapping of the enzyme leads to a depletion of the cellular pool of DNMTs and subsequent DNA hypomethylation over time (cf. Section 15.3). This in turn results in activation, particularly reactivation, of silenced genes. Additionally, the covalently trapped DNMT may inhibit RNA and DNA polymerases, which leads to an inhibition of protein biosynthesis and DNA strand breaks. This can lead to apoptosis and, hence, cytotoxicity. Thus, it is not easy to dissect the reasons for the clinical efficacy of these inhibitors in terms of real epigenetic and plain cytotoxic effects [128].

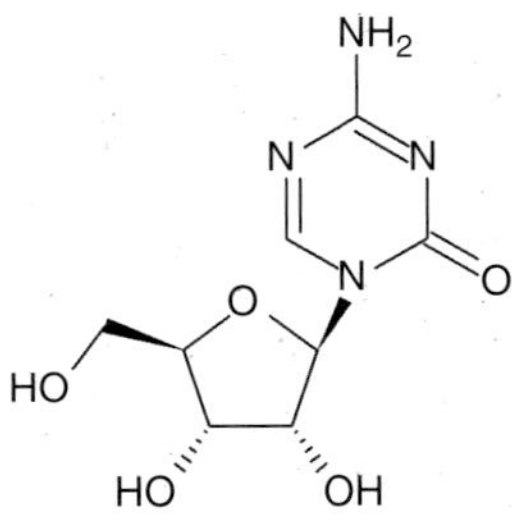

Azacitidine **1**

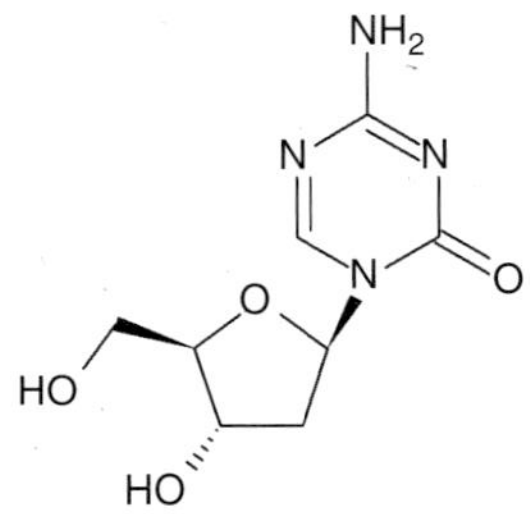

Decitabine **2**

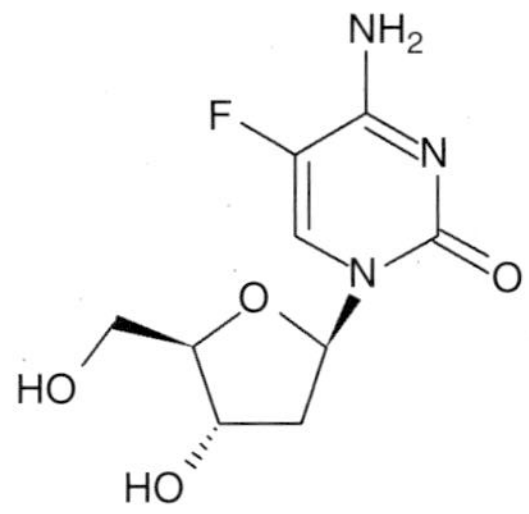

2′-deoxy-5-fluoro-cytidine **3**

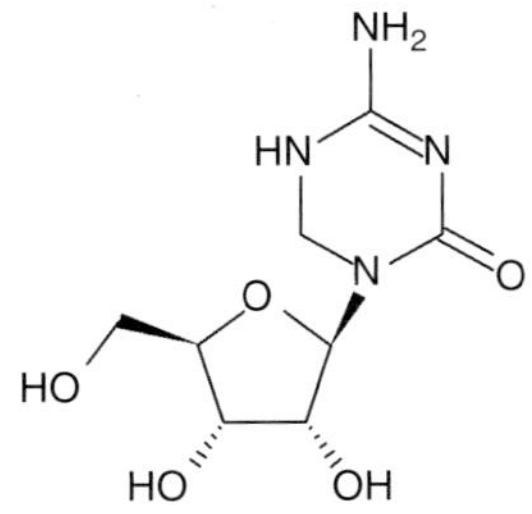

5,6-dihydro-azacitidine **4**

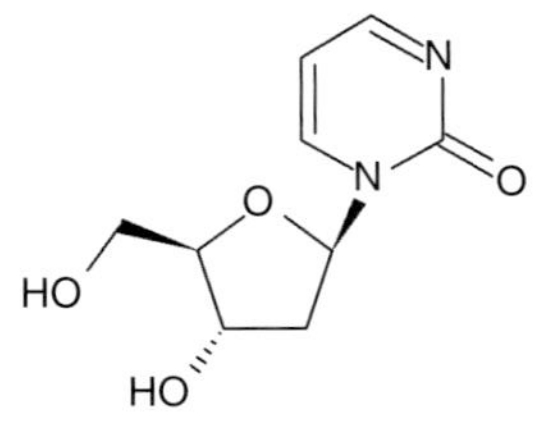

Zebularine **5**

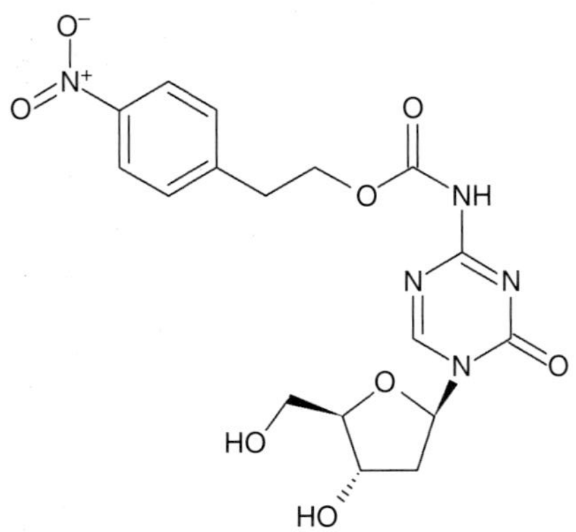

NPEOC-DAC **6**

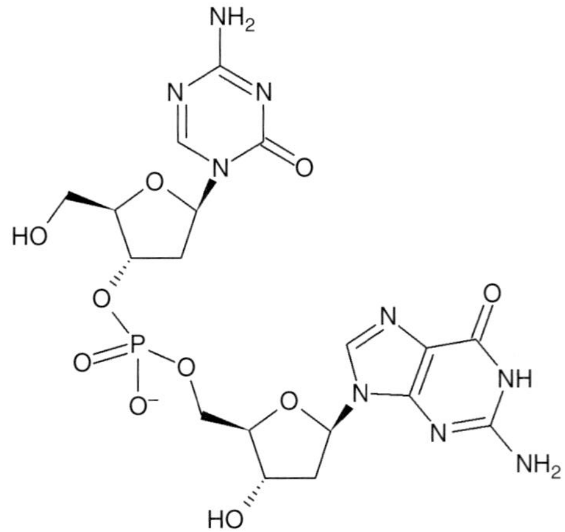

SGI-110 **7**

Figure 15.10 Chemical structures of nucleoside-mimicking DNMT inhibitors. Azacitidine **1** and decitabine **2** are licensed drugs, SGI-110 **7** is undergoing phase II and phase III clinical trials, while **3**–**6** were not developed further.

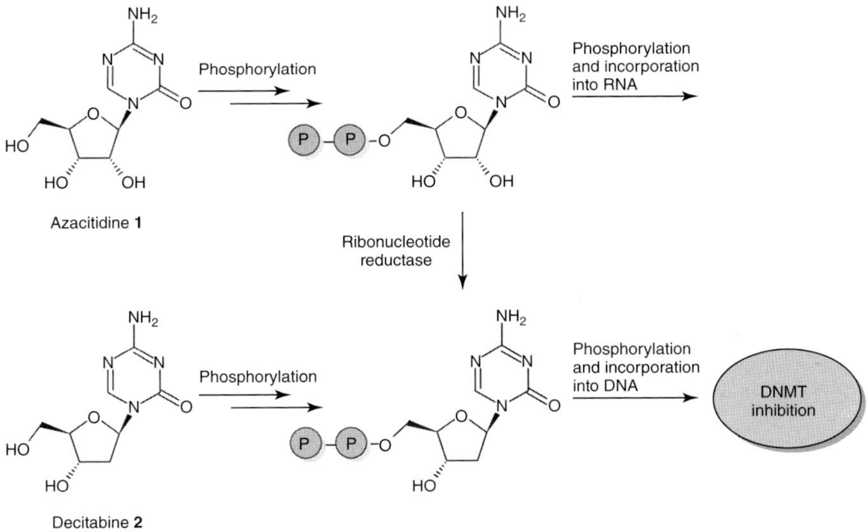

Figure 15.11 Metabolic activation of nucleoside-mimicking prodrug inhibitors *in vivo*.

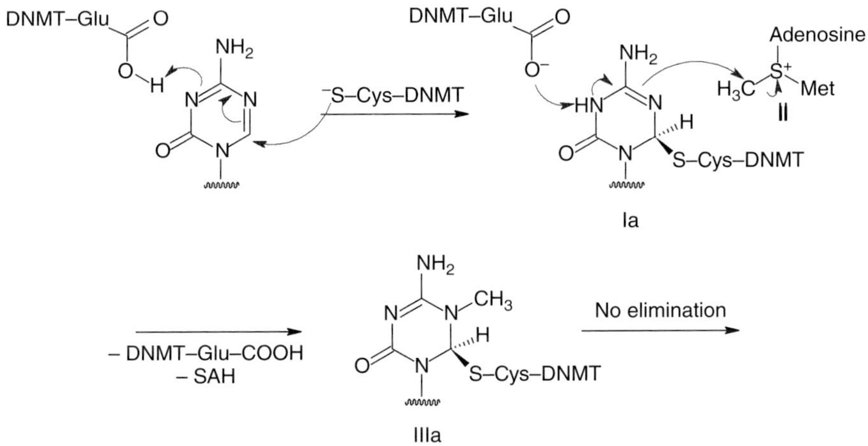

Figure 15.12 Mode of action of covalent inhibition of DNMTs by 5-azanucleotides in the DNA. After formation of the covalent methylated adduct **IIIa**, elimination is no longer possible, and the enzyme is irreversibly trapped.

Based on this mode of action, it follows that these inhibitors show no selectivity for any of the DNMT enzymes and inhibit *de novo* as well as maintenance methylation across the entire genome. However, it is unknown whether selective inhibition of any one enzyme or pan-inhibition is required for the therapeutically beneficial effect. However, in terms of probe molecules to investigate the physiological role of any given enzyme, inhibitors with some degree of selectivity would be desirable.

Compounds of this class generally exhibit poor bioavailability and metabolic stability in physiological media. Rapid metabolic degradation includes opening of the triazine ring, deformylation, and anomerization. The plasma half-life for **1** and **2** is less than one hour, and subcutaneous injection or application by infusion is required [129, 130]. In addition, resistance to these drugs in myelodysplastic syndrome (MDS) is also known [131, 132].

In an effort to overcome some of these issues, different analogues of **1** and **2** have been prepared that are assumed to function by a similar mechanism (Figure 15.10). This includes a 5-fluoro derivative **3** of deoxycytidine [122, 133] and the 5,6-dihydro derivative **4** of azacitidine **1** [122], with little improvement in their cellular effects, however. The derivative zebularine **5** was developed in 2002 [79] and shown to function via a slightly different mechanism by forming a reversible bond with the DNMT enzymes, which exhibits very slow dissociation kinetics, however [134]. About 10-fold higher concentrations of **5** are required to reach the same treatment effect as with **1** and **2**, presumably due to less efficient metabolic activation [125]. Zebularine **5** also inhibits cytidine deaminase [135], which is involved in nucleoside catabolism and deactivates azacitidine **1** and its deoxy analogue **2** [136]. Thus, it increases the concentrations of azanucleoside triphosphates for incorporation into the DNA, the efficacy of DNA methylation inhibition, and ultimately the anticancer activity of **1** and **2** [137, 138]. Zebularine **5** is metabolized by aldehyde oxidase, and it has been shown that its activity can be increased if an inhibitor of that enzyme, e.g. raloxifene, is given in combination [139].

In terms of bioavailability, two prodrugs of decitabine **2** have been proposed (Figure 15.10). A nitrophenylethoxycarbamoyl-protected DAC (NPEOC-DAC **6**), however, showed no improvement in *in vivo* activity compared with its parent compound **2** [140]. Since azanucleosides are prodrugs themselves that require metabolic activation, this can be considered a prodrug of a prodrug. The dinucleotide SGI-110 **7** (formerly S110 and guadecitabine, Astex Pharmaceuticals) is a prodrug of decitabine, in which **2** is coupled through a phosphodiester linker to deoxyguanosine, mimicking the CpG island substructure. It is more resistant to cleavage by cytidine deaminase, and thus, longer plasma half-lives can be obtained, facilitating application regimens [141–143]. Compound **7** is currently being tested in several clinical trials against naïve and refractory acute myeloid leukemias (AMLs) (both phase III), MDS (phase III), and solid tumors such as ovarian, hepatocellular, and colorectal cancer (phase II) [144].

15.8.2 Non-nucleosidic DNMT Inhibitors

These inhibitors act as traditional small-molecule reversible binders directly on the enzyme without the necessity for incorporation into the DNA. In theory, this should shift the dual activity of reactivation of silenced genes and cytotoxicity toward the epigenetic effects, which may result in reduced side effects. Over the last decade and a half, quite a number of such molecules have been published from screening campaigns as reviewed below and in the literature [31, 96]. However, none of them have progressed into the clinic so far. One reason may be that these compounds often exhibit only limited potency *in vitro* and/or *in vivo*, which is often also difficult to independently reproduce in different assays [96], or are derived from very general natural products and assay-interfering compounds that have been reported to possess all kinds of biological activity. As discussed by Erdmann et al. in their seminal perspective article, there is also a need for better screening assays to obtain more meaningful lead structures [31, 145]. The compounds published so far (Figure 15.13) share no common pharmacophore or clear structure–activity relationship.

The phthalimide derivative RG108 (NSC401077) **8** was identified by a virtual screening approach of the NCI compound library against the structure of human DNMT1, based on a homology model to bacterial DNMTs, and is active *in vitro* against bacterial DNMTs between 10 and 100 μM and reduced DNA methylation in leukemia cells [146, 147]. In the same study, NSC303530 **9** was also identified as a somewhat weaker inhibitor (Figure 15.13).

Hydralazine **10** is a vasodilating drug used against hypertension and inhibition of DNA methyltransferases in the range of 10–20 μM has been reported in cellular assays, but not *in vitro* [148]. In addition, hypomethylation in cell culture has also been shown [149]. Moreover, the local anesthetic procaine **11a** [150] and its amide analogue procainamide **11b** [149] have been identified as DNA methyltransferase inhibitors. Constrained synthetic analogues of procaine with oxazoline or isoxazoline linkers like **12** have shown only weak activity around 500 μM on DNMT1 in biochemical and cellular assays [151] (Figure 15.13). Hybrid molecules based on the procaine template coupled to phthalimides like RG108 **8** have also been reported with activity on the murine DNMT3A/3L complex and human DNMT1 at least 50 times greater than that of the parent compounds [152].

A screening campaign using a nonradioactive *in vitro* assay identified several hits including the pesticide dichlone **13** with an IC_{50} value of 460 nM, which was able to reactivate expression of the reporter gene YFP in HEK293 cells by demethylation of its promoter [153] but contains a reactive quinone structure that makes an unspecific mode of action very likely.

Another report identified quinoline derivative SGI-1027 **14** as an inhibitor of bacterial M.*Sss*I and mammalian DNMT1, DNMT3A, and DNMT3B with IC_{50} values of around 10 μM [154]. It was initially based on quaternary quinolinium salts known to bind to the minor groove of DNA. The mode of enzyme inhibition remained unclear, as **14** was found to be noncompetitive with respect to SAM and did not interact with DNA. However, in cells, it led to selective degradation of DNMT1, leaving DNMT3A/B levels unaffected. Compound **14** induced hypomethylation and re-expression of tumor suppressor genes; however, no global hypomethylation was observed as is typically found with **1** and **2** [154]. The lead structure **14** has become the starting point for a number of chemical optimization programs elucidating the structure–activity relationship, optimizing compounds of this class, and rationalizing its mode of action [155–157]. This lead structure class probably represents the most promising family of non-nucleosidic DNMT inhibitors to date.

In addition to these small-molecule chemical inhibitors, some natural products were also identified as inhibitors of DNMTs [158]. Certain disulfide bromotyrosine derivatives, such as psammaplin A, isolated from the sponge *Pseudoceratina purpurea*, were not only found to be potent inhibitors of DNMT1 but also HDACs [159]. (−)-Epigallocatechin-3-gallate (EGCG), a major polyphenol from

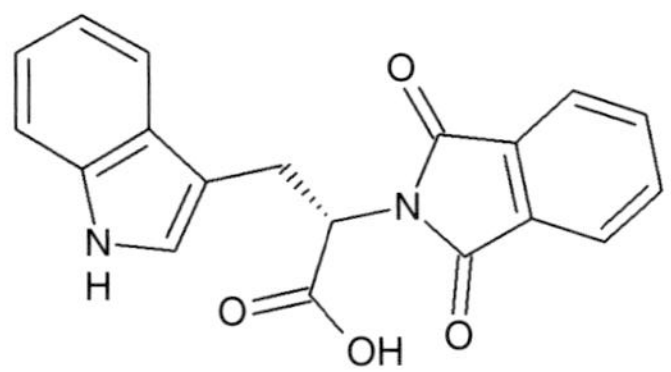

RG108 **8**

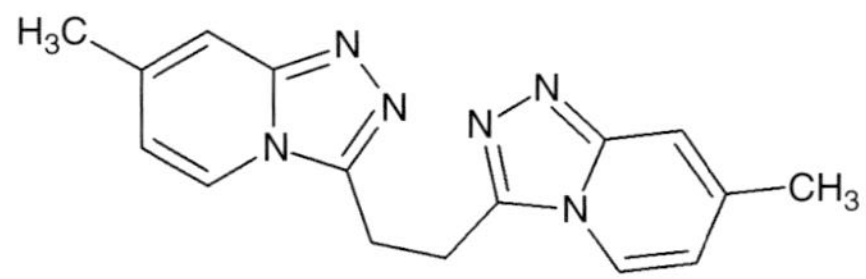

NSC303530 **9**

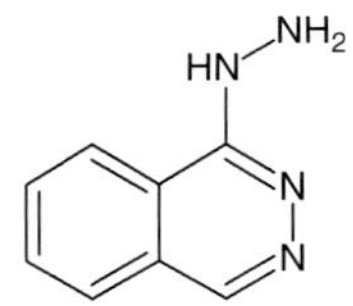

Hydralazine **10**

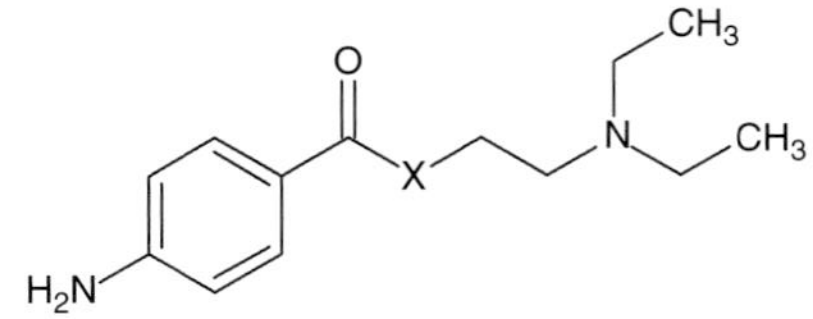

X = O procaine **11a**
X = NH procainamide **11b**

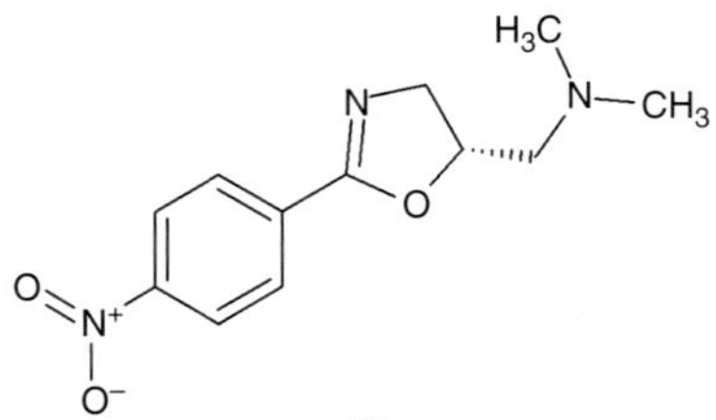

12

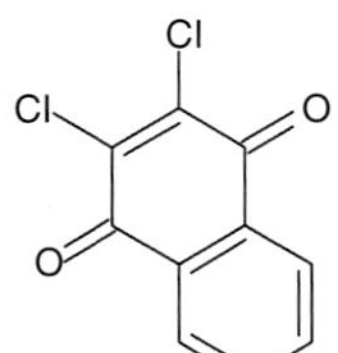

Dichlone **12**

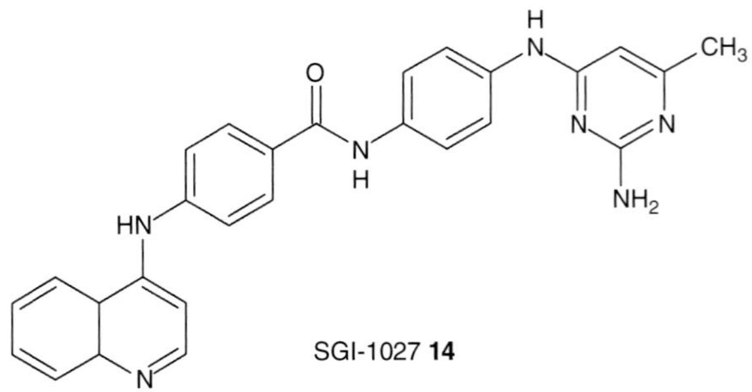

SGI-1027 **14**

Figure 15.13 Chemical structures of reported non-nucleosidic DNMT inhibitors.

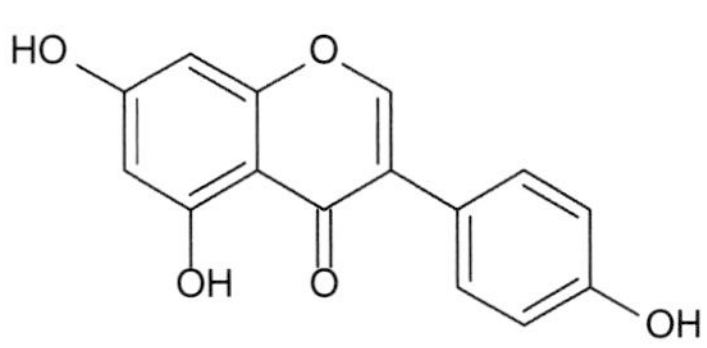

Genistein **15**

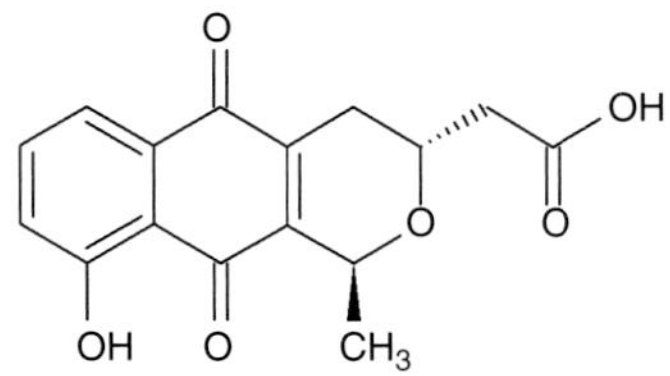

Nanaomycin A **16**

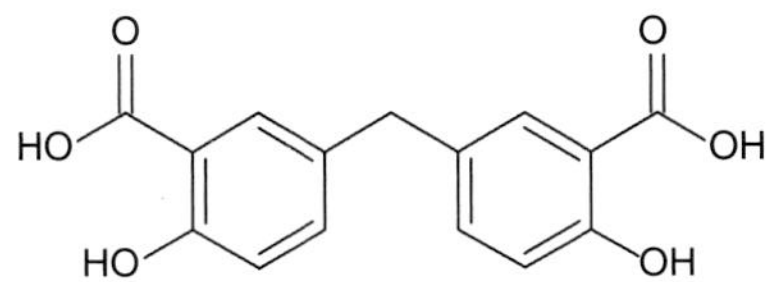

NSC14778 **17**

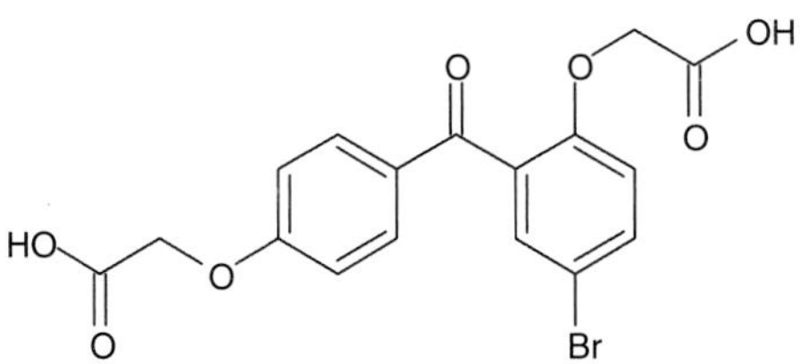

NSC106084 **15**

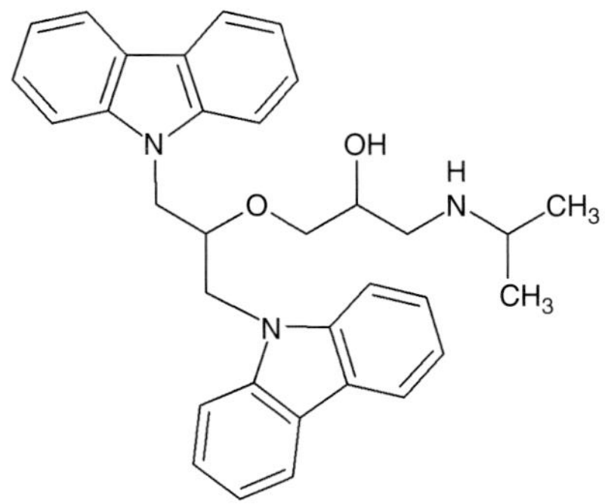

DC_517 **19**

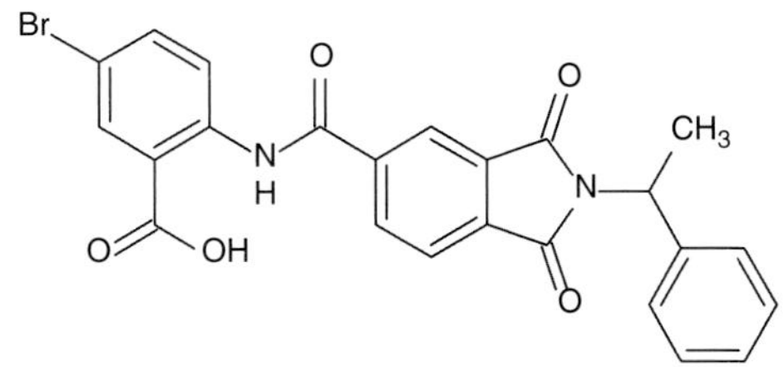

20

Figure 15.13 (*Continued*)

green tea leaves, was reported to inhibit DNA methyltransferases and to thereby be able to reactivate silenced genes like *p16*^{INK4a} and *hMLH1* in tumor cells [160]. However, EGCG is a pleiotropic inhibitor that has received a lot of attention especially for cancer chemoprevention. It remains to be determined which of its inhibitory or stimulatory activities are the most important for its net effect in biological systems [161]. Two polyphenols from coffee, caffeic acid, and chlorogenic acid have also been reported to inhibit DNMT1 in the low micromolar range [162], and also polyphenols from apples have been described to have a similar activity [163]. An isoflavone from soy beans, genistein **15**, inhibits methylation of DNA and can lead to re-expression of methylation-silenced genes, such as *RARβ*, *p16*^{INK4}, and *MGMT* in the esophageal squamous cell carcinoma cell line KYSE-510 [164] (Figure 15.13). As histone acetylation levels are also known to have an impact on the re-expression of these genes, the weak inhibition of HDAC activity observed after treatment of the cells with genistein **15** possibly contributes to the gene reactivation as well. Biochanin A and daidzein, two other isoflavones, are weaker inhibitors of DNMTs and also less effective in reactivating the *RARβ* gene. Thus, a direct correlation between the inhibition of DNA methyltransferases and the reactivation of the silenced genes was assumed [164].

A further natural product inhibitor is mithramycin A, a structurally complex anticancer antibiotic. Mithramycin A is a member of a group of aureolic acid-type polyketides that are produced by the soil bacterium *Streptomyces argillaceus* [165]. Mithramycin A binds to GC-rich DNA sequences and blocks DNMT methylation activity. Docking of mithramycin A into the DNMT1 catalytic domain indicated that the trisaccharide moiety of the inhibitor could fit into the putative cytosine pocket and the aglycon core is bound between the two arms that fasten up the hemimethylated DNA. It has also been reported to deplete lung cancer cells of DNMT1, possibly by proteasomal degradation of the enzyme.

Furthermore, the microbial antibiotic nanaomycin A **16** was found to selectively inhibit DNMT3B, leading to DNA demethylation in cells and reactivation of the tumor suppressor gene *RASSF1A* [166].

The insect-derived natural product laccaic acid A was discovered as a DNMT1 inhibitor in an *in vitro* high-throughput screening campaign using a highly active truncated variant of DNMT1 lacking the auto-inhibitory domain. It competes with the DNA substrate with a K_i of 0.3 μM and specifically induced expression of tumor suppressor genes in MCF7 breast cancer cells at high concentrations [167].

A very different approach to therapeutically downregulate DNA methylation is represented by the DNMT1 antisense oligonucleotide MG98, which has been used in a number of clinical studies, however, so far with only limited success [168–172].

Lastly, the recently growing availability of high-quality structural data for human DNMTs from X-ray crystallography (cf. Section 15.4) has enabled the use of virtual screening methods for the identification of novel lead structures. This has led to the identification of the structurally unrelated hit compounds NSC14778 **17** and NSC106084 **18**, which were identified by docking-based virtual screening of the NCI database and their activity confirmed in *in vitro* assays [74]. Compound **17** was active at around 20 μM and exhibited fivefold selectively for DNMT3B over DNMT1 *in vitro*, while **18** was selective for DNMT1 but also much less potent. The binding mode from docking suggests substrate-competitive binding of **17** in the deoxycytidine binding pocket. Cellular effects were not disclosed, however [74].

More recently, Chen et al. used a docking approach to virtually screen the SPECS database containing ~200 000 compounds for DNMT1 inhibitors [75]. Two hits active in the low micromolar range were identified, which were used for further chemical optimization. The most potent DNMT1 inhibitor DC_517 **19** showed an IC_{50} value of 1.7 μM in a radioactive enzyme activity assay and a K_d value of 0.9 μM in a surface plasmon resonance binding experiment. Even though the inhibitor is postulated to block the SAM cofactor site of DNMT1, it has shown remarkable selectivity for this enzyme over other SAM-dependent protein methyltransferases (DNMT3A, DNMT3B, G9a, SUV39H1, MLL1, SET7/9, and PRMT1). Compound **19** was shown to induce dose-dependent apoptotic cell death in HCT116 cells [75].

The same group recently also reported on the virtual screening for DNMT3A applying a similar approach as in the case of DNMT1 [76]. Virtual screening of the SPECS database resulted in a DNMT3A inhibitor of the lead structure **20** (Figure 15.13) with an IC_{50} value of 46.5 μM in an *in vitro* radioactive methylation assay. A number of derivatives were synthesized and also tested. The hits were further tested in different tumor cell lines and showed growth inhibition at around 100 μM.

15.9 Therapeutic Applications of DNMT Inhibitors

The two DNMT inhibitors licensed for human use, azacitidine **1** and decitabine **2**, have been approved for the treatment of MDS [173, 174]. MDS summarizes a set of different conditions that affect the maturation of blood cells. It is a group of bone marrow stem cell malignancies that have a pathogenic overlap with AML and show peripheral blood cytopenias and, in more advanced subtypes, varied degrees of maturation arrest [175]. Both drugs are approved for all subtypes of MDS. Response rates are usually around 30%. The question, whether the clinical benefit results more from epigenetic effects and reactivation of silenced maturation factors or more from cytotoxic effects on the immature hyperproliferative cells, remains open.

AML of older, medically nonfit patients is a major therapeutic challenge, since aggressive, standard chemotherapy usually is much too toxic and thus not feasible for these patients. Therefore the recent approval of decitabine **2** and azacitidine **1** as the first standard first-line treatment for these patients constituted a breakthrough. As demonstrated in the pivotal studies [176, 177], the treatment was overall well tolerated and provided a survival benefit over conventional treatment that was not dramatic but nonetheless clinically meaningful. Particularly the activity of these hypomethylating agents (HMAs) even in patients with the most adverse genetics, i.e. those with multiple chromosomal monosomies

[178–180], provides clear proof of an underlying mechanism of action that is different from low-dose cytarabine or other conventional treatments.

Further development of single-agent DNA-HMA treatment is focused on the dinucleotide guadecitabine **7**, a molecule showing more favorable pharmacokinetics than its parent compound decitabine [141], and is currently in phase III testing in newly diagnosed older AML/MDS patients. Combination therapy approaches building on approved HMAs are plentiful, with many rational combinations between HMAs and other drugs such as histone deacetylase inhibitors (HDACis) of different classes and structures, inhibitors of histone demethylases and methyltransferases, and differentiation-inducing drugs (e.g. retinoic acid), chemotherapeutic agents, and immune checkpoint inhibitors, to name just a few. In solid tumors, HMAs have until now only limited activity, here also prompting combination therapies and biomarker development in order to identify subgroups of patients that may show a superior response [181].

Several studies have shown the ability of DNMT inhibitors to prevent cancer using different pathways [182–185]. Genistein **15** demonstrated cancer chemopreventive activities in animal models [186]. This has been studied in different cancer cell lines, and inhibition of cancer cell growth could be shown [187]. However, several mechanisms, such as anti-hormonal effects, inhibition of tyrosine kinases, and lack of selectivity of **15** for DNMTs, are discussed to explain this activity, and the exact contribution of DNMT inhibition to *in vivo* activity in cancer chemoprevention remains to be determined.

Many *in vitro* investigations have shown a cooperation of DNA methylation and histone deacetylation in the repression of transcription [90, 188, 189]. Thus, it seems logical that the combination of DNMT and HDAC inhibition has been shown to induce differentiation, apoptosis, and growth arrest in cancer cell lines [190]. The re-expression of hypermethylated genes such as *MLH1, TIMP3, CDKN2B,* and *CDKN2A* was increased when HDAC inhibitors were combined with inhibitors of DNA methyltransferases. HDAC inhibition alone, on the contrary, was not able to reactivate transcription [90]. The silencing of the *COX-2* gene, whose promoter was shown to be hypermethylated in gastric cancer, can be relieved with a synergistic treatment of DNMT and HDAC inhibitors [191]. More connections and hence potential for combined therapeutic approaches between inhibitors targeting histone-modifying enzymes and DNA methyltransferase inhibitors are becoming evident [192, 193]. These observations now form the basis for the more advanced combination therapies in epigenetics [11, 14, 91, 95].

15.10 Conclusion

Besides modifications to amino acid residues in histone tails, chemical modifications to DNA bases themselves are one cornerstone of epigenetic regulation. While the biological role of the enzymes involved in cytosine 5-methylation is now relatively well understood, this is much less the case for the more complex and rarer oxidized cytosine derivatives and other base methylations. DNA methyltransferases have been recognized as vital, but challenging, drug targets, and this has spurred an interest in the development of potent inhibitors of these enzymes.

Two DNMT inhibitors have been approved for clinical use, azacitidine (5-azacytidine) **1** and decitabine **2**, which are both nucleoside mimics and rely on incorporation into the DNA. Consequently, they suffer from poor pharmacokinetic properties and metabolic stability. Attempts at overcoming this by prodrug strategies and better oral formulations are ongoing. As for non-nucleosidic small-molecule inhibitors, some literature has been published in recent years. However, potent and selective high-quality probes are still elusive as the published compounds are often natural products with additional *in vivo* effects, not very potent, or lack rigorous screening controls. It has even been suggested that DNMTs may generally be "poorly druggable" by small molecules [31].

As a result, the field has mostly shifted from the discovery of new lead structures to creating innovative applications for the ones already known. This relates to combination therapies of DNMT inhibitors with HDAC inhibitors, with inhibitors of other epigenetic modifiers, or with other cytotoxic chemotherapeutics [11]. Another future avenue is to use DNMT inhibitors to prime patients for immune checkpoint therapy, a novel paradigm in cancer treatment, which has already shown some promising results [194–196]. Moreover, the applicability of the DNMT inhibitors to other diseases than the ones they were originally licensed for is constantly being expanded in clinical trials.

Acknowledgment

M. R. and M. J. thank the Deutsche Forschungsgemeinschaft for funding within CRC992 (Project A04).

References

1 Kouzarides, T. (2007). Chromatin modifications and their function. *Cell* 128: 693–705.
2 Allis, C.D., Jenuwein, T., and Reinberg, D. (2007). Overview and concepts. In: *Epigenetics* (ed. C.D. Allis, T. Jenuwein and D. Reinberg). New York, NY: Cold Spring Harbor Laboratory Press.
3 Jenuwein, T. and Allis, C.D. (2001). Translating the histone code. *Science* 293: 1074–1080.
4 Berger, S.L. (2007). The complex language of chromatin regulation during transcription. *Nature* 447: 407–412.
5 Su, Z. and Denu, J.M. (2016). Reading the combinatorial histone language. *ACS Chem. Biol.* 11: 564–574.
6 Arrowsmith, C.H., Bountra, C., Fish, P.V. et al. (2012). Epigenetic protein families: a new frontier for drug discovery. *Nat. Rev. Drug Discovery* 11: 384–400.

7 Huston, A., Arrowsmith, C.H., Knapp, S., and Schapira, M. (2015). Probing the epigenome. *Nat. Chem. Biol.* 11: 542–545.

8 Chi, P., Allis, C.D., and Wang, G.G. (2010). Covalent histone modifications – miswritten, misinterpreted and mis-erased in human cancers. *Nat. Rev. Cancer* 10: 457–469.

9 Hoffmann, I., Roatsch, M., Schmitt, M.L. et al. (2012). The role of histone demethylases in cancer therapy. *Mol. Oncol.* 6: 683–703.

10 Carey, N. (ed.) (2016). *Epigenetics for Drug Discovery*. Cambridge, UK: The Royal Society of Chemistry.

11 Jones, P.A., Issa, J.P., and Baylin, S. (2016). Targeting the cancer epigenome for therapy. *Nat. Rev. Genet.* 17: 630–641.

12 Nebbioso, A., Carafa, V., Benedetti, R., and Altucci, L. (2012). Trials with 'epigenetic' drugs: an update. *Mol. Oncol.* 6: 657–682.

13 Morera, L., Lübbert, M., and Jung, M. (2016). Targeting histone methyltransferases and demethylases in clinical trials for cancer therapy. *Clin. Epigenetics* 8: 57.

14 Schiffmann, I., Greve, G., Jung, M., and Lübbert, M. (2016). Epigenetic therapy approaches in non-small cell lung cancer: update and perspectives. *Epigenetics* 11: 858–870.

15 Bestor, T.H. (2000). The DNA methyltransferases of mammals. *Hum. Mol. Genet.* 9: 2395–2402.

16 Yoo, C.B., Valente, R., Congiatu, C. et al. (2008). Activation of *p16* gene silenced by DNA methylation in cancer cells by phosphoramidate derivatives of 2′-deoxyzebularine. *J. Med. Chem.* 51: 7593–7601.

17 Wolffe, A.P. and Matzke, M.A. (1999). Epigenetics: regulation through repression. *Science* 286: 481–486.

18 Lopez-Serra, L. and Esteller, M. (2008). Proteins that bind methylated DNA and human cancer: reading the wrong words. *Br. J. Cancer* 98: 1881–1885.

19 Schmutte, C. and Jones, P.A. (1998). Involvement of DNA methylation in human carcinogenesis. *Biol. Chem.* 379: 377–388.

20 Morgan, H.D., Dean, W., Coker, H.A. et al. (2004). Activation-induced cytidine deaminase deaminates 5-methylcytosine in DNA and is expressed in pluripotent tissues: implications for epigenetic reprogramming. *J. Biol. Chem.* 279: 52353–52360.

21 Petersen-Mahrt, S. (2005). DNA deamination in immunity. *Immunol. Rev.* 203: 80–97.

22 Bird, A. (2002). DNA methylation patterns and epigenetic memory. *Genes Dev.* 16: 6–21.

23 McCabe, D.C. and Caudill, M.A. (2005). DNA methylation, genomic silencing, and links to nutrition and cancer. *Nutr. Rev.* 63: 183–195.

24 Ehrlich, M., Gama-Sosa, M.A., Huang, L.-H. et al. (1982). Amount and distribution of 5-methylcytosine in human DNA from different types of tissues of cells. *Nucleic Acids Res.* 10: 2709–2721.

25 Gardiner-Garden, M. and Frommer, M. (1987). CpG islands in vertebrate genomes. *J. Mol. Biol.* 196: 261–282.

26 Takai, D. and Jones, P.A. (2002). Comprehensive analysis of CpG islands in human chromosomes 21 and 22. *Proc. Natl. Acad. Sci. U.S.A.* 99: 3740–3745.
27 Dunn, B.K. (2003). Hypomethylation: one side of a larger picture. *Ann. N.Y. Acad. Sci.* 983: 28–42.
28 Ooi, S.K.T. and Bestor, T.H. (2008). The colorful history of active DNA demethylation. *Cell* 133: 1145–1148.
29 Métivier, R., Gallais, R., Tiffoche, C. et al. (2008). Cyclical DNA methylation of a transcriptionally active promoter. *Nature* 452: 45–50.
30 Kangaspeska, S., Stride, B., Métivier, R. et al. (2008). Transient cyclical methylation of promoter DNA. *Nature* 452: 112–115.
31 Erdmann, A., Halby, L., Fahy, J., and Arimondo, P.B. (2015). Targeting DNA methylation with small molecules: what's next? *J. Med. Chem.* 58: 2569–2583.
32 Tahiliani, M., Koh, K.P., Shen, Y. et al. (2009). Conversion of 5-methylcytosine to 5-hydroxymethylcytosine in mammalian DNA by MLL partner TET1. *Science* 324: 930–935.
33 Tan, L. and Shi, Y.G. (2012). Tet family proteins and 5-hydroxymethylcytosine in development and disease. *Development* 139: 1895–1902.
34 Hausinger, R.P. (2004). Fe(II)/α-ketoglutarate-dependent hydroxylases and related enzymes. *Crit. Rev. Biochem. Mol. Biol.* 39: 21–68.
35 Franchini, D.-M., Schmitz, K.-M., and Petersen-Mahrt, S.K. (2012). 5-Methylcytosine DNA demethylation: more than losing a methyl group. *Annu. Rev. Genet.* 46: 419–441.
36 Maiti, A. and Drohat, A.C. (2011). Thymine DNA glycosylase can rapidly excise 5-formylcytosine and 5-carboxylcytosine: potential implications for active demethylation of CpG sites. *J. Biol. Chem.* 286: 35334–35338.
37 Hu, X., Zhang, L., Mao, S.-Q. et al. (2014). Tet and TDG mediate DNA demethylation essential for mesenchymal-to-epithelial transition in somatic cell reprogramming. *Cell Stem Cell* 14: 512–522.
38 Pfaffeneder, T., Spada, F., Wagner, M. et al. (2014). Tet oxidizes thymine to 5-hydroxymethyluracil in mouse embryonic stem cell DNA. *Nat. Chem. Biol.* 10: 574–581.
39 Hu, L., Li, Z., Cheng, J. et al. (2013). Crystal structure of TET2-DNA complex: insight into TET-mediated 5mC oxidation. *Cell* 155: 1545–1555.
40 Hauser, A.-T., Roatsch, M., Schulz-Fincke, J. et al. (2015). Discovery of histone demethylase inhibitors. In: *Epigenetic Technological Applications* (ed. Y.G. Zheng). Amsterdam: Academic Press.
41 McAllister, T.E., England, K.S., Hopkinson, R.J. et al. (2016). Recent progress in histone demethylase inhibitors. *J. Med. Chem.* 59: 1308–1329.
42 Kaniskan, H.Ü., Martini, M.L., and Jin, J. (2017). Inhibitors of protein methyltransferases and demethylases. *Chem. Rev.* doi: 10.1021/acs.chemrev.1026b00801.
43 Okano, M., Xie, S., and Li, E. (1998). Cloning and characterization of a family of novel mammalian DNA (cytosine-5) methyltransferases. *Nat. Genet.* 19: 219–220.

44 Bestor, T., Laudano, A., Mattaliano, R., and Ingram, V. (1988). Cloning and sequencing of a cDNA encoding DNA methyltransferase of mouse cells. The carboxyl-terminal domain of the mammalian enzymes is related to bacterial restriction methyltransferases. *J. Mol. Biol.* 203: 971–983.
45 Yoder, J.A. and Bestor, T.H. (1998). A candidate mammalian DNA methyltransferase related to pmt1p of fission yeast. *Hum. Mol. Genet.* 7: 279–284.
46 Goll, M.G. and Bestor, T.H. (2005). Eukaryotic cytosine methyltransferases. *Annu. Rev. Biochem.* 74: 481–514.
47 Aapola, U., Shibuya, K., Scott, H.S. et al. (2000). Isolation and initial characterization of a novel zinc finger gene, *DNMT3L*, on 21q22.3, related to the cytosine-5-methyltransferase 3 gene family. *Genomics* 65: 293–298.
48 Deplus, R., Brenner, C., Burgers, W.A. et al. (2002). Dnmt3L is a transcriptional repressor that recruits histone deacetylase. *Nucleic Acids Res.* 30: 3831–3838.
49 Goll, M.G., Kirpekar, F., Maggert, K.A. et al. (2006). Methylation of $tRNA^{Asp}$ by the DNA methyltransferase homolog Dnmt2. *Science* 311: 395–398.
50 Barau, J., Teissandier, A., Zamudio, N. et al. (2016). The DNA methyltransferase DNMT3C protects male germ cells from transposon activity. *Science* 354: 909–912.
51 Pósfai, J., Bhagwat, A.S., Pósfai, G., and Roberts, R.J. (1989). Predictive motifs derived from cytosine methyltransferases. *Nucleic Acids Res.* 17: 2421–2435.
52 Jeltsch, A. (2002). Beyond Watson and Crick: DNA methylation and molecular enzymology of DNA methyltransferases. *ChemBioChem* 3: 274–293.
53 Lauster, R., Trautner, T.A., and Noyer-Weidner, M. (1989). Cytosine-specific type II DNA methyltransferases. A conserved enzyme core with variable target-recognizing domains. *J. Mol. Biol.* 206: 305–312.
54 Trautner, T.A., Balganesh, T.S., and Pawlek, B. (1988). Chimeric multispecific DNA methyltransferases with novel combinations of target recognition. *Nucleic Acids Res.* 16: 6649–6658.
55 Wilke, K., Rauhut, E., Noyer-Weidner, M. et al. (1988). Sequential order of target-recognizing domains in multispecific DNA-methyltransferases. *EMBO J.* 7: 2601–2609.
56 Cheng, X., Kumar, S., Posfai, J. et al. (1993). Crystal structure of the HhaI DNA methyltransferase complexed with S-adenosyl-L-methionine. *Cell* 74: 299–307.
57 O'Gara, M., Klimašauskas, S., Roberts, R.J., and Cheng, X. (1996). Enzymatic C5-cytosine methylation of DNA: mechanistic implications of new crystal structures for *Hha*I methyltransferase-DNA-AdoHcy complexes. *J. Mol. Biol.* 261: 634–645.
58 O'Gara, M., Horton, J.R., Roberts, R.J., and Cheng, X. (1998). Structures of *Hha*I methyltransferase complexed with substrates containing mismatches at the target base. *Nat. Struct. Biol.* 5: 872–877.
59 Reinisch, K.M., Chen, L., Verdine, G.L., and Lipscomb, W.N. (1995). The crystal structure of HaeIII methyltransferase covalently complexed to DNA: an extrahelical cytosine and rearranged base pairing. *Cell* 82: 143–153.

60 Tajima, S., Suetake, I., Takeshita, K. et al. (2016). Domain structure of the Dnmt1, Dnmt3a, and Dnmt3b DNA methyltransferases. In: *DNA Methyltransferases – Role and Function* (ed. A. Jeltsch and R.Z. Jurkowska). Cham, Switzerland: Springer International.
61 Cheng, X. and Blumenthal, R.M. (2008). Mammalian DNA methyltransferases: a structural perspective. *Structure* 16: 341–350.
62 Dong, A., Yoder, J.A., Zhang, X. et al. (2001). Structure of human DNMT2, an enigmatic DNA methyltransferase homolog that displays denaturant-resistant binding to DNA. *Nucleic Acids Res.* 29: 439–448.
63 Jia, D., Jurkowska, R.Z., Zhang, X. et al. (2007). Structure of Dnmt3a bound to Dnmt3L suggests a model for *de novo* DNA methylation. *Nature* 449: 248–251.
64 Guo, X., Wang, L., Li, J. et al. (2015). Structural insight into autoinhibition and histone H3-induced activation of DNMT3A. *Nature* 517: 640–644.
65 Song, J., Rechkoblit, O., Bestor, T.H., and Patel, D.J. (2011). Structure of DNMT1-DNA complex reveals a role for autoinhibition in maintenance DNA methylation. *Science* 331: 1036–1040.
66 Song, J., Teplova, M., Ishibe-Murakami, S., and Patel, D.J. (2012). Structure-based mechanistic insights into DNMT1-mediated maintenance DNA methylation. *Science* 335: 709–712.
67 Cheng, J., Yang, H., Fang, J. et al. (2015). Molecular mechanism for USP7-mediated DNMT1 stabilization by acetylation. *Nat. Commun.* 6: 7023.
68 Zhang, Z.-M., Liu, S., Lin, K. et al. (2015). Crystal structure of human DNA methyltransferase 1. *J. Mol. Biol.* 427: 2520–2531.
69 Ooi, S.K.T., Qiu, C., Bernstein, E. et al. (2007). DNMT3L connects unmethylated lysine 4 of histone H3 to *de novo* methylation of DNA. *Nature* 448: 714–717.
70 Wu, H., Zeng, H., Lam, R. et al. (2011). Structural and histone binding ability characterizations of human PWWP domains. *PLoS One* 6: e18919.
71 Qiu, C., Sawada, K., Zhang, X., and Cheng, X. (2002). The PWWP domain of mammalian DNA methyltransferase Dnmt3b defines a new family of DNA-binding folds. *Nat. Struct. Biol.* 9: 217–224.
72 Otani, J., Nankumo, T., Arita, K. et al. (2009). Structural basis for recognition of H3K4 methylation status by the DNA methyltransferase 3A ATRX-DNMT3-DNMT3L domain. *EMBO Rep.* 10: 1235–1241.
73 Syeda, F., Fagan, R.L., Wean, M. et al. (2011). The replication focus targeting sequence (RFTS) domain is a DNA-competitive inhibitor of Dnmt1. *J. Biol. Chem.* 286: 15344–15351.
74 Kuck, D., Singh, N., Lyko, F., and Medina-Franco, J.L. (2010). Novel and selective DNA methyltransferase inhibitors: docking-based virtual screening and experimental evaluation. *Bioorg. Med. Chem.* 18: 822–829.
75 Chen, S., Wang, Y., Zhou, W. et al. (2014). Identifying novel selective non-nucleoside DNA methyltransferase 1 inhibitors through docking-based virtual screening. *J. Med. Chem.* 57: 9028–9041.

76 Shao, Z., Xu, P., Xu, W. et al. (2017). Discovery of novel DNA methyltransferase 3A inhibitors via structure-based virtual screening and biological assays. *Bioorg. Med. Chem. Lett.* 27: 342–346.
77 Vilkaitis, G., Merkienė, E., Serva, S. et al. (2001). The mechanism of DNA cytosine-5 methylation. Kinetic and mutational dissection of *Hha*I methyltransferase. *J. Biol. Chem.* 276: 20924–20934.
78 Klimasauskas, S., Kumar, S., Roberts, R.J., and Cheng, X. (1994). HhaI methyltransferase flips its target base out of the DNA helix. *Cell* 76: 357–369.
79 Zhou, L., Cheng, X., Connolly, B.A. et al. (2002). Zebularine: a novel DNA methylation inhibitor that forms a covalent complex with DNA methyltransferases. *J. Mol. Biol.* 321: 591–599.
80 Brannan, C.I. and Bartolomei, M.S. (1999). Mechanisms of genomic imprinting. *Curr. Opin. Genet. Dev.* 9: 164–170.
81 Hansen, R.S. (2003). X inactivation-specific methylation of LINE-1 elements by DNMT3B: implications for the Lyon repeat hypothesis. *Hum. Mol. Genet.* 12: 2559–2567.
82 Watt, F. and Molloy, P.L. (1988). Cytosine methylation prevents binding to DNA of a HeLa cell transcription factor required for optimal expression of the adenovirus major late promoter. *Genes Dev.* 2: 1136–1143.
83 Wade, P.A. (2001). Methyl CpG binding proteins: coupling chromatin architecture to gene regulation. *Oncogene* 20: 3166–3173.
84 Lopez-Serra, L., Ballestar, E., Ropero, S. et al. (2008). Unmasking of epigenetically silenced candidate tumor suppressor genes by removal of methyl-CpG-binding domain proteins. *Oncogene* 27: 3556–3566.
85 Fahrner, J.A., Eguchi, S., Herman, J.G., and Baylin, S.B. (2002). Dependence of histone modifications and gene expression on DNA hypermethylation in cancer. *Cancer Res.* 62: 7213–7218.
86 Kondo, Y., Shen, L., and Issa, J.-P.J. (2003). Critical role of histone methylation in tumor suppressor gene silencing in colorectal cancer. *Mol. Cell. Biol.* 23: 206–215.
87 Griffiths, E.A. and Gore, S.D. (2008). DNA methyltransferase and histone deacetylase inhibitors in the treatment of myelodysplastic syndromes. *Semin. Hematol.* 45: 23–30.
88 Walton, T.J., Li, G., Seth, R. et al. (2008). DNA demethylation and histone deacetylation inhibition co-operate to re-express estrogen receptor beta and induce apoptosis in prostate cancer cell-lines. *Prostate* 68: 210–222.
89 Venturelli, S., Armeanu, S., Pathil, A. et al. (2007). Epigenetic combination therapy as a tumor-selective treatment approach for hepatocellular carcinoma. *Cancer* 109: 2132–2141.
90 Cameron, E.E., Bachman, K.E., Myöhänen, S. et al. (1999). Synergy of demethylation and histone deacetylase inhibition in the re-expression of genes silenced in cancer. *Nat. Genet.* 21: 103–107.
91 Azad, N., Zahnow, C.A., Rudin, C.M., and Baylin, S.B. (2013). The future of epigenetic therapy in solid tumours – lessons from the past. *Nat. Rev. Clin. Oncol.* 10: 256–266.

92 Grishina, O., Schmoor, C., Döhner, K. et al. (2015). DECIDER: prospective randomized multicenter phase II trial of low-dose decitabine (DAC) administered alone or in combination with the histone deacetylase inhibitor valproic acid (VPA) and all-*trans* retinoic acid (ATRA) in patients >60 years with acute myeloid leukemia who are ineligible for induction chemotherapy. *BMC Cancer* 15: 430.

93 Silverman, L.R., Verma, A., Odchimar-Reissig, R. et al. (2008). A phase I trial of the epigenetic modulators Vorinostat, in combination with azacitidine (azaC) in patients with the myelodysplastic syndrome (MDS) and acute myeloid leukemia (AML): a study of the New York cancer consortium. *Blood* 112: 1252–1252.

94 Prebet, T., Sun, Z., Figueroa, M.E. et al. (2014). Prolonged administration of azacitidine with or without entinostat for myelodysplastic syndrome and acute myeloid leukemia with myelodysplasia-related changes: results of the US leukemia intergroup trial E1905. *J. Clin. Oncol.* 32: 1242–1248.

95 Juergens, R.A., Wrangle, J., Vendetti, F.P. et al. (2011). Combination epigenetic therapy has efficacy in patients with refractory advanced non-small cell lung cancer. *Cancer Discov.* 1: 598–607.

96 Heightman, T.D. and McCullar, M. (2016). Targeting DNA methylation. In: *Epigenetics for Drug Discovery* (ed. N. Carey). Cambridge, UK: The Royal Society of Chemistry.

97 Li, E., Bestor, T.H., and Jaenisch, R. (1992). Targeted mutation of the DNA methyltransferase gene results in embryonic lethality. *Cell* 69: 915–926.

98 Ateeq, B., Unterberger, A., Szyf, M., and Rabbani, S.A. (2008). Pharmacological inhibition of DNA methylation induces proinvasive and prometastatic genes *in vitro* and *in vivo*. *Neoplasia* 10: 266–278.

99 Choi, J.K. and Kim, S.C. (2007). Environmental effects on gene expression phenotype have regional biases in the human genome. *Genetics* 175: 1607–1613.

100 Weidman, J.R., Dolinoy, D.C., Murphy, S.K., and Jirtle, R.L. (2007). Cancer susceptibility: epigenetic manifestation of environmental exposures. *Cancer J.* 13: 9–16.

101 Jones, P.A. and Baylin, S.B. (2007). The epigenomics of cancer. *Cell* 128: 683–692.

102 Kuratomi, G., Iwamoto, K., Bundo, M. et al. (2008). Aberrant DNA methylation associated with bipolar disorder identified from discordant monozygotic twins. *Mol. Psychiatry* 13: 429–441.

103 Kaminsky, Z., Petronis, A., Wang, S.-C. et al. (2008). Epigenetics of personality traits: an illustrative study of identical twins discordant for risk-taking behavior. *Twin Res. Hum. Genet.* 11: 1–11.

104 Walton, E., Liu, J., Hass, J. et al. (2014). *MB-COMT* promoter DNA methylation is associated with working-memory processing in schizophrenia patients and healthy controls. *Epigenetics* 9: 1101–1107.

105 Fraga, M.F., Ballestar, E., Paz, M.F. et al. (2005). Epigenetic differences arise during the lifetime of monozygotic twins. *Proc. Natl. Acad. Sci. U.S.A.* 102: 10604–10609.

106 Esteller, M. (2006). The necessity of a human epigenome project. *Carcinogenesis* 27: 1121–1125.

107 Wang, S.S., Smiraglia, D.J., Wu, Y.-Z. et al. (2008). Identification of novel methylation markers in cervical cancer using restriction landmark genomic scanning. *Cancer Res.* 68: 2489–2497.

108 Yu, L., Liu, C., Vandeusen, J. et al. (2005). Global assessment of promoter methylation in a mouse model of cancer identifies *ID4* as a putative tumor-suppressor gene in human leukemia. *Nat. Genet.* 37: 265–274.

109 Mulero-Navarro, S. and Esteller, M. (2008). Epigenetic biomarkers for human cancer: the time is now. *Crit. Rev. Oncol. Hematol.* 68: 1–11.

110 Baylin, S.B. and Herman, J.G. (2000). DNA hypermethylation in tumorigenesis: epigenetics joins genetics. *Trends Genet.* 16: 168–174.

111 Costello, J.F. and Plass, C. (2001). Methylation matters. *J. Med. Genet.* 38: 285–303.

112 Esteller, M. (2002). CpG island hypermethylation and tumor suppressor genes: a booming present, a brighter future. *Oncogene* 21: 5427–5440.

113 Butcher, D.T. and Rodenhiser, D.I. (2007). Epigenetic inactivation of *BRCA1* is associated with aberrant expression of *CTCF* and DNA methyltransferase (*DNMT3B*) in some sporadic breast tumours. *Eur. J. Cancer* 43: 210–219.

114 Lin, R.-K., Hsu, H.-S., Chang, J.-W. et al. (2007). Alteration of DNA methyltransferases contributes to 5′CpG methylation and poor prognosis in lung cancer. *Lung Cancer* 55: 205–213.

115 Liu, S., Shen, T., Huynh, L. et al. (2005). Interplay of RUNX1/MTG8 and DNA methyltransferase 1 in acute myeloid leukemia. *Cancer Res.* 65: 1277–1284.

116 Ostler, K.R., Davis, E.M., Payne, S.L. et al. (2007). Cancer cells express aberrant *DNMT3B* transcripts encoding truncated proteins. *Oncogene* 26: 5553–5563.

117 Sproul, D., Kitchen, R.R., Nestor, C.E. et al. (2012). Tissue of origin determines cancer-associated CpG island promoter hypermethylation patterns. *Genome Biol.* 13: R84.

118 Esteller, M. (2008). Epigenetics in cancer. *N. Engl. J. Med.* 358: 1148–1159.

119 Lujambio, A., Ropero, S., Ballestar, E. et al. (2007). Genetic unmasking of an epigenetically silenced microRNA in human cancer cells. *Cancer Res.* 67: 1424–1429.

120 Zeller, C. and Brown, R. (2010). Therapeutic modulation of epigenetic drivers of drug resistance in ovarian cancer. *Ther. Adv. Med. Oncol.* 2: 319–329.

121 Christman, J.K. (2002). 5-Azacytidine and 5-aza-2′-deoxycytidine as inhibitors of DNA methylation: mechanistic studies and their implications for cancer therapy. *Oncogene* 21: 5483–5495.

122 Jones, P.A. and Taylor, S.M. (1980). Cellular differentiation, cytidine analogs and DNA methylation. *Cell* 20: 85–93.

123 Stresemann, C., Brueckner, B., Musch, T. et al. (2006). Functional diversity of DNA methyltransferase inhibitors in human cancer cell lines. *Cancer Res.* 66: 2794–2800.

124 Plagemann, P.G.W., Behrens, M., and Abraham, D. (1978). Metabolism and cytotoxicity of 5-azacytidine in cultured Novikoff rat hepatoma and P388 mouse leukemia cells and their enhancement by preincubation with pyrazofurin. *Cancer Res.* 38: 2458–2466.

125 Brueckner, B., Kuck, D., and Lyko, F. (2007). DNA methyltransferase inhibitors for cancer therapy. *Cancer J.* 13: 17–22.

126 Santi, D.V., Garrett, C.E., and Barr, P.J. (1983). On the mechanism of inhibition of DNA-cytosine methyltransferases by cytosine analogs. *Cell* 33: 9–10.

127 Santi, D.V., Norment, A., and Garrett, C.E. (1984). Covalent bond formation between a DNA-cytosine methyltransferase and DNA containing 5-azacytosine. *Proc. Natl. Acad. Sci. U.S.A.* 81: 6993–6997.

128 Stresemann, C. and Lyko, F. (2008). Modes of action of the DNA methyltransferase inhibitors azacytidine and decitabine. *Int. J. Cancer* 123: 8–13.

129 Derissen, E.J.B., Beijnen, J.H., and Schellens, J.H.M. (2013). Concise drug review: azacitidine and decitabine. *Oncologist* 18: 619–624.

130 Karahoca, M. and Momparler, R.L. (2013). Pharmacokinetic and pharmacodynamic analysis of 5-aza-2′-deoxycytidine (decitabine) in the design of its dose-schedule for cancer therapy. *Clin. Epigenetics* 5: 3.

131 Qin, T., Castoro, R., El Ahdab, S. et al. (2011). Mechanisms of resistance to decitabine in the myelodysplastic syndrome. *PLoS One* 6: e23372.

132 Prébet, T., Gore, S.D., Esterni, B. et al. (2011). Outcome of high-risk myelodysplastic syndrome after azacitidine treatment failure. *J. Clin. Oncol.* 29: 3322–3327.

133 Osterman, D.G., DePillis, G.D., Wu, J.C. et al. (1988). 5-Fluorocytosine in DNA is a mechanism-based inhibitor of *Hha*I methylase. *Biochemistry* 27: 5204–5210.

134 Champion, C., Guianvarc'h, D., Sénamaud-Beaufort, C. et al. (2010). Mechanistic insights on the inhibition of C5 DNA methyltransferases by zebularine. *PLoS One* 5: e12388.

135 Xiang, S.B., Short, S.A., Wolfenden, R., and Carter, C.W. Jr., (1995). Transition-state selectivity for a single hydroxyl group during catalysis by cytidine deaminase. *Biochemistry* 34: 4516–4523.

136 Laliberté, J., Marquez, V.E., and Momparler, R.L. (1992). Potent inhibitors for the deamination of cytosine arabinoside and 5-aza-2′-deoxycytidine by human cytidine deaminase. *Cancer Chemother. Pharmacol.* 30: 7–11.

137 Lemaire, M., Momparler, L.F., Bernstein, M.L. et al. (2005). Enhancement of antineoplastic action of 5-aza-2′-deoxycytidine by zebularine on L1210 leukemia. *Anti-Cancer Drug Des.* 16: 301–308.

138 Lemaire, M., Momparler, L.F., Raynal, N.J.-M. et al. (2009). Inhibition of cytidine deaminase by zebularine enhances the antineoplastic action of 5-aza-2′-deoxycytidine. *Cancer Chemother. Pharmacol.* 63: 411–416.

139 Klecker, R.W., Cysyk, R.L., and Collins, J.M. (2006). Zebularine metabolism by aldehyde oxidase in hepatic cytosol from humans, monkeys, dogs, rats, and mice: influence of sex and inhibitors. *Bioorg. Med. Chem.* 14: 62–66.

140 Byun, H.-M., Choi, S.H., Laird, P.W. et al. (2008). 2′-Deoxy-N4-[2-(4-nitrophenyl)ethoxycarbonyl]-5-azacytidine: a novel

inhibitor of DNA methyltransferase that requires activation by human carboxylesterase 1. *Cancer Lett.* 266: 238–248.

141 Issa, J.-P.J., Roboz, G., Rizzieri, D. et al. (2015). Safety and tolerability of guadecitabine (SGI-110) in patients with myelodysplastic syndrome and acute myeloid leukaemia: a multicentre, randomised, dose-escalation phase 1 study. *Lancet Oncol.* 16: 1099–1110.

142 Yoo, C.B., Jeong, S., Egger, G. et al. (2007). Delivery of 5-aza-2′-deoxycytidine to cells using oligodeoxynucleotides. *Cancer Res.* 67: 6400–6408.

143 Chuang, J.C., Warner, S.L., Vollmer, D. et al. (2010). S110, a 5-aza-2′-deoxycytidine-containing dinucleotide, is an effective DNA methylation inhibitor *in vivo* and can reduce tumor growth. *Mol. Cancer Ther.* 9: 1443–1450.

144 ASTX Pharmaceuticals. http://www.astx.com/research-development/clinical-pipeline (accessed 28 March 2017).

145 Eglen, R.M. and Reisine, T. (2011). Screening for compounds that modulate epigenetic regulation of the transcriptome: an overview. *J. Biomol. Screening* 16: 1137–1152.

146 Brueckner, B., Boy, R.G., Siedlecki, P. et al. (2005). Epigenetic reactivation of tumor suppressor genes by a novel small-molecule inhibitor of human DNA methyltransferases. *Cancer Res.* 65: 6305–6311.

147 Siedlecki, P., Boy, R.G., Musch, T. et al. (2006). Discovery of two novel, small-molecule inhibitors of DNA methylation. *J. Med. Chem.* 49: 678–683.

148 Arce, C., Segura-Pacheco, B., Perez-Cardenas, E. et al. (2006). Hydralazine target: from blood vessels to the epigenome. *J. Transl. Med.* 4: 10.

149 Segura-Pacheco, B., Trejo-Becerril, C., Perez-Cardenas, E. et al. (2003). Reactivation of tumor suppressor genes by the cardiovascular drugs hydralazine and procainamide and their potential use in cancer therapy. *Clin. Cancer Res.* 9: 1596–1603.

150 Villar-Garea, A., Fraga, M.F., Espada, J., and Esteller, M. (2003). Procaine is a DNA-demethylating agent with growth-inhibitory effects in human cancer cells. *Cancer Res.* 63: 4984–4989.

151 Castellano, S., Kuck, D., Sala, M. et al. (2008). Constrained analogues of procaine as novel small molecule inhibitors of DNA methyltransferase-1. *J. Med. Chem.* 51: 2321–2325.

152 Halby, L., Champion, C., Sénamaud-Beaufort, C. et al. (2012). Rapid synthesis of new DNMT inhibitors derivatives of procainamide. *ChemBioChem* 13: 157–165.

153 Ceccaldi, A., Rajavelu, A., Ragozin, S. et al. (2013). Identification of novel inhibitors of DNA methylation by screening of a chemical library. *ACS Chem. Biol.* 8: 543–548.

154 Datta, J., Ghoshal, K., Denny, W.A. et al. (2009). A new class of quinoline-based DNA hypomethylating agents reactivates tumor suppressor genes by blocking DNA methyltransferase 1 activity and inducing its degradation. *Cancer Res.* 69: 4277–4285.

155 Gamage, S.A., Brooke, D.G., Redkar, S. et al. (2013). Structure–activity relationships for 4-anilinoquinoline derivatives as inhibitors of the DNA methyltransferase enzyme DNMT1. *Bioorg. Med. Chem.* 21: 3147–3153.

156 Rilova, E., Erdmann, A., Gros, C. et al. (2014). Design, synthesis and biological evaluation of 4-amino-*N*-(4-aminophenyl)benzamide analogues of quinoline-based SGI-1027 as inhibitors of DNA methylation. *ChemMedChem* 9: 590–601.

157 Valente, S., Liu, Y., Schnekenburger, M. et al. (2014). Selective non-nucleoside inhibitors of human DNA methyltransferases active in cancer including in cancer stem cells. *J. Med. Chem.* 57: 701–713.

158 Hauser, A.-T. and Jung, M. (2008). Targeting epigenetic mechanisms: potential of natural products in cancer chemoprevention. *Planta Med.* 74: 1593–1601.

159 Piña, I.C., Gautschi, J.T., Wang, G.-Y.-S. et al. (2003). Psammaplins from the sponge *Pseudoceratina purpurea*: inhibition of both histone deacetylase and DNA methyltransferase. *J. Org. Chem.* 68: 3866–3873.

160 Fang, M.Z., Wang, Y., Ai, N. et al. (2003). Tea polyphenol (−)-epigallocatechin-3-gallate inhibits DNA methyltransferase and reactivates methylation-silenced genes in cancer cell lines. *Cancer Res.* 63: 7563–7570.

161 Narayanan, B.A. (2006). Chemopreventive agents alters global gene expression pattern: predicting their mode of action and targets. *Curr. Cancer Drug Targets* 6: 711–727.

162 Lee, W.J. and Zhu, B.T. (2006). Inhibition of DNA methylation by caffeic acid and chlorogenic acid, two common catechol-containing coffee polyphenols. *Carcinogenesis* 27: 269–277.

163 Fini, L., Selgrad, M., Fogliano, V. et al. (2007). Annurca apple polyphenols have potent demethylating activity and can reactivate silenced tumor suppressor genes in colorectal cancer cells. *J. Nutr.* 137: 2622–2628.

164 Fang, M.Z., Chen, D., Sun, Y. et al. (2005). Reversal of hypermethylation and reactivation of $p16^{INK4a}$, *RARβ*, and *MGMT* genes by genistein and other isoflavones from soy. *Clin. Cancer Res.* 11: 7033–7041.

165 Lin, R.-K., Hsu, C.-H., and Wang, Y.-C. (2007). Mithramycin a inhibits DNA methyltransferase and metastasis potential of lung cancer cells. *Anti-Cancer Drug Des.* 18: 1157–1164.

166 Kuck, D., Caulfield, T., Lyko, F., and Medina-Franco, J.L. (2010). Nanaomycin a selectively inhibits DNMT3B and reactivates silenced tumor suppressor genes in human cancer cells. *Mol. Cancer Ther.* 9: 3015–3023.

167 Fagan, R.L., Cryderman, D.E., Kopelovich, L. et al. (2013). Laccaic acid a is a direct, DNA-competitive inhibitor of DNA methyltransferase 1. *J. Biol. Chem.* 288: 23858–23867.

168 Davis, A.J., Gelmon, K.A., Siu, L.L. et al. (2003). Phase I and pharmacologic study of the human DNA methyltransferase antisense oligodeoxynucleotide MG98 given as a 21-day continuous infusion every 4 weeks. *Invest. New Drugs* 21: 85–97.

169 Winquist, E., Knox, J., Ayoub, J.-P. et al. (2006). Phase II trial of DNA methyltransferase 1 inhibition with the antisense oligonucleotide MG98

in patients with metastatic renal carcinoma: a National Cancer Institute of Canada clinical trials group investigational new drug study. *Invest. New Drugs* 24: 159–167.
170 Klisovic, R.B., Stock, W., Cataland, S. et al. (2008). A phase I biological study of MG98, an oligodeoxynucleotide antisense to DNA methyltransferase 1, in patients with high-risk myelodysplasia and acute myeloid leukemia. *Clin. Cancer Res.* 14: 2444–2449.
171 Amato, R.J. (2007). Inhibition of DNA methylation by antisense oligonucleotide MG98 as cancer therapy. *Clin. Genitourin. Cancer* 5: 422–426.
172 Amato, R.J., Stephenson, J., Hotte, S. et al. (2012). MG98, a second-generation DNMT1 inhibitor, in the treatment of advanced renal cell carcinoma. *Cancer Invest.* 30: 415–421.
173 Melchert, M. and List, A. (2008). Targeted therapies in myelodysplastic syndrome. *Semin. Hematol.* 45: 31–38.
174 Issa, J.-P.J. (2007). DNA methylation as a therapeutic target in cancer. *Clin. Cancer Res.* 13: 1634–1637.
175 Sekeres, M.A. (2007). The myelodysplastic syndromes. *Expert Opin. Biol. Ther.* 7: 369–377.
176 Dombret, H., Seymour, J.F., Butrym, A. et al. (2015). International phase 3 study of azacitidine vs conventional care regimens in older patients with newly diagnosed AML with >30% blasts. *Blood* 126: 291–299.
177 Kantarjian, H.M., Thomas, X.G., Dmoszynska, A. et al. (2012). Multicenter, randomized, open-label, phase III trial of decitabine versus patient choice, with physician advice, of either supportive care or low-dose cytarabine for the treatment of older patients with newly diagnosed acute myeloid leukemia. *J. Clin. Oncol.* 30: 2670–2677.
178 Lubbert, M., Ruter, B.H., Claus, R. et al. (2012). A multicenter phase II trial of decitabine as first-line treatment for older patients with acute myeloid leukemia judged unfit for induction chemotherapy. *Haematologica* 97: 393–401.
179 Lubbert, M., Suciu, S., Hagemeijer, A. et al. (2016). Decitabine improves progression-free survival in older high-risk MDS patients with multiple autosomal monosomies: results of a subgroup analysis of the randomized phase III study 06011 of the EORTC leukemia cooperative group and German MDS study group. *Ann. Hematol.* 95: 191–199.
180 Welch, J.S., Petti, A.A., Miller, C.A. et al. (2016). TP53 and decitabine in acute myeloid leukemia and myelodysplastic syndromes. *N. Engl. J. Med.* 375: 2023–2036.
181 Linnekamp, J.F., Butter, R., Spijker, R. et al. (2017). Clinical and biological effects of demethylating agents on solid tumours – a systematic review. *Cancer Treat. Rev.* 54: 10–23.
182 Laird, P.W., Jackson-Grusby, L., Fazeli, A. et al. (1995). Suppression of intestinal neoplasia by DNA hypomethylation. *Cell* 81: 197–205.
183 Davis, C.D. and Uthus, E.O. (2002). Dietary selenite and azadeoxycytidine treatments affect dimethylhydrazine-induced aberrant crypt formation in rat colon and DNA methylation in HT-29 cells. *J. Nutr.* 132: 292–297.

184 Lantry, L.E., Zhang, Z., Crist, K.A. et al. (1999). 5-Aza-2′-deoxycytidine is chemopreventive in a 4-(methyl-nitrosamino)-1-(3-pyridyl)-1-butanone-induced primary mouse lung tumor model. *Carcinogenesis* 20: 343–346.

185 McGregor, F., Muntoni, A., Fleming, J. et al. (2002). Molecular changes associated with oral dysplasia progression and acquisition of immortality: potential for its reversal by 5-azacytidine. *Cancer Res.* 62: 4757–4766.

186 Dixon, R.A. and Ferreira, D. (2002). Genistein. *Phytochemistry* 60: 205–211.

187 Dean, N.M., Kanemitsu, M., and Boynton, A.L. (1989). Effects of the tyrosine-kinase inhibitor genistein on DNA synthesis and phospholipid-derived second messenger generation in mouse 10T1/2 fibroblasts and rat liver T51B cells. *Biochem. Biophys. Res. Commun.* 165: 795–801.

188 Jones, P.L., Veenstra, G.C.J., Wade, P.A. et al. (1998). Methylated DNA and MeCP2 recruit histone deacetylase to repress transcription. *Nat. Genet.* 19: 187–191.

189 Nan, X., Ng, H.-H., Johnson, C.A. et al. (1998). Transcriptional repression by the methyl-CpG-binding protein MeCP2 involves a histone deacetylase complex. *Nature* 393: 386–389.

190 Zhu, W.-G. and Otterson, G.A. (2003). The interaction of histone deacetylase inhibitors and DNA methyltransferase inhibitors in the treatment of human cancer cells. *Curr. Med. Chem. Anticancer Agents* 3: 187–199.

191 Kikuchi, T., Itoh, F., Toyota, M. et al. (2002). Aberrant methylation and histone deacetylation of cyclooxygenase 2 in gastric cancer. *Int. J. Cancer* 97: 272–277.

192 Jones, P.A. and Baylin, S.B. (2002). The fundamental role of epigenetic events in cancer. *Nat. Rev. Genet.* 3: 415–428.

193 Yoo, C.B. and Jones, P.A. (2006). Epigenetic therapy of cancer: past, present and future. *Nat. Rev. Drug Discovery* 5: 37–50.

194 Wrangle, J., Wang, W., Koch, A. et al. (2013). Alterations of immune response of non-small cell lung cancer with azacytidine. *Oncotarget* 4: 2067–2079.

195 Héninger, E., Krueger, T.E., and Lang, J.M. (2015). Augmenting antitumor immune responses with epigenetic modifying agents. *Front. Immunol.* 6: 29.

196 Chiappinelli, K.B., Zahnow, C.A., Ahuja, N., and Baylin, S.B. (2016). Combining epigenetic and immunotherapy to combat cancer. *Cancer Res.* 76: 1683–1689.

16

Parasite Epigenetic Targets

Raymond J. Pierce and Jamal Khalife

Université de Lille, CNRS, Inserm, CHU Lille, Institut Pasteur de Lille, U1019-UMR-8204-CIIL (Centre d'Infection et d'Immunité de Lille), 59000 Lille, France

16.1 Introduction: The Global Problem of Parasitic Diseases and the Need for New Drugs

Despite continuing progress and improvements in the diagnosis, prevention, and treatment of parasitic diseases, they continue to rank high among global health problems. Diseases caused by parasites affect more than 1 billion people worldwide and kill more than one million each year, but their true impact is much more profound: these diseases are disablers that affect children's growth and development and school attendance and have a wide economic impact. They are major contributors to maintaining at-risk populations in the poverty trap [1].

The 2012 London Declaration (http://unitingtocombatntds.org/london-declaration-neglected-tropical-diseases) and a World Health Assembly resolution in 2013 (www.who.int/mediacentre/events/2013/wha66/en/) focused attention on neglected tropical diseases (NTDs) in general and the parasitic diseases in particular, with ambitious targets for elimination and eradication, or advanced control, to be met by 2020. Priorities for these targets include the need for the development of new drug treatments, which remains in the front line of control strategies and is a major priority for parasitic diseases. Among the latter, leishmaniasis, sleeping sickness, and Chagas disease, caused by kinetoplastid protozoan parasites (*Leishmania* spp., *Trypanosoma brucei*, and *Trypanosoma cruzi*), and schistosomiasis, caused by flatworm parasites of the genus *Schistosoma*, have been particularly prioritized, due both to their global impact [2] and the urgent need for new treatments. Although not, *stricto sensu*, a neglected disease, malaria remains a major health threat and also a priority for drug development [3].

In the case of malaria, parasite resistance to the current recommended treatment, artemisinin combination therapy (ACT) in Southeast Asia is of serious concern [4]. Current treatments for Chagas disease involve long dosing schedules, cause significant side effects [5], and are ineffective against the chronic phase of the disease. Treatments for leishmaniasis and in particular the only oral treatment, miltefosine, are also encountering problems of resistance

Epigenetic Drug Discovery, First Edition. Edited by Wolfgang Sippl and Manfred Jung.

and/or cause severe side effects [6, 7]. For schistosomiasis, praziquantel (PZQ) is the only available treatment, and although well tolerated and cheap, its use in mass-treatment control strategies in sub-Saharan Africa is a source for concerns about the development of resistance [8]. Moreover, recent surveys show that while generally successful in reducing disease morbidity, prevalence, and intensity, this is not the case in all regions [8]. In addition, lower rates of cure and reduction in parasite egg excretion have been observed where multiple annual treatments have been carried out [9]. PZQ reduces the fecundity of worms that survive treatment, probably leading to an overestimation of drug efficacy, measured by fecal or urinary egg counts [8].

Although several promising compounds are in the pipeline for the development of new drugs against these parasitic diseases (see, for example, www.dndi.org/diseases-projects/portfolio/; www.mmv.org/), the high attrition rate during clinical trials means that new candidates are constantly needed to feed into the testing process. Furthermore, the rapid and constant development of drug resistance by these parasites underlines that in this arena the Red Queen reigns supreme [10], and we need to progress very rapidly merely to maintain the effectiveness of treatments, let alone develop better ones. Many different strategies for drug discovery against parasites are currently pursued, including the repurposing of drugs used in other pathologies [11], target-agnostic phenotypic screening with compound libraries, and target-based approaches. Among the latter, the screening of epigenetic targets in parasites is particularly attractive since it affords a unified strategy that can be applied to different disease organisms. Moreover, genome sequencing of the main parasite species has been followed up by studies of transcription factors, histone posttranslational modifications (PTMs), and enzymes that write or erase these marks, as well as reader proteins that interact with them. The latter three categories of actors in epigenetic processes are the main targets for drug discovery that have been exploited so far and will be considered in this book chapter.

16.2 Parasite Epigenetic Mechanisms

The parasites causing the major parasitic diseases have complex life cycles involving an invertebrate host in addition to the human host and two modes of reproduction, sexual and asexual, and their development within the human host is complex (Figure 16.1). All these characteristics indicate that they tightly control transcription and that this is achieved via epigenetic mechanisms. These mechanisms, defined as structural modifications in chromatin without any alteration to the DNA sequence, which cause heritable changes in gene expression, include DNA methylation, reversible posttranslational histone modifications, variant histones, factors that remodel chromatin, and noncoding RNAs. Although parasites, such as eukaryotes, exhibit all these mechanisms, they have been more or less studied depending on the organism. For example, microRNAs (miRNAs) have been extensively characterized in schistosomes (reviewed in Ref. [12]), and their roles in ovary development [13] and the overall pattern of transcription during

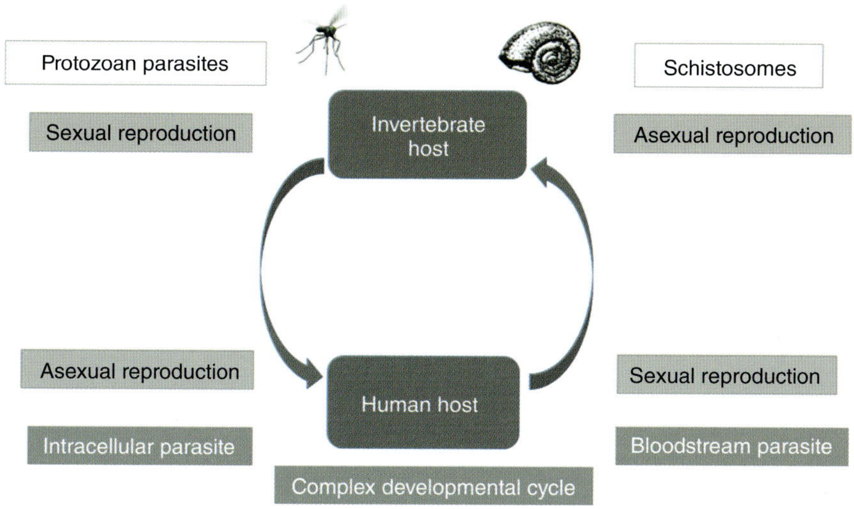

Figure 16.1 Schematic representation of the life cycles of parasites causing neglected tropical diseases. All involve an invertebrate host (an insect for protozoan parasites, a freshwater snail in the case of schistosomes) as well as the human host, both asexual and sexual reproductive cycles, and a complex developmental cycle within the human host. These characteristics indicate the necessity for tight control of gene transcription via epigenetic mechanisms.

the transition from juvenile to adult worms [14] have recently been described. However, the therapeutic potential of targeting their function in schistosomes has yet to be exploited [12]. In the case of protozoan parasites, attention has mainly focused on the modulation of host responses by miRNA, generated by both the parasite and the host. Intracellular protozoan parasites can secrete extracellular vesicles containing miRNA that transfer the latter to the host cell, modulating its functions (reviewed in Ref. [15]). Here again, this mechanism has not been targeted so far for drug development.

16.2.1 DNA Methylation

DNA methylation, which encompasses the methylation or hydroxymethylation of cytosine residues within CpG dinucleotides [16], is a modification associated with gene repression in vertebrates. Both its presence and potential role in invertebrate organisms in general and in parasites in particular remain to be fully elucidated, but evidence exists for its importance in both schistosomes and in the malaria parasite *Plasmodium falciparum*. A genome-wide analysis of cytosine methylation in *P. falciparum* showed an asymmetrical pattern with features common to undifferentiated plant and mammalian cells [17]. Hypomethylation of core promoters, along with methylation transitions at nucleosomes and exon–intron boundaries, as well as the correlation of transcript levels with intraexonic methylation suggests a role for this modification in the control of transcription, particularly of the virulence-associated genes. In schistosomes evidence for cytosine methylation, although still controversial (reviewed in Ref. [12]), was provided by several methods [18], and hypermethylation of a repetitive intron demonstrated. Moreover, in accordance with the suggested conservation of DNA methylation throughout the phylum Platyhelminthes [19], approximately 2% of cytosine residues were found to be methylated in the *Schistosoma mansoni* intermediate snail host, *Biomphalaria glabrata* [20], demonstrating that this modification is present in the lophotrochozoan invertebrates. The DNA methyltransferases (DNMTs) that effect this modification have consequently been suggested as therapeutic targets in both schistosomes and *P. falciparum*.

16.2.2 Histone Posttranslational Modifications

PTMs of histones and the proteins that write, erase, or read them have come under the most attention as potential therapeutic targets in parasites. An increasing number of possible modifications include phosphorylation, ubiquitinylation, and SUMOylation as well as the most studied acetylation and methylation. However, in addition to acetylation, histone "acetyltransferases" and in particular p300/CBP are able to effect different types of fatty and dicarboxylic acid transfer reactions of histone lysines, such as propionylation, butyrylation, succinylation, or malonylation [21].

A global analysis of histone modifications in *P. falciparum* mapped 44 different marks including lysine acetylation, ubiquitinylation, and mono-, di-, or trimethylation, as well as arginine mono- or dimethylation [22]. Histone serine/threonine phosphorylation was analyzed separately [23], and 14 sites were detected. Overall the *P. falciparum* histones showed conserved modification sites, but a different pattern compared to human [22]. The functional importance of most of these modifications is still unknown, but phenotypic variation based on differential gene expression has been shown to be governed by the reversible formation of trimethylated histone 3 lysine 9 (H3K9me3)-based heterochromatin [24]. Acetylation of H3K9 (H3K9ac) and di- or trimethylation of H3K4 (H3K4me2, H3K4me3) are associated with the single transcriptionally active *var* gene locus [25] whereas repressive marks, such as H3K9me3, H3K36me3, and H3K20me3, are present in repressive clusters at the promoters of inactive virulence gene family members including *var* but also *rifin, stevor,* and *pfmc-2tm* (reviewed in Ref. [26]). However, the repressive mark H3K27me3 is absent from the parasite.

Despite the unconventional transcriptional mechanisms of most genes in kinetoplastid parasites, functional histone marks are present. Kinetoplastid gene transcription is characterized by constitutive expression from polycistronic transcription units (PTUs), which contain functionally unrelated genes. Exceptions are the variable surface glycoprotein (VSG) and procyclin genes in *T. brucei*, which are transcriptionally regulated [27]. PTUs, on the other hand, are transcribed bidirectionally from divergent strand switching regions (SSRs). Initiation loci are enriched in the histone marks H4K10ac and H3K4me3, in addition to the histone variants H2A.Z and H2B.V and the bromodomain protein *Tb*BDF3 (reviewed in Ref. [27]). Histone acetylations were also found to mark the divergent SSR in *Leishmania major* [28]. An analysis of the repertoire of histone marks in *T. brucei* [29] showed methylation and acetylation of lysine residues of canonical histones, as well as threonine phosphorylation. However, the lack of conservation of the N-terminal tail sequences of kinetoplastid histones complicates the comparison with mammalian histone marks. For example, H3K9 and H3K27 are absent in *T. brucei*, and it is not certain that H3K10 and H3K23, which are both methylated, are their equivalents [27]. Nevertheless, the stage dependency of some histone marks, notably blood-stage lysine acetylations on histones H2A and H2B, suggests that these marks have a distinct function.

Histone modifications are also stage dependent in *T. cruzi* [30] in which the replicative epimastigote stage shows more modifications than trypomastigotes. For example, H3K76 mono- or dimethylation was found in cells during mitosis and cytokinesis. A large number and variety of histone modifications have been catalogued in *T. cruzi* [30, 31]. As well as PTMs that are conserved in eukaryotes, some are unique to *T. cruzi* or trypanosomatids. Moreover, some unusual modifications were found [31] including lysine acylations, serine/threonine acetylation, or N-terminal methylation. Moreover, individual residues were found to be substrates of different marks. For example, H3K61 is modified, alone or in combination, by acetylation, crotonylation, succinylation, glutarylation, and ubiquitination. This argues for the existence of a complex mechanism for the regulation of transcription via the modulation of these PTMs and that individual histone residues may have key roles. This contradicts earlier interpretations of the repertoire of trypanosomatid histone marks and histone-modifying enzymes (HMEs) that emphasized a perceived lack of complexity and a limited role in chromatin regulation (reviewed in Ref. [32]). On the contrary, epigenetic control in kinetoplastid parasites has the potential to be as complex as in other eukaryotes.

The description of the full complement of histone gene orthologues encoded in the *S. mansoni* genome [33] shows that the core histones, and in particular their N-terminal tails, are highly conserved compared to those of other metazoans, including humans. Therefore, although no systematic study of schistosome histone PTMs has been done, it is reasonable to suppose that their nature and functions will also resemble those in other metazoan species. This is further borne out by the conservation of the HMEs in schistosomes (see Section 16.2.4). Nevertheless, an expanding body of work confirms the importance of critical histone PTMs in schistosome sexual differentiation, adaptation to the freshwater snail intermediate host, and changes in gene expression on invasion of the mammalian definitive host. Sexual differentiation is associated with changes in the expression of 12 miRNAs and in changes in the enrichment profiles of the repressive H3K27me3 mark across the genome between cercariae (infective larvae) and adult worms, particularly in the case of males [34]. Epigenetic changes are a major cause of the expression of adaptive phenotypic variants. The expression of variant mucins (*Sm*PoMuc) that determine parasite compatibility with different snail host strains is governed by changes in the H3K9me3/ac ratio at the transcription start sites (TSSs) [35]. Moreover, different snail strains induce transient changes in the epigenome of schistosomes [36]. Finally, the comparison of genome-wide patterns of histone modifications at the TSS in cercariae and adult worms shows that the former maintain a bivalent transcriptional state at 121 loci. The concomitant presence of the transcription activating H3K4me3 and repressive H3K27me3 marks at these loci is characteristic of poised transcription and is similar to the pattern found in embryonic stem cells [37].

16.2.3 Histone-modifying Enzymes in Parasites

The enzymes that "write" or "erase" histone PTMs, and increasingly proteins that "read" these marks, are the principal targets for the development of drugs impacting parasite epigenetic processes: the so-called "epi-drugs." However, these enzymes and proteins have conserved functions and are generally members of protein families that are conserved in humans. Therefore the key to developing viable epi-drugs against parasites is to ensure their selectivity. Unusual features of the malaria epigenome and, by extension, those of other parasites have been proposed to represent a weakness, making the parasites uniquely vulnerable to epi-drug treatment [38]. An example is the singular importance of a particular histone mark, H3K9me3, in *var* gene repression in *P. falciparum*. Nevertheless, it is preferable to develop both enzyme isoform- and species-selective inhibitors of HMEs to ensure selectivity and minimize off-target effects of epi-drugs. Ideally, the protein targeted should be essential to the survival and development of the parasite, as well as exhibiting specific structural features that can be exploited to develop species selectivity. So far attention has focused on the enzymes involved in acetylation/deacetylation and methylation/demethylation, as well as bromodomain proteins that interact with acetylated lysine residues.

16.2.4 HMEs Validated as Therapeutic Targets

Gene knockout and transcript knockdown methods have been used to show that individual parasite HMEs can be essential to survival within the mammalian host. Four Zn^{2+}-dependent histone deacetylases (HDACs) are present in *T. brucei*, of which one class I HDAC (DAC1) and one class II enzyme (DAC3) were essential [39]. Similarly two of the three MYST class histone acetyltransferases (HATs), HAT1 and HAT2, were essential [40], the latter being responsible for H4K10 acetylation, present at the transcription initiation loci [27]. The essentiality of other trypanosome HME can be inferred from their functions. The histone methyltransferase (HMT) DOT1B methylates H3K76 in *T. brucei* and is required both for rapid silencing of VSG gene expression and for the mitotic checkpoint control in bloodstream forms. The essential DAC1 and DAC3 impact subtelomeric gene silencing along with the bromodomain proteins BDF2 and 3 (reviewed in Ref. [41]).

The roles of the sirtuins (class III, NAD^{+}-dependent HDACs) in kinetoplastid parasites are less well known, but seem to be distinct from those of the Zn^{2+}-dependent HDACs. Three sirtuins are encoded in the genomes of *T. brucei* and *Leishmania* sp. [42], one class I enzyme (homologue of human Sirt1, Sirt2, and Sirt3), one class II enzyme (orthologue of human Sirt4), and one class III enzyme (orthologue of human Sirt5). In *T. brucei* the class I sirtuin *Tb*Sir2 is involved in telomeric silencing, but not VSG silencing [41]. In *Leishmania infantum*, in contrast, the essential class I sirtuin *Li*SIR2RP1 is cytosolic, deacetylates α-tubulin, and is associated with the microtubule network [43].

In *P. falciparum* three Zn^{2+}-dependent HDACs and two sirtuins are present [44]. *Pf*HDAC1 is a class I enzyme, while *Pf*HDAC2 (also called *Pf*Hda1) and *Pf*HDAC3 (also called *Pf*Hda2) are class II HDACs [45, 46]. *Pf*HDAC1 is the only one available in recombinant form [47], and *in silico* modeling predicts differences in the catalytic pocket structure compared to human HDACs that should allow the development of selective inhibitors [48]. Knockdown of *Pf*HDAC3/*Pf*Hda2 disrupts intraerythrocytic stage growth and the expression of both the virulence-associated *var* genes and *Pf*ApiAP2, a transcription factor controlling conversion of asexual to sexual stages [49]. Both the *P. falciparum* sirtuins *Pf*Sir2a and *Pf*Sir2b regulated *var* gene transcription, but neither was found to be essential for parasite survival *in vitro* [50] and has not been considered as therapeutic targets.

The enzymes effecting histone methylation and demethylation in *P. falciparum* are under increasing scrutiny as drug targets. Ten SET-domain histone lysine methyltransferases (HMTs), three arginine methyltransferases (PRMT), and two JmjC-domain histone demethylases are encoded in the *P. falciparum* genome [42, 51]. Gene knockout studies showed that six of the HMTs seem to be essential for parasite survival [52]. Methylation of H3K36 by PfSET2 was found to silence *var* genes. Knockout of *Pf*SET2 resulted in the transcription of almost all the *var* genes in a single parasite as well as their expression on the surface of infected erythrocytes [52]. In contrast, methylation of H3K4 by *Pf*SET10 is essential for the maintenance of the active *var* gene in a poised state during cell division before its reactivation in daughter parasites [53].

Among the proteins that read histone PTMs, those containing bromodomains have received most scrutiny as potential drug targets, particularly for protozoan parasitic diseases (reviewed in Ref. [54]), although in most cases their precise functions are as yet unknown. In *P. falciparum Pf*BDP1 knockdown caused defects in parasite invasion of red blood cells and growth, resulting from the downregulation of invasion-related genes [55]. Moreover, in addition to binding to acetylated H3, *Pf*BDP1 interacts with a second bromodomain protein, *Pf*BDP2. In *T. cruzi* overexpression of bromodomain factor 3 (*Tc*BDF3), which binds to acetylated α-tubulin, led to resistance of the parasite to bromodomain inhibitors. In contrast, the expression of a mutated version of *Tc*BDF3 led to growth defects in cultured epimastigotes [56]. The effects of bromodomain protein transcript knockdown using inducible RNAi or of gene knockout have been extensively studied in *T. brucei* [57, 58]. Notably, the knockdown of *Tb*BDF2 and *Tb*BDF3 led to growth and cell cycle defects. Transcriptome analyses of these parasites showed that, for example, knockdown of *Tb*BDF3 led to transcription of VSG genes located at silent bloodstream expression sites being increased. The effects of bromodomain protein knockdown were reiterated by the use of an inhibitor, I-BET151 [58].

The *S. mansoni* genome encodes 55 HMEs involved in protein acetylation/deacetylation or methylation/demethylation [59]. Most attention has so far focused on the deacetylases: the Zn^{2+}-dependent HDACs and the sirtuins. Genes encoding five sirtuins are present [42, 60] including nuclear, mitochondrial, and cytoplasmic forms [59, 61]. So far, transcript knockdown has only been carried out for the class I sirtuin, *Sm*Sirt1 [60], in adult worms. This led to marked effects on female worm reproductive organs including an increase in mature oocytes in the ovaries. These effects mirrored those seen after treatment with an inhibitor of human sirtuins, salermide.

Three class I and four class II HDACs are encoded in the *S. mansoni* genome [42, 62]. Of the class I HDACs, *Sm*HDAC8 was initially identified as a possible target due to its high expression levels and its lower level of sequence conservation with respect to its human orthologue compared to either *Sm*HDAC1 or *Sm*HDAC3 [62] including substitutions in and around the catalytic pocket. Its importance to parasite growth and development in the mammalian host was confirmed by transcript knockdown using RNAi [63], which led to markedly reduced worm survival in infected mice.

Of the nine HATs present in schistosomes [59], two orthologues, human GCN5 and CBP/p300, have been shown to be essential for normal egg production by adult female worms [64]. Both HATs were found to be recruited to the promoter of the major eggshell protein *Sm*14 by nuclear receptors and RNAi transcript knockdown of either led to changes in the reproductive system of female worms, reduced egg laying, and altered egg morphology. These effects were reiterated by treatment with an HAT inhibitor, PU139, which was also shown to cause chromatin condensation at the *Sm*14 promoter, which correlated with an increase in the repressive H3K27me3 mark.

16.2.5 Structure-based Approaches for Defining Therapeutic Targets

The determination of crystal structures of parasite HMEs and the use of computer-generated homology models have been used only sparingly so far, whether for the comparison of catalytic pocket structures or for computer-aided drug design. Nevertheless, these approaches are extremely informative for the identification of structural differences that can be exploited for the development of selective inhibitors and for the choice of enzymes to target. Homology modeling of *Pf*HDAC1 highlighted differences compared to human class I HDACs, particularly at the entrance to the active site tunnel, and suggested that selective inhibitors could be developed [65]. Homology modeling of *Sm*HDAC8 also pointed out some differences in its structure compared to the human enzyme and allowed *in silico* screening in order to detect potential inhibitors [66]. However, the generation of crystal structures of *Sm*HDAC8 was decisive in defining these structural differences [63]. Notably, in addition to the substitution of a methionine residue by a histidine (H292), introducing a charge difference, the "flipped-out" conformation of F151 in the schistosome enzyme leads to catalytic pocket that is much broader and shallower than in human HDAC8. This conformation is favored by the absence of residues causing steric hindrance in loops around the catalytic pocket. The presence of such residues in the human enzyme leads to steric constraint of the flipped-in conformation of F152 (reviewed in Ref. [67]). These differences, together with the solved crystal structures of the enzyme in apo form or complexed with different inhibitors [68], have allowed the investigation of the flexibility of the catalytic pocket and structure-based drug design for *Sm*HDAC8.

In a further study, homology modeling of HDAC8 proteins from the major flatworm parasites, including other schistosome species, other trematodes, and cestodes [69], was carried out. These models showed that the flipped-out configuration of the F151 side chain in *Sm*HDAC8 is also characteristic for

Table 16.1 Validated parasite epigenetic targets for drug development.

Parasite	Target class	Validated target(s)	Validation method	Reference
T. brucei	HDAC	DAC1, DAC3	Gene KO	[39]
	HAT	HAT1, HAT2	Gene KO	[40]
	HMT	DOT1B	Gene KO	[41]
	Bromodomain protein	*Tb*BDF2, *Tb*BDF3	Gene KO, conditional transcript knockdown	[58]
L. infantum	Sirtuin	*Li*SIR2RP1	Gene KO	[43]
P. falciparum	HDAC	*Pf*HDAC1, *Pf*HDAC3/*Pf*Hda2	Gene KO, conditional protein depletion	[48, 49]
	HMT	Six SET-domain HMTs	Gene KO	[52]
	Bromodomain protein	*Pf*BDP1	Conditional protein depletion	[55]
S. mansoni	HDAC	*Sm*HDAC8	RNA interference	[63]
	HAT	*Sm*GCN5, *Sm*CBP	RNA interference	[64]

the orthologues in other species of flatworm. In addition, conformational changes induced by the flipped-out F151 and the H292 substitution for M274 in SmHDAC8 lead to the formation of a "side pocket" that can also be exploited for drug design [69]. The conservation of these characteristics in flatworm HDAC8 suggests that SmHDAC8 inhibitors active against *S. mansoni* should not only be active against other schistosomes, but should in fact be "*pan*-flatworm infection" drugs. This study was also extended to the HDACs of protozoan parasites, and a catalogue of conserved and variable amino acid residues both within the catalytic pocket, in the walls of the pocket and the side-pocket entrance was compiled [69]. Overall this study allowed the identification of promising targets and suggested a drug design strategy based on human isotype-selective inhibitors as a starting point for chemical optimization of their inhibitory efficacy.

The various epigenetic actors that have been validated as therapeutic targets are summarized in Table 16.1.

16.3 Development of Epi-drugs for Parasitic Diseases

A significant advantage of targeting epigenetic processes in parasites is that these mechanisms have been very actively studied in other pathologies, particularly in cancer. In addition to the four existing drugs (all HDAC inhibitors) that have been approved for use in humans by the Federal Drug Administration (United States), many more compounds are currently in the drug development pipeline [70, 71]. This means that in addition to *de novo* strategies for drug discovery, repurposing of existing molecules and their optimization to develop parasite-selective inhibitors are both valid approaches.

16.3.1 Repurposing of Existing Epi-drugs

Four clinically approved anticancer drugs that target HDACs (Vorinostat, Belinostat, Panobinostat, and Romidepsin – all *pan*-HDAC inhibitors) have been tested *in vitro* against *Plasmodium knowlesi, Leishmania amazonensis, Leishmania donovani,* and *S. mansoni* [72]. None was active against *Leishmania*, whether tested against the intracellular amastigote form or against extracellular promastigotes. In the case of *S. mansoni*, Romidepsin was extremely active against adult worms, but not against schistosomula, and none of the other drugs had any activity. In contrast, all four compounds were potent (IC_{50} in the nanomolar range) inhibitors of *P. knowlesi* growth, and all except Romidepsin showed selectivity for the parasite with respect to human cell lines. Moreover, a moderate effect of both Vorinostat and Panobinostat was observed after oral dosing of the drugs to *Plasmodium berghei*-infected mice. A newer, hydroxamate-based orally active drug with improved pharmacokinetic properties compared to Vorinostat, SB939, had similar *in vitro* properties, but when administered to mice infected with *P. berghei*, ANKA was able to both inhibit parasite growth and to prevent cerebral malaria-like symptoms [73]. This suggests that improved efficacy and selectivity is achievable, even with *pan*-HDAC inhibitors.

A panel of HDAC inhibitors was also tested against *T. brucei* [74]. Although the FDA-approved drugs Belinostat and Panobinostat, and others, such as Quisinostat, still in clinical trials, proved effective, none was able to kill parasites at concentrations achievable at tolerated oral doses. Nevertheless, these drugs may be viable starting points for further development.

The importance of methylation of H3K9 in the repression of inactive *P. falciparum var* genes [26] suggested the study of inhibitors of human G9a/GLP (responsible for H3K9 methylation) as starting points for the development of antimalarial drugs [75]. Using the G9a inhibitor BIX-01294 [76] as a starting point, a focused library of synthesized derivatives yielded a compound, TM2-115, that killed parasites in culture with low nanomolar IC_{50}, was selective, and was also active *in vivo*. Interestingly, this compound reduced methylation of H3K4. This approach was pursued, and even more effective and selective compounds developed [77], confirming the interest of the diaminoquinazoline scaffold as a lead for the development of new drugs for malaria [78].

Preliminary work in *S. mansoni* also suggests the interest of HMT inhibitors as potential drugs. A transcriptomic study to determine the targets of pan-HDAC inhibition in this parasite using trichostatin A (TSA) identified a component of the PRC2 repressive complex, EED, as being significantly downregulated [79]. The EED protein regulates the H3K27 methyltransferase EZH2, and only a single orthologue with this activity is present in the *S. mansoni* genome. The EZH2 inhibitor GSK343 [80] was active against schistosome larvae *in vitro* and showed strong synergistic activity with TSA. This suggests both that *Sm*EZH2 is a promising therapeutic target and that GSK343 could serve as a starting point for drug development.

Bromodomain inhibitors also show promise as anticancer agents, and several are in clinical trials [81]. Seven inhibitors were tested on *T. cruzi*, three of which – JQ1, I-BET151, and GSC-CBP30 – inhibited parasite replication [56] and interacted with the bromodomain of *Tc*BDF3. I-BET151 also impacted VSG gene expression in *T. brucei*, and its effects reiterated the effects of bromodomain protein knockdown [58]. This compound also inhibits *Toxoplasma gondii* tachyzoite replication [82], showing that bromodomain inhibitors are likely to be effective against other apicomplexan parasites, including *P. falciparum*.

16.3.2 Candidates from Phenotypic or High-throughput Screens

No results derived from high-throughput screening of large compound libraries using individual parasite HMEs as targets have so far been published, but these have been carried out, and compounds derived from them are under study [83]. Similarly, the identification of inhibitors of epigenetic processes via unbiased phenotypic screens of compound libraries on parasites has rarely been undertaken. Bisnaphthalimidopropyl derivatives were initially shown to have activity against *L. infantum* intracellular amastigotes [84] and were subsequently shown to inhibit the sirtuin LiSIR2RP1. More generally, focused libraries of compounds with the characteristics of, for example, HDAC inhibitors have been used in phenotypic screens against the parasites. This approach was used to screen HDAC inhibitors for activity against *T. brucei* and led to the identification of a sulfonepiperazine compound with an IC_{50} of 34 nM [85]. Numerous similar approaches have been undertaken to identify HDAC inhibitors active against *P. falciparum* (reviewed in Ref. [86]). Screening of a panel of phenyl-thiazolyl-hydroxamate-based compounds yielded a compound (WR301801) that was extremely active *in vitro* [87] but was less effective *in vivo*, despite having a plasma half-life of 3.6 hours and achieving concentrations well above the IC_{90} value after an oral dose. Other compound classes tested include HDAC inhibitors with an alkoxyamide connecting unit linker region [88] and terephthalic acid-based compounds [89]. Both of these compound classes were active both against gametocytes and asexual blood stages and showed low cytotoxicity against human cell lines.

16.3.3 Structure-based Development of Selective Inhibitors

Understanding the binding mode of HME inhibitors to the catalytic pocket of the target enzyme is key to the reasoned development of increased isotype and species selectivity. Methods based on homology modeling and/or solution of the crystal structure of the enzyme–inhibitor complex are extensively utilized in drug development in general and increasingly in studies of parasite HME inhibitors. Homology modeling of *T. cruzi* sirtuin *Tc*Sir2RP1 in complex with the NAD^+ co-substrate allowed the molecular docking of phytochemical inhibitors and identified four compounds that interact both with this enzyme and with *Tc*Sir2PR2 [90]. Co-crystallization of the *T. brucei* bromodomain protein *Tb*Bdf2 with the inhibitor I-BET151 revealed that the latter is flipped by about 180° compared to its complex with human BRD4-BD1 [58] and consequently adopts a very different binding mode. The observed differences suggest ways in which the inhibitor can be modified to produce a trypanosome-specific, high-affinity ligand. The co-crystallization of the PCAF bromodomain chemical probe L-45 with the bromodomain from the *P. falciparum* PCAF orthologue provides similar opportunities [91].

The most complete study of inhibitor binding to a parasite HME, together with bioguided optimization of inhibitor structures in order to generate increasingly selective and effective compounds, has been carried out with *Sm*HDAC8. The resolution of the crystal structure of *Sm*HDAC8 [63] and *in silico* screening of an extensive compound library [63, 66] identified two hydroxamate-based inhibitors, J1038 and J1075, which, although not more effective inhibitors of the schistosome enzyme than of the human orthologue, showed significant isotype selectivity, e.g. over human HDAC1. Co-crystallization studies showed that each exploited different specific characteristics of the catalytic pocket of *Sm*HDAC8 and suggested a strategy for the generation of inhibitors with greater selectivity. The synthesis of a series of benzamidohydroxamate derivatives [92] and their co-crystallization with *Sm*HDAC8 in an iterative process revealed the presence of a schistosome-specific clamp binding selective inhibitors in the pocket. The best compounds showed up to eightfold selectivity for *Sm*HDAC8 over the human HDAC8. Moreover, compounds with low nanomolar IC_{50} values (enzyme inhibition), effective against both adult and larval schistosomes, and that have very low toxicity toward human cell lines have been generated. Co-crystallization studies with *Sm*HDAC8 were also employed with other inhibitor classes, mercaptoacetamides [68], and hydroxamates connected with alkoxyamide or hydrazide groups [93] and offer a rationale for further development.

16.4 Conclusions

Many of the enzymes catalyzing the PTM of histones, particularly those that acetylate/deacetylate or methylate/demethylate lysine residues in histone N-terminal tails, as well as the proteins that read these marks, are essential to parasite survival in the human host. As such they are excellent targets for drug development, and the fact that their human counterparts are also targets for cancer therapy means that developers of inhibitors for parasite HMEs can "piggyback" on the available expertise, approaches, and inhibitor libraries.

Nevertheless there are drawbacks in targeting highly conserved enzymes that are vital to all organisms. HDAC inhibitors that target both class I and II enzymes, like the approved drugs Romidepsin, Panobinostat, and Vorinostat, are associated with serious side effects such as cardiotoxicity (reviewed in Ref. [94]). In the case of bromodomain inhibitors, concerns have been raised about the potential of these compounds to reactivate viral infections including HIV [81]. These examples underline the general need for the development of both isotype- and species-specific inhibitors. Therefore, while the existing approved drugs and those currently in clinical trials will probably not be directly used for treating parasitic diseases, they and compounds like the bromodomain inhibitor I-BET151 (for *T. brucei*) [58] or GSK343 the inhibitor of the methyltransferase EZH2 (for *S. mansoni*) [79] may prove to be useful starting points for the development of more selective inhibitors, in addition to the other strategies we have outlined.

Finally, in the context of the treatment of diseases caused by parasites that readily develop resistance to all drugs so far developed, consideration should be given to testing HME inhibitors in combination with existing drugs or drugs currently in development. This approach is currently being extensively trialed in cancer therapy [94] where combining drugs that target different signaling pathways can lessen adverse side effects while increasing treatment efficacy [95].

Acknowledgments

This work and the authors of this manuscript received funding from the European Union's Seventh Framework Programme for research, technological development, and demonstration under grant agreement nos. 241865 (SEtTReND) and 602080 (A-ParaDDisE). RJP and JK were also supported by institutional funds from the CNRS, the Institut Pasteur de Lille, and the Université de Lille.

References

1 Herricks, J.R., Hotez, P.J., Wanga, V. et al. (2017). The global burden of disease study 2013: what does it mean for the NTDs? *PLoS Negl.Trop. Dis.* 11 (8): e0005424.

2 Hotez, P.J., Pecoul, B., Rijal, S. et al. (2016). Eliminating the neglected tropical diseases: translational science and new technologies. *PLoS Negl.Trop. Dis.* 10 (3): e0003895.

3 Burrows, J.N., Duparc, S., Gutteridge, W.E. et al. (2017). New developments in anti-malarial target candidate and product profiles. *Malar. J.* 16: 26.

4 Dondorp, A.M., Smithuis, F.M., Woodrow, C., and Seidlein, L.V. (2017). How to contain artemisinin- and multidrug-resistant falciparum malaria. *Trends Parasitol.* 33: 353–363.

5 Olivera, M.J., Cucunuba, Z.M., Alvarez, C.A., and Nicholls, R.S. (2015). Safety profile and treatment interruption for chronic Chagas disease in Colombian adults. *Am. J. Trop. Med. Hyg.* 93: 1224–1230.

6 Deep, D.K., Singh, R., Bhandari, V. et al. (2017). Increased miltefosine tolerance in clinical isolates of *Leishmania donovani* is associated with reduced drug accumulation, increased infectivity and resistance to oxidative stress. *PLoS Negl.Trop. Dis.* 11 (6): e0005641.

7 Garcia-Hernandez, R., Manzano, J.I., Castanys, S., and Gamarro, F. (2012). *Leishmania donovani* develops resistance to drug combinations. *PLoS Negl.Trop. Dis.* 6 (12): e1974.

8 Lamberton, P.H.L., Faust, C.L., and Webster, J.P. (2017). Praziquantel decreases fecundity in *Schistosoma mansoni* adult worms that survive treatment: evidence from a laboratory life-history trade-offs selection study. *Infect. Dis. Poverty* 6 (1): 110.
9 Crellen, T., Walker, M., Lamberton, P.H. et al. (2016). Reduced efficacy of praziquantel against *Schistosoma mansoni* is associated with multiple rounds of mass drug administration. *Clin. Infect. Dis.* 63: 1151–1159.
10 Van Valen, L. (1973). A new evolutionary law. *Evol. Theory* 1: 1–30.
11 Panic, G., Vargas, M., Scandale, I., and Keiser, J. (2015). Activity profile of an FDA-approved compound library against *Schistosoma mansoni*. *PLoS Negl.Trop. Dis.* 9 (7): e0003962.
12 Cabezas-Cruz, A., Lancelot, J., Caby, S. et al. (2014). Epigenetic control of gene function in schistosomes: a source of therapeutic targets? *Front Genet.* 5: 317.
13 Zhu, L., Zhao, J., Wang, J. et al. (2016). MicroRNAs are involved in the regulation of ovary development in the pathogenic blood fluke *Schistosoma japonicum*. *PLoS Pathog.* 12 (4): e1005582.
14 Protasio, A.V., van Dongen, S., Collins, J. et al. (2017). MiR-277/4989 regulate transcriptional landscape during juvenile to adult transition in the parasitic helminth *Schistosoma mansoni*. *PLoS Negl.Trop. Dis.* 11 (5): e0005559.
15 Bayer-Santos, E., Marini, M.M., and da Silveira, J.F. (2017). Non-coding RNAs in host–pathogen interactions: subversion of mammalian cell functions by protozoan parasites. *Front Microbiol.* 8: 474.
16 Baubec, T. and Schübeler, D. (2014). Genomic patterns and context-specific interpretation of DNA methylation. *Curr. Opin. Genet. Dev.* 25: 85–92.
17 Ponts, N., Fu, L., Harris, E.Y. et al. (2013). Genome-wide mapping of DNA methylation in the human malaria parasite *Plasmodium falciparum*. *Cell Host Microbe* 14: 696–706.
18 Geyer, K.K., Rodriguez-Lopez, C.M., Chalmers, I.W. et al. (2011). Cytosine methylation regulates oviposition in the pathogenic blood fluke *Schistosoma mansoni*. *Nat. Commun.* 2: 424.
19 Geyer, K.K., Chalmers, I.W., Mackintosh, N. et al. (2013). Cytosine methylation is a conserved epigenetic feature found throughout the phylum Platyhelminthes. *BMC Genomics* 14: 462.
20 Fneich, S., Dheilly, N., Adema, C. et al. (2013). 5-Methylcytosine and 5-hydroxymethylcytosine in the genome of *Biomphalaria glabrata*, a snail intermediate host of *Schistosoma mansoni*. *Parasit. Vectors* 6: 167.
21 Sabari, B.R., Zhang, D., Allis, C.D., and Zhao, Y. (2017). Metabolic regulation of gene expression through histone acylations. *Nat. Rev. Mol. Cell Biol.* 18: 90–101.
22 Trelle, M.B., Salcedo-Amaya, A.M., Cohen, A.M. et al. (2009). Global histone analysis by mass spectrometry reveals a high content of acetylated lysine residues in the malaria parasite *Plasmodium falciparum*. *J. Proteome Res.* 8: 3439–3450.
23 Dastidar, E.G., Dzeyk, K., Krijgsveld, J. et al. (2013). Identification of Pf14-3-3 protein as a histone H3 phosphorylation reader in malaria parasites. *PLoS One* 8 (1): e0053179.

24 Rovira-Graells, N., Gupta, A.P., Planet, E. et al. (2012). Transcriptional variation in the malaria parasite *Plasmodium falciparum*. *Genome Res.* 22: 925–938.

25 Lopez-Rubio, J.J., Gontijo, A.M., Nunes, M.C. et al. (2007). 5′ Flanking region of *var* genes nuclear histone modification patterns linked to phenotypic inheritance of virulence traits in malaria parasites. *Mol. Microbiol.* 66: 1296–1305.

26 Ay, F., Bunnik, E.M., Varoquaux, N. et al. (2015). Multiple dimensions of epigenetic gene regulation in the malaria parasite *Plasmodium falciparum*. *Bioessays* 37: 182–194.

27 Maree, J.P. and Patterton, H.G. (2014). The epigenome of *Trypanosoma brucei*: a regulatory interface to an unconventional transcriptional machine. *Biochem. Biophys. Acta* 1839: 743–750.

28 Thomas, S., Green, A., Sturm, N.R. et al. (2009). Histone acetylations mark origins of polycistronic transcription in *Leishmania major*. *BMC Genom.* 10: 152.

29 Mandava, V., Fernandez, J.P., Deng, H. et al. (2007). Histone modifications in *Trypanosoma brucei*. *Mol. Biochem. Parasitol.* 156: 41–50.

30 Leandro de Jesus, T.C., Nunes, V.S., Lopes, M.C. et al. (2016). Chromatin proteomics reveals variable histone modifications during the life-cycle of *Trypanosoma cruzi*. *J. Proteome Res.* 15: 2039–2051.

31 Picchi, G.F.A., Zulkievicz, V., Krieger, M.A. et al. (2017). Post-translational modifications of *Trypanosoma cruzi* canonical and variant histones. *J. Proteome Res.* 16: 1167–1179.

32 Figueiredo, L.M., Cross, G.A.M., and Janzen, C.J. (2009). Epigenetic regulation in African trypanosomes: a new kid on the block. *Nat. Rev. Microbiol.* 7: 504–513.

33 Anderson, L., Pierce, R.J., and Verjovski-Almeida, S. (2012). *Schistosoma mansoni* histones: from transcription to chromatin regulation; an *in silico* analysis. *Mol. Biochem. Parasitol.* 183: 105–114.

34 Picard, M.A.L., Boissier, J., Roquis, D. et al. (2016). Sex-biased transcriptome of *Schistosoma mansoni*: host-parasite interaction, genetic determinants and epigenetic regulators are associated with sexual differentiation. *PLoS Negl.Trop. Dis.* 10 (9): e0004930.

35 Roquis, D., Rognon, A., Chaparro, C. et al. (2016). Frequency and mitotic heritability of epimutations in *Schistosoma mansoni*. *Mol. Ecol.* 25: 1741–1758.

36 Fneich, S., Théron, A., Cosseau, C. et al. (2016). Epigenetic origin of adaptive phenotypic variants in the human blood fluke *Schistosoma mansoni*. *Epigenetics Chromatin* 9: 27.

37 Roquis, D., Lepesant, J.M.J., Picard, M.A.L. et al. (2015). The epigenome of *Schistosoma mansoni* provides insight about how cercariae poise transcription until infection. *PLoS Negl.Trop. Dis.* 9 (8): e0003853.

38 Salcedo-Amaya, A.M., Hoeijmakers, W.A.M., Bartfai, R., and Stunnenberg, H. (2010). Malaria: could its unusual epigenome be the weak spot? *Int. J. Biochem. Cell Biol.* 42: 781–784.

39 Ingram, A.K. and Horn, D. (2002). Histone deacetylases in *Trypanosoma brucei*: two are essential and another is required for normal cell cycle progression. *Mol. Microbiol.* 45: 89–97.
40 Kawahara, T., Siegel, T.N., Ingram, A.K. et al. (2008). Two essential MYST-family proteins display distinct roles in histone H4K10 acetylation and telomeric silencing in trypanosomes. *Mol. Microbiol.* 69: 1054–1068.
41 Duraisingh, M.T. and Horn, D. (2016). Epigenetic regulation of virulence gene expression in parasitic protozoa. *Cell Host Microbe* 19: 629–640.
42 Scholte, L.S.S., Mourao, M.M., Sviatopolk-Mirsky Pais, F. et al. (2017). Evolutionary relationships among protein lysine deacetylases of parasites causing neglected diseases. *Infect. Genet. Evol.* 53: 175–188.
43 Tavares, J., Ouaissi, A., Santarem, N. et al. (2008). The *Leishmania infantum* cytosolic SIR2-related protein 1 (LiSIR2RP1) is an NAD^+-dependent deacetylase and ADP-ribosyltransferase. *Biochem. J.* 415: 377–386.
44 Aurrecoechea, C., Brestelli, J., Brunk, B.P. et al. (2009). PlasmoDB: a functional genomic database for malaria parasites. *Nucleic Acids Res.* 37 (Database issue): D539–D543.
45 Darkin-Rattray, S.J., Gurnett, A.M., Myers, R.W. et al. (1996). Apicidin: a novel antiprotozoal agent that inhibits parasite histone deacetylase. *Proc. Natl. Acad. Sci. U.S.A.* 93: 13143–13147.
46 Horrocks, P., Wong, E., Russell, K., and Emes, R.D. (2009). Control of gene expression in *Plasmodium falciparum* – ten years on. *Mol. Biochem. Parasitol.* 164: 9–25.
47 Patel, V., Mazitschek, R., Coleman, B. et al. (2009). Identification and characterization of small molecule inhibitors of a class I histone deacetylase from *Plasmodium falciparum*. *J. Med. Chem.* 52: 2185–2187.
48 Andrews, K.T., Tran, T.N., Lucke, A.J. et al. (2008). Potent antimalarial activity of histone deacetylase inhibitor analogues. *Antimicrob. Agents Chemother.* 52: 1454–1461.
49 Coleman, B.I., Skillman, K.M., Jiang, R.H.Y. et al. (2014). A *Plasmodium falciparum* histone deacetylase regulates antigenic variation and gametocyte conversion. *Cell Host Microbe* 16: 177–186.
50 Tonkin, C.J., Carret, C.K., Duraisingh, M.T. et al. (2009). Sir2 paralogues cooperate to regulate virulence genes and antigenic variation in *Plasmodium falciparum*. *PLoS Biol.* 7 (4): e84.
51 Cui, L., Fan, Q., Cui, L., and Miao, J. (2008). Histone lysine methyltransferases and demethylases in *Plasmodium falciparum*. *Int. J. Parasitol.* 38: 1083–1097.
52 Jiang, L., Mu, J., Zhang, Q. et al. (2013). PfSETvs methylation of histone H3K36 represses virulence genes in *Plasmodium falciparum*. *Nature* 499: 223–227.
53 Volz, J.C., Bartfai, R., Petter, M. et al. (2012). PfSET10, a *Plasmodium falciparum* methyltransferase, maintains the active *var* gene in a poised state during parasite division. *Cell Host Microbe* 11: 7–18.
54 Jeffers, V., Yang, C., Huang, S., and Sullivan, W.J. Jr., (2017). Bromodomains in protozoan parasites: evolution, function and opportunities for drug development. *Microbiol. Mol. Biol. Rev.* 81: e00047–e00016.

55 Josling, G.A., Petter, M., Oehring, S.C. et al. (2015). A *Plasmodium falciparum* bromodomain protein regulates invasion gene expression. *Cell Host Microbe* 17: 741–751.

56 Alonso, V.L., Ritagliati, C., Cribb, P. et al. (2016). Overexpression of bromodomain factor 3 in *Trypanosoma cruzi* (TcBDF3) affects differentiation of the parasite and protects it against bromodomain inhibitors. *FEBS J.* 283: 2051–2066.

57 Alsford, S. and Horn, D. (2012). Cell-cycle-regulated control of VSG expression sited silencing by histones and histone chaperones ASF1A and CAF-1b in *Trypanosoma brucei*. *Nucl. Acids Res.* 40: 10150–10160.

58 Schulz, D., Mugnier, M.R., Paulsen, E.M. et al. (2015). Bromodomain proteins contribute to maintenance of bloodstream form stage identity in the African trypanosome. *PLoS Biol.* 13 (12): e1002316.

59 Pierce, R.J., Dubois-Abdesselem, F., Lancelot, J. et al. (2012). Targeting schistosome histone modifying enzymes for drug development. *Curr. Pharm. Des.* 18: 3567–3578.

60 Lancelot, J., Caby, S., Dubois-Abdesselem, F. et al. (2013). *Schistosoma mansoni* sirtuins: characterization and potential as chemotherapeutic targets. *PLoS Negl.Trop. Dis.* 7 (9): e2428.

61 Lancelot, J., Cabezas-Cruz, A., Caby, S. et al. (2015). Schistosome sirtuins as drug targets. *Future Med. Chem.* 7: 765–782.

62 Oger, F., Dubois, F., Caby, S. et al. (2008). The class I histone deacetylases of the platyhelminth parasite *Schistosoma mansoni*. *Biochem. Biophys. Res. Commun.* 377: 1079–1084.

63 Marek, M., Kannan, S., Hauser, A.T. et al. (2013). Structural basis for the inhibition of histone deacetylase 8 (HDAC8), a key epigenetic player in the blood fluke *Schistosoma mansoni*. *PLoS Pathog.* 9 (9): e1003645.

64 Carneiro, V., de Abreu da Silva, I.C., Torres, E.J. et al. (2014). Epigenetic changes modulate schistosome egg formation and are a novel target for reducing transmission of schistosomiasis. *PLoS Pathog.* 10 (5): e1004116.

65 Mukherjee, P., Pradhan, A., Shah, F. et al. (2008). Structural insights into the *Plasmodium falciparum* histone deacetylase 1 (PfHDAC-1): A novel target for the development of antimalarial therapy. *Bioorg. Med. Chem.* 16: 5254–5265.

66 Kannan, S., Melesina, J., Hauser, A.T. et al. (2014). Discovery of inhibitors of *Schistosoma mansoni* HDAC8 by combining homology modeling, virtual screening and in vitro validation. *J. Chem. Inf. Model.* 54: 3005–3019.

67 Marek, M., Oliveira, G., Pierce, R.J. et al. (2015). Drugging the schistosome zinc-dependent HDACs: current progress and future perspectives. *Future Med. Chem.* 7: 783–800.

68 Stolfa, D.A., Marek, M., Lancelot, J. et al. (2014). Molecular basis for the antiparasitic activity of a mercaptoamide derivative that inhibits histone deacetylase 8 (HDAC8) from the human pathogen *Schistosoma mansoni*. *J. Mol. Biol.* 426: 3442–3453.

69 Melesina, J., Robaa, D., Pierce, R.J. et al. Homology modeling of parasite histone deacetylases to guide the structure-based design of selective inhibitors. *J. Mol. Graph. Model.* 62: 342–361.

70 Arrowsmith, C.H., Bountra, C., Fish, P.V. et al. (2012). Epigenetic protein families: a new frontier for drug discovery. *Nat. Rev. Drug Discovery* 11: 384–400.

71 Pfister, S.X. and Ashworth, A. (2017). Marked for death: targeting epigenetic changes in cancer. *Nat. Rev. Drug Discovery* 16: 241–263.

72 Chua, M.J., Arnold, M.S.J., Xu, W. et al. (2017). Effect of clinically approved HDAC inhibitors on *Plasmodium*, *Leishmania* and *Schistosoma* parasite growth. *Int. J. Parasitol. Drugs Drug Resist.* 7: 42–50.

73 Sumanadasa, S.D.M., Goodman, C.D., Lucke, A.J. et al. (2012). Antimalarial activity of the anticancer histone deacetylase inhibitor SB939. *Antimicrob. Agents Chemother.* 56: 3849–3856.

74 Carillo, A.K., Guiguemde, W.A., and Guy, R.K. (2015). Evaluation of histone deacetylase inhibitors (HDACi) as therapeutic leads for African trypanosomiasis (HAT). *Bioorg. Med. Chem.* 23: 5151–5155.

75 Malmquist, N.A., Moss, T.A., Mecheri, S. et al. (2012). Small-molecule histone methyltransferase inhibitors display rapid antimalarial activity against all blood stage forms in *Plasmodium falciparum*. *Proc. Natl. Acad. Sci. U.S.A.* 109: 16708–16713.

76 Kubicek, S., O'Sullivan, R.J., August, E.M. et al. (2007). Reversal of H3K9me2 by a small-molecule inhibitor for the G9a histone methyltransferase. *Mol. Cell* 25: 473–481.

77 Sundriyal, S., Malmquist, N.A., Caron, J. et al. (2014). Development of diaminoquinazoline histone lysine methyltransferase inhibitors as potent blood-stage anti-malarial compounds. *ChemMedChem* 9: 2360–2373.

78 Malmquist, N.A., Sundriyal, S., Caron, J. et al. (2017). Histone methyltransferase inhibitors are orally bioavailable, fast-acting molecules with activity against different species causing malaria in humans. *Antimicrob. Agents Chemother.* 59: 950–959.

79 Anderson, L., Gomes, M.R., da Silva, L.F. et al. (2017). Histone deacetylase inhibition modulates histone acetylation at gene promoter regions and affects genome-wide gene transcription in *Schistosoma mansoni*. *PLoS Negl.Trop. Dis.* 11 (4): e0005539.

80 Verma, S.K., Tian, X., LaFrance, L.V. et al. (2012). Identification of potent, selective, cell-active inhibitors of the histone lysine methyltransferase EZH2. *ACS Med. Chem. Lett.* 3: 1091–1096.

81 Andrieu, G., Belkina, A.C., and Denis, G.V. (2016). Clinical trials for BET inhibitors run ahead of the science. *Drug Discov. Today Technol.* 19: 45–50.

82 Jeffers, V., Kamau, E., Srinivasan, A. et al. (2017). TgPRELID, a novel mitochondrial protein linked to multi-drug resistance in the parasite *Toxoplasma gondii*. *mSphere* 2 (1): e00229.

83 Pierce, R.J., MacDougall, J., Leurs, R., and Costi, M.P. (2017). The future of drug development for neglected tropical diseases: how the European Commission can continue to make a difference. *Trends Parasitol.* 33: 581–583.

84 Tavares, J., Ouaissi, A., Kong Thoo Lin, P. et al. (2010). Bisnaphthalimidopropyl derivatives as inhibitors of *Leishmania* SIR2 related protein 1. *ChemMedChem* 5: 140–147.

85 Kelly, J.M., Taylor, M.C., Horn, D. et al. (2012). Inhibitors of human histone deacetylase with potent activity against the African trypanosome *Trypanosoma brucei*. *Bioorg. Med. Chem. Lett.* 22: 1886–1890.
86 Andrews, K.T., Tran, T.N., and Fairlie, D.P. (2012). Towards histone deacetylase inhibitors as new malarial drugs. *Curr. Pharm. Des.* 18: 3467–3479.
87 Dow, G.S., Chen, Y., Andrews, K.T. et al. (2008). Antimalarial activity of phenylthiazolyl-bearing hydroxamate-based histone deacetylase inhibitors. *Antimicrob. Agents Chemother.* 52: 3467–3477.
88 Hansen, F.K., Sumanadasa, S.D.M., Stenzel, K. et al. (2014). Discovery of HDAC inhibitors with potent activity against multiple malaria parasite life-cycle stages. *Eur. J. Med. Chem.* 82: 204–213.
89 Stenzel, K., Chua, M.J., Duffy, S. et al. (2017). Design and synthesis of terephthalic acid-based histone deacetylase inhibitors with dual-stage anti-plasmodium activity. *ChemMedChem* 12: 1627–1636.
90 Sacconnay, L., Angleviel, M., Randazzo, G.M. et al. (2014). Computational studies on sirtuins from *Trypanosoma cruzi*: structures, conformations and interactions with phytochemicals. *PLoS Negl.Trop. Dis.* 8 (2): e2689.
91 Moustakim, M., Clark, P.G.K., Trulli, L. et al. (2017). Discovery of a PCAF bromodomain chemical probe. *Angew. Chem. Int. Ed.* 56: 827–831.
92 Heimburg, T., Chakrabarti, A., Lancelot, J. et al. (2016). Structure-based design and synthesis of novel inhibitors targeting HDAC8 from *Schistosoma mansoni* for the treatment of schistosomiasis. *J. Med. Chem.* 59: 2423–2435.
93 Stenzel, K., Chakrabarti, A., Melesina, J. et al. (2017). Isophthalic acid-based HDAC inhibitors as potent inhibitors of HDAC8 from *Schistosoma mansoni*. *Arch. Pharm. (Weinheim)* 350 (8): doi: 10.1002/ardp.201700096.
94 Benedetti, R., Conte, M., and Altucci, L. (2015). Targeting histone deacetylases in diseases: where are we ? *Antioxid. Redox Signaling* 23: 99–126.
95 Sato, T., Suzuki, M., Sato, Y. et al. (2006). Sequence-dependent interaction between cisplatin and histone deacetylase inhibitors in human oral squamous cell carcinoma cells. *Int. J. Oncol.* 28: 1233–1241.

Index

Epigenetic Drug Discovery, First Edition. Edited by Wolfgang Sippl and Manfred Jung.

c

i